Lecture Notes in Mechanical Engineering

Series Editors

Fakher Chaari, National School of Engineers, University of Sfax, Sfax, Tunisia

Mohamed Haddar, National School of Engineers of Sfax (ENIS), Sfax, Tunisia

Young W. Kwon, Department of Manufacturing Engineering and Aerospace Engineering, Graduate School of Engineering and Applied Science, Monterey, CA, USA

Francesco Gherardini, Dipartimento Di Ingegneria, Edificio 25, Università Di Modena E Reggio Emilia, Modena, Modena, Italy

Vitalii Ivanov, Department of Manufacturing Engineering Machine and Tools, Sumy State University, Sumy, Ukraine

AF805630

Lecture Notes in Mechanical Engineering (LNME) publishes the latest developments in Mechanical Engineering—quickly, informally and with high quality. Original research reported in proceedings and post-proceedings represents the core of LNME. Volumes published in LNME embrace all aspects, subfields and new challenges of mechanical engineering. Topics in the series include:

- Engineering Design
- Machinery and Machine Elements
- Mechanical Structures and Stress Analysis
- Automotive Engineering
- Engine Technology
- Aerospace Technology and Astronautics
- Nanotechnology and Microengineering
- Control, Robotics, Mechatronics
- MEMS
- Theoretical and Applied Mechanics
- Dynamical Systems, Control
- Fluid Mechanics
- Engineering Thermodynamics, Heat and Mass Transfer
- Manufacturing
- Precision Engineering, Instrumentation, Measurement
- Materials Engineering
- Tribology and Surface Technology

To submit a proposal or request further information, please contact the Springer Editor of your location:

China: Dr. Mengchu Huang at mengchu.huang@springer.com
India: Priya Vyas at priya.vyas@springer.com
Rest of Asia, Australia, New Zealand: Swati Meherishi at swati.meherishi@springer.com
All other countries: Dr. Leontina Di Cecco at Leontina.dicecco@springer.com

To submit a proposal for a monograph, please check our Springer Tracts in Mechanical Engineering at http://www.springer.com/series/11693 or contact Leontina.dicecco@springer.com

Indexed by SCOPUS. The books of the series are submitted for indexing to Web of Science.

More information about this series at http://www.springer.com/series/11236

Hari Vasudevan · Vijaya Kumar N. Kottur ·
Amool A. Raina
Editors

Proceedings of International Conference on Intelligent Manufacturing and Automation

ICIMA 2020

Springer

Editors
Hari Vasudevan
Dwarkadas J. Sanghvi College
of Engineering
Mumbai, India

Vijaya Kumar N. Kottur
Dwarkadas J. Sanghvi College
of Engineering
Mumbai, India

Amool A. Raina
RWTH Aachen University
Aachen, Germany

ISSN 2195-4356 ISSN 2195-4364 (electronic)
Lecture Notes in Mechanical Engineering
ISBN 978-981-15-4484-2 ISBN 978-981-15-4485-9 (eBook)
https://doi.org/10.1007/978-981-15-4485-9

© Springer Nature Singapore Pte Ltd. 2020
This work is subject to copyright. All rights are reserved by the Publisher, whether the whole or part of the material is concerned, specifically the rights of translation, reprinting, reuse of illustrations, recitation, broadcasting, reproduction on microfilms or in any other physical way, and transmission or information storage and retrieval, electronic adaptation, computer software, or by similar or dissimilar methodology now known or hereafter developed.
The use of general descriptive names, registered names, trademarks, service marks, etc. in this publication does not imply, even in the absence of a specific statement, that such names are exempt from the relevant protective laws and regulations and therefore free for general use.
The publisher, the authors and the editors are safe to assume that the advice and information in this book are believed to be true and accurate at the date of publication. Neither the publisher nor the authors or the editors give a warranty, expressed or implied, with respect to the material contained herein or for any errors or omissions that may have been made. The publisher remains neutral with regard to jurisdictional claims in published maps and institutional affiliations.

This Springer imprint is published by the registered company Springer Nature Singapore Pte Ltd.
The registered company address is: 152 Beach Road, #21-01/04 Gateway East, Singapore 189721, Singapore

ICIMA 2020

Chief Patron

Shri Amrish R. Patel, Chief Patron, President, SVKM

Patrons

Shri Bhupesh R. Patel, Joint President, SVKM
Shri Bharat M. Sanghvi, Vice President and Trustee, SVKM and I/C DJSCE
Shri Chintan A. Patel, Vice President, SVKM
Shri Sunandan R. Divatia, Hon. Secretary, SVKM
Shri Harshad H. Shah, Hon. Treasurer, SVKM
Shri Shalin S. Divatia, Hon. Joint Secretary, SVKM
Shri Jayant P. Gandhi, Hon. Joint Secretary, SVKM
Shri Harit H. Chitalia, Hon. Joint Treasurer, SVKM
Shri Jagdish B. Parikh, Hon. Joint Secretary, SVKM

International Advisory Committee

Dr. Huynh T. Luong, Asian Institute of Technology, Thailand
Dr. Amit S. Jariwala, Georgia Institute of Technology, USA
Dr. Michel Fillon, University of Poitiers, France
Dr. Amool A. Raina, RWTH Aachen University, Germany
Dr. Konstantinos Salonitis, Cranfield University, UK
Dr. Rohan A. Shirwaiker, North Carolina State University, USA

National Advisory Committee

Dr. S. S. Mantha, Former Chairman, AICTE, New Delhi
Dr. D. N. Malkhede, Adviser–I(RIFD), AICTE
Dr. S. K. Mahajan, Joint Director, DTE, Maharashtra
Dr. S. G. Deshmukh, ABV–IIITM, Gwalior
Dr. S. K. Ukarande, Dean, FOST, University of Mumbai
Dr. M. K. Tiwari, IIT Kharagpur
Dr. K. P. Karunakaran, IIT Bombay
Dr. S. M. Khot, FCRIT, Navi Mumbai
Dr. Sushil Kumar, IIT Lucknow
Dr. Tusar Desai, NIT Surat
Dr. V. R. Kalamkar, VNIT Nagpur
Dr. C. M. Venkateswaran, Aker Solution
Dr. L. Ganapathy, NITIE, Mumbai
Dr. K. Madduley, SPJSGM, Mumbai
Dr. Vivek Sunnapwar, LTCE, Navi Mumbai
Dr. P. Sakthivel, VIT Vellore
Dr. Suhash Deshmukh, GCE, Karad
Dr. S. R. Chandak, Ex. V.P./Advisor (HR), Bharat Forge Ltd.

Organising Committee

Dr. Hari Vasudevan, General Chair and Convenor, Principal, DJSCE
Dr. A. C. Daptardar, General Co-Chair, Vice Principal (Admin.), DJSCE
Dr. M. J. Godse, General Co-Chair, Vice Principal (Acad.), DJSCE
Dr. Vijaya Kumar N. Kottur, Joint Convenor, Professor and Head, Department of Mechanical Engineering, DJSCE
Dr. Rajendra S. Khavekar, Co-Convenor, Training and Placement Officer, DJSCE

Members

Dr. S. N. Takalikar
Dr. Atul Dhale
Dr. Vinayak H. Khatawate
Dr. Sanjeev Thool
Dr. Meeta N. Gandhi
Mr. E. Narayanan
Mr. Sandeep R. Vaity
Mr. Prasad S. Shirodkar

Mr. Vyankatesh U. Bagal
Mr. P. Frank Crasta
Mr. Prashant P. Patankar
Mr. Avdhut Samant
Mr. Rajnarayan M. Yadav
Mr. Bronin Cyriac
Mr. Greegory Mathew
Mr. Dharam Ranka
Mr. Amit Chaudhari
Mr. Ramesh R. Rajguru
Mrs. Trupti Markose
Mr. Rohit K. Chaurasia
Mr. Mehul S. Prajapati
Mr. Vinit R. Katira
Mr. Pavan R. Rayar
Mr. Kartik M. Ajugia
Mr. Dhaval J. Birajdar
Mr. Sandip H. Mane
Mr. Sanket D. Parab
Mr. Shashikant M. Auti
Mr. N. C. Deshpande
Mr. Dhanjay Shukla
Mr. Ravikant Hattale
Mr. Sachin Kamble
Mr. Dhiraj Nigade

Preface

Advanced technologies in the current era ensure that we are in turbulent and challenging times as we attempt for wealth creation and business in the fields of manufacturing and automation. Industries are currently looking up to manufacturing and automation engineers for their assistance in increasing the overall productivity in their organizations. It is also the time when policymakers across the globe have started to focus more of their attention on the manufacturing sector due to the presence of disruptive technologies in manufacturing. The International Conference on Intelligent Manufacturing and Automation 2020 (ICIMA 2020) was therefore designed to encourage discussions and research on advancements and applications in the areas of manufacturing and automation. The primary focus of this conference was to bring together academicians, researchers and scientists for knowledge sharing in various areas of manufacturing, automation and other allied domains. This conference covered topics encompassing automation, mechatronics, robotics, manufacturing processes, management and other related areas, such as product design and development, green manufacturing and smart materials with the objective of brainstorming and with specific emphasis on the applications in the field of intelligent manufacturing and automation. The response to the call for papers was overwhelming with 181 articles submitted for this conference. Finally, 82 articles, covering a wide spectrum of topics related to the theme of this conference, were accepted after a thorough review process. We express our sincere appreciation to the authors for their contribution to this book. We would also like to express our sincere gratitude to all the experts and referees for their valuable comments and support extended during the review process. Thank you everyone once again!

Mumbai, India — Hari Vasudevan
Mumbai, India — Vijaya Kumar N. Kottur
Aachen, Germany — Amool A. Raina

Contents

About the Editors

Dr. Hari Vasudevan has his Masters in Production Engineering as well as Postgraduate Diploma in Industrial Engineering from VJTI (University of Mumbai) and Ph.D. from IIT Bombay. He has also done a 3 months full time certificate programme (ERP-BaaN) from S.P.Jain Institute of Management & Research, Mumbai, under the University Synergy Programme of BaaN Institute, Netherlands. His areas of interest include manufacturing engineering, manufacturing systems & strategy, market orientation of manufacturing firms and world class manufacturing. He is an approved Ph.D. guide at the University of Mumbai and NMIMS (Deemed to be University) and has so far guided 7 Ph.D. students for their degree. He is the President of the Indian Society of Manufacturing Engineering (ISME), life member of ISTE, New Delhi, Fellow of the Institution of Engineers India, Fellow of ISME and a senior member of IEDRC. He has 27 years of experience in teaching and 2 years of experience in industry. Presently, he is working as the Principal of Dwarkadas J. Sanghvi College of Engineering, Mumbai. He has published over 115 papers in international conferences and journals as well as in national conferences and journals.

Dr. Vijaya Kumar N. Kottur heads the Department of Mechanical Engineering at Dwarkadas J. Sanghvi College of Engineering, Mumbai. He completed his Master's degree in Engineering Management from SJCE Mysore, Master's degree in Mechanical Engineering with Machine Design as specialization from SPCE, Mumbai, and Ph.D. from the University of Mumbai. He has 30 years of teaching experience and published 52 papers in national and international journals and conferences. He is an approved Ph.D. guide at Mumbai University, Pune University and JJTU University, and has so far guided 3 Ph.D. students. He worked as a Guest Faculty at NITIE Mumbai. His areas of interest are quality engineering, world class manufacturing, supply chain management and system dynamics. He is a life member of professional bodies like ISNDT, IIIE, ISTE, and ISME.

Dr. Amool A. Raina currently coordinates and heads the aerospace programme at the Institut für Textiltechnik (ITA) of RWTH Aachen University. His doctoral thesis (highest distinction received) majoring in Aerospace Engineering at the University of Kansas presented solutions for improvement in wind turbine design and manufacturing. He is considered as an expert in wind turbine design and engineering and has a work experience of over 6 years in the aerospace and renewable industry prior to joining ITA. He has designed and engineered over 35 wind turbine blades ranging from 2 m to 105 m in length with over 16,000 blades successfully flying worldwide. Dr. Raina has also been involved in projects relating to design and optimization of aircraft engines and other components. Dr. Raina is currently developing and promoting textile-based solutions for several sectors including aerospace, automotive, traditional textiles as well as digital solutions as per Industry 4.0 norms for the above industries. He has contributed and led engineering and product development teams for manufacturing reflector antennas for communication satellites, composite airframe parts and high temperature aircraft engine components. Apart from his technical responsibilities, he also heads all affairs and activities pertaining to the European Union and European Space Agency at ITA.

Manufacturing

Exploring Ideal Process Parameters to Enhance the Surface Integrity Using Grey Fuzzy Integrated Technique

Ramesh R. Rajguru and Hari Vasudevan

Abstract Inconel 625 belongs to the category of austenitic nickel chromium-based super alloys and is extensively used in sea water applications, including propeller blades in aerospace as well as oil and gas industries. However, the precision machining of Inconel 625 alloy is still a challenge due to high rate of work hardening, which hinders further applications in industries. In this context, this research study was carried out, involving the end milling of Inconel 625 nickel-based alloy, using process parameters, such as cutting speed, feed per tooth, radial depth of cut and radial rake angle. The study explored the ideal process parameters to enhance the surface integrity of machined surface under a dry cutting environment. The grey fuzzy integrated approach followed gave the best experimental performance, using grey relational coefficient and multi-performance criteria index. Combination of cutting and tool geometry parameters at the feed per tooth of 0.1 mm/tooth, cutting speed 90 m/min, radial depth of cut 0.4 mm and radial rake angle of 11° are recommended as part of the result. It minimizes work hardening of the machined surface and could ensure induction of superior surface integrity in the machined surfaces, by inducing compressive residual stress 158 MPa, minimum average roughness 0.906 μm and microhardness 379 HV.

Keywords Grey fuzzy · Inconel 625 · Surface integrity · PVD TiAlSiN coating

R. R. Rajguru (✉)
Department of Mechanical Engineering, Dwarkadas J. Sanghvi College of Engineering, Vile Parle, Mumbai, India
e-mail: ramesh.rajguru@djsce.ac.in

H. Vasudevan
Dwarkadas J. Sanghvi College of Engineering, Vile Parle, Mumbai, India
e-mail: harivasudevan@iitb.ac.in

© Springer Nature Singapore Pte Ltd. 2020

H. Vasudevan et al. (eds.), *Proceedings of International Conference on Intelligent Manufacturing and Automation*, Lecture Notes in Mechanical Engineering,
https://doi.org/10.1007/978-981-15-4485-9_1

1 Introduction

Surface integrity of machined components demands more attention at the manufacturing stage, since it directly affects the performance of the machined component during its intended application. The functionality and durability of machined components depend on induced residual stresses, surface roughness as well as microhardness that occur due to machining. Hence, these are all important aspects of surface integrity that affects the overall machined components for its superiority.

In the experimental investigation of tool coating and machining parameters during turning operation on Inconel 825, Thakur et al. [1] developed a predictive model with fuzzy logic. They observed that the experimental values and predicted result of fuzzy logic matched closely with the average errors in microhardness that were 1.07 and 1.18 for uncoated and CVD coated carbide insert tool, respectively. According to Rajyalakshmi et al. [2], fuzzy grey relational analysis is a hybrid method of process parameters optimization for response characteristics in machining. They optimized multiple responses of nonconventional machining process in wire electric discharge machining on Inconel 825, applying grey fuzzy. In a study on turning operation on Hadfield steel, Horng and Chiang [3] used fast and effective hybrid algorithms of grey fuzzy logic to find the optimal machining parameters in terms of surface roughness and tool wear. They employed cutting speed, feed rate, depth of cut and nose radius as independent variables and pointed out that the grey coupled with fuzzy logic approach significantly improved the performance characteristics in the turning operation.

Rajmohan et al. [4] applied grey fuzzy techniques to find out the optimal level of multiple responses during the drilling of metal matrix composite, a difficult to cut situation, through the setting of machining parameters. They affirmed that the responses including thrust force, surface roughness and burr height were improved by using this integrated method.

In the electric discharge machining (EDM) of AISI P20 steel with RSM-based experimental design, Dewangan et al. [5] utilized a hybrid technique grey fuzzy logic to optimize the integrity of surface aspects, such as surface roughness, surface crack density and white layer thickness. They concluded that the grey fuzzy method has good potential to achieve minimum surface integrity during EDM of AISI P20 steel. Vasudevan et al. [6] used an integrated multi-objective optimization method involving grey and fuzzy rule-based approach to optimize the chosen output responses, such as surface roughness, material removal rate and tangential cutting force during CNC turning of GFRP/E composites. They pointed out that the grey fuzzy method is a simple and effective technique for robust finish turning operation for GFRP/E composites. Vetri Velmurgan et al. [7] investigated experimentally the effect of cutting parameters during the turning of hybrid metal matrix composites and optimized machining parameters, using a grey fuzzy approach coupled with Taguchi loss function. They confirmed the effectiveness of the grey fuzzy method for optimization of cutting parameters with more than one quality criteria response. None of the studies reported so far in the literature have brought out conclusive results on the ideal process parameters on surface integrity of Inconel 625. In this context, this study was carried out involving the end milling of Inconel 625 nickel-based alloy, using cutting and tool geometry parameters to explore the ideal process parameters that enhance the surface integrity with a grey fuzzy integrated approach.

2 Grey Relational Analysis and Fuzzy Logic Approach

If there are many responses for the same experimental conditions with input research design variables, the statistical technique provides the best combination of experimental settings for each response, but these settings could be dissimilar from each other. Milling being a complex operation, the relation between research design variables is complicated, and it produces uncertainty. Such systems are called grey; hence, the grey approach is adopted. Grey fuzzy integrated approach is a combination of grey relational analysis and fuzzy logic theory. To remove the uncertainty of grey response due to imprecision and lack of exactness of information in grey relational coefficient, fuzzy inference system was used. Figure 1 depicts the methodology applied in the grey fuzzy method.

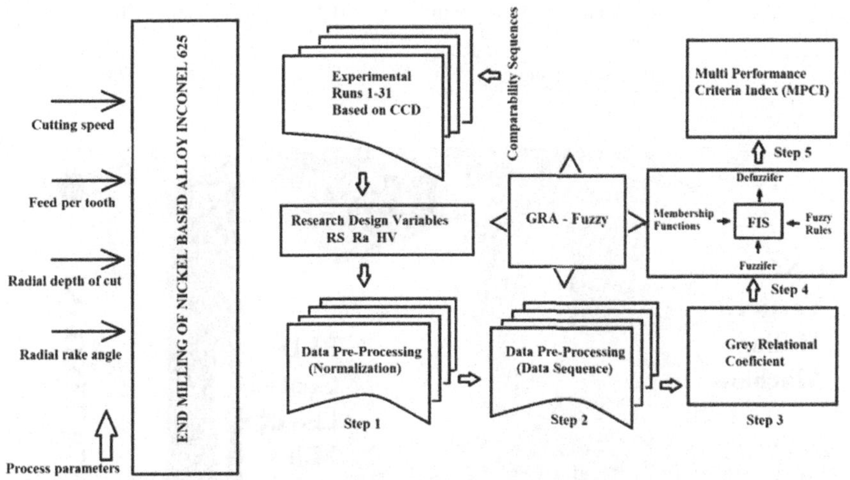

Fig. 1 Grey fuzzy integrated approach

3 Experimental Work

In this study, cutting speed, feed per tooth, radial depth of cut and radial rake angle were selected as input variables with five levels, while the axial depth of cut was kept constant (Table 1). Figure 2 depicts the experimental setup and details. The levels of machining parameters were selected based on pilot experiment results. The experiments were conducted with various cutting conditions, as per the Central Composite Design (CCD) matrix that included full factorial design with axial points and centre points. A four-flute end mill of 12 mm diameter PVD coated with TiAlSiN was used as the cutting tool. The measurement of surface roughness was done on Taylor Hobson Taly surf 4. Microhardness of the machine surface was measured using Vickers indenter with square-based pyramid. Residual stress measurement was done using Panalytical by X-ray diffraction method at IIT Bombay.

Table 1 Cutting parameters and their levels

Cutting parameters designation	Process parameters	Units	Levels				
			−2	−1	0	1	2
V_c	Cutting speed	m/min	60	70	80	90	100
f_z	Feed per tooth	mm/tooth	0.04	0.06	0.08	0.10	0.12
a_e	Radial depth of cut	mm	0.2	0.4	0.6	0.8	1.0
γ	Radial rake angle	°	5	7	9	11	13

Fig. 2 Experimental set up with work material Inconel 625 and end mill cutter

4 Analysis Using Grey Fuzzy Method

This section discusses the steps taken to arrive at ideal process parameters using the grey fuzzy integrated approach.

4.1 Grey Relational Analysis

Initially, sequence x_0 and x_i in k point, using grey relational theory was determined [8].

$$x_i^0(k) = \text{Comparability sequence and } x_o^0(k) = \text{Referance sequence}$$

where $i = 1, \ldots, m$, $k = 1, \ldots, n$. 'm' is the number of experimental data items, and 'n' is the number of parameters.

$$\begin{aligned} x_0 &= x_0(1), x_0(2) \ldots x_0(k) \\ x_1 &= x_1(1), x_1(2) \ldots x_1(k) \\ &\vdots \\ x_m &= x_m(1), x_m(2) \ldots x_m(k) \end{aligned}$$

The deviation sequence $\Delta oi(k)$ is the absolute difference between the reference sequence $x_o^*(k)$ and the comparability sequence $x_i^*(k)$ after normalization. Normalization of response data is presented in Table 2.

$$\Delta oi(k) = \left|x_o^*(k) - x_i^*(k)\right|, i = 1, 2, 3, \ldots, 31 \text{ and } k = 1, 2, 3, 4$$

$$\text{For } i = 1, 2, 3 \ldots 31, \text{ there are}$$

$$\begin{aligned} \Delta o1(1) &= \left|x_o^*(k) - x_1^*(1)\right| \\ &\vdots \\ \Delta o31(31) &= \left|x_o^*(k) - x_{31}^*(31)\right| \end{aligned}$$

The grey relational coefficient is calculated from the deviation sequence, and the same is presented as in Table 3.

Table 2 Input variables in coded form and normalization of response data

Expt. No.	Research design variables in coded form				Normalization of response data $x_i^*(k) = \frac{x_i^o(k) - \min x_i^o(k)}{\max x_i^o(k) - \min x_i^o(k)}$		
	V_c	f_z	a_e	γ	Surface roughness (R_a)	Residual stress (RS)	Microhardness (HV)
1	−1	−1	−1	−1	0.913	0.756	0.461
2	1	−1	−1	−1	0.953	0.683	0.292
3	−1	1	−1	−1	0.517	0.399	0.449
4	1	1	−1	−1	0.339	0.358	0.652
5	−1	−1	1	−1	0.852	0.419	0.382
6	1	−1	1	−1	0.926	0.188	0.472
7	−1	1	1	−1	0.222	0.537	0.494
8	1	1	1	−1	0.249	0.423	0.551
9	−1	−1	−1	1	0.923	0.611	0.000
10	1	−1	−1	1	0.826	0.000	0.663
11	−1	1	−1	1	0.657	0.495	0.348
12	1	1	−1	1	0.455	1.000	0.337
13	−1	−1	1	1	0.841	0.277	0.382
14	1	−1	1	1	0.825	0.791	0.618
15	−1	1	1	1	0.387	0.938	0.685
16	1	1	1	1	0.297	0.849	0.225
17	−2	0	0	0	0.660	0.572	1.000
18	2	0	0	0	0.630	0.541	0.831
19	0	−2	0	0	1.000	0.882	0.562
20	0	2	0	0	0.000	0.183	0.112
21	0	0	−2	0	0.7971	0.636	0.685
22	0	0	2	0	0.731	0.566	0.157
23	0	0	0	−2	0.570	0.113	0.135
24	0	0	0	2	0.540	0.894	0.292
25	0	0	0	0	0.617	0.613	0.415
26	0	0	0	0	0.560	0.581	0.090
27	0	0	0	0	0.709	0.637	0.124
28	0	0	0	0	0.720	0.577	0.551
29	0	0	0	0	0.652	0.613	0.494
30	0	0	0	0	0.663	0.624	0.112
31	0	0	0	0	0.606	0.621	0.045

Table 3 Individual grey relational coefficient values of the responses and MPCI

Expt. No.	Deviation sequence $\Delta oi(k) = \lvert x_o^*(k) - x_i^*(k) \rvert$			Grey relational coefficient $\gamma(x_o^*(k), x_i^*(k)) = \frac{\Delta\min+\xi\Delta\max}{\Delta_{ik}+\xi\Delta\max}$			MPCI
	R_a	RS	HV	R_a	RS	HV	
1	0.087	0.244	0.539	0.852	0.672	0.481	0.682
2	0.047	0.317	0.708	0.915	0.612	0.414	0.643
3	0.482	0.601	0.551	0.508	0.454	0.476	0.436
4	0.660	0.642	0.348	0.430	0.438	0.589	0.419
5	0.148	0.581	0.618	0.772	0.463	0.447	0.452
6	0.074	0.812	0.528	0.871	0.381	0.486	0.379
7	0.778	0.463	0.506	0.391	0.519	0.497	0.532
8	0.751	0.577	0.449	0.400	0.464	0.527	0.448
9	0.077	0.389	1.000	0.866	0.562	0.333	0.596
10	0.174	1.000	0.337	0.742	0.333	0.597	0.455
11	0.343	0.505	0.652	0.593	0.498	0.434	0.496
12	0.544	0.000	0.663	0.478	1.000	0.430	0.811
13	0.158	0.723	0.618	0.759	0.409	0.447	0.452
14	0.175	0.209	0.382	0.741	0.705	0.567	0.693
15	0.613	0.062	0.315	0.449	0.889	0.614	0.789
16	0.703	0.151	0.775	0.416	0.768	0.392	0.64
17	0.339	0.428	0.000	0.595	0.539	1.000	0.75
18	0.370	0.459	0.169	0.575	0.521	0.748	0.583
19	0.000	0.118	0.438	1.000	0.808	0.533	0.792
20	1.000	0.817	0.888	0.333	0.380	0.360	0.379
21	0.414	0.364	0.315	0.711	0.578	0.614	0.595
22	0.269	0.434	0.843	0.650	0.535	0.372	0.556
23	0.430	0.887	0.865	0.538	0.361	0.366	0.365
24	0.459	0.106	0.708	0.521	0.825	0.414	0.67
25	0.383	0.387	0.584	0.566	0.564	0.461	0.632
26	0.440	0.419	0.910	0.532	0.544	0.355	0.562
27	0.291	0.363	0.876	0.632	0.579	0.363	0.535
28	0.280	0.423	0.449	0.641	0.541	0.527	0.61
29	0.348	0.387	0.506	0.589	0.563	0.497	0.535
30	0.337	0.376	0.888	0.597	0.571	0.360	0.598
31	0.394	0.379	0.955	0.560	0.569	0.344	0.368

4.2 Fuzzy Logic

In this study, a fuzzy inference system (Fig. 3) was used to estimate the MPCI, when values of the grey relational coefficient were given as inputs to the system. Gaussian-type membership functions were used for research design variables input, such as residual stress, surface roughness and microhardness as well as output (MPCI), as shown in Figs. 4 and 5. Total 125 fuzzy rules based on Mamdani types were derived to generate the fuzzy value in terms of MPCI, based on higher the better of grey relational coefficients.

Rule 1: IF GRC-Ra is very low and GRC-RS is very low and GRC-HV is very low, THEN MPCI is very low.
ELSE … Continuing up to Rule 125.

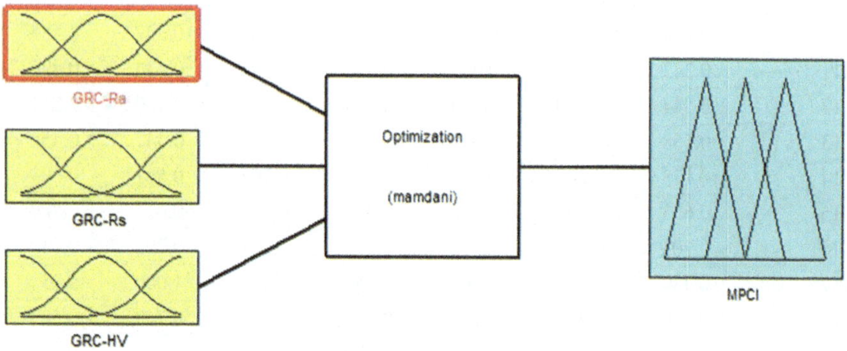

Fig. 3 Fuzzy inference system (FIS)

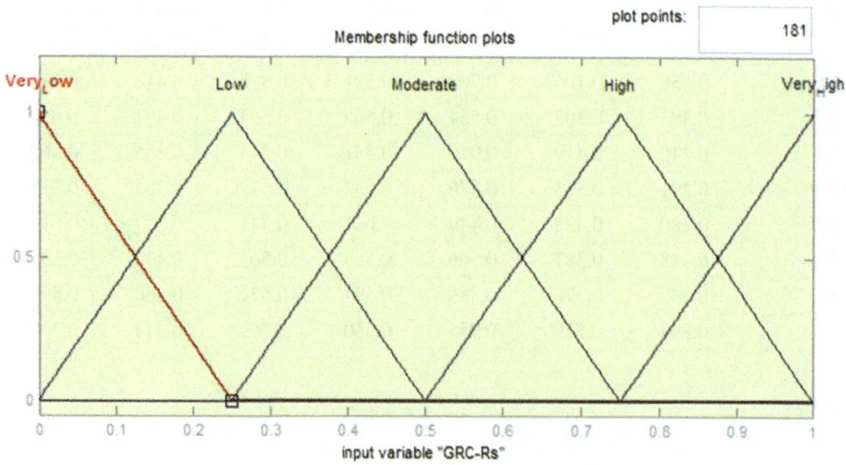

Fig. 4 Membership functions of the input (residual stress)

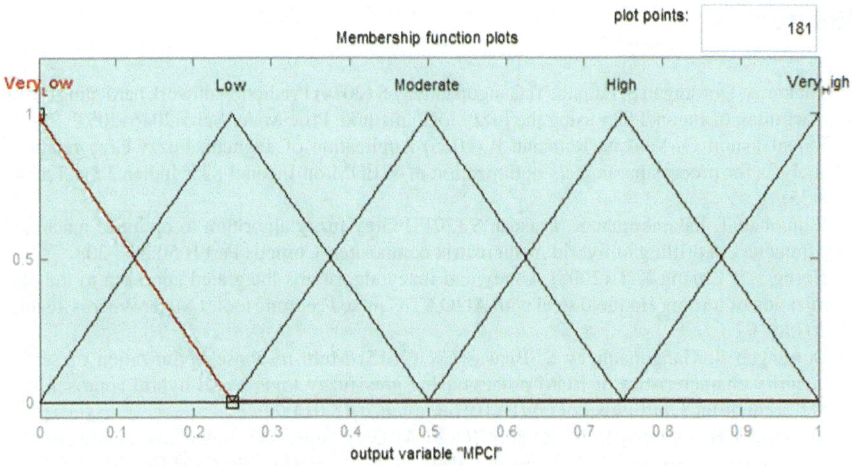

Fig. 5 Membership functions of the output (MPCI)

Rule 125: IF GRC-Ra is very high and GRC-RS is high and GRC-HV is low, THEN MPCI is high.

Individual grey relational coefficient values of the responses and the calculated MPCI are as given in Table 3. The larger value of MPCI, the better was the multiple process response. Hence, cutting speed at level (90 m/min), feed per tooth at level (0.1 mm/tooth), radial depth of cut at level (0.4 mm) and radial rake angle (11°) at level were the ideal process parameters to enhance the surface integrity of machined surface.

5 Conclusion

In this research work, an experimental investigation into the end milling of nickel-based super alloy Inconel 625 was conducted in order to explore the ideal process parameters with grey fuzzy integrated approach. Results show that superior surface integrity would lead to improved fatigue performance, which could be achieved by implementing best performance cutting and tool geometry parameters, as obtained using grey fuzzy method. Basically, the larger the grey fuzzy grade, the better are the multiple performance characteristics. From the grey fuzzy grade, the ideal process parameters, such as cutting speed at level (90 m/min), feed per tooth at level (0.1 mm/tooth), radial depth of cut at level (0.4 mm) and radial rake angle at level (11), were determined. Research approach developed in this study could be applied in case of other difficult to cut materials or different grade of nickel-based alloys as well.

References

1. Thakur A, Dewangan S, Patnaik Y, Gangopadhyay S (2014) Prediction of work hardening during machining of Inconel 825 using the fuzzy logic method. Proc Mater Sci 5:2046–2053
2. Rajyalakshmi G, Venkata Ramaiah P (2015) Application of Taguchi, Fuzzy-Grey relational analysis for process parameters optimization of WEDM on Inconel 825. Indian J Sci Technol 8(35)
3. Rajmohan T, Palanikumar K, Prakash S (2013) Grey-fuzzy algorithm to optimize machining parameters in drilling of hybrid metal matrix composites. Compos Part B 50:297–308
4. Horng J-T, Chiang K-T (2008) A grey and fuzzy algorithms integrated approach to the optimization of turning Hadfield steel with Al2O3/TiC mixed ceramic tool. J Mater Process Technol 207:89–97
5. Dewangan S, Gangopadhyay S, Biswas CK (2015) Multi-response optimization of surface integrity characteristics of EDM process using grey-fuzzy logic-based hybrid approach. Eng Sci Technol Int J. http://dx.doi.org/10.1016/j.jestch.2015.01.009
6. Vasudevan H, Deshpande N, Rajguru R (2014) Grey fuzzy multi-objective optimization of process parameters for CNC turning of GFRP/epoxy composites. Procedia Eng 97:85–94
7. Vetri Velmurgan K, Venkatesan K (2016) An investigation of the parametric effects of cutting parameters on quality characteristics during the dry turning of Inconel 718 alloy. Indian J Sci Technol 9(44)
8. Deng JL (1989) Introduction to grey system. J Grey Syst 1:1–24

Multi-objective Optimization of Dry EDM with Inconel 718 Using Grey Relational Analysis

A. S. Bhandare and U. A. Dabade

Abstract In dry EDM machining, hollow electrodes are applied to supply gaseous medium at the machining zone. In this paper, an attempt has been made with L_{27} orthogonal array using compressed air as a dielectric medium, for Inconel 718 workpiece material and with copper tool electrode, to study and investigate the effects of machining parameters like as gas pressure, discharge current, pulse on time and gap voltage to optimize the output variables such as material removal rate (MRR), surface roughness (SR) and tool wear rate (TWR) based on Taguchi grey relational array. Analysis of variance (ANOVA) was carried out to find the significance of process parameters on grey relational grade. Gas pressure at level 2 (2 bar), current at level 1 (13 A), pulse on time at level 3 (200 μs) and gap voltage at level 3 (60 V) were found to be the optimum parameters for this process.

Keywords Dry electric discharge machining (Dry EDM) · Material removal rate (MRR) · Surface roughness (SR) · Tool wear rate (TWR) · Grey relational analysis · Analysis of variance

1 Introduction

Electric discharge machining (EDM) is a non-traditional machining process in which material removal takes place through the process of controlled spark generation. This process generates a perfect replica of the tool shape on the workpiece by using controlled spark energy. By using thermal energy, workpiece is melted and vaporized, and the amount of removed material can be effectively controlled to produce complex and precise machine components. The electrical discharges generate impulsive pressure in combination with dielectric explosion to remove the melted material. By

A. S. Bhandare (✉) · U. A. Dabade
Department of Mechanical Engineering, Walchand College of Engineering Sangli, Sangli, India
e-mail: bhandareamar@gmail.com

© Springer Nature Singapore Pte Ltd. 2020
H. Vasudevan et al. (eds.), *Proceedings of International Conference on Intelligent Manufacturing and Automation*, Lecture Notes in Mechanical Engineering,
https://doi.org/10.1007/978-981-15-4485-9_2

considering process advantages, it is also necessary to consider its impact on environmental issues which limits the versatility of process [1]. Different oil-based dielectric fluids used in EDM reflected as a primary source of pollution from the process. New eco-friendly dry EDM techniques get developed by replacing liquid dielectric using gases [1]. Hollow tool electrodes are used, and through which, defined velocity gas is supplied in the machining zone. Due to defined velocity gas in the machining region, it removes debris of machined material and also avoids extreme heating of electrode and workpiece material in the machining region which reduces thermal loading. It provides advantages compared to conventional EDM process are lesser tool wear, minimum discharge gap, lesser residual stresses, lower white layer and heat affected zone [2]. Dry EDM holds the potential for complex and precision-oriented machining applications with its unique identity.

2 Literature Review

Chakravorty et al. [3] tried with four simpler methods for multi-response optimization compared to GRA method for determination of optimal process conditions in EDM processes. They found that WSN and utility theory method provided effective optimization compared to GRA and other multi-response optimization techniques for EDM processes. Vikasa et al. [4] studied the influence of the input process parameters during EDM machining like pulse on time, off time, current and voltage over the surface roughness for an EN41 material by using Grey-Taguchi method, and they found that current provided most significance and after that voltage for surface roughness value. Mishra and Routara [5] compared optimization of the EDM process using Taguchi methodology and grey relational approach on EN-24 alloy steel with pulse on and off time dielectric pressure and current for material removal rate (MRR) coupled with tool wear rate (TWR) as response variables. They found for MRR pulse on time and input current provided most significance and after that pulse off time and for TWR pulse on time and current provided most significance and after that Pulse off time. For GRA approach, they found optimal parameters setting with pulse on time level 3, pulse off time level 3, current level 2 and dielectric pressure level 1. Purohit et al. [6] executed optimization of EDM machining of tool steel M2 with process parameters as tool rotation speed, voltage and spark time for metal removal rate (MRR), electrode wear ratio (EWR) and over cut (OC) as a response variables using grey relational approach with a L_9 orthogonal array. For GRA approach, they found electrode rotation speed was most significant parameter succeeded by voltage and the spark time. Khundrakpam et al. [7] investigated on near dry electrical discharge machining with pulse on time, off time, discharge current, gap voltage and tool rotating speed as process parameters for surface roughness as output variable with air plus deionized water as a dielectric mixture on EN-8 material under Grey-Taguchi approach. Under GRA approach, they observed that discharge current and pulse on time were the significant parameters in sequence. Priyadarshini and Pal [8] demonstrated parametric optimization of EDM for machining of Ti–6Al–4V alloy with copper as tool using L_{25} orthogonal array. After GRA, they found optimal parameter settings were current 10 A, pulse on time 10 μs, duty factor 9 and gap voltage 8 V.

Here, an attempt is made with an objective to optimize the process variables for dry EDM machining of Inconel-718 for the multiple responses. In order to obtain optimum MRR, TWR, and SR work was attempted to determine the suitable testing parameters, with the application of statistical-based ANOVA coupled with GRA. Experiments were conducted to verify the combinations of optimal test parameters.

3 Steps for Grey Relational Analysis

In following section, procedure of grey relational analysis with corresponding equations to process the data is discussed.

Step 1: Normalization

Smaller-the-better

$$x_i^*(k) = \frac{\max x_i^{(o)}(k) - x_i^{(o)}(k)}{\max x_i^{(o)}(k) - \min x_i^{(o)}(k)} \tag{1}$$

Larger the better

$$x_i^*(k) = \frac{x_i^*(k) - \min x_i^0\ (k)}{\max x_i^0(k) - \min x_i^0\ (k)} \tag{2}$$

Step 2: Deviation sequence

$$\Delta 0_i(k) = \left| x_0^*(k) - x_i^*(k) \right| \tag{3}$$

Step 3: Grey Relational Coefficient

$$\gamma(x_0(k), x_i(k)) = \frac{\Delta_{\min} + \zeta\,\Delta_{\max}}{\Delta_{0i}(k) + \zeta\,\Delta_{\max}} \tag{4}$$

ζ is grey relational constant, here it is taken as 0.5.

Step 4: Grey Relational Grade

$$\gamma(x_0, x_i) = \frac{1}{m}\sum_{i=1}^{m}\gamma(x_0(k), x_i(k)) \quad (5)$$

Step 5: Determination of Optimal parameters
Here, grey relational grades are grouped by the factor level for each column in the orthogonal array, and its average is taken to judge optimal parameters
Step 6: Prediction of grey relational grade under optimum parameters

$$\hat{\gamma} = \gamma_m + \sum_{i=1}^{0}(\gamma^{-} - \gamma_m) \quad (6)$$

4 Experimental Set-up and Methodology

As it is dry EDM machining, a new attachment was developed on the existing CNC EDM machine for machining of Inconel 718 workpiece material with copper as electrode material and compressed air as dielectric medium.

Table 1 shows details of machining parameters selected for investigation, and Table 2 shows L_{27} orthogonal array with machining parameters and response variables. Experimental set-up is shown in Fig. 1.

Table 1 Levels of process parameters

Parameter	Levels		
	A	B	C
Gas pressure, GP (Bar)	1	2	3
Current (I) A	13	16	19
Pulse on time (POT) μs	100	150	200
Gap voltage (V)	40	50	60

Table 2 Taguchi L_{27} array with experimental parameters and evaluated response variables

Gas pressure (Bar)	Current (I) A	Pulse on time (POT) μs	Gap voltage (V)	MRR (mm^3/min)	TWR (mm^3/min)	R_a (μm)
1	13	100	40	0.0625	0.0019	1.2430
1	13	150	50	0.0726	0.0003	3.6160
1	13	200	60	0.0851	0.0006	2.0860
1	16	100	50	0.1231	0.0001	3.1760
1	16	150	60	0.1138	0.0017	3.2800
1	16	200	40	0.1347	0.0036	3.1860
1	19	100	60	0.1345	0.0135	4.9880
1	19	150	40	0.1429	0.0198	5.1040
1	19	200	50	0.1425	0.0019	3.8020
2	13	100	40	0.2103	0.0060	5.1440
2	13	150	50	0.1997	0.0094	3.7140
2	13	200	60	0.2001	0.0005	1.3560
2	16	100	50	0.2643	0.0002	4.0960
2	16	150	60	0.2851	0.0002	6.2250
2	16	200	40	0.2991	0.0043	2.2400
2	19	100	60	0.2841	0.0038	4.0100
2	19	150	40	0.3115	0.0037	2.7620
2	19	200	50	0.3721	0.0052	1.3300
3	13	100	40	0.2273	0.0003	3.4280
3	13	150	50	0.2136	0.0008	3.9420
3	13	200	60	0.3460	0.0011	2.2020
3	16	100	50	0.2143	0.0028	3.0700
3	16	150	60	0.2620	0.0002	3.4100
3	16	200	40	0.2752	0.0103	3.3600
3	19	100	60	0.2664	0.0031	1.5840
3	19	150	40	0.2756	0.0304	3.2560
3	19	200	50	0.3162	0.0140	1.8440

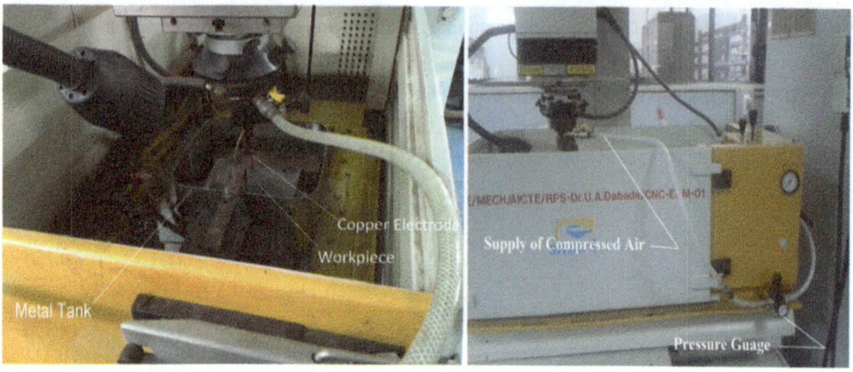

Fig. 1 Experimental set-up

5 Results and Discussion

The measured values of material removal rate (MRR), tool wear rate (TWR) and surface roughness are as shown in Tables 2, 3 and 4 show all the calculations derived from equations of grey relational analysis.

From the response and ANOVA Table 6, it is visible that pulse on time is most significant parameter for all response variables. It is pursued by gas pressure, gap voltage and current. All machining parameters have significant effect on GRG. The optimal factor setting is G2C1P3V3, i.e. gas pressure at level 2 (2 bar) current at level 1 (13 A), pulse on time at level 3 (200 μs), gap voltage at level 3 (60 V). The sequence of significance for all parameters is shown in Table 5.

Predicted (0.8054) and experimental value (0.8767) of GRG show good agreement as observed in Table 7. The improvement of 0.0742, i.e. 7.42% in grey relational grade is found from initial parameter combination, i.e. P3I3T1G3 to optimal parameter combination, i.e. P2I1T3G3.

6 Conclusion

- Multi-objective optimization of dry EDM process of Inconel 718 using grey relational analysis has been presented in this paper. Gas pressure, discharge current, pulse on time and gap voltage are the process parameters considered for investigation. Material removal rate (MRR), surface roughness (SR) and tool wear rate (TWR) are the response variables.

Table 3 Normalized data and deviation sequences

Normalized data			Deviation sequence		
MRR (mm^3/min)	TWR (mm^3/min)	R_a (μm)	MRR (mm^3/min)	TWR (mm^3/min)	R_a (μm)
0.0000	0.9416	1.0000	1.0000	0.0584	0.0000
0.0327	0.9948	0.5237	0.9673	0.0052	0.4763
0.0730	0.9827	0.8308	0.9270	0.0173	0.1692
0.1958	1.0000	0.6120	0.8042	0.0000	0.3880
0.1658	0.9462	0.5911	0.8342	0.0538	0.4089
0.2333	0.8850	0.6100	0.7667	0.1150	0.3900
0.2326	0.5567	0.2483	0.7674	0.4433	0.7517
0.2596	0.3513	0.2250	0.7404	0.6487	0.7750
0.2584	0.9391	0.4864	0.7416	0.0609	0.5136
0.4773	0.8069	0.2170	0.5227	0.1931	0.7830
0.4431	0.6946	0.5040	0.5569	0.3054	0.4960
0.4443	0.9884	0.9773	0.5557	0.0116	0.0227
0.6517	0.9964	0.4273	0.3483	0.0036	0.5727
0.7189	0.9964	0.0000	0.2811	0.0036	1.0000
0.7642	0.8603	0.7999	0.2358	0.1397	0.2001
0.7158	0.8783	0.4446	0.2842	0.1217	0.5554
0.8043	0.8825	0.6951	0.1957	0.1175	0.3049
1.0000	0.8324	0.9825	0.0000	0.1676	0.0175
0.5324	0.9918	0.5614	0.4676	0.0082	0.4386
0.4882	0.9764	0.4582	0.5118	0.0236	0.5418
0.9158	0.9664	0.8075	0.0842	0.0336	0.1925
0.4905	0.9123	0.6333	0.5095	0.0877	0.3667
0.6444	0.9963	0.5650	0.3556	0.0037	0.4350
0.6869	0.6638	0.5751	0.3131	0.3362	0.4249
0.6586	0.8994	0.9316	0.3414	0.1006	0.0684
0.6882	0.0005	0.5959	0.3118	0.9995	0.4041
0.8193	0.5400	0.8794	0.1807	0.4600	0.1206

Table 4 Grey relational grade

Grey relational coefficients				Rank
MRR (mm^3/min)	TWR (mm^3/min)	R_a (μm)	GRG	
0.3333	0.8953	1.0000	0.7429	5
0.3408	0.9897	0.5121	0.6142	18
0.3504	0.9665	0.7472	0.6880	9
0.3834	1.0000	0.5631	0.6488	14
0.3748	0.9028	0.5501	0.6092	19
0.3947	0.8129	0.5618	0.5898	21
0.3945	0.5299	0.3995	0.4413	26
0.4031	0.4351	0.3922	0.4101	27
0.4027	0.8914	0.4933	0.5958	20
0.4889	0.7213	0.3897	0.5333	23
0.4731	0.6207	0.5020	0.5319	24
0.4736	0.9773	0.9566	0.8025	3
0.5894	0.9928	0.4661	0.6828	11
0.6401	0.9928	0.3333	0.6554	13
0.6796	0.7816	0.7142	0.7251	6
0.6376	0.8042	0.4738	0.6385	17
0.7187	0.8097	0.6212	0.7165	7
1.0001	0.7489	0.9663	0.9051	1
0.5167	0.9838	0.5327	0.6778	12
0.4942	0.9548	0.4800	0.6430	15
0.8559	0.9370	0.7220	0.8383	2
0.4953	0.8507	0.5769	0.6410	16
0.5844	0.9927	0.5348	0.7039	8
0.6149	0.5978	0.5406	0.5844	22
0.5942	0.8325	0.8796	0.7688	4
0.6159	0.3333	0.5531	0.5008	25
0.7346	0.5207	0.8056	0.6870	10

Table 5 Sequence of significance

Factors	Level 1	Level 2	Level 3	Max–Min	Rank
Gas pressure	0.5934	**0.6879**	0.6717	0.0945	2
Current	**0.6747**	0.6489	0.6293	0.0454	4
Pulse on time	0.6417	0.5983	**0.7129**	0.1146	1
Gap voltage	0.6090	0.6611	**0.6829**	0.0739	3
Mean grey relational grade = 0.6510					

Bold values in the table indicates optimal parameter setting is G2C1P3V3, i.e., Gas pressure at level 2 (2 bar), current at level 1 (13 amp), pulse on time at level 3(200 μs), gap voltage at level 3 (60 V)

Table 6 *F* and *P* values of ANOVA for parameters

Source	DF	Adj SS	Adj MS	F	P
Gas pressure (Bar)	2	0.046003	0.023002	2.21	0.139
Current (I) A	2	0.009309	0.004654	0.45	0.647
Pulse on time (POT) μs	2	0.060206	0.030103	2.89	0.082
Gap voltage (V)	2	0.025960	0.012980	1.25	0.312
Error	18	0.187661	0.010426		
Total	26	0.329139			

Table 7 Confirmation table

	Exp. No. 4	Optimal	
	Initial (P3I3T1G3)	Predication (P2I1T3G3)	Experimental (P2I1T3G3)
MRR	0.2644		0.3119
TWR	0.0031		0.0009
SR	1.5840		1.3459
GRG	0.8025	0.8054	0.8767

- For multi-objective optimization, grey relational analysis is used because of its simplicity to solve complex problems with simple mathematical equations. Gas pressure at level 2 (2 bar), current at level 1 (13 A), pulse on time at level 3 (200 μs) and gap voltage at level 3 (60 V) were found as the optimum parameters for the process.
- Good agreement between predicted (0.8054) and experimental values (0.8767) of grey relational grade was found during the investigation.

References

1. Leao FN, Pashby IR (2004) A review on the use of environmentally-friendly dielectric fluids in electrical discharge machining. J Mater Process Technol 149(1–3):341–346
2. Saha SK, Choudhury SK (2009) Multi-objective optimization of the dry electric discharge machining process, pp 1–4
3. Chakravorty R, Gauri SK, Chakraborty S (2013) A study on the multi-response optimisation of EDM processes. Int J Mach Mach Mater 13(1):91–109
4. Vikasa AKR, Kumarb K (2014) Effect and optimization of various machine process parameters on the MRR, over-cut and surface roughness in EDM for an EN41 material using grey-taguchi approach. Int J Appl Eng. Res 9(26:8963–8966
5. Mishra BP, Routara BC (2017) An experimental investigation and optimisation of performance characteristics in EDM of EN-24 alloy steel using Taguchi Method and Grey Relational Analysis. Mater Today Proc 4(8):7438–7447

6. Purohit R, Rana RS, Dwivedi RK, Banoriya D, Singh SK (2015) Optimization of electric discharge machining of M2 tool steel using grey relational analysis. Mater Today Proc 2(4–5):3378–3387
7. Singh Khundrakpam N, Singh Brar G, Deepak D (2018) Grey-Taguchi optimization of near dry EDM process parameters on the surface roughness. Mater Today Proc 5(2):4445–4451
8. Priyadarshini M, Pal K (2015) Grey-taguchi based optimization of EDM process for titanium alloy. Mater Today Proc 2(4–5):2472–2481

Predictive Modeling of Surface Roughness in the Machining of Inconel 625 Using Artificial Neural Network

Hari Vasudevan and Ramesh R. Rajguru

Abstract Inconel 625 belongs to the category of nickel-based alloy and it has wide applications in aerospace, defense, marine, nuclear, and oil and gas industries, due to its superior properties at elevated temperatures. However, machinability of Inconel 625 is very poor, because of high strength at higher temperature as well as high rate of tool wear and work hardening. In this context, this research work developed a predictive model using the artificial neural network, during the turning operation on nickel-based alloy Inconel 625, with chemical vapor deposition-coated inserts. Cutting speed, feed rate, and depth of cut were selected as input variables and surface roughness was selected as the response variable. Totally, twenty-seven experiments were performed, based on the design of experiments, using L_{27} orthogonal array and the measured result was analyzed using statistical software MATLAB. Mean absolute percentage error between predicted results and experimental values was 4.90% for surface roughness. The overall correlation coefficient between the artificial neural network-predicted result and experimental values was 0.97734, which showed good agreement between experimental values and predicted results and was nearly accurate.

Keywords Artificial neural network · Inconel 625 · Surface roughness · CVD coating · Turning

1 Introduction

Nickel-based alloy has been used widely in aerospace, defense, marine, nuclear, and oil and gas industries, because they constitute properties, such as great strength,

H. Vasudevan
Dwarkadas J. Sanghvi College of Engineering, Vile Parle, Mumbai, India
e-mail: harivasudevan@iitb.ac.in

R. R. Rajguru (✉)
Department of Mechanical Engineering, Dwarkadas J. Sanghvi College of Engineering, Vile Parle, Mumbai, India
e-mail: ramesh.rajguru@djsce.ac.in

© Springer Nature Singapore Pte Ltd. 2020
H. Vasudevan et al. (eds.), *Proceedings of International Conference on Intelligent Manufacturing and Automation*, Lecture Notes in Mechanical Engineering,
https://doi.org/10.1007/978-981-15-4485-9_3

sturdy corrosion resistance, and exceptional thermal stability [1–4]. However, they are among the most difficult to cut materials, because of high strength at elevated temperature and low thermal conductivity. Various researchers have carried out machining operation on nickel-based alloy. Waghmode and Dabade [5] investigated the effect of process parameters on machinability of Inconel 625 and optimized them by the Taguchi method. The study pointed out that the feed rate and depth of cut have significant effect on surface roughness and material removal rate, respectively. Further, ANOVA result showed maximum contribution of 55.05% on cutting force. Venkatesan et al. [6] analyzed the effect of PVD AlTiN-coated carbide inserts on machinability aspects, such as cutting forces and surface roughness in the turning of Inconel 625. Their study showed that the surface roughness is most significantly affected by the feed rate, followed by the cutting speed. Dabhade et al. [7] investigated on the experimental effect of sialon ceramic insert during dry turning of Inconel 625 on surface roughness and material removal rate (MRR). The results revealed that the feed rate and depth of cut show significant effect on surface roughness and MRR, respectively. Further, the process parameters were optimized using the response surface method. A study by Vasudevan et al. [8] investigated the optimization of process parameters using gray relational analysis for surface roughness in dry turning operation of Inconel 625. They found the optimal process parameters, namely the feed rate of 0.35 mm/rev, cutting speed of 90 m/min, and the depth of cut 0.2 mm. None of the studies reported so far in the literature has been about predictive modeling of surface roughness of Inconel 625 during the turning operation. Therefore, this study was carried out to develop a predictive model of surface roughness in the machining of Inconel 625, using artificial neural network.

2 Experimental Details

The material used during the machining was a round bar of Inconel 625 (Fig. 1) of ten samples. The length of the round bars was 60 mm and the diameter was 30 mm. Turning operation was performed on a CNC machine using CVD-coated carbide inserts. The input parameters chosen for the experiments were cutting speed (40, 60, and 80 m/min), feed rate (0.04, 0.08, and 0.12 mm/rev), and depth of cut (0.25, 0.35, and 0.45 mm). Totally, twenty-seven experiments were performed according to the Taguchi method, using L_{27} orthogonal array. Each experiment was performed for 10 mm length with four experiments performed on each sample. Thus, eight samples were machined. Surface roughness was measured using Taylor Hobson Talysurf 4 instrument (Fig. 2). Table 1 presents the experimental design, using orthogonal array with factors and responses.

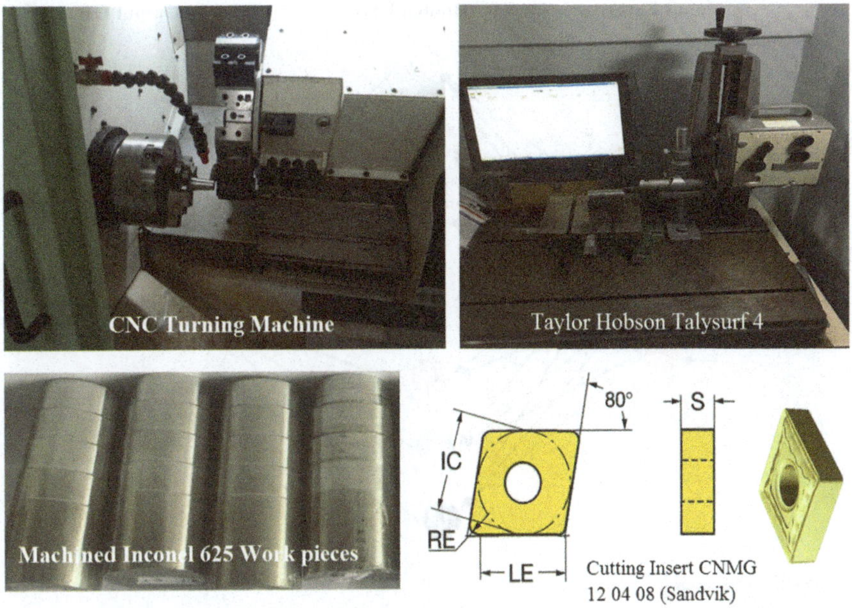

Fig. 1 Experimental setup

3 Artificial Neural Network

Artificial neural network (ANN)-based method is one of the novel approaches for predictive modeling of machining operations [9, 10]. ANN-based predictive model developing process includes data acquisition and preparation, system analysis, network structures, network training, and testing and validation [11, 12]. Feed-forward neural networks are characterized by simple structure and easy numerical explanation [13]. This network allows the signals from the input layer to the output layer in one direction only. Multilayer feed-forward artificial neural network is the first choice for modeling of nonlinear behavior of industrial systems [12]. In this study, three layers such as input, hidden, and output were used as shown in Fig. 2.

4 Analysis Using Artificial Neural Network

The ANN model trained and developed using eight steps and normalized values of trained and predicted data are presented in Table 2 with mean absolute percentage error. It was observed that the network structure of type 3-7-1 gave best results for the prediction of surface roughness. The mean absolute percentage error (MAPE) in the prediction of response performance by predictive model was 4.9082 for training the model. The proposed predictive ANN model could successfully predict the surface

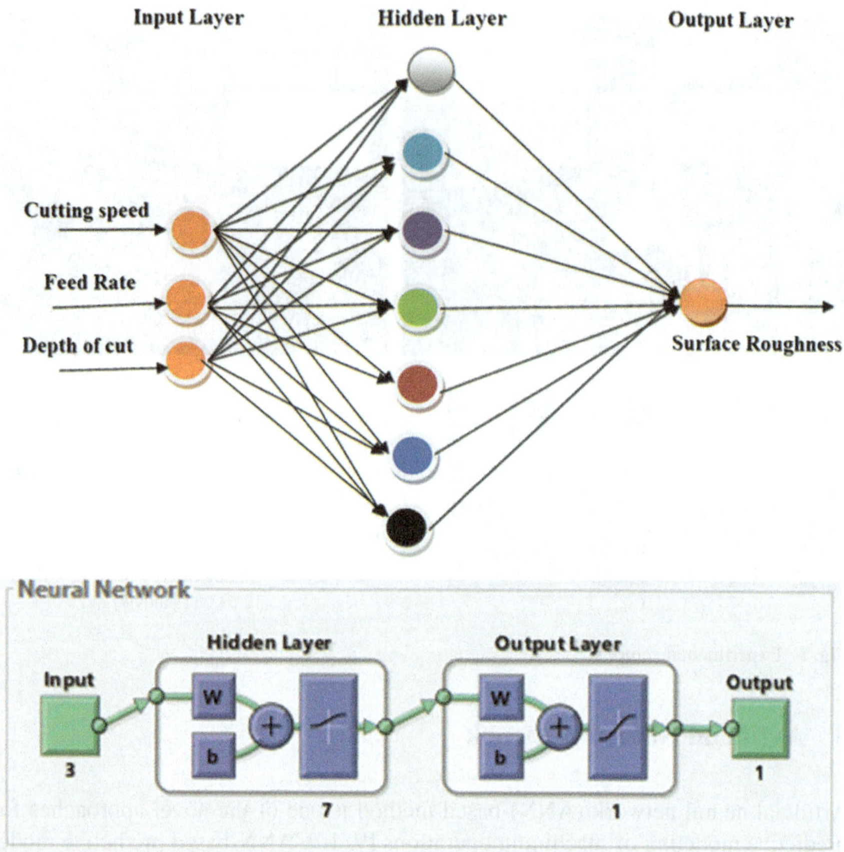

Fig. 2 ANN 3-7-1 network

roughness within the experimental limits. The overall correlation coefficient between the ANN-predicted result and experimental values was 0.97734, as depicted in Fig. 3. Further, giving very good predictions during the turning operation leads to saving money and the time taken for conducting the experiments.

5 Conclusion

Feed-forward back propagation with one hidden layer, having seven neurons with transfer function (trainlm), was used as the network architecture for developing the ANN model and for comparison between ANN models predicted results and the experimental values. The average relative error between ANN-predicted results and experimental values was 4.9082% for surface roughness. The overall correlation

Table 1 Experimental design using orthogonal array with factors and responses

Expt. No.	Cutting speed		Feed rate		Depth of cut		R_a (μm)	R_a ANN predicted
	Actual	Coded	Actual	Coded	Actual	Coded		
1	40	1	0.04	1	0.25	1	0.805	0.847
2	40	1	0.04	1	0.35	2	0.653	0.697
3	40	1	0.04	1	0.45	3	0.362	0.402
4	40	1	0.08	2	0.25	1	0.892	1.003
5	40	1	0.08	2	0.35	2	0.501	0.516
6	40	1	0.08	2	0.45	3	0.693	0.616
7	40	1	0.12	3	0.25	1	0.982	1.022
8	40	1	0.12	3	0.35	2	0.695	0.739
9	40	1	0.12	3	0.45	3	0.853	0.895
10	60	2	0.04	1	0.25	1	0.362	0.412
11	60	2	0.04	1	0.35	2	0.759	0.805
12	60	2	0.04	1	0.45	3	0.486	0.533
13	60	2	0.08	2	0.25	1	0.604	0.666
14	60	2	0.08	2	0.35	2	1.396	1.435
15	60	2	0.08	2	0.45	3	0.734	0.777
16	60	2	0.12	3	0.25	1	0.618	0.652
17	60	2	0.12	3	0.35	2	0.354	0.403
18	60	2	0.12	3	0.45	3	0.814	0.858
19	80	3	0.04	1	0.25	1	0.47	0.488
20	80	3	0.04	1	0.35	2	0.494	0.590
21	80	3	0.04	1	0.45	3	1.743	1.780
22	80	3	0.08	2	0.25	1	0.563	0.604
23	80	3	0.08	2	0.35	2	0.853	0.899
24	80	3	0.08	2	0.45	3	1.239	1.280
25	80	3	0.12	3	0.25	1	0.398	0.463
26	80	3	0.12	3	0.35	2	0.477	0.588
27	80	3	0.12	3	0.45	3	0.567	0.614

coefficient between ANN predicted result and experimental values was 0.97734, which showed that the result between experimental values and predicted results was accurate. ANN-predicted model could be used further for improving the performance by increasing the number of training data. In small- and medium-scale precision machining industry, operator could use an ANN predictive model to set up the optimum levels of process parameters for turning of Inconel 625. This would help in reducing the machining process time and accordingly, increasing the production rate.

Table 2 Normalization of network input and ANN-predicted values along with their percentage error w.r.t. experimental values

S. No.	Normalization of process parameters			ANN predicted and actual value of surface roughness		Error %
	Cutting speed	Feed rate	Depth of cut	R_a	R_a	
1	0.1	0.1	0.1	0.339	0.338	4.468
2	0.1	0.1	0.5	0.252	0.252	2.052
3	0.1	0.1	0.9	0.084	0.080	2.968
4	0.1	0.5	0.1	0.389	0.429	10.333
5	0.1	0.5	0.5	0.164	0.146	10.701
6	0.1	0.5	0.9	0.275	0.204	10.768
7	0.1	0.9	0.1	0.441	0.440	0.657
8	0.1	0.9	0.5	0.276	0.276	25.365
9	0.1	0.9	0.9	0.367	0.366	18.421
10	0.5	0.1	0.1	0.084	0.087	1.219
11	0.5	0.1	0.5	0.313	0.314	0.534
12	0.5	0.1	0.9	0.156	0.157	25.555
13	0.5	0.5	0.1	0.223	0.234	2.554
14	0.5	0.5	0.5	0.680	0.679	4.602
15	0.5	0.5	0.9	0.298	0.298	0.028
16	0.5	0.9	0.1	0.232	0.226	0.123
17	0.5	0.9	0.5	0.080	0.081	0.174
18	0.5	0.9	0.9	0.344	0.345	0.367
19	0.9	0.1	0.1	0.146	0.131	0.33
20	0.9	0.1	0.5	0.160	0.190	0.102
21	0.9	0.1	0.9	0.880	0.879	0.223
22	0.9	0.5	0.1	0.200	0.197	0.437
23	0.9	0.5	0.5	0.367	0.369	10.07
24	0.9	0.5	0.9	0.589	0.589	0.273
25	0.9	0.9	0.1	0.105	0.116	0.014
26	0.9	0.9	0.5	0.150	0.189	0.116
27	0.9	0.9	0.9	0.202	0.203	0.043
	MAPE					4.902

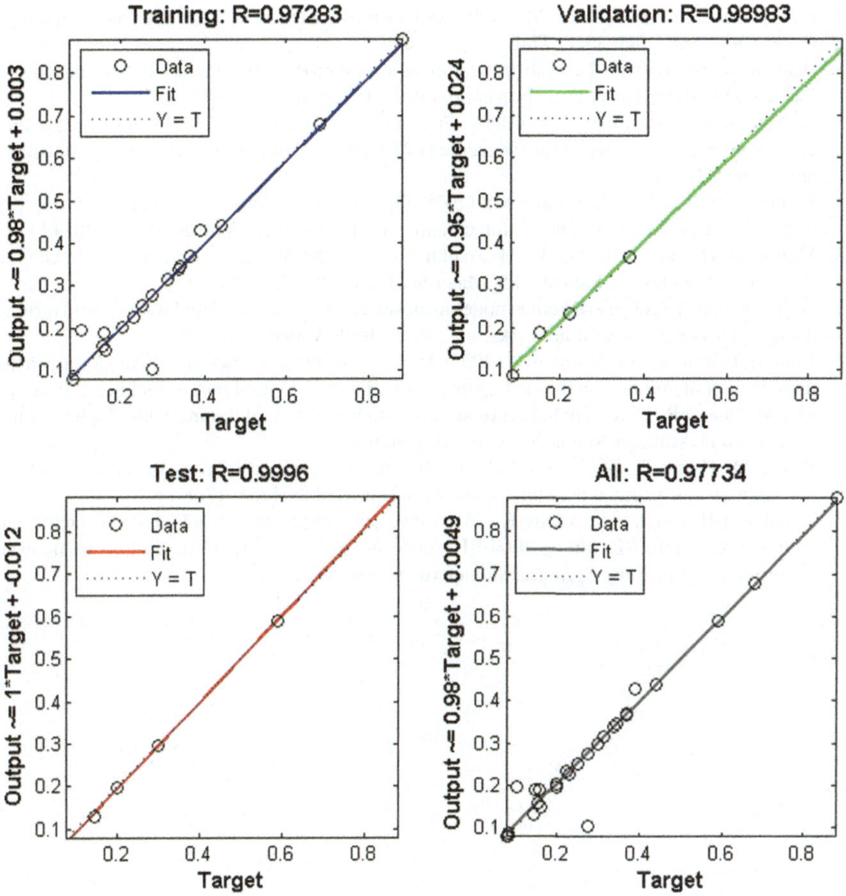

Fig. 3 Correlation value for the training, validation, testing, and overall correlation coefficient

Acknowledgements The authors would like to thank the University of Mumbai for research support given under the minor research grant (Ref. APD/ICD/2018-19/593, dated 16th March 2019, Sr. No. 195).

References

1. Ezugwu EO, Wang ZM, Machado AR (1999) The machinability of nickel-based alloys: a review. J Mater Process Technol 86:1–16
2. Smith GD, Tillack DJ, Patel SJ (2001) Alloy 625—impressive past/significant presence/awesome future. Super Alloys 718:633–706
3. Arunachalam R, Mannan MA (2007) Machinability of nickel-based high temperature alloys. Mach Sci Technol Int J 4(1):127–168

4. Hanasaki S, Fajiwara J, Touse M (1990) Tool wear of coated tools when machining and high nickel alloy. Ann CIRP 39(1):77–80
5. Waghmode SP, Dabade UA (2019) Optimization of process parameters during turning of Inconel 625. Mater Today Proc. https://doi.org/10.1016/j.matpr.2019.08.138
6. Venkatesan K, Ramanujam R, Saxena V, Chawdhury N, Choudhary V (2014) Influence of cutting parameters on dry machining of Inconel 625 alloy with coated carbide insert—a statistical approach. ARPN J
7. Dabhade DS, Kathar RA, Kadam MS (2018) Optimization of process parameters during dry turning of Inconel 625 by using sialon ceramic insert. Int J Eng Sci Invent 07(11):38–44
8. Vasudevan H, Rajguru R, Shaikh M, Shaikh A (2019) Optimization of process parameters in the turning operation of Inconel 625. Mater Sci Forum 969:756–761
9. Ozel T, Karpat Y (2005) Predictive modeling of surface roughness and tool wear in hard turning using regression and neural networks. Int J Mach Tools Manuf 45:467–479
10. Ktem HO, Erzurumlu T, Kurtaran H (2005) Application of response surface methodology in the optimization of cutting conditions for surface roughness. J Mater Process Technol 170:11–16
11. Mia M, Dhar NR (2016) Prediction of surface roughness in hard turning under high pressure coolant using Artificial Neural Network. Measurement
12. Zain AM, Haron H, Sharif S (2010) Prediction of surface roughness in the end milling machining using artificial neural network. Exp Syst Appl 37:1755–1768
13. Vasudevan H, Rajguru R, Yadav R (2018) Predictive modelling of delamination factor and cutting forces in the Machining of GFRP composite material using ANN. In: Lecture notes in mechanical engineering. Springer, Singapore, pp 301–313

Measurement of Spur Gear Parameters Using Machine Vision

Ketaki Joshi and Bhushan Patil

Abstract This paper presents a machine vision approach as a non-contact, automated method of measuring all critical dimensions of a spur gear, using image processing algorithms. Spur gear images are captured using image acquisition devices under backlight illumination and acquired in image processing software (MATLAB). Critical spur gear dimensions, such as addendum circle radius, dedendum circle radius, pitch circle radius, module, number of teeth, pressure angle, tooth thickness, and circular pitch, were extracted using various image processing and feature extraction algorithms. The values obtained using the machine vision approach are found to be in good agreement with those obtained using traditional metrology instruments and standard spur gear formulae. The approach can be adopted for developing an automated inspection system using image data.

Keywords Machine vision · Measurement · Spur gear inspection

1 Introduction

Gears are one of the effective modes of power transmission having applications in wide spectrum of industries. Their performance in terms of quality of operations, level of dynamic forces, noise, and vibrations can be significantly affected due to manufacturing defects. Therefore, accurate measurement of all critical quality parameters is very crucial in their inspection and quality control [1]. Machine vision offers an automated solution for 100% inspection of spur gears by facilitating the measurement of all individual errors using image processing algorithms. On account of improved technology affordances, enhanced image acquisition techniques, accuracy in processing hardware and software; the technique provides an

K. Joshi (✉) · B. Patil
Fr. Conceicao Rodrigues College of Engineering, Affiliated to University of Mumbai, Bandra, Mumbai, Maharashtra, India
e-mail: ketaki.joshi@fragnel.edu.in

B. Patil
e-mail: bhushan.patil@fragnel.edu.in

© Springer Nature Singapore Pte Ltd. 2020

H. Vasudevan et al. (eds.), *Proceedings of International Conference on Intelligent Manufacturing and Automation*, Lecture Notes in Mechanical Engineering,
https://doi.org/10.1007/978-981-15-4485-9_4

opportunity for its applicability in industrial metrology. Machine vision approach is used by researchers for different purposes including gear measurement, gear sorting, analysis of gear tooth profile, and detection of manufacturing defects. A machine vision-based inspection system typically uses an image acquisition setup, involving an inspection table illuminated with backlight, image acquisition device (camera) mounted on camera holder for capturing spur gear images, frame grabber video card for transferring images to hardware typically a computer and image acquisition and processing software [2].

Various approaches are used by researchers to process gear image data and extract its critical dimensions of interest. Firstly, color image acquired in the software is preprocessed using a sequence of steps typically including conversion from red-green-blue (RGB) image into grayscale image, binarization using threshold value, and image segmentation for extraction of gear tooth profile [2–4]. Then outer edges are detected by comparing the color/grayscale value of a pixel with four or eight neighboring pixels, labeling the pixels as foreground or background pixel based on the comparison, edge thinning by removing two consecutive pixels having same pixel values and morphological operations to fill the image [2, 4, 5]. Gear center can be located directly from center of gravity of the gear tooth profile image [4]. Image erosion method is used for extracting outer and root diameters from the gear tooth profile image. Different approaches used for measurement of diameters include detection of outermost and innermost pixels followed by measurement of diameters using distance formula [2], measurement after transforming polar coordinates into rectangular coordinates for improving calculation ease and accuracy [4], location of outermost circle based on convex polygon encompassing the image, and circle equation for that polygon as an alternative method [6]. Various approaches are explored by researchers for calculating number of teeth. The most logical method is the extraction of connected components due to its simplicity and efficiency [3, 5], whereas calculation of number of teeth based on peak distances from image center [7] and calculation based on distance between consecutive gear tooth profile pixels [2] are some other approaches. Other gear parameters such as module, base circle, pitch circle diameter, pressure angle are calculated using standard spur gear equations using number of teeth, addendum, and root circle diameters extracted earlier [2]. Points on gear tooth profile intersecting pitch circle are used to compute tooth thickness and circular pitch [6].

2 Problem Definition

The paper aims at developing a machine vision system for measuring critical dimensions of standard spur gears using image data. The images of spur gear need to be captured using image acquisition devices and processed further in image processing software to extract parameters of interest. The values obtained using machine vision are validated against those obtained using traditional metrology instruments such as gear tooth caliper, micrometer, and standard spur gear formulae.

3 Methodology

Methodology adopted in the current paper extracting important spur gear dimensions using various image processing and feature extraction algorithms is elaborated below.

RGB image of spur gear under inspection is captured by 12 MP camera and acquired in image processing software (MATLAB) in computer. Various image processing algorithms are used for extracting major dimensions of spur gear under inspection [8].

Image preprocessing steps include conversion from RGB to grayscale based on a threshold value, conversion of grayscale image to B/W image using a threshold value, cleaning the image by its morphological opening and filling the holes to extract the gear teeth contour using MATLAB functions [8]. Rgb2gray ( ) function converts RGB images to grayscale image by eliminating hue and saturation in the image while retaining its luminance. Imcomplement ( ) function complements grayscale image by subtracting each pixel value from maximum pixel value supported by the data class and using this difference in the output image to make dark areas lighter and vice versa. Graythresh ( ) and Im2bw ( ) are used for converting grayscale image into binary image. For thresholding, global threshold value is calculated using Otsu's method [9] based on the selection of threshold that minimizes intraclass variance of thresholded black and white pixels. Then, grayscale image is converted into B/W image by changing pixel value of all pixels having luminance greater than threshold to 1 (White) and that for others to 0 (Black). Bwareaopen ( ) is used to remove all connected objects in the image hawing pixels fewer than the specified target value for reducing noise in the B/W image. Then imfill ( ) is used to fill holes in the image identified by set of background pixels that are unreachable by filling background from edges in the image. The sequence of operation mentioned above helps to extract a B/W image of gear tooth profile without a hole for gear shaft and keyway.

Furthermore, regionprops ( ) calculates area of the image based on number of pixels, X–Y co-ordinates of centroid and bounding box. Then convexhull ( ) finds the smallest convex polygon containing the region and returns an array of its vertices. Coordinates of these vertices (X_i, Y_i) are used to calculate addendum circle radius using distance formula from the centroid (X_c, Y_c) using the following Eq. (1) and averaging all the values.

$$\text{addrad}_{(i)} = \sqrt{(X_i - X_c)^2 + (Y_i - Y_c)^2} \tag{1}$$

The value is then converted from pixel scale to millimeter scale using conversion ratio obtained through image calibration.

For measuring number of teeth, image erosion method is used. Poly2mask ( ) function converts region of interest (ROI) polygon to region mask by setting pixels inside the polygon to 1 and outside the polygon to 0. Then imerode ( ) erodes the image using a flat disk-shaped morphological structuring element, which is equal to local-minimum operator. Regionprops ( ) finds properties of eroded image. Pixel values of eroded region are set to 0 for subtracting this eroded image from the original image. Number of connected components is counted using bwlabel ( ) that essentially provides the number of teeth in the gear image.

Then module is calculated from addendum circle radius and number of teeth using formula given in Eq. (2) and rounding off the value to the nearest standard module.

$$\text{Module} = \frac{2 * \text{Addendum Radius}}{(\text{Number of Teeth} + 2)} \tag{2}$$

For measuring dedendum circle radius, erosion of spur gear image and calculation of number of teeth as mentioned above are continued in a loop till the number of teeth calculated now becomes less than the number of teeth calculated earlier. This helps in obtaining the position of bottomland, i.e., dedendum circle in the gear image. Then the radius is calculated using convex polygon properties similar to the procedure mentioned for addendum circle radius. Pitch circle radius is calculated using formula given in Eq. (3).

$$\text{Pitch Circle Radius} = \frac{\text{Dedendum Circle Radius} * \text{Number of Teeth}}{(\text{Number of Teeth} + 2)} \tag{3}$$

For measuring pressure angle, base circle is positioned by adding fillet radius to dedendum circle radius for calculation of base circle radius. Imfindcircle ( ) is used to get the fillet radius that finds circles having a specified range of radii using circular Hough transform.

Using the value of average fillet radius obtained by circular Hough transform, base circle is positioned in the original image space. Then the pressure angle is obtained by marking a point on the pitch circle, plotting a tangent to the base circle passing through this pitch point, and calculating pressure angle using the formula given in Eq. (4).

$$\alpha = \tan^{-1}\frac{Y_{\text{vct}}}{X_{\text{vct}}} - \tan^{-1}\frac{\text{Tngt}}{X_{\text{vct}}} \tag{4}$$

In order to determine circular pitch and tooth thickness, points on pitch circle are extracted. The convex polygon is obtained enclosing the image that is eroded using pitch circle radius value, and for each labeled element (tooth), its extreme points are obtained using distance formula.

In order to determine circular pitch and tooth thickness, points on pitch circle are extracted. The convex polygon is obtained enclosing the image that is eroded using pitch circle radius value, and for each labeled element (tooth), its extreme points are obtained using distance formula.

4 Results and Discussion

Spur gear image is captured using 12 MP camera and transferred to MATLAB for extracting its critical dimensions. The image is processed using image processing algorithms mentioned above. Figure 1 provides sample images of gear features extracted using machine vision algorithms. Figure 1 (a) indicates spur gear contour extracted by image preprocessing, (b) shows spur gear teeth extracted by erosion, and (c) shows addendum circle, pitch circle, base circle, tangent to base circle, and X-vectors for pressure angle measurement.

The values obtained using machine vision approach compared with actual values measured using traditional metrology instruments such as micrometer, gear tooth caliper and related standard spur gear formulae. Accuracy of the machine vision inspection system is evaluated in terms of percent error in the measurement of critical quality parameters using the formula given in Eq. (5).

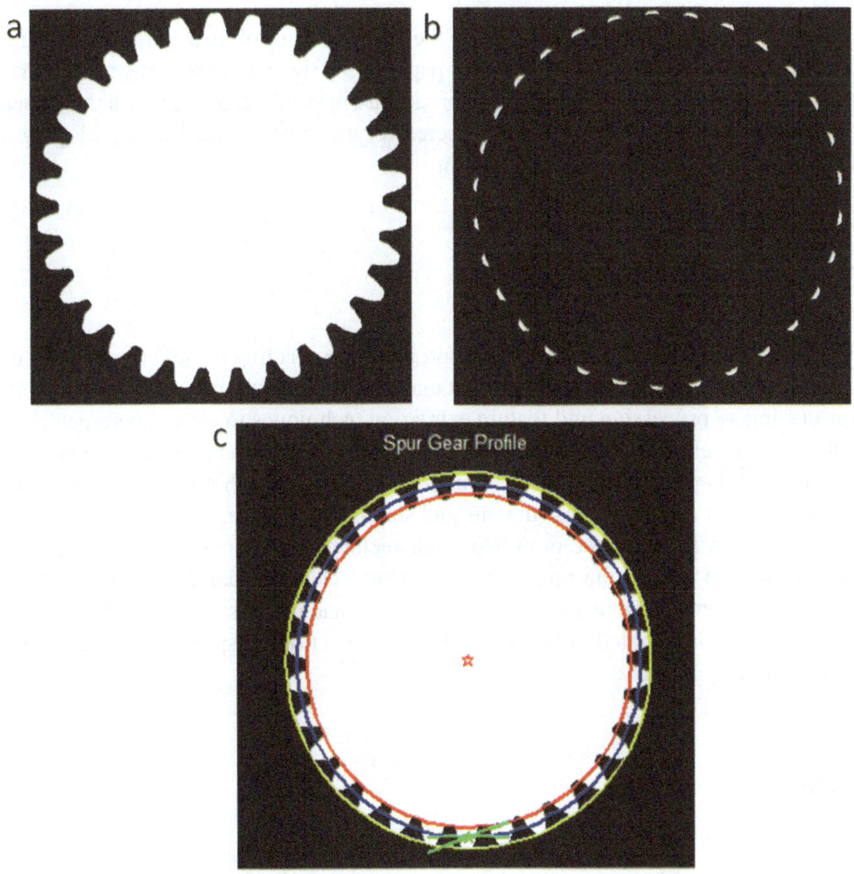

Fig. 1 **a** Spur gear profile, **b** number of teeth extracted, **c** pressure angle

Table 1 Comparison between gear parameter values obtained using machine vision and traditional approach

Parameter	Machine vision value	Actual value	Percent error
Dedendum circle radius (mm)	22.29	22.14	0.67
Pitch circle radius (mm)	23.89	23.72	0.72
Base circle radius (mm)	22.45	22.29	0.72
Tooth thickness (mm)	2.22	2.36	5.93
Circular pitch (mm)	4.69	4.71	0.42
Number of teeth	30	30	0
Module (mm)	1.5	1.5	0
Pressure angle (°)	20	20	0

$$\%\ \text{Error} = \frac{\text{Machine Vision Value} - \text{Actual Value}}{\text{Actual Value}} \times 100 \tag{5}$$

Table 1 indicates maximum error in the measurement of gear parameters using machine vision approach adopted in this paper as 5.93%. The values of gear parameters calculated using machine vision approach are in good agreement with the values obtained using traditional approach. Therefore, it can be adopted for developing a vision-based spur gear inspection system.

5 Conclusion

The paper presents a machine vision approach for extracting critical dimensions of spur gears using image data. Image of spur gear under inspection was processed using various image processing and feature extraction techniques to extract parameters of interest. The values of critical dimensions obtained using machine vision approach adopted in this paper are found to be in good agreement with those obtained using traditional gear metrology instruments and standard formulae.

There lies a future scope in utilizing this technique for incorporating more gear parameters and calculating manufacturing errors. The approach can be adopted for developing a spur gear inspection system using machine vision. It has a potential of offering an automated solution for 100% inspection of spur gears to the gear manufacturing industry.

References

1. Goch G (2003) Gear metrology. CIRP Ann Manuf Technol 52(2):659–695
2. Gadelmawla ES (2011) Computer vision algorithms for measurement and inspection of spur gears. Measurement 44(9):1669–1678

3. Mavi A, Kaur M (2012) Identify defects in gears using digital image processing. Int J Eng Res Dev 1:49–55
4. Huang Y, Gu J, Wang S, Xiao H, Yuan K (2016) Vision based embedded tiny spur gear inspection and measurement system. In: 2016 IEEE international conference on mechatronics and automation (ICMA). IEEE, pp 1679–1684
5. Wang W, Guan F, Ma S, Li J (2015) Measurement system of gear parameters based on machine vision. Meas Control 48(8):242–248
6. Ge DY, Yao XF, Lian ZT, Wang HZ, Tao T (2017) Applications of computer vision in measuring total cumulative pitch deviation of a gear. Tehničkivjesnik 24(1):71–78
7. Qingmin CXWLL (2004) Research on measuring system for gear parameters based on CCD. Tool Eng 5:022
8. Mathworks Documentation (2019) Available at: https://in.mathworks.com/help/matlab. Accessed 26 Mar 2019
9. Otsu N (1979) A threshold selection method from gray-level histograms. IEEE Trans Syst Man, Cybern 9(1):62–66

Modeling and Optimization of Cutting Temperature in Hard Turning of AISI 52100 Hardened Alloy Steel Using Response Surface Methodology

Sandip Mane and Sanjay Kumar

Abstract This paper analyses the effect of cutting parameters, such as cutting speed, feed and depth of cut on the cutting temperature in dry turning of AISI hardened 52100 alloy steel of 58 HRC using multilayer coated carbide tool insert. The heat generation and temperature at the cutting zone have a significant impact on tool wear, tool life and surface integrity. The cutting temperature is strongly influenced by cutting parameters and increases with their levels. Therefore, it was imperative to optimize the cutting parameters and develop a model for the accurate prediction of cutting temperature. Response surface methodology based on central composite design (CCD) was used to investigate and optimize the cutting parameters on cutting temperature response. The diagnostic tests were carried out to check its validity. The analysis of variance (ANOVA) was performed to analyse the effect of process parameters and their interactions on cutting temperature response. The quadratic regression model in terms of cutting speed, feed and depth of cut for cutting temperature was developed. The predicted values of cutting temperature response are in good agreement with the experimental results.

Keywords Hard turning · Surface integrity · Response surface methodology · Central composite design

S. Mane (✉)
Department of Mechanical Engineering, Dwarkadas J. Sanghvi College of Engineering, Mumbai, Maharashtra, India
e-mail: sandip.mane@djsce.ac.in

S. Kumar
Department of Mechanical Engineering, Thakur College of Engineering and Technology, Mumbai, Maharashtra, India
e-mail: sanjay.kumar@thakureducation.org

© Springer Nature Singapore Pte Ltd. 2020
H. Vasudevan et al. (eds.), *Proceedings of International Conference on Intelligent Manufacturing and Automation*, Lecture Notes in Mechanical Engineering,
https://doi.org/10.1007/978-981-15-4485-9_5

1 Introduction

The heat generation and temperature at the cutting zone in hard machining affect the tool wear, tool life and machined surface integrity and can be improved by reducing the heat generation and temperature at tool–chip and tool–workpiece interface. Aouici et al. [1] developed the parametric relationship between the cutting speed, feed, depth of cut and the response cutting temperature using response surface methodology in dry hard turning of AISI H11 steel using CBN insert. The cutting speed, feed rate and depth of cut are found to be the most significant and influencing parameter on cutting temperature as it increases with an increase in their values. Tönshoff et al. [2] reported that the hard turning has been gaining the importance over conventional grinding process due to its benefits such as lesser set-up time, fewer process steps, process flexibility, greater part geometry and higher material removal rate. Huang et al. [3] stated that the hard turning is used to machine the complex parts, and the cost can be reduced up to 30% of the total manufacturing costs. Suresh et al. [4] and Saboo et al. [5] stated that the cubic boron nitride (CBN) and ceramic tool inserts are widely acceptable as it performs well in the machining of hardened alloy steels, but these tool inserts are very costlier. The development of new coating materials and coating deposition techniques leads the researchers to work in the area of machining of hardened alloy steel using coated carbide tools. Sutter et al. [6] studied the effects of the cutting speed and depth of cut on the temperature profile at the chip during orthogonal machining of 42CrMo4 steel using standard carbide TiCN coated tools. They performed the machining with a gas gun. It was found from their results that the temperature at the chip increases with the increase in both cutting speed and the depth of cut. Ren et al. [7] determined the cutting temperatures during hard turning of high chromium hard facing materials using PCBN tools. They found that the average cutting temperatures ranged from 600 to 700 °C, which increased with higher cutting speed and feed rate. Fnides et al. [8] found that the cutting speed has more significant impact on cutting temperature than the feed and depth of cut in turning of AISI H11 hardened alloy steel of hardness 50 HRC using ceramic tool insert at the dry condition. Lin et al. [9] reported that the cutting speed is the most dominant factor, which raises the cutting temperature significantly in hard turning of hardened AISI 4340 alloy steel by cubic boron nitride (CBN) tool inserts. Bouchelaghem et al. [10] showed that the cutting speed, feed and depth of cut are most influential factors on cutting temperature, and the increase in the levels of cutting parameters raises the cutting temperature in hard turning of AISI D3 steel at elevated hardness of 60 HRC using CBN tool inserts. However, very few publications can be found in the literature that discusses the effect of cutting parameters on cutting temperature in turning of alloy steel at elevated hardness of 58–60 HRC. The modeling and optimization of the cutting temperature in hard turning of alloy steels using multilayer coated carbide tools at elevated hardness are rarely reported in the literature. In view of this, an attempt has been made to investigate the effect of cutting parameters on cutting temperature in turning of AISI 52100 hardened

Table 1 Chemical composition of AISI 52100 hardened alloy steel (weight percentage)

C %	Si %	Mn %	P %	S %	Cr %	Ni %	Cu %	Fe %
1.04	0.18	0.35	0.007	0.004	1.35	0.076	0.058	Balance

steel at elevated hardness of hardness 58 HRC, which has wide applications in the automotive and allied industries.

2 Experimentation

2.1 Selection of Workpiece Material

In this study, AISI 52100 hardened alloy steel having a hardness of 58 HRC was selected as workpiece material. AISI 52100 hardened alloy steel has wide applications and is being used in automotive and allied industries such as bearings, forming rolls, spindles, tools and precision instrument parts. Table 1 shows the chemical composition of the workpiece material.

2.2 Selection of Tool

The cutting tool inserts and the tool holder were selected based on the literature review and the tool manufacturer's recommendation. The MTCVD multilayer coated carbide (TiN/TiCN/Al_2O_3)–[HK150, K-type] cutting tool insert having specification CNMG120408 and the tool holder with PCLNR 2020 K12 specification were selected for experimentation. The experiments were carried out on a rigid high precision HMT NH-18 lathe machine.

2.3 Selection of Cutting Parameters

Based on the previous research carried out and the tool manufacturer's recommendations, the cutting parameters were selected. Table 2 presents the cutting parameters and their coded and actual levels based on the central composite design of response surface methodology.

Table 2 Cutting parameters and their coded and actual levels using central composite design

Process parameters	Units	Limits				
		−1.682	−1	0	1	1.682
Cutting speed	m/min	90	100	115	130	140
Feed rate	mm/rev	0.05	0.075	0.10	0.125	0.150
Depth of cut	mm	0.1	0.2	0.3	0.4	0.5

2.4 Design of Experiment

The response surface methodology uses the statistical experimental design technique and least square fitting method for model generation. The central composite design (CCD) was used for experimental design. The prediction of the cutting temperature was carried out in terms of cutting parameters under dry environment. Cutting temperature values were measured using infrared thermometer and embedded thermocouple technique. Table 3 shows the experimental design and result for cutting temperature.

Table 3 Experimental design and result for cutting temperature

S. No.	Cutting speed (m/min)	Feed rate (mm/rev)	Depth of cut (mm)	Cutting temperature (°C)
1	115	0.1	0.3	640
2	130	0.08	0.4	702
3	115	0.1	0.3	635
4	90	0.1	0.3	651
5	140	0.1	0.3	748
6	115	0.1	0.1	597
7	115	0.05	0.3	631
8	115	0.1	0.3	635
9	115	0.1	0.3	642
10	100	0.08	0.2	610
11	130	0.12	0.4	728
12	115	0.1	0.5	667
13	130	0.12	0.2	658
14	115	0.1	0.3	643
15	130	0.08	0.2	643
16	100	0.08	0.4	638
17	100	0.12	0.4	657
18	100	0.12	0.2	621
19	115	0.1	0.3	644
20	115	0.15	0.3	662

2.5 *Measurement of Cutting Temperature*

The cutting temperature at the cutting edge of the insert was measured with the help of embedded K-type thermocouple of range 50–1370 °C and −58–2498 °F and infrared thermometer (HTC IRX-68) of range 50–1850 °C and −58–3362 °F. Table 3 shows the experimental design and result for cutting temperature.

3 Result and Discussion

The MINITAB software was used for the analysis. Table 4 shows the ANOVA table for the quadratic model of cutting temperature, and it is clear from the ANOVA table that the *P*-values less than 0.0500 indicate the model terms are significant and the values greater than 0.1000 indicate the model terms are not significant. The cutting speed (*V*), depth of cut (*D*) and quadratic value of cutting velocity (V^2) have the most significant effect; while the feed (*F*) and the interaction between cutting speed and depth of cut (*VD*) have a less significant effect on cutting temperature. But the quadratic value of feed rate (F^2), the quadratic value of depth of cut (D^2), the interaction between cutting speed and feed rate (*VF*) and the interaction between feed rate and depth of cut (*FD*) all have no significant effect on cutting temperature. The below equation is in terms of coded factors can be used to make predictions of cutting temperature for given levels of each factor.

Table 4 ANOVA for quadratic model of cutting temperature

Source	DF	Seq SS	Contribution (%)	Adj SS	Adj MS	*F*-value	*P*-value
Model	9	24,932.0	98.52	24,932.0	2770.22	73.91	0.000
V	1	9918.0	39.19	9918.0	9918.03	264.62	0.000
F	1	1075.7	4.25	1075.7	1075.72	28.70	0.000
D	1	6930.6	27.39	6930.6	6930.56	184.91	0.000
V * *V*	1	6182.3	24.43	5709.5	5709.52	152.33	0.000
F * *F*	1	109.3	0.43	56.4	56.43	1.51	0.248
D * *D*	1	127.7	0.50	127.7	127.72	3.41	0.095
V * *F*	1	15.1	0.06	15.1	15.12	0.40	0.540
V * *D*	1	528.1	2.09	528.1	528.13	14.09	0.004
F * *D*	1	45.1	0.18	45.1	45.12	1.20	0.298
Error	10	374.8	1.48	374.8	37.48		
Lack of fit	5	296.0	1.17	296.0	59.19	3.75	0.086
Pure error	5	78.8	0.31	78.8	15.77		0.000
Total	19	25,306.8	100.00				

$$\text{Cutting temperature} = 639.31 + 27.05 * V + 7.24 * F + 20.81 * D \\ + 1.38 * V * F + 8.13 * V * D + 2.38 * F * D \\ + 20.79 * V * V + 0.9761 * F * F - 2.25 * D * D.$$

The model F-value of 73.91 implies the model is significant and could be used to predict the cutting temperature accurately. There is only a 0.01% chance that an F-value this large could occur due to noise. P-values less than 0.0500 indicate model terms are significant. The lack of fit F-value of 3.75 implies that the lack of fit is not significant relative to the pure error and can fit the model well (Table 5).

The predicted R^2 of 0.8966 is in reasonable agreement with the Adjusted R^2 of 0.9719; i.e. the difference is less than 0.2. Adeq precision measures the signal-to-noise ratio. A ratio greater than 4 is desirable, and ratio of 35.4493 indicates an adequate signal. This model can be used to navigate the design space.

Figure 1a shows the normal probability plot of the residuals of cutting temperature. The normal probability plot indicates whether the residuals follow a normal distribution or not. The points on the normal probability plots of the residuals fall on a straight line implying that the errors are distributed normally. It can be seen from Fig. 1b that all the actual values are following the predicted values. Figure 1c represents residuals versus the predicted response plot for cutting temperature. The plot shows the random scatter and contains no obvious pattern. This implies that the model proposed is adequate, and there is no reason to suspect any violation of the independence or constant variance assumptions.

Figure 2 shows contour plots of cutting temperature. Figure 2a shows the cutting temperature versus cutting speed and feed rate. From this graph, lower cutting temperature value of 586 °C has arrived between a feed rate of 0.05–0.12 mm/rev

Table 5 Model summary of cutting temperature

Std. dev.	R^2	R^2 (adj)	Press	R^2 (pred)	Adeq precision
6.12213	98.52%	97.19%	2617.60	89.66%	35.4493

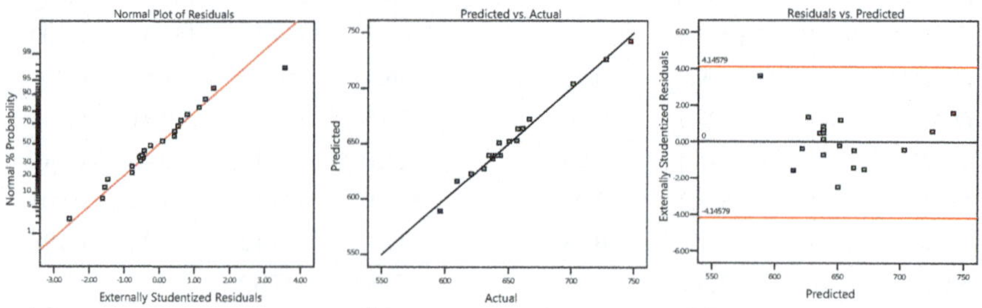

(a) (b) (c)

Fig. 1 **a** Plot of residuals versus normal probability, **b** plot of predicted versus actual values, **c** plot of residuals versus predicted values for cutting temperature

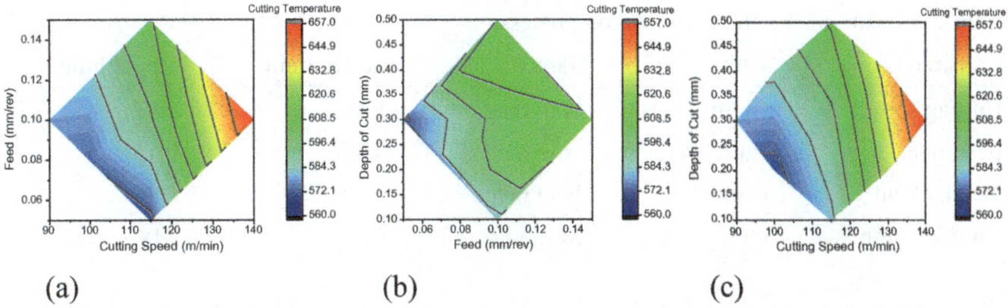

Fig. 2 Three-dimensional contour plots for cutting temperature **a** cutting speed versus feed rate, **b** cutting speed versus depth of cut and **c** feed rate versus depth of cut

and a cutting speed of 100–115 m/min. Figure 2b shows cutting temperature against the depth of cut and feed rate. The cutting temperature value of 586 °C has arrived between a feed rate of 0.05–0.12 mm/rev and depth of cut of 0.1–0.3 mm. Figure 2c shows cutting temperature against the depth of cut and cutting speed. The cutting temperature value of 586 °C has arrived between a depth of cut of 0.1–0.3 mm and cutting speed of 100–120 m/min. Figure 3 shows the surface plot for better visualization of the functional relationship between the dependant and two independent variables.

3.1 *Optimization of Cutting Condition*

The optimum values of the cutting factors selected for the minimization of cutting temperature were obtained by numerical optimization applying desirability function. Table 6 shows the constraint for minimizing the cutting temperature. Table 7 shows the optimized values of cutting parameters as cutting speed: 102.318 m/min, feed

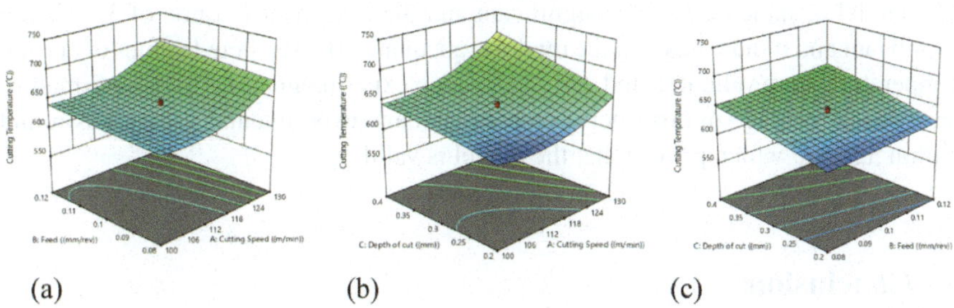

Fig. 3 Three-dimensional surface plots for cutting temperature **a** cutting speed versus feed rate, **b** cutting speed versus depth of cut and **c** feed rate versus depth of cut

Table 6 Constraints for the optimization process

Constraints	Unit	Target	Lower limit	Upper limit
Cutting speed	(m/min)	Is in range	100	140
Feed rate	(mm/rev)	Is in range	0.05	0.15
Depth of cut	(mm)	Is in range	0.10	0.50
Cutting Temp.	(°C)	Minimize	597	748

Table 7 Optimization results

Number	Cutting speed	Feed rate	Depth of cut	Cutting temperature	Desirability	
1	**102.318**	**0.051**	**0.100**	**596.987**	**1.000**	**Selected**
2	101.650	0.069	0.101	596.047	1.000	
3	105.185	0.054	0.117	595.152	1.000	
4	104.257	0.061	0.106	593.570	1.000	
5	102.484	0.056	0.102	596.172	1.000	
6	103.081	0.063	0.110	595.901	1.000	
7	103.332	0.053	0.101	595.045	1.000	
8	103.632	0.053	0.114	596.911	1.000	
10	107.574	0.055	0.101	589.442	1.000	

rate: 0.051 mm/rev and depth of cut: 0.10 mm and the corresponding value of cutting temperature response is 596 °C.

3.2 *Validation of the Model*

The model *F*-value of 73.91 is significant and the lack of fit *F*-value of 3.75 is not significant and it indicates that the model is adequate. The predicted values of cutting temperature were validated and verified through experimental runs and confirmatory test. The percentage of error between predicted and experimental values was found within ±5–8% which proved that the model is valid.

4 Conclusions

In this investigation, a central composite design (CCD)-based response surface methodology (RSM) was used to design the experiment and model the cutting temperature in terms of cutting parameters within the constraint in dry hard turning of AISI 52100 hardened alloy steel at elevated hardness of 58 HRC using multilayer coated carbide tool. The multilayer coated carbide (TiN/TiCN/Al_2O_3) insert performed well in hard turning and found to be more economical compared to costlier CBN insert. It was found that the cutting speed is the most significant parameter on cutting temperature followed by the depth of cut. The quadratic terms of *F, D* and interactions *VF* and *FD* had no influence on cutting temperature. The percentage contribution of cutting parameters, such as cutting speed, feed rate, depth of cut and quadratic term of cutting speed for cutting temperature is 39.19%, 4.25%, 27.39% and 24.43% respectively. The developed predictive model could be successfully used for the prediction of cutting temperature for different combination of cutting parameters within the constraint defined.

References

1. Aouici H, Yallese MA, Fnides B, Mabrouki T (2010) Machinability investigation in hard turning of AISI H11 hot work steel with CBN tool. Mechanika 86:71–77
2. Tönshoff HK, Arendt C, Ben Amor R (2000) Cutting of hardened steel. CIRP Ann Manuf Technol 49(2):547–566
3. Huang Y, Chou YK, Liang SY (2007) CBN tool wear in hard turning: a survey on research progresses. Int J AdvManufTechnol 35:443–453
4. Suresh R, Basavarajappa S, Gaitonde VN, Samuel GL (2012) Machinability investigations on hardened AISI 4340 steel using coated carbide insert. Int. J Refract Met Hard Mater 33:75–86
5. Sahoo AK., Sahoo B (2012) Experimental investigations on machinability aspects in finish hard turning of AISI 4340 steel using uncoated and multilayer coated carbide inserts. Measurement 45:2153–2165
6. Sutter G, Faure L, Molinari A, Ranc N, Pina V (2003) An experimental technique for the measurement of temperature fields for the orthogonal cutting in high speed machining. Int J Mach Tools Manuf 43:671–678
7. Ren XJ, Yang QX, James RD, Wang L (2004) Cutting temperatures in hard turning chromium hard facings with PCBN tooling. J Mater Process Technol 147:38–44
8. Fnides B, Yallese MA, Aouici H (2008) Hard turning of hot work steel AISI H11: Evaluation of cutting pressures, resulting force and temperature. Mechanika 72:59–63
9. Lin HM, Liao YS, Wei CC (2008) Wear behavior in turning high hardness alloy steel by CBN tool. Wear 264:679–684
10. Bouchelaghem H, Yallese MA, Mabrouki T, Amirat A, Rigal JF (2010) Experimental investigation and performance analyses of CBN insert in hard turning of cold work tool steel (D3). Int J Mach Sci Technol 14:471–501

Optimization of Surface Roughness During High Speed Milling of Inconel 825 Using Grey Relation Analysis

Balagopal Unnikrishnan, Armaan Valjee, Vyankatesh Bagal, and Prashantkumar Patankar

Abstract Inconel 825 is extensively used in harsh environment due to outstanding mechanical and chemical properties. However, Inconel 825 belongs to difficult to cut material due to the presence of hard carbide in microstructure that leads to poor machinability. In this context, this study experimentally investigated and optimized machining parameters during end milling using PVD TiAlN coated carbide tool to enhance surface quality and material removal rate. For this, three input variables such as cutting speed, feed per tooth and radial rake angle with three levels are used. L_{27} orthogonal array-based Taguchi's design of experiments is used. Responses are analysed using statistical software MiniTab 2017. Grey relation analysis is used to convert multi-objective optimization into single response and is optimized using Taguchi method. It is observed that, during end milling of Inconel 825, the most optimal conditions are the cutting speed of 90 m/min, feed rate of 0.03 mm/rev and radial rake angle of 7°. Using the response table for mean, it was concluded that the feed rate has substantial influence on various performance characters over other milling parameters.

Keywords Grey relation analysis · Inconel 825 · Surface roughness · PVD TiAlN coating · Milling

1 Introduction

Inconel 825 has good impact strength at room temperature and retains its strength at cryogenic temperatures. The main drawback of Inconel 825 is that work hardening occurs on machining and it is basically in a category of difficult to cut type metals affecting the flutes of the cutting tools. Inconel 825 provides a serious challenge as a work material during machining, due to their unique combination of properties, such as high temperature strength, hardness and chemical wear resistance [1]. Although

B. Unnikrishnan (✉) · A. Valjee · V. Bagal · P. Patankar
Mechanical Engineering Department, Dwarkadas J. Sanghvi College of Engineering, Mumbai, India
e-mail: vishnubuk2013@yahoo.com

© Springer Nature Singapore Pte Ltd. 2020

H. Vasudevan et al. (eds.), *Proceedings of International Conference on Intelligent Manufacturing and Automation*, Lecture Notes in Mechanical Engineering,
https://doi.org/10.1007/978-981-15-4485-9_6

these properties are the desirable design requirements, they pose a greater challenge to manufacturing engineers, due to high temperature and stresses generated during machining. End milling is commonly used in manufacturing industry for finish milling of this advanced material.

Ozcelik et al. [2] conducted an experiment of optimization of surface roughness in end milling of Inconel 718. The input parameters used were cutting speed, feed, axial depth of cut and radial rake angle. These cutting parameters of Inconel 718 were determined to enable minimum surface roughness under the constraints of several roughness and material removal rates (MRR). They concluded that surface roughness values are highly influenced by MRR and it is not possible to reduce the surface roughness values without considering any sacrifice on the MRR. Rajguru and Vasudevan [3] conducted experiment on Inconel 625 to investigate the impact of process parameters on surface integrity aspects such as surface roughness, residual stress and microhardness. They found that the feed per tooth, cutting speed, radial depth of cut and radial rake angle significantly affect the surface integrity aspects.

Xavior et al. [4] carried out an experiment of evaluation of surface integrity during turning of Inconel 718. The inputs used were cutting speed, tool material (coated carbide, ceramic, CBN, etc.) and cutting conditions (dry, MQL, flood cooling, etc.) to obtain surface roughness and surface residual stress. They concluded that a cutting speed of 100 m/min and flood cooling can be considered as the optimum parameters for machining Inconel 718.

Vasudevan et al. [5] carried out an experimental investigation and optimization of end milling parameters in the machining of Inconel 825 using carbide coated tool. The cutting tools used were two fluted TiAlN carbide coated cutting tool. The input parameters such as feed per tooth, cutting speed, radial depth of cut were used to obtain surface roughness and material removal rate. It was concluded that the most optimal conditions were found when the speed of cutting was 55 m/min, feed per tooth is 0.09 mm/tooth and the radial depth of cut of 1.2 mm. Jahanbakhsh et al. [6] experimented on optimizing the flank wear on turning of Inconel 625 using ceramic tool. The input parameters used were cutting speed, feed and depth of cut to obtain output parameters such as flank wear and material removal rate. They concluded that the size of flank wear decreases with the increase in feed rate, while the increase in depth of cut intensifies flank wear, especially, when it is associated with the increase in cutting speed. The flank wear propagation could be moderated by the increase in feed rate and decrease in depth of cut. However, the depth of cut at the middle range should be used to satisfy the MRR when equal weight is given to the tool wear.

So far, there is few research work related to the machining of Inconel 825 considering cutting and tool geometry parameters to optimize process parameters in dry end milling. Therefore, there is need to investigate multi-objective optimization of machining parameters under dry end milling with coated carbide tool.

2 Experimental Details

The work material used for this experiment is Inconel 825 nickel chromium iron-based super alloy, having a dimension of 150 mm length and 50 mm breadth and a thickness of 6 mm. The cutting tools used are four fluted solid TiALN coated carbide end mill. The detail specification of end mill is depicted in Table 1.

The cutting parameters selected were cutting speed, feed per tooth and radial rake angle, and their levels are presented in Table 2. The response parameters considered were material removal rate (MRR) and surface roughness parameters such as mean roughness (roughness average R_a), mean roughness depth (R_z) and maximum height of the profile (R_t). Other parameters are kept constant such as axial depth of cut, dry working environment and tool holder. The end milling process is carried out in vertical three-axis milling CNC machine. The model is BMV 45 TC++. It has a traversing distance of 600 mm on X-axis, 450 mm on Y-axis and 500 mm on Z-axis. The table dimensions are 900 mm by 450 mm. The output power is 11 kW/30 min rating. L_{27} orthogonal array based on design of experiments was used to conduct 27 experiments. The measurement for roughness was done on a Taylor Hobson Talysurf-4 set-up with data acquisition by a diamond tip stylus profiler using Se-surf software. Material removal rate is calculated using Eq. 1, and same is presented in Table 3. Detail of the experimental set-up, work piece and cutting tool as well as surface roughness profile is depicted in Fig. 1.

The material removal rate is calculated using the formula given below

$$\text{MRR} = \frac{f_Z \times Z \times V_C \times a_e \times a_p}{1000} \quad (1)$$

Table 1 Tool Specifications

Specifications	Unit	End mill 1	End mill 2	End mill 3
Cutter diameter	mm	12	12	12
Overall length	mm	108	108	108
Flute length	mm	38	38	38
Helix angle		43	43	43
Radial rake angle		5	7	9

Table 2 Process parameters with their levels

Process parameters	Units	Factor levels		
		1	2	3
Cutting speed	m/min	70	80	90
Feed rate	mm/tooth	0.03	0.06	0.09
Radial rake angle		5	7	9

Table 3 Input variable and measured responses

Runs	Actual process parameters			MRR	R_a	R_t	R_z
	Cutting speed	Feed rate	Radial rake angle	cm^3/min	μm	μm	μm
1	70	0.03	5	0.670	0.447	3.498	2.294
2	80	0.03	5	0.764	0.550	4.312	2.780
3	90	0.03	5	0.859	0.326	2.385	1.678
4	70	0.06	5	1.339	0.231	2.016	1.222
5	80	0.06	5	1.527	0.572	3.808	2.860
6	90	0.06	5	1.719	0.560	2.977	2.638
7	70	0.09	5	2.009	1.688	8.274	7.444
8	80	0.09	5	2.292	1.023	4.279	4.108
9	90	0.09	5	2.579	0.615	3.510	3.213
10	80	0.03	7	0.764	0.274	1.635	1.346
11	90	0.03	7	0.859	0.206	1.379	1.102
12	70	0.03	7	0.669	0.343	1.930	1.660
13	80	0.06	7	1.527	0.476	2.602	2.346
14	90	0.06	7	1.719	0.463	2.38	2.128
15	70	0.06	7	1.339	0.250	3.724	1.710
16	80	0.09	7	2.292	1.381	6.142	4.994
17	90	0.09	7	2.579	0.776	3.994	3.583
18	70	0.09	7	2.079	1.991	8.332	7.758
19	90	0.03	9	0.859	0.258	1.945	1.506
20	70	0.03	9	0.669	0.405	2.532	2.068
21	80	0.03	9	0.764	0.371	2.76	1.972
22	90	0.06	9	1.719	0.705	3.634	3.124
23	70	0.06	9	1.339	0.346	2.634	1.876
24	80	0.06	9	1.578	0.560	4.572	2.978
25	90	0.09	9	2.579	0.552	4.08	2.816
26	70	0.09	9	2.009	1.236	8.278	5.610
27	80	0.09	9	2.292	1.211	5.921	4.509

f_z = feed per tooth (mm/tooth), N = spindle speed, Z = number of effective teeth on tool, V_c = cutting speed.

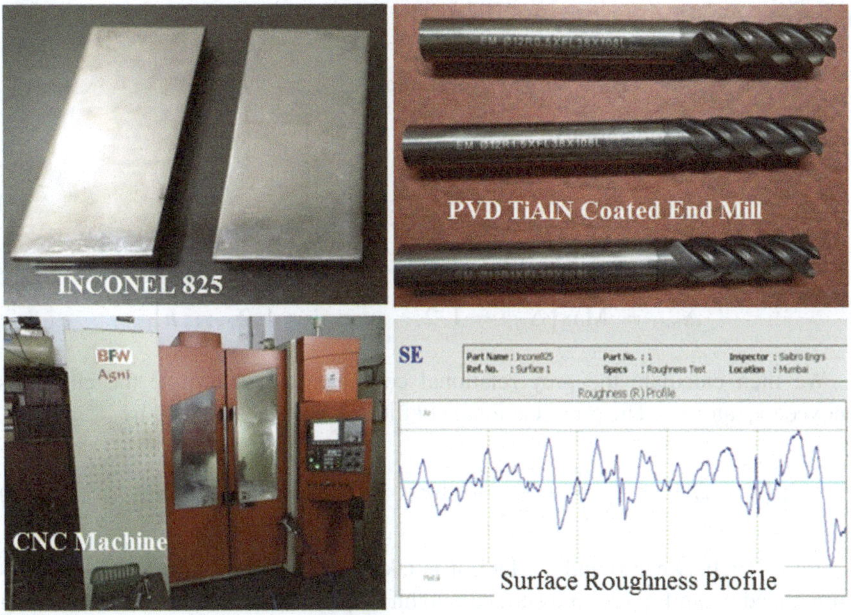

Fig. 1 Experimental set-up, work material, coated end mill and profile of surface roughness

3 Grey Relation Analysis

Grey systems are the mathematical tool which is used to deal with incomplete and uncertain information [7]. It is also known as Deng's grey incidence analysis model. It is discovered by Chinese Professor Julong Deng of Huazhong University of Science and Technology. It is one of the most widely used models of grey system theory. In grey systems theory, white denotes having all the information, and black denotes complete absence of information. Grey is the intermediate level of information. This theory is applicable in different domains like grey relational analysis, grey modelling, grey programming, grey control and grey clustering.

$$X_{ij} = \frac{Y_{ij} - \text{Min}\{Y_{ij}, i = 1, 2, \ldots, m\}}{\text{Max}\{Y_{ij} = 1, 2, \ldots, m\} - \text{Min}\{Y_{ij}, i = 1, 2, \ldots, m\}} \tag{2}$$

For parameters such as R_a, R_z and $R_{t,}$ the normalization formulae is given by

$$X_{ij} = \frac{\text{Max}\{Y_{ij}, i = 1, 2, \ldots, m\} - Y_{ij}}{\text{Max}\{Y_{ij} = 1, 2, \ldots, m\} - \text{Min}\{Y_{ij}, i = 1, 2, \ldots, m\}} \tag{3}$$

Y_{ij} a parameter (e.g. R_a, R_z, R_t and MRR)
i each of the parameters
j reading numbers from 1 to 27 experiments.

We find out the deviation

$$\Delta_{ij} = \left|X_{oj} - X_{ij}\right| \tag{4}$$

$$\Delta_{\min} = \text{Min}\{\Delta_{ij,}i = 1, 2, \ldots, m;\ j = 1, 2, \ldots, n\}$$
$$\Delta_{\max} = \text{Max}\{\Delta_{ij,}i = 1, 2, \ldots, m;\ j = 1, 2, \ldots, n\}$$

We then calculate the grey relational coefficient to determine the closeness between x_{ij} and x_{0j}. The grey relational coefficient can be calculated by

$$\gamma\left(X_{oj,}X_{ij}\right) = \frac{\Delta\min + \zeta\Delta\max}{\Delta ij + \zeta\Delta\max} \text{ for } i = 1, 2, \ldots, m,\ j = 1, 2, \ldots, n \tag{5}$$

where ζ lies between 0 and 1. Here, it is taken as 0.5. Material removal rate was normalized using Eq. (2), and surface roughness parameters were normalized using Eq. (3) and presented in Table 4. Further, the deviation sequence was calculated using Eqs. (4) and (5) was used to calculate the grey relation coefficient. Then, the grey relational grade is computed by using Eq. (6)

$$\Gamma(X_o, X_i) = \frac{1}{n}\sum_{j=1}^{n} wj\gamma\left(X_{o,j}, X_{ij}\right) \text{ for } i = 1, 2, \ldots, m. \tag{6}$$

where $\Gamma(X_o, X_i)$ is the grey relational grade between X_o and X_i. Calculated grade is presented in Table 5. This grade is treated as a single quality index, and the alternative having maximum grade is usually chosen as optimum. Many scholars have used this approach for multicriteria optimization.

4 Results and Discussion

The weighted grey relational grade calculated for each sequence was taken as a response for further analyses.

The larger-the-better quality characteristic is used for analysing the grey relational grade (GRG), since a larger value indicates the better performance of the process. The grey relational grade is further analysed by using Taguchi method. From response Table 6, it was observed that the deviation between the maximum and minimum value of the GRG for end milling parameters is 0.1004 for cutting speed, 0.1410 for feed per tooth and 0.0396 for radial rake angle. By comparison of values given in

Table 4 Measured values of responses and data preprocessing

Expt. No.	Measured responses				Normalized				Deviation sequence			
	R_a	R_t	R_z	MRR	R_a	R_t	R_z	MRR	R_a	R_t	R_z	MRR
1	0.447	3.498	2.294	0.670	0.865	0.695	0.821	0.000	0.135	0.305	0.179	1.000
2	0.550	4.312	2.780	0.764	0.807	0.578	0.748	0.049	0.193	0.422	0.252	0.951
3	0.326	2.385	1.678	0.860	0.933	0.855	0.913	0.099	0.067	0.145	0.087	0.901
4	0.231	2.016	1.222	1.340	0.986	0.908	0.982	0.351	0.014	0.092	0.018	0.649
5	0.572	3.808	2.860	1.528	0.795	0.651	0.736	0.449	0.205	0.349	0.264	0.551
6	0.560	2.977	2.638	1.719	0.802	0.770	0.769	0.550	0.198	0.230	0.231	0.450
7	1.688	8.274	7.444	2.009	0.170	0.008	0.047	0.702	0.830	0.992	0.953	0.298
8	1.023	4.279	4.108	2.292	0.542	0.583	0.548	0.850	0.458	0.417	0.452	0.150
9	0.615	3.510	3.213	2.579	0.771	0.694	0.683	1.000	0.229	0.306	0.317	0.000
10	0.274	1.635	1.346	0.764	0.962	0.963	0.963	0.049	0.038	0.037	0.037	0.951
11	0.206	1.379	1.102	0.860	1.000	1.000	1.000	0.099	0.000	0.000	0.000	0.901
12	0.343	1.930	1.660	0.670	0.923	0.921	0.916	0.000	0.077	0.079	0.084	1.000
13	0.476	2.602	2.346	1.528	0.849	0.824	0.813	0.449	0.151	0.176	0.187	0.551
14	0.463	2.380	2.128	1.719	0.856	0.856	0.846	0.550	0.144	0.144	0.154	0.450
15	0.250	3.724	1.710	1.340	0.975	0.663	0.909	0.351	0.025	0.337	0.091	0.649
16	1.381	6.142	4.994	2.292	0.342	0.315	0.415	0.850	0.658	0.685	0.585	0.150
17	0.776	3.994	3.583	2.579	0.681	0.624	0.627	1.000	0.319	0.376	0.373	0.000
18	1.991	8.332	7.758	2.010	0.000	0.000	0.000	0.702	1.000	1.000	1.000	0.298
19	0.258	1.945	1.506	0.860	0.971	0.919	0.939	0.099	0.029	0.081	0.061	0.901
20	0.405	2.532	2.068	0.670	0.889	0.834	0.855	0.000	0.111	0.166	0.145	1.000

(continued)

Table 4 (continued)

Expt. No.	Measured responses				Normalized				Deviation sequence			
	R_a	R_t	R_z	MRR	R_a	R_t	R_z	MRR	R_a	R_t	R_z	MRR
21	0.371	2.760	1.972	0.764	0.908	0.801	0.869	0.049	0.092	0.199	0.131	0.951
22	0.705	3.634	3.124	1.719	0.720	0.676	0.696	0.550	0.280	0.324	0.304	0.450
23	0.346	2.634	1.876	1.340	0.922	0.820	0.884	0.351	0.078	0.180	0.116	0.649
24	0.560	4.572	2.978	1.528	0.802	0.541	0.718	0.449	0.198	0.459	0.282	0.551
25	0.552	4.080	2.816	2.579	0.806	0.612	0.742	1.000	0.194	0.388	0.258	0.000
26	1.236	8.278	5.610	2.010	0.423	0.008	0.323	0.702	0.577	0.992	0.677	0.298
27	1.211	5.921	4.509	2.292	0.437	0.347	0.488	0.850	0.563	0.653	0.512	0.150

Table 5 Grey relation grade with rank

Expt. No.	Grey relation coefficient				Grey relation grade	Rank
	R_a	R_t	R_z	MRR	GRG	
1	0.787	0.621	0.736	0.333	0.620	17
2	0.722	0.542	0.665	0.345	0.568	22
3	0.881	0.776	0.852	0.357	0.717	8
4	0.973	0.845	0.965	0.435	0.805	2
5	0.709	0.589	0.654	0.476	0.607	18
6	0.716	0.685	0.684	0.526	0.653	16
7	0.376	0.335	0.344	0.626	0.420	26
8	0.522	0.545	0.525	0.769	0.590	20
9	0.686	0.620	0.612	1.000	0.729	7
10	0.929	0.931	0.932	0.345	0.784	3
11	**1.000**	**1.000**	**1.000**	**0.357**	**0.839**	**1**
12	0.867	0.863	0.856	0.333	0.730	6
13	0.768	0.740	0.728	0.476	0.678	13
14	0.776	0.776	0.764	0.526	0.711	10
15	0.953	0.597	0.846	0.435	0.708	11
16	0.432	0.422	0.461	0.769	0.521	24
17	0.610	0.571	0.573	1.000	0.688	12
18	0.333	0.333	0.333	0.626	0.407	27
19	0.945	0.860	0.892	0.357	0.763	4
20	0.818	0.751	0.775	0.333	0.669	15
21	0.844	0.716	0.793	0.345	0.674	14
22	0.641	0.607	0.622	0.526	0.599	19
23	0.864	0.735	0.811	0.435	0.711	9
24	0.716	0.521	0.640	0.476	0.588	21
25	0.721	0.563	0.660	1.000	0.736	5
26	0.464	0.335	0.425	0.626	0.463	25
27	0.470	0.434	0.494	0.769	0.542	23

Experiment 11 indicates best machining condition to achive superior surfae roughness but not the optimum condition

Table 6 Response table for mean

Factors	Level 1	Level 2	Level 3	Max.–Min.	Rank
Cutting speed	0.6147	0.6170	0.7151	0.1004	2
Feed rate	0.7072	0.6733	0.5662	0.1410	1
Radial rake angle	0.6344	0.6740	0.6384	0.0396	3

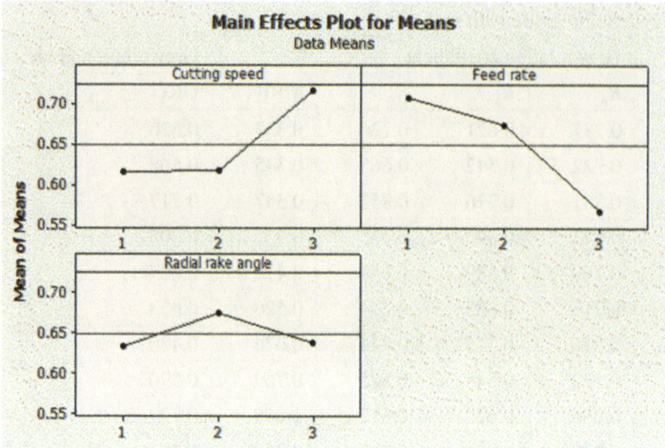

Fig. 2 Main effect plots for grey relation grade

the above table, the most compelling factor affecting multiple performance characteristics was found out. This comparison leads to identify the level of importance of the controllable factors over the multiple performance characteristics. The maximum value of deviation is 0.1410. Hence, feed rate has the greatest influence or substantial effect on the various performance characters over other milling factors.

From mean effect plot, it is observed that the optimum cutting condition to enhance surface roughness as well as material removal rate is as follows:

Cutting speed at level 3 (90 m/min), feed per tooth at level 1 (0.03 mm/tooth) and radial rake angle at level 2(7°).

5 Conclusion

In this study, the process parameters are optimized successfully using grey relation analysis in dry end milling operation on Inconel 825. From mean effect plot and response table, it is observed that, during end milling of Inconel 825, the most optimal conditions are the cutting speed of 90 m/min, feed rate of 0.03 mm/tooth and radial rake angle of 7°. These parameters can ensure better surface roughness and maximum material removal rate. Using the grey relation grade, we have found that feed rate has the substantial influence on various performance characters over other milling parameters (radial rake angle and cutting speed) when the minimization of surface roughness and maximization of material removal rate are considered.

Acknowledgements The authors would like to thank Mr. Ramesh and R. Rajguru for guidance and help in designing experiments and support rendered towards grey relation analysis.

References

1. Thakur A, Gangopadhyay S (2016) State-of-the-art in surface integrity in machining of nickel-based super alloys. Int J Mach Tools Manuf 100:25–54
2. Ozcelik B, Oktem H, Kurtaran H (2005) Optimum surface roughness in end milling Inconel 718 by coupling neural network model and genetic algorithm. Int J Adv Manuf Technol 27. https://doi.org/10.1007/s00170-004-2175-7
3. Rajguru R, Vasudevan H (2019) Investigation of the impact of cutting parameters on surface integrity in the end milling of Inconel 625. Mater Sci Forum 969:762–767
4. Anthony Xavior M, Manohar M, Patil MP, Jeyapandiarajan P (2017) Investigation of surface integrity during turning Inconel 718. Trans Can Soc Mech Eng 41(3)
5. Vasudevan H, Rajguru R, Dave G, Alva A, Punjani V, Bhurke D (2019) Experimental investigation and optimization of end milling parameters in the machining of Inconel 825 using carbide coated tool. In: Lecture notes in mechanical engineering. Published by Springer, Singapore. https://doi.org/10.1007/978-981-13-2490-1, pp 401–412
6. Jahanbakhsh M, Akhavan Farid A, Lotfi M (2016) Optimal flank wear in turning of Inconel 625 super-alloy using ceramic tool. Proc Inst Mech Eng Part B: J Eng Manuf. https://doi.org/10.1177/0954405416640698
7. Vasudevan H, Rajguru R, Tank K, Shetty N (2018) Optimization of multi-performance characteristics in the turning of GFRP(E) composites using principle component analysis combined with grey relational analysis. Mater Today Proc 5(2, Part 1):5955–5967

Role of Technological Innovativeness in the Manufacturing Performance of Indian SMEs

Anup N. Chawan and Hari Vasudevan

Abstract Technological innovativeness is one of the most important dimensions of a firm's overall innovativeness, as they look for global competitiveness. This study deployed a survey questionnaire to investigate the role of technological innovativeness in the manufacturing performance of the Indian SMEs. The study also explored various parameters associated with customer orientation, technology orientation, technological alliances, product and process innovativeness of the firms. It was found that SMEs in India have started exploring the latest developments related to technology in their particular sectors, which is one of the good indicators of them having acquired technological innovativeness. Technological innovativeness in this context has in fact increased significantly. The study also found that the majority of the SMEs have reported incremental innovations. The study further revealed that the Indian SMEs give topmost priority to quality in the context of manufacturing performance, product uniqueness and customer orientation.

Keywords Technological innovativeness · Technology orientation · Manufacturing performance

1 Introduction

Innovativeness is the key enabler for survival and growth of any firm involved in manufacturing and related activities [1]. Extant literature in the area shows that it is important to understand the current scenario of technological innovativeness and manufacturing performance in India, especially as its presence in the SMEs still remains unexplored to a large extent. It is also important to understand and

A. N. Chawan (✉)
Department of Mechanical Engineering, Dwarkadas J. Sanghvi College of Engineering, Vile Parle (W), Mumbai, India
e-mail: chawananup@gmail.com

H. Vasudevan
Dwarkadas J. Sanghvi College of Engineering, Vile Parle (W), Mumbai, India
e-mail: harivasudevan@iitb.ac.in

© Springer Nature Singapore Pte Ltd. 2020

H. Vasudevan et al. (eds.), *Proceedings of International Conference on Intelligent Manufacturing and Automation*, Lecture Notes in Mechanical Engineering,
https://doi.org/10.1007/978-981-15-4485-9_7

explore various enablers of technological innovativeness, which include customer orientation, technology orientation and technological alliances. Product and process innovativeness are also very important and are needed for a better manufacturing performance. "Make in India" initiative at the national level is expected to improve technological innovativeness of SMEs and many Indian SMEs have also started making entries into new markets. A deeper understanding is therefore required to know as to how the SMEs reposition themselves in competitive markets, by making changes in their technological capabilities, in spite of their financial constraints. SMEs now have been given much more importance by the governments and the research institutions, because of their technological capabilities. It is also known that many special schemes have been launched recently for the benefit of SMEs in India.

2 Literature Review

Technological innovativeness is the extent to which, the new technologies are incorporated in a new product [2]. There are few studies that have been conducted on two basic dimensions of innovativeness, i.e., organizational innovativeness and technological innovativeness [3, 4].

Strategic alliances are defined as voluntary agreements between firms involved in any exchange, sharing or co-development of products, technologies or services [5]. Alliances are the need of hour, as many OEMs and SMEs in India are in the process of making alliances, considering the current industrial and economic scenario.

A technology-oriented firm can be defined as a firm, with the ability and the will to acquire a substantial technological background and then use it in the development of new products [6]. Technology monitoring and technology adoption are two important dimensions of technology orientation [7]. Technology orientation is more important than the customer orientation in explaining product newness to customers [8]. Technology-orientated firms promote openness to new ideas that use the state of art technologies [9]. Technology orientation plays an important role in the development of innovation capabilities [10].

Customer orientation is one of the most important dimensions of what is broadly the market orientation of firms. Customer orientation is defined as the set of beliefs, that give priority to customer's interest first, but keeping in mind not to exclude all other stakeholders of the firm [11]. Studies have found that the customer orientation has a positive effect on SME's performance [12].

There are three important dimensions of product innovativeness, which include newness to the firm, newness to the customer and new product uniqueness [13]. Product innovativeness consists of having incrementally new products, moderately innovative products and really new products [14]. Process innovativeness captures the introduction of new production methods, new management approaches and new technologies [15]. This study has considered the Manufacturing aspect of the firm performance, especially in the context of Indian SMEs. Manufacturing performance is a key determinant in the firm's success [16]. The important dimensions of manufacturing performance are quality, cost, flexibility and delivery [17–20].

3 Research Methodology

In order to achieve the basic research objectives, a survey instrument was developed with the help of relevant literature review and expert's guidance. The first part consisted of the general profile of the SMEs, which included name, year of establishment, number of employees and sector type. The respondents' profile included name, designation, qualification, experience and contact information. The second part consisted of parameters associated with the technological innovativeness and firm performance. There were a total of nine constructs considered in the survey instrument and a total of 42 items, measuring different constructs. The measuring items for the constructs were identified with the help of an elaborate literature review. Previously tested and established scales were used in the research.

Likert-type scales were used to measure technological innovativeness, technological orientation, process innovativeness, product newness to customer, product uniqueness and customer orientation. Technological alliances and manufacturing performance were measured using the ordinal scale. Product newness to firm was measured using the binary scale.

In this study, SMEs were randomly selected from the database available in the business directory. Prior appointment was taken as per the availability and convenience of the respondents. A covering letter, addressing the purpose and importance of the study, was provided to the respondents. Around 78 SMEs were selected for the study, of which 28 refused to answer, because of their busy schedule. The responses were received from 50 manufacturing SMEs, ensuring a response rate of 64%.

The responses collected were analyzed for descriptive statistics and reliability analysis for the constructs under consideration.

4 Results and Discussion

The study reported respondent's profile, which included their designation, qualifications and SMEs profile. The details of the respondent's designation are given in Table 1. The respondents with different designation in the study consisted of 46% managing directors, 10% CEOs, 28% design heads/engineers and 16% marketing/sales head. Table 2 describes the qualification of the respondents. Majority of the respondents were 34% graduate engineers, 28% are postgraduates, 30% are diploma holders, 6% are SSC and 2% under school level. Table 3 describes the profile of the SMEs. The study sample comprised of 34% plastic mold manufacturers, 28% machine manufacturers, 8% plastic products, 8% equipment manufacturers, 4% rubber, 4% chemical, 4% metal and others 10%.

Table 1 Respondents designation

Designation	No.	%
C.E.O.	05	10
Managing director	23	46
Design/head engineer	14	28
Marketing/sales head	08	16

Table 2 Respondents qualification

Qualification	No.	%
Postgraduate	14	28
Graduate	17	34
Diploma	15	30
SSC	03	06
Under SSC	01	02

Table 3 SMEs profile

Type	No.	%
Plastic molding	17	34
Machine manufacturers	14	28
Plastic products	04	08
Equipment manufacturers	04	08
Rubber products	02	04
Chemical	02	04
Metal	02	04
Others	05	10

The constructs under consideration were analyzed using SPSS software for its internal consistency. Further, it was observed that the Cronbach alpha value for the technological alliances construct was less than 0.6. This indicated that the scale for this construct was not reliable. The value of Cronbach alpha for other constructs ranged from 0.632 to 0.891, indicating that the items representing the constructs were highly reliable. The reliability measures of the constructs are shown in Table 4. The values of the corrected item to total correlation for all the items of the reliable constructs were more than 0.3, indicating the acceptability level. Among the constructs under the study, product uniqueness was the most important aspect, having the mean value of 37.06, which ranked among the first. Process innovativeness stood at the second position with the mean value of 28.04. All other constructs followed the next positions in that order. Also, the subcriteria of the constructs under consideration were studied in detail. Table 5 shows the item statistics for product uniqueness. The maximum mean value was 6.47 for higher quality. This indicated that the product's higher quality was being considered as most important dimension in the context of product uniqueness.

Table 4 Reliability measures

Sr. No.	Construct	Items	Cronbach alpha	Mean	Standard deviation
1	Product uniqueness	06	0.89	37.06	4.57
2	Process innovativeness	05	0.88	28.30	5.15
3	Technological orientation	05	0.80	28.04	4.21
4	Customer orientation	04	0.78	24.63	3.31
5	Technological innovativeness	05	0.79	19.92	2.66
6	Manufacturing performance	04	0.63	16.19	2.18
7	Product newness to customer	04	0.87	7.50	5.59
8	Product newness to firm	01	NA	0.14	0.35

Table 5 Item statistics for PU

Subcriteria	Mean	Std. deviation
MPP	5.97	0.97
UNI	6.10	1.03
MU	6.12	0.91
HQ	6.47	0.74
ST	6.10	1.01
MCN	6.27	0.96

Table 6 shows the item statistics for technology orientation with the highest mean value of 5.90, indicating that the policy of Indian manufacturers has been to consider the most updated technology available in the market. Table 7 shows that the highest mean value for the subcriteria of process innovativeness was 5.92, indicating that the technology of the main machinery in use was updated. As shown in Table 9, the highest mean value of the subcriteria for the technological innovativeness was 4.30, indicating that the SMEs gave priority on exploring the technological developments

Table 6 Item statistics for TO

Subcriteria	Mean	Std. deviation
UPT	5.90	1.03
NME	5.78	1.01
NPD	5.40	1.26
BQS	5.34	1.13
CC	5.62	1.17

Table 7 Item statistics for PI

Subcriteria	Mean	Std. deviation
MM	5.92	1.17
MP	5.52	1.38
PM	5.50	1.19
FIP	5.68	1.20
FIM	5.68	1.23

in their respective sector. The maximum mean value of the subcriteria for customer orientation was 6.48 as shown in Table 8. It indicated that most of the firms use information from customers to improve their product quality.

The maximum mean value of the subcriteria of manufacturing performance was 4.38, which indicated that the manufacturers gave more priority to the quality of product compared to their competitors (see Table 10). The maximum mean value of the subcriteria for product newness to customer was 2.16, as shown in Table 11, indicating that the customer does not require much effort in using the product. 62% of the SMEs are having products of higher quality. 40% of the SMEs have technology of their core machinery updated. 70% of them uses information from customer to enhance product quality. 56% agreed that the customers do not require major learning

Table 8 Item statistics for CO

Subcriteria	Mean	Std. deviation
OP	6.32	1.00
PQQ	6.48	0.86
DNP	6.14	1.04
CCC	5.67	1.28

Table 9 Item statistics for TI

Subcriteria	Mean	Std. deviation
INV	3.72	0.78
EMPT	3.86	0.80
ATD	3.76	0.74
ETI	4.28	0.67
ETDI	4.30	0.58

Table 10 Item statistics for MP

Subcriteria	Mean	Std. deviation
PQ	4.38	0.63
CEE	3.92	0.87
PDPP	4.30	0.73
PPF	4.32	0.86

Table 11 Item statistics for PNC

Subcriteria	Mean	Std. deviation
LEE	2.16	1.77
PA	2.04	1.80
PC	1.76	1.49
PNN	1.54	1.43

effort in using their product. 46% of the SMEs' product quality was better than the average compared to that of their competitors. 40% of the firms investigated new trends and technologies. 58% of them were seen to be exploring technological developments in their respective industries. 52% of them have adopted technologies available in the market. 54% of them gave importance to technological innovation. 86% of them reported incremental innovation and 14% radical innovation. Table 12 shows the KMO and Bartlett's test of Sphericity of the constructs. The Bartlett's test indicates that all the constructs under consideration are significant ($p < 0.05$). The KMO value ranged from 0.693–0.849 and was greater than 0.5, for all the constructs mentioned above. Both the tests indicated suitability of the constructs for further process.

As discussed above, there are four dimensions of manufacturing performance, and in each of these dimensions, technological innovativeness has a different role to perform. Study shows that the Indian SMEs improve their product quality by using information from the customer. Customer orientation is one of the enablers for technological innovativeness. As the study shows, the role of technological innovativeness here was by adopting a technology, for example, an Advanced CNC Milling machine, which could produce a precision machined component. Here, the technology adoption rate of precision manufacturing machine was one of the indicators

Table 12 KMO and Bartlett's test of sphericity

Sr. No.	Constructs	KMO	Bartlett's significance value (p)
1	Product uniqueness	0.849	0.000
2	Process innovativeness	0.763	0.000
3	Technological orientation	0.707	0.000
4	Customer orientation	0.721	0.000
5	Technological innovativeness	0.773	0.000
6	Manufacturing performance	0.693	0.001
7	Product newness to customer	0.774	0.000

for technological innovativeness. Product uniqueness in terms of different aspects of quality would also lead to a better performance. However, design innovativeness was also very important as to how many times firms make changes in the design, compared to their competitors to improve the quality of the product.

Cost is considered one of the important aspects of manufacturing performance and is always compared with that of the competitors. Cost of the product varies according to the type of market targeted to sell the product. In certain cases, it was observed that there is no need to invest in proprietary technology, not even adopt technologies, SMEs could outsource some of the parts. Cost also differs according to the quality and design of the product.

Technological innovativeness plays a very important role, when it comes to delivering the product on time. Firm has to identify as to which activity takes more time. For example, how frequently the firm makes changes in the quality inspection techniques to reduce time? Another way of addressing this is by knowing how frequently the firm make changes in the design for assembly to reduce the time of assembly.

Volume flexibility is the context in which, technological innovativeness can be addressed by measuring how frequently the firm makes changes in their machineries to vary production volumes. As far as the product flexibility is concerned, firm's emphasis on technological innovation plays a very important role. It can be measured by knowing, how frequently the firm makes changes in the product mix for better performance.

5 Conclusion

In this study, an attempt was made to explain the role of technological innovativeness in the manufacturing performance of the SMEs. Technological innovativeness is quality driven in Indian SMEs. Majority of the SMEs under study were recognized by the OEMs for better product quality. It was observed that the product innovation is incremental in nature but contributes considerably to the manufacturing performance.

SMEs are seen investigating new trend and technologies, which would definitely help to improve their technological innovativeness. The study also found that the Indian SMEs give first priority to quality and this is a change widely observed. This change is very important from a global competitiveness point of view and is in line with government initiatives in specific sectors.

The results obtained are based on a small sample size; however, it was important that the current scenario of technological innovativeness was explored, which could form the basis for further investigation. Further research could also focus on sector-specific role of technological innovativeness in the context of manufacturing performance.

References

1. Simon H (2009) Hidden champions of the twenty-first century, the success strategies of unknown world market leaders. Springer, New York
2. Garcia R, Calantone R (2002) A critical look at technological innovation typology and innovativeness terminology: a literature review. J Prod Innov Manag 19(2):110–132
3. Avlonthis GJ, Kouremenos A, Tzokas N (1994) Assessing innovativeness of organization and its antecedents: project innovstrat. Eur J Mark 28(11):5–28
4. Laforet S, Tann J (2006) Innovative characteristics of small manufacturing firms. J Small Bus Enterp Dev 13(3):363–380
5. Gulati R (1998) Alliances and networks. Strateg Manag J 19(4):293–317
6. Gatignon H, Xuereb J-M (1997) Strategic orientation of the firm and new product performance. J Market Res 34:77–90
7. Dvir D, Segev E, Shenhar A (1993) Technology's varying impact on the success of strategic business units within the miles and snow typology. Strateg Manag J 14(2):155–161
8. Salavou H (2005) Do customer and technology orientations influence product innovativeness in SMEs? Some new evidence from Greece. J Market Manag 21(3–4):307–338
9. Zhou KZ, Yim CK, Tse D (2005) The effects of strategic orientations on technology and market-based breakthrough innovations. J Market 69(2):42–60
10. Hortinha P, Lages C, Lages LF (2011) The trade-off between customer and technology orientations: impact on innovation capabilities and export performance. J Int Market 19(3):36–58
11. Deshpande R, Farley J, Webster F (1993) Corporate culture, customer orientation, and innovativeness in japanese firms: a quadrad analysis. J Market 57(1):23–37
12. Appiah-Adu K, Singh S (1998) Customer orientation and performance: a study of SMEs. Manag Decis 36(6):385–394
13. Salavou H (2004) The concept of innovativeness should we need to focus. Eur J Innov Manag 7(1):33–44
14. Ansoff HI (1957) Strategies for diversification. Harvard Bus Rev 35(5):113–124
15. Wang CL, Ahmed PK (2004) The development and validation of the organisational innovativeness construct using confirmatory factor analysis. Eur J Innov Manag 7(4):303–313
16. Richardson PR, Taylor AJ, Gordon JRM (1985) A strategic approach to evaluating manufacturing performance. Interfaces 15(6):15–27
17. Skinner W (1969) Manufacturing-missing link in corporate strategy. Harvard Bus Rev 47(3):136–145
18. Klassen RD, Whybark DC (1999) The impact of environmental technologies on manufacturing performance. Acad Manag J 42(6):599–615
19. Gaur S, Vasudevan H, Gaur A (2011) Market orientation and manufacturing performance of Indian SMEs: moderating role of firm resources and environmental factors. Eur J Market 45(7/8):1172–1193
20. Dangayach GS, Deshmukh SG (2001) Manufacturing strategy: literature review and some issues. Int J Oper Prod Manag 21(7):884–932

Implementation of 6R Paradigm in the Life Cycle of Automobiles for Sustainability

Neel Sanghvi, Jash Patel, Dhairya Vora, Ujwal Sutaria, and Satish Takalikar

Abstract Meeting the needs of the present without compromising those of future is sustainability. Sustainability has become a critical function as natural resources are depleting at an alarming rate. Sustainability is closely associated with economic, social, and ecological aspects of human life and culture. It inspires businesses to look far beyond the balance sheets of current years and make decisions. The 6R paradigm is a key principle of sustainable design, fabrication, and use. Its use has gained much importance recently in refining the life cycle of various products. In this paper, the same is implemented to automobiles throughout its stages of life to boost sustainable practices.

Keywords Sustainability · Automobile · Manufacturing · Life cycle

1 Introduction

The word sustainability means the ability to occur and exist constantly. In the broader scope of this paper, this word also means the ability of the biosphere and human civilization to coexist without causing damage to the biosphere, while keeping in perspective the technological advances that the human civilization is so vigorously pursuing at this point of time. To attain the required level goals of sustainability, to secure a good future for the future generations, the sustainable development goals (SDG's) were adopted by all the UN member states in 2015 to put an end to poverty, protect the earth, and ensure peace and prosperity for all by 2030 [1]. The SDG's are 17 in number, each having their own targets and indicators. The strategies developed in this paper can directly contribute to the following SDGs: # 12: Responsible Consumption, # 13: Climate Action, # 15: Life on Land. In this paper, an effort is being made to implement the 6R paradigm in the life cycle of an automobile, with an intent of promoting sustainability in terms of its production and use [1].

N. Sanghvi · J. Patel · D. Vora (✉) · U. Sutaria · S. Takalikar
Department of Mechanical Engineering, Dwarkadas J. Sanghvi College of Engineering, Mumbai, India
e-mail: dhairyakvora2018@gmail.com

© Springer Nature Singapore Pte Ltd. 2020

H. Vasudevan et al. (eds.), *Proceedings of International Conference on Intelligent Manufacturing and Automation*, Lecture Notes in Mechanical Engineering,
https://doi.org/10.1007/978-981-15-4485-9_8

2 Evolution of Manufacturing

The driving factor for the economy of the manufacturing industry is the desire to produce saleable finished products at a very reasonable price while still maintaining suitable standards of quality, timeliness, and functionality of products [2]. Keeping this aspect and various other factors in consideration, manufacturing techniques have been broadly categorized as follows:

1. Traditional manufacturing
2. Lean manufacturing
3. Green manufacturing
4. Sustainable manufacturing.

Traditional Manufacturing

During the period between 1760 and 1840, i.e., during the Industrial Revolution or now known as the First Industrial Revolution, there was a transition to formal manufacturing systems in Europe and the USA. During this period textile industry was dominant in terms of capital invested, value output, and employment opportunities created [3]. Several new manufacturing techniques were developed which served the iron, wool, silk, etc. industries apart from textile sector. During the days when traditional manufacturing was dominant, it was dangerous for humans due to, handling of toxic materials and monitoring of manufacturing processes without adequate security.

As a result, there has been more generation of waste materials that have caused more landfills and pollution and no automation techniques were implemented in that period. As the demand for manufacturing was more, waste generation increased exponentially and caused more and more landfilling and gave a threat to environment. To overcome these troubles of excessive waste generation and landfilling, lean manufacturing method was developed further.

Lean Manufacturing

Lean manufacturing is a fairly new concept developed in the middle of twentieth century with an objective of efficient manufacturing and operations. In the other word, it is a revolutionized manufacturing philosophy compared to the traditional mass production standard that has been used for nearly a century [4]. However, when we hear the name "lean," the first name that strikes our mind is Toyota. The philosophy of lean manufacturing originated in Japan in the 1940s when Toyota Motor Company reassessed their manufacturing techniques to survive in the market [5]. Henry Ford first tried and conducted different experiments for integrating the concept of lean into manufacturing to achieve the standardization of work. But the technique developed by him lacked variety and was applicable in only a single specification. The product system developed by Toyota aimed at meeting the dynamic customer needs. The aims of the system included; reduction of cost of production, enhancing product quality, and increasing the throughput times.

Green Manufacturing
The term green manufacturing is related to the methodology of 3R (*Reduce, Reuse, and Recycle*). Green manufacturing was developed to not only reduce the waste generation, but it also aimed to reuse the product manufactured by segregating the different parts of the product and using them in other/different products. It also aims to recycle those products, which will eventually reduce the need for extracting, refining and processing of the raw materials, and thus decrease the negative environmental effects. Applying the concept of green manufacturing in the production unit of an organization helps them in improving economically environmentally and socially [6].

Sustainable Manufacturing
Sustainable manufacturing is regarded as a new manufacturing paradigm that provides an abstract representation of a manufacturing system. The concept is becoming more and more mature due to being the focus of more research projects. It based on the 6R methodology (*Reduce, Reuse, Recycle, Remanufacture, Redesign, Recover*) There is no crystal clear definition that could define sustainable manufacturing, but for the ease of understanding, we may deploy the definition given by the US Department of Commerce "The creation of manufactured products that use process to minimize negative environmental impacts, conserve energy and natural resources, and also are safe for employees, communities, and consumers and are economically sound" [6]. It helps in integrating the product and product design issues with the issues in manufacturing/production planning and control in such a manner as to identify, quantify, assess, and manage the flow of environmental waste with an objective of reducing negative environmental impacts and cost cutting on various parameters very effectively which helps in increasing the profits.

3 Implementation of 6R Methodology in the Life Cycle of Automobiles

In this paper, an action matrix has been proposed which represents the organized form of the various steps taken to implement the 6R methodology throughout the life cycle of automobiles. Well-identified actions or strategies aiming to achieve our goal can be referred to as $X \cdot Y \cdot Z$ where $X = \{A, B, C, D\}$, $Y = \{1, 2, 3, 4, 5, 6\}$ and Z refers to the serial number of the strategy within the concerned subcategory, i.e., the combination of values of X and Y.
where,

A Pre-manufacturing
B Manufacturing
C Use
D Post-use.

And,

1 Reuse
2 Reduce
3 Recycle
4 Recover
5 Remanufacture
6 Redesign.

The various actions/strategies mapped with phases of life cycle of an automobile are discussed here. Summary matrix is given in Table 1.

A. **Pre-Manufacturing**

1. **Reuse**
   Material reuse instead of extraction: Manufacture dashboard with thermoplastic polymers instead of thermosetting polymer composites, so that the dashboard materials can be reused.
2. **Reduce**
   Reduce energy-intensive material processing methods: General processes involved in material manufacturing involve high amount of energy requirement. These processes can be reduced by opting for alternate energy-efficient methods.
3. **Recycle**
   Recycle materials involved in accessories: In end of life vehicles, materials of some accessories are in good condition and can undergo recycling to get utilized in another product, ex. floor mats and upholstery.
4. **Recover**
   Spent chemical agents can be recovered for use in different processes: In certain chemical processes used in material manufacturing the secondary by-products or catalysts used can be recovered directly or indirectly (by chemical activity) and can be used in other processes.
5. **Remanufacture**
   Develop framework to use remanufactured parts: Develop a methodology to support remanufacturing of ancillary and spare components. Example: A headlight with a scratched glass can be polished and used as a fresh product with a new life cycle.
6. **Redesign**
   a. Redesign component materials for circular economy: Ancillary and spare components should be redesigned with those materials for which circular economy is being implemented.
   b. Redesign processing methods for increased efficiency and safety: Several raw materials may undergo some hazardous operations to be completely transformed into a finished product. These hazardous operations cause life risk for the workers and also efficiency of these processes may be low. Alternate operations may be inculcated which have greater efficiency and are also safe for humans working on the operation.

Table 1 Summarized matrix for 6R application

	A. Pre-manufacturing	B. Manufacturing	C. Use	D. Post-use
1. Reuse	1. Material reuse instead of extraction	1. Coolant reuse till its life ends 2. Reuse tools by regrinding them	–	–
2. Reduce	1. Reduce energy-intensive material processing methods	1. Use mist cooling instead of flood cooling wherever possible 2. Reduce scrap and power usage by taking minimal required dimension for the stock used for machining	1. Do not use cars for very short distances 2. Instead of hose, use bucket for washing purpose 3. Do not use car for less number of occupants	–
3. Recycle	1. Recycle materials involved in accessories, e.g., floor mats and upholstery	1. Recycling of scrap metal obtained from machining	–	–
4. Recover	1. Spent chemical agents can be recovered for use in same or different processes	–	–	1. Recover materials that can be reused/recycled 2. Recover parts which can be reconditioned/remanufactured 3. Recover automobile fluids for recycle/reuse or safe disposal 4. E-waste
5. Remanufacture	1. Develop framework to use remanufactured parts	–	–	–
6. Redesign	1. Redesign component materials for circular economy 2. Redesign processing methods for increased efficiency and safety 3. Digital paperwork	1. Redesign quality control and safety checks to streamline process 2. Redesign manufacturing/assembly processes for better efficiency and safety 3. Digital paperwork	1. Redesign for higher operation efficiency 2. Digital paperwork	1. Redesign for easy disassembly 2. Redesign for remanufacture 3. Digital paperwork

c. Digital paperwork: During pre-manufacturing phase of an automobile, there is a lot of paperwork involved right from the beginning. Example: A correct set of documents has to be prepared for the procurement of materials, transportation of materials, bills of commodities purchased, etc. A lot of paper is used and sometimes wasted for maintaining these records, which creates a huge pessimistic impact on environment. Redesigning this whole concept into an electronic form would help in maintaining large amount of records easily and would also cause less harm on the environment by using less paper.

B. **Manufacturing**

1. **Reuse**
   a. Coolant reuse till its life ends: In CNC machining systems, coolants should be reused till their end of life.
   b. Reuse tools by regrinding them: As the tool loses its sharpness, instead of using a new tool, and scrapping the old one, the old tool can be reused again by sharpening it through regrinding for the machining process.
2. **Reduce**
   a. Use mist cooling instead of flood cooling wherever possible: Spraying of coolant with a mixture of air on the raw material during a machining process instead of using continuous flow of coolant would help in reducing the amount of coolant being used.
   b. Reduce scrap and power usage by taking minimal required dimension for stock used for machining. The dimension for stock used to manufacture the required product should be optimum, so that the scrap generated would be less and also the power consumption would be reduced.
3. **Recycle**
   Recycling of scrap metal obtained from machining: The scrap obtained from machining can be recycled by into a raw material again by performing certain chemical processes.
4. **Redesign**
   a. Redesign quality control and safety checks to streamline processes to reduce idle time and redundant checks.
   b. Redesign manufacturing/assembly processes for better efficiency and safety: Reengineer manufacturing and assembly technology to improve quality efficiency and safety.
   c. Digital paperwork: Redesigning this whole concept into an electronic form would help in maintaining large amount of records easily and would also cause less harm on the environment by using less paper.

C. **Use**

1. **Reduce**
   a. Do not use cars for very short distances
   b. Instead of hose, use bucket for washing purpose

   c. Do not use car for less number of occupants.
2. **Redesign**
   a. Redesign for higher operation efficiency.
   b. Opting for various methods which would improve the efficiency of engine; hence, the fuel consumption would be reduced.
   c. Digital paperwork: Redesigning this whole concept into an electronic form would help in maintaining large amount of records easily and would also cause less harm on the environment by using less paper.

D. **Post-use**

1. **Recover**
   a. Recover materials that can be reused/recycled: After the vehicle has reached its end of life, several materials from it can be recovered and then recycled which can further be used in the manufacturing of new vehicles.
   b. Recover parts which can be reconditioned/remanufactured.
   c. Recover automobile fluids for recycle/reuse or safe disposal: Toxic fluids inside the vehicle should be recovered so that they can be disposed safely and do not cause any harm to the environment.
   d. Recover electronic components instead of throwing them away as e-waste: Several electronic components of an automobile such as battery, lights, sensors, etc. can be recovered and severe land and water contamination can be prevented.
2. **Redesign**
   a. Redesign for easy disassembly: In the post-use phase of an automobile, the process for the disassembling the automobile should be easy so that parts can be disassembled easily without any fractures or strains in them.
   b. Redesign for remanufacture: Components should be redesigned in such a way that they can be used easily for remanufacturing
   c. Digital paperwork: Redesigning this whole concept into an electronic form would help in maintaining large amount of records easily and would also cause less harm on the environment by using less paper.

4 Conclusion

The literature study reveals that the 6R paradigm can be effectively applied to automobiles for sustainable manufacturing and use. The life cycle of an automobile is divided in four phases: pre-manufacturing, manufacturing, use, and post-use. The strategies under various processes of 6R methodology are identified and then mapped with various phases of life cycle of an automobile. The work documented in this paper provides a clear framework for effective management of life cycle of automobiles for implementing sustainability.

References

1. Resolution adopted by the General Assembly on 25 September 2015, 14/35, A/RES/70/1
2. "2. Manufacturing: context, content, and history", information technology for manufacturing: a research agenda. National Research Council Washington, DC; The National Academies Press (1995) https://doi.org/10.17226/4815
3. Landes DS (1969) The unbound prometheus. Press Syndicate of the University of Cambridge. ISBN 978-0-521-09418-4
4. Sundar R, Balaji AN, Satheesh Kumar RM (2014) A review on lean manufacturing implementation techniques. Procedia Eng 97:1875–1885. ISSN 1877-7058
5. Manzouri M, Ab Rahman MZ, Che Mohd Zain CR, Jamsari EA (2014) Increasing production and eliminating waste through lean tools and techniques for halal food companies. Sustainability 6:9179–9204
6. Houshyar AN, Sulaiman Rb (2014) Review paper on sustainability in manufacturing system. J Appl Environ Biol Sci 4(4):7–11. ISSN: 2090-4274

Evaluating GSCM Practice–Performance Relationship in Chemical, Textile and Rubber/Plastic SMEs in India

Meeta Gandhi and Hari Vasudevan

Abstract Green supply chain management (GSCM) practice–performance relationship has gained a lot of research interest in recent years. Though this practice–performance relationship has been studied in developed and developing countries, there is little empirical evidence about this relationship for small- and medium-scale manufacturing industries in India. This research work presents one of the earliest studies on GSCM practice–performance relationship for SMEs in the chemical, textile and rubber/plastics manufacturing firms in India. A survey questionnaire was developed based on extant literature and feedback from academics and industry experts. Data collected was analyzed using descriptive statistics, one-way ANOVA and Post hoc Scheffe test. From the results of the data analysis, it was concluded that the SMEs are still in the early stage of adoption of GSCM practices. The environmental performance of textile firms and the economic performance of chemical firms are significantly higher than the other two in respective cases, whereas the operational performance is seen similar among the three sectors. This can be attributed to the lack of knowledge, experience and tools available to deploy so as to meritoriously and proficiently improve their business performance through the implementation of GSCM practices.

Keywords Green supply chain management practices · Business performance · Chemical · Textile · Rubber/plastic · One-way ANOVA · Post hoc Scheffe test

M. Gandhi (✉)
Department of Production Engineering, Dwarkadas J. Sanghvi College of Engineering, University of Mumbai, Mumbai, India
e-mail: meeta.gandhi@djsce.ac.in

H. Vasudevan
Dwarkadas J. Sanghvi College of Engineering, University of Mumbai, Mumbai, India
e-mail: principaldjs@gmail.com

© Springer Nature Singapore Pte Ltd. 2020

H. Vasudevan et al. (eds.), *Proceedings of International Conference on Intelligent Manufacturing and Automation*, Lecture Notes in Mechanical Engineering,
https://doi.org/10.1007/978-981-15-4485-9_9

1 Introduction

One of the key challenges being faced by the world today is environmental sustainability. In the current scenario, environmental sustainability has become the world's focal point. At the United Nations Ambassadors Conference on April 20, 2006, the prime minister of UK delivered a speech in this context, and according to him, the environmental sustainability is no longer a choice but a necessity. He can be quoted here: "For the economies to flourish, for global poverty to be banished, for the well-being of the world's people to be enhanced—not just in this generation, but in the succeeding generations—we have a compelling and ever more urgent duty of stewardship to take care of the natural environment and resources on which our economic activity and social fabric depends. A new paradigm that sees economic growth, social justice and environmental care, advancing together can become the common sense of our age".

Developing a sustainable global economy is a futuristic need [1]. Sustainability is also important in the context of supply chain management (SCM). Competitions in markets have risen as consumers demand environmentally friendly products. Increase in fuel cost and high energy security have led various regulatory bodies like NGOs, governments, etc. put a lot of pressure on firms to adopt and effectively implement green business practices [2]. Moreover, according to Klassen and Vachon [3], for genuine long-term care and future of the planet, green practices are important, as it creates winning opportunities in environmental and organizational performance, leading to a competitive advantage and higher profitability for an organization. To reduce the environmental pollution, manufacturing industries in particular are required to incorporate the concept of green in their supply chain. Green is the process of using environmentally friendly inputs and transforming these inputs through change agents—whose by-products could improve or be recycled within the existing environment. Thus, effective development of new and environmentally improved, i.e. greener products will clearly be crucial in creating successful environmental strategies and this will help companies and economies to move towards environmental sustainability.

Asian Emerging Economies (AEE) have seen a rise in manufacturing bases and production facilities from developed nations. The principal reason behind this is availability of cheap labour and lower material costs [4]. This swelling/heightened global awareness about the environmental impact of production processes puts unsurmountable pressure on manufacturing industries in the developed as well as developing economies like AEE. Success for any industry in today's global scenario depends on monitoring, anticipating as well as adopting quick responsiveness to the change.

In this context, this study has covered the Small and Medium Enterprises (SMEs) in India from chemical, textile and rubber/plastic industry sectors. SMEs in the manufacturing sector were chosen, because they are the invisible link in the supply chain, and therefore, their lack of environmental management capability poses a danger of having a negative impact on the business performance and brand image of the customer company [5–7].

SMEs act as a catalyst in driving the economic growth of any country, in terms of production, employment generation and contribution to exports. They are a vital link between the large, local and international supply chains. The Government of India through its various initiatives like Skill India, Make in India and Startup India plan to boost the sector in the long run, both economically and socially. According to the MSME Annual Report 2015–16, India has the second-largest number of SMEs in the world. Thus, it is very crucial to evaluate the impact of the GSCM practice–performance relationship in industries in the SME sector in India.

2 Literature Review

Common environmental aspects like relationship between supply chain operations and activities undertaken by managers, suppliers and consumers are the main focus areas of research on green supply chains. There exists a direct link between GSCM and environmental protection, as environmental protection requires a long-term planning and strategy. At the national level, green supply chains can help change the market's orientation to become more "green", together with the creation of incentives for small and medium-sized enterprises to implement right practices to improve environmental protection [8].

GSCM is a powerful way to differentiate a firm from its competitors and it can greatly influence the planned success.

For the overall growth of a nation's economy, firms use different internal as well as external GSCM practices [9] has highlighted three major types of these practices: internal environment management practices comprising of reduced power and paper consumption: external GSCM practice as training suppliers to practices green purchasing and cooperating with customers for environmental requirements, reduction in waste and investment recovery. Thus, any energy-saving measure can be termed as a green measure, and hence, such energy-saving measures in a supply chain can also be defined as GSCM. Manufacturers not only play a crucial role in the entire product life cycle right from its origin to its disposal but are equally responsible for using waste and emission-free energy sources to cater to green consumers in order to reduce the impact on the environment [10]. In order to eliminate the waste in all forms, energy used and material consumption, manufacturing firms are driven towards implementation of green quality standards like ISO 14000 and also ISO 14001 [11]. Previous studies by Ann et al. [12] and Rao and Diane [13] indicated that ISO certified companies are more likely to have adopted GSCM practices. The process of adopting ISO certification provides high level of awareness and experience with environmental issues for companies, which facilitates the adoption of GSCM practices [12, 14] conducted a study on Chinese firms and concluded about the positive relationship of organizational learning between ISO 14001 certification and the adoption of GSCM practices.

According to the study performed by Lefebvre [15] on SMEs in Canada, implementing green strategies into an organization could help in improving the organizational innovativeness, i.e. product, process and managerial innovation could be improved and also the organizational competitiveness (cost containment, liability management and export performance) could be developed [16]. Concluded that Environment Management Systems (EMS), such as GSCM, have a positive relationship on the firm's economic performance.

A review of the research studies that are based on GSCM and that have focused on manufacturing in the Indian context has brought out the following. Various researchers [17–21] analyzed and identified the contextual relationship between drivers and barriers in implementing green practices for Indian manufacturing firms. It is concluded from their studies that major drivers are pressure from regulatory bodies and customers, while major barriers are the lack of knowledge, finance and incentives for adoption [22]. Designed a GSCM strategy for Indian manufacturing firms. They emphasized that the GSCM can be a source of competitive advantage for organizations through improvement in their environmental performance [23]. Developed a conceptual framework showing various indices for performance measurement of an enterprise towards GSCM implementation. However, in Indian context, analysis with different industry perspectives such as in SME firms is missing. Clearly, there is very little work on the analysis of GSCM implementation as well as in exploring its effect on business performance in the Indian scenario in the SME sector.

Green has moved from being perceived as a "necessary evil" to being seen as "good business". Companies that undertake green initiatives stand to be advantaged on brand enhancement, political traction and regulatory compliance, greater ability to attract and retain talent, enhanced customer retention and potential cost savings. However, these benefits require a long-term commitment and willingness to make trade-offs against short term objectives, as the economics of green manufacturing are not well understood yet.

Performance measurement can be difficult due to multidimensionality, non-standardized data, poor technological integration, geographical and culture differences, difference in organizations policy and lack of agreed-upon metrics or poor understanding of the need. According to Walton et al. [24] and Zhu and Lai [25] corporations, adopting GSCM practices may generate positive environmental and business performances.

Based on the reported studies and extant literature as well as inputs received from experts in the industry and academia, various parameters for exploring the impact of GSCM practices on business performance were selected in this study. They are as follows. Environmental performance parameters represent reduction in the following such as, air emissions, waste water, solid waste, energy consumption, consumption of hazardous/harmful/toxic materials, total flow quantity of scrap, frequency of environmental accidents and improvement in a company's environmental image [9, 13, 16, 26–30]. According to these researchers, economic performance parameters are cost per operating hour, manufacturing cost, operating expenses, cost of purchasing environment-friendly materials, cost of scrap/rework, disposal cost, recycling cost, transportation cost, cost of avoidance from environmental actions, increased revenue from green products, fine and penalties and cash rewards/subsidy for using renewable/alternate energy sources. They also indicated operational performance parameters as increased amount of goods delivered on time, reduced rates of customer complaints, better after-sales service efficiency, better responsiveness to urgent delivery, decreased inventory level, enhanced product quality, reduced customer reject rate, inplant defect fallout rate, reduced cycle time, delivery lead time, increased product line and improved capacity utilization.

3 Research Instrument Design

A structured questionnaire was developed with emphasis on environmental performance, economic performance and operational performance. The questionnaire was divided into three sections: Sect. 1 captured the general information of the industry, Sect. 2 included the importance of GSCM practices in the industry and Sect. 3 measured the business performance in light of GSCM practices. All the questions were rated on a five-point Likert scale. Questions were rated as follows: 1 = No, 2 = little, 3 = Moderate, 4 = Very Much and 5 = Greatly.

3.1 Sample Selection

A pilot study was conducted on a sample of 25 SMEs (seven chemical industries, eight textile industries and ten rubber/plastic industries) to check on the reliability and validity of the questionnaire. Overall, goals and objectives of the research were communicated through personal interactions to the target respondents to ensure that they fully understood all the items in the questionnaire. The Cronbach Alpha value obtained was 0.860 and Kaiser-Meyer-Olkin (KMO) test value obtained was 0.811, which indicated high internal consistency among the various items, sampling adequacy and suitability of the data for factor analysis.

3.2 Data Collection

After analyzing the reliability and validity of the research instrument, an extensive survey was conducted covering 256 manufacturing firms in sectors such as chemical, textile and rubber/plastic. A total of 300 firms were targeted for the survey, and finally, 272 responses were obtained. Among these 272 responses, 256 of them were found meeting all the completeness of proper data for further analysis. The finally fully usable responses of sample size of 256 met the statistical suitability as well, considering the survey instrument items and the type of research context. The questionnaire was administered directly to the director/proprietor/chief operating officer/chief executive officer/chief financial officer of these SMEs. They were chosen, because they are in "a critical position" to influence the size of overall environmental footprint of the industry. Data collected was collated in Microsoft Excel and then transferred to SPSS and analyzed using descriptive statistics, one-way ANOVA and Post hoc tests to evaluate, if a significant difference in practice–performance relationship exists among the three sectors of SMEs.

4 Data Analyses and Results

Out of the 256 respondent firms, 22% were from textile, 35% were from chemical and 43% were from rubber/plastics industries. Majority of the respondents were managing directors/proprietors (58%), followed by 26% managers and 16% chief executive officers. 29% of the respondent firms had employees less than 10, 38% had 10–50, 9% had 51–100 and 24% had 101–200. Thus, they fall in the category of SMEs.

As seen from Table 1, chemical, textile and rubber/plastic SMEs have a significant difference in their environmental, economic and operational performance. The textile firms have better environmental performance, whereas chemical firms have better economic and operational performance. However, the mean of all performance is less than the median value, and hence, it is concluded that the firms covered here are in their early stage of adoption of green practices.

One-way ANOVA and Post hoc Scheffe tests were then performed to compare the performance across the three sectors. The results are given in Tables 2 and 3.

Table 1 Results of descriptive statistics

		N	Minimum	Maximum	Mean	Median	Std. deviation
Environmental performance	Chemicals	89	13.00	36.00	3.15	3.13	4.88
	Rubber/plastics	110	15.00	33.00	2.98	2.89	4.58
	Textiles	57	20.00	39.00	3.37	3.38	5.52
	Total	256	13.00	39.00	3.13	3.13	5.03
Economic performance	Chemicals	89	18.00	56.00	3.18	3.23	5.56
	Rubber/plastics	110	22.00	56.00	3.17	3.23	5.41
	Textiles	57	15.00	50.00	2.99	3.0	4.91
	Total	256	15.00	56.00	3.14	3.15	5.43
Operational performance	Chemicals	89	18.00	55.00	3.45	3.5	6.72
	Rubber/plastics	110	20.00	53.00	3.33	3.42	5.94
	Textiles	57	13.00	51.00	3.25	3.5	8.04
	Total	256	13.00	55.00	3.35	3.47	6.76

Table 2 Results of one-way ANOVA test

		Sum of squares	df	Mean square	F	Sig.
Environmental performance	Between groups	372.804	2	186.402	7.748	0.001
	Within groups	6086.806	253	24.059		
	Total	6459.609	255			
Economic performance	Between groups	255.728	2	127.864	4.453	0.013
	Within groups	7264.737	253	28.714		
	Total	7520.465	255			
Operational performance	Between groups	198.161	2	99.080	2.191	0.114
	Within groups	11,442.148	253	45.226		
	Total	11,640.309	255			

Table 3 Results of post hoc Scheffe test

Dependent Variable			Mean difference $(I - J)$	Std. error	Sig.
Environmental performance	Chemicals	Rubber/Plastics	1.32043	0.69931	0.170
		Textiles	−1.81530	0.83211	0.095
	Rubber/plastics	Chemicals	−1.32043	0.69931	0.170
		Textiles	−3.13573*	0.80050	0.001
	Textiles	Chemicals	1.81530	0.83211	0.095
		Rubber/plastics	3.13573*	0.80050	0.001
Economic performance	Chemicals	Rubber/plastics	0.05312	0.76399	0.998
		Textiles	2.43111*	0.90906	0.029
	Rubber/plastics	Chemicals	−0.05312	0.76399	0.998
		Textiles	2.37799*	0.87453	0.026
	Textiles	Chemicals	−2.43111*	0.90906	0.029
		Rubber/plastics	−2.37799*	0.87453	0.026
Operational performance	Chemicals	Rubber/plastics	1.40981	0.95880	0.341
		Textiles	2.28445	1.14087	0.137
	Rubber/plastics	Chemicals	−1.40981	0.95880	0.341
		Textiles	0.87464	1.09753	0.728
	Textiles	Chemicals	−2.28445	1.14087	0.137
		Rubber/plastics	−0.87464	1.09753	0.728

*The mean difference is significant at the 0.05 level

Results of the test show that there was a statistically significant difference in environmental performance (F (2.253) = 7.748, p = 0.001) and economic performance (F (2.253) = 4.453, p = 0.013) between groups as determined by one-way ANOVA. There was no statistically significant difference in the operational performance between groups ($p = 0.114$).

The Post hoc Scheffe test revealed that the environmental performance of textile industry (27.02 ± 5.52 min, $p = 0.001$) is statistically significantly higher than that of rubber/plastic industry (23.88 ± 4.583.2 min, p = 0.001) and chemical industry (25.2 ± 4.88 min). The economic performance of chemical industry (41.33 ± 5.56 min, $p = 0.029$) is statistically significantly higher than textile industry (38.89 ± 4.91 min) and rubber/plastic industry (41.27 ± 5.41 min). However, there is no statistically significant difference in the operational performance of chemical, textile and rubber/plastic industries.

5 Conclusion

The findings of this study show that the GSCM practices are in its early stage of adoption in the SME firms in India. Coercive pressures are responsible for GSCM adoption and implementation in chemical sector, while mimetic and normative forces are responsible in textile and rubber/plastic sectors. The environmental performance of textile firms and the economic performance of chemical firms are significantly higher than the other two in respective cases, whereas the operational performance is seen similar among the three sectors. Thus, we can conclude that the chemical, textile and rubber/plastic firms covered in the study have differing drivers and pressures, and therefore have resulted in varying levels of GSCM adoption and implementation and hence have different potential for improvements. Literature has highlighted significant environmental and economic benefits arising out of implementation of GSCM practices. In order to improve business performance, Indian SMEs require knowledge, expertise and tools to effectively and efficiently implement GSCM practices.

6 Future Scope

Limitations in this study indicate directions for future research. The motivation and barriers on why organizations do or do not implement typical GSCM practices can be explored in the future. Understanding such reasons could help policy makers to establish suitable regulations and laws to promote environmental practices in industries and help manufacturers address the barriers. To gain and keep their competitiveness, it is important for Indian manufacturers to improve their environmental image through environmental tools, such as the adoption of ISO 9000, ISO 14000 or Environment Management Systems. As the Indian manufacturing SMEs are in their early stages of GSCM adoption, a longitudinal analysis on contribution to business performance can be done to suggest adoption of advanced GSCM practices, like reverse logistics. This study has not considered business ethics, institutional pressures, leadership, organizational culture or human resource issues. Future studies could incorporate these factors to build a more comprehensive model.

References

1. Hart SL, Ahuja G (1996) Business strategy and the Environment 5:30–37
2. Wolf C, Seuring S (2010) Environmental impacts as buying criteria for third party logistical services. Int J Phys Distrib Logistics Manag 40(1):84–91
3. Klassen RD, Vachon S (2003) Collaboration and evaluation in the supply chain: their impact on plant-level environmental investment. Prod Oper Manag 12:336–352
4. Lai K, Wong CWY (2012) Green logistic management and performance: some empirical evidence from Chinese manufacturing exporters. Omega 40(3):267–282
5. Lee SY, Klassen RD (2008) Drivers and enablers that foster environmental management capabilities in small- and medium-sized suppliers in supply chains. Prod Oper Manag 17(6):573–586
6. Sarkis J, Dijkshoorn J (2007) Relationships between solid waste management performance and environmental practice adoption in Welsh small and medium-sized enterprises (SMEs). Int J Prod Res 45(21):4989–5015
7. Shearlock C, Hooper P, Millington S (2000) Environmental improvement in small and medium-sized enterprises: a role for the business-support network. Greener Manag Int 30:50–60
8. Gilbert S (2001) Greening supply chain: enhancing competitiveness through green productivity 1–6
9. Kudroli K (2014) Green supply chain management: operation and environmental impact at different stages of the supply chain. Int J Conceptions Manag Soc Sci 2(2):128–132
10. Sambarani V, Pol N (2016) Green supply chain management: a literature review. IUP J Supply Chain Manag 13(4)
11. Srivastava SK (2007) Green supply-chain management: a state of the art literature review. Int J Manag Rev 9(1):53–80
12. Ann GE, Zailani S, Wahid NA (2006) A study on the impact of environmental system (EMS) certification towards firms performance in Malayasia. Int J Manag Environ Qual 17(1):73–93
13. Rao P, Diane H (2005) Do green supply chains lead to competitiveness and economic performance? Int J Oper Product Manag 25(9):898–916
14. Zhu Q, Sarkis A, Corderio JJ, Lai K-H (2008) Firm-level correlates of emergent green supply chain management practices in the Chinese context. Omega 36(4):577–591
15. Lefebvre E, Lefebvre LA, Talbot S (2000) Environmental initiatives, innovativeness and competiveness: some empirical evidence. In: Proceedings of the IEEE EMS international engineering management conference, pp 674–679
16. Zhu S, Lai K (2007) Green supply chain management: pressures, practices and performance within the Chinese automobile industry. J Clean Prod 15:1041–1052
17. Diabat A, Govindan K (2011) An analysis of drivers affecting the implementation of green supply chain management. Resour Conserv Recycl 55(6):659–667
18. Luthra S, Kumar V, Kumar S, Haleem A (2011) Barriers to implement green supply chain management in automobile industry using interpretive structural modeling technique-an Indian perspective. J Ind Eng Manag 4(2):231–257
19. Mathiyazhagan K, Govindan K, Noorul Haq A, Geng Y (2013) An ISM approach for the barrier analysis in implementing green supply chain management. J Clean Prod 47:283–297

20. Mudgal RK, Shankar R, Talib P, Raj T (2010) Modeling the barriers of green supply chain practices: an Indian perspective. Int J Logistics Syst Manag 7(1):81–107
21. Toke LK, Gupta RC, Dandekar M (2012) An empirical study of green supply chain management in Indian perspective. Int J Appl Sci Eng Res 1(2):372–383
22. Bag S (2013) Designing the green supply chain strategy for Indian manufacturing firm J Supply Chain Manag 2(1):8–18
23. Sekhar PS, Kumar KS (2013) Green technology performance measurement using BSC-DEA approach. IUP J Knowl Manag 11(4):20–35
24. Walton SV, Handfield RB, Melynk ST (1998) The green supply chain: integrating suppliers into environmental management process. J Supply Chain Manag 34(2):2–11
25. Zhu Q, Cote RP (2004) Integrating green supply chain management into an embryonic eco-industrial development: a case study of Guotang group. J Clean Prod 12(8–10):1025–1035
26. Kumar BA, Rajesh B, Sarbjit S, Anish S (2011) Study of green supply chain management in the Indian manufacturing industries: a literature review cum an analytical approach for the measurement of performance. Int J 13:84–89
27. Kenneth W, Jr Green et al (2012) Green supply chain management practices: impact on performance. Supply Chain Manag Int J 17(3):290–305
28. Vijayvargy L, Agarwal G (2013) A comparative study of green supply chain management practices in Indian, Japanese and Chinese companies. IUP J Supply Chain 10(3):7–18
29. Dey PK, Cheffi W (2013) Green supply chain performance measurement using the analytic hierarchy process: a comparative analysis of manufacturing organizations. Prod Plan Control 24(8):702–720
30. Shaw S, Grant DB, Mangan J (2010) Developing environmental supply chain performance measures. Benchmarking Int J 17(3):320–339

Investigations on Effect of Cutting and Cutting Fluid Application Parameters on Surface Roughness and Microhardness in Hard Turning of AISI 52100 Alloy Steel

Sandip Mane and Sanjay Kumar

Abstract The cost, health, and environment concerns associated with the use of cutting fluid calls for minimizing its usage in machining. This research work is aimed to investigate machining performance in turning of AISI 52100 hardened alloy steel with multilayer-coated carbide tool under dry and minimal cutting fluid environments. A Taguchi's L9 orthogonal array was used to design the experiments. The aim was to identify the optimal combination of the cutting and cutting fluid application parameters. The response measured was the surface roughness and microhardness under different cutting environment. The experimental result showed that the hard turning with minimal cutting fluid application improves surface roughness and reduces the microhardness variation at machines surface which in turn improves fatigue life of the machined components.

Keywords Minimal cutting fluid application · Surface roughness · Microhardness

1 Introduction

Hardened alloy steels are being widely used in automotive and allied industries due to their high compressive strength and wear resistance. Attanasio et al. [1] reported that the turning of hardened alloy steels above 45 HRC is considered to be hard turning; but in actual practice, the hard turning is carried out at elevated hardness of 45–68 HRC. Huang et al. [2] reported that the manufacturing cost can be reduced by 30%, if the hard turning process is employed to machine the complex parts. Bartarya and Choudhury [3] stated that the hard turning can replace conventional grinding process,

S. Mane (✉)
Department of Mechanical Engineering, Dwarkadas J. Sanghvi College of Engineering, Mumbai, Maharashtra, India
e-mail: sandip.mane@djsce.ac.in

S. Kumar
Department of Mechanical Engineering, Thakur College of Engineering and Technology, Mumbai, Maharashtra, India
e-mail: sanjay.kumar@thakureducation.org

© Springer Nature Singapore Pte Ltd. 2020
H. Vasudevan et al. (eds.), *Proceedings of International Conference on Intelligent Manufacturing and Automation*, Lecture Notes in Mechanical Engineering,
https://doi.org/10.1007/978-981-15-4485-9_10

and surface finish of Ra 0.4–0.6 μm can be achieved with proper selection of process parameter. Krishna et al. [4]; Tzeng et al. [5]; Sahin et al. [6]; Sharma et al. [7] reported that the turning of component in hardened state can offer high accuracy but the major concern is about the surface quality. The surface integrity is an important aspect in determining the functional performance such as fatigue life and tribological properties of machined components; in turn, it affects the product quality. Khan et al. [8]; Grzesik and Wanat [9]; Ezugwu et al. [10]; Dhar et al. [11] reported that the surface roughness and microhardness are the major performance indicators of the surface integrity, which must be studied in depth for enhancing the product performance. Fernandes et al. [12] showed that the increase in value of surface roughness leads to poor surface finish and thus reduces the fatigue life of the machined components. Many researchers studied the effect of cutting parameters on surface roughness and microhardness (surface integrity) and attempted to develop the relationships between cutting parameters and the surface integrity in hard turning. In this work, an attempt has been made to investigate the effect of cutting parameters and cutting fluid application parameters on the surface characteristics, such as roughness and microhardness in turning of AISI 52100 hardened alloy steel of 58 HRC with multilayer coated carbide tool in dry turning and turning with minimal cutting fluid application (MCFA). The effects of the cutting and cutting fluid application parameters on surface roughness and microhardness were investigated, and optimal conditions were determined to achieve the better surface integrity of the machined component.

2 Experimentation

2.1 Selection of Workpiece Material

In this study, AISI 52100 hardened alloy steel having hardness of 58 HRC was selected as workpiece material. AISI 52100 hardened alloy steel has wide applications and is being used in automotive and allied industries such as bearings, forming rolls, spindles, tools, and precision instrument parts. Table 1 shows the chemical composition of the workpiece material.

Table 1 Chemical composition of AISI 52100 hardened alloy steel (weight percentage)

C %	Si %	Mn %	P %	S %	Cr %	Ni %	Cu %	Fe %
1.04	0.18	0.35	0.007	0.004	1.35	0.076	0.058	Balance

2.2 Selection of Tool

The cutting tool inserts and the tool holder were selected based on the literature review and the tool manufacturer's recommendation. The MTCVD multilayer-coated carbide (TiN/TiCN/Al_2O_3)—[HK150, K-type] cutting tool insert having specification CNMG120408 and the tool holder with PCLNR 2020 K12 specification were selected for experimentation. The experiments were carried out on a rigid high precision HMT NH-18 lathe machine.

2.3 Selection of Cutting Fluid

The quantity of cutting fluid delivered per pulse was extremely small, and a commercially available SUN Cut ECO-33 high-performance eco-friendly semi-synthetic cutting fluid was used in this investigation.

2.4 Selection of Cutting and Cutting Fluid Application Parameters

Based on the previous research carried out and the tool manufacturer's recommendations, the cutting parameters were selected. Table 2 presents the cutting parameters and their levels.

Similarly, based on the previous studies conducted, the cutting fluid application parameters were selected. Table 3 presents the cutting fluid application parameters and their levels.

Table 2 Process parameters and their levels

Cutting parameters	Units	Levels		
Cutting speed	m/min	80	110	140
Feed rate	mm/rev	0.04	0.08	0.12
Depth of cut	mm	0.2	0.3	0.4

Table 3 Cutting fluid parameters and their levels

Cutting fluid application parameters	Units	Levels		
Pressure	Bar	60	80	100
Frequency of pulsing	Pulses/min	200	300	400
Flow rate	mL/min	4	8	12
Nozzle standoff distance	mm	20	30	40

Table 4 Experimental result for surface roughness and microhardness in dry turning

Exp. No.	Cutting speed (m/min)	Feed rate (mm/rev)	Depth of cut (mm)	Surface roughness (μm)	Microhardness (HV)
1	80.000	0.04	0.15	0.671	751
2	80.000	0.08	0.30	0.701	795
3	80.000	0.12	0.45	0.752	781
4	110.00	0.04	0.30	0.575	746
5	110.00	0.08	0.45	0.657	759
6	110.00	0.12	0.15	0.827	775
7	140.00	0.04	0.45	0.667	809
8	140.00	0.08	0.15	0.922	816
9	140.00	0.12	0.30	1.170	822

2.5 Design of Experiment

The experiments were carried out with Taguchi's L9 orthogonal array to reduce the number of experimentation without losing the significance of each input parameter in turning of hardened AISI 52100 alloy steel under dry cutting condition and with minimal cutting fluid application method. The optimized values of cutting parameters were obtained under dry cutting condition. These optimized cutting parameters were kept constant, and cutting fluid jet application parameters were varied in three levels. Talysurf surface roughness measuring machine was used to measure the surface roughness and microhardness tester (FM-300e) to measure microhardness of the machined surface. Table 4 shows the experimental design and results for surface roughness and microhardness of machined surface in dry turning.

The cutting parameters that were optimized in dry turning were kept constant, and fluid application parameters were varied in three levels. Table 5 shows the design of experiment based on Taguchi's L9 orthogonal array and results for surface roughness and microhardness of machined surface in turning of hardened AISI 52100 alloy steel under minimal cutting fluid application (MCFA) environment.

3 Result and Discussion

Table 6 shows the percentile contribution effect of the cutting parameters on the surface roughness and microhardness in dry turning through analysis of variance (ANOVA).

It was observed that the feed rate and cutting speed are the most significant cutting parameters affecting the surface roughness, whereas the depth of cut has no considerable effect on the surface roughness. On the other hand, the microhardness

Table 5 Experimental result for surface roughness and microhardness in MCFA turning

Pressure (bar)	Frequency (pulses/min)	Quantity (mL/min)	Nozzle standoff distance (mm)	Surface roughness (μm)	Microhardness (HV)
60	200	4	20	0.668	774
60	300	8	30	0.446	765
60	400	12	40	0.522	772
80	200	12	30	0.453	771
80	300	4	40	0.428	768
80	400	8	20	0.401	763
100	200	8	40	0.433	759
100	300	12	20	0.310	745
100	400	4	30	0.413	752

Table 6 Percentile contribution of cutting parameters on responses in dry turning

Factors	Effect on surface roughness (%)	Effect on microhardness (%)
Cutting speed	37.97	76.52
Feed	44.57	16.07
Depth of cut	10.83	1.18
Other error	6.63	6.24
Total %	100.00	100.00

is significantly affected by the cutting speed. The feed rate has less significant effect, and the depth of cut has no effect on the microhardness of machined surface.

Figure 1 shows the column effect graph for surface roughness and microhardness. Figure 1a shows that the cutting speed V2 = 110 m/min, feed f2 = 0.04 mm/rev, and

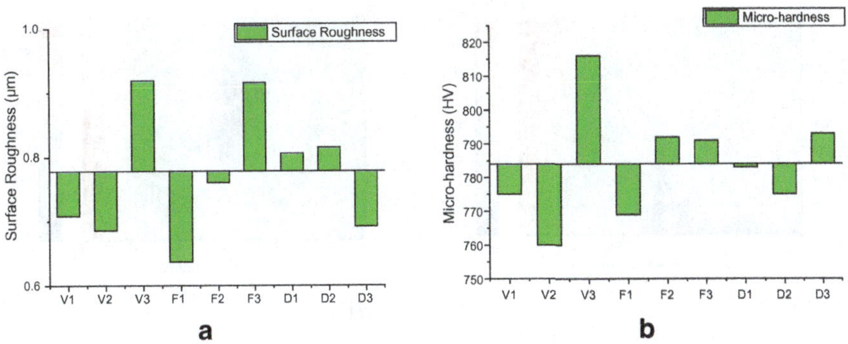

Fig. 1 Column effect graphs of **a** surface roughness, **b** microhardness in dry turning

depth of cut d3 = 0.45 mm contributed most in minimizing the surface roughness. Figure 1b presents the cutting speed V2 = 110 m/min, feed f2 = 0.04 mm/rev, and depth of cut d3 = 0.30 mm contributed more on reduction in variation of microhardness at surface and subsurface layer of machined surface.

Figure 2 shows the contour plot of surface roughness versus cutting parameters. The minimum value of surface roughness 0.5–0.6 μm observed between a cutting speed of 100–110 m/min, feed rate of 0.04–0.08 m/min, and depth of cut 0.20–0.45 mm. Figure 2a, b shows that the feed rate is the most influential parameter on surface roughness followed by the cutting speed and depth of cut have less significant effect on surface roughness.

Figure 3 shows the contour plot of microhardness versus cutting parameters. The microhardness value of 745–750 HV observed between a cutting speed of 100–110 m/min, feed rate of 0.04–0.08 m/min, and depth of cut 0.15–0.45 mm. Figure 3a, b shows that the feed has less significant effect and depth of cut has no effect on microhardness, whereas the cutting speed is the most influential parameter on microhardness.

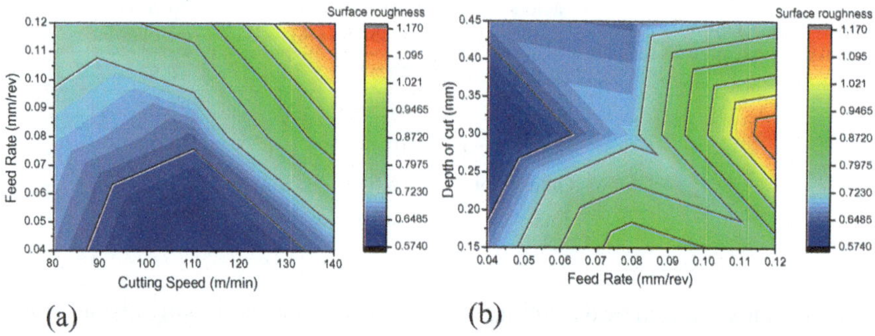

Fig. 2 Contour plot for surface roughness **a** cutting speed versus feed, **b** feed versus depth of cut in dry turning

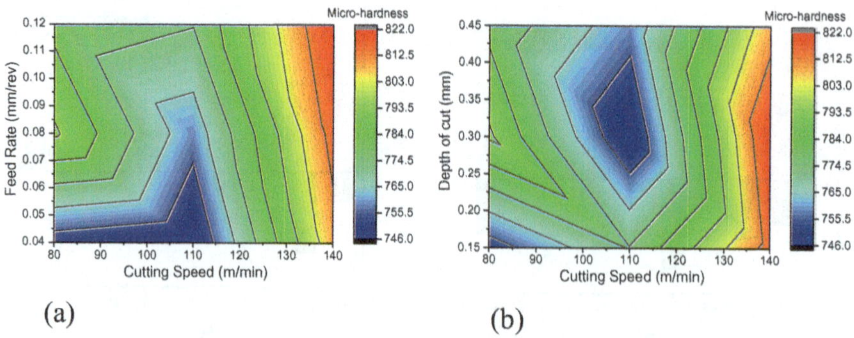

Fig. 3 Contour plot for microhardness **a** cutting speed versus feed, **b** cutting speed versus depth of cut, **c** FEED versus depth of cut in dry turning

Table 7 Percentile contribution of fluid application parameters on responses in MCFA turning

Factors	Effect on surface roughness (%)	Effect on microhardness (%)
Pressure	64.65	72.65
Frequency of pulsing	17.45	9.34
Flow rate	8.71	8.28
Nozzle standoff distance	8.10	7.61
Other error	1.10	2.11
Total	100.00	100.00

Table 7 shows the percentile contribution effect of the cutting fluid application parameters on the surface roughness and microhardness in turning under MCFA environment through analysis of variance (ANOVA). It has been observed that the cutting fluid pressure is the most influential cutting fluid application parameter affecting the surface roughness and microhardness followed by the frequency of pulsing. The flow rate and nozzle standoff distance have less significant effect on surface roughness and microhardness.

Figure 4 shows column effect graph of surface roughness and microhardness. The cutting fluid pressure at level-3 (100 bar), frequency at level-2 (300 pulses/min), quantity at level-3 (12 mL), and nozzle standoff distance at level-1 (20 mm) contributes more on the reduction of surface roughness and less variation in the microhardness values at machined surface.

Figures 5 and 6 show the comparison of effect of cutting speed on microhardness and surface roughness in dry turning and turning with MCFA respectively. The results obtained validated the optimized cutting fluid application parameters. The considerable reduction in the variation of microhardness value in the range of 20–100 HV compared to base hardness of material and significant reduction of 40–50% in the surface roughness have been observed in turning with MCFA when compared to dry turning.

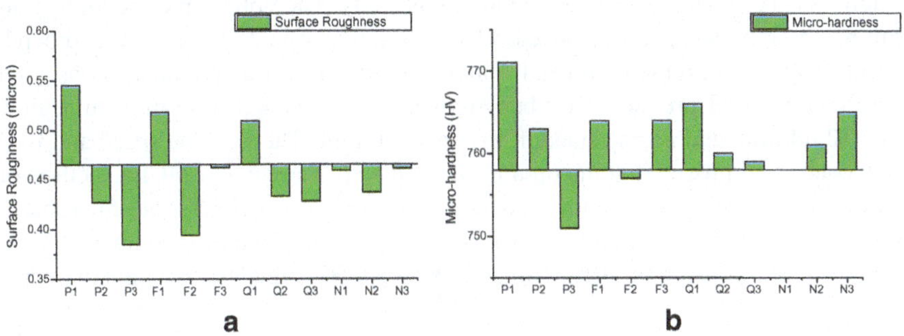

Fig. 4 Column effect graphs of **a** surface roughness, **b** microhardness in turning with MCFA

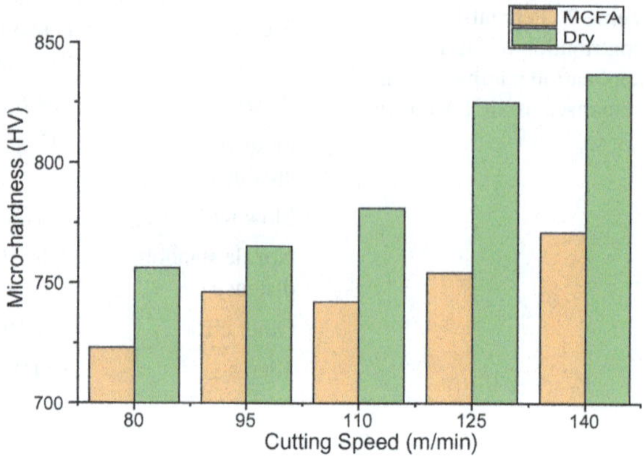

Fig. 5 Cutting speed versus (μH)

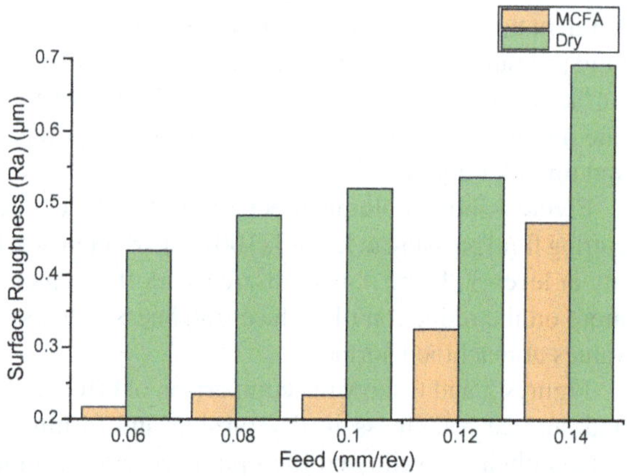

Fig. 6 Feed rate versus Ra

4 Conclusion

The surface integrity of a machined component is greatly influenced by the surface roughness and microhardness as found in this study. The optimum dry cutting conditions observed were cutting speed: 110 m/min, feed rate: 0.04 mm/rev, and depth of cut: 0.30 mm, and the optimized values of fluid application parameters observed were cutting fluid pressure: 100 bar, frequency of pulsing: 300 pulses/min, flow rate: 12 ml/min, and nozzle standoff distance: 20 mm. The variable speed and feed tests conducted proved that the optimized results are correct and MCFA method of cutting fluid application is more efficient. The surface roughness and microhardness variation decreased significantly under MCFA environment, as compared to dry turning . The reduction in surface roughness by 40–50% and less variation in microhardness values in the range of 20–100 HV compared to base hardness of material were observed in turning with MCFA compared to dry turning. Minimal cutting fluid application method helped in minimizing the usage of cutting fluid and to overcome the problems associated with cutting fluid such as its cost, storage, disposal, health, and environmental concerns.

References

1. Attanasio A, Umbrello D, Cappellini C, Rotella G, M'Saoubi R (2012) Tools wear effects on white and dark layer formation in hard turning of AISI 52100 steel. Wear 286:98–107
2. Huang Y, Chou YK, Liang SY (2007) CBN tool wear in hard turning: a survey on research progresses. Int J Adv Manuf Technol 35(2007):443–453
3. Bartarya G, Choudhury SK (2012) Effect of cutting parameters on cutting force and surface roughness during finish hard turning AISI52100 grade steel. Procedia CIRP 1:651–656
4. Krishna PV, Srikant RR, Rao DN (2010) Experimental investigation on the performance of nanoboric acid suspensions in SAE-40 and coconut oil during turning of AISI 1040 steel. Int J Mach Tools Manuf 50:911–916
5. Tzeng C-J, Lin YH, Yang YK, Jeng MC (2009) Optimization of turning operations with multiple performance characteristics using the Taguchi method and Grey relational analysis. J Mater Process Technol 209:2753–2759
6. Sahin Y, Motorcu AR (2008) Surface roughness model in machining hardened steel with cubic boron nitride cutting tool. Int J Refract Metals Hard Mater 26:84–90
7. Sharma VS, Dhiman S, Sehgal R, Sharma SK (2008) Estimation of cutting forces and surface roughness for hard turning using neural networks. J Intell Manuf 19:473–483
8. Dhar NR, Paul S, Chattopadhyay AB (2002) The influence of cryogenic cooling on tool wear, dimensional accuracy and surface finish in turning AISI 1040 and E4340C steels. Wear 249:932–942
9. Ezugwu EO, Silva RB, Da Bonney J, Machado AR (2005) Evaluation of the performance of CBN tools when turning Ti–6Al–4V alloy with high-pressure coolant supplies. Int J Mach Tools Manuf 45:1009–1014
10. Grzesik W, Wanat T (2006) Surface finish generated in hard turning of quenched alloy steel parts using conventional and wiper ceramic inserts. Int J Mach Tools Manuf 46:1988–1995
11. Khan MMA, Mithu MAH, Dhar NR (2009) Effects of minimum quantity lubrication on turning AISI 9310 alloy steel using vegetable oil-based cutting fluid. J Mater Process Technol 209:5573–5583
12. Fernandes FAP, Gallego J, Picon CA, Tremiliosi Filho G, Casteletti LC (2015) Wear and corrosion of niobium carbide coated AISI 52100 bearing steel. Surf Coat Technol 279:112–117

Optimization of Cutting Parameters in Dry Turning of AISI 4340 Hardened Alloy Steel with Multilayered Coated Carbide Tool

Sandip Mane, Anjali Mishra, and Vaidehi Kannawar

Abstract AISI 4340 has been widely used in automobile and aerospace industries due to their adequate mechanical and chemical properties. This paper investigated the effect of cutting parameters such as cutting speed, feed and depth of cut on cutting force and cutting temperature in dry turning of AISI 4340 hardened alloy steel with a multilayer coated carbide tool. The optimization of cutting conditions is highly important that helps the manufacturing technocrats in process planning as the economy of machining operation defines the competitive advantage. The study investigated the optimum cutting parameters that could produce a significant reduction in cutting temperature and cutting forces leading to improve tool life as well as the surface finish of the machined surface in hard turning. Taguchi's L9 orthogonal array was used to design the experiment and analysis of variance (ANOVA) was performed to identify the significant factors affecting the cutting forces and cutting temperatures. The experimental results showed that the feed is the most influencing factor on cutting force and cutting temperature followed by the depth of cut and cutting speed.

Keywords Tool life · Surface finish · Analysis of variance · Signal-to-noise ratio

1 Introduction

The machining is a major and critical sector of the manufacturing process, which has major influences on the final product quality. The heat generation due to plastic deformation, friction between tool-chip interface and tool-workpiece interface affects the tool life, tool wear and surface integrity of the machined surface. Cutting force and cutting temperature are the performance indicator in the machining process.

S. Mane (✉)
Department of Production Engineering, Dwarkadas. J. Sanghvi College of Engineering, Mumbai 400056, India
e-mail: sandip.mane@djsce.ac.in

A. Mishra · V. Kannawar
Dwarkadas. J. Sanghvi College of Engineering, Mumbai 400056, India

© Springer Nature Singapore Pte Ltd. 2020
H. Vasudevan et al. (eds.), *Proceedings of International Conference on Intelligent Manufacturing and Automation*, Lecture Notes in Mechanical Engineering,
https://doi.org/10.1007/978-981-15-4485-9_11

The proper selection of cutting conditions could reduce the cutting temperature and cutting force significantly, and in turn, it would improve the tool life and the surface quality of the machined component. Suresh et al. [1] analyzed the influence of cutting parameters on resultant cutting force in turning of hardened alloy steel AISI 4340 of 48 HRC using full factorial design. The results revealed that the combination of low feed rate, low depth of cut, and high cutting speed is beneficial for minimizing the machining force. Das et al. [2] studied the effect of cutting parameters during dry turning of hardened AISI 4340 steel (47 HRC) and showed that the feed is a prime factor, cutting speed second prime and depth of cut having no significance on surface roughness and cutting force. Aouici et al. [3] investigated the effect of cutting parameters and workpiece hardness on surface roughness and cutting force components in hard turning of AISI H11 steel by using cubic boron nitride insert. The results showed that the cutting force is most influenced by the depth of cut and surface roughness by feed rate. Qasima et al. [4] described the numerical method of optimization of machining parameters with multiple cutting tools. It showed that the optimization helps to reduce the cutting forces and cutting temperature while machining AISI 1045 steel. It was inferred that the feed rate and depth of cut have the most dominant effect on cutting force and the cutting temperature is highly influenced by cutting speed and rake angle in hard turning process. Gupta et al. [5] showed that the cutting condition such as cutting speed and feed rate have a remarkable effect on the tool wear, surface roughness, and cutting force in dry turning of AISI 4340. Das et al. [6] studied the effect of cutting speed, feed rate, and nose radius on cutting force in dry turning of AISI 4340 alloy steel with coated carbide tool and optimized the parameters to get the best response. In a similar class of materials, Singh et al. [7] performed the cutting operation on D2 steel and found that the optimum condition of cutting parameters such as feed rate and depth of cut, which influences the cutting forces and surface roughness significantly. Motorcu et al. [8] investigated the effect of cutting conditions on AISI 4140 on tool–chip interface temperature, tool temperature, and surface roughness using Taguchi analysis. Aslan et al. [9] reported that the cutting speed, depth of cut, and feed rate have a considerable effect on cutting force, tool life, and surface roughness in turning of AISI 4140 alloy steel. Bouacha et al. [10] investigated the effects of cutting parameters on cutting forces during machining of bearing steel with CBN tool and found that the cutting force is greatly influenced by the depth of cut, whereas the feed rate and cutting speed have a less significant effect on cutting force.

1.1 Taguchi Approach

The word “optimization” in the Taguchi method implies determining the best level of control factors, which maximize the signal-to-noise ratio. The desired output properties are achieved by calculating the log function of the signal-to-noise ratio. The determination of these levels is conducted by experiments based on “Orthogonal Arrays,” which are less in number still have a balance for all control factors. This, in

Table 1 % chemical composition of AISI 4340

Fe	Ni	Cr	Mn	C	Mo	Si	S	P
95.195	1.8	0.9	0.7	0.4	0.25	0.15	0.040	0.035

turn, means that the material and time required for the experiments also become lesser. Taguchi approach uses the signal-to-noise ratio to measure the response variation with respect to target value under different noise conditions. Depending upon the aim of the experimentation, one can select the signal-to-noise ratio as "smaller-to-better," "larger-to-better," and "nominal-to-best." The percentage contribution of each response parameter is calculated using variance analysis (ANOVA). According to experimental data, the highest value of the signal-to-noise ratio is the optimal level for process parameters. The objective of this study was to investigate the influence of the cutting speed, feed, and depth of cut on cutting temperature and cutting force in dry turning of AISI 4140 hardened alloy steel by applying Taguchi's technique and analysis of variance (ANOVA).

2 Experimental Procedure

2.1 Equipment and Material

The cutting parameters such as cutting speed, feed, and depth of cut were selected as process parameters, and AISI 4340 hardened alloy steel of 55 HRC was selected as a workpiece material for this study. AISI 4340 alloy steel is widely used in the automotive and aircraft industries. The typical applications include components such as connecting rods, collets, conveyor pins, gears, stem assemblies, pump shafts, and tool holders. The chemical configuration of AISI 4340 is listed in Table 1. NH-18 lathe machine (HMT Make) was used for turning experiments; it has a maximum spindle speed of 1800 rpm and a maximum power of 7.5 kW. Multilayer-coated (TiN/TiCN/Al_2O_3) carbide tool insert with −6° rake angle, 6° clearance angle, and 0.08 mm nose radius was used for cutting operation. IEICOS lathe tool dynamometer (model 652) with digital multicomponent force indicator with three independent digital panel meters were used to measure *XYZ* force. Also, IEICOS handheld digital infrared thermometer (Range: 50–1850 °C) and K-type embedded thermocouple were used for temperature measurement.

2.2 Design of Experiments

Taguchi's optimization technique was used in deriving an orthogonal array to examine the quality aspect through minimum practicable trials. The results obtained through experimentation based on the orthogonal array (OA) were transformed in the form of signal-to-noise ratios to analyze the effects of cutting speed, depth of cut, and feed on the responses. An orthogonal array for three parameters and for each parameter three different levels were planned according to the design of experiments method. The cutting parameters were selected based on the literature review and tool manufacturer's recommendation. Table 2 shows the cutting parameters and their levels.

To perform the parametric optimization, there has to be a minimum number of experiments to be carried out which is determined as:

Total number of runs " $= [(L - 1) \times P] + 1 = [(3 - 1) \times 3] + 1 = 7 \approx$ L9"

L9 orthogonal array (OA) was a sufficient array to optimize the cutting parameters, on the basis of which experiments were performed and output parameters were determined, i.e., cutting force and cutting temperature. The signal-to-noise ratio is used to investigate the influence of turning parameters on the response parameters. For examination, the following quality characteristics were selected, "smaller-the-better", "larger-the-better", and "nominal-the-better" (Tables 3 and 4).

"Smaller-the-better (minimize): signal-to-noise $= -10 * \log(\Sigma(Y2)/N)$"
"Larger-the-better (maximize): signal-to-noise $= -10 * \log(\Sigma(1/Y\hat{}2)/n)$"
"Nominal-the-best: signal-to-noise $= -10 * \log(S2)$".

Table 2 Factors and their levels

Parameters	Units	Levels		
Cutting speed	(m/min)	80	110	140
Feed	(mm/rev)	0.08	0.10	0.12
Depth of cut	(mm)	0.10	0.20	0.30

Table 3 Orthogonal array L9 of Taguchi

Experiment No.	Cutting speed (m/min)	Feed (mm/rev)	Depth of cut (mm)
1	80.0	0.08	0.10
2	80.0	0.10	0.20
3	80.0	0.12	0.30
4	110.0	0.08	0.20
5	110.0	0.10	0.30
6	110.0	0.12	0.10
7	140.0	0.08	0.30
8	140.0	0.10	0.10
9	140.0	0.12	0.20

Table 4 Experimental design and results

Exp. No.	Cutting speed (m/min)	Feed (mm/rev)	Depth of cut (mm)	Force in *X* (N)	Force in *Y* (N)	Cutting Temp. (°C)
1	80.0	0.08	0.10	107.2	68.6	645.0
2	80.0	0.10	0.20	45.35	25.0	745.0
3	80.0	0.12	0.30	194.7	100.2	760.0
4	110.0	0.08	0.20	90.42	51.2	656.0
5	110.0	0.10	0.30	161.8	89.1	680.0
6	110.0	0.12	0.10	66.9	30.1	687.0
7	140.0	0.08	0.30	126.2	73.3	661.0
8	140.0	0.10	0.10	153.6	84.6	677.0
9	140.0	0.12	0.20	113.8	54.94	802.0

3 Results and Discussion

For statistical analysis, Minitab 17 software was employed. The smaller-the-better technique of signal-to-noise ratio was used. The table of response for cutting speed, feed rate, and depth of cut is expressed based on the determined combined signal-to-noise ratio by averaging this ratio for each level of input parameters as presented in Table 5. From the response table, the main effects plot of *S/N* ratio is plotted for cutting speed, feed rate, and depth of cut as shown in Fig. 1.

On basis of response table and main effects plot of the combined signal-to-noise ratio, the level of optimum parameters was identified as 110 m/min cutting speed, 0.08 mm/rev feed rate, and 0.1 mm depth of cut (A2B1C1). Table 6 shows optimal cutting conditions. To analyze the impact on categorical factors on responses, the analysis of variance was used and it proved to be an effective method. Factors having the highest contribution to the quality characteristics are deduced through analysis of variance. Table 7 shows the analysis of variance for combined responses. It was observed that the feed rate has a significant effect on responses followed by cutting speed and depth of cut (Table 6).

From the ANOVA table, it is observed that the feed rate is the most significant factor contributing by 58.30%, followed by the depth of cut by 26.64% and cutting

Table 5 Response table for the signal-to-noise ratios (smaller-the-better)

Level	Cutting speed	Feed	Depth of cut
1	−54.24	−53.47	−53.67
2	−53.72	−54.09	−54.36
3	−54.23	−54.63	−54.15
Delta	0.53	1.16	0.69
Rank	3	1	2

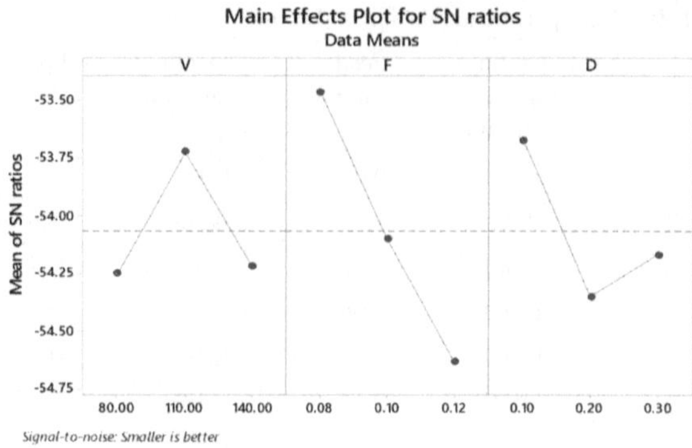

Fig. 1 Main effect plot of signal-to-noise ratio is shown on three different levels of cutting speed, depth of cut, and feed

Table 6 Optimal cutting condition

Cutting speed (m/min)	Feed (mm/rev)	Depth of cut (mm)
110.0	0.08	0.10
A2	B1	C1

Table 7 Analysis of variance

Source	DF	Seq SS	Cont. %	Adj MS	F-value	P-value
Cutting speed	2	3307.20	14.01	1653.6	13.3	0.070
Feed	2	13,764.4	58.30	6882.2	55.6	0.018
Depth of cut	2	6289.00	26.64	3144.5	25.4	0.038
Error	2	247.50	1.05	123.8		
Total	8	23,608.20	100.00			

Table 8 Model summary

S	R-sq	R-sq (adj)	Press	R-sq (pred)
11.1254	98.95%	95.81%	501.86	78.77%

Table 9 Responses at optimum cutting parameters

Cutting temperature (°C)	Cutting force (N)		
	F_x	F_y	$F_{rs/}$
574.63	64.29	37.91	74.63

speed by 14.01% (Tables 8 shows the model summary and Table 9 shows the cutting force and cutting temperatures at optimal conditions).

4 Conclusion

This paper evaluated the merits of the Taguchi method, as it aids in making decision for selecting the most optimal parameters to minimize the cutting force and cutting temperature in turning AISI 4340 hardened alloy steel. It is also evident that proper selection of process parameters appraises the quality attributes in turning of hard-to-cut material. The optimal cutting conditions were as cutting speed 110 m/min, feed rate 0.8 mm/rev, and depth of cut 0.10 mm. From the ANOVA table, it has been observed that the feed is the most significant factor contributing 58.03%, followed by the depth of cut 26.64% and cutting speed by 14.01%. It was observed that the hard turning at optimum cutting parameters decreases the cutting force and cutting temperature significantly, which improves the tool life and surface finish thereby reducing machining cost and increasing the economy of the operation.

References

1. Suresh R, Basavarajappa S, Gaitonde VN, Samuel GL (2012) Machinability investigations on hardened AISI 4340 steel using coated carbide insert. Int J Refract Metals Hard Mater 75–86
2. Das SR, Kumar A, Dhupal D, Mohapatra SK (2013) Optimization of surface roughness in hard turning of AISI 4340 steel using coated carbide inserts. Int J Inf Comput Technol 871–880
3. Aouici H, Yallese MA, Chaoui K (2012) Analysis of surface roughness and cutting force components in hard turning with CBN tool: prediction model and cutting conditions optimization. Measurement 344–353
4. Qasima A, Nisara S, Shaha A, Mohammed MSK, Sheikhb A (2015) Optimization of process parameters for machining of AISI-1045 steel using Taguchi design and ANOVA. Simul Model Pract Theory 36–51
5. Gupta MK, Sood PK (2015) Optimization of machining parameters for turning AISI 4340 steel using Taguchi based grey relational analysis. Indian J Eng Mater Sci 679–685
6. Das S, Mandal B (2016) Exploring machinability of AISI 4340 steel with coated carbide inserts. Adv Intell Syst Res 137:165–170
7. Singh P, Singh R (2014) Optimization of machining characteristics of D2 steel under different in turning conditions. Int J Innov Sci Eng Technol 219–223
8. Motorcu AR, Isik Y, Kus A, Cakir MC (2015) Analysis of the cutting temperature and surface roughness during the orthogonal machining of AISI 4140 Alloy Steel via the Taguchi method. Izvirni znanstveni lane 343–351
9. Aslan E, Camuşcu N, Birgören B (2007) Design optimization of cutting parameters when turning hardened AISI 4140 steel (63 HRC) with Aluminum oxide + TiCN mixed ceramic tool. Mater Des 1618–1622
10. Bouacha K, Yallese MA, Mabrouki T, Rigal JF (2010) Statistical analysis of surface roughness and cutting forces using response surface methodology in hard turning of AISI 52100 bearing steel with CBN tool. Int J Refract Metals Hard Mater 349–361

Optimization of Cutting and Cutting Fluid Application Parameters in Turning of AISI 52100 Hardened Alloy Steel Under Minimal Cutting Fluid Application

Sandip Mane and Sanjay Kumar

Abstract Hardened alloy steel materials are in great demand in automotive and allied manufacturing industries due to their better strength, wear resistance and high thermal stability. The hard turning offers many potential benefits over conventional grinding operations. During hard turning, the excessive heat generation and temperature affects the tool wear, tool life and surface integrity of the machined components. The cutting force and cutting temperature are the important technological parameter to control the machining process. The cutting fluid removes and reduces the heat from the cutting zone through cooling and lubrication action, but the use of conventional cutting fluid is being restricted due to the cost, health and environmental concern associated with its use in machining processes. The machining performance and cost-effectiveness can be improved by minimizing the cutting fluid. The main objective of this research work was to investigate the effect of cutting parameters and cutting fluid application parameters on cutting force and cutting temperature in turning of AISI 52100 hardened alloy steel with multilayer coated carbide insert under minimal cutting fluid application. The result showed that the minimal cutting fluid application (MCFA) technique significantly reduces the cutting force and cutting temperature compared to dry turning.

Keywords Hard turning · Minimal cutting fluid application · Tool life · Surface integrity

S. Mane (✉)
Department of Mechanical Engineering, Dwarkadas J. Sanghvi College of Engineering, Mumbai, Maharashtra, India
e-mail: sandip.mane@djsce.ac.in

S. Kumar
Department of Mechanical Engineering, Thakur College of Engineering and Technology, Mumbai, Maharashtra, India
e-mail: sanjay.kumar@thakureducation.org

© Springer Nature Singapore Pte Ltd. 2020
H. Vasudevan et al. (eds.), *Proceedings of International Conference on Intelligent Manufacturing and Automation*, Lecture Notes in Mechanical Engineering,
https://doi.org/10.1007/978-981-15-4485-9_12

1 Introduction

The hard turning is the viable alternative for the conventional grinding operation. Huang et al. [1] reported that the manufacturing cost can be reduced by 30%, if the hard turning process is employed to machine the complex parts. The AISI 52100 hardened alloy steel is being widely used in manufacturing of automotive components such as gear, excel, shaft, bearing, tool and die. Pal et al. [2] stated that the cutting forces increase drastically when machining materials with hardness higher than about 45 HRC. Bartarya et al. [3, 4] observed that the cutting force is mainly affected by depth of cut rather than cutting speed and feed rate in turning of AISI 52100 hardened alloy steel. Fnides et al. [5] studied the effect of cutting parameters on cutting temperature in turning of AISI H11 steel and showed that the cutting speed has more significant impact on cutting temperature followed by feed rate, whereas the depth of cut has no influence on cutting temperature in turning of AISI H11 hardened alloy steel of hardness 50 HRC using ceramic tool insert at dry condition. Lin et al. [6] have concluded that the cutting speed is most dominant factor, which raises the cutting temperature significantly in hard turning of hardened AISI 4340 alloy steel by cubic boron nitride (CBN) tool inserts. Bouchelaghem et al. [7] have explored the effect of cutting conditions on cutting temperature and reported that the cutting speed, feed and depth of cut are the most influential factors on cutting temperature; the increase in the levels of cutting parameters raises the cutting temperature in hard turning of AISI D3 steel at elevated hardness of 60 HRC using CBN tool inserts. Varadarajan et al. and Vikas Kumar et al. [8–10] reported that the minimal cutting fluid application technique uses very small quantity of cutting fluid in the form of pulsating jet at very high velocity is impinged at tool–chip interface. Sam Paul et al. [11] concluded that the turning of hardened alloy steel with semi-solid lubricants under MCFA environment showed significant reduction in tool wear, surface roughness, cutting force and cutting temperature. Raj et al. [12] analysed the effect of cutting and cutting fluid application parameters on cutting temperature and surface roughness in hard turning of OHNS steel of 45 HRC; the significant reduction in cutting temperature and surface roughness was observed. The minimal usage of cutting fluid makes this method resembles to dry machining and reduces the cutting force and cutting temperature significantly, and in turn, it improves the machinability characteristics remarkably.

2 Experimentation

2.1 Selection of Workpiece Material

In this study, AISI 52100 hardened alloy steel having hardness of 58 HRC was selected as workpiece material. AISI 52100 hardened alloy steel has wide applications and is being used in automotive and allied industries such as bearings, forming

Table 1 Chemical composition of AISI 52100 hardened alloy steel (weight percentage)

C %	Si %	Mn %	P %	S %	Cr %	Ni %	Cu %	Fe %
1.04	0.18	0.35	0.007	0.004	1.35	0.076	0.058	Balance

rolls, spindles, tools and precision instrument parts. Table 1 shows the chemical composition of the workpiece material.

2.2 *Selection of Tool*

The cutting tool inserts and the tool holder were selected based on the literature review and the tool manufacturer's recommendation. The MTCVD multilayer coated carbide ($TiN/TiCN/Al_2O_3$)—[HK150, K-type] cutting tool insert having specification CNMG120408 and the tool holder with PCLNR 2020 K12 specification were selected for experimentation. The experiments were carried out on a rigid high precision HMT NH-18 lathe machine.

2.3 *Selection of Cutting Fluid*

The quantity of cutting fluid delivered per pulse is extremely small; a commercially available SUN Cut ECO-33 high performance eco-friendly semi-synthetic cutting fluid was used in this investigation.

2.4 *Selection of Cutting and Cutting Fluid Application Parameters*

Based on the previous research carried out and the tool manufacturer's recommendations, the cutting parameters were selected. Table 2 presents the cutting parameters and their levels.

Table 2 Cutting parameters and their levels

Cutting parameters	Units	Levels		
Cutting speed	m/min	80	110	140
Feed rate	mm/rev	0.04	0.08	0.12
Depth of cut	mm	0.15	0.30	0.45

Table 3 Cutting fluid application parameters and their levels

Cutting fluid application parameters	Units	Levels		
Pressure	Bar	60	80	100
Frequency of pulsing	Pulses/min	200	300	400
Flow rate	ml/min	4	8	12
Nozzle stand-off distance	mm	20	30	40

Similarly, based on the previous studies conducted, the cutting fluid application parameters were selected. Table 3 presents the cutting fluid application parameters and their levels.

2.5 *Design of Experiment*

The experiments were carried out with Taguchi's L9 orthogonal array to reduce the number of experimentation without losing significance of each input parameter in turning of hardened AISI 52100 alloy steel under dry cutting condition and with minimal cutting fluid application method. The optimized values of cutting parameters were obtained under dry cutting condition. These optimized cutting parameters were kept constant, and cutting fluid jet application parameters were varied in three levels. Table 4 shows the experimental design and results for the responses in dry turning.

Table 4 Experimental result for cutting force and cutting temperature in dry turning

Exp. No.	Cutting speed (m/min)	Feed rate (mm/rev)	Depth of cut (mm)	Cutting force (N)	Cutting Temp. (°C)
1	80.000	0.04	0.15	157.194	556
2	80.000	0.08	0.30	186.830	619
3	80.000	0.12	0.45	253.313	643
4	110.00	0.04	0.30	182.700	526
5	110.00	0.08	0.45	217.817	585
6	110.00	0.12	0.15	152.720	621
7	140.00	0.04	0.45	265.470	680
8	140.00	0.08	0.15	184.470	654
9	140.00	0.12	0.30	224.149	749

Table 5 Experimental result for cutting force and cutting temperature in MCFA turning

Pressure (bar)	Frequency (pulses/min)	Quantity (mL/min)	Nozzle stand-off (mm)	Cutting force (N)	Cutting Temp. (°C)
60	200	4	20	223.05	607
60	300	8	30	196.65	615
60	400	12	40	211.60	566
80	200	12	30	201.70	538
80	300	4	40	202.39	561
80	400	8	20	191.72	527
100	200	8	40	184.42	504
100	300	12	20	131.26	418
100	400	4	30	173.43	100

2.6 Measurement of Cutting Force and Cutting Temperature

The cutting force was measured with the help of IEICOS Tool Lathe Dynamometer. The cutting temperature at the cutting edge of insert was measured with the help of embedded K-type thermocouple of range 50–1370 °C/−58–2498 °F and infrared thermometer (HTC IRX-68) of range 50–1850 °C/−58–3362 °F.

The cutting parameters that were optimized in dry turning were kept constant, and fluid application parameters were varied in three levels. Table 5 shows the design of experiment based on Taguchi's L9 orthogonal array and response parameters in turning of hardened AISI 52100 alloy steel under minimal cutting fluid application (MCFA) environment.

3 Result and Discussion

Table 6 shows the percentile contribution effect of the cutting parameters on the cutting force and cutting temperature in dry turning through analysis of variance (ANOVA).

Table 6 Percentile contribution of cutting parameters on responses in dry turning

Factors	Effect on cutting force (%)	Effect on cutting temperature (%)
Cutting speed	19.57	62.38
Feed	2.24	29.89
Depth of cut	77.56	3.14
Other error	0.63	4.60
Total %	100.00	100.00

It was observed that the depth of cut is the most significant cutting parameters affecting the cutting force followed by the cutting speed. The feed rate has no considerable effect on cutting force. On other hand, the cutting temperature is significantly affected by the cutting speed followed by the feed rate, whereas the depth of cut has no impact on the cutting temperature.

Figure 1 shows the column effect graph for cutting force and cutting temperature. Figure 1a shows that the cutting speed $V2 = 110$ m/min, feed $F2 = 0.08$ mm/rev and depth of cut $D3 = 0.15$ mm contributed most in minimizing the cutting force. Figure 1b presents the cutting speed $V2 = 110$ m/min, feed $F1 = 0.04$ mm/rev and depth of cut $D1 = 0.15$ mm which contributes more on reduction of cutting temperature.

Figure 2 shows the contour plot of cutting force versus cutting parameters. Figure 2a shows that the minimum value of cutting force 160–180 N observed between a cutting speed of 100–110 m/min, feed rate of 0.04–0.08 m/min and depth of cut 0.15–0.20 mm. Figure 2b, c shows that the cutting speed has less significant effect and feed has no effect on cutting force, whereas the depth of cut is most influential parameter on cutting force.

Figure 3 shows the contour plot of cutting temperature versus cutting parameters. Figure 3a shows that the minimum value of cutting temperature 520–550 °C observed

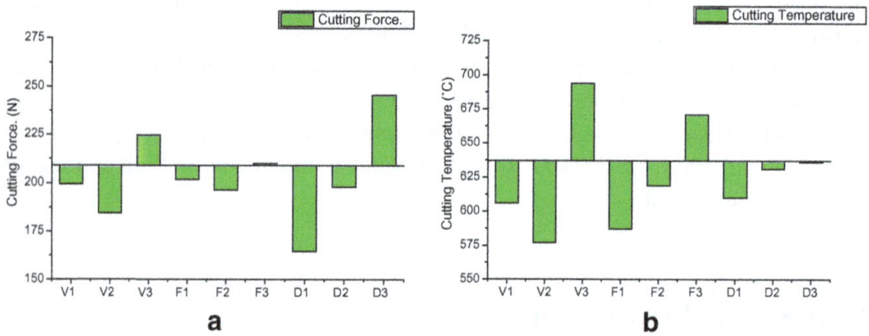

Fig. 1 Column effect graphs of **a** cutting force, **b** cutting temperature in dry turning

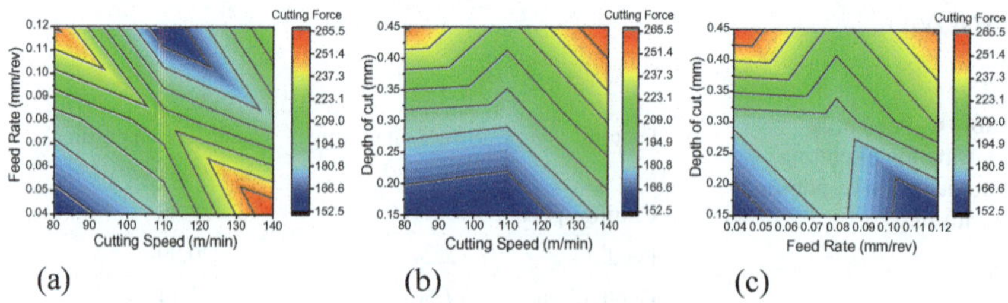

Fig. 2 Contour plot for cutting force **a** cutting speed versus feed, **b** cutting speed versus depth of cut, **c** feed versus depth of cut in dry turning

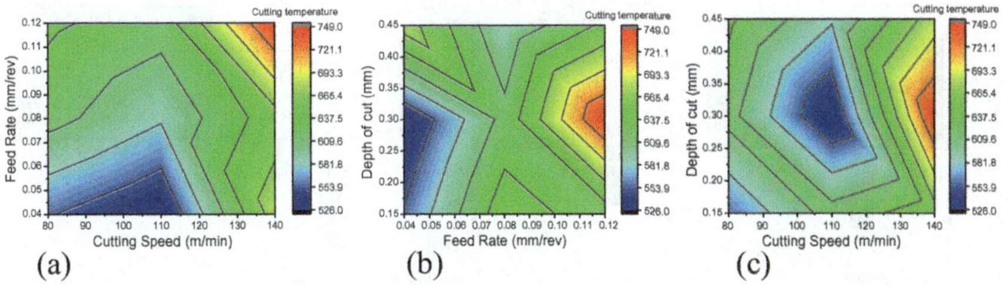

Fig. 3 Contour plot for cutting temperature **a** cutting speed versus feed, **b** cutting speed versus depth of cut, **c** feed versus depth of cut in dry turning

between a cutting speed of 100–110 m/min, feed rate of 0.04–0.08 m/min and depth of cut 0.15–0.30 mm. Figure 3b, c shows that the feed has less significant effect and depth of cut has no effect on cutting temperature, whereas the cutting speed is the most influential parameter on the cutting temperature.

Table 7 shows the percentile contribution effect of the cutting fluid application parameters on the cutting force and cutting temperature in turning with MCFA through analysis of variance (ANOVA). It has been observed that the cutting fluid pressure is the most significant fluid parameters affecting the cutting force and cutting temperature, whereas the frequency of pulsing, flow rate and nozzle and stand-off distance have less considerable impact on cutting force and cutting temperature.

Figure 4 shows column effects graph of various output parameters. The cutting fluid injected pressure at level-3 (100 bar), frequency at level-2 (300 pulses/min), quantity at level-3 (12 mL) and nozzle stand-off distance at level-1 (20 mm) contribute more on the reduction of cutting force and cutting temperature.

Variable Speed Test on AISI 52100 Hardened Alloy Steel

The results obtained during optimization were considered for conducting variable speed and feed tests. The cutting fluid parameters were kept constant as pressure: 100 bar, frequency of pulsing: 300 pulses/min, flow rate: 12 mL/min and nozzle stand-off distance: 20 mm. In variable speed test, the cutting speed was varied at five

Table 7 Percentile contribution of fluid application parameters on responses in MCFA turning

Factors	Effect on cutting force (%)	Effect on cutting temperature (%)
Pressure	64.65	72.65
Frequency of pulsing	17.45	9.34
Flow rate	8.71	8.28
Nozzle stand-off distance	8.10	7.61
Other error	1.10	2.11
Total	100.00	100.00

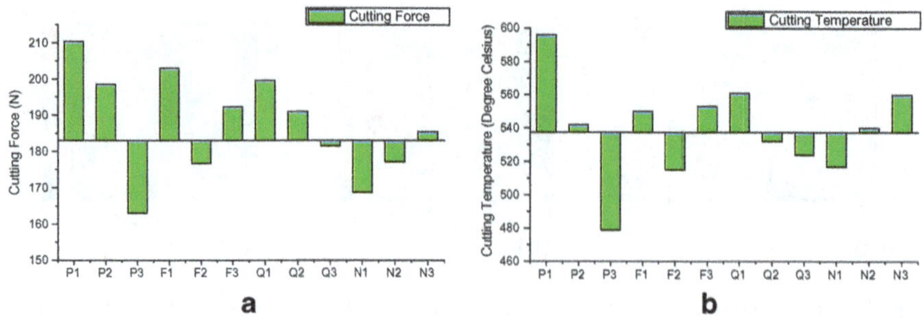

Fig. 4 Column effect graphs of **a** cutting force, **b** cutting temperature in turning with MCFA

Fig. 5 Cutting speed versus cutting temperature

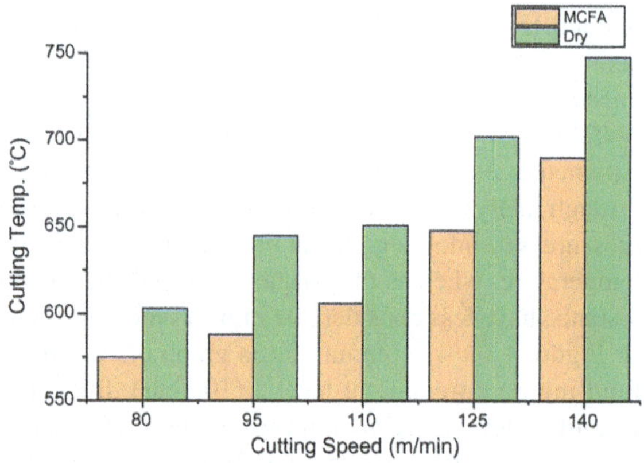

levels (80, 95, 110, 125 and 140 m/min) while the feed and depth of cut were kept constant at 0.08 mm/rev and 0.4 mm, respectively.

Figure 5 shows the comparison of effect of cutting speed on cutting temperature at constant feed rate and depth of cut in dry turning and turning with MCFA. The results obtained validated the optimized cutting fluid application parameters and showed the 15–20% reduction in the cutting temperature in turning with MCFA in comparison with dry turning.

4 Conclusion

The optimization of cutting parameters and cutting fluid application parameters helped in minimizing the usage of cutting fluid and other related issues. The optimum dry cutting conditions were cutting speed: 110 m/min, feed rate: 0.08 mm/rev and depth of cut: 0.15 mm. The optimized values of fluid application parameters were cutting fluid pressure: 100 bar, frequency of pulsing: 300 pulses/min, flow rate: 12 ml /min and nozzle stand-off distance: 20 mm. The variable speed test conducted proved that the optimized results are correct. The cutting force was reduced significantly by 25–30% and cutting temperature by 15–20% under (MCFA) environment, as compared to dry turning.

References

1. Huang Y, Chou YK, Liang SY (2007) CBN tool wear in hard turning: a survey on research progresses. Int J Adv Manuf Technol 35:443–453
2. Pal A, Choudhury SK, Chinchanikar S (2014) Machinability assessment through experimental investigation during hard and soft turning of hardened steel. Procedia Mater Sci 6:80 –91
3. Bartarya G, Choudhury SK (2012) Effect of cutting parameters on cutting force and surface roughness during finish hard turning AISI52100 grade steel. Procedia CIRP 1:651–656
4. Bartarya G, Choudhury SK (2014) Influence of machining parameters on forces and surface roughness during finish hard turning of EN 31 steel. Proc IMechE Part B: J Eng Manuf 228(9):1068–1080
5. Fnides B, Yallese MA, Aouici H (2008) Hard turning of hot work steel AISI H11: evaluation of cutting pressures, resulting force and temperature. Mechanika 72:59–63
6. Lin HM, Liao YS, Wei CC (2008) Wear behavior in turning high hardness alloy steel by CBN tool. Wear 264:679–684
7. Bouchelaghem H, Yallese MA, Mabrouki T, Amirat A, Rigal JF (2010) Experimental investigation and performance analyses of CBN insert in hard turning of cold work tool steel (D3). Int J Mach Sci Technol 14:471–501
8. Philip PK, Varadarajan AS, Ramamoorthy B (2000) Influence of cutting fluid composition and delivery variables on performance in hard turning using minimal fluid in pulsed jet form. J Inst Eng (India) 68–72
9. Vikram Kumar CHR, Ramamoorthy B (2007) Performance of coated tools during hard turning under minimum fluid application. J Mater Process Technol 185:210–216
10. Varadarajan AS, Philip PK, Ramamoorthy B (2002) Investigations on hard turning with minimal cutting fluid application (HTMF) and its comparison with dry and wet turning. Int J Mach Tools Manuf 42:193–200
11. Sam Paul P, Varadarajan AS (2013) Performance evaluation of hard turning of AISI 4340 steel with minimal fluid application in the presence of semi-solid lubricants. Proc IMechE Part J: J Eng Tribol 227(7):738–748
12. Raj A, Leo Dev Wins K, Varadarajan AS (2016) Evaluation of the performance during hard turning of OHNS steel with minimal cutting fluid application and its comparison with minimum quantity lubrication. In: IOP conference series: materials science and engineering, vol 149, pp 012021. https://doi.org/10.1088/1757-899X/149/1/012021

Supplier Selection in MSME Gear Manufacturing Industries Using MCDM Technique

Ashish J. Deshmukh and Hari Vasudevan

Abstract This study is an attempt to find out, whether the manufacturing industries follow only the traditional criteria in their supplier selection or they also include green criteria in the supplier selection. Data for the study were collected from 50 different gear manufacturing micro, small, and medium enterprises (MSME) sector and all the criteria were tested using a multi-criteria decision-making technique, such as the analytical hierarchy process (AHP). Results concerning the main criteria followed in the study show that cost, quality, delivery, and environmental manufacturing management are considered prime criteria in the supplier selection. Gear manufacturing MSMEs using various criteria concerning different supplier selection are discussed in this paper. The study is expected to act as a future guide for manufacturing industries, especially in the MSME sector in their evaluation and selection of suppliers, as they adopt various supplier selection criteria.

Keywords Traditional supply chain · Green supply chain · AHP · MSME · Sensitivity analysis

1 Introduction

Although, industries in the micro-small and medium enterprises (MSME) sector contribute remarkably to the economy of India, so far only a few research studies have been reported on exploring various aspects related to manufacturing firms in the sector, especially in the gear manufacturing MSMEs. The Indian gear manufacturing MSMEs are dependent on heavy engineering industries, such as textile, cement, sugar, etc., and light engineering industries, such as process control instruments, steel forging, bicycle, etc., for manufacturing varieties of gears. In MSMEs, different

A. J. Deshmukh
SVKMs NMIMS MPSTME, Mumbai, India
e-mail: ashish.deshmukh@nmims.edu

H. Vasudevan (✉)
Dwarkadas J. Sanghvi College of Engineering, University of Mumbai, Mumbai, India
e-mail: principaldjs@gmail.com

© Springer Nature Singapore Pte Ltd. 2020

H. Vasudevan et al. (eds.), *Proceedings of International Conference on Intelligent Manufacturing and Automation*, Lecture Notes in Mechanical Engineering,
https://doi.org/10.1007/978-981-15-4485-9_13

types of gears and gearboxes are produced according to the conditions laid down by the different heavy and light manufacturing industries, who are their customers.

Supply chain management (SCM) is about having an end-to-end view of operations from the point of production to delivery to the customer. With this approach, industries can service their customers through a better trade-off between the costs and service level. SCM requirements of gear manufacturing MSMEs diverge in different ways, including the availability of resources such as manpower, raw materials, energy, information, and finance to obtain the desired output. Supplier selection is a very important aspect to be addressed in the context of the management of supply chain [5]. The authors in this study gathered various research papers on criteria for supplier selection (traditional and green), published from 1992 to 2007.

In the above backdrop, the current study is an attempt to analyze the use of analytical hierarchy process (AHP), a multi-criteria decision-making (MCDM) method and determine its applicability in the supplier selection process. A comprehensive literature review was first carried out to arrive at an understanding of various studies that have used AHP as a multi-criteria decision-making method as well as to understand various criteria adopted by the industries. Subsequently, an in-depth interview with senior industry personnel, such as purchase managers, proprietors, and general managers of different gear manufacturing MSMEs was conducted. Cost, quality, risk, environmental performance assessment, service performance, innovation and learning, deliver and environmental manufacturing management are the total eight criteria that were identified along with 40 sub-criteria to be adopted as part of the study. Data for the study were collected from 50 different gear manufacturing MSMEs in India and the relationship suggested in the framework was verified and analyzed using AHP, and the conclusions were drawn accordingly.

2 Literature Review

AHP is a multi-criterion decision-making process that is exclusively useful while dealing with multifaceted problems by using a nine-point scale [10]. AHP is a perfect technique for providing the preferences to the alternatives when more number of alternatives (criteria) and sub-alternatives (sub-criteria) are existing in the process of decision-making [12]. AHP uses a process to prefer alternatives and due to this reason, AHP is suitable for the process of supplier selection [2]. AHP and preference ranking organization method for enrichment evaluation (PROMETHEE) is used as a tool for information system outsourcing decisions. The AHP is used to determine the weights and criteria for outsourcing problems and PROMETHEE is used for final ranking [17]. AHP and TOPSIS were used to find out the faculty performance in engineering education [18]. AHP is one of the best tools for deciding the complex criteria at different levels and fuzzy AHP is the extension of AHP for to consider the fuzziness of the decision makers. AHP and FAHP are used for supplier selection in the Indian gear manufacturing MSME sector and to find the traditional and green manufacturing factors [8]. Decision makers and researchers have most extensively used the AHP approach as an MCDM process since its invention [13]. AHP was used to evaluate various transport policies to reduce climate change impact [3].

Green supply chain management (GSCM) is considered a major part of the companies in their effort to maintain environmental friendly policies to meet the demand of the customers [6]. Green and traditional supplier selection criteria identified and considered together for supplier selection in plastic manufacturing MSMEs India [9]. Multi-criteria approach was also used for evaluating the environmental performance of suppliers [11]. First time the combined traditional and green supplier selection criteria checked in the case of Indian manufacturing MSMEs [7]. Reverse logistics, green marketing, purchasing (green), and manufacturing (green) are four major activities that are covered in GSCM in the Indian context [4].

3 Methodology

For analyzing and organizing a complex decision, analytic hierarchy process (AHP) is used as a structured technique [1]. It was developed by Thomas L. Saaty in 1970 [14]. In AHP, the pair-wise comparison is used to determine the preferences between the alternatives [16]. AHP uses a 1–9 numerical scale for preferences, where 1 is used for "equal importance" and 9 is used for "extreme importance". If one criterion is extremely important than the other criterion, then it is indicated by the value 9 and the other criterion which is extremely more important than the first is indicated by the value 1/9 in the pair-wise comparison [15]. The advantage of structuring the problem is better understanding of the decision is to be achieved and all alternatives are to be evaluated. Another advantage is that it allows tangible factors (e.g., cost with cost) as well as intangible factors (e.g., cost with quality) as criteria in the decision. It is to be noted that the data obtained for AHP have been collected from industry experts of gear manufacturing MSMEs by using the direct interviews and questionnaires. Hierarchy for main criteria, sub-criteria, and alternatives is formed with the goal as selection of best supplier.

4 Analysis and Discussion

This study illustrates the application and use of AHP for supplier selection and the problem consisted of prioritizing three suppliers, from whom material had to be purchase based on eight major criteria, deemed to be important for the organization. Gear manufacturing MSMEs are the main suppliers for heavy engineering (textile, cement, sugar, material handling, etc.) and light engineering (process control instruments, steel forging, bicycle, etc.) industries because most of these industries subcontract MSMEs for the gears. Nowadays, because of the increasing concern over environmental safety, gear manufacturing MSMEs is answerable for their final products and overall sustainability. This research paper has been an attempt to find out whether the companies are following only the traditional way for supplier selection or they are including the green criteria also. This study has also explored major factors included in the green supply chain along with the traditional criteria for the supplier selection process [6]. In this research paper, decision hierarchy was formed for supplier selection problem and the weights assigned using the AHP method for the evaluation process. Industry experts helped in pair-wise comparison using the Saaty's scale.

An opinion given by industry experts for the major criteria in the form of the pair-wise comparison matrix is in Table 1. After getting the pair-wise comparison from industry experts, the next step in AHP was the calculation of local weights of individual component in the matrix. Algebra is used for calculating the local weights of the individual component in the matrix to find the total weights in matrix calculation. In the next step, to normalize each column of an individual component to the sum to 1, the element of the column was divided by the total of the column and then summed up. Finally, all the elements of each row are added and the sum of that row was divided by the total number of elements in that row to find the average shown in the last column (Local weights) of Table 1.

At last, the consistency ratio (CR) and the consistency index (CI) are calculated. Inconsistent matrix, the eigenvalue value calculated has to be greater than or equal to the number criteria, i.e., $\lambda_{max} \geq n$ (where λ_{max} is the maximum eigenvalue and n is the number of criteria), and hence the judgments are decided to be consistent. In the next step, the consistency index is to be calculated from $(\lambda_{max} - n)/(n - 1)$, the need for that to assess the judgment made completely at random. Finally, Saaty suggested that CR go beyond 0.1, the judgment given by the respondent is invalid and if the CR is less than nor equal to 0.1, then the judgments are accepted. Table 1 shows eigenvalue $\lambda_{max} \geq 8$, CI is 0.0437, and CR is 0.0300. This means for the pair-wise comparison calculations for the main criteria is valid and was accepted.

Final results for the main criteria are as shown in Fig. 1 and it was observed that the local weights of the criterion, cost, quality, and delivery were high compared to others. Therefore, these criteria are ranked as first, second, and third, whereas the others are showing less importance in the process of supplier selection.

Similarly, for all sub-criteria, the opinion given by the industry experts and matrix for pair-wise comparison was calculated. Figure 2 shows that in sub-criteria of cost, the purchase cost is preferred even today (with its weight 0.496) for supplier selection by industries. Freight cost is the cost incurred for moving goods and recycling cost shows equal importance (with its weight 0.158). Cost for recycling generally comes in green supplier selection criterion and from the result, and it was observed that it was also taken into account in gear manufacturing MSMEs. In quality criterion, the reliability of component (the ability of the component to function under the given condition for a specified period) shows the highest importance (with the weight of 0.420). Supplier capacity (the amount that the supplier could supply), supply variety, and buffer stock (supply of inputs held as a reserve to safeguard against unforeseen shortages) with its weights 0.486, 0.151, and 0.131 are at first three positions.

Table 1 Direct and indirect effect of major criteria

Goal	C	Q	S	E	R	D	EP	I&L	Local weights
C	1	3.0000	5.0000	4.0000	6.0000	3.0000	6.0000	5.0000	0.4660
Q	0.3333	1	2.0000	1.0000	3.0000	1.0000	3.0000	2.0000	0.1120
S	0.2000	0.5000	1	0.5000	1.0000	0.5000	1.0000	1.0000	0.0790
E	0.2500	1.0000	2.0000	1	2.0000	0.5000	0.5000	1.0000	0.0660
R	0.1667	0.3333	1.0000	0.5000	1	0.3333	1.0000	0.5000	0.0660
D	0.3333	1.0000	2.0000	2.0000	3.0000	1	3.0000	2.0000	0.0910
EP	0.1667	0.3333	1.0000	2.0000	1.0000	0.3333	1	0.5000	0.0770
I&L	0.2000	0.5000	1.0000	1.0000	2.0000	0.5000	2.0000	1	0.0440

C cost, *Q* quality, *S* supplier performance, *E* environmental manufacturing management, *R* risk, *D* delivery, *EP* environmental performance, *I&L* innovation and learning

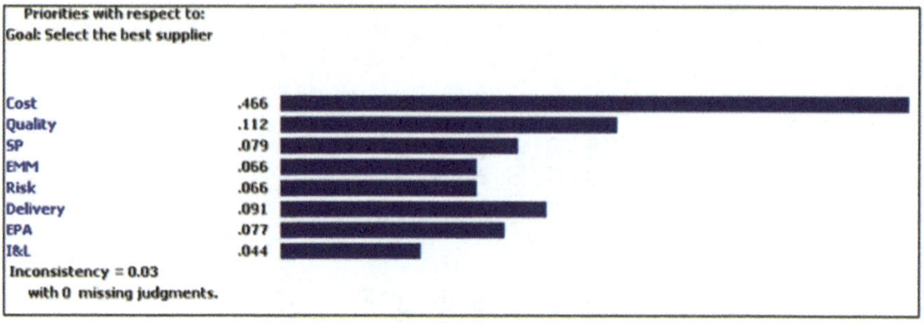

Fig. 1 Results of main criteria for gear manufacturing MSMEs

Fig. 2 Results of cost criteria

For all 8 main criteria and 40 sub-criteria, a matrix of pair-wise comparison was formed and their weights calculated. To find the overall rating of the suppliers, finally their global weights are calculated. The results and their global weights for each supplier are shown in Fig. 3. It can be noted that among the three given suppliers, supplier "1" is better than others. Therefore, supplier "1" satisfies the objectives and goals of one of the gear manufacturing MSMEs. As can be seen, the supplier 1s' score of (0.5973) is greater than the other suppliers' supplier 2 (0.3033) and supplier 3 (0.1883). Similarly, for all the 50 gear manufacturing, MSMEs best supplier is selected for this research.

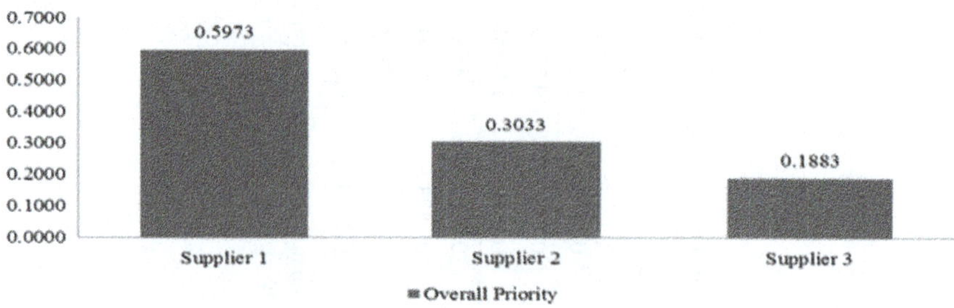

Fig. 3 Final ranking and scores of the suppliers

5 Sensitivity Analysis

Sensitivity analysis recognizes the effect of changes in the priority of criteria on the supplier's performance and order quantities. In this research study, dynamic sensitivity analyses were carried out to find the overall response of alternatives and the changes in the relative importance of each criterion. Figure 4 shows the prioritization of alternatives and their dynamic sensitivity concerning suppliers. Prioritization of alternatives on the left-hand side of the figure shows that for the cost criterion in alternative rankings and on right-hand side of the figure found that supplier 3 is at first position followed by supplier 1 and supplier 2. Using the software expert choice, it is possible to move the criteria forward and backward and see the effect of the same on the suppliers.

It is observed from Table 2 and Fig. 5 (which is for service performance criterion) that if we vary the weight 0–20% then supplier 3 is performing better than supplier 1 and 2. Increasing the weight of service performance criterion from 40 to 60% shows that supplier 1 and supplier 3 are coming closer for their weights and it is a very crucial decision for the manager as to how much of the order will be allocated to

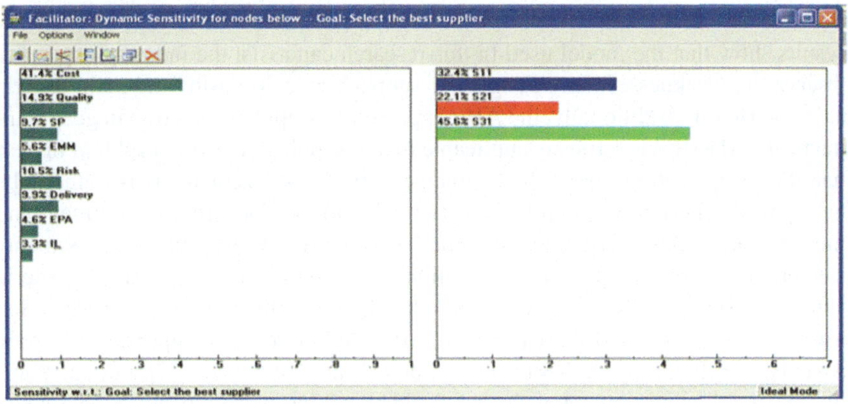

Fig. 4 Dynamic sensitivity analysis

Table 2 Suppliers ranking for service performance criterion

Area	Service performance criterion (%)	Suppliers' ranking
1	00.00–20.00	S3 > S1 > S2
2	20.00–40.00	S3 > S1 > S2
3	40.00–60.00	S1 > S3 > S2
4	60.00–80.00	S1 > S3 > S2
5	80.00–100	S1 > S3 > S2

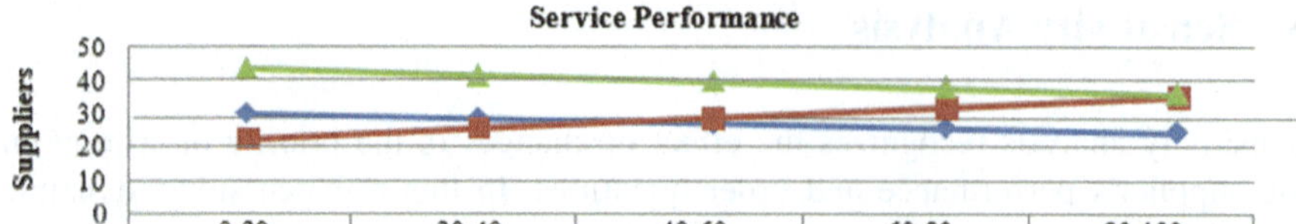

	0-20	20-40	40-60	60-80	80-100
S1	31.4	30	28.8	27.4	26.1
S2	24.2	27.4	30.4	33.4	36.4
S3	44.3	42.5	40.8	39.1	37.4

Fig. 5 Results of variation in trends for supplier selection with changes in the percentage of weights for a service performance criterion

supplier 1 and supplier 2 if supplier 3 does not satisfy all the demand. Further, if we increase the weights from 60 to 80%, then supplier 2 is found performing better than supplier 1.

6 Conclusion

Results show that the model used in this research can assist the industry experts to observe the weaknesses and strengths of supplier selection using the main criteria and sub-criteria. Additionally, the AHP approach for supplier selection in gear manufacturing MSME industries is applicable to any supplier selection problem in gear manufacturing sector in India. In the main criteria, "cost" as on today is found to be the topmost criterion for supplier selection. Due to pressure from government and international market environmental norms "environmental performance assessment" is emerging as criteria for supplier selection. In assessing these suppliers, 40 criteria were considered, so that each criterion adds their weight in the supplier selection process and it was found that "purchase cost," "reliability of component," "supply capacity," "technology and R&D management," "production risk," "production volume changes," "achievement of cleaner production," "product innovation" are ranked first with respect to all main criteria. The study also found that supplier 1s' score of (0.5973) is greater than the other suppliers', supplier 2 (0.3033) and supplier 3 (0.1883). If the industries are willing to assess and compare for their next suppliers, the same work that has been carried out in this paper can be reused. The projected model using AHP is remarkably successful in the process of decision-making. But in further research, researches could use the matrix with fuzzy numbers for pair-wise comparison and this could further improve the proposed method.

References

1. Albayrak E, Erensal YC (2009) Using analytic hierarchy process (AHP) to improve human performance: an application of multiple criteria decision-making problem. J Intell Manuf 15:491–503
2. Awasthi A, Chauhan SS, Goyal S (2010) A fuzzy multi-criteria approach for evaluating the environmental performance of suppliers. Int J Prod Econ 126(2):370–378
3. Buyukozkan G, Cifci G (2012) A novel hybrid MCDM approach based on fuzzy DEMATEL, fuzzy ANP, and fuzzy TOPSIS to evaluate green suppliers. Expert Syst Appl 39(3):3000–3011
4. Chatterjee D, Mukherjee B (2010) Study of the fuzzy-AHP model to search the criterion in the evaluation of the best technical institutions: a case study. Int J Eng Sci Technol 2(7):2499–2510
5. Deshmukh A, Chaudhari A (2011) A review for supplier selection criteria and methods. Syst Manag 145:283–291
6. Deshmukh A, Vasudevan H (2014) Emerging supplier selection criterion in the context of traditional and green supply chain management. Int J Manag Value Supply Chain (IJMVSC) 5(1):19–33
7. Deshmukh A, Vasudevan H (2016) Analysis of supplier selection criteria in traditional as well as green supply chain management in Indian MSMEs. Int J Bus Quant Econ Appl Manag Res 3(3):73–85
8. Deshmukh AJ, Vasudevan H (2018) A combined approach for supplier selection using AHP and fuzzy AHP in Indian gear manufacturing MSMEs. Mater Sci Eng IOP Ser 376:1–7
9. Deshmukh AJ, Vasudevan H (2018) Supplier selection in plastic products manufacturing MSMEs using a combined traditional and green criteria based on AHP and fuzzy AHP. ICIMA, Lecture notes in mechanical engineering. Springer, Singapore, pp 593–600
10. Ertugrul I, Karakasoglu N (2009) Performance evaluation of Turkish cement firms with fuzzy analytic hierarchy process and TOPSIS methods. Expert Syst Appl 36(1):702–715
11. Kannan G, Sarkis J, Sivakumar R, Palaniappan M (2012) Multi-criteria decision making approaches for green supplier evaluation and selection: a literature review. In: Conference on the greening of industry network, GIN 2012 (Linkoping, Sweden, 2012)
12. Noci G (1997) Designing green vendor rating systems for the assessment of a suppliers environmental performance. Eur J Purch Supply Manag 3(2):103–114
13. Omkarprasad SV, Kumar S (2006) Analytic hierarchy process: an overview of applications. Eur J Oper Res 169:1–29
14. Saaty TL (1980) The analytic hierarchy process. McGraw-Hill, New York
15. Sarkis J, Talluri S (2004) Evaluating and selecting e-commerce software and communication systems for a supply chain. Eur J Oper Res 159(2):318–329
16. Taha HA (2003) Operations research. Pearson Education Inc., Fayetteville
17. Wang JJ, Yang DL (2007) Using a hybrid multi-criteria decision aid method for information systems outsourcing. Comput Oper Res 34(12):3691–3700
18. Wu GC, Ding JH, Chen PS (2012) The effects of GSCM drivers and institutional pressures on GSCM practices in Taiwan's textile and apparel industry. Int J Prod Econ 135:618–636

Agricultural Supply Chain Using Blockchain

Ahan Fernandez, Ashriel Waghmare, and Shweta Tripathi

Abstract A constant rise in population has led to exponential growth in the food requirements all over the world. Not only has the demand increased over time, but concerns regarding food quality and safety have also proliferated. Many cases of food contamination like the Sudan Red food colouring and horse meat scandal have caused havoc in the food supply industry. This is not limited to the production sector only, but even the agricultural supply chain has felt the added scrutiny for the safety of crops. Hence, under such circumstances, it is essential to ensure food safety. Customers question the handling of crops during transportation and packaging. Quality of food also becomes priority. Concerns arise regarding, if the farmers get paid adequately. This brings the need of a system that can enable not only the grocery markets, but also the consumers to track the origin of the food they buy and all the other aspects involved in the agricultural supply chain. The current farmer-to-table supply chain model in India is archaic and requires immediate reforms to improve the plight of Indian farmers. Lack of structured transactions has allowed corruption to creep into every stage of food reaching from the farms to a consumer's table. The current model has almost seven levels of intermediates that affect the transactions. The blockchain model developed here aims to restrict the middleman's intrusive involvement and aims to improve farmer's profits. Every transaction will be authenticated and traceable, improving transparency for customers and removing unnecessary monetary leaks, thus optimizing the income of the farmers.

Keywords Supply chain · Blockchain · Traceable · Transaction

A. Fernandez · A. Waghmare (✉) · S. Tripathi
Computer Department, Fr.C. Rodrigues Institute of Technology, Agnel Technical Complex, Vashi, Navi Mumbai, India
e-mail: ashriel31@gmail.com

A. Fernandez
e-mail: fernandez.ahan@gmail.com

© Springer Nature Singapore Pte Ltd. 2020
H. Vasudevan et al. (eds.), *Proceedings of International Conference on Intelligent Manufacturing and Automation*, Lecture Notes in Mechanical Engineering,
https://doi.org/10.1007/978-981-15-4485-9_14

1 Introduction

Blockchain is an efficient technology that has pioneered solutions in many domains. India is an agrarian country and yet the conditions and arduous struggles of our farmers required immediate solutions. As of 2018, in Maharashtra alone, more than 60,000 suicides had taken place, with an average of 10 suicides every day [1]. This catastrophe needs rectification. Blockchain can really improve the tracking of crops and eliminate any wastage of resources (monetary or crops), while transferring from farmer to customers. The system is like a ledger record, having every transaction, monetary and process, at every stage of movement of the crop. It begins with recording, which farmer harvests and sends the crop to a storage facility including a timestamp. The blockchain is public, allowing new farmers to assimilate easily. The system would be operating in real time and accessible to all. Currently, the project aims to tackle only one crop, i.e. grapes. The attributes it factors in are, cost, time and to some extent quality. This can be scaled up to crop management, soil management and extend to other crops as well. The input data used is primarily from the state of Maharashtra, since grapes are predominantly cultivated in the state of Maharashtra. The system can be used for the entire country. The ease of utility for farmers, so that they can independently get on the blockchain and utilize tools to understand the market prices, mark up on their crops and the profit margins they can get. This can be achieved using a mobile application.

2 Literature Survey

The study is aimed to guarantee optimized income for farmers from a supply chain management perspective. The key issue is building a decentralized information system which is transparent for the whole agricultural supply chain. By using blockchain technology, this new decentralized information system could prove to be an innovation, providing openness, transparency, neutrality, reliability and security [2]. The proposed an information system for food safety monitoring in supply chains, based on HACCP for China's food supply chain market. He suggested a system using blockchain and Internet of things technologies, which establishes a food supply chain traceability system for real-time food tracing, in order to realize the traceability with trusted information in the entire food supply chain and effectively guarantee food safety by gathering, transferring and sharing authentic data of food products in production, processing, warehousing and distribution. Finally, the logistics capability of the cold-chain logistics enterprise could influence the quality and safety of food products directly. All of these will ultimately enhance the safety assurance of a food supply chain. The author also discussed how this new information system could improve the performance of a cold-chain logistics enterprise and demonstrated a performance evaluation research for a Chinese agri-food cold-chain logistics enterprise by using improved analytic hierarchy process (AHP) and fuzzy comprehensive evaluation methods. His research provided some ideas, methods and management suggestions from a supply chain management perspective for improving the quality and safety of food products [3] and discussed the future challenges on the use of blockchain for food traceability. The steady increase in food falsification, which has caused large economic losses and eroded consumers' trust, has become a pressing issue for producers, researchers, governments, consumers and other stakeholders. Tracking and authenticating the food supply chain to understand provenance are critical with a view to identifying and addressing sources of contamination in the food supply chain worldwide. One way of solving traceability issues and ensuring transparency is by using blockchain technology to store data from chemical analysis in chronological order so that they are impossible to manipulate afterwards. This review examines the potential of blockchain technology for assuring traceability and authenticity in the food supply chain. It is thus considered to be a relevant approach to assure the quality of the third step of the analytical processes: data acquisition and management.

3 Study of Agricultural Supply Chain Using Blockchain

3.1 Supply Chain

A processing-based and organized agri-supply chain functions as a part of a very complex network. Figure 1 shows a generic supply chain at the organization level

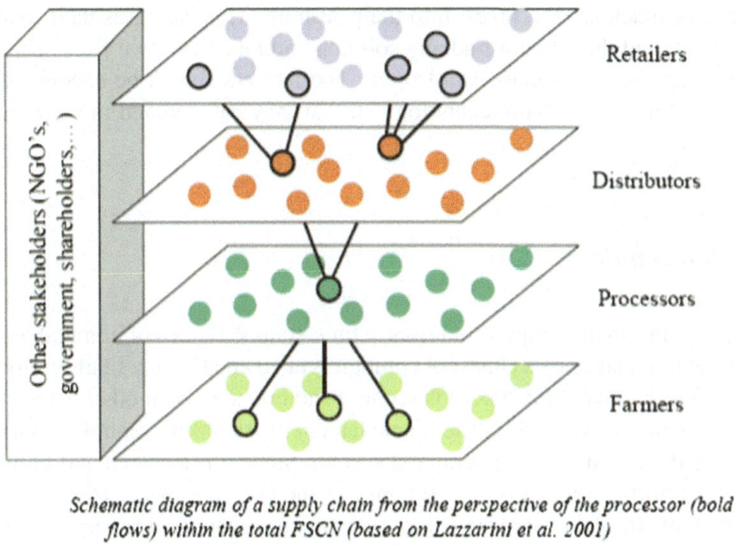

Fig. 1 Schematic diagram of supply chain

within the context of a complete supply chain network. Each firm is positioned in a network layer and belongs to at least one supply chain, i.e. it usually has multiple (varying) suppliers and customers at the same time and overtime.

3.2 The Traditional Route

The current "Mandi" route is that, first a farmer harvests his crops. Most of them do not have facilities to refrigerate; hence, they have a limited time frame to sell these crops. They loan transportation and pack their crops on, to send to APMCs. APMC is designated centres by the government that are used to assimilate fruits and vegetables from a plethora of farmers. APMCs are self-governed and determine the price of crops depending on various factors. The current price of the crop is then sent across to the farmers. The farmers send forward their crops depending on these prices. These prices, however, may fluctuate due to the demand and supply ratio. During transportation, octroi is usually charged for heavy vehicles while transporting across borders. This marks up the original cost incurred to the farmer. The precise and detailed changes in these prices are unknown to farmers, and this lack of clarity is targeted by the middlemen to avail profits for themselves. The trucks unload at APMC and begin selling. The crops are perishable, and if not sold within a certain time limit, they will rot. Hence in the unfortunate situation that the crop does not fully sell, the crop is sold at dearth cheap prices, which again incurs losses to the farmer. This whole model is flawed since the farmers do not get their deserved profits. They work hard, grow crops and harvest them, and after that they cannot track how they can maximize their profits. They make meagre money; in most cases, they do not make as much as they invest into their farming. This increases their loans and after a point, and they do not find any solution and take their own lives. This helps in preventing the unfortunate catastrophe. The farmers should be encouraged and supported to use blockchain technology so that they can demand more money for their labour.

3.3 Blockchain

A blockchain is, in the simplest of terms, a time-stamped series of immutable record of data that is managed by a cluster of computers not owned by any single entity. Each of these blocks of data (i.e. block) is secured and bound to each other using cryptographic principles (i.e. chain). The blockchain network has no central authority—it is the very definition of a democratized system. Since it is a shared and immutable ledger, the information in it is open for anyone and everyone to see. Hence, anything that is built on the blockchain is by its very nature transparent and everyone involved is accountable for their actions [4]. A blockchain carries no transaction cost (an infrastructure cost yes, but no transaction cost). The blockchain is a simple yet ingenious way of passing information from A to B in a fully automated and safe manner. One party to a transaction initiates the process by creating a block. This block is verified by thousands, perhaps millions of computers distributed around the net. The verified block is added to a chain, which is stored across the net, creating not just a unique record, but a unique record with a unique history. Falsifying a single record would mean falsifying the entire chain in millions of instances. That is virtually impossible. Bitcoin uses this model for monetary transactions, but it can be deployed in many other ways. Picture a spreadsheet that is duplicated thousands of times across a network of computers. Then imagine that this network is designed to regularly update this spreadsheet and you have a basic understanding of the blockchain. Information held on a blockchain exists as a shared and continually reconciled database. This is a way of using the network that has obvious benefits. The blockchain database isn't stored in any single location, meaning the records it keeps are truly public and easily verifiable. No centralized version of this information exists for a hacker to corrupt. Hosted by millions of computers simultaneously, its data is accessible to anyone on the internet.

The farmer joins the blockchain, and his farming techniques and processes are authenticated using either digital certificates or physical site visits. Once this process is done, the farmer can easily get on the blockchain and start communication with the respective parties. Each farmer inputs the type of crop harvested, date of harvest, quality of the crop (organic or non-organic) and total cost incurred. If the crop further goes to a warehouse for refrigeration and storage, the distributor inputs the date of arrival, storage costs and miscellaneous costs. The transporter inputs transportation costs, octroi charges, date of departure and date of arrival. Like this, every stage that a crop goes through, to reach a customer's table is documented and can be seen transparently by every participant.

4 System Designs

4.1 Block Diagram

The design aims at tracking three factors, viz. source, cost and quality:

- Current source of the crops, from which region is they exported, or which places they had been stored.
- Cost of the crops as it moves through the supply chain from farmer to vendor, how much profit or loss is incurred by each participant.
- Quality of the commodity, whether they have been grown through organic or inorganic methods, whether they are fresh from harvest or not. As blockchain keeps track of data fed into it with assured integrity, all the above factors can be looked through in the blockchain as and when demanded. Further, a Hyperledger framework for the development of the blockchain application is chosen.

Some of the features provided by Hyperledger framework are:

- Provisions for ledger control
- Privacy of ledgers through channels
- Security provided through membership services
- Consensus of transaction added to the blockchain.

The traditional agricultural supply chain model involves participants, the business network storing all factors of the supply chain and the transactions. When implemented on Hyperledger, the model involves the agricultural blockchain network executing on the run-time environment and the channels to each ledger. A brief model of the same is described in Fig. 2. Each participant can get a view of the data relevant to their roles and access rights. As the ledgers are separated through channels, an additional level of data privacy is obtained.

The basic components of this architecture are:

1. **Participants**: The participants involve all actors who are interacting with the blockchain right from the farmer who produces and sells his crops to the distributor, right up to the vendor who sells those crops to the customers, such as supermarkets and vegetable/fruit vendors in markets. Each of these participants interacts with the blockchain for adding transactions on to the ledger as well as tracing those transactions back whenever required.
2. **Assets**: The assets define the commodities that are going to be exchanged among the participants. These not only include the crops in our project but also the money that is being handed over to the seller. In every transaction, the exchange or transfer of assets takes place, which is then recorded into the ledger and a block

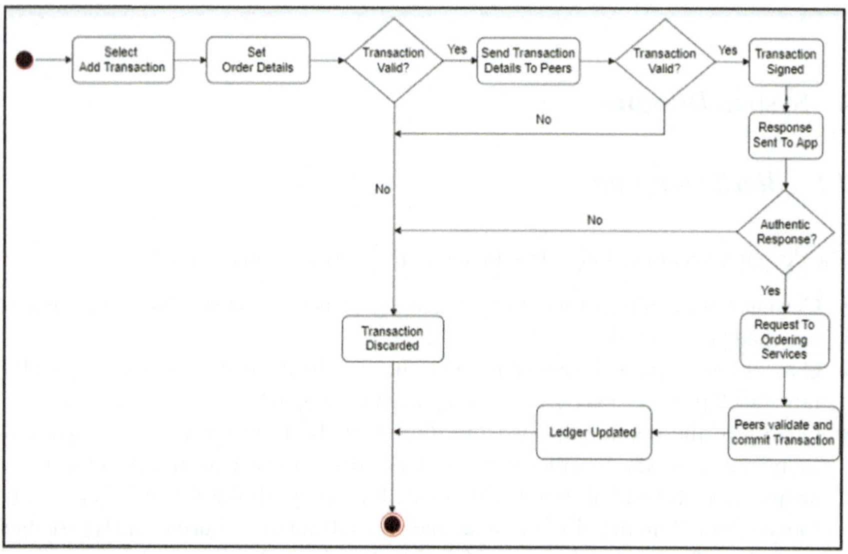

Fig. 2 Activity diagram for transaction in Hyperledger

gets added in the blockchain. Initially, a certain amount of the assets would be made available to the users. Using these initial asset values, they can perform transactions and exchange the ownership of assets.

3. **Transactions**: The transactions are not just restricted towards monetary exchanges but also product ownership exchange. In simpler words, when the ownership of a crop is handed over from one participant to another, we do consider that as a transaction; as we need to track when the commodity was handed over to a particular person in a specific point in time. This would not only help us trace the origin of any problems like product defect or mishandling, but would also help us to mitigate those weak and faulty points in the supply chain.
4. **Blockchain Network**: The agricultural blockchain network involves the main business logic which runs the entire supply chain blockchain. It authenticates users and the transactions, maintains the integrity of those transactions, separates the ledgers through channels and performs and adds transactions on those ledgers. The blockchain network runs on a runtimes environment, which in this case is Hyperledger Fabric. It is accessible to all the participants of the supply chain through a variety of interfaces including desktop and mobile applications.

5 Hyperledger Environment

Hyperledger provides a run-time environment for executing or deploying our blockchain projects. The run-time environment includes a REST server environment, embedded Node.js runtime and the Hyperledger framework. We would be able to access all these through the terminal in CLI mode. It can be manually switched on and off the server environments, add business cards or import the business network to the Hyperledger framework [5].

Activity Diagram

Figure 2 shows the activity diagram for the transactions in hyperledger.

Transactions in Hyperledger take place in the following eight steps:

1. Transaction proposal is created and submitted by the users
2. Transaction proposal is sent by the application to the endorsing peers
3. Endorsing peers send a response by verifying and signing the proposal
4. Verification and integrity check of response takes place
5. Request invocated to blockchain service to add transaction to ledger
6. Invocation response sent to peers
7. Peers validate and commit transactions
8. Ledger is updated to show up-to-date copy.

6 Conclusion and Scope

Complexities of the various stages of an agricultural supply chain were studied. By scaling the supply chain activities, an attempt was made here to improve the output primitives of a supply chain. To enhance transparency in the supply chain, focus was laid on the cost and on tracking its transfer from hand to hand. This promises to give an impetus to the credibility of crops and allow the farmers to understand, how to get right benefit. There are already existing systems that use artificial intelligence to predict prices of crops daily. There are companies using Ethereum tokens to handle transactions. This system aims to improve the transparency of the supply chain, which in turn will improve profit margins for farmers.

References

1. Maharashtra crosses 60,000 farm suicides. People's archive of rural India (PARI). www.ruralindiaonline.org. Retrieved 25 Mar 2019
2. Feng T (2018) An information system for food safety monitoring in supply chains based on HACCP, blockchain and internet of things. Doctoral thesis, WU Vienna University of Economics and Business
3. Galvez JF, Mejuto JC, Simal-Gandara J (2018) Future challenges on the use of blockchain for food traceability analysis. Trends Anal Chem. https://doi.org/10.1016/j.trac.2018.08.011
4. Rosic A (2016) 17 blockchain applications that are transforming society. Blockgeeks
5. Blummer T et al An introduction to Hyperledger—V1.1. https://www.hyperledger.org/wp-content/uploads/2018/08/HLWhitepaperIntroductiontoHyperledger.pdf

An Empirical Analysis of the Strategic, Organizational, Financial and Technological Issues in the Implementation of Knowledge Management in Indian Automotive SMEs

Ibrahim Shaikh and Hari Vasudevan

Abstract Knowledge is considered as the currency in the current knowledge economy. It is a key business asset and is considered vital to creating a sustainable competitive advantage. The growing importance of knowledge-intensive organizations makes knowledge management (KM) an essential strategic tool and a potent competitive weapon for gaining competitive advantage and improved business performance. However, many small and medium-sized enterprises (SMEs) do not realize these benefits, because of an uncertain and unclear idea of knowledge management and most of them lack innovative approaches to knowledge management. Hence, the SMEs still lag in the implementation of KM. SMEs are required to be fully aware and be perceptive of the issues that would hinder the implementation of knowledge management in SMEs. This study investigated the influence of strategic issues, organizational issues, financial issues, and technological issues in the implementation of KM in automotive component manufacturing SMEs across three tiers, i.e., Tier 1, Tier 2 and Tier 3, in India. The overall results of the study show that the strategic issues and financial issues are significant, whereas organizational issues and technological issues are insignificant across the three tiers of SMEs. Moreover, Tier 3 SMEs are found to have several strategic and financial issues in comparison with Tier 1.

Keywords Knowledge Management (KM) · KM issues · Automotive small and medium-sized enterprises

---

I. Shaikh (✉)
Department of Mechanical Engineering, Dwarkadas J. Sanghvi College of Engineering, Mumbai, India
e-mail: ibrahim.shaikh@mhssce.ac.in

H. Vasudevan
Dwarkadas J. Sanghvi College of Engineering, University of Mumbai, Mumbai, India
e-mail: harivasudevan@iitb.ac.in

© Springer Nature Singapore Pte Ltd. 2020

H. Vasudevan et al. (eds.), *Proceedings of International Conference on Intelligent Manufacturing and Automation*, Lecture Notes in Mechanical Engineering,
https://doi.org/10.1007/978-981-15-4485-9_15

1 Introduction

The contribution of small and medium-sized enterprises (SMEs) can never be overlooked in the growth of a national economy, especially in the current globalized era [1]. According to the Organization for Economic Cooperation and Development (OECD), over 95% of OECD enterprises are SMEs and that account for 60–70% of employment in most countries. Though it is widely known that the SMEs seldom conduct R&D activities as compared to larger firms, they have often proved themselves innovative by creating or re-engineering products and services and many times, introducing new organizational approaches to enhance productivity. Baporikar [2] concluded that many innovative ideas and technological breakthroughs in society have been contributed by small businesses.

Among the Indian manufacturing sectors, the auto-component industry has seen remarkable growth in the recent past and is mostly dominated by SMEs [3]. This industry has come a long way from being a small supplier of the auto-components to the local market into becoming a key performer in the local and international markets. The Indian auto-component SMEs provide robust support to the Indian automotive industry as well as several other industries that contribute to the Indian economy. These include machine tools, metals, rubber and plastics industries as well as various forging and machining industries, etc. Due to the make in India initiative of the Government of India, the Indian auto-component industry is currently transforming itself to be very competitive. The Indian auto-component business is poised for a big leap by 2026 and is presently undergoing a new phase of global exposure and technology adoption. In order to keep pace with the ever-changing market and for being competitive, the auto-component industries are advised to invest in technology and R&D with the focus on creating a strong knowledge base [4].

In the context of such a scenario, this study was carried out to investigate upon and further compare the strategic, organizational, financial and technological issues, which impede the implementation of knowledge management (KM) in different tiers across the Indian auto-components SMEs.

1.1 Definition of Knowledge and Knowledge Management

A number of definitions and opinions about knowledge exist in the literature. According to Bhanumathi [5], knowledge plays the role of a critical differentiator in today's competitive world, which has embarked on a new era as compared to a period, which was governed by man, material and money. Ragab and Arisha [6] regarded "knowledge as the coinage, a valuable organizational strategic resource and a foundation for competitive advantage in the current knowledge economy". According to them, knowledge is a critical business asset and is vital to creating a sustainable competitive advantage. According to Saini [7], knowledge is the state of consciousness, understanding and accumulated thoughts gained through know-how and edification.

According to Pillania [3], "knowledge is defined as a whole set of intuition, reasoning, insights and experiences related to technology, products, processes, customers, markets, competition and so on, that enables effective action".

Typically, knowledge management (KM) can be defined as the management of knowledge or the management of organizational knowledge for creating business value and generating a competitive advantage [8]. Jennex et al. [9] defined "KM", as the practice of selectively applying knowledge from previous experiences of decision making to current and future decision-making activities with the express purpose of improving the organization's effectiveness. Whereas, according to Swan et al. [10] "KM is a process through which knowledge creation, acquisition and sharing takes place". According to them, knowledge management is used to enhance learning as well as the performance of an organization. According to Alavi and Leidner [11], "KM" is about defining and implementing an organization-wide process for enhancing organizational performance and ensuring that it remains competitive. "KM" is defined as a systematic, structured, distinct and cautious ongoing process of contriving, spreading, implementing, recreating and amending the knowledge for accomplishing organizational goals [12].

2 Literature Review

Handzic [13] advocated that in order to preserve or create value, an organization must implement knowledge management. She advocated that knowledge management must be aligned with the strategic requirements of the organization. According to her, implementing knowledge management can certainly enhance the knowledge base and further add value to the organization. McAdam and Reid [14] carried out a survey to investigate the perceptions of KM dimensions in large-scale companies and SMEs. The outcome of their research indicated that the SME sector lack in KM understanding and implementation. The SME sector was found to be less developed and having a mechanistic approach to knowledge and lacking investment in KM approaches and systems.

Wong and Aspinwall [15] investigated the various critical success factors (CSFs) for implementing knowledge management (KM) in small and medium-sized enterprises. They criticized that the implementation of KM has received very little attention in the literature. They further emphasized that small and medium-sized enterprises lack a thorough and comprehensive investigation of critical success factors as far as knowledge management implementation is concerned. Based on the review, the authors proposed an extensive model with eleven critical factors for knowledge management in SMEs like culture, information technology, leadership and top management support, strategy and purpose, KM measurement, motivational aids, organizational set-up, training and education and so on.

According to Hasanali [16], the success of a knowledge management initiative depends on a number of factors. She underlined five categories of factors, namely leadership, culture, structure, roles and responsibilities, IT infrastructure and measurement. Although the factors mentioned are incredibly reasonable, there are few more aspects, which dictate the success of KM.

Desouza and Awazu [17] concluded that there is a considerable difference in the knowledge management approach of SMEs and larger organizations. They found it inappropriate to view SMEs' knowledge management practices as scaled-back forms of the practices found in larger organizations. Their research elaborated on five key peculiarities in knowledge management practices found in the SMEs.

Migdadi [18] carried out perhaps one of the first systematic research studies to ascertain the critical success factors (CSFs) for KM implementation in SMEs. According to him, the outcome of the research would embolden small and medium-sized enterprises to practice knowledge management from the right viewpoint, so as to realize better outcomes from knowledge management initiatives. Further, Nunes et al. [19] emphasized that implementing knowledge management activities could lead to more significant innovation and productivity within SMEs.

Singh et al. [20] conducted a study to assess the impact of KM practices in Indian manufacturing industries. They observed that the culture and financial constraints are among the highest-rated barriers in the implementation of KM in Indian manufacturing industries. They further added that quality, cost reduction, improvement in efficiency, improved delivery, flexibility and innovation are some of the competitive priorities, for which Indian organizations are looking for KM.

Pillania [3] carried out a study to explore the knowledge creation and categorization among Indian automotive components manufacturing SMEs. He claimed that a better discernment about knowledge creation is found among the international automotive components manufacturers as compared to their Indian counterparts. In his firm-level comparison, the researcher asserted that the international automotive components manufacturers have more regard and value for knowledge in contrast with Indian automotive ancillaries.

Another study by Pillania [12] focused on knowledge management strategies. This research work by the author was an endeavour to explore strategic issues in knowledge management within the Indian SMEs with a particular reference to the automotive component manufacturing industries. The research revealed that the most important strategy by the Indian automotive component SMEs is to give more importance to customer-focused knowledge. Furthermore, the author suggested that the Indian SMEs should focus more on the strategic issues to realize the premiums of KM for sustainable competitiveness.

Edvardsson and Durst [21] suggested that the SMEs need to use the available resources with caution as they face resource constraints. Consequently, any flawed decisions would be a grimmer on the existence of SMEs than they would have in large businesses. According to the authors, lack of financial resources and expertise would result in most knowledge being kept in the minds of a few key employees, rather than disseminated throughout the organization.

3 Conceptual Model

The following sections give a brief analysis of various issues that are involved in the implementation of KM, particularly in the SMEs. The conceptual model is shown in Fig. 1.

3.1 Strategic Issues

Developing a comprehensive enterprise-wide KM strategy is one of the crucial elements for the success of KM in any organization [22]. The strategy specifies the precise objectives, purposes and goals that need to be set and understood by everyone involved in achieving the objectives of an organization [23]. Organizations need a well-defined KM strategy to enable themselves to employ its capabilities and resources to achieve their KM goals [12]. Active top management participation and a well-perfected and aligned KM strategy are two of the critical strategic issues in knowledge management. However, M. H. Zack [24] and Ribière and Calabrese [25] emphasized that there is a rarity of focus in KM literature on strategic aspects of knowledge. Although numerous strategies have been suggested in the extant literature for implementing KM, the most common is the one, which links KM strategy with the overall business strategy [8]. Companies that succeed over the long-term need to align their KM processes with their strategy. Researchers, such as Ribière and Calabrese [25] and Edvardsson and Durst [26], argued that the SMEs do not

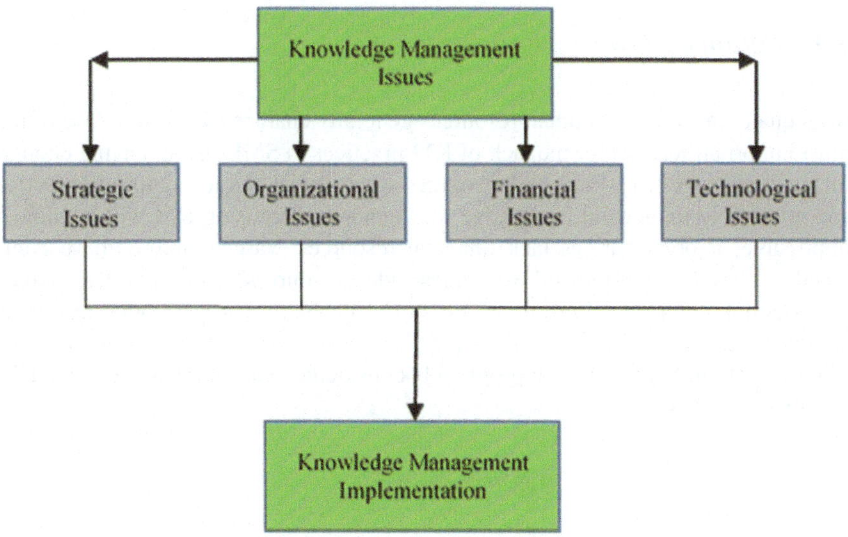

Fig. 1 Conceptual model of KM issues

follow an organizational learning approach. According to Pillania [12], there is a dearth of research on strategic issues concerning KM in SMEs in general and in the Indian automotive components SMEs in particular.

3.2 Organizational Issue

The successful implementation of knowledge management in SMEs requires an in-depth analysis of the organizational issues, as they can adversely influence the perception and attitudes towards KM success. Rajesh K. Pillania [3] asserted that the industrious participation and backing of top management is one of the most critical requirements for successful KM implementation. However, knowledge management efforts in SMEs lack support from top management [7, 15]. Moreover, extant literature has criticized that the SMEs lack a thorough understanding of knowledge management and its implementation [15]. SMEs are also found to have no unified and validated process for implementing knowledge management [27, 15]. A well-developed knowledge management system cannot be expected to deliver the anticipated profitability, unless the employee in organizations has the skills required and learning capability to use knowledge resourcefully [28]. However, several researchers have argued that the SMEs lack in training and education for employees [15]. A few more organizational issues that could be investigated in the context of Indian automotive SMEs include knowledge audit [29] and knowledge loss/leakages, which are associated with a missing or improper succession planning [30].

3.3 Financial Issues

Inadequate financial and human resources generally characterize SMEs. One of the most important reasons for the lack of KM practices in SMEs is not having enough financial resource [31]. Financial resources are essential and mandatory to practice and effectively implement knowledge management. Practising KM will be almost impossible, if organizations lack financial resources. Various authors have combined financial and non-financial resources under "resources". Although all resources are essential, the financial resources are more valuable, because all other resources are dependent on financial resources [15]. Since the availability of resources is of prime importance, it has to be appropriately considered, when implementing the KM initiative in SMEs.

3.4 Technological Issues

Information technology undoubtedly is considered as one of the significant enablers in the implementation of knowledge management [23]. However, knowledge management in SMEs is seen as more human-oriented and that is the reason, as to why the role of technology in knowledge management in SMEs is considered to be a divisive issue. In SMEs, core knowledge is possessed by the manager or key personnel and the knowledge is conveyed mostly through the informal way. Use of technology in SMEs is given less importance, and they manage knowledge in a humanistic way. Moreover, technology solutions are also considered not affordable due to financial constraints. Significant issues that need to be considered in the implementation of KM in SMEs include lack of investment in IT system, technological lucidness, ease of usage, pertinence to users' needs, the applicability of knowledge content and standardization.

4 Data Analyses and Results

After analyzing the extant literature, the following hypotheses were formulated as mentioned below. Hypotheses were framed in null as well as alternate forms. Null hypotheses were used for testing purposes and alternate ones retained, in case the null hypotheses were rejected.

The following null hypotheses have been conceptualized as:

$H_0$1: There is no significant difference in the strategic issues faced by Tier 1, Tier 2 and Tier 3 SMEs in the implementation of knowledge management.
$H_0$2: There is no significant difference in the organizational issues faced by Tier 1, Tier 2 and Tier 3 SMEs in the implementation of knowledge management.
$H_0$3: There is no significant difference in the financial issues faced by Tier 1, Tier 2 and Tier 3 SMEs in the implementation of knowledge management.
$H_0$4: There is no significant difference in the technological issues faced by Tier 1, Tier 2 and Tier 3 SMEs in the implementation of knowledge management.

The following alternate hypotheses have been conceptualized as:

H_A1: There is a significant difference in the strategic issues faced by Tier 1, Tier 2 and Tier 3 SMEs in the implementation of knowledge management.
H_A2: There is a significant difference in the organizational issues faced by Tier 1, Tier 2 and Tier 3 SMEs in the implementation of knowledge management.
H_A3: There is a significant difference in the financial issues faced by Tier 1, Tier 2 and Tier 3 SMEs in the implementation of knowledge management.
H_A4: There is a significant difference in the technological issues faced by Tier 1, Tier 2 and Tier 3 SMEs in the implementation of knowledge management.

Data collected were collated in Microsoft Excel, transferred to SPSS and then analysed using descriptive statistics. One-way ANOVA and post hoc test were conducted. One-way ANOVA comparisons were computed for each type of issues in the questionnaire to compare the Tier 1, Tier 2 and Tier 3 industries. The following conclusions were drawn from ANOVA and post hoc tests.

The F ratio (3.624), as shown in Table 1, was significant for strategic issues, and the p-values for construct were 0.028 ($p < 0.05$). Hence, it was concluded that there is a significant difference among Tier 1, Tier 2 and Tier 3 SMEs, as far as the strategic issues are concerned. The post hoc analysis shown in Table 2 suggests that for strategic issues, the p-value for Tiers 1 and 3 was 0.020. The mean of Tier 3 (22.00) was higher than that for Tier 1 (20.42). It was concluded that Tier 3 SMEs have a significant number of strategic issues as compared to Tier 1. The other pairs, viz. Tier 1 with Tier 2 and Tier 2 with Tier 3, do not show a significant difference for strategic issues. Hence, the null hypothesis was rejected.

The F ratio (2.612), as shown in Table 3, was not significant for organizational issues, and the p-values for construct were 0.075 ($p > 0.05$). Hence, it was concluded that there is no significant difference among Tier 1, Tier 2 and Tier 3 SMEs, as far as the organizational issues are concerned. They face similar challenges for these types of issues. Hence, the null hypothesis was retained.

The F ratio (3.549), as shown in Table 4, was significant for financial issues, and the p-values for construct were 0.030 ($p < 0.05$). Hence, it was concluded that there is a significant difference among Tier 1, Tier 2 and Tier 3 SMEs, as far as the

Table 1 ANOVA—strategic issues

	Sum of squares	df	Mean square	F	Sig.
Between groups	114.324	2	57.162	3.624	0.028
Within groups	5252.673	333	15.774	F ratio is significant. Hence, the null hypothesis is rejected	
Total	5366.997	335			

Table 2 Descriptives—strategic issues

	N	Mean	Std. deviation	Std. error	95% confidence interval for mean		Minimum	Maximum
					Lower bound	Upper bound		
Tier 1	122	20.4180	4.48107	0.40570	19.6148	21.2212	10.00	30.00
Tier 2	141	20.9929	3.88862	0.32748	20.3455	21.6404	11.00	30.00
Tier 3	73	22.0000	3.13138	0.36650	21.2694	22.7306	13.00	30.00
Total	336	21.0030	4.00261	0.21836	20.5734	21.4325	10.00	30.00

Table 3 ANOVA—organizational issues

	Sum of squares	df	Mean square	F	Sig.
Between groups	226.237	2	113.119	2.612	0.075
Within groups	14,423.188	333	43.313	F ratio is not significant. Hence, the null hypothesis is retained	
Total	14,649.426	335			

Table 4 ANOVA—financial issues

	Sum of squares	df	Mean square	F	Sig.
Between groups	129.353	2	64.676	3.549	0.030
Within groups	6069.287	333	18.226	F ratio is significant. Hence, the null hypothesis is rejected	
Total	6198.640	335			

financial issues are concerned. The post hoc analysis shown in Table 5 suggests that for strategic issues, the p-value for Tiers 1 and 3 was 0.022. The mean of Tier 3 (21.33) was greater than that for Tier 1 (19.65). It was, therefore, concluded that Tier 3 SMEs have a significant number of financial issues as compared to Tier 1. The other pairs, viz. Tier 1 with Tier 2 and Tier 2 with Tier 3, do not show a significant difference for financial issues. Hence, the null hypothesis was rejected.

The F ratio (1.779), as shown in Table 6, was not significant for technological issues and the p-values for construct were 0.170 ($p > 0.05$). Hence, it was concluded that there is no significant difference among Tier 1, Tier 2 and Tier 3 SMEs, as far as the technological issues are concerned. They face similar challenges for these types of issues. Hence, the null hypothesis was retained.

Table 5 Descriptives—financial issues

	N	Mean	Std. deviation	Std. error	95% confidence interval for mean		Minimum	Maximum
					Lower bound	Upper bound		
Tier 1	122	19.6475	4.25999	0.38568	18.8840	20.4111	10.00	27.00
Tier 2	141	20.3333	4.48118	0.37738	19.5872	21.0794	10.00	28.00
Tier 3	73	21.3288	3.84077	0.44953	20.4326	22.2249	10.00	30.00
Total	336	20.3006	4.30156	0.23467	19.8390	20.7622	10.00	30.00

Table 6 ANOVA—technological issues

	Sum of squares	df	Mean square	F	Sig.
Between groups	108.736	2	54.368	1.779	0.170
Within groups	10178.967	333	30.567	F ratio is not significant. Hence, the null hypothesis is retained	

5 Conclusion

The Indian auto-component SMEs provide robust support to the Indian as well as the global automotive industries. In order to keep pace with the ever-changing market and for being competitive, the auto-component industries must invest in technology and R&D with a firm focus on creating knowledge.

This research was carried out to investigate and compare the strategic, organizational, financial and technological issues, which impede the implementation of knowledge management (KM) in different tiers across Indian auto-components SMEs.

The empirical analyses of research suggest that there is a significant difference among Tier 1, Tier 2 and Tier 3 SMEs, as far as the strategic issues and financial issues are concerned. However, Tier 3 SMEs have a significant number of strategic issues as well as financial issues, as compared to Tier 1 SMEs. Whereas there is no significant difference seen among Tier 1, Tier 2 and Tier 3 SMEs, as far as the organizational issues and technological issues are concerned. All of them face similar challenges with these types of issues. The outcome of this research suggests that all the issues are essential, as far as knowledge management implementation across the SMEs is concerned. However, Tier 3 SMEs are found to have several strategic as well as financial issues, as compared to Tier 1 and Tier 2 SMEs. Tier 3 SMEs also need to consider strategic and financial issues more seriously, so as to implement KM successfully in the SMEs in the Indian context.

References

1. OECD (2017) Enhancing the contributions of SMEs in a global and digitalised economy. In: Meeting of the OECD Council at ministerial level, 7–8 Jun 2017
2. Baporikar N (2016) Understanding knowledge management spectrum for SMEs in global scenario. Int J Soc Organ Dyn IT 5(1):1–15
3. Pillania RK (2008) Creation and categorization of knowledge in automotive components SMEs in India. Manag Decis 46(10):1452–1464
4. ACMA (2018) ACMA AutoNews. ACMA
5. Bhanumathi P (2015) Knowledge management enablers, processes and organizational performance: a case study of select SMEs in Bangalore (Chapter 1). Dr Philos Manag Sci 1–13

6. Ragab MAF, Arisha A (2013) Knowledge management and measurement: a critical review. J Knowl Manag 17(6):873–901
7. Saini R (2013) Model development for key enablers in the implementation of knowledge management. IUP J Knowl Manag XI(2):46–63
8. Pillania RK (2008) Knowledge management in SMEs in India: a study of the automotive components sector. Int J Electr Hybrid Veh 1(3):308–318
9. Jennex ME et al (2008) Towards measuring knowledge management success. In: Proceedings of 41st Hawaii international conference on systems science, pp 1–8
10. Swan J et al (1999) Emerald Article: Knowledge management and innovation: networks and networking. J Knowl Manag 3(4):262–275
11. Alavi M, Leidner D (1999) Knowledge management and knowledge management systems. J Strateg Inf Syst 9(2–3):101–105
12. Pillania RK (2008) Strategic issues in knowledge management in small and medium enterprises. Knowl Manag Res Pract 6(4):334–338
13. Handzic M (2006) Knowledge management in SMEs. CACCI J 1:1–11
14. McAdam R, Reid R (2001) SME and large organisation perceptions of knowledge management : comparisons and contrasts. J Knowl Manag 5:231–241
15. Wong KY, Aspinwall E (2005) An empirical study of the important factors for knowledge-management adoption in the SME sector. J Knowl Manag 9(3):64–82
16. Hasanali BF (2002) Critical success factors of knowledge management, pp 1–4
17. Desouza KC, Awazu Y (2006) Knowledge management at SMEs: five peculiarities. J Knowl Manag 10(1):32–43
18. Migdadi M (2008) Knowledge management enablers and outcomes in the small-and-medium sized enterprises. Ind Manag Data Syst 109(6):840–858
19. Nunes MB, Annansingh F et al (2005) Knowledge management issues in knowledge-intensive SMEs. J Doc. 62(1):101–119
20. Singh MD et al (2006) Survey of knowledge management practices in Indian manufacturing industries. J Knowl Manag 10(6):110–128
21. Edvardsson IR, Durst S (2013) The benefits of knowledge management in small and medium-sized enterprises. Procedia Soc Behav Sci 81:351–354
22. Liebowitz J (1999) Key ingredients to the success of an organization's knowledge management strategy. Knowl Process Manag 6(1):37–40
23. Wong KY (2005) Critical success factors for implementing knowledge management in small and medium enterprises. Ind Manag Data Syst 105(3):261–279
24. Zack MH (1999) Developing a knowledge strategy. Calif Manag Rev 41(3):125–145
25. Ribière V, Calabrese FA (2015) The state of the art of knowledge management. In: Knowledge management, pp 53–70
26. Edvardsson IR, Durst S (2013) Does knowledge management deliver the goods in SMEs? Bus Manag Res 2(2):52–60
27. Chan I, Chao C-K (2008) Knowledge management in small and medium-sized enterprises. Commun ACM 51:83–88
28. Kim S et al (2003) Building the knowledge map: an industrial case study. J Knowl Manag 7(2):34–45
29. Handzic M (2004) Knowledge management in SMEs. Tech Monitor, Sydney, Australia, pp 29–34
30. Durst S, Wilhelm S (2012) Knowledge management and succession planning in SMEs. J Knowl Manag 16(4):637–649
31. OECD (2003) Measuring knowledge management in the business sector

Optimization of CNC Die-Sinking EDM Process Parameters Based on MRR and EWR by Taguchi Method Using Copper Electrode on P20 Tool Steel

Mehul Prajapati and Sowmin Trivedi

Abstract Optimization using Taguchi methodology is one of the techniques for industries by which manufacturing for quality products at lower cost is achieved. The process parameters, such as current, pulse-ON and pulse-OFF time in electrical discharge machining (EDM) process, give variations in the performance characteristics, such as material removal rate (MRR) on workpiece and electrode wear rate (EWR) on tool while machining P20 tool steel using copper electrode. By conducting Taguchi design of experiments using L9 orthogonal array, the analysis had been carried out. Using Minitab, its response tables and graphs were observed to find out the optimal levels of parameters in the EDM process. Thus, the process parameters for EDM were optimized during machining of P20 steel for achieving the combined objectives of higher rate of material removal and lower wear rate on tool. The obtained results helped to identify the major and minor parameters affecting MRR and EWR.

Keywords EDM process · MRR · EWR · Taguchi method

1 Introduction

Electrical discharge machining is a type of an unconventional machining technique where the electrical energy is directly used to remove or cut the metals. It is also called as spark erosion machining or electro-erosion machining [4]. The metal is removed by electrical spark discharge between electrode tool (cathode) and workpiece (anode). Electrical discharge machining is used in mould and die-making industries, automobile industries and also making of aerospace components. In die and mould making, a die-sinking EDM is especially used for machining intricate and unique patterns and shapes which otherwise are difficult to machine or time-consuming using conventional CNC milling machines. The paper describes an investigation of EDM process

M. Prajapati (✉) · S. Trivedi
Department of Production Engineering, Dwarkadas J. Sanghvi College of Engineering, Vile Parle, Mumbai, India
e-mail: mehul.prajapati@djsce.ac.in

© Springer Nature Singapore Pte Ltd. 2020

H. Vasudevan et al. (eds.), *Proceedings of International Conference on Intelligent Manufacturing and Automation*, Lecture Notes in Mechanical Engineering,
https://doi.org/10.1007/978-981-15-4485-9_16

parameters and its optimization using Taguchi method and determining the major and minor parameters affecting MRR and EWR as well as finding optimized values based on the values considered for this experiment [1].

2 Literature Survey

Optimization of EDM process parameters using Taguchi method with copper electrode by Niraj Kumar Ohdar, Babuli Kumar Jena, Saumya Kanta Sethi discusses optimizing process parameters that are current, pulse-ON time, pulse-OFF time and flushing pressure using Taguchi method with copper electrode based on MRR and EWR with corresponding S/N ratio and then finding out mean of their S/N ratios with 'larger the better' for MRR and 'smaller the better' for EWR and then finding their residual plots and respectively the major and minor parameters were identified for both MRR and EWR.

The implementation of Taguchi method on EDM process of tungsten carbide by Mohd Amri Lajis, H. C. D. Mohd Radzi and A. K. M. Nurul Amin discusses using Taguchi method with graphite as an electrode and tungsten carbide as workpiece and accordingly finding optimized values for the parameters, i.e. current, pulse-ON time, pulse-OFF time and voltage, and similarly, the major and minor parameters based on its effects on MRR, EWR and surface roughness are determined with its optimized values from the graph plotted.

Optimization of EDM process parameters using Taguchi method with graphite electrode by Vishal J. Nadpara and Prof. Ashok Choudhary discusses optimizing process parameters that are current, pulse-ON time and pulse-OFF time by Taguchi method with graphite as electrode and AISI D3 Steel as workpiece. The MRR and EWR are respectively found based on machining time and weight loss, and the S/N ratio is founded for each values. The mean of S/N ratio graph is plotted based on 'smaller the better' for EWR and 'larger the better' for MRR for finding again the major and minor parameters affecting them and similarly finding the optimized values.

3 Experimental Set-Up

The experiments were conducted using a CNC die-sinking EDM machine manufactured by HCM Taiwan as shown in Fig. 1. The copper electrode is fed downwards into the workpiece under servo control in this EDM machine. The workpiece material used for this experiment was P20 tool steel. In this experiment, Taguchi design of experiments with L9 orthogonal array, the analysis had been carried out [2]. Using this, the proposed process parameters were arranged accordingly and corresponding MRR and EWR along with their signal-to-noise ratio based on 'higher the better'

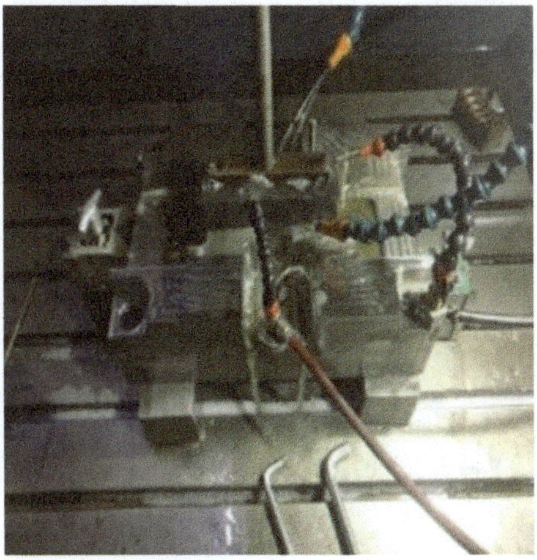

Fig. 1 CNC die-sinking EDM machine setup

and ‘smaller the better’, respectively. With the help of Minitab software, the mean of S/N ratio graph is plotted, and accordingly, the values and the ranking of each parameter based on importance are found.

4 Material Properties

The workpiece is P20 tool steel for this experiment with the following physical properties (Table 2) and Chemical Composition (Table 1) [5].

5 Experiment Data

The parameters considered for this experiment are current, pulse-ON time and pulse-OFF time [3]. The duration of time (in μs) when the current is passed gives pulse-ON time (td) to flow per cycle. The duration of time (in μs) in between two consecutive sparks and current (I) is current flowing through the whole cycle (in AMP) is pulse-OFF time (to).

5.1 Experiment Procedure

During the experiment, 3 levels (Table 3) were taken for each parameter and then arranged in a L9 orthogonal array (Table 4) in Taguchi method. During the readings, all the weight loss and machining time are taken for calculating MRR and EWR for

Table 1 Chemical composition of 'P20'

Components	Percentages (%)
C	0.378
Mn	1.412
Si	0.300
P	0.014
S	0.0001
Cr	1.8600
Ni	0.230
Mo	0.188
Al	0.020
Cu	0.220
Nb	0.0030
V	0.005
Ti	0.003
N (ppm)	83

Table 2 Physical properties of 'P20'

Properties of 'P20'	Metric
Hardness, Brinell (typical)	300
Hardness, Rockwell C (typical)	30
Tensile strength, ultimate	965–1030 MPa
Tensile strength, yield	827–862 MPa
Compressive strength	862 MPa
Elastic modulus	190–210 GPa
Thermal expansion	12.8×10^{-6}/°C at 20–425 °C

Table 3 Levels for process parameters

Parameters	Levels		
	1	2	3
Current (*I*)	4	5	6
Pulse-ON time (T-ON)	80	100	120
Pulse-ON time (T-OFF)	40	50	60

each of the 9 readings. These values are input in Minitab 19 software and then under DOE and analysis of Taguchi method, S/N ratio and mean table were calculated, and based on the values, graph was plotted accordingly for MRR and EWR, respectively, where the optimized values were then determined from the graph [2].

For calculating MRR, the formula:

Table 4 L9 orthogonal array for Taguchi method

Experiment levels	Current (I)	Pulse-ON time (T-ON)	Pulse-OFF time (T-OFF)	MRR	EWR
1	4	80	40	2.11	2.05
2	4	100	50	2.72	2.64
3	4	120	60	3.24	3.13
4	5	80	50	4.50	4.85
5	5	100	60	5.18	5.58
6	5	120	40	2.92	9.13
7	6	80	60	3.22	11.16
8	6	100	40	9.60	8.58
9	6	120	50	6.66	10.30

$$\frac{W_b - W_a}{d * t}\ \text{mm}^3/\text{min}$$

where

W_b—Workpiece weight before machining (g)
W_a—Workpiece weight after machining (g)
d—Density of P20 tool steel (g/cc)
t—Machining Time (in mins)
The density of P20 tool steel is 7.85 g/cc.

For calculating EWR, the formula:

$$\frac{E_b - E_a}{d * t}\ \text{mm}^3/\text{min}$$

where

E_b—Electrode weight before machining (g)
E_a—Electrode weight after machining (g)
d—Density of electrode (g/cc)
t—Machining Time (in mins)
The density of copper electrode is 8.96 g/cc.

5.2 *Results and Analysis*

After putting the MRR values in Minitab software, for 'larger the better' option to get S/N ratio values, Table 5 represents the S/N ratio for MRR. Accordingly, the response table (Table 6) and graph (Figs. 2 and 3) for mean of S/N ratio along with

Table 5 S/N ratio

MRR	S/N ratio
2.11	6.4856
2.72	8.6914
3.24	10.2109
4.50	13.0643
5.18	14.2866
2.92	9.3077
3.22	10.1571
9.60	19.6454
6.66	16.4695

Table 6 Response table for signal-to-noise ratios

Level	*I*	T-ON	T-OFF
1	8.463	9.902	11.813
2	12.220	14.208	12.742
3	15.424	11.996	11.552
Delta	6.961	4.305	1.190
Rank	1	2	3

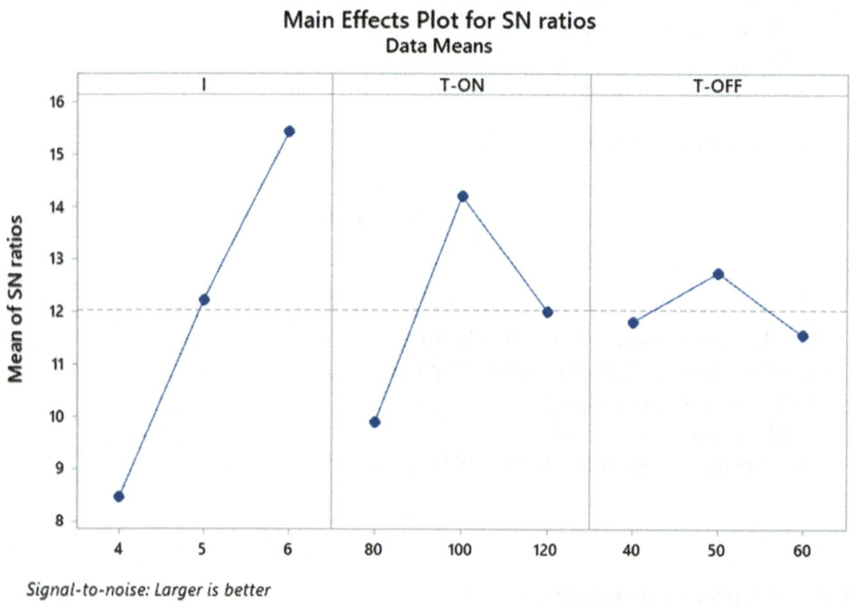

Fig. 2 Signal to noise ratios for MRR

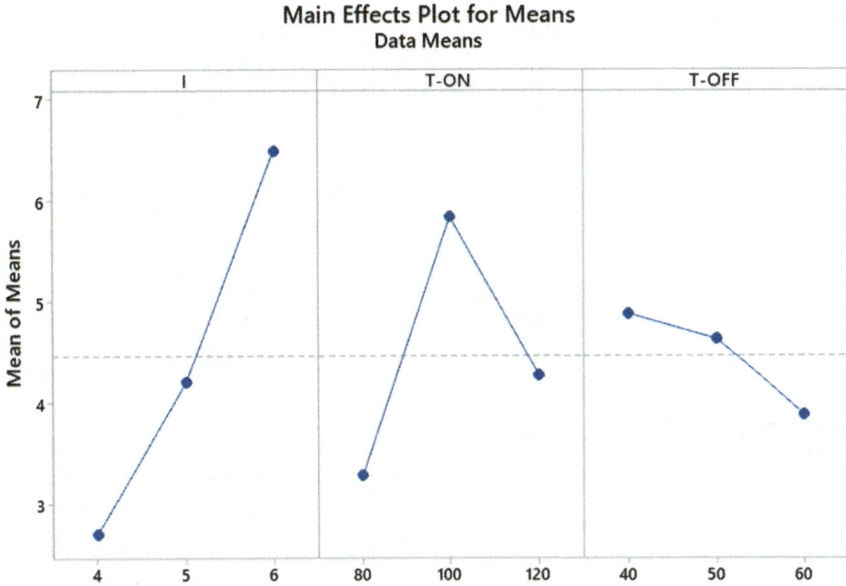

Fig. 3 Mean for MRR

a significance of each of the process parameters were ranked based on delta value from high to low (Table 7).

So accordingly, MRR is most affected by current followed by current, pulse-ON time and then pulse-OFF time. The parameters where MRR is said to be maximum are: current (6 A), pulse-ON time (100 μs) and pulse-OFF time (40 μs).

Similarly, for EWR for 'smaller the better' option is selected in Minitab software for obtaining S/N values, Table 8 represents the S/N ratio for EWR. Accordingly, the response table (Table 9) and graph (Figs. 4 and 5) for mean of S/N ratio along with significance of each of the process parameters were ranked based on delta value from high to low (Table 10).

So accordingly, the EWR is most affected by current followed by current, pulse-ON time and then pulse-OFF time. The parameters where EWR is said to be minimum are: current (4 A), pulse-ON time (100 μs) and pulse-OFF time (50 μs).

Table 7 Response table for means

Level	*I*	T-ON	T-OFF
1	2.690	3.277	4.877
2	4.200	5.833	4.627
3	6.493	4.273	3.880
Delta	3.803	2.557	0.997
Rank	1	2	3

Table 8 S/N ratio

EWR	S/N ratio
2.05	−6.2351
2.64	−8.4321
3.13	−9.9109
4.85	−13.7148
5.58	−14.9327
9.13	−19.2094
11.16	−20.9533
8.58	−18.6697
10.30	−20.2567

Table 9 Response table for signal-to-noise ratios

Level	*I*	T-ON	T-OFF
1	−8.193	−13.634	−14.705
2	−15.952	−14.012	−14.135
3	−19.960	−16.459	−15.266
Delta	11.767	2.825	1.131
Rank	1	2	3

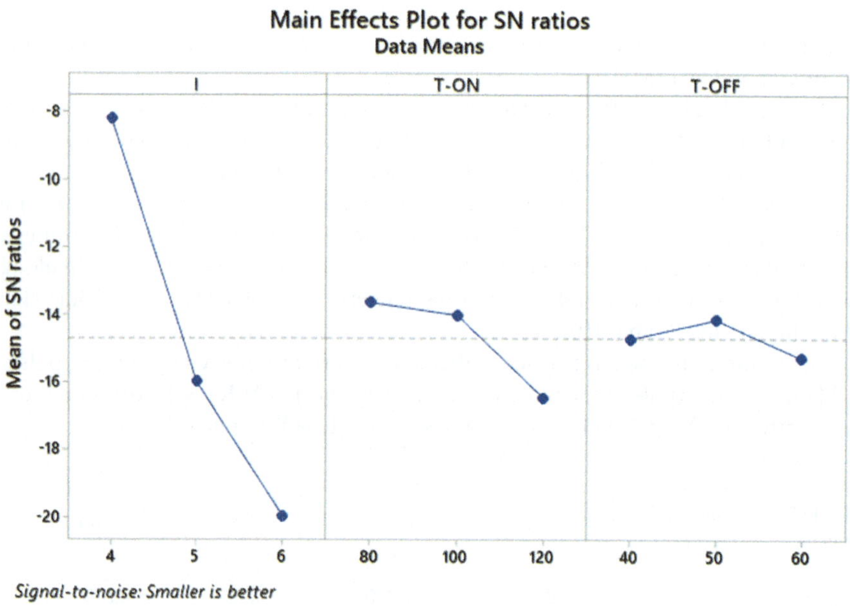

Fig. 4 Signal to noise ratios for EWR

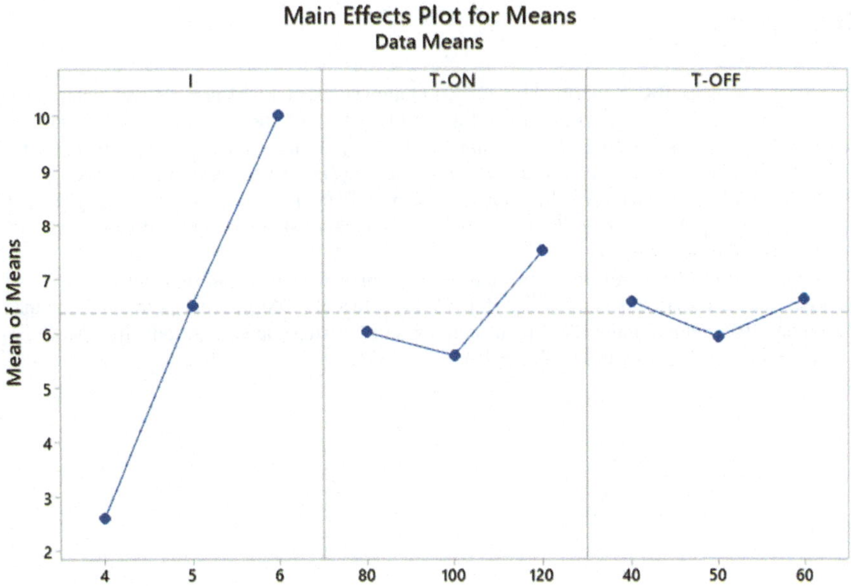

Fig. 5 Mean for EWR

Table 10 Response table for means

Level	I	T-ON	T-OFF
1	2.607	6.020	6.587
2	6.520	5.600	5.930
3	10.013	7.520	6.623
Delta	7.407	1.920	0.693
Rank	1	2	3

6 Conclusion

It was found that MRR is most affected by current followed by pulse-ON time and pulse-OFF time and is maximized at current (6 A); pulse-ON time (100 μs) and pulse-OFF time (40 μs).

As far as EWR is concerned, it is most affected again by current followed by pulse-ON time and pulse-OFF time, and further, EWR is minimized at current (4 A); pulse-ON time (100 μs) and pulse-OFF time (50 μs).

References

1. Ohdar NK, Jena BK, Kanta Sethi S (2017) Optimization of EDM process parameters using Taguchi method with copper electrode. Int Res J Eng Technol (IRJET) 4(4). ISSN: 2395-0072
2. Lajis MA, Mohd Radzi HCD, Nurul Amin AKM (2009) The implementation of Taguchi method on EDM process of tungsten carbide. Eur J Sci Res 26(4):609–617. ISSN 1450-216X
3. Nadpara VJ, Choudhary A (2014) Optimization of EDM process parameters using Taguchi method with graphite electrode. Int J Eng Trends Technol (IJETT). https://doi.org/10.14445/22315381/ijett-v7p239
4. McGeough JA (1998) Advanced methods of machining. Chapman and Hall, New York
5. Gaikwad A, Tiwari A, Kumar A, Singh D (2014) Effect of EDM parameters in obtaining maximum MRR and minimum EWR by machining SS 316 using copper electrode. Int J Mech Eng (IJMET) 5(6):101–109. ISSN 0976-6340 (print)/ISSN 0976-6359 (online)

Prediction of Surface Roughness and Optimisation of Cutting Parameters in Hard Turning of AISI 52100 Steel Based on Response Surface Methodology

Sandip Mane and Sanjay Kumar

Abstract The surface roughness is one of the most important surface characteristics, which affects the quality and its service life of the product. This paper aimed to analyse the effect of cutting parameters, such as cutting speed, feed and depth of cut on surface roughness (Ra) in turning of hardened AISI 52100 steel of 58 HRC with multilayer coated carbide insert under dry environment. Response surface methodology based on central composite design (CCD) was used for the optimisation and development of mathematical models for surface roughness. The analysis of variance (ANOVA) was performed to explore the effect of cutting parameters and their interactions on surface roughness. The RSM-based quadratic regression model in terms of cutting speed, feed and depth of cut for surface roughness (Ra) was developed with 95% confidence level. The adequacy and validity of the developed model were checked on the basis of *P*-value, *F*-value and R^2-value. The results showed that feed is the most influencing cutting parameter on surface roughness, followed by cutting speed and depth of cut which has no significant effect on surface roughness.

Keywords Hard turning · Surface roughness · Response surface methodology · Central composite design

1 Introduction

Hardened alloy steels are being widely used in automotive and allied industries due to their high compressive strength and wear resistance. Tönshoff et al. [1] reported that the hard turning is a viable and cost-effective machining process. In practice, hard

S. Mane (✉)
Department of Mechanical Engineering, Dwarkadas J. Sanghvi College of Engineering, Mumbai, Maharashtra 400056, India
e-mail: sandip.mane@djsce.ac.in

S. Kumar
Department of Mechanical Engineering, Thakur College of Engineering and Technology, Mumbai, Maharashtra, India
e-mail: sanjay.kumar@thakureducation.org

© Springer Nature Singapore Pte Ltd. 2020

H. Vasudevan et al. (eds.), *Proceedings of International Conference on Intelligent Manufacturing and Automation*, Lecture Notes in Mechanical Engineering,
https://doi.org/10.1007/978-981-15-4485-9_17

turning is carried out in the range of 58–68 HRC. König et al. [2] stated that the hard turning has been gaining the importance over conventional grinding process due to its benefits such as lesser set-up time, fewer process steps, process flexibility, greater part geometry, higher material removal rate, compatible surface roughness, less health and environment issues. The excessive heat generation at cutting zone during hard turning raises the temperature significantly and affects the tool wear, tool life and surface integrity of the machined component. The surface finish can be improved by reducing the heat generation and temperature at tool-chip-workpiece interface. This objective can be achieved by proper selection of cutting parameters. Aouici et al. [3] investigated the effect of ceramic and CBN insert on surface roughness Ra and Rt during hard turning of AISI H11 steel using ANOVA and RSM. The results showed that the surface roughness is mainly influenced by feed rate, followed by the cutting speed and the depth of cut. Rashid et al. [4] have used Taguchi for experimental design and ANOVA to determine the implication of cutting parameters on surface roughness (Ra). For the prediction of surface roughness (Ra) in terms of the cutting parameter, the regression analysis was used. The analysis revealed that lower feed rate produces good surface finish. Shihab et al. [5] studied the effect of cutting speed, feed and depth of cut on surface roughness and microhardness during dry hard turning of hardened AISI 52100 alloy steel with coated carbide inserts using RSM based on CCD. The results revealed that the combination of lower feed rate and depth of cut and higher cutting speed produces improved surface integrity. Asiltürk and Akkus [6] reported that the feed rate is the most significant cutting parameter on surface roughness (Ra and Rz) in turning of hardened AISI 4140 steel at 51 HRC with coated carbide inserts. Hessainia et al. [7] explored the influences of cutting parameters and cutting tool vibrations on surface roughness parameters (Ra and Rz) in the machining of 42CrMo4 steel of 56 HRC with ceramic tool and developed the response surface-based model for surface roughness and tool vibration. The results obtained indicate that the feed rate is the most influential parameter on the surface roughness, whereas the tool vibrations have no significant influence on surface roughness. Das et al. [8] analysed the effect of cutting speed, feed and depth of cut on surface roughness in machining of EN24 hardened steel of 50 HRC with the coated carbide insert by Grey–Taguchi approach and showed that the feed is a most important parameter, which significantly influences the surface roughness (Ra). Elbah et al. [9] performed the optimisation study based on response surface methodology (RSM) and ANOVA in turning of hardened AISI 4140 steel at 60 HRC by ceramic inserts. The results revealed that the surface roughness can be reduced by the combination of lower feed rate and depth of cut. Singh and Rao [10] studied the influence of cutting parameters and tool geometry on surface roughness in turning of AISI 52100 at 60 HRC with the mixed ceramic inserts. The results indicated that the surface finish is highly influenced by feed rate, followed by nose radius and cutting speed. Sahin and Motorcu [11] developed response surface methodology-based mathematical models for surface roughness in terms of cutting speed, feed and depth of cut. The predictions were in good agreement with the experiment result. Arbizu and Perez [12] reported that the predictive model of surface roughness based on response surface methodology gives better and accurate predictions. However, very few publications are available in the

Table 1 Chemical composition of AISI 52100 hardened alloy steel (weight percentage)

C %	Si %	Mn %	P %	S %	Cr %	Ni %	Cu %	Fe %
1.04	0.18	0.35	0.007	0.004	1.35	0.076	0.058	Balance

literature that discusses the influence of cutting parameters on surface roughness in turning of hardened alloy steel. A key limitation of previous research is that it does not address the effect of cutting parameters on surface integrity aspect at elevated hardness of 58–68 HRC in turning of hardened alloy steels. Thus, there was a need to explore the machinability aspect of various hardened alloy steel with respect to surface characteristic of machined surface in hardened steel at elevated hardness.

2 Experimentation

2.1 Selection of Workpiece Material

In this study, AISI 52100 hardened alloy steel having hardness of 58 HRC was selected as workpiece material. AISI 52100 hardened alloy steel has wide applications and is being used in automotive and allied industries such as bearings, forming rolls, spindles, tools and precision instrument parts. Table 1 shows the chemical composition of the workpiece material.

2.2 Selection of Tool

The cutting tool inserts and the tool holder were selected based on the literature review and the tool manufacturer's recommendation. The MTCVD multilayer coated carbide (TiN/TiCN/Al_2O_3)–[HK150, K-type] cutting tool insert having specification CNMG120408 and the tool holder with PCLNR 2020 K12 specification were selected for experimentation. The experiments were carried out on a rigid high precision HMT NH-18 lathe machine.

2.3 Selection of Cutting Parameters

Based on the previous research carried out and the tool manufacturer's recommendations, the cutting parameters were selected. Table 2 presents the cutting parameters and their coded and actual levels based on the central composite design of response surface methodology.

Table 2 Cutting parameters and their coded and actual levels using central composite design

Process parameters	Units	Limits				
		−1.682	−1	0	1	1.682
Cutting speed	m/min	90	100	115	130	140
Feed rate	mm/rev	0.05	0.075	0.10	0.125	0.15
Depth of cut	mm	0.1	0.2	0.3	0.4	0.5

2.4 Design of Experiment

The response surface methodology uses the statistical experimental design technique and the least square fitting method for model generation. The central composite design (CCD) was used for experimental design. The prediction of the surface roughness was carried out in terms of cutting parameters under dry environment. Surface roughness was measured using Talysurf4 surface roughness measuring machine. Table 3 shows the experimental design and result for surface roughness.

Table 3 Experimental design and result for surface roughness

S. No.	Cutting speed (m/min)	Feed rate (mm/rev)	Depth of cut (mm)	Surface roughness (μm)
1	115	0.10	0.3	0.872
2	130	0.08	0.4	0.822
3	115	0.10	0.3	0.863
4	90	0.10	0.3	0.667
5	140	0.10	0.3	0.926
6	115	0.10	0.1	0.870
7	115	0.05	0.3	0.436
8	115	0.10	0.3	0.857
9	115	0.10	0.3	0.805
10	100	0.08	0.2	0.657
11	130	0.12	0.4	0.910
12	115	0.10	0.5	0.963
13	130	0.12	0.2	0.870
14	115	0.10	0.3	0.850
15	130	0.08	0.2	0.681
16	100	0.08	0.4	0.670
17	100	0.12	0.4	0.752
18	100	0.12	0.2	0.718
19	115	0.10	0.3	0.863
20	115	0.15	0.3	0.636

2.5 *Measurement of Surface Roughness*

To measure the surface finish of the machined component, Talysurf4 surface roughness measuring machine was used. Figure 1 shows the Talysurf4 surface roughness measuring machine and surface roughness profile.

3 Result and Discussion

It is clear from ANOVA table that the *P*-values less than 0.0500 indicate model terms are significant and the values greater than 0.1000 indicate the model terms are not significant. The quadratic value of feed (F^2) has the most significant effect, followed by feed (F) and cutting speed (V). The depth of cut (D) has a minor effect on surface roughness. The quadratic value of cutting speed (V^2), the quadratic value of depth of cut (D^2), the interaction between cutting speed and feed rate (VF), the interaction between cutting speed and depth of cut (VD) and the interaction between feed rate and depth of cut (FD) all have no significant effect on surface roughness. We have selected the highest order polynomial where the additional terms are significant and the model is not aliased. The below equation is in terms of coded and actual factors can be used to make predictions about the response for given levels of each factor. Table 4 shows ANOVA for quadratic model of surface roughness and table 5 shows the model summary of surface roughness.
Final equation in terms of coded factors:

$$\begin{aligned}\text{Surface roughness} = {} & +0.8457 + 0.0677 * V + 0.0449 * F + 0.0259 * D \\ & + 0.0168 * V * F + 0.0168 * V * D - 0.0100 * F * D \\ & - 0.0277 * V * V - 0.0515 * F * F + 0.0129 * D * D\end{aligned}$$

Final equation in terms of actual factors:

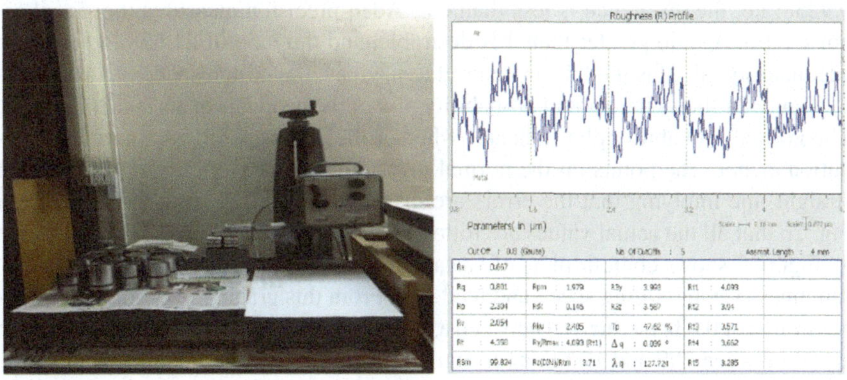

Fig. 1 Talysurf4 surface roughness measuring machine and surface roughness profile

Table 4 ANOVA for quadratic model of surface roughness

Source	DF	Sum of square	Contr. (%)	Mean square	*F*-value	*P*-value	
Regression	9	0.304125	96.10	0.033792	27.36	<0.0001	Significant
V	1	0.041288	13.05	0.041288	33.43	<0.0001	Significant
F	1	0.062123	19.63	0.062123	50.29	<0.0001	Significant
D	1	0.010712	3.38	0.010712	8.67	0.015	
V * *V*	1	0.001150	0.36	0.010170	8.23	0.017	
F * *F*	1	0.179401	56.69	0.157318	127.36	<0.0001	Significant
D * *D*	1	0.004162	1.32	0.004162	3.37	0.096	
V * *F*	1	0.002245	0.71	0.002245	1.82	0.207	
V * *D*	1	0.002244	0.71	0.002244	1.82	0.207	
F * *D*	1	0.000800	0.25	0.000800	0.65	0.440	
Error	10	0.012352	3.90	0.001235			
Lack-of-fit	5	0.009472	2.99	0.001894	3.29	0.109	Not significant
Pure error	5	0.002879	0.91	0.000576			
Cor total	19	0.316477	100.00				

Table 5 Model summary of surface roughness

Std. dev.	R^2	R^2 (adj)	PRESS	R^2 (pred)	Adeq precision
0.0351452	96.10%	92.58%	0.0762256	75.91%	21.6282

$$\begin{aligned}\text{Surface roughness} = {} & -1.901 + 0.023 * V + 23.091 * F - 1.296 * D \\ & + 0.055 * V * F + 0.011 * V * D - 5.000 * F * D \\ & - 0.000 * V * V - 128.842 * F * F + 1.285 * D * D\end{aligned}$$

The predicted R^2 of 0.7591 is in reasonable agreement with the adjusted R^2 of 0.9258; i.e. the difference is less than 0.2. Adeq precision measures the signal-to-noise ratio. A ratio greater than 4 is desirable, and the ratio of 21.628 indicates an adequate signal. This model can be used to navigate the design space.

Figure 2a shows the normal probability plot of the residuals of surface roughness. The normal probability plot indicates whether the residuals follow a normal distribution or not. The points on the normal probability plots of the residuals fall on a straight line implying that the errors are distributed normally. It can be seen from Fig. 2b that all the actual values are following the predicted values.

Figure 3 shows contour plots of surface roughness. Figure 3a shows the surface roughness versus cutting speed and feed rate. From this graph, lower surface roughness value of 0.634 has arrived between a feed rate of 0.08–0.10 mm/rev and a cutting speed between 100 and 120 m/min. Figure 3b shows surface roughness against the

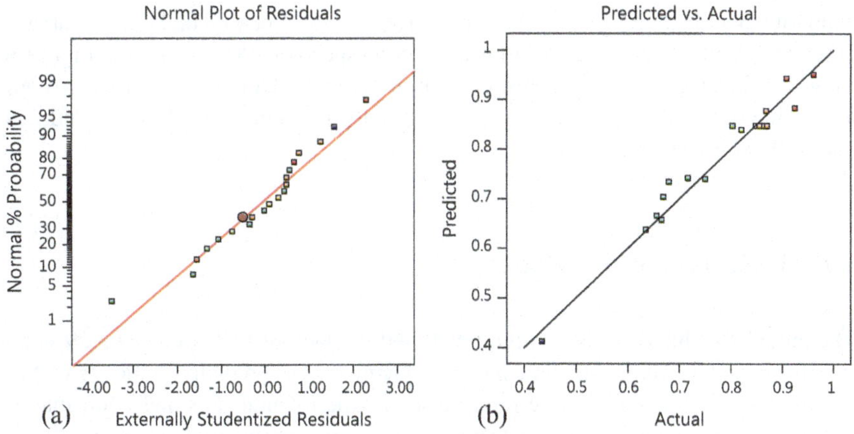

Fig. 2 **a** Plot of residuals versus normal probability for Ra. **b** Plot of predicted versus actual values for Ra. **c** Plot of residuals versus predicted values for Ra

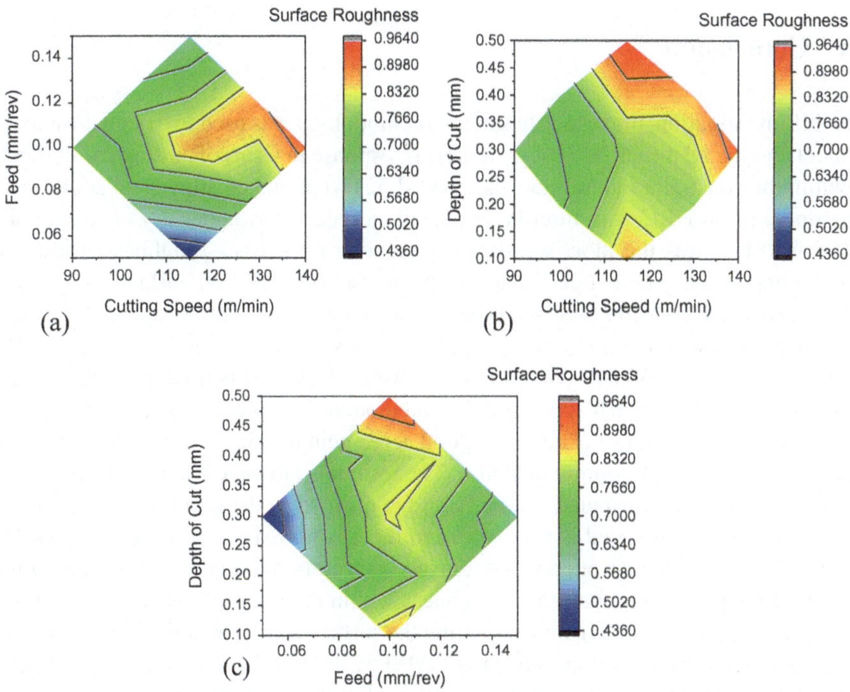

Fig. 3 Three-dimensional contour plots for surface roughness (Ra) **a** cutting speed versus feed rate, **b** cutting speed versus depth of cut and **c** feed rate versus depth of cut

depth of cut and cutting speed. The surface roughness value of 0.634 μm has arrived between a depth of cut of 0.2–0.3 mm and cutting speed of 100–120 m/min. Figure 3c shows surface roughness against the depth of cut and feed rate. The surface roughness value of 0.634 μm has arrived between a depth of cut of 0.2–0.3 mm and feed rate of 0.08–0.10 mm/rev.

3.1 Validation of the Model

The model F-value of 27.36 is significant and the lack of fit F-value of 3.29 is not significant and it indicates that the model is adequate. The predicted values of surface response were validated and verified through experimental runs and confirmatory test. The percentage of error between predicted and experimental values was found within ±4–6% which proved that the model is valid.

4 Conclusion

In this investigation, the response surface methodology was used to design the experiment and model the surface roughness (Ra) response in terms of cutting parameters, within the constraint in hard turning of AISI 52100 hardened alloy steel at elevated hardness of 58 HRC with multilayer coated carbide insert under dry environment. The feed rate was the most significant parameter on surface roughness, followed by cutting speed. The depth of cut was having no influence on surface roughness. The percentage contributions of cutting parameters, such as cutting speed, feed rate, depth of cut and quadratic term of feed rate for surface roughness, were 13.05%, 19.63%, 3.38% and 56.69%, respectively. The developed mathematical model of surface roughness based on response surface methodology was adequate and considered statistically significant. It exhibited a good agreement between the experimental and predicted values of the output response. The residuals fall on a straight line implying that the errors are distributed normally and confirmed to the normal distribution of data. The model proposed is adequate, and there is no reason to suspect any violation of the independence or constant variance assumptions and could be successfully used for the prediction of surface roughness within the range of cutting parameters. This investigation helped in understanding the surface characterisation under various cutting conditions in hard dry turning of AISI 52100 of 58 HRC, which would help the manufacturer technocrats in the selection of cutting parameters.

References

1. Tönshoff HK, Arendt C, Ben Amor R (2000) Cutting of hardened steel. CIRP Ann Manuf Technol 49(2):547–566
2. König W, Hochschule T, Komanduri R, Schenectady D, Tönshoff HK (1984) Machining of hard materials. CIRP Ann 33(2):417–427
3. Aouici H, Fnides B, Elbah M, Benlahmidi S, Bensouilah H, Yallese MA (2016) Surface roughness evaluation of various cutting materials in hard turning of AISI H11. Int J Ind Eng Comput 7:339–352
4. Rashid WB, Goel S, Davim JP, Joshi SN (2016) Parametric design optimization of hard turning of AISI 4340 steel (69 HRC). Int J Adv Manuf Technol 82(1):451–462
5. Shihab SK, Khan ZA, Mohammad A, Siddiquee AN (2014) Optimization of surface integrity in dry hard turning using RSM. Sadhana 39:1035–1053
6. Asiltürk I, Akkus H (2011) Determining the effect of cutting parameters on surface roughness in hard turning using the Taguchi method. Measurement 44:1697–1704
7. Hessainia Z, Belbah A, Yallese MA, Mabrouki T, Rigal JF (2013) On the prediction of surface roughness in the hard turning based on cutting parameters and tool vibrations. Measurement 46:1671–1681
8. Das DK, Sakho AK, Das R, Routara BC (2014) Investigations on hard turning using coated carbide insert: Grey based Taguchi and regression methodology. Procedia Mater Sci 6:1351–1358
9. Elbah M, Yallese MA, Aouici H, Mabrouki T, Rigal JF (2013) Comparative assessment of wiper and conventional ceramic tools on surface roughness in hard turning AISI 4140 steel. Measurement 46:3041–3056
10. Singh D, Rao PV (2007) A surface roughness prediction model for hard turning process. Int J Adv Manuf Technol 32:1115–1124
11. Sahin Y, Motorcu AR (2008) Surface roughness model in machining hardened steel with cubic boron nitride cutting tool. Int J Refract Metal Hard Mater 26:84–90
12. Arbizu IP, Perez CJL (2003) Surface roughness prediction by factorial design of experiments in turning processes. J Mater Process Technol 143–144:390–396

Productivity Improvement in Blow Molding Process Through Energy Savings

Hari Vasudevan, Rajendra Khavekar, and Nida Sayed

Abstract Rapid urbanization has resulted in enormous demand for energy in all sectors. There is a need to break the gap between demand and supply by devising energy-efficient approach. Through this study, an attempt has been made to bring down the power and to save energy utilized for cooling in the blow molding process. Cooling of the products was the biggest hurdle in plastics processing technology, which consumes the maximum amount of energy. Geothermal cooling system was explored for cooling of water used for blow molding process in lieu of the prevailing system using a chiller. Underground soil was used as a cooling medium for cooling the water emerging out from the molds. The system was thus analyzed with software simulation using ANSYS. The results have indicated a substantial reduction in energy utilization. The percentages of saving energy amounted to 15.28 and 24.27% when the geothermal system was implemented partially and fully. In the future, there is a scope to implement the geothermal cooling system in the plant in order to get a clear picture of the energy benefits.

Keywords Blow molding · Energy saving · Geothermal cooling system · Kasauda formula

1 Introduction

The ongoing technological advancements and developmental activities happening globally on various fronts have resulted in an ever-increasing demand for energy. Energy has become a precious commodity, necessitating its optimal use through

H. Vasudevan · R. Khavekar (✉) · N. Sayed
Dwarkadas J. Sanghvi College of Engineering, University of Mumbai, Mumbai, India
e-mail: khrajendra@rediffmail.com

H. Vasudevan
e-mail: principaldjs@gmail.com

N. Sayed
e-mail: sayednida23@gmail.com

© Springer Nature Singapore Pte Ltd. 2020
H. Vasudevan et al. (eds.), *Proceedings of International Conference on Intelligent Manufacturing and Automation*, Lecture Notes in Mechanical Engineering,
https://doi.org/10.1007/978-981-15-4485-9_18

selective and efficient approaches. Even, as far as manufacturing activities are concerned, energy management has a vital role to play, such that it is made more sustainable. Renewable energy resources have been widely explored these days and successfully used to achieve energy efficiency. It has been found that underground soil temperature has less temperature value in comparison with the ambient temperature. Thus, the difference between both the temperatures can be used as an efficient source for cooling purpose. Geothermal energy is a sustainable source of energy that uses the underground surface as a heat sink. This research study is an attempt to explore the saving of energy in the cooling system employed for the plastic blow molding process. Cooling of molds influences the productivity to a considerable extent. An efficient system has been conceived for cooling of water used in blow molding, using the concept of geothermal energy.

2 Literature Review

Ralegaonkar et al. [1] in their research study designed a geothermal cooling system and conducted a case study at the VNIT Nagpur Campus. It was observed that, at a depth of 2.5 m, the underground soil temperature was 20 °C in comparison with the ambient temperature that was 40 °C. Matsuda and Mino [2] implemented a geothermal cooling system at a datacenter located in Japan. They arrived at the conclusion that the developed geothermal air-conditioning system reduces electrical power consumption by 21%. Urchueguia et al. [3] in their research carried out an experimental study at Valencia, Spain. A comparison was made between HVAC system and geothermal cooling system, and they concluded that energy saving by the geothermal cooling system was $37 \pm 18\%$, in comparison with the HVAC system. De Nardin et al. [4] designed a geothermal air-conditioning system in Brazil to reduce the energy consumed in the office. Alam et al. [5] in their research stated that the underground soil can be used as a cooling medium. Cui et al. [6] carried out a study on temperature variation of underground soil and found that the temperature of soil beneath the surface is always lower than the ambient temperature. There is a temperature fluctuation up to the depth of 3 m, and beyond that, the temperature remains constant throughout the year. At a depth of 4 m, the average underground soil temperature was found to be 17 °C. Hence, from the literature review, it was concluded that by the application of soil temperature below the earth surface, which is lower in comparison with the ambient temperature, a geothermal cooling system could be built, which is an efficient and sustainable source of energy, and with that the electricity consumption can be reduced, hence saving energy.

3 Research Methodology

A flowchart depicting the methodology employed for developing the geothermal cooling system is given below.

1. Study of the existing system.
2. Calculation of underground soil temperature.
3. Design of the new system.
4. Computational fluid dynamics using ANSYS fluent.
5. Calculation of electricity unit.

3.1 Study of the Existing System

There are total of 8 machines in the plant and at one time, minimum of 5 machines work. In Fig. 1, the distance of each machine from the chiller is being specified. To meet the cooling requirements of blow molding machines, water which is being supplied to the mold was being chilled by the chiller (as shown in Fig. 1). The data, which was collected from the company about the electricity unit consumption, is being presented in the graph as indicated. In Fig. 2 the *X*-axis and *Y*-axis describe the electricity unit consumed per month, and from the graph, it can be concluded that the electricity unit consumption was in the range between 30,000 units and 40,000 units. The breakup of power consumption is furnished in the pie chart, as shown in Fig. 3. The chiller consumes a large amount of energy (shown in Fig. 3). In order to reduce the electrical power consumed by the chiller, in this research study, the use of naturally existing geothermal energy has been explored.

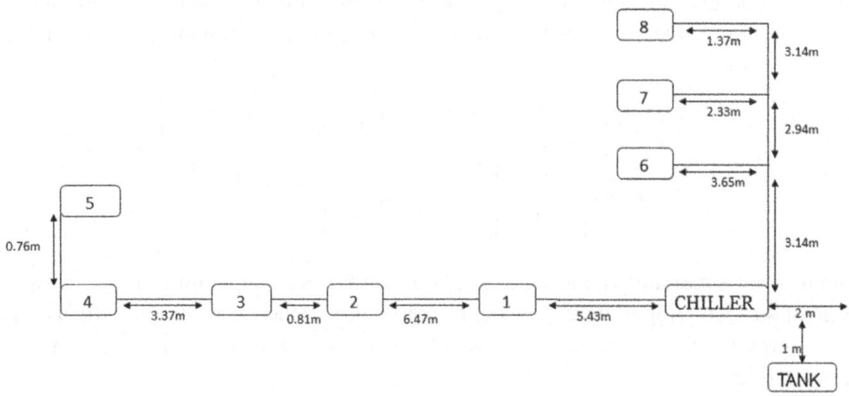

Fig. 1 Plan view of the studied plan

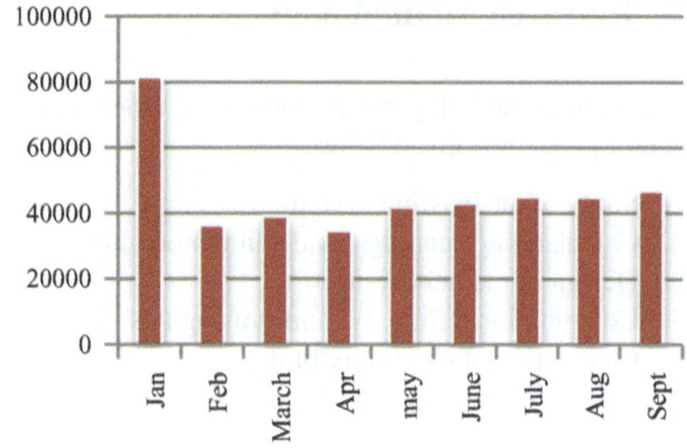

Fig. 2 Electricity unit consumption of the year-2019

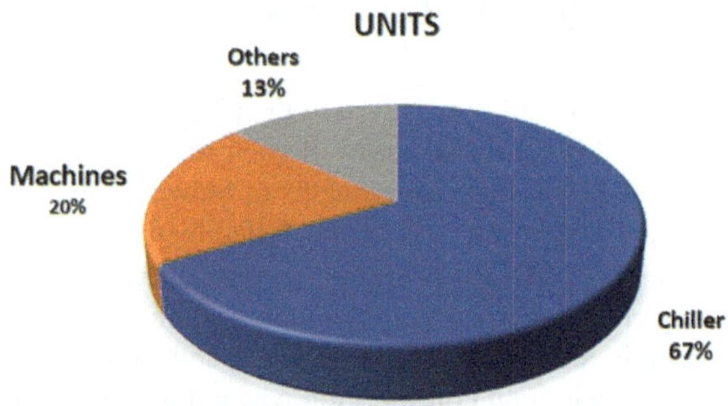

Fig. 3 Consumption of energy

3.2 Underground Temperature Calculation

The underground temperature was calculated by the analytical method. It was found that the underground temperature is a function of the depth beneath the ground surface and time of the year. The same could be described by the following correlation.

$$T = T_{\text{mean}} - T_{\text{amp}} * \exp\left(-z * \sqrt{\frac{\pi}{365 * a}}\right) * \cos\left[\frac{2\pi}{365} * \left\{T_{\text{year}} - T_{\text{shift}} - \frac{z}{2} * \sqrt{\frac{365}{\pi * a}}\right\}\right]$$

where T = temperature of the soil, T_{mean} = mean surface temperature, T_{amp} = amplitude of surface temperature, Z = depth below the surface, a = thermal diffusivity of the ground, T_{year} = current time (day) and T_{shift} = day of the minimum surface temperature.

For the calculation, the weather variables for Mumbai were needed. For this purpose, the temperature data recorded and published by the meteorological department for the year 2019 was collected. In the month of January, the minimum temperature of 14 °C was recorded on 2nd January. Therefore, the time shift was 38 h for the entire month of January. According to the data, the time shift was of 38 h, the mean surface temperature, 21.5 °C, the amplitude temperature, 15 °C with maximum temperature as 29 °C and minimum temperature as 14 °C. Therefore, at a depth of 1 m, the underground temperature value calculated for 2nd January was 13 °C. Similarly, the value was calculated for each day and the average temperature value computed for the month of January was 12.86 °C. The calculated temperature values obtained using the Kasauda formula are furnished in Tables 1 and 2.

Thus, from the above data, it was observed that as the depth increases, the temperature was getting higher. Therefore, for this research it was favorable to implement the geothermal cooling system at a depth of 1 m below the ground surface. The inlet water for the mold was required to be at a temperature of 17 °C. The temperature value obtained from Kasauda was a theoretical value, and in reality, it could vary depending on the physical properties of the ground (shown in Fig. 4) and the layout of the same is as given in Fig. 5. In the blow molding machine, the water which was being circulated in the molds had an inlet temperature of 17 °C and the outlet temperature of the water flowing out from the molds varied between 21 and 25 °C. In the existing system, the water emerging out of the molds had a temperature of around 25 °C and was subsequently supplied back to the chiller for cooling. From the chiller, the cold water having a temperature of 17 °C was supplied to the molds. The chiller used in the plant was having a cooling capacity of 15,000 K cal/h, and

Table 1 Calculated temperature value obtained from the Kasauda formula from the month of Jan–June 2018

Depth (m)	January (°C)	February (°C)	March (°C)	April (°C)	May (°C)	June (°C)
1	12.86	14.24	18.27	20.59	24.36	23.09
2	14.81	16.29	19.90	21.99	25.40	24.15
3	16.45	18.00	21.26	23.16	26.27	25.02
4	17.82	19.43	22.40	24.13	27.00	25.74
5	18.98	20.61	23.34	24.94	27.60	26.34

Table 2 Calculated temperature value obtained from the Kasauda formula from the month of July–December 2018

Depth (m)	July (°C)	August (°C)	September (°C)	October (°C)	November (°C)	December (°C)
1	23.98	23.71	22.41	20.49	17.24	13.95
2	24.63	24.38	23.42	22.18	19.23	16.00
3	25.17	24.94	24.26	23.58	20.86	17.69
4	25.63	25.40	24.95	24.73	22.21	19.08
5	26.00	25.79	25.52	25.69	23.33	20.23

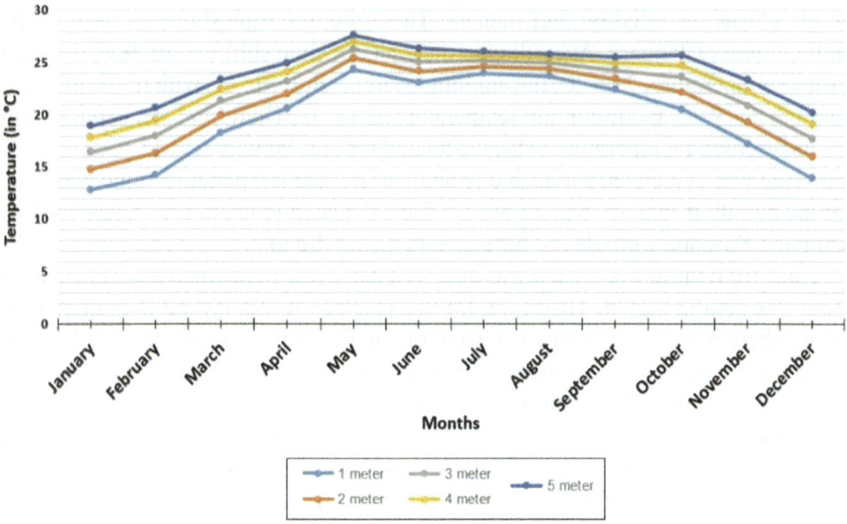

Fig. 4 Results obtained for a typical year 2018, by using the Kasauda formula at various depths

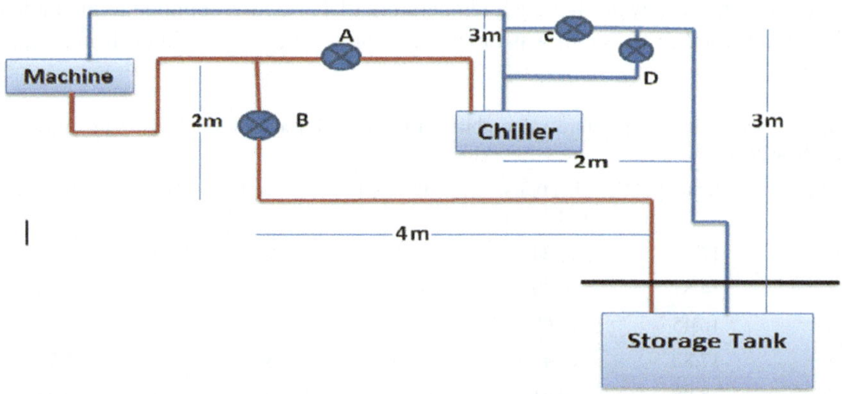

Fig. 5 Layout of the new system

the electricity consumption per month was 10,821 units. The chiller was consuming almost 40% of the total electricity consumption of the plant.

3.3 Layout of the New System

In the devised system, the components that were going to be used are storage tank, chipset, solenoid valves and temperature sensors. The storage tank was installed 1 m below the ground surface. There were two temperature sensors that were used for sensing the temperatures of the ground surface and were sending the input values to the chipset. The chipset was used for monitoring the solenoid valves A, B, C and D according to the input values received from the temperature sensors.

The system can be perceived as under:

i. The system is working on geothermal cooling
ii. The system is working on partial geothermal cooling
iii. The system is working by the use of chiller.

Case 1 (*The system is working on geothermal cooling*) In this case, the solenoid valves A and D will be closed, whereas the solenoid valves B and C will be open. The water will flow from the molds to the storage tank, and from the storage tank, water will be supplied back to the molds for cooling.

Case 2 (*The system is working on partial geothermal cooling*) In this case, the solenoid valves A and C will be closed, whereas the solenoid valves B and D will be opened. The water will flow from the molds to the storage tank, and from the storage tank, water will flow to the chiller for further cooling. After that, the water will be supplied back to the molds for cooling.

Case 3 (*The mold cooling is done by the chiller*) In this case, the solenoid valve B, C and D will be closed, whereas the solenoid valve A will be open. The water will flow from the molds to the chiller, and from the chiller, the water will be supplied back to the molds.

3.4 Computational Fluid Dynamics Using ANSYS Fluent Under Different Boundary Conditions

Analysis software was employed (shown in Figs. 6, 7, 8 and 9) to examine the role of underground soil in cooling the hot water emerging from the molds. Three-dimensional parametric model was made in SOLIDWORKS software. In ANSYS, solid body was meshed using 3D hexahedron. In order to perform computational fluid dynamics, ANSYS fluent was used. 50 mm element size was maintained. K-Epsilon model was used as in the tank region flow might become turbulent. Energy equation was turned on, in order to find heat exchange. Default mathematical model was used for convergence. The storage tank wall, which is in contact with the underground soil, was assumed to have the same temperature as that of the soil temperature. The boundary condition given was the discharge rate that was 3 L/min and the inlet temperature of water entering the storage tank was 25 °C. The above two boundary conditions were the same, whereas the cold wall temperature values were changed. From the analysis, it was concluded that underground soil is very effective in transferring the heat from the water stored in the storage tank to the underground soil. The obtained result through the ANSYS analysis gave an indication of the temperature drop in water as the water flows in the downward direction.

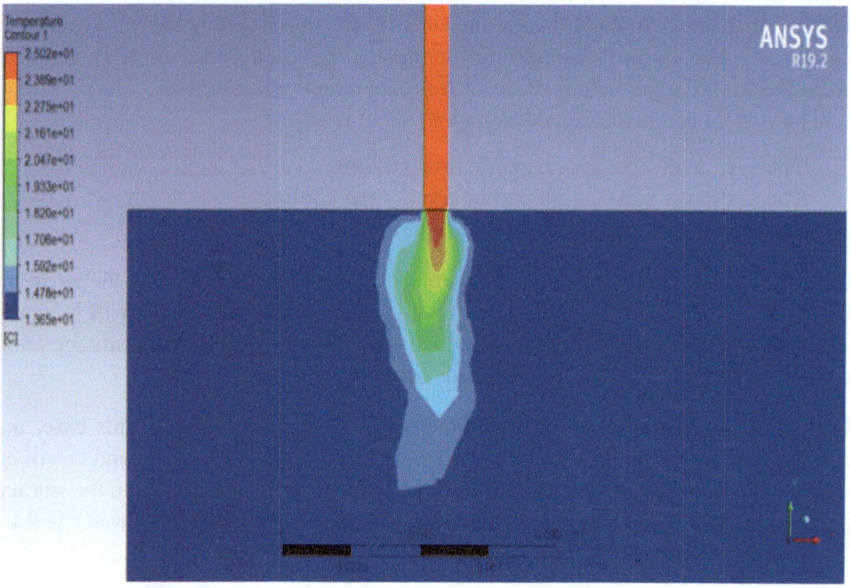

Fig. 6 ANSYS result at 14 °C temperature

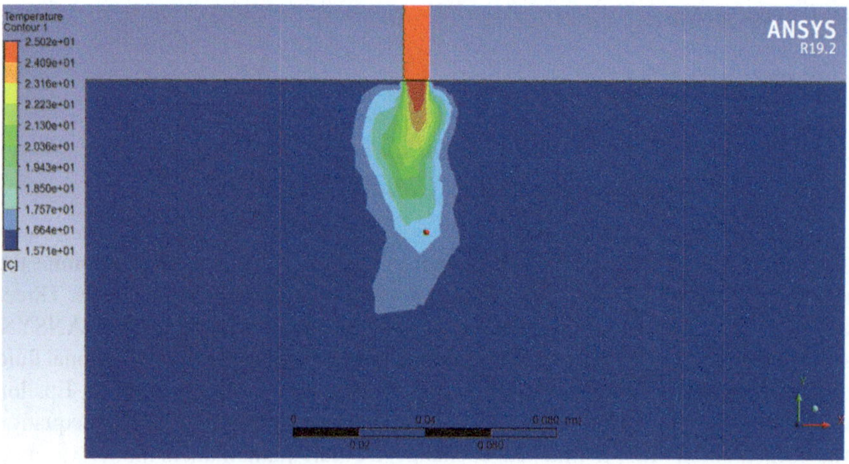

Fig. 7 ANSYS result at 16 °C temperature

3.5 Electricity Unit Calculation

Case 1 (*Electricity consumption calculation of the existing system*) (1) 3 HP pump: 2238 W, (2) Chiller: 17,580 W, (3) Total power consumed: 19,818 W, (4) Unit

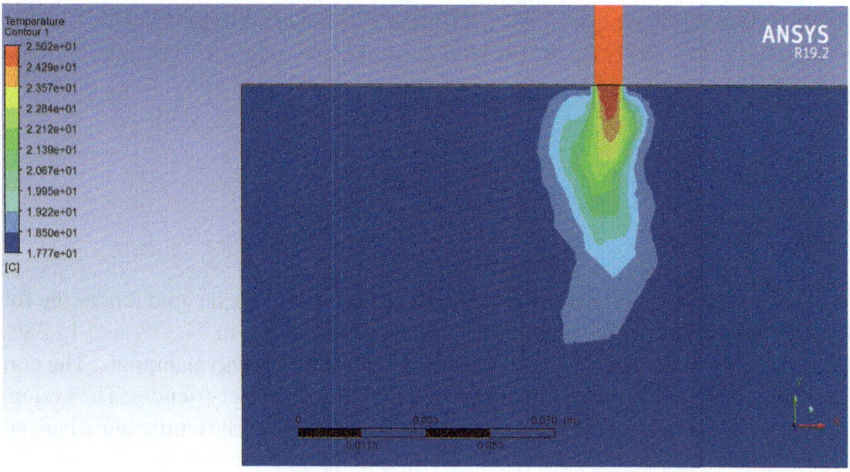

Fig. 8 ANSYS result at 18 °C temperature

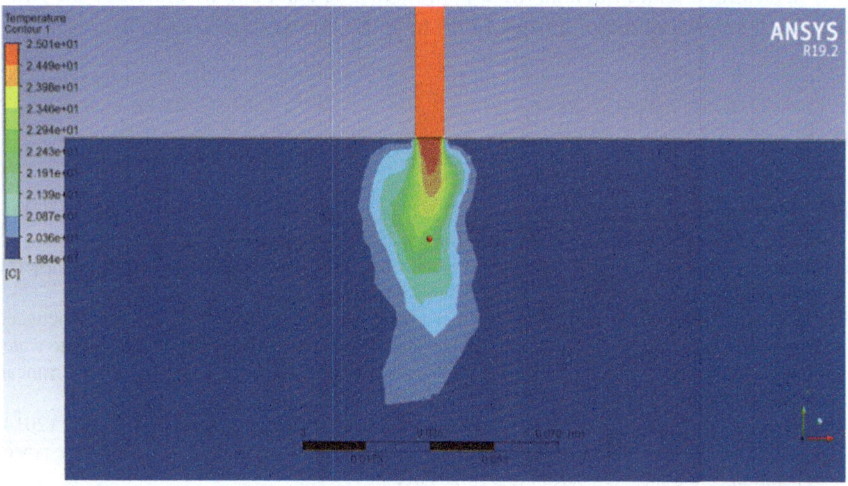

Fig. 9 ANSYS result at 20 °C temperature

consumed per day: 416.178 units, (5) Unit consumed in a month: 10,821 units, (6). Unit consumed per month: 10,821 units.

Case 2 (*Electricity consumption of the geothermal cooling system*) (1) 3 HP pump: 2238 W, (2) 1 HP pump: 746 W, (3) Solenoidal valves: 44 W, (4) Temperature sensor: 0.044 W, (5) Chipset: 1 W, (6) Total power consumed: 3029.044 W, (7) Unit consumed per day: 63.609 Units, (8) Unit Consumed in a month: 1654 units.

Case 3 (*Electricity consumption with partial geothermal cooling system*) (1) 3 HP pump: 2238 W, (2) 1 HP pump: 746 W, (3) Solenoidal valves: 44 W, (4) Temperature sensor: 0.044 W, (5) Chipset: 1 W, (6) Chiller (¼ load condition): 1782 W, (7) Total power consumed: 4811.044 W, (8) Unit consumed per day: 101.031 units, (9) Unit consumed in a month: 2627 units.

4 Conclusion

The new system devised with the geothermal cooling approach could render the following benefits. The percentage saving of energy amounted to 24.27% and 15.28%, respectively, while operating entirely and partially with geothermal inputs. The constructional and design feature of the system is simple and user-friendly. The system, however, has been designed with the following constraints. The temperature beneath the soil could vary from place to place, depending upon the climate situations and characteristics of the soil. Hence, the actual data pertaining to the soil temperature at the place, where the factory was located could not be obtained and the same was determined using the analytical approach. The calculations indicated the most feasible temperature ranging from 12.86 to 20.49 °C over the season.

References

1. Ralegaonkar R, Kamath MV, Dakwale VA (2014) Design and development of geothermal cooling system for composite climatic zone in India. J Inst Eng India Ser A 95(3):179–183
2. Matsuda C, Mino Y (2016) Study on power-saving effects in direct use of geothermal energy for datacenter cooling system. IEEE. ISBN: 978-1-5090-1877-2/16
3. Urchueguıa J, Zacarés M, Corberán J, Montero A, Martos J, Witte H (2008) Comparison between the energy performance of a ground coupled water to water heat pump system and an air to water heat pump system for heating and cooling in typical conditions of the European Mediterranean coast. Energy Convers Manag 49(10):2917–2923
4. De Nardin CR, Fernandes FT, Longo AJ, Farret FA, De Nardin AC, Lima L, Cruz DR (2014) Underground geothermal energy to reduce the electricity consumption. In: 3rd renewable power generation conference. https://doi.org/10.1049/cp.2014.0933
5. Alam MR, Zain MFM, Kaish ABMA, Jamil M (2013) Underground soil and thermal conductivity materials based heat reduction for energy-efficient building in tropical environment. SAGE J 24(2):185–200
6. Cui W, Liao Q, Chang G, Chen G, Peng Q, Jen TC (2011) Measurement and prediction of undisturbed underground temperature distribution. In: Proceedings of the ASME—2011 international mechanical engineering congress and exposition, pp 671–676

Effect of Cutting Parameters on Microhardness in Turning of AISI 52100 Hardened Alloy Steel with Multilayer Coated Carbide Insert

Sandip Mane and Sanjay Kumar

Abstract The microhardness (μH) and surface roughness (Ra) constitute the surface integrity aspect, which affects the quality and service life of a product. This paper aimed to analyse the effect of cutting parameters, such as cutting speed, feed and depth of cut on microhardness (μH) in hard turning of AISI 52100 steel of 58 HRC, with multilayer coated carbide tool insert under dry environment. Response surface methodology based on central composite design (CCD) was employed to design the experiment and develop the mathematical models for microhardness (μH). The analysis of variance (ANOVA) was used to explore the main and interaction effect of cutting parameters on microhardness (μH). The (ANOVA) results revealed that the microhardness of the machined surface is significantly influenced by the cutting speed, feed rate and depth. A quadratic regression model in terms of cutting speed, feed and depth of cut for microhardness (μH) was developed with 95% confidence level. The adequacy and validity of the developed model were checked on the basis of *P*-value, *F*-value and R^2-value.

Keywords Microhardness · Surface roughness · Fatigue life · Surface integrity

1 Introduction

Tönshoff et al. [1] reported that the hardened alloy steels are being widely used in automotive and allied industries due to their high compressive strength and wear resistance. Hard turning is a viable and cost-effective machining process. In practice, hard turning is carried out in the range of 58–68 HRC. König et al. [2] stated

S. Mane (✉)
Department of Mechanical Engineering, Dwarkadas J. Sanghvi College of Engineering, Mumbai, Maharashtra 400056, India
e-mail: sandip.mane@djsce.ac.in

S. Kumar
Department of Mechanical Engineering, Thakur College of Engineering and Technology, Mumbai, Maharashtra, India
e-mail: sanjay.kumar@thakureducation.org

© Springer Nature Singapore Pte Ltd. 2020

H. Vasudevan et al. (eds.), *Proceedings of International Conference on Intelligent Manufacturing and Automation*, Lecture Notes in Mechanical Engineering,
https://doi.org/10.1007/978-981-15-4485-9_19

that the hard turning has been gaining the importance over conventional grinding process due to its benefits such as lesser set-up time, fewer process steps, process flexibility, greater part geometry, higher material removal rate, compatible surface roughness, less health and environment issues. Guo et al. [3] pointed out, the excessive heat generation and cutting temperature at the cutting zone during hard turning significantly affect the surface integrity of the machined component. The surface finish can be improved by reducing the heat generation and temperature at the tool-chip-workpiece interface. This objective can be achieved by proper selection of cutting parameter. Kundrak et al. [4] and Saini et al. [5] reported that the surface integrity aspect investigates the functional behaviour, fatigue life, reliability and the wear resistance of the machined components. The factors such as surface roughness, microhardness, residual stresses and microstructural changes are the performance indicators of the surface integrity and need to be studied in depth to understand and set the adequate guideline in the machining process. Goel et al. [6] reported that the cutting parameters such as cutting speed, feed rate and depth of cut are found to be very important in metal machining, which affects the surface characteristics like surface roughness, microhardness, residual stresses and microstructure. These parameters are the indeed constituents of the surface integrity, and thus, it is imperative to select an optimum combination of these parameters. The surface roughness (Ra) and the microhardness (μH) are very important parameters amongst all the other surface integrity parameters, as they correlate with the surface profiles in order to better characterize the different machining processes. Liu et al. [7] investigated the effects of cutting parameters on microhardness in the machining of Inconel 718, and the results revealed that the microhardness is significantly influenced by the cutting followed by the feed rate, while the depth of cut has less influence. Ndaliman et al. [8] analysed the effects of process parameters on the microhardness and reported that the machining parameters have considerable effects on the microhardness. The variation of the microhardness at the machined surface can be reduced by proper selection of cutting parameters. Cai et al. [9] showed that the increase in the level of cutting speed increases the microhardness of the machined surface in high-speed milling of Inconel 718 due to a high strain rate. Rafai et al. [10] studied the influence of cutting speed and depth of cut on microhardness in milling of hardened AISI D2 steel and showed that the cutting speed and the radial depth of cut are the important parameters affecting the microhardness. Bhopale et al. [11] reported that the microhardness is a surface characteristic, which affects the service life of the component and significantly influenced by the cutting parameters. Hassanpour et al. [12] investigated and developed the relationship between cutting parameters such as cutting speed, feed rate, and depth of cut and microhardness of the machined surface in the milling of hardened 4340 alloy steel. However, the previous studies have not investigated the effects of cutting parameters on microhardness in turning of hardened steel at elevated hardness in the range of 58–65 HRC, which is being widely used in automotive and allied industries. In this study, an attempt has been made to study and develop parametric relationship between the cutting parameters and the microhardness of machined surface in turning of hardened AISI 52100

Table 1 Chemical composition of AISI 52100 hardened alloy steel (weight percentage)

C %	Si %	Mn %	P %	S %	Cr %	Ni %	Cu %	Fe %
1.04	0.18	0.35	0.007	0.004	1.35	0.076	0.058	Balance

alloy steel of 58 HRC with multilayer coated carbide tool insert in dry environment. The result provides a reference for the selection of cutting parameters in practice.

2 Experimentation

2.1 Selection of Workpiece Material

In this study, AISI 52100 hardened alloy steel having a hardness of 58 HRC was selected as workpiece material. AISI 52100 hardened alloy steel has wide applications and is being used in automotive and allied industries such as bearings, forming rolls, spindles, tools and precision instrument parts. Table 1 shows the chemical composition of the workpiece material.

2.2 Selection of Tool

The cutting tool inserts and the tool holder were selected based on the literature review and the tool manufacturer's recommendation. The MTCVD multilayer coated carbide (TiN/TiCN/Al_2O_3)–[HK150, K-type] cutting tool insert having specification CNMG120408 and the tool holder with PCLNR 2020 K12 specification were selected for experimentation. The experiments were carried out on a rigid high precision HMT NH-18 lathe machine.

2.3 Selection of Cutting Parameters

Based on the previous research carried out and the tool manufacturer's recommendations, the cutting parameters were selected. Table 2 presents the cutting parameters and their coded and actual levels based on the central composite design of response surface methodology.

Table 2 Cutting parameters and their coded and actual levels using central composite design

Process parameters	Units	Limits				
		−1.682	−1	0	1	1.682
Cutting speed	m/min	90	100	115	130	140
Feed rate	mm/rev	0.05	0.075	0.10	0.125	0.15
Depth of cut	mm	0.10	0.20	0.30	0.40	0.50

2.4 *Design of Experiment*

The response surface methodology uses the statistical experimental design technique and the least square fitting method for model generation. The central composite design (CCD) was used for experimental design. The prediction of the microhardness was carried out in terms of cutting parameters under dry environment. Microhardness values were measured using microhardness tester. Table 3 shows the experimental design and result for microhardness.

Table 3 Experimental design and result for microhardness

S. No.	Cutting speed (m/min)	Feed rate (mm/rev)	Depth of cut (mm)	Microhardness (HV)
1	115	0.10	0.3	778
2	130	0.08	0.4	811
3	115	0.10	0.3	780
4	90	0.10	0.3	756
5	140	0.10	0.3	837
6	115	0.10	0.1	776
7	115	0.05	0.3	759
8	115	0.10	0.3	775
9	115	0.10	0.3	783
10	100	0.08	0.2	752
11	130	0.12	0.4	825
12	115	0.10	0.5	801
13	130	0.12	0.2	796
14	115	0.10	0.3	780
15	130	0.08	0.2	781
16	100	0.08	0.4	751
17	100	0.12	0.4	778
18	100	0.12	0.2	765
19	115	0.10	0.3	785
20	115	0.15	0.3	781

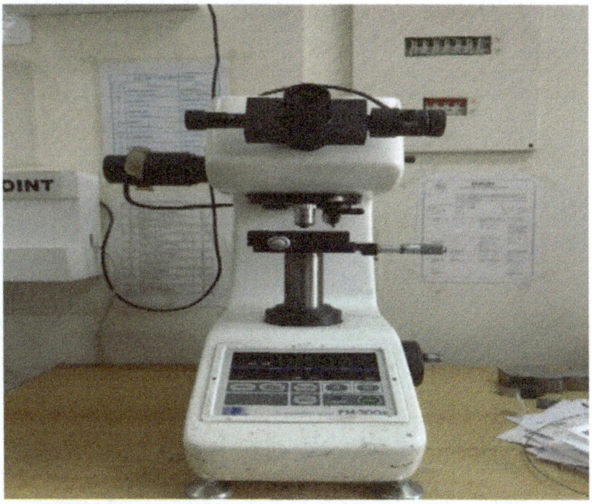

Fig. 1 Microhardness tester (FM-300e)

2.5 *Measurement of Microhardness*

To measure the microhardness of the machined component, the microhardness tester (FM-300e) was used. Figure 1 shows the microhardness tester (FM-300e).

3 Result and Discussion

It is clear from ANOVA table that the *P*-values less than 0.0500 indicate model terms are significant, and the values greater than 0.1000 indicate the model terms are not significant. The cutting speed (V) has the most significant effect, while the feed rate (F), depth of cut (D), the quadratic term of cutting speed (V), the quadratic term of feed rate (F^2) and the interaction between cutting speed and depth of cut (VD) have less significant effect on the microhardness. But the quadratic term of the depth of cut (D^2), the interaction between cutting speed and feed rate (VF) and the interaction between feed rate and depth of cut (FD) all have no significant effect. The below equation is in terms of coded and actual factors can be used to make the predictions. Table 4 shows the ANOVA for quadratic model of microhardness and table 5 shows the model summary of microhardness.

Final equation in terms of coded factors:

$$\begin{aligned}\text{Microhardness} = {} & +779.51 + 6.05 * F + 7.56 * D - 1.37 * V * F + 5.88 * V * D \\ & + 1.63 * F * D + 5.01 * V * V - 0.74 * F * F + 1.71 * D * D\end{aligned}$$

Final equation in terms of actual factors:

$$\begin{aligned}\text{Microhardness} = {} & +929.117 - 4.355 * V + 1455.813 * F - 558.925 * D \\ & - 583 * V * F + 3.916 * V * D + 812.50 * F * D + 0.022 * V * V \\ & - 4350.206 * F * F + 171.473 * D * D\end{aligned}$$

Table 4 ANOVA for quadratic model of microhardness

Source	DF	Seq SS	Cont. (%)	Adj MS	*F*-value	*P*-value	
Regression	9	9444.71	96.78	1049.41	33.39	0.000	Significant
V	1	6728.16	68.94	6728.16	214.07	0.000	Significant
F	1	750.05	7.69	750.05	23.86	0.001	Significant
D	1	915.06	9.38	915.06	29.12	0.000	Significant
$V * V$	1	413.83	4.24	331.97	10.56	0.009	
$F * F$	1	251.22	2.57	179.34	5.71	0.038	
$D * D$	1	74.00	0.76	74.00	2.35	0.156	
$V * F$	1	15.12	0.15	15.13	0.48	0.504	
$V * D$	1	276.13	2.83	276.12	8.79	0.014	
$F * D$	1	21.13	0.22	21.13	0.67	0.431	
Error	10	314.29	3.22	31.43			Not
Lack of fit	5	251.46	2.58	50.29	4.00	0.077	significant
Pure error	5	62.83	0.64	12.57			
Total	19	9759.00	100.00				

The model *F*-value of 33.39 implies the model is significant and could be used to predict the microhardness accurately. The lack of fit *F*-value of 4.00 implies that the lack of fit is not significant relative to the pure error and can fit the model well (Table 5).

The predicted R^2 of 0.7604 is in reasonable agreement with the adjusted R^2 of 0.9388; i.e. the difference is less than 0.2. Adeq precision measures the signal to noise ratio. A ratio greater than 4 is desirable, and ratio of 19.4371 indicates an adequate signal. This model can be used to navigate the design space.

Figure 2a shows the normal probability plot of the residuals for microhardness. The normal probability plot indicates whether the residuals follow a normal distribution or not. The points on the normal probability plots of the residuals fall on a straight line implying that the errors are distributed normally. It can be seen from Fig. 2b that all the actual values are following the predicted values. Figure 2c represents residuals versus the predicted response plot for surface roughness. The plot shows the random scatter and contains no obvious pattern. This implies that the model proposed is adequate, and there is no reason to suspect any violation of the independence.

Figure 3 shows contour plots of microhardness. Figure 3a shows the microhardness versus cutting speed and feed rate. From this graph, lower microhardness

Table 5 Model summary of microhardness

Std. Dev.	R^2	R^2 (adj)	PRESS	R^2 (pred)	Adeq precision
5.60616	96.78%	93.88%	2338.70	76.04%	19.4371

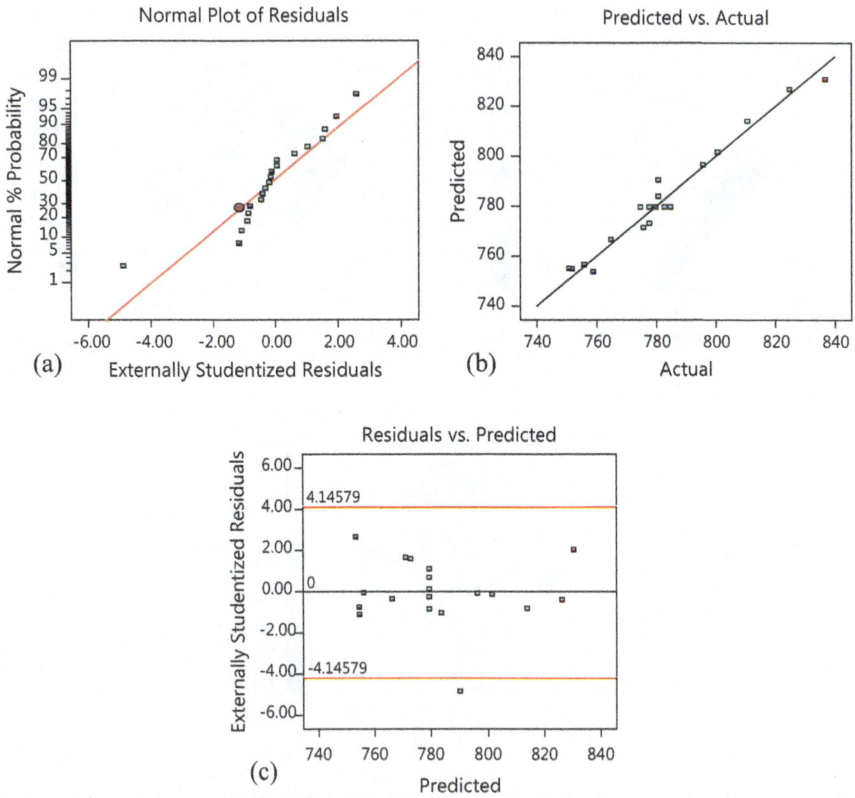

Fig. 2 **a** Plot of residuals versus normal probability. **b** Plot of predicted versus actual values. **c** Plot of residuals versus predicted values for microhardness

value of 761 μm was arrived between a feed rate of 0.05–0.10 mm/rev and a cutting speed between 90 and 115 m/min. Figure 3b shows microhardness against the depth of cut and cutting speed. An optimum surface roughness value of 761 μm has arrived between a depth of cut of 0.15–0.35 mm and cutting speed of 90–110 m/min. Figure 3c shows microhardness against the depth of cut and feed rate. An optimum microhardness value of 754 μm has arrived between a depth of cut of 0.15–0.35 mm and feed rate of 0.05–0.10 mm/rev. Figure 4 shows the surface plot for better visualization of the functional relationship between the dependant and two independent variables.

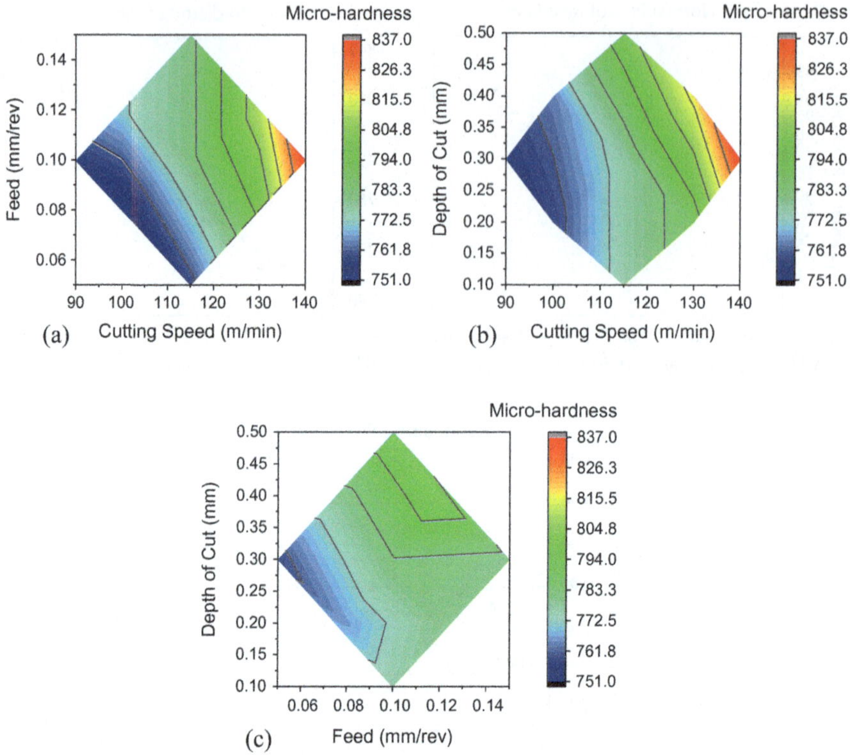

Fig. 3 Three-dimensional contour plots for microhardness **a** cutting speed versus feed rate, **b** cutting speed versus depth of cut, **c** feed rate versus depth of cut

4 Conclusion

In this paper, the response surface methodology was used to design the experiment and model the microhardness of machined surface in terms of cutting parameters within the constraint in hard turning of AISI 52100 hardened alloy steel at elevated hardness of 58 HRC with multilayer coated carbide insert under dry environment. The cutting speed was the most significant parameter on microhardness. The feed rate and depth of cut were having less significant effect on surface roughness. The quadratic terms of cutting parameters and their interactions had no effects on microhardness. The percentage contribution of cutting parameters, such as cutting speed, feed rate and depth of cut for microhardness were 68.94, 7.69% and 9.38%, respectively. The developed mathematical model of microhardness based on response surface methodology is adequate and statistically significant. It exhibited good agreement between the experimental and predicted values of the output response. This investigation would help in understanding the surface characterization under various cutting conditions in turning of hardened alloy steel.

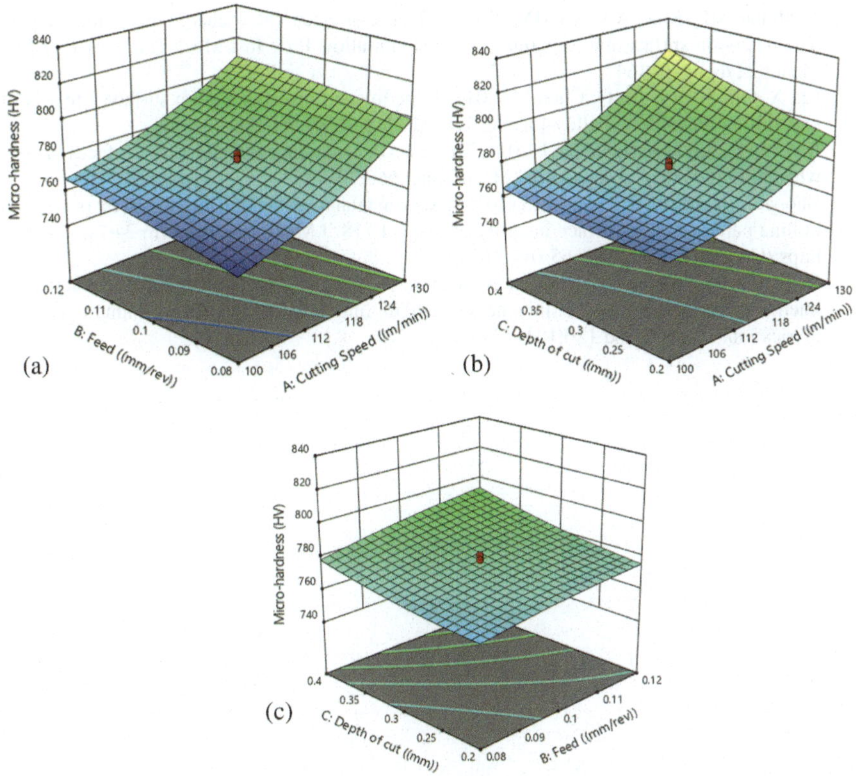

Fig. 4 Three-dimensional surface plots for microhardness **a** cutting speed versus feed rate, **b** cutting speed versus depth of cut, **c** feed rate versus depth of cut

References

1. Tönshoff HK, Arendt C, Ben Amor R (2000) Cutting of hardened steel, CIRP Ann Manuf Technol 49(2):547–566
2. König W, Hochschule T, Komanduri R, Schenectady D, Tönshoff HK (1984) Machining of hard materials. CIRP Ann 33(2):417–427
3. Guo YB, Li W, Jawahir IS (2009) Surface integrity characterization and prediction in machining of hardened and difficult-to-machine alloys: a state-of-art research review and analysis. Mach Sci Technol 13:437–470
4. Kundrak J, Mamalis AG, Gyani K, Bana V (2011) Surface layer microhardness changes with high-speed turning of hardened steels. Int J Adv Manuf Technol 53:105–112
5. Saini S, Ahuja IS, Sharma VS (2012) Residual stresses, surface roughness, and tool wear in hard turning: a comprehensive review. Mater Manufact Proc 27:583–598
6. Goel P, Khan ZA, Siddiquee AN, Kamaruddin S, Gupta RK (2012) Influence of slab milling process parameters on surface integrity of HSLA: a multi-performance characteristics optimization. Int J Adv Manuf Technol 61:859–871
7. Liu WW, Li XY, Wan XS, Li F, Shan CW (2013) The effects of turning parameters on machining surface integrity in high speed turning Inconel 718. Mech Sci Technol Aerosp Eng 32(8)

8. Ndaliman MB, Khan AA, Ali MY (2013) Influence of electrical discharge machining process parameters on surface micro-hardness of titanium alloy. Proc Inst Mech Eng Part B: J Eng Manuf 227(3):460–464
9. Cai X, Qin S, Li J, An Q, Chen M (2014) Experimental investigation on surface integrity of end milling nickel-based alloy-Inconel 718. Mach Sci Technol 18(1):31–46
10. Rafai NH, Lajis MA, Hosni NAJ (2014) The effect of machining parameters on surface integrity when milling hardened steel. Appl Mech Mater 660:70–73
11. Bhopale NN, Joshi SS, Pawade RS (2015) Experimental investigation into the effect of ball end milling parameters on surface integrity of Inconel 718. J Mater Eng Perform 24(2):986–998. https://doi.org/10.1007/s11665-014-1323-y
12. Hassanpour H, Sadeghi MH, Rasti A, Shajari S (2016) Investigation of surface roughness, microhardness and white layer thickness in hard milling of AISI 4340 using minimum quantity lubrication. J Clean Prod 120:124–134

Analysis and Manufacturing of Aerodynamic Components

Rajnarayan Yadav, Vinit Katira, Ruchit Doshi, Shakshi Himmatramka, Parshva Mehta, and Harshil Mody

Abstract Aerodynamic components are extensively used in motorsport, such as Formula One to increase the downforce, thereby ensuring traction of the tires, which increases the performance of the race car during cornering. However, these aerodynamic components that provide the downforce also increase drag. An ideal solution, therefore, is dynamically adjustable aerodynamic components, which have the ability to change the amount of downforce and drag. Such components have been used as aerodynamic components for an FSAE car as part of this study. After completing an extensive research on current car design without aerodynamic components, preliminary design concepts were formulated and full car solid model was designed on SOLIDWORKS software. These designs were then compared against each other based on the analysis performed on STAR CCM+ software on the basis of downforce, and drag was then generated. The software analysis was used to compare and conclude the final design from each iteration to get an optimum aerodynamic package which consists of all components for the car. The product was manufactured using some advanced manufacturing techniques to save cost, time as well as to provide it for further development.

Keywords Formula one · Lap simulations · Aerodynamics package · Finite elements analysis · Composite materials

1 Introduction

A car designed for Formula Student/SAE competition must be lightweight, quick, and enduring [1, 2]. If the components cannot be utilized in improving the performance of race car, it becomes dead weight resulting in design degradation [3, 4]. In order to find out the need for aerodynamic components, extensive research was carried

R. Yadav (✉) · V. Katira · R. Doshi · S. Himmatramka · P. Mehta · H. Mody
Department of Mechanical Engineering, Dwarkadas J. Sanghvi College of Engineering, Vile Parle West, Mumbai 400056, India
e-mail: rajnarayan.yadav@djsce.ac.in

© Springer Nature Singapore Pte Ltd. 2020

H. Vasudevan et al. (eds.), *Proceedings of International Conference on Intelligent Manufacturing and Automation*, Lecture Notes in Mechanical Engineering,
https://doi.org/10.1007/978-981-15-4485-9_20

out by comparing the results on optimum lap simulations software with a car having aerodynamic components and car without aerodynamic components [5–7]. Also, we had done a manual calculation based on the previous car data to find out improvement in lap times.

2 Methodology

The test track was modeled and simulated on the optimum lap simulation software. After this, manual calculations considering the engine to be KTM RC 390 were carried out. As per the rules, the engine intake system is restricted to 20 mm to limit the power of the engine. Accordingly, the engine power was taken as 18 kW. The engine's top speed is calculated without the aerodynamic components and then the amount of power that will be used by the drag force of the aerodynamic package is decided.

To find out C_{dmax} value for rear wings, manual calculations were carried out.

$$P_{max} = 18\,\text{kW}$$
$$V_{max} = 2 * P_{max}/(\rho * A * C_d)$$
$$A = 0.65 \quad \text{(without wings)}$$
$$C_d = 1.025 \quad \text{(without wings)}$$

Therefore, $V_{max} = 35.38$ m/s $= 127.37$ kmph (theoretical top speed).

Now assuming speed that we want to achieve to be 110 kmph = 30.556 m/s.

Therefore, power used at this speed is given by

$$P = 0.5 * 1.22 * 0.65 * 1.025 * (30.556)3 = 11.595\,\text{kW}$$

Therefore, power left to overcome the drag produced by wings is $\Delta P = 6.405$ kW.

Hence, we get $C_{dmax} = 1.338$.

Now, for this C_{dmax} value, C_l value for multielement wing can be found out by simulation at particular angle of attack.

The following observations are drawn from the lap simulations:

- The car with wings has lower lap times compared to the car without wings.
- The top speed is lower in the car with aerodynamics but the time for which it stays at that speed is high.
- The maximum lateral acceleration increases in the car with aerodynamics.
- The fuel consumption is more in the car with aerodynamics because it produces more drag.
- The percentage of braking reduces in the car with aerodynamics.

The aerodynamics package development consists of several stages. The first stage is to develop an aerofoil capable of producing high downforce. It is expected that the under-tray will have a negligible effect on the load distribution of the vehicle, and thus it is designed after the aerofoil configuration before the front and rear wing is introduced, as these will affect the balance. When the under-tray design is established, the strategy is to design the front wing to produce as much downforce as possible while complying with the competition rules for wing dimensions and clearance to the wheels and the ground.

3 Formula One Car Modeling and Analysis

Using a 3D modeling software, SOLIDWORKS, a solid model of the car was created out of surfaces, as shown in Fig. 1. The computers used for the Star CCM+ solver had 8 cores at 2.4 GHz each and 16 GB of ram. The geometry was simplified to remove sharp corners and small radii, which require increased cell densities to mesh.

A half car CAD model as shown in Fig. 2 was used for simulation which allows

Fig. 1 Full CAD model

Fig. 2 Symmetry setup of the car

Fig. 3 Velocity contour on the symmetric plane

for higher cell density on one half of the car. A plane was created to replicate the symmetry for the right side of the car, and the results are mirrored for the right half. The car was meshed with a trimmer mesh of 2 million cells around the car. An unstructured mesh is required around the car to limit skewed cells due to the complex geometry, which requires both high-density and low-density cell regions around the car. Since flow propagates both upstream and downstream, it is important to have a sufficiently large volume to capture the full effect of the flow. As a result, the domain was designed for a distance of two car lengths in front, two-car heights above, one car width out from the cut plane and four-car lengths behind the car. Several turbulence models are available in Star CCM+. Through a literature search, the *k-omega* realizable viscous solver was selected due to its high correlation with experimental data. In order to aid convergence and to improve the boundary layer formation, the "velocity-inlet" was initialized at 20 m/s. A "pressure-outlet" was set for the exit conditions behind the car. The model was run with double precision to improve the accuracy of the results. The simulation results include the velocity contour and the pressure distribution over the wing surfaces and they are shown in Figs. 3 and 4, respectively.

4 Mount Design

Sprung aero mounting: In this method, most of the aerodynamic components are connected to chassis, where downforce generated by these components is transferred to chassis.

Unsprung aerodynamic mountings: In this method, all the mountings used for the aerodynamic components are directly connected to suspension, in order to maximize the efficiency of downforce and directly transferring it to tires. It was decided to go

Fig. 4 Pressure distribution contour

with sprung aero mounting because of the simplicity in analyzing the effects of downforce on the vehicle dynamics of the car.

The mount must provide rigid mounting for all the components when different types of forces act on it but it should be light in weight. It should be designed considering axial and lateral force which may act throughout the operation.

After analyzing the different materials and their weight, it was decided to use composite panel which consists of hard density foam used as a core material in sandwich panel of carbon fire composite layer.

Mounts have to be designed according to track and worst possible environment to have a safe and reliable design. Car has to be driven at different places and different types of tracks. After taking weight, bump force, weather conditions, and other criteria, we were able to formulate the loads which will act on the front and rear wings (Table 1).

$$\begin{aligned}\text{Force } Z &= \text{Downforce by } V_{\max}(80\,\text{km/s}) + \text{Weight of wings}\\ &\quad + \text{Bump}(10\,\text{g}) + \text{Gust of winds}\\ \text{Force } Z &= Z1 + Z2 + Z3 + Z4\end{aligned}$$

- $Z1 = ½ \times$ Frontal area (m^2) $\times$ Density of air (kg/m^3) $\times$ Velocity of car $\times$ (-2.55)
- $Z2 =$ Weight of the wings (kg) $\times$ ($-$Acceleration due to gravity) (m/s^2)
- $Z3 =$ Weight of the wings (kg) $\times$ 10 $\times$ ($-$Acceleration due to gravity) (m/s^2)

Table 1 Details of front and rear wing

Component	Frontal area (m^2)	Mass of the component (kg)
Front wing	0.298	4
Rear wing	0.331	5

- Z4 = ½ × Frontal area (m^2) × Density of air (kg/m^3) × Velocity of Air Gust × (−2.55)
- Factor of safety = 1.5.

Force on front wing mount = 1300 N.
Force on rear wing mount = 1500 N.

Front mounting will be used to mount the front wings on chassis. Initially, a rough plate was designed without having any weight reduction to understand the stressed region. Ansys ACP was used to generate the layer model having a different number of carbon fiber layers and foam layers and transferred to Ansys structural, where two points which will be bolted with chassis are kept fixed and loads acting on the single mount were applied on other two holes.

The simulation results are shown in Figs. 5 and 6. We are able to get region were stressed induced were minimum. As the total deformation and induced stress was very less, we had a scope of weight reduction. Removing the material from such a region and performing analysis with multiple iterations was carried out.

As a designed constraint, it was decided to mount using two carbon fiber sandwich panels, each bolted with 2 M6 bolt on chassis and 2 M6 bolt on rear wing main element with the help of aluminum mount bonded on wing element. At each end plate, carbon fiber tubes with aluminum rod ends were used to provide flexural stiffness.

Initially, a rough plate was designed without having any weight reduction to understand the stressed region. Ansys ACP was used to generate the layer model having a different number of carbon fiber layers and foam layers and transferred to Ansys structural, where two points which will be bolted with chassis are kept fixed and loads acting on the single mount were applied on other two holes. The design of the mount was iterated keeping in mind the stress concentrations and weight of the component. The designs are shown in Figs. 7, 8, 9 and 10.

Fig. 5 Stress induced

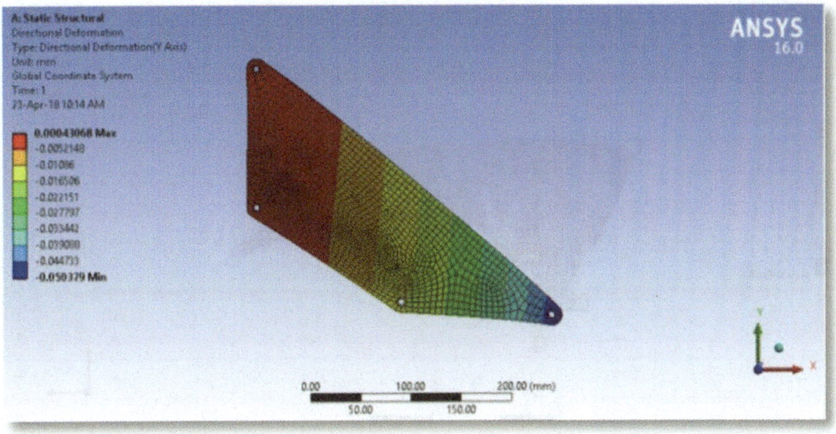

Fig. 6 Total deformation

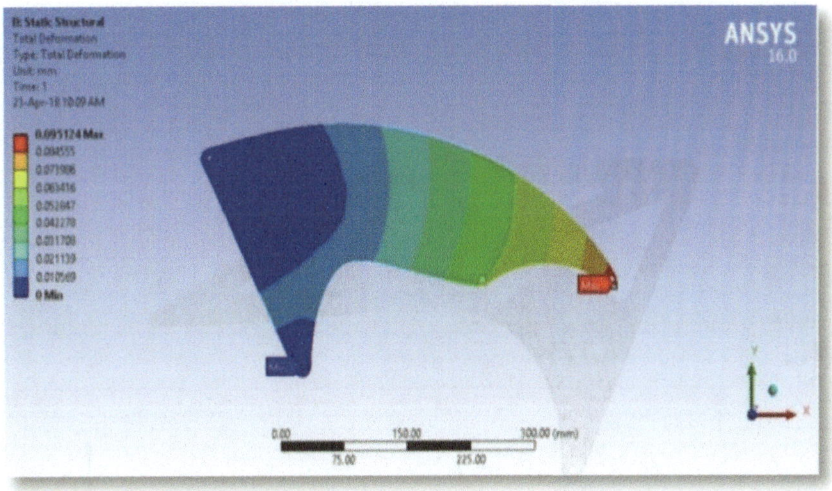

Fig. 7 Initial model: total deformation

The comparison of initial and final model on the basis of deformation and stress induced is shown in Table 2.

The mounts were manufactured using water-jet cutting process. First, the sandwich panel was created using vacuum bagging, and then, it was brought into a required shape from flat panel to below shape as shown in Figs. 11 and 12.

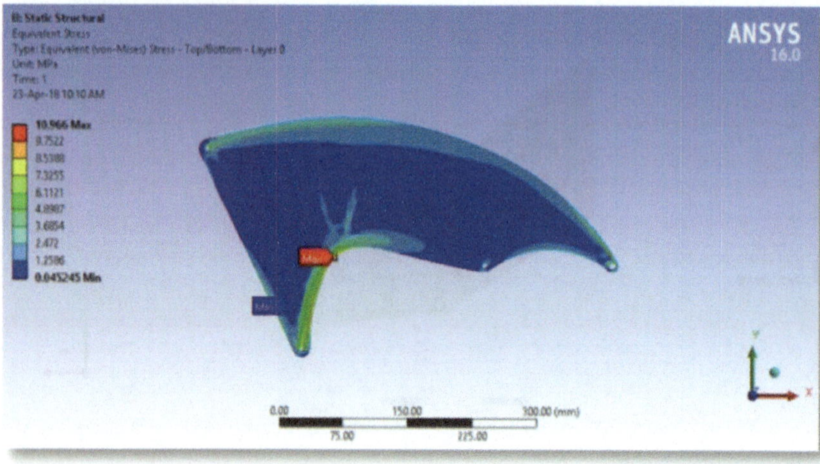

Fig. 8 Initial design: stress induced

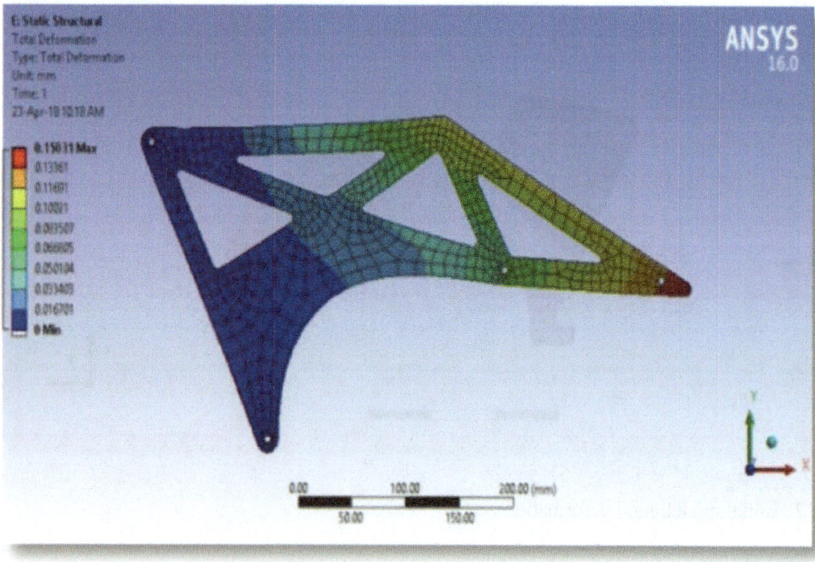

Fig. 9 Final iteration: deformation

5 Carbon Fiber Selection

Carbon fiber-reinforced plastic (CFRP) was used for its higher specific strength and also for its low density. The parts made from CFRP are comparatively lighter than those made from aluminum; since it is a fiber, it is also able to conform to different

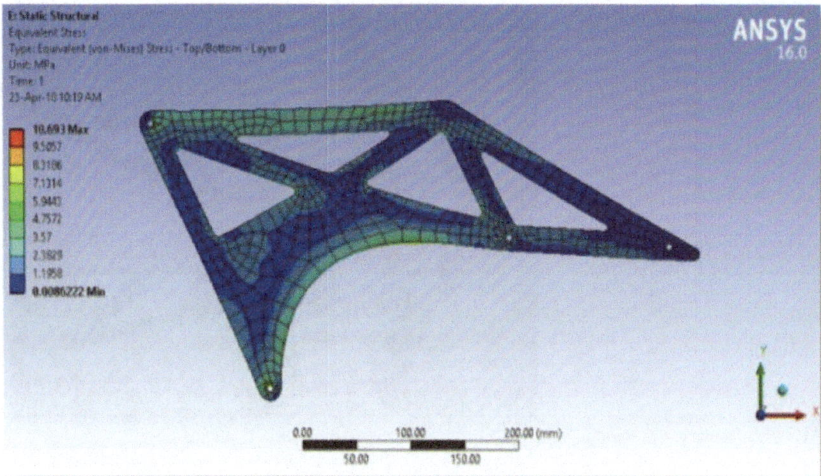

Fig. 10 Final iteration: induced stress

Table 2 Results

Model	Deformation (mm)	Stress induced (MPa)
Initial model	0.0951	10.96
Final model	0.1503	10.69

Fig. 11 Front wing mount

shapes and thus parts having complex curves can be made [8]. Carbon fiber is selected on the basis of its thickness. The thickness of carbon fiber is measured with gram per square meter (gsm). The thickness of carbon fiber was selected as 200 gsm because it is flexible enough to fit into tight curves.

Fig. 12 Rear wing mount

The manufacturing of carbon fiber composite parts is done mainly by 3 processes: wet layup, vacuum bagging, and vacuum infusion. These processes mainly defer on the equipment that is used for them and the method of resin application. The equipment necessary for all the processes is molds and mold release agent.

The main function of the mold is to give the shape of the part to carbon fiber that is laid on the mold. The mold also provides a base for laying up the fiber reinforcement.

Before laying up carbon fiber on the mold, 3–4 layers of a mold release agent have to be applied to the mold. This ensures that the part is easily released from the mold and that it does not get stuck to the mold and thus damages both the part and the mold.

6 Manufacturing of Components

The vacuum bagging process involves wet layup initially and the excess epoxy is removed after that. The equipment required exclusively for vacuum bagging is peel ply, sealant tape, and vacuum bag. The peel ply film is placed on the laid material to prevent it from adhering to the rest of the vacuum bagging material as shown in Fig. 13. The peel ply is porous and allows the excess resin to flow through when vacuum bagging is done. The peel ply can be easily peeled off from the part after it has been cured. On the peel ply, the breather fabric is placed, as shown in Fig. 14. The breather fabric absorbs most of the excess resin extracted by the vacuum.

The sealant tape makes the area around vacuum bag airtight preventing air leaks and thus creating the vacuum. The vacuum bag covers the entire cavity of the mold. It is attached around the mold with the sealant tape. The vacuum bag helps in creating the vacuum that helps extract the excess resin from the part. The vacuum pump is connected to the vacuum bag via a connector. The vacuum pump must be of a high enough CFM so that all the excess resin can be removed effectively form the part.

Fig. 13 Peel ply

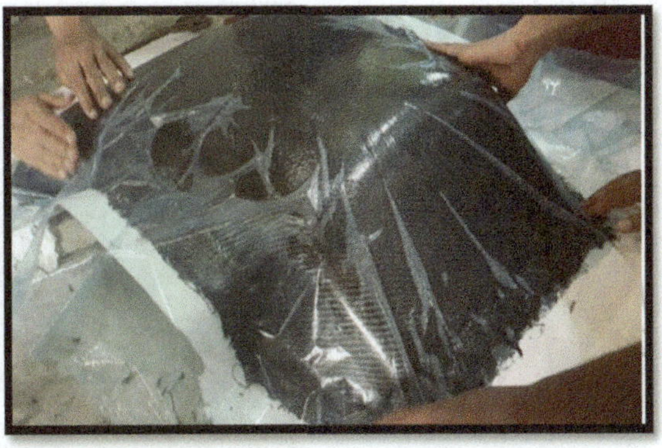

Fig. 14 Breather fabric

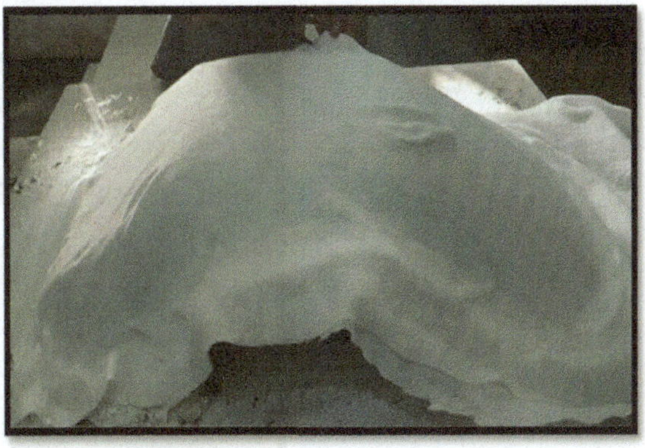

The vacuum pump must be connected before the resin begins to gel; otherwise, the resin becomes too viscous to flow out of the part.

The side plates were made to cover the side impact structure of the chassis. It was decided to use medium density XPS foam patterns for the side plates. It was decided to lay up carbon fiber directly onto the pattern and do vacuum bagging. Four layers of carbon fiber were added to the side plates as shown in Fig. 15.

Resin was applied using plastic rollers and an aluminum roller was used to evenly spread the resin along the surface of the part. The vacuum bagging equipment was then placed on the pattern, and then, the vacuum pump was started. After completion, the vacuum bagging equipment was removed and the part was allowed to cure for 24 h after that the side plates were released, cut, and then fitted to the chassis.

The wings were designed to increase the downforce of the car. The wing profiles were manufactured from carbon fiber keeping XPS foam as core. The foam core profiles were machined in hot wire cutting machine. After that, carbon fiber layup was done on it by vacuum bagging process as shown in Fig. 16.

Fig. 15 Carbon fiber layup over the mold

Fig. 16 Wing vacuum bagging after hot wire cutting process

7 Conclusion

This report has summarized the development of an aerodynamic package in an FSAE car. The main objectives of the report were to design and manufacture the aerodynamic package that includes components, namely wings, under-tray, and side pods. The carbon fiber sandwiched XPS foam was the preferred material for manufacturing of the components. Structural as well as aerodynamic analyses were carried out on Ansys, and StarCCM+ was used to predict their performance.

References

1. SAE Formula (2017) SAE rules hand book Society of Automotive Engineers
2. McBeath S (1998) Competition car downforce. Haynes Publishers, Newbury Park
3. Katz J (1995) race car aerodynamics. Bentley Publishers, USA

4. Wind tunnel tests of DRS on a Formula SAE car report (2011) Monash Motorsport
5. Design report (2016) Munich Motorsport
6. Prasanth A, Biswal S, Gupta A, Barodawala A (2016) Complete design and optimization of aerodynamics of a FSAE car using Solid works ANSYS & XFLR5. In: WCE 2016, world congress on engineering, proceedings, London, GB, Jun 29–Jul 1
7. Oxyzoglou I (2017) Design and development of an aerodynamic package for race car. Thesis for: M.Eng. in Mechanical Engineering University of Southampton
8. Padmaban P, Aruna M Design data: data book of engineers. PSG College—Coimbatore Kalaikathir Achchagam

Design and Manufacturing of Compact and Portable Smart CNC Machine

Amit Choudhari, Shamir Talkar, Pavan Rayar, and Aditya Rane

Abstract Nowadays, due to the advent of technology, there is huge importance given to compact machinery. World is moving towards the compact, portable and multi-purpose machinery. A large number of SMEs are getting engaged in high-precision manufacturing jobs. They need high-end CNC machines. The cost of such machines is very high. In a small-scale industry, it is very difficult to install a computerized or intelligent machine-like CNC because they are very costly and also requires scheduled maintenance. We are aiming to provide an option for compact and portable machines. So, by using this projected machine, an option can be given to all those small-scale industries who aim to machine the job economically and also want to save their time with a significant increase in production and accuracy to contribute to Indian GDP in the manufacturing sector.

Keywords Compact machinery · Demand · Constraint of available space · Costly · Portable · Small-scale industries · Increase in production

1 Introduction

This research paper is about design and manufacturing of microcontroller-based CNC machine (drilling only) which is a kind of special-purpose machine. Aim of the project was to provide compact, user-friendly and cost-effective machines to small-scale industries. In any machine, complexity goes on increasing as the degree of

A. Choudhari (✉)
Worley, Mumbai, India
e-mail: arc201296@gmail.com

S. Talkar
Arkk Consulting, MEP, Mumbai, India

P. Rayar
Department of Production Engineering, Dwarkadas J. Sanghvi College of Engineering, Mumbai, India

A. Rane
Rajiv Gandhi Institute of Technology, Mumbai, India

© Springer Nature Singapore Pte Ltd. 2020

H. Vasudevan et al. (eds.), *Proceedings of International Conference on Intelligent Manufacturing and Automation*, Lecture Notes in Mechanical Engineering,
https://doi.org/10.1007/978-981-15-4485-9_21

freedom increased, and therefore, to make it simple and effective, degree of freedom was kept as one. In this research paper, design procedure along with the construction details is explained. During the design and selection of the different components, emphasis on its cost and durability is given so that the machine can be used by the small-scale industry and hence they can atomize their processes like centre punching, drilling, etc., operations with this machine and save their time with a significant increase in production.

2 Constructional Details

As shown in Fig. 1, primary motor is connected to the chuck through the driving shaft. A driving shaft connects the faceplate on which chuck is mounted to the shaft of the motor [1]. To provide smooth rotation and to minimize the load on the motor shaft, a bearing is used between motor and chuck. Bearing is placed inside a bearing housing which is completely packed and keep all the contaminants away from it. To hold the tool, a tool holder is connected to a rigid plate that is mounted over the linear motion guide. To get the feed, a rack and pinion mechanism is used. The rack is connected to the linear motion guide, and the pinion is connected to the secondary motor. To have precise control over the motions of the motor, a microcontroller is used which is programmed according to the requirement of the customer.

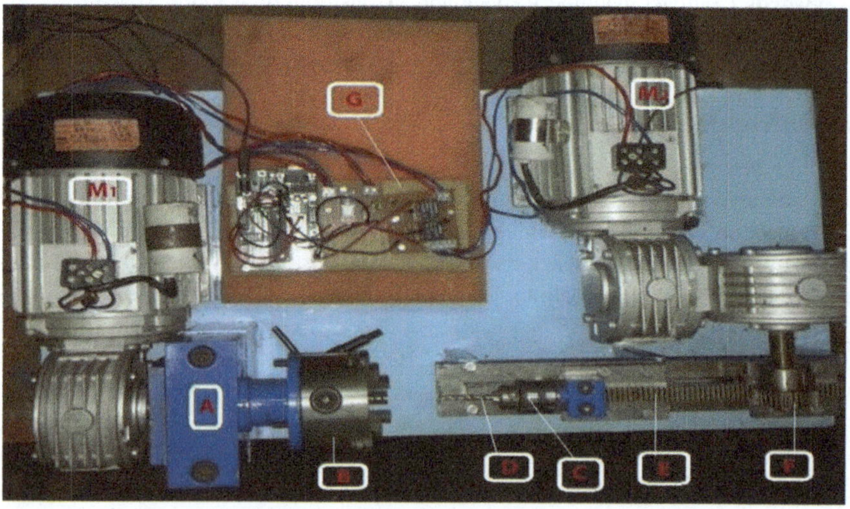

Fig. 1 This figure shows the constructional details of the machine. Different abbreviations used are as follows: M_1—primary motor connected to the chuck; M_2—secondary motor connected to the pinion to get feed with rack and pinion mechanism; A—bearing housing; B—three-jaw self-aligned chuck; C—tool holder; D—drill bit; E—rack; F—pinion; G—microcontroller

Fig. 2 This figure shows the constructional details of the primary motor (on left side) and secondary motor (on right side)

2.1 This Machine Has the Following Main Component

2.1.1 Primary Motor (M1) Connected to Chuck

It is a single-phase induction motor having a rating of 230 V and 0.5 HP. The initial speed of this motor was 1440 rpm but using gearbox with 4.8 gear ratio, the speed is stepped down to 300 RPM. This motor output shaft is connected to the spindle. This motor is having an aluminium frame body, due to which it is lighter when compared to the same grade motor. Therefore, it has also very good convection of heat phenomenon mechanism. The type of gearbox used is of worm and worm wheel type (Fig. 2).

2.1.2 Secondary Motor Connected to Chuck

It is also a single-phase induction motor having a rating of 230 V and 0.5 HP which is as shown in Fig. 1 (on right side). The initial speed of this motor was 1440 RPM but using gearbox speed is stepped down to 1 RPM. We selected 1 RPM motor because during feed drill bit should move very slowly to get better finishing. Motor speed is reduced from 1440 to 1 RPM by using a two-stage gearbox. It is controlled in such a way that it rotates in both directions, i.e. clockwise during the cutting stroke and anticlockwise during the return stroke.

2.1.3 Chuck

In this machine, three-jaw chuck is used to get self-alignment which is as shown in Fig. 1(B) and its size is 3 in. Since the chuck is coupled with the driving shaft using the faceplate, therefore, when driving shaft rotated, the chuck also starts rotating.

Fig. 3 Left figure shows the arrangement of rack and pinion and right figure shows the details of linear motion guide

2.1.4 Rack and Pinion

As shown in Fig. 3 (on left side), the pinion is directly mounted on the secondary motor which rotates at 1 RPM. Since the pinion is connected to the rack, as the pinion rotates, the rack moves in a linear direction [2]. During rotation of the pinion, some of the load also goes to the base through the linear motion guide and the support, and hence, it is designed in such a way that it should not deflect otherwise there are chances of errors.

2.1.5 Linear Motion Guide

A linear motion guide is provided with a rail and guide block. Guide block having internal ball bearing, due to which they can slide on the rail guide in a linear direction. In this machine, linear motion guide serves as a sliding bed. The rack is fitted on to this block by screws. These linear motion guides are efficient and can take load also. Figure 3 (on right side) shows linear motion guide.

2.1.6 Microcontroller

Figure 4 shows an ATMEGA16 microcontroller by which the complete operation is controlled intelligently. It is made up from a double-sided PHT PCB board to provide extra strength and increase the strength of connectors. This microcontroller can work on 7–15 V AC or DC power supply. Built-in reverse polarity protection is given to it. We are using a 9 V external power supply for its operation [2]. The program is embedded in it with the help of a cable through the computer. It contains a read–write memory; therefore, we can easily change and store the program with great accuracy [3]. It comes with 1 KB of static RAM which stores information for short period of time, whereas 16 KB of flash memory, also known as ROM, is also incorporated in the device which is non-volatile in nature and can store information for long period of time. It is connected to both motors using their microcontrollers and relays circuit and the command on the motor is given by this microcontroller.

Fig. 4 Left figure shows the details of microcontroller and right figure shows the drilling of workpiece just before the tool enters into the workpiece

3 Working of Machine

Figure 4 (on right side) shows a picture just before the drilling starts. In the first step, workpiece is fitted in the chuck and the required diameter of the drill bit into the tool holder. Now, connect the power supply of the machine and press the start button. When the start button is pressed, then microcontroller gives the command to the primary motors to rotate in a particular direction. Since this machine is controlled by a microcontroller, there is no need to operate this machine manually. Initially, chuck starts rotating in the anticlockwise direction and pinion rotates in a clockwise direction. After research and by some practical data, we found that the most suitable speed for drilling a hole ranging from 1 to 10 mm is 300 RPM.

The secondary motor is controlled by the microcontroller; hence, it also starts rotating with the primary motor. As per the length of the workpiece and the depth of hole, a travelling distance is calculated and fed into the microcontroller with respect to the initial fixed point. Hence, the entire tool holder moves towards workpiece and the drilling of the workpiece starts. As per the design of machine, total tool movement in forwarding direction is 300 mm; hence, workpiece ranging from 10 to 250 mm can be used for drilling. As the one loop gets completed, it gets automatically stopped, and for the next cycle, the same procedure is to be repeated. This machine is designed to operate in cycle to perform the same operation as per the coding embedded in microcontroller. Therefore, before the drilling, the required travel length is calculated manually and entered into the coding. The program is designed in such a way that when the required length has been achieved the program runs in reverse so that the tool again reaches to its initial position.

4 Design Calculations

4.1 Torque and Power Calculation for Primary and Secondary Motor

In this project, there are some criteria, as the drilling range is from 1 to 6 mm diameter.

So, we designed for maximum diameter as, i.e. $d = 6$ mm.

From the design data book, properties of aluminium are as follows,

Ultimate strength = 340 N/mm^2.

The yield strength of aluminium = 241 N/mm^2.

Some constant factors selected from the book Workshop Technology written by Hazara Choudhary.

$C = 80$, KS = 1050, KC = $\frac{\text{KS}}{2}$ = 1025. As the diameter of the pinion is 43 mm.

So, circumference = $\pi \times d$.

Feed = $\frac{\pi \times D}{60} = \frac{\pi \times 43}{60} = 2.25$ mm/s.

Therefore, from the standard equation,

Torque = $C \times d^{1.9} \times s^{0.8}$ $\therefore T = 80 \times 6^{1.9} \times 2.25^{0.8}$ $\therefore T = 4605.99$ N mm $\therefore T =$ 4.606 N m. While rotating the workpiece in the chuck, drill touches the workpiece. There are chances of stopping the rotation, due to the thrust/force of the drill bit.

Therefore, also considering axial thrust,

$$B = kc \times d \times s^{0.78}, \quad B = 525 \times 6 \times 2.25^{0.8}, \quad B = 5929.42\ \text{N mm} = 5.92\ \text{N mm}$$

where B = Thrust due to drilling. Total horsepower develops due to drilling a hole in an aluminium workpiece is given by,

Horsepower due to torque + Horsepower due to thrust

$$\therefore \text{HPC} = \frac{2\pi NT}{60 \times 1000 \times 102} + \frac{\text{BSN}}{60 \times 1000 \times 102}$$
$$\therefore \text{HPC} = \frac{2\pi \times 300 \times 4.627}{60 \times 1000 \times 102} + \frac{5.92 \times 2.25 \times 1}{60 \times 1000 \times 102}$$
$$\therefore \text{HPC} = 0.2236\ \text{HP} = 166.80\ \text{W}$$

It is theoretical torque required to drill a hole.

$$P = \frac{2\pi NT}{60} = \frac{2\pi \times 300 \times T}{60}$$

$T = 5.28$ N m.

By considering the factor of safety, we selected a motor of 0.5 HP.

Therefore, as per the selection of motor, the original torque is [4],

$$373 = \frac{2\pi \times 300 \times T}{60} \quad \therefore T = 11.87\ \text{N m} > 5.28\ \text{N m}$$

Motor torque is greater than the designed torque, and hence, our design is safe with a factor of safety of 2.

4.2 Torque Calculation for Secondary Motor

As we know that yield tensile strength of aluminium = 241 N/mm^2, according to shear stress theory, $(\tau)_{\text{Max}} = 0.5 \times \sigma_{yt}$ $(\tau)_{\text{Max}} = 0.5 \times 241 = 120.50$ N m.

$\therefore$ Force $(F_t) = 3407.05$ N $\therefore$ $F_t = 3.407$ kN but in practical as per our motor selection, Power = 373 W, $N = 1$ RPM $\therefore$ $T = 3561.88$ N m $\therefore$ Torque = Force $\times$ Radius $\therefore$ $T = F_t \times R_{\text{pinion}}$ $\therefore$ $3561.88 = F_t \times (21.5)$.

$\therefore$ $F_t = 165.66$ kN > 3.40 kN. Therefore, design is safe.

4.3 Design Procedure to Fix the Position of the Linear Motion Guide

[5] On linear motion guide (LMG), there are mainly two downward forces were acting, one is due to tool holder weight with the drill bit and the other is due to pinion during rotation. Therefore, fixing the position of LMG was very critical. During the preliminary design stage, we come up with the different positions of LMG and for each case, we find out the bending moment and shear stress produced in the rack. Figure 5 shows a sample shear force diagram (SFD) and bending moment diagram (BMD) for positioning of LMG over the rack. Out of different options, we selected the below case in which minimum bending stress is induced. Distance between the LMG is 100 mm.

Total force exerted by the pinion on the rack is given by,

$$F = 0.3945 \times 9.81 = 3.7\,\text{N}.$$

A sample calculation for shear force diagram is as below:

$$\sum \text{FY} = 0 \ldots \text{(Sum of all vertical forces)} \quad \therefore -1.5 + R_{\text{A}} + R_{\text{B}} - 3.87 = 0$$

$$\therefore R_A + R_B = 5.37$$

Taking moment about point A, we get, $R_{\text{B}} \times 80–125 \times 3.87 + 1.54 \times 175 = 0$

$\therefore R_{\text{B}} = 2.67$ N $\therefore R_{\text{A}} + 2.67 = 5.37$ $R_{\text{A}} = 2.69$ N…. Where R_{A} and R_{B} are the reactions at the supports. A sample calculation for bending moment is as below:

$M_{\text{A}} = 1.54 \times 100$ N mm = 154 N mm; $M_{\text{B}} = 3.87 \times 18$ N mm = 69.66 N mm.

$M_{\text{C}} = 0$; $M_{\text{D}} = 1.54 \times 25 = 38.5$ N mm; $M_{\text{E}} = 0$, where $M_{\text{A}}, M_{\text{B}}, M_{\text{C}}, M_{\text{D}}$ are the bending moment at respective points as shown in Fig. 5.

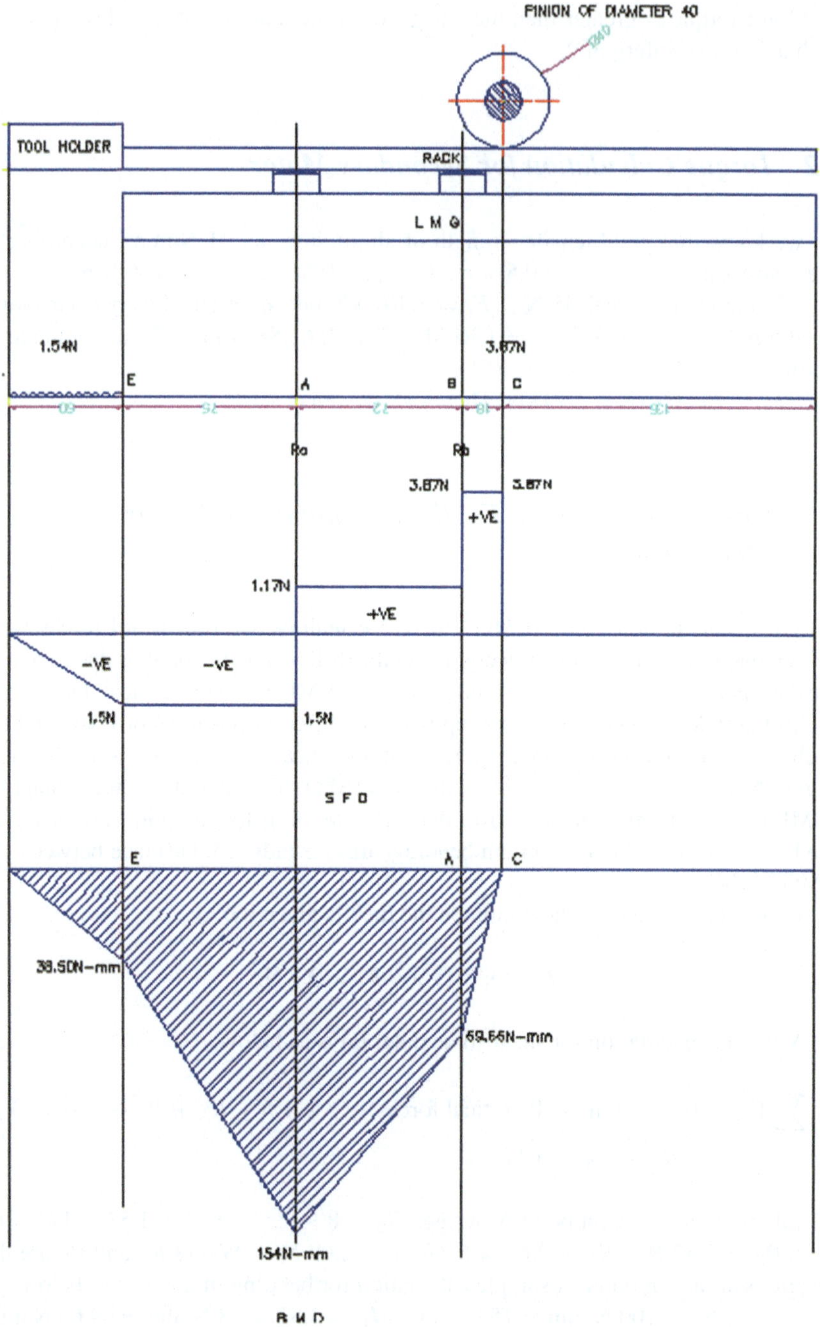

Fig. 5 This figure shows shear force and bending moment diagram for LMG

5 Unique Features of the Machine

1. The part program can be input to the controller unit through the computer.
2. The part program once entered into the microcontroller can be used again and again.
3. The part program can be edited and optimized at the machine tool itself.
4. The input information can be reduced to a great extent with the use of special subprograms.
5. NC control unit allows compensation for any changes in the dimension of cutting tools.
6. The combined characteristics of the machine tool and the control determine the precision of positioning [6]. Three critical measures of precision are resolution, accuracy and repeatability, which is possible by using the NC control.
7. It is compact in size and having lesser weight.
8. It requires less space due to less number of parts hence it is compact in size.
9. It gives greater accuracy of the job as compared to manual operations.
10. This machine requires less skilled full workers for its operation.
11. This machine reduces the in-process inventory of the parts in the process.
12. Tool set-up time is minimum and better machine utilization hence reduced cycle time.

6 Application of the Machine

Present model can be used:

1. For drilling in holes in shafts.
2. For making drills in the small circular component.
3. For drilling in soft materials.
4. For the centre drilling purpose.
5. For drilling in nut and bolts industry.
6. In nozzle industry.
7. In watch-making industry.

7 Conclusion and Future Scope

A cross-slide can be added from which it will have four-axis. A tool post can be mounted in XY-direction from which it can be work as a four-axis CNC machine. An automatic coolant can be provided with the help of a motor. A separator can be used to separate the scrap and oil, and oil can be reused. The controller system can replace as PLC and the CNC control system, which will make more user-friendly to this machine. From varying motor speed, it will be able to machine M.S. materials.

Machining is an important manufacturing process that is used in a wide range of applications. From aerospace applications to the manufacturing of energy systems and medical robots. In this research paper, the more emphasis was given on the design and development of the machine starting from analysing the requirements of the customers to the delivery of the product to the needful customers. A successful machine for those small-scale industries who cannot afford a special-purpose machine for doing their work with the small tolerance zones is successfully built. Detailed engineering analysis of each component was made keeping their cost, reliability and durability as parameters which helped to attain machining without any breakdown or failures. Tool selection, speeds feed, selection of linear motion guide and positioning it, centring of the tool and chuck to have zero eccentricity for great accuracy, selection of the motor & torque calculation and material selection—these were the main relevant factors which were examined. Many other parameters like ergonomics, portability and ease of handling should be taken care while designing commercial CNC machine.

References

1. Jodh G et al (2014) Design of low cost CNC drilling machine. Int J Eng Res Gen Sci 2(2):189–196
2. Hasan MM et al (2019) Design and implementation of a microcontroller based low cost computer numerical control (CNC) plotter using motor driver controller. In: 2019 international conference on electrical, computer and communication engineering (ECCE). IEEE
3. Daadoo M, Daraghmi YA (2016) Design and implementation of low cost computer numerical control-printed circuit boards drilling machine
4. Sawant PR, Barawade RA (2012) Design and development of SPM—a case study in multi drilling and tapping machine. Int J Adv Eng Res Stud (IJAERS) 1(2):55–57
5. Bate Andy (2005) Drilling machine cuts through traditional design principles. Eng Technol 8(9):38–41
6. Kumar M, Puttige V Low cost automation for CNC machining center. IJMET 3. Pahole I, Rataj L, Ficko M, Klancnik S, Brezovnik S, Brezocnik M, Balic J (2009) Construction and evaluation of low-cost table CNC milling machine. Sci Bull Ser C Mech Tribol Mach Manuf Technol XXIII:143–148

Effect of Plunging and Dwelling Period on Temperature Profile and Energy Dissipation in FSSW and Its Relevance in FSW

Niyati N. Raut, Vivek Yakkundi, Akshay Vartak, and S. N. Teli

Abstract FSW is the welding technique, where filler material for welding is not required resulting into no additional mass being added, which proves very useful in the industry where the addition of small material plays a significant difference on the performance of the working unit. Energy generation and its utilization during welding are crucial in understanding the physics behind the process and further extended to know the effect of process parameters on a way of effective welding to achieve sound weld. FSW experiments are costly as well as tedious where the requirement of the accurate plasticized stage should be reached so that the tool can travel flawlessly in traverse direction to achieve sound weld. Computational modelling and numerical simulation are the solution to the above problem. In this research paper, numerical simulation on FSSW is carried out first for better understanding of energy generation and its utilization at plunge and dwell stage before going further for next step in FSW that is traverse movement of the tool in the direction of the weld line. During the plunging stage, the friction causes the metal around the contiguity area to soften. It transforms the material into the plasticized stage during dwell stage, so the temperature profile obtained during those stages is validated with experimental results for the accuracy of the numerical model. Also, energy generation and its utilizations are confirmed with the experimental values for accurate validation of the numerical model. CEL approach is used for simulation of FSS welding of the aluminium sheet. Plunge speed and dwell time were chosen as different parameters. Simulation results showed that thermal response, along with energy history, varies with simulation parameters used in FSSW. To conclude, this numerical model can be effectively used to predict the FSW process for any other material.

Keywords Friction stir welding · FSSW · Peak temperature · Frictional energy

N. N. Raut (✉) · V. Yakkundi · A. Vartak · S. N. Teli
Mumbai University, Mumbai, India
e-mail: niyatiraut@gmail.com

© Springer Nature Singapore Pte Ltd. 2020

H. Vasudevan et al. (eds.), *Proceedings of International Conference on Intelligent Manufacturing and Automation*, Lecture Notes in Mechanical Engineering,
https://doi.org/10.1007/978-981-15-4485-9_22

1 Introduction

The friction stir welding process was developed by TWI, Abington, UK, in 1991 for joining light alloys such as aluminium [1]. Friction stir spot welding (FSSW) is a subdivision of friction welding. It is generally used in the automotive industry as a partial replacement of resistance spot welding. The friction spot stir welding process improved the assembly operations of the automotive and aerospace industry using lightweight material for construction [2].

The various stages of friction stir spot welding process are shown in Fig. 1:

- The process starts with the tool rotating at high speed and then with specified plunge rate, it plunges into a workpiece.
- After the tool shoulder comes in contact with the top surface of the plate, the pressure and temperature rise causing inter-diffusion of the material across the interface joining.
- Finally, when a specified plunge depth is reached, the rotating tool dwells for a specified dwell period. After dwell, the tool retracts from the workpiece.

In this research, the impact of the plunge rate variation on the process of friction stir spot welding (FSSW) is the primary topic of interest. This methodical approach helps us understand that the dwell time of the welding tool and the plunge rate has a significant influence on the frictional energy generation during FSSW process.

This model allows the prediction of the temperature profile, energy dissipation and stress and strain rate during FSSW. Experimental and simulation results have been studied and collated to validate the model.

This research paper aims to carry out a numerical simulation on FSSW before the FSW process for a better understanding of energy generation and its utilization at plunge and dwell stage before going any further for the next step in FSW, i.e. traverse movement of the tool in the direction of the weld line. The simulation gives a clear idea up to the dwelling stage of the FSW process. If the process up to dwell stage is wrong in anyway, we are alerted at dwell stage itself and do not have to wait until the whole FSW simulation is completed.

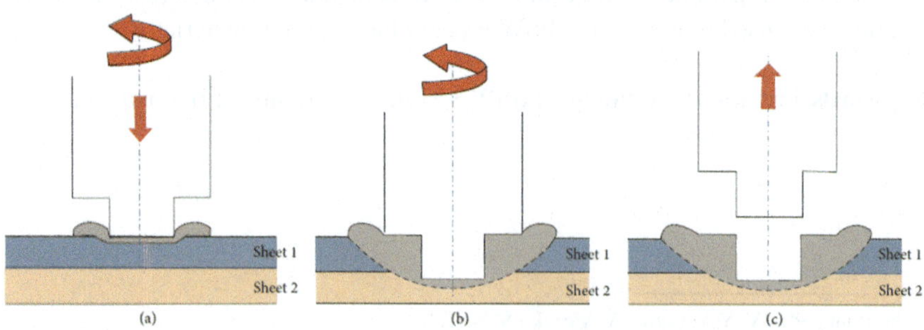

Fig. 1 Schematic of FSSW process; **a** plunging, **b** dwell and **c** retraction of the tool [3]

During the plunging phase, the metal around the contact area softens due to friction [4, 5]. It transforms the material into the plasticized stage during the dwell stage, so the temperature profile obtained during those stages is validated with experimental results for the accuracy of the numerical model. Also, energy generation and its utilization are cross-checked with experimental values for precise validation of the numerical model. CEL approach is used for commencing the simulation of FSS welding of the aluminium sheet. Plunge speed and dwell time used in FSSW were chosen as different parameters.

While creating the input file, steps similar to experiments are followed for the material Al 6061-T6. Created input file is executed, and the results of simulation and experimental results are compared. If it holds well, then it is implemented for our aimed high-strength material like titanium and nickel for further FSW simulation.

2 Numerical Model of FSSW

Abaqus/Explicit software is used to develop a 3D dynamic, fully coupled thermal displacement model of FSSW analysis. The thermo-mechanical analysis was carried out to obtain thermo-mechanical and frictional energy dissipation responses of FSSW process.

The FSW process involves coupled temperature-displacement response of the rotating tool-fixed workpiece system. In this analysis, the Eulerian model of the material is used to produce both energy and temperature distribution as the tool plunges in and dwell forming the weld. Heat generation during FSSW welding is primarily affected by plunge rate, dwell time and rotational speed [6]. To conduct finite elemental analysis, specific plunge rate and tool rotational speed were selected and implemented by applying appropriate boundary conditions. The Coupled Eulerian-Lagrangian (CEL) method used in this study comprises of a domain with an Eulerian mesh which represents volume through which material flows and interacts with Lagrangian part [7].

Plunge rate was changed in three different models to study its effects on various parameters of welding. For the first model, plunge rate was 2.5 mm/s and it varies from 5 to 10 mm/s for further models. The model with 2.5 mm/s plunge rate is used to verify thermal response with experimental results performed by previous researchers [8].

2.1 *Geometry and Mesh*

Abaqus/Explicit solver code was utilized to solve the frictional energy conservation and heat transfer equations of the FSSW process. Due to the low computational efficiency, only a localized region of the workpiece is included in the analysis. The shape of the Eulerian domain was cubical with a size of $25 \times 25 \times 10$ mm. The tool

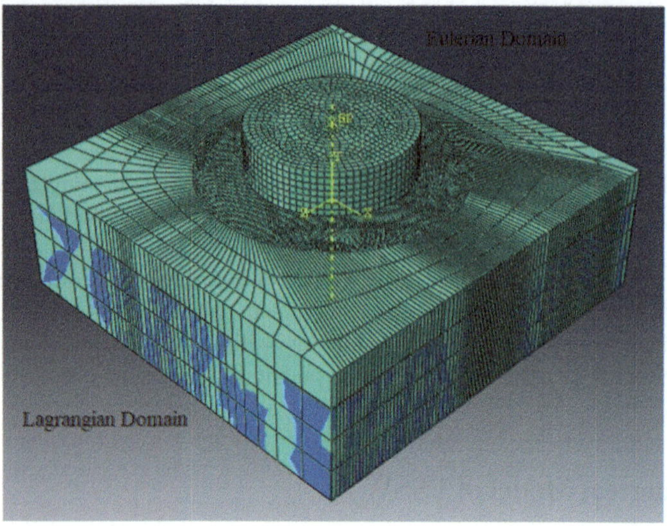

Fig. 2 Meshing of FSSW CEL mode

with shoulder diameter 10 mm and pin diameter 4 mm with a pin height of 2.2 mm is used in the current simulation.

In order to increase the computational efficiency, sweep mesh technique with the middle axis is applied to create fine mesh at centre zone where tool and workpiece interact as shown in Fig. 2.

2.2 *Material Model Used and Related Properties*

In the FSS welding nugget zone (NZ) and the thermo-mechanically affected zone (TMAZ) are areas where considerably great extent of plastic deformation occurs. The interaction knowledge of flow stress with temperature, plastic strain and strain rate is crucial for numerical analysis modelling of the FSS welding process [9]. For this reason, Johnson-Cook material law was employed to describe the plasticity of workpiece material, and the flow stress is defined as follows:

$$\overline{\sigma} = \left[A + B \cdot \left(\overline{\varepsilon}^{\mathrm{pl}}\right)^{n}\right]\left[1 + C \cdot \ln\left(\frac{\dot{\overline{\varepsilon}}_{\mathrm{pl}}}{\dot{\varepsilon}_{0}}\right)\right]\left[1 - \left(\frac{T - T_{\mathrm{ref}}}{T_{\mathrm{melt}} - T_{\mathrm{ref}}}\right)^{m}\right] \text{[GPa]}$$

where

σ_{y}—is the yield stress, $\overline{\varepsilon}^{\mathrm{pl}}$—the effective plastic strain;
$\dot{\overline{\varepsilon}}_{\mathrm{pl}}$—is the effective plastic strain rate, $\dot{\varepsilon}_{0}$—is the normalizing strain rate;
n and m are material constants, C—represents strain rate sensitivity;

Table 1 Johnson-cook material Model Al 6061-T6 [12]

Material	A (MPa)	B (MPa)	C	n	m	$T_{melting}$ (°C)	$T_{transition}$ (°C)
AL 6061-T6	293.4	121.26	0.002	0.23	1.34	582	25

T_{ref}—is the temperature at which we determine the parameters A, B, n;
T_{solid}—is the material's solidification temperature.

Aluminium alloy 6061-T6 was chosen as a material for the workpiece. Alloy aluminium 6061-T6 material's properties are: heat capacity −896 J/kg K, convection coefficient −20 W/m^2 K, emissivity −700 W/m^2 K4 and thermal conductivity −167 W/m K [10, 11]. In this study, the value of the coefficient of friction is taken as 0.4 (Table 1).

3 Results and Discussions

The numerical simulation is carried out in Abaqus/Explicit using Eulerian domain for the workpiece and commercial FEM software by analysing a three-dimensional FEA model of the welding process. Figure 3 presents the temperature contours after

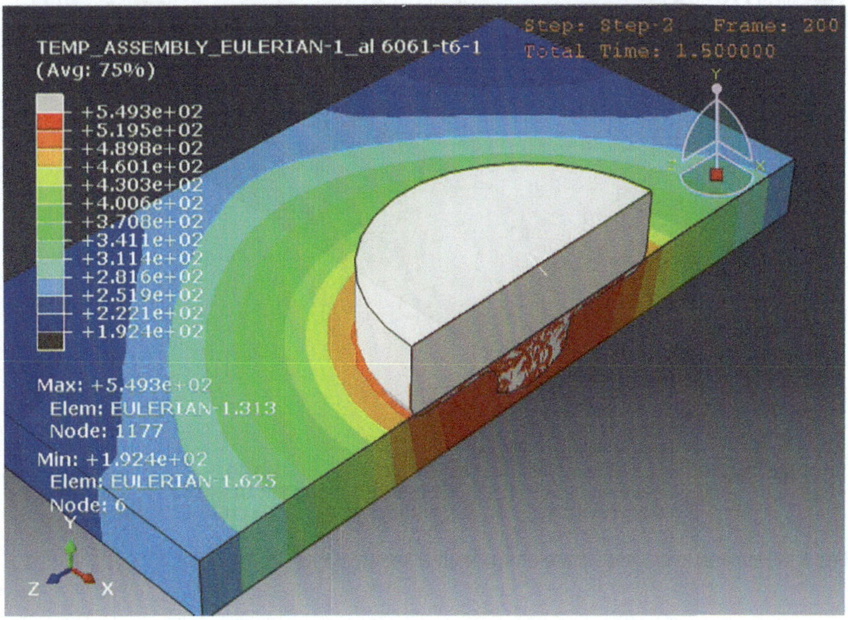

Fig. 3 Temperature contours after 1.5 s

1.5 s run. The maximal temperature accomplished at weld zone (for this case) is 549.3 °C, which is sufficiently high enough to reach metal's plasticized stage for smooth and even rotation of the tool to produce sound spot weld without defects. In friction stir welding, the temperature at the weld zone can be controlled by three process variables, namely tool rotation velocity, plunge rate and tool pressure [13].

Figure 4a and b shows the comparison of temperature history between the FEM experimental results reported by Su et al. and the simulation results. The analysis was performed using material Al 6061-T6, rotational tool speed of 3000 rpm and plunge rate of 2.5 mm/s. The tool geometry and the thickness of the workpiece are kept similar to that of the experimentation. Numerical simulation results show that maximal temperature that occurs at the tooltip is 549.3 °C, whereas experimental results show peak temperature at the tip of the tool is about 542 °C. From the comparison, we can conclude that both results are somewhat in agreement with each other with a 1.32% difference.

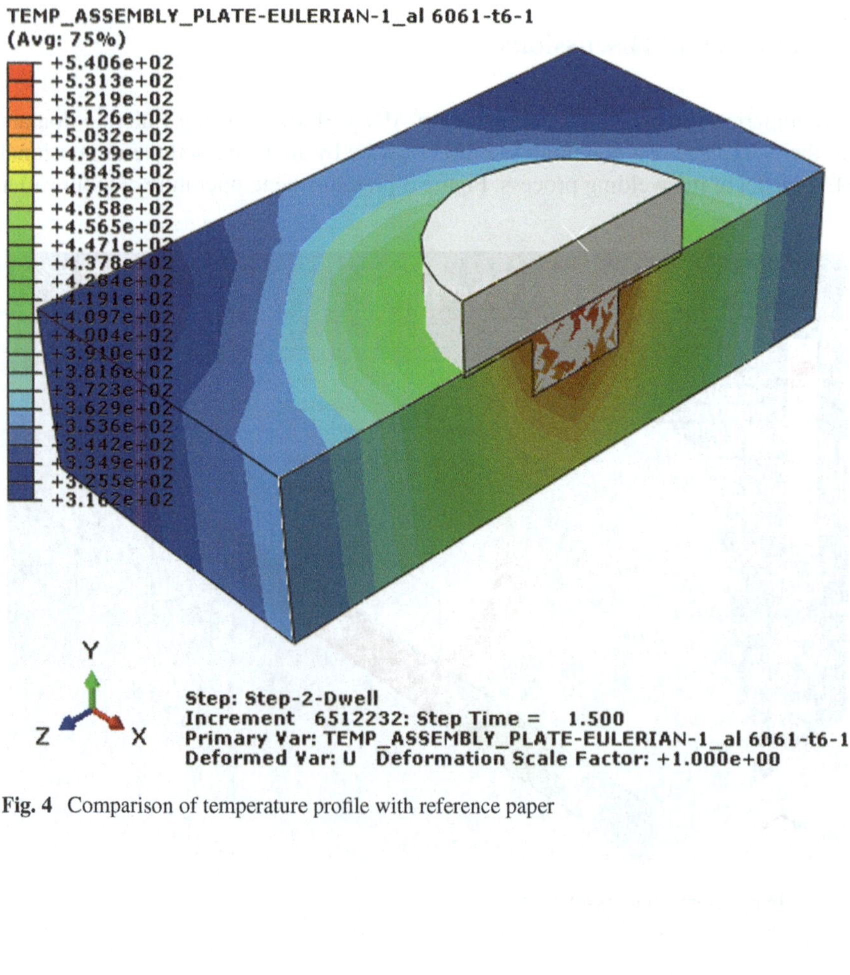

Fig. 4 Comparison of temperature profile with reference paper

3.1 *Effect of Plunge Rate on Friction Energy and Plastic Energy with Plunge Depth*

In the FSSW process, the heat generated due to frictional energy is utilized to form the weld and join two workpieces together [14, 15]. Frictional heat generation occurs due to two main criteria during FSSW process, i.e. workpiece material plastic deformation and friction work at the tool and top workpiece interface. In this research, the frictional and plastic energy dissipation has been explored using .obd history outputs.

The angular velocity of the tool is same for all three models, i.e. 3000 rpm and plunge rates of the tool were 2.5 mm/s, 5 mm/s and 10 mm/s, respectively, which is kept same as the experimental study with a 6.3 mm thick plate.

The graphs shown in Fig. 5a–c based on Table 3 (energy values validated with reference paper Su et al.) are drawn at plunge depth of 1.5 mm and show total energy dissipation, frictional energy dissipation and plastic energy dissipation taken out from obd history output for three different plunge rates. Based on the curves, total energy increases as the plunge rate decreases. This is because, at a low plunge

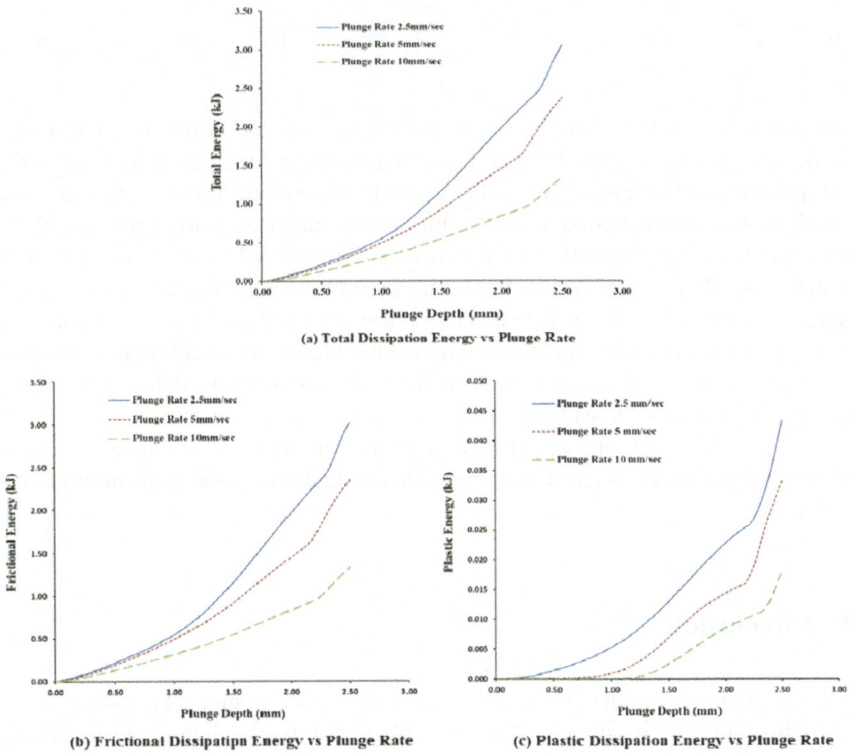

Fig. 5 a–c Graphical representation of energy contribution ratio at different plunge depths (3000 rpm, 2.5 mm/s plunge rate, 6.3 mm thick Al 6061-T6)

Table 2 Pin energy contribution ratio at different plunge depths (3000 rpm, 2.5 mm/s plunge rate, 6.3 mm thick Al 6061-T6)

Plunge depth (mm)	Energy input (kJ) Smooth pin (4 mm diameter, 2.2 mm long) Tool with shoulder (10 mm diameter)		% difference
	Experimental [13]	Analysis	
1.71	1.24	1.31	5.34
1.9	1.51	1.62	6.79
2.09	1.89	2.03	6.90
2.28	2.35	2.35	0.00
2.47	2.83	2.89	2.08

Table 3 Summary of energies for different tools' plunge rate

Plunge rate (mm/s)	Total energy (kJ)	% diff.	Frictional energy (kJ)	% diff.	Plastic energy (kJ)	% diff.
2.5	3.064	100	3.020	100	0.043	100
5	2.377	22	2.344	22	0.033	23
10	1.349	43	1.331	43	0.018	46

rate, adequate time is available to create friction between tool–workpiece interfaces. Evidently, frictional energy dissipation and plastic energy dissipation increase as the tool plunge rate decreases [16]. Table 2 shows energies at different plunge rates. The frictional energy dissipation of 5 mm/s plunge rate reduces to about 22% of the 2.5 mm/s frictional dissipation energy. It is further reduced by 43% with 10 mm/s plunge rate. This decrement in energies is caused because when the tool is at high penetration rate contact time between tool pin surface and workpiece decreases and this causes less time to create friction and in turn, plastic deformation. Due to lower heat generation, no plasticized stage is achieved and leads to defect generation. Parameters taken into consideration.

Welding should be carefully chosen to get a sound weld. Tool plunge rate is one of the significant parameters which plays an essential role in the generation of heat (Table 3).

4 Conclusion

A detailed numerical analysis is reported in this study, to understand the process of friction stir spot welding to show the effect of different plunge rate on energy generation and temperature distribution at tool–workpiece interface during friction stir welding of A1 6061 plates.

A fully developed 3D thermo-mechanical FEM with coupled Eulerian-Lagrangian domain with an Eulerian mesh which represents volume through which material flows and interacts with Lagrangian part was designed to simulate the FSSW process. The model is used in Abaqus/Explicit environment and validated by matching the simulated temperature with experimental results published by other researchers. The effects of different welding parameters are investigated and the conclusions of importance are drawn as follows:

1. The numerical simulation analysis results acquired were validated with experimental results within the variation of 1.32%.
2. Maximum temperature reached during simulation, which is equivalent to 0.95 of its solidus temperature, is fairly matched with finding reported by Su et al.
3. Major percentage of the energy generated within the model is due to frictional work at the interface between tool and workpiece which is about 99.44%, for the FSSW process.
4. Tool plunge rate effect is also considered, from which we can conclude that lower plunge rates produce higher frictional energy. The frictional energy dissipation of 5 mm/s plunge rate reduces to about 22% of the 2.5 mm/s frictional dissipation energy. It is further decreased with 10 mm/s plunge rate (43% decrement).

References

1. Thomas WM, Nicholas ED, Needham JC, Murch MG, Temple-Smith P, Dawes CJ (1991) Friction stir butt welding. GB Patent No. 9125978· 8, UK
2. Zhu L, Li N, Childs PRN (2018) Light-weighting in aerospace component and system design. Propuls Power Res 7(2):103–119
3. Chen Y (2015) Refill friction stir spot welding of dissimilar alloys. Master's thesis, University of Waterloo
4. Assidi M, Fourment L, Guerdoux S, Nelson T (2010) Friction model for friction stir welding process simulation: calibrations from welding experiments. Int J Mach Tools Manuf 50(2):143–155
5. Zhang Z, Wu Q (2015) Numerical studies of tool diameter on strain rates, temperature rises and grain sizes in friction stir welding. J Mech Sci Technol 29(10):4121–4128
6. Lomholt TC, Pantleon K, Somers MA (2011) Microstructure evolution during friction stir spot welding of TRIP steel. DTU Mekanik. http://www.mek.dtu.dk/Forskning/Projekter-phd/phd_trinecoldinglomholt.aspx
7. Dassault Systèmes (2014) Abaqus analysis user's guide, version 6.14
8. Su P, Gerlich A, North TH, Bendzsak GJ (2006) Energy utilisation and generation during friction stir spot welding. Sci Technol Weld Join 11(2):163–169
9. Abbasi M, Bagheri B, Keivani R (2015) Thermal analysis of friction stir welding process and investigation into affective parameters using simulation. J Mech Sci Technol 29(2):861–866
10. Chao YJ, Qi X (1998) Thermal and thermo-mechanical modeling of friction stir welding of aluminum alloy 6061-T6. J Mater Process Manuf Sci 7:215–233
11. Chao YJ, Qi X (1999) Heat transfer and thermo-mechanical modeling of friction stir joining of AA6061-T6 plates. In: Proceedings of the first international symposium on friction stir welding. Thousand Oaks, CA, USA, pp 14–16

12. Cao JY, Wang M, Kong L, Yin YH, Guo LJ (2017) Numerical modeling and experimental investigation of material flow in friction spot welding of Al 6061-T6. Int J Adv Manuf Technol 89(5–8):2129–2139
13. Meran C, ErselCanyurt O (2011) The effects of tool rotation speed and traverse speed on friction stir welding of AISI 304 austenitic stainless steel. Int J Mater Res 102(4):420–428
14. Pradhan SP (2012) An investigation into the friction stir welding of aluminium pipe with stainless steel plate. Doctoral dissertation
15. Chao YJ, Qi X, Tang W (2003) Heat transfer in friction stir welding—experimental and numerical studies. J Manuf Sci Eng 125(1):138–145
16. Veljić DM, Rakin MP, Perović MM, Medjo BI, Radaković ZJ, Todorović PM, Pavičić MN (2013) Heat generation during plunge stage in friction stir welding. Therm Sci 17(2):489–496

Prediction of Optimum Sheet Metal Blanking Clearance for IS513CR Steel Using Artificial Neural Network

Pradip P. Patil, Vijaya P. Patil, and R. Ramaswamy

Abstract In a quest for higher productivity, sheet metal manufacturing industry is undergoing significant development in the field of sensing and automation. One of the sheet metal operations is blanking, which is affected by an uneven crack which leads to a loss in productivity. In present work, an experiment is carried using the uni-punch tool on power press for varied punch penetration, to observe crack initiation and to find optimum clearance for IS 513 cold-rolled steel. The crack initiation is measured using shear angle, fracture angle and punch penetration. As the blanking process is complex and nonlinear, artificial neural network (ANN) is employed to predict clearance for input parameters. The predicted values are well within the experimental values.

Keywords Sheet metal blanking · Optimum punch–die clearance · Artificial neural network

1 Introduction

The metal-forming industry has been undergoing the requirement for an automated manufacturing system. One of the metal-forming processes is sheet metal blanking, which needs an accurate assessment of the parameters. In perspective on the advancement in the field of robotized fabricating, it is discovered that the present modern status is experiencing quick advancement in detecting framework equipped for recognizing and consequently replacing broken or worn segments, and absence

P. P. Patil (✉) · V. P. Patil
SIES Graduate School of Technology, Navi Mumbai, India
e-mail: pradip.patil@siesgst.ac.in

V. P. Patil
e-mail: vijaya.patil2017@nitie.ac.in

V. P. Patil · R. Ramaswamy
National Institute of Industrial Engineering, Mumbai, India
e-mail: ramaswamy@nitie.ac.in

© Springer Nature Singapore Pte Ltd. 2020

H. Vasudevan et al. (eds.), *Proceedings of International Conference on Intelligent Manufacturing and Automation*, Lecture Notes in Mechanical Engineering,
https://doi.org/10.1007/978-981-15-4485-9_23

of sufficient procedure model fits for discussing the multifaceted nature of the assembling framework. In any case, this advancement is seen through adaptable assembling frameworks, direct numerical control, shrewd assembling frameworks and PC incorporated assembling.

ANN modelling has been employed in developing an intelligent system in the field of machining for enhanced accuracy of prediction [1, 2]. Maiti et al. [3] modelled the sheet metal blanking and studied mechanical and geometrical parameters which include the wear state of the tool, thickness of the sheet and blanking clearance.

The aim of this work is to build up a predictive model for the optimum sheet metal blanking conditions. A feed-forward back propagation ANN system model is utilized for developing a prediction model. As customarily embraced means are frequently intelligent and compelling, the answer for the multifaceted nature of blanking parameter requires ANN modelling. A portion of the springiness and capacity of the human brain can be mirrored by the premise of ANN. The role of ANN is coming into the picture when the nonlinear plotting from inputs to outputs and the other way is not achievable by the conventional and regression methods. The following section explains literature review followed by result and analysis. The conclusion is presented at the end.

2 Literature Review

2.1 Blanking Process

In blanking processes, punch penetrates the sheet metal pressed by the blank holder as shown in Fig. 1. The process separates blank from sheet metal.

The quality of sheared edge of the cut workpiece/blank is defined by geometrical attributes (fracture depth, burr height, roll-over depth and smooth-sheared depth) through the blanking operation for a given material as shown in Fig. 2.

Fig. 1 Sheet metal blanking operation

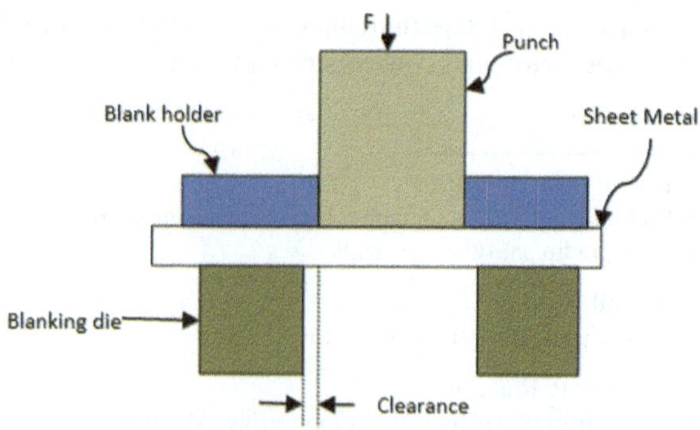

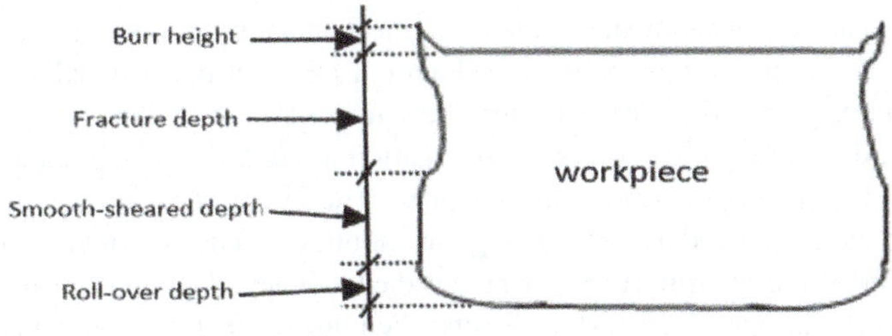

Fig. 2 Parameters of sheet metal blanking

Fig. 3 Process of crack propagation

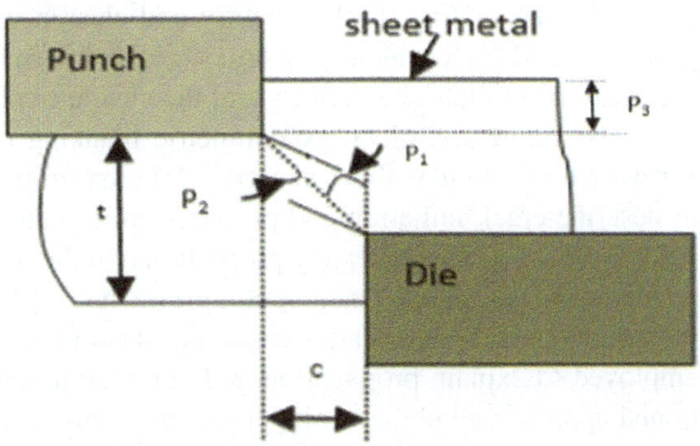

The process of finding optimum clearance is expressed by shear diagonal angle (P1), fracture angle (P2) and punch penetration (P3) when crack initiates, as illustrated in Fig. 3. The quality of sheared edge will be optimum when a crack propagates along diagonal line joining the punch and die corners. However, the clearance of the process is expressed as c.

$$c = 100\frac{D_m - D_p}{2t}(\%)$$

where t, D_m and D_p, and are, respectively, sheet thickness, the die diameter and the punch diameter. Since the primary material variable illustrating the fracture at beginning and propagation conditions is the strain at the crack, the fracture angles which is the input data for the ANN is determined by material elongation. The networks are utilized as digital devices for switching the experimental values required for the optimum clearance.

2.2 *State of Art*

Fang et al. [4] applied the Cockcroft and Latham fracture criterion and finite element method (FEM) to optimize clearance values for specified sheet material and thickness. They have established a shearing mechanism by mimicking the blanking operation of an aluminium alloy 2024. Faura et al. [5] propositioned a method to find optimum punch–die clearance for sheet material and thickness, applying the FEM. The diagonal angle and crack propagation angle for various clearances were calculated to uncover the optimum clearance. The effect of clearance on an angle of crack propagation and a diagonal angle is monitored. The diagonal angle and the clearance are directly proportional, whereas the tendency of crack propagation stays almost invariable. The clean surface forms when line emanating from punch and die coincide as shown in Fig. 2. This process leads the optimum clearance and punch incursion increases as per the ratio increases of clearance to metal thickness.

Hambli and Guerin [6] illustrated the blanking process and blanked surfaces shaped by the tooling (tool geometry and clearance) and properties of the workpiece material (mechanical properties, blank thickness and microstructure). For a specified material, tool geometry and the clearance configurations essential parameters, the simulation of an axisymmetric blanking process with Abaqus–Explicit software for particular sheet material. A Lemaitre-type damage model is deployed to describe crack initiation and propagation into the sheet. They explored that the material elongation is inversely proportional to the optimum clearance.

Hambli [7] proposed a methodology with the FEM and NN simulation to envisage the optimal punch–die clearance through sheet metal blanking. A damage model is employed to explain propagation and crack initiation into the sheet. The propositioned approach combines finite projecting element and NN modelling of the principal blanking parameters. As researched by previous researchers, the ANN is a powerful means for dealing with the complicated nature of the turning process [8]. Thus, the ANN was preferred to develop a predictive model to foresee optimum punch–die clearance. A backpropagation NN depleted in the prediction of optimum clearance with the help of backpropagation NN modelling [2, 9–11]. They carried out the experimentation on four materials with discrete ductility. The data of investigational fracture angles were operated to train the established simulation environment built on backpropagation NN modelling. The established prediction system achieved the capability of accurate clearance classification for the trained range. Mucha and Tutak [12] experimented the influences of the clearance on the deflection, initial values of burr and the bending radius of the hook and measured influence the wear of the punch on the geometric values of the cut. The outcome shows proper selection clearance and new tool materials.

The material selected for this work is ‘IS 513 CR’, suitable for stamping operations which includes a variety of processes, such as punching, blanking, embossing, bending, flanging and coining. The applications involve forming precision tubes, railway coaches, corrosion resistance applications, industrial storage systems, drum and barrel. Variety of shapes are developed (simple or complex) with high tooling and equipment costs at high production rates with low labour cost.

2.3 Artificial Neural Network

NN is made out of straightforward components working in parallel which are propelled by organic sensory systems. As in nature, the system work is acquired by the associations between parts. We can prepare a NN to play out capacity by adjusting the estimations of the associations (loads) between elements. By and large, NN is balanced or trained with the goal that information prompts the specific objective yield. The log-sigmoid capacity is utilized to standardize the information as shown in Fig. 4.

This transfer function takes the input (which may have any value between plus and minus infinity) and squashes the output in the range 0–1 according to the expression.

$$a = \frac{1}{1+e^{-n}}$$

This shows the neuron with R inputs and a single output, so it is called multiple input neurons, as shown in Fig. 5.

$$n = Wp + b$$

$$n = w_{1,1}p_1 + w_{1,2}p_2 + \cdots + w_{1,R}p_R + b$$

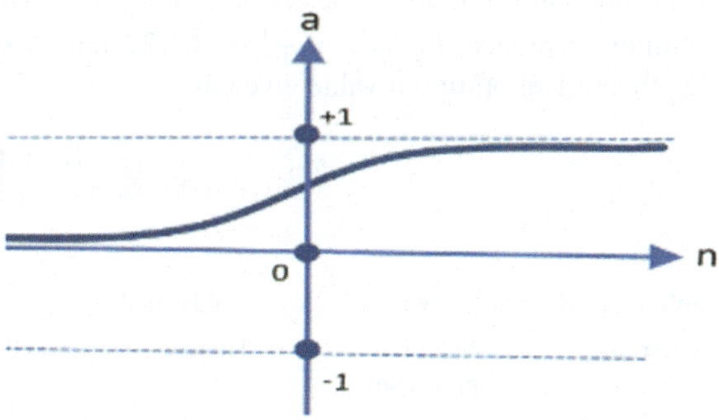

Fig. 4 Log-sigmoid transfer function

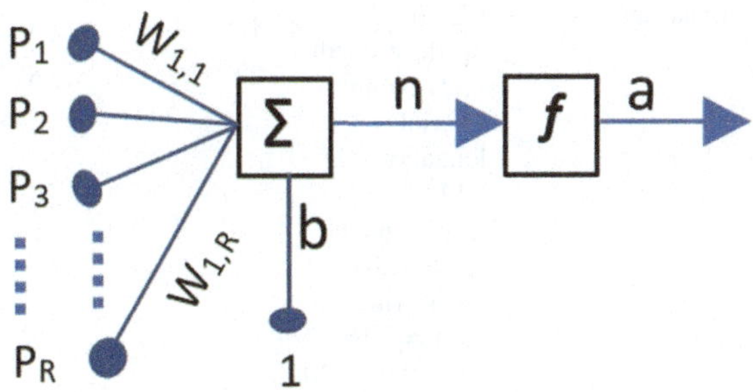

Fig. 5 Neural network model

where R = inputs, W = weight matrix.

The backpropagation algorithm with Levenberg–Marquardt training is used to train the neural network [13].

3 Result and Analysis

Experiments are carried out using the power press of 100-ton capacity for blanking operation. The uni-punch tooling system is used, and tonnage varies from 10 to 100 tons. The blank holder on the machine table holds the sheet and sheet is subjected to blanking. The material properties and input parameters for the experiment of 'IS 513 Cold Rolled' are shown in Table 1.

Experimentally, the values for engineering strain, true strain, optimum clearance C_0 and punch penetration are obtained [14].

Engineering strain $e_f = 7.16$, true strain $\varepsilon_f = (e_f + 1) = 2.1$

$$\frac{t}{C_0} = 1.36 \times e^{\varepsilon_f}\left[\frac{2.3\times e^{(\varepsilon_f)}-1}{2\times e^{(\varepsilon_f)}-1}\right]$$

$$\frac{t}{C_0} = 1.36 \times e^{1.75}\left[\frac{2.3\times e^{(1.75)}-1}{2\times e^{(1.75)}-1}\right]$$

Theoretical optimum value, $C_0 = \frac{2}{12.88} = 0.155\,\text{mm}$, from experimental study, optimum clearance, $C_0 = 3 \times \frac{6.1}{100} = 0.122\,\text{mm}$. Also, punch penetration to clear slug theoretical optimum value given as

$$\frac{\Delta + C_0}{t} = \frac{1}{2.45}\left[\frac{1.9\times e^{(\varepsilon_f)}-1}{2.56\times e^{(\varepsilon_f)}-1}\right]$$

Table 1 Mechanical properties of IS 513 cold rolled

Material	Material properties	Thickness (mm)	Clearance (mm)	Percent clearance (% of sheet thickness)
IS 513 cold rolled steel	Yield stress = 195 N/mm^2 Tensile strength = 321 N/mm^2 Material elongation (%) = 49.25 Reduction area (%) = 62.64 Engineering strain = 7.16 True strain = 2.1	0.8	0.2	25
		1.5	0.2	13.33
		2.0	0.2	10

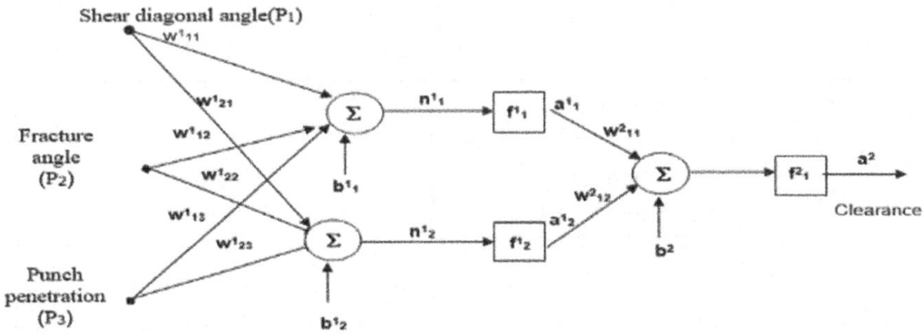

Fig. 6 Neural network for predicting optimum clearance

Punch penetration, $\Delta + C_0 = 0.28 \times t = 0.28 \times 2 = 0.58\,\text{mm}$.

Punch penetration from experimentation when crack initiates at Up = 0.31 mm.

The input to the network as shown in Fig. 6 is shear diagonal angle, fracture angle, punch penetration and output is clearance. The angles are determined from slug removed by the blanking process. The network uses three neurons and corresponding three transfer functions (2-log sigmoid and 1-purelin) Weights and biases are selected randomly at the beginning. A set of known parameters from experiments is selected for error minimization. At the beginning was calculated as $e = 0.075281$, which is brought down to $e = -0.0004325$ and took 4 iterations at a learning rate of $\alpha = 0.4$. Backpropagation iterations: following steps are followed in combination with LM training to solve the neural network (Fig. 7).

Results of the simulation are shown in Table 2 and the plot of input versus output and goal versus training error, as shown in Fig. 8. The graph shows improvement in approximation as shown in the graphs.

The inference of optimum clearance is derived where the line of shear diagonal angle intersects with the fracture angle. Theoretically, the shear line and fracture line shall lie on one line to avoid secondary fracture. The variation of fracture and shear angle verses per cent clearance and optimum clearance is observed to be 7% as shown in Fig. 7, and this outcome is confirmed by following earlier work [10]. The intersection of these two linear fitting of angles denotes the optimum clearance for IS513CR. The results of the simulation are plotted for input versus output and goal versus training error graph shows improvement in approximation.

The results of the simulation are validated using the analytical methodology of the neural network.

4 Conclusion and Future Scope

The purpose of this work is to find optimum clearance for IS 513 CR steel. The experimental values are validated by ANN simulation. The plot of the intersection of shear angle and fracture angle shows optimum clearance. The optimum clearance

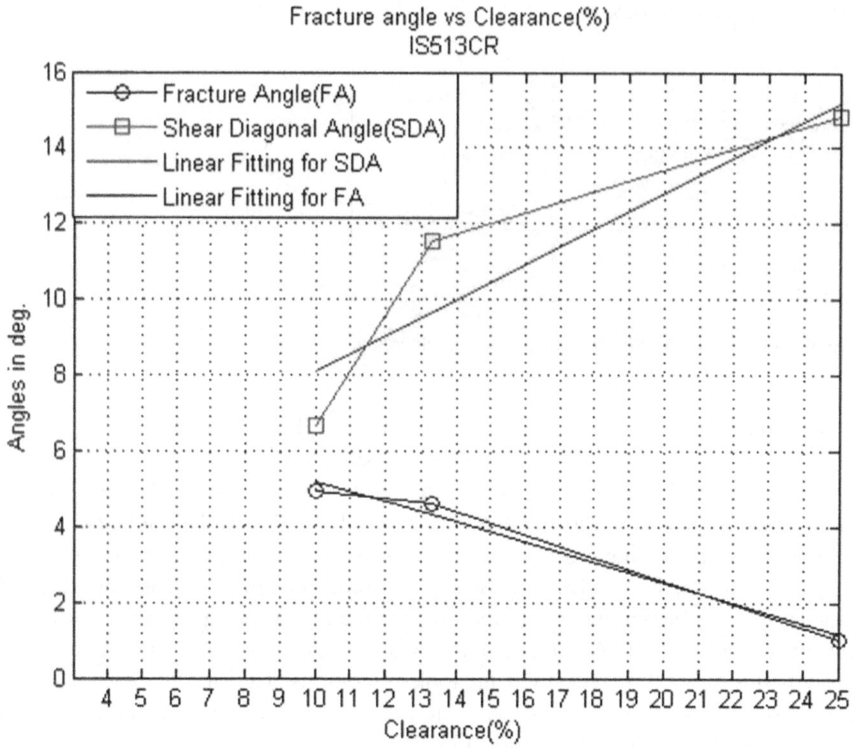

Fig. 7 Plot of angles vs clearance for IS513CR

Table 2 Results of neural network simulation for IS513CR for 2 mm sheet

S. No.	Punch penetration (mm)	Fracture angle	Shear angle	Clearance
1	0.004	0.7162	14.74	0.13
2	0.005	0.7162	14.74	0.15
3	0.004	0.7162	14.74	0.15
4	0.34	5.57	6.87	0.14
5	0.34	4.86	6.87	0.13
6	0.34	5.28	6.87	0.14
7	0.52	6.4659	11.76	0.16
8	0.52	6.4659	11.30	0.15
9	0.52	6.4659	11.53	0.14

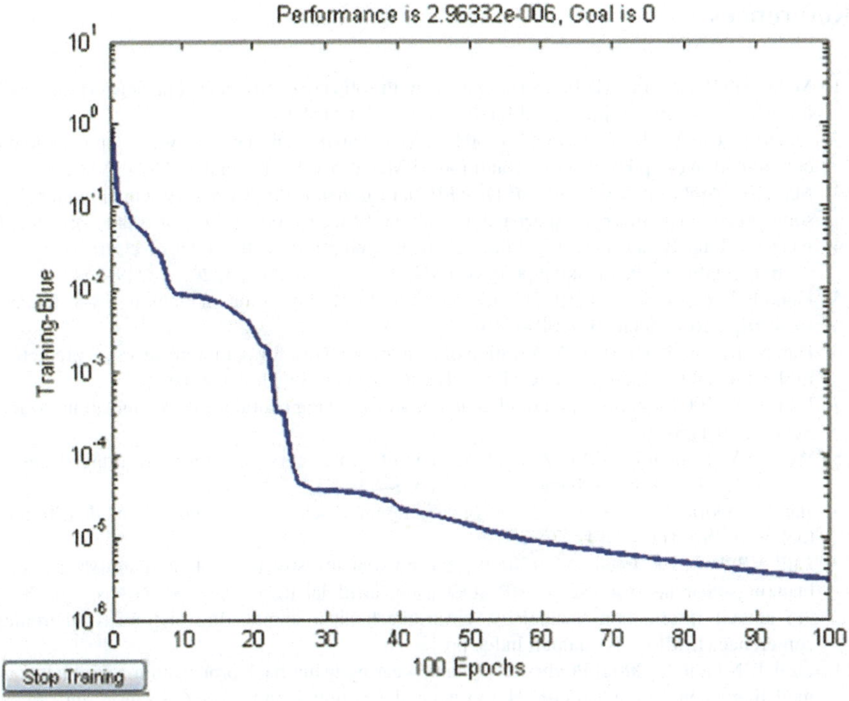

Fig. 8 Error graph

for IS513CR experimentally is found to be 6.1% of sheet thickness. The simulation using the neural network result gives optimum clearance of 7%. The backpropagation algorithm with LM training is used to solve the network. The backpropagation algorithm is an extension of least mean square algorithm that can be used to train the multilayer network. The backpropagation is an approximate steepest descent algorithm that minimizes squared error. After the four iterations performed, the error in the desired target and obtained value by the neural network approached to zero (0.0004428). The corresponding system is the optimum network and used to verify the clearance for different input conditions. However, the accuracy of prediction can be improved by considering more layers.

This study demonstrates the optimum clearance for IS513CR, a widely used material for a variety of in industrial applications.

The punch-load penetration, tool and die wear characteristics, sheet metal surface roughness will improve modelling of blanking process. The variation of clearance influences punch load because strains are larger when applying smaller clearance, and the stresses are also higher. The little difference in experimental and numerical results could be caused by other factors such as wear state of the tool and the thickness of the sheet.

References

1. Malakooti B, Raman V (2000) An interactive multi-objective artificial neural network approach for machine setup optimization. J Intell Manuf 11(1):41–50
2. Zuperl U, Cus F, Mursec B, Ploj T (2004) A hybrid analytical-neural network approach to the determination of optimal cutting conditions. J Mater Process Technol 157:82–90
3. Maiti SK, Ambekar AA, Singh UP, Date PP, Narasimhan K (2000) Assessment of influence of some process parameters on sheet metal blanking. J Mater Process Technol 102(1–3):249–256
4. Fang G, Zeng P, Lou L (2002) Finite element simulation of the effect of clearance on the forming quality in the blanking process. J Mater Process Technol 122(2–3):249–254
5. Faura F, Lopez J, Sanes J (1997) Criterion for tool wear limitation on blanking 18-8 stainless steel strips. Rev Metal 33(5):304–310
6. Hambli R, Guerin F (2003) Application of a neural network for optimum clearance prediction in sheet metal blanking processes. Finite Elem Anal Des 39(11):1039–1052
7. Hambli R (2005) Optimization of blanking processes using neural network simulation. Arab J Sci Eng 30(1):3–16
8. Wong SV, Hamouda AMS (2003) Machinability data representation with artificial neural network. J Mater Process Technol 138(1–3):538–544
9. Jiaa CL, Dornfeld DA (1998) A self-organizing approach to the prediction and detection of tool wear. ISA Trans 37(4):239–255
10. Patil VP, Patil PP, Ingale NE (2019) Experimental investigations of optimum sheet metal blanking clearance for IS2062 HR steel using artificial neural network (ANN). In: 2019 9th annual information technology, electromechanical engineering and microelectronics conference (IEMECON), Jaipur, India, pp 12–16
11. Özel T, Nadgir A (2002) Prediction of flank wear by using back propagation neural network modeling when cutting hardened H-13 steel with chamfered and honed CBN tools. Int J Mach Tools Manuf 42(2):287–297
12. Mucha J, Tutak J (2019) Analysis of the influence of blanking clearance on the wear of the punch, the change of the burr size and the geometry of the hook blanked in the hardened steel sheet. Materials 12(8):1261
13. Hagan MT, Demuth HB, Beale MH, De Jesús O (1996) Neural network design, vol 20. PWS Pub, Boston
14. Ghosh A, Mallik Manufacturing science. IIT Kanpur, pp 148–156

Improving the On-Time Delivery of Projects in a Complex Industrial Environment

Hari Vasudevan, Rajendra Khavekar, and Krishnan Kaushik

Abstract In companies, where a large number of projects are undertaken, the environment starts depicting characteristics of projects and batch manufacturing. Thus, traditional models which solve problems in a singular environment prove to be insufficient, when applied on environments of such nature. In this study, a similar case is presented, wherein the company works in a multi-project environment. Due to the repetitive nature of the product portfolio, the aforementioned complexity was seen to be present. Thus, an attempt has been made to solve this, by creating a generalized process flow, which holds true for any product that is dispatched. The results of the study showed that for the PERT chart, each singular flow finishes with the same three or four activities at the end. Thus, by using the backward planning approach, the due dates for each flow could be accurately determined, thereby shifting the emphasis on how much is left to be done rather than how much is finished.

Keywords Theory of constraints · Critical chain · Project management · On-time delivery

1 Introduction

This study presents the case of a company dealing in heavy manufacturing in India. Being in a project-based environment and dealing with engineered-to-order type orders, the company has a seemingly large product portfolio. However, on a deeper investigation, an aggregation in the product types became apparent, which served as an explanation for the discrepancy in the functions carried out by the company. Each project was treated and tracked as separate, but due to a plethora of the same being

H. Vasudevan · R. Khavekar (✉) · K. Kaushik
Dwarkadas J. Sanghvi College of Engineering, University of Mumbai, Mumbai, India
e-mail: khrajendra@rediffmail.com

H. Vasudevan
e-mail: principaldjs@gmail.com

K. Kaushik
e-mail: krishnan.kaushik1@gmail.com

© Springer Nature Singapore Pte Ltd. 2020

H. Vasudevan et al. (eds.), *Proceedings of International Conference on Intelligent Manufacturing and Automation*, Lecture Notes in Mechanical Engineering,
https://doi.org/10.1007/978-981-15-4485-9_24

undertaken, there were certain operations being carried out in a way synonymous to how a batch manufacturing environment would actually function. This disparity resulted in various complexities, arising in planning and tracking, thus resulting in a majority of the orders getting delayed for dispatch. Thus, through this study, a methodology is proposed using the concepts of critical chain project management to help improve the on-time delivery of projects in general.

2 Literature Review

Hall [1] concluded in his study that an increased competition, shorter product and service life cycles, tighter budgets, unfamiliar and more complex applications, globally distributed and multicultural project teams are the new trends in the domain of project management. Goh and Sim [2] focussed on a linear optimization problem with uncertainties, having expectations in the objective and in the set of constraints. They presented a modular framework to obtain an approximate solution to the problem, that is distributional robust and more flexible than the standard technique of using linear rules. Pich et al. [3] developed a model for planning, coordination, incentives and monitoring of the projects. Bertsimas and Sim [4] developed a mathematical model, which could be applied in discrete optimization problems. Kwak and Anbari [5] discussed regarding the project management and its future in their study. Munns and Bjeirmi [6] highlighted the overlap that existed between projects and project management with respect to their objectives.

3 Objective and Scope

Being in a project-based environment dealing in heavy manufacturing, the nature of the projects that the company delivers has long lead times. Long lead times result in many uncertainties arising during the course of the project. Inherently, projects of such nature are dependent on a multitude of small activities for its completion. Tracking these activities, while ensuring that they follow the overall project plan set in the beginning proves to be a near-impossible task. Tracking all of these activities, while ensuring that the critical path of the project is not disturbed, is a very tedious process and therein lies the scope of the study. The objective of this study has been to develop a mechanism, which ensures that the major tasks of the project are governed, while maintaining an ease in tracking daily activities, thereby ensuring on-time delivery.

4 Analysis of Current Reality

Projects, the company is involved in, have a lead time of around one year. Once a project is undertaken, an initial plan is created with the help of the planning department, which is followed by the assigning of tasks to each individual department. This initial plan is further updated and refined as the project progresses, based on the updates from each department. This results in each individual department becoming insular from the overall picture of the project, with only the planning department being aware of the actual overall status of the project. This situation coupled with the plethora of projects undertaken leads to constant fire-fighting, with each department getting focussed on completing only its daily tasks, resulting in loss of overall priority in work.

5 Current Challenges

The company faced the following challenges in its daily execution of work:

- Plans never seem realistic in execution.
- Constant shifts in priorities, leading to re-scheduling and confusion in execution.
- Constant fire-fighting between departments.
- Long lead times coupled with pressure from the market to be aggressive in accepting new projects.
- Too many changes coming from customers leading to frequent changes.
- Customer dissatisfaction and disruption due to unpredictable deliveries.
- Often delays in projects.

At a company level, this meant increased costs, reduced profitability and adverse impact on customer satisfaction.

6 Analysis of Problem

For a project-based environment, one would expect that the functions revolving around the project would work closely with that of the critical path. The complexity in the environment was arising from the repetitive nature of the projects, which resulted in a multitude of them being undertaken. This coupled with the limited capacity of the departments, resulted in delays as each department had to cater to the needs of multiple projects at a given time. This ultimately resulted in last-minute changes in priorities defined on the basis of which project's critical path would mostly get affected. Hence, each project when looked at individually resembled that of a pure project-based environment, but, when all of them were taken together, the environment resembled that of batch manufacturing.

7 Direction of Solution

To derive the tasks and assign responsibilities, the first requirement was to derive a PERT chart. The challenge, in this case, was to balance the need for visibility vis-a-vis the details. Two conflicting needs were observed, while detailing a plan. On one hand, execution needed details, while on the other hand, managing priorities needed a broader view. By creating one plan, this would never be achieved. Hence, one of the first decisions made was to restrict the PERT to a "planning" level, so that the focus could be made on timely delivery and tracking. The details could always be added at a functional level, which was not really needed for the planner. The pre-manufacturing stage mainly comprised of activities such as procurement, subcontracting and making of plant layouts. The activities in this phase functioned almost parallel in a way, such that any delay in one of the major activities resulted in the delay of the entire project. The manufacturing phase could start only when each of the activities in the pre-manufacturing phase would be completed. The manufacturing phase would comprise of activities to fabricate the project on-hand involving manufacturing processes such as welding, heat treatment, rolling, marking and cutting. Thus, the PERT chart to be made also had to be made in two parts. The first part of the process flow which depicted the pre-manufacturing phase was obtained by studying the "primavera" plans of the various projects undertaken. By doing this, it was able to derive a standard work breakdown structure (WBS), which served as the basis for making the PERT chart. The process flow obtained is shown in Fig. 1. Once the top half was defined, the next step was to elaborate on the activities in the manufacturing phase.

Thus, the first step was to generalize the product portfolio offered by the department. Their product line could be aggregated to three main skews being either a reactor, column or a heat exchanger. The next step was to understand the manufacturing activities necessary to fabricate each of these products, while also taking into account the variations in processes that might occur depending on its nature. The flow thus obtained is shown in Fig. 2. This flow accounted for all possible variations that could occur across product lines. Thus, for a given project, only a part of the chart would be valid by eliminating the processes relevant for the other two products and any other variations that do not pertain to it. Once a rough process flow was obtained for the pre-manufacturing and manufacturing phase, it was possible to arrive at the final PERT chart. A few changes were made after combining the two phases and reviewing and studying a few more projects, where the chart had failed. The final PERT thus obtained is shown in Fig. 3.

8 Depicting the Environment

A "Matrix" format was envisioned, wherein each project with its PERT chart followed a "row", while the departments represented "columns". Each participating task of a PERT that needed to be done by a department would then be a "cell" of the matrix,

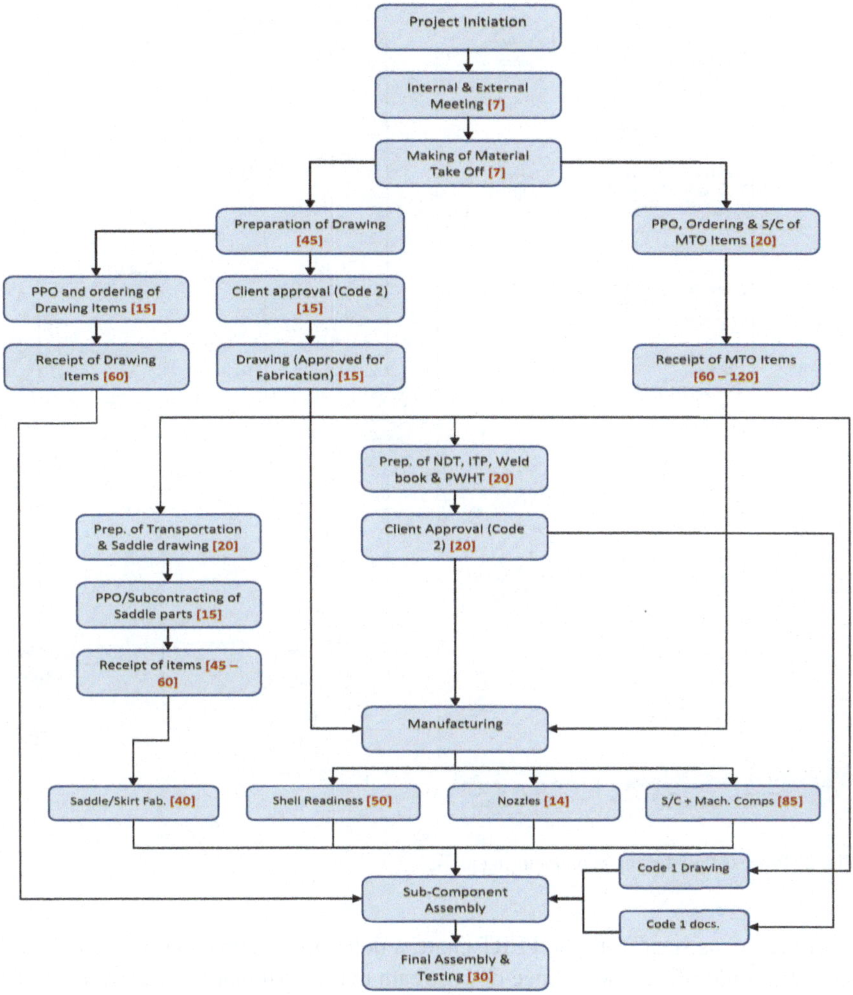

Fig. 1 Process flow for the pre-manufacturing phase

which had a scope and time assigned to it. This step was critical, as it helped visualize the system and the overall structure, which was lost in the myriad of software, reports and charts. The envisioned matrix is as shown in Fig. 4.

Thus, each project would flow individually in accordance with its critical path, while keeping in mind its goal of achieving its on-time delivery. Parallelly, the department would be dealing with tasks derived from the PERT chart and would carry out their tasks, while keeping in mind their goal of achieving maximum efficiency or productivity. However, it would be clear that the overall company priorities would supersede departmental priorities, without creating conflicts for the teams. This method would solve the problems present between the departments and the planning team.

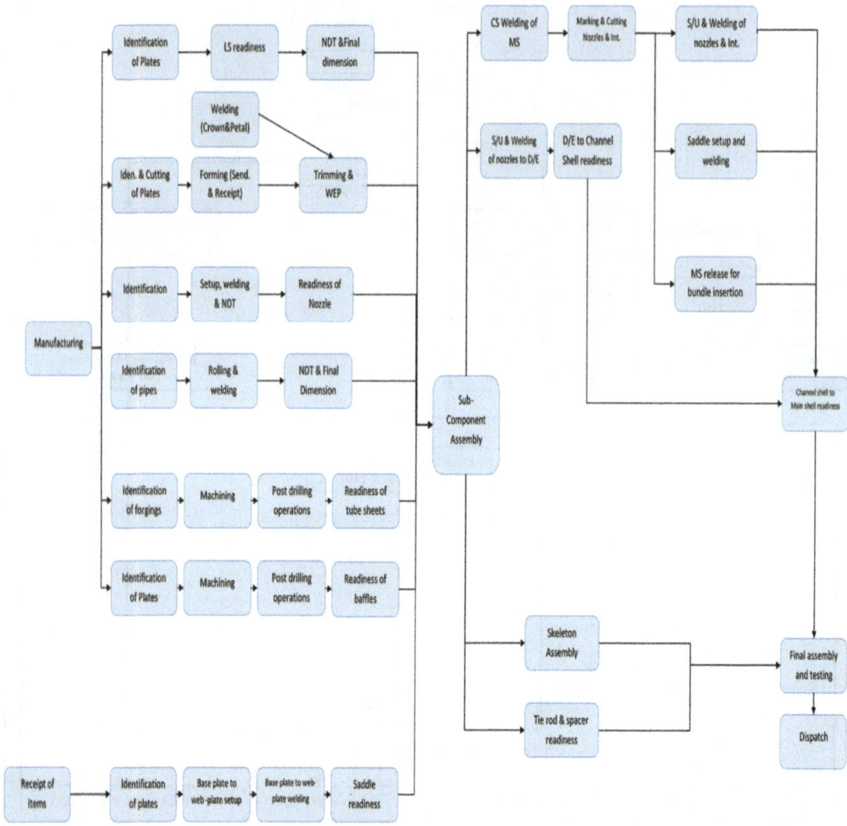

Fig. 2 Process flow for the manufacturing phase

However, as seen from the PERT chart, a delay in procurement of a small consumable could often lead to huge downstream delays. Similarly, a relatively newly arrived part in manufacturing could be lost in the queue, while a crucial activity of phase 2 would be waiting for it. With planning needed to chase, this would lead to departments breaking queues, which is usually not welcomed. In the new approach, if selectively, departments could know "urgent" tasks as different from "normal" tasks, wherein the urgency is derived from the overall project, then it becomes easy to follow priorities without chasing, follow-ups or inter-departmental fires.

9 Planning Algorithm

It was proposed to use the concepts of "Theory of Constraints" and "Critical Chain Project Management" to derive the necessary tool for tracking. Specifically, the project priorities to be derived were based on the buffers added to the PERT chart

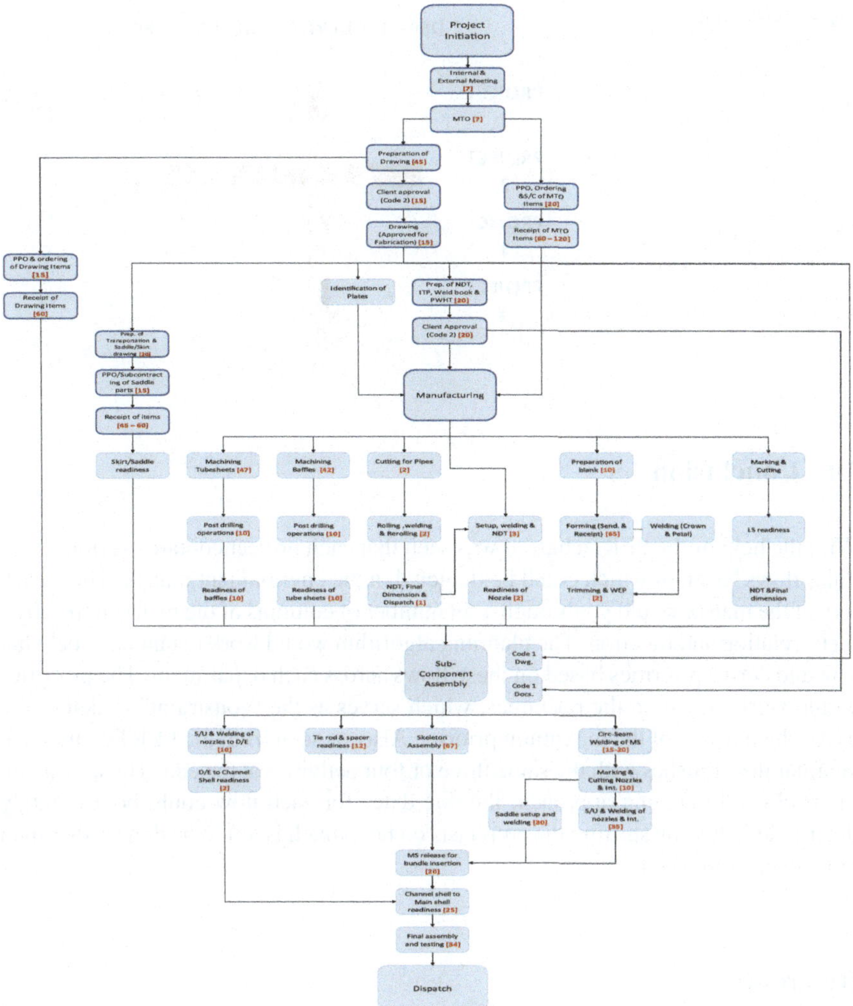

Fig. 3 Final PERT of the process

in accordance with the critical chain. This would serve as the backward planning approach. Then, to factor for the manufacturing process and the ability to deal with discrete items, the functionality would be added to arrive at release mechanisms for each function.

The tool would comprise of the following features:

(a) Modelling the environment. (b) Templatization of the plan. (c) Determination of direction of flow. (d) Ability to track discrete items as part of a task. (e) Deriving project priorities using buffers. (f) Backward planning to arrive release dates. (g) Customized mechanism to update daily progress and to derive reports to review and track.

Fig. 4 Environment matrix

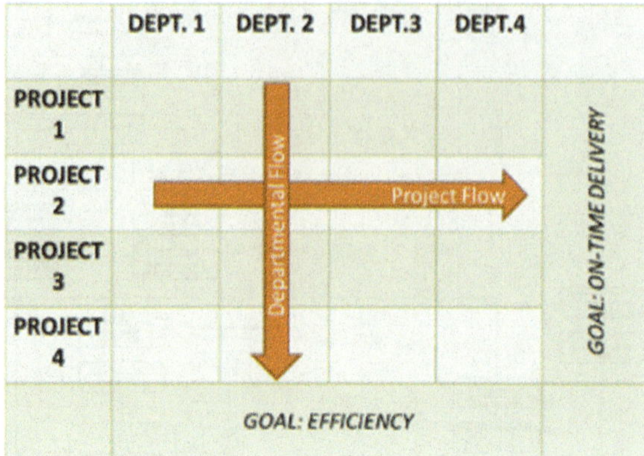

10 Conclusion

With the help of the PERT chart, it was seen that each project contains various "singular flows" each of which could be depicted in the environment matrix. Thus, each row of the matrix would go through a set number of columns of the matrix depending on its relation and function. The planning algorithm would work in tandem, such that it would derive priorities based on these flows across each department. The priorities would work, based on the resources, which serves as the "constraint" to determine as to which row is of the maximum priority. Also, as seen from the PERT chart, each singular flow finishes with the same three or four activities at the end. Thus, by using the backward planning approach, the due dates for each flow could be accurately determined, thereby shifting the emphasis on how much is left to be done rather than how much is finished.

References

1. Hall NG (2012) Project management: recent developments and research. J Sci Syst Eng 21(2):129–143
2. Goh J, Sim M (2010) Distributionally robust optimization and its tractable approximations. Oper Res 58:902–917
3. Pich MT, Loch CH, De Meyer A (2002) On uncertainty, ambiguity, and complexity in project management. Manag Sci 48(8):1008–1023
4. Bertsimas D, Sim M (2004) The price of robustness. Oper Res 52:35–53
5. Kwak YH, Anbari FT (2009) Analyzing project management research: perspectives from top management. Int J Project Manag 27:435–446
6. Munns AK, Bjeirmi BF (1996) The role of project management in achieving project success. Int J Project Manag 14(2):81–87

Investigating the Influence of Infill Pattern on the Compressive Strength of Fused Deposition Modelled PLA Parts

Sanket Parab and Nilay Zaveri

Abstract The research work presents an analysis of the optimal infill pattern of 3D-printed Polylactic Acid (PLA) based on its compressive strength. Samples with varying infill percentages and types are loaded in different directions. The samples are built using fused deposition modelling process (FDM) on a Creality Ender-3 3D printing machine. Strength is calculated based on the results of testing and feasibility of infill type and pattern according to application is analysed. The testing shows triangular infill to be superior to the default line infill pattern in every condition. Gyroid infill shows nearly isotropic properties, thus being a useful pattern to select when the part is subjected to multiple loading directions.

Keywords Fused deposition modelling (FDM) · Additive manufacturing (AM) · Rapid prototyping · Polylactic acid (PLA) · Process parameters · Compressive strength testing

1 Introduction

Additive manufacturing permits programmed creation of complex shapes with a critical decrease in assembling cost, contrasted with conventional subtractive manufacturing methods [1].

Fused deposition modelling (FDM®) is a material extrusion process used to make thermoplastic parts through heated extrusion and deposition of materials layer by layer [2].

The process begins with the creation of a CAD model of the part to be printed. This is then exported to a .stl format, which simplifies the part by reducing it to its most basic components. The disadvantage is that the part loses some resolution, because the geometry is now represented by triangles, and not by features like true arcs and splines [3]. However, the errors introduced by these approximations are

S. Parab (✉) · N. Zaveri
Production Engineering Department, Dwarkadas J. Sanghvi College of Engineering, Mumbai, India
e-mail: sanket.parab@djsce.ac.in

© Springer Nature Singapore Pte Ltd. 2020

H. Vasudevan et al. (eds.), *Proceedings of International Conference on Intelligent Manufacturing and Automation*, Lecture Notes in Mechanical Engineering,
https://doi.org/10.1007/978-981-15-4485-9_25

acceptable as long as they are less than the inaccuracy inherent in the manufacturing process [4].

The file is then loaded into slicing software, which slices the part into multiple horizontal layers. A G-code is generated that determines the tool path and other printing parameters of the 3D printer.

A filament of a thermoplastic material such as acrylonitrile butadiene styrene (ABS) or polylactic acid (PLA) is pushed through a heated nozzle. The resulting fused filament is then deposited onto a heated bed. After one layer of semi-molten material is deposited, the nozzle moves up and deposits another layer on top of the previous one, fusing them together. This process is repeated until the part is formed.

FDM parts are generally not printed solid. To save on material and decrease build time, parts are printed with an internal, low-density structure known as infill. Infill percentage is a parameter that can be varied based on the application of a part. The geometry of the infill also impacts the performance of an FDM part [5].

Alvarez et al. [6] investigated the influence of infill percentage on the tensile strength of ABS parts using a honeycomb infill pattern. Their results show a trend in increase of strength with infill percentage. Mohammed Raffic and Ganesh Babu [7] and Ahn et al. [8] are some of the researchers who have studied the effects of layer thickness, air gap, raster orientation, bead width, raster width and model temperature on the mechanical properties of 3D printed plastics. Most of these papers present conclusions of the effects of these parameters by either keeping the infill percentage or the types of infill constant.

Infill type is a very simple parameter to tweak. This paper attempts to fill the gaps in these previous studies by analysing the effect of different infill structures, as it is largely an independent parameter, on the strength of a part, and thus comment upon the optimal infill type to utilise according to the situation.

2 Methodology

2.1 Selection of Material

PLA, ABS and nylon are three of the most popular 3D printing materials. All can be extruded on basic 3D printers and they are among the most affordable filaments available today [9]. However, since Nylon is a flexible material, it cannot be used for compression testing.

PLA has a larger strength and lower ductility than the traditional acrylonitrile butadiene styrene (ABS) material [10]. PLA has a lower coefficient of thermal expansion, which reduces the effects of warping, not adhering to the printed surface, and large parts cracking as they are printed.

PLA will give a more defined graph and will reduce the possibility of manufacturing defects due to warping. Considering these factors, the material chosen is 1.75 mm diameter PLA filament.

2.2 Selection of Infill Pattern and Density

The default infill option in most software is the line infill. It is a simple infill structure as compared to other infills. Triangle infill was chosen as the second infill structure, as it has comparable weight and print time to line infill. These infill structures are two-dimensional infill structures, i.e. they repeat the same pattern in every layer.

To counteract the anisotropic property of FDM parts, a 3D infill structure had to be chosen next. Gyroid was selected as the infill structure since the nozzle head does not have to constantly change direction as compared to other structures.

Infill percentages of 20, 50 and 80 are chosen as they are very common infill percentages for low, medium and high strength applications, respectively, as shown in Figs. 1 and 2.

Fig. 1 Line, triangle and gyroid infill at 20% density with internal structure below

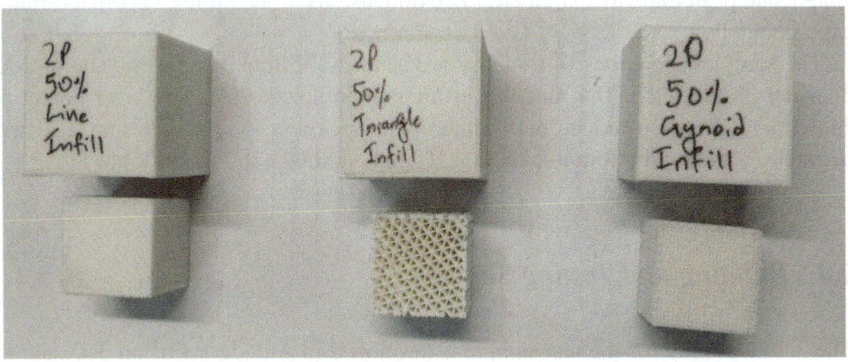

Fig. 2 Line, triangle and gyroid infill at 50% density with internal structure below

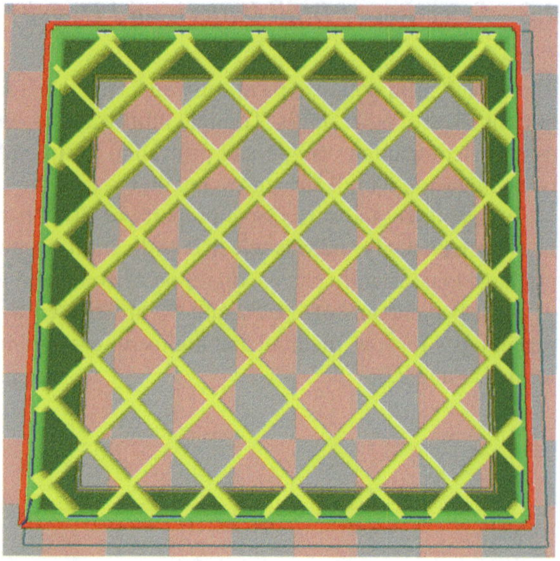

Fig. 3 Internal structure of 20% line infill sample

2.3 *Identifying Print Parameters for Test Samples*

The printed test samples are to be subjected to compression under an Aimil (AIM UT 100 DG) Universal Testing Machine (UTM). Its working load range is 1–1000 kN. To attain more accurate results, a part must sustain at least the first 2% of the load capacity of a machine, i.e. 20 kN. The test samples have to be cubes as they will be loaded in different orientations to account for their anisotropic properties. Taking these factors into consideration, along with practicality regarding sample weight and print time, the parameters and dimensions of the samples are determined experimentally.

The samples are cubes of dimension 35 mm × 35 mm × 35 mm. The nozzle diameter is selected as 0.4 mm as it is a common nozzle size. The layer height is chosen as 0.3 mm to reduce build time. Shell thickness is set to 0.8 mm, i.e. two perimeters, and top/bottom thickness is 0.9 mm as shown in Fig. 3 (Table 1).

2.4 *Building and Testing Samples*

The 3D model file for the test sample is developed in Autodesk Fusion 360, and is imported to the slicing software, Ultimaker Cura. This specific slicing software is chosen as it is user friendly and has the required selection of infill types. The variable parameters for the test are given in Table 2.

Two samples of each combination are built to reduce the effect of error, totalling at 36 samples.

Table 1 Control parameters

Fixed parameters	
Cube dimensions	35 mm × 35 mm × 35 mm
Nozzle diameter	0.4 mm
Layer height	0.3 mm
Shell thickness	0.8 mm (2 perimeters)
Top/bottom thickness	0.9 mm (3 layers)
Print speed	50 mm/s
Nozzle temperature	205 °C
Bed temperature	55 °C

Table 2 Control parameters

Variable parameters	
Infill type	• Line • Triangle • Gyroid
Infill percentage	• 20% • 50% • 80%
Loading direction	• Along build direction • Perpendicular to build direction

2.5 Test Procedure

The testing process is done according to ASTM standard—ASTM D695-15 (Standard Test Method for Compressive Properties of Rigid Plastics) [11]. A test sample is loaded onto the machine, ensuring that it is centred. Compressive load is increased at a fixed rate and load vs displacement graph is generated by the UTM Machine. Two identical samples are tested for each permutation, and the average value is taken. Figures 4 and 5 show samples loaded both along and perpendicular to build direction.

Due to the limitations such as immaturity of FDM technology, there are still many problems to be resolved in the production process. Many products built by FDM have the defects such as insufficient precision, warping, short filling, string shearing, or nozzle failure, which seriously restrict the application of FDM [12]. These defects can cause the printed samples to prematurely fail as shown in Fig. 6.

The generated graphs for identical samples are analysed. In case of a defect, the shapes of the graphs are dissimilar, and the samples exhibit different points or modes of failure. Mechanical defects were observed in two such samples, which were then reprinted. Samples with similar graphs were selected for the dataset, thus significantly reducing the error due to mechanical defects.

Fig. 4 Sample loaded in build direction

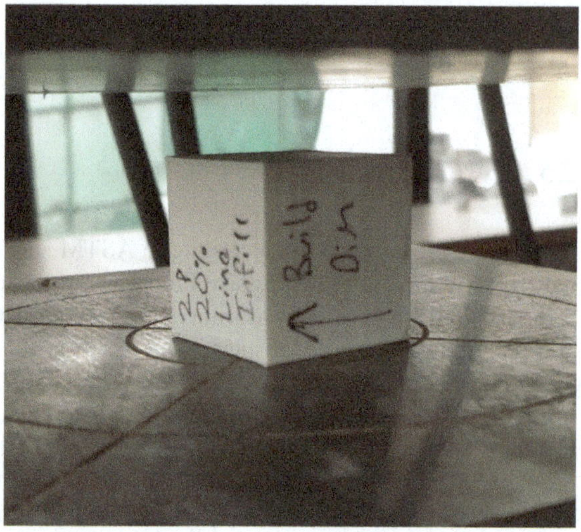

Fig. 5 Sample perpendicular to build direction

3 Results

3.1 Load Orientation

Figure 7 shows average values of peak yield load, extracted and plotted into a graph.

Line and gyroid patterns exhibit similar properties in this region, with gyroid sustaining slightly higher load values than line for 20 and 50% infill, and slightly lower values for 80% infill.

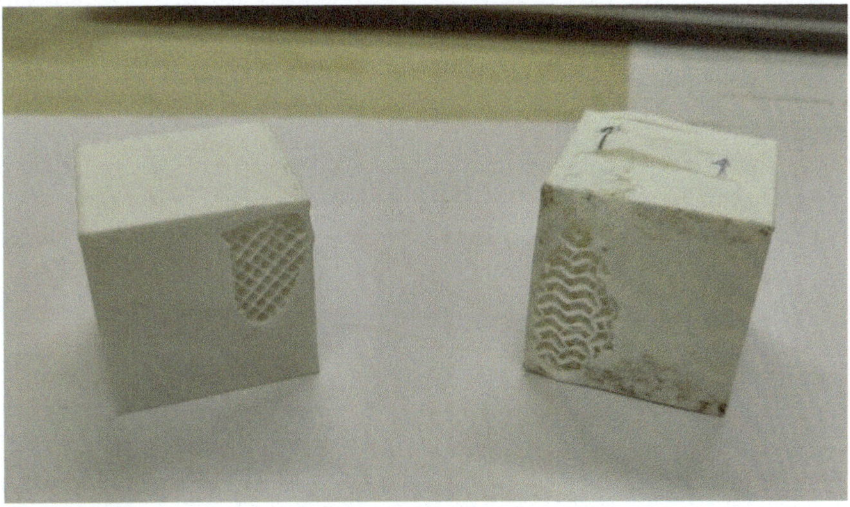

Fig. 6 Buckling of walls, common failure mode for sample

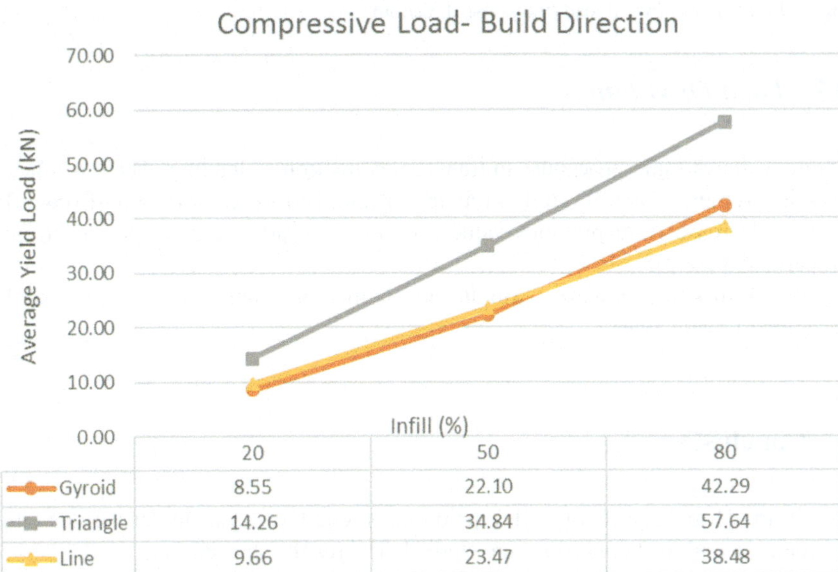

	20	50	80
Gyroid	8.55	22.10	42.29
Triangle	14.26	34.84	57.64
Line	9.66	23.47	38.48

Fig. 7 Peak values of yield load (build direction)

Triangle infill performs significantly better in this condition, sustaining between 4 and 15 kN more load than its counterparts.

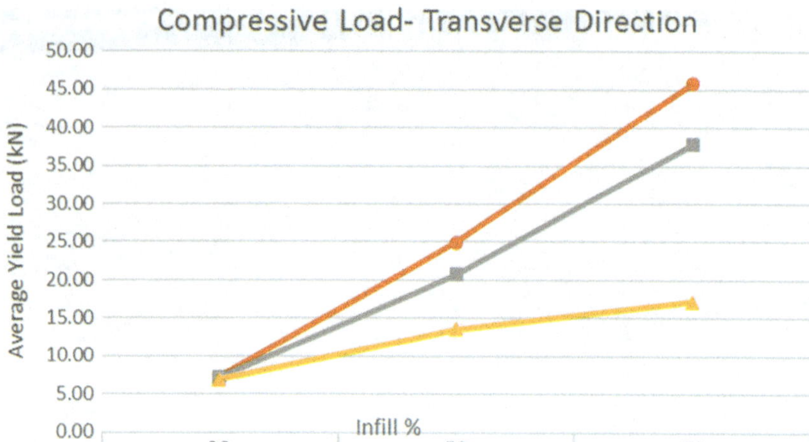

	20	50	80
Gyroid	7.25	24.93	45.79
Triangle	7.22	20.76	37.85
Line	6.92	13.64	17.30

Fig. 8 Peak values of yield load (transverse direction)

3.2 Load Orientation

Figure 8 shows same procedure in transverse direction of loading. There is a large drop in strength of the 2D infill structures of line and triangle, ranging from a 35 to 55% decrease from previous values. This can be attributed to the anisotropic properties of the parts.

Gyroid structure however exhibits no significant change apart from a slight variation in load values.

4 Conclusion

In this article, the effects of infill type on the strength of a part under various infill percentages and load conditions are studied. The results showed that:

1. 2D infill structures exhibit a sharp decrease in strength when loaded in the transverse direction.
2. Triangle infill is extremely strong in build direction.
3. Triangle infill sustains higher loads than line infill in all conditions and infill percentages. It is beneficial to make triangular infill the default infill structure, as it offers better mechanical properties for the same weight and printing time as line infill.

4. Gyroid infill should be a preferred infill pattern for when the direction of loads to be applied on the parts is unknown or the part is going to be subjected to transverse loads.
5. This information will help manufacturer's waste of minimum material and build time to create a useful product.

5 Future Scope

More 2D structures such as hexagonal, tri-hexagonal and 3D structures such as cubic, 3D honeycomb can be tested at two more infill percentages, leading to more comprehensive data. Time required to print can also be analysed to find the right fit of infill. Some of the fixed parameters such as shell thickness can also be varied in conjunction with infill to find the optimal mix of weight, time and strength.

References

1. Dimitrov D, van Wijck W, Schreve K, de Beer N (2006) Investigating the achievable accuracy of three dimensional printing. Rapid Prototyp J 12(1):42–52
2. ASTM (2012) Standard terminology for additive manufacturing technologies. Standard F2792-12a, ASTM International. West Conshohocken, PA. https://doi.org/10.1520/F2792-10
3. Wright PK (2001) 21st Century Manufacturing. Prentice Hall, New Jersey
4. Ahn S-H, Montero M, Odell D, Roundy S, Wright PK (2002) Anisotropic material properties of fused deposition modeling ABS. Rapid Prototyp 8(4):248–257
5. Redwood B, Schoffer F, Garret B (2017) The 3D Printing Handbook: Technologies, design and applications. 3D Hubs B.V. Amsterdam, The Netherlands, p 40
6. Alvarez KL, Lagos RF, Aizpun M (2016) Investigating the influence of infill percentage on the mechanical properties of fused deposition modelled ABS parts. Ingeniería e Investigación 36(3):110–116. https://doi.org/10.15446/ing.investig.v36n3.56610
7. Mohammed Raffic N, Ganesh Babu K, Palaniappan PL, RajeshKannan P, Santhanabharathi NS, Venkatramanan S (2017) Effect of Fdm process parameters in Abs plastic material. Int J Mech Prod Eng 3(1):14–23. www.aetsjournal.com
8. Ahn SH, Montero M, Odell D et al (2002) Anisotropic material properties of fused deposition modeling ABS. Rapid Prototyp J 8(4):248–257
9. https://markforged.com/blog/pla-abs-nylon/
10. Letcher T (2014) Material property testing of 3D-printed specimen in PLA on an entry-level 3D printer. In: International mechanical engineering congress and exposition 2014, IMECE, vol 2, pp 1–8. Montreal
11. ASTM (2014) Standard test method for compressive properties of rigid plastics. In: D695-02a, United States
12. He K, Wang H, Hu H (2018) Approach to online defect monitoring in fused deposition modeling based on the variation of the temperature field. Complexity 2018(Article ID 3426928), 13 pages. https://doi.org/10.1155/2018/3426928

Analysis of Compound Column-Based Supporting Structures Used in Suburban Railway Transport System: Use of Stiffener Plates

Herin Savla, Neel Sanghvi, Saurabh Rasal, and Vinayak H. Khatawate

Abstract This paper documents a comparative numerical analysis aimed at verifying the application of stiffener plates in a compound column, which is being used as a supporting member at multiple railway stations in the Mumbai Suburban Railway Network. Stiffener plate is a term used for a group of plate-like structural elements that are used to increase the rigidity of a beam or column. These stiffener plates are installed at specific locations in the column to decrease mainly the total deformation, equivalent stress, equivalent elastic strain, and the shear stress. Linear and non-linear analyses were performed for the compound column, both with and without the presence of the stiffener plates. It was concluded that with a 3% increase in the weight of the structure, the total deformation decreases by 8% and the equivalent stress decreases by 28%.

Keywords Stiffener · Safety · FEA · Column · Structural engineering

1 Introduction

The design of a railway station is a complex process, primarily because major repairs can incur very large costs, and also because these areas witness a large flux of people [1]. The capability of the railway station to safely handle a large volume of people during technical snags and natural calamities is also an aspect which has to be considered during railway station design [2]. In cities with high population and complex railway transit systems, e.g. Mumbai, New York City, London [3, 4], etc., railway stations often have to accommodate a very large number of people at a given time, especially during the peak hours. And the land area available for construction is often limited.

Thus, railway stations accommodate certain features in their design which help to accommodate a large number of people at a given time. One such common feature

H. Savla (✉) · N. Sanghvi · S. Rasal · V. H. Khatawate
Mechanical Engineering Department, Dwarkadas J. Sanghvi College of Engineering, Mumbai 400056, India
e-mail: h.savla@outlook.com

© Springer Nature Singapore Pte Ltd. 2020

H. Vasudevan et al. (eds.), *Proceedings of International Conference on Intelligent Manufacturing and Automation*, Lecture Notes in Mechanical Engineering,
https://doi.org/10.1007/978-981-15-4485-9_26

is the construction of an upper level above the ground level to place features like cafeterias, ticket counters and other places of interest. This structural feature prevents overcrowding at the railway stations [5]. In the Mumbai suburban railway network, which is made of approximately 430 km of railway track, this feature is observed at multiple railway stations—Borivali, Andheri and Goregaon to name a few. The provision of a second level helps to keep the population density low.

This second or upper level can be described as a large slab of concrete with columns passing through it. Given the average density of masonry [6], the mass and hence weight of this slab will be quite high. The columns or stationary vertical members used to support this second level will have to be strong enough to successfully bear this load without large deformations. Moreover, this loading will be of a complex and dynamic nature due to movement of people and the presence of wind and other such factors [7].

The supporting columns that were observed at Andheri and Goregaon railway stations were of absolute metal type, using no concrete or other such material. The columns were essentially two I-beams fused together such that their longitudinal axes were coinciding and they had their end faces in the same horizontal plane but the cross-sections were perpendicular to each other. The feature of the column which is the major interest in this experiment is the use of stiffener plates in the columns [8, 9]. Stiffeners are secondary plates or sections which are attached to beam webs or flanges to stiffen them against out of plane deformations [9, 10]. An image of a part of the above-described compound column is shown in Fig. 1. Stiffener plate is basically an example of a structural element. The fundamental goal of a structural element is to function as a part of a given structure; although in many cases, structural elements are used to increase the strength of a structure and its inherent safety. Two interesting features about these plates are that the thickness is always much smaller compared to the length and width and that the forces applied on them generate stresses whose resultants lie exclusively in a plane normal to the plate thickness. Stiffener plates are used widely in marine applications especially for hull girder construction [11, 12].

The use of stiffeners contributes additional strength to the web by ensuring that the web does not laterally buckle or cripple due to application of heavy concentrated load [13]. The horizontal intermediate stiffeners are used to increase the buckling strength of the web and these are located in the compression zone [14]. They prevent the web from bending laterally [15, 16].

2 Objectives

The authors's principle objective is to perform a comparative study aimed at verifying the application of stiffener plates in a compound column while applying basic concepts of FEA to solve this real-life engineering problem. Their aim is to also understand the application and importance of FEA in structural engineering while studying and analysing an example of practical application of structural engineering.

Fig. 1 Image of lower portion of column observed at Andheri station

3 Methodology

The software used for designing the compound columns both with and without stiffener plates is SOLIDWORKS 2018. The subsequent analysis of these columns was done by using FEA. To create the 3D CAD model of this column, the authors paid a visit to the Andheri railway station to measure and note down the required dimensions. For performing the analysis, structural steel was assigned as the material to this column design. A suitable mesh was applied, fixed supports were selected and a remote force was added. Then, the analysis to calculate the values of total deformation, directional deformation, equivalent elastic strain, shear elastic strain, shear stress, equivalent stress, and maximum shear stress was performed on both columns(with and without stiffener plates). Figures 2 and 3 given below show the meshed CAD model of the compound columns with and without the stiffener plates.

4 Design Calculations

In this comparative study, two versions of the column are considered. The first version is the basic compound column. The second column has stiffener plates installed in

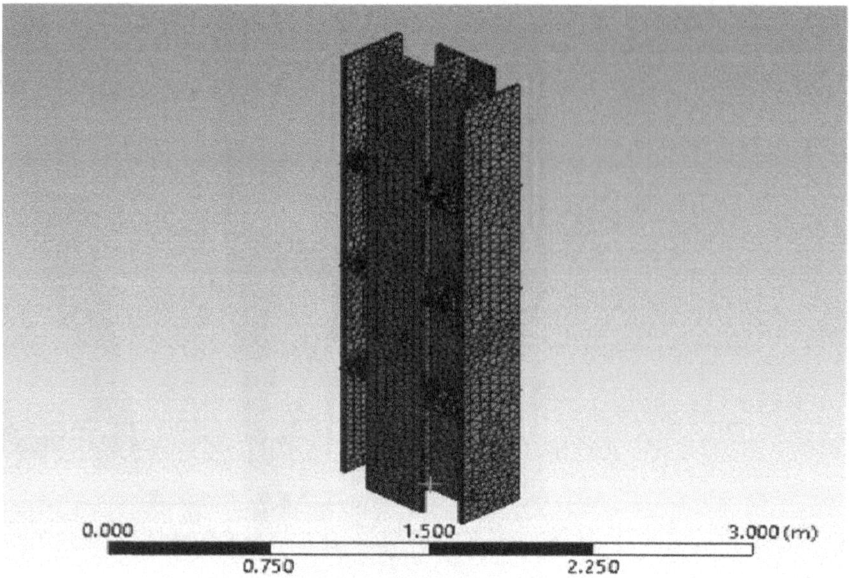

Fig. 2 Meshed body for compound column with stiffener plates

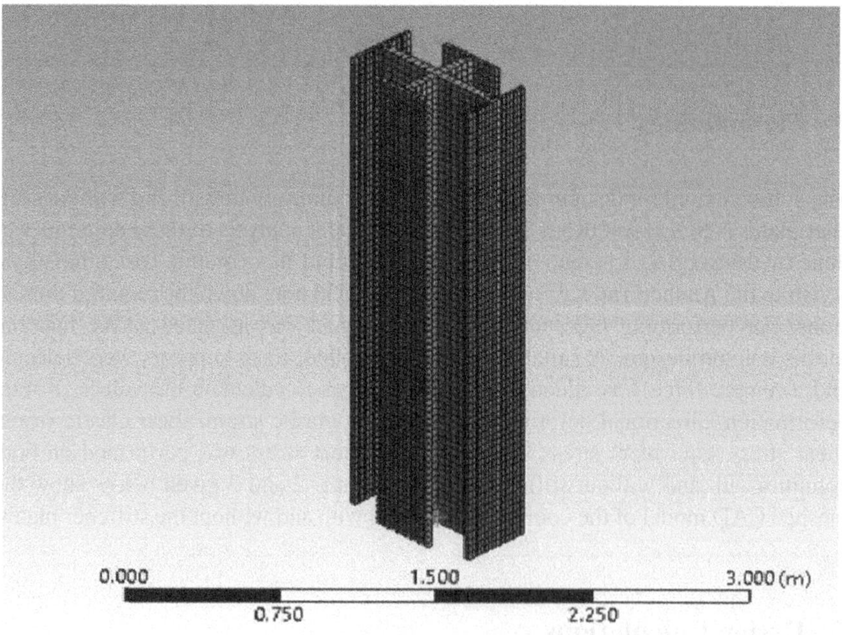

Fig. 3 Meshed body for compound column without stiffener plates

the plane perpendicular to the longitudinal axis of the column. Four stiffener plates are installed in one horizontal plane. The stiffener plates are taken to be manufactured with the same material as that of the column.

After a basic survey of Andheri railway station, the initial conditions are decided such that a representative computer model of the system can be developed. By means of measurement and suitable estimates, it is estimated that one compound column is responsible for bearing the weight of a part of the first level having an area of 3 m × 8 m. Slab thickness of the first level is considered to be 25 cm.

The column observed at the before-mentioned location consists of 11 segments (one segment implies one stiffener plate group and a certain length of continuous column). Due to increased computational cost, the analyses are restricted to representative model of the column consisting of three such segments. This gives rise to a scaling factor of 3/11 in the design calculations. The average density of masonry is taken as 2000 kg/m^3 [6].

By making use of the appropriate mathematical formulae, the load exerted by a part of the first level on the column is calculated to be 117.72 kN. This is multiplied by the scaling factor to give a value of 32.105 kN which is increased to 40 kN to account for the presence of people on the bridge and also the presence of various structures like ticket counters, ATMs, and fast food joints on the first level.

5 Results and Discussion

The authors have performed a non-linear analysis in 20 steps of 2 KN each, while keeping large deflection mode ON. The authors have taken *Y*-axis as the axis along the height of the column, *Z*-axis as the axis perpendicular to the railway track, and *X*-axis as the axis parallel to railway track. A total of 40 kN load was applied.

It is clear from Table 1 that the addition of stiffener plates to the compound column has the desired result of decrease in the values of stress and deformation experienced by the column. The linear analysis shows that there is 8.209% decrease in total deformation and 28.316% decrease in equivalent stress on addition of stiffener plates. The non-linear analysis shows that there is 8.84% decrease in total deformation and 28.36% decrease in equivalent stress on addition of stiffener plates. In Figs. 4 and 5, the authors have shown the results for von-Mises stress and total deformation of the linear analyses for both types of columns.

6 Conclusion

Linear and non-linear analyses were performed to study the effect of stiffener plate on the rigidity of the compound column. Linear analysis was performed by applying a static load of 40 KN whereas the non-linear analysis was performed by applying the same load but in 20 steps of 2 KN each. The force was applied at a calculated point

Table 1 Results of linear and non-linear analyses (DEC indicates decrease in corresponding quantity)

	Linear without stiffener	Linear with stiffener	Change (%)	Non-linear without stiffener	Non-linear with stiffener	Change (%)
Total deformation (m)	3.1816e−5	2.9204e−5	−8.209 DEC	3.2035e−5	2.9203e−5	−8.84 DEC
Deformation X (m)	6.1183e−6	2.5481e−6	−58.35 DEC	6.1185e−6	2.539e−6	−58.50 DEC
Deformation Y (m)	3.1735e−5	2.9098e−5	−8.31 DEC	3.1953e−5	2.9158e−5	−8.747 DEC
Equivalent elastic strain	5.602e−5	4.0156e−5	−28.32 DEC	5.605e−5	4.015e−5	−28.36 DEC
Shear elastic strain	2.102e−5	1.896e−5	−9.844 DEC	2.105e−5	1.907e−5	−9.36 DEC
Equivalent stress (Pa)	1.120e7	8.0313e6	−28.32 DEC	1.121e7	8.031e6	−28.36 DEC
Shear stress (Pa)	1.617e6	1.458e6	−9.84 DEC	1.619e6	1.467e6	−9.36 DEC
Maximum shear stress (Pa)	5.875e6	4.246e6	−27.74 DEC	5.879e6	4.246e6	−27.78 DEC

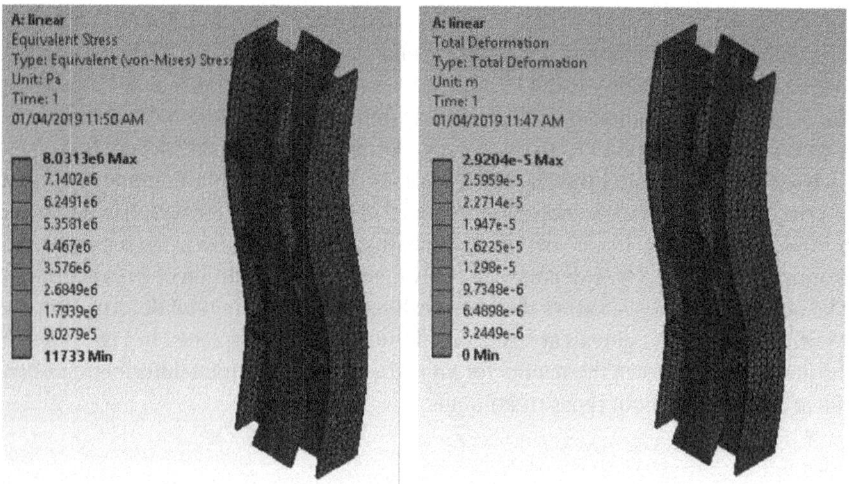

Fig. 4 Results for von-Mises stress and total deformation of linear analyses for columns with stiffener plates

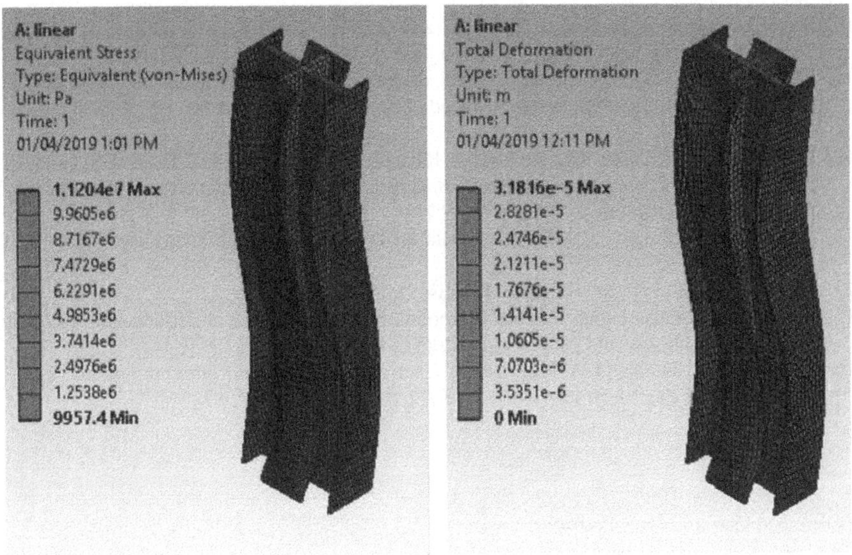

Fig. 5 Results for von-Mises stress and total deformation of linear analyses for columns without stiffener plates

as a remote force. The results were analysed and it was observed that the addition of stiffener plates increases the mass of the column by 3.04% but decreases the von-Mises stress by more than 28% and decreases the total deformation by more than 8%. Thus, it is effectively concluded that the addition of stiffener plates in the design of these compound columns is justified.

References

1. Guoqiang C, Limin J, Liming Z, Yu L, Xi L (2010) Research on rail safety security system. Int J Econ Manage Eng 4(8):1938–1943
2. Choudhary S, Pipralia S (2017) Architectural perception for redevelopment of railway termini. Int J Emerg Technol 8(1):575–585
3. Fitzsimmons E, Fessenden F, Lai KR (2017) Every New York City subway line is getting worse. Here's why. The New York Times, 28 June 2017
4. Crerar P (2017) Revealed: 17 Tube stations that face chronic overcrowding if Crossrail 2 is stopped. Evening Standard, 18 July 2017
5. Press Information Bureau, Government of India, Ministry of Railways. Measures to control and regulate crowd at railway stations
6. Technical staff, Masonry—densities. The Concrete society
7. Takabatake H, Kusumoto S, Inoue T (1991) Lateral buckling behaviour of *I* beams stiffened with stiffeners. J Struct Eng 117(11):3203–3215
8. Building Construction and Design Viewpoint. Density related properties of concrete masonry. 1(1) (2007)
9. Bhat A (2016) Effect of Stiffeners on natural frequencies of plate. IRJET 3(1):1048–1053

10. Serror M, Bahaa M, Hassan A, Sobhy H (2004) Effect of vertical web stiffeners on Lateral Torsional buckling behaviour of Cantilever steel I—beams. J Appl Mech 7(11):233–246
11. Bai Y, Jin W-L (2016) Ultimate strength of plates and stiffened plates. In Bai Y, Jin W-L (eds) Marine structural design, 2nd edn. Butterworth-Heinemann, pp 339–352. ISBN 9780080999975
12. De Queiroz JPTP, Cunha ML, Pavlovic A, Rocha LAO, Dos Santos ED, Troina GS, Isoldi LA (2019) Geometric evaluation of stiffened steel plates subjected to transverse loading for naval and offshore applications. J Mar Sci Eng 7:7
13. Kumar A, Arakerimath R (2015) A review and buckling analysis of stiffened plate. IOSR-JMCE 4(11):192–198
14. Ding X, Yamazaki K (2004) Struct Multidisc Optim 26(1):99
15. Rahmzadeh A, Ghassemieh M, Park Y, Abolmaali A (2016) Effect of stiffeners on steel plate shear wall systems. Steel Compos Struct 20(3):545–569
16. Sherbourne A, Bedair O (1993) Plate stiffener assemblies in uniform compression. Part 2: postbuckling. J Eng Mech 119(10):1956–1972

Drilling Process Quality Improvement by Grey Relation Analysis

Janak Suthar, S. N. Teli, Siddesh Lad, and Vijaya Kumar N. Kottur

Abstract It is very important to take an optimum process parameter to improve the quality of machined products with a maximum tool life and less defects. Production of hole by drilling operation is a key process in composite manufacturing because drilling operation is being performed after cure/solidification process. Poor hole quality leads to rejections up to 60% of all products manufactured. Carbon fibre-reinforced plastic (CFRP) composite have large range of application in aircraft, automotive and medical industries. This work deals with a selection of optimum process parameters in drilling of CFRP to minimize defect, thrust and torque by using grey relational analysis (GRA). After conducting the study using multi-objective GRA, gave optimum process parameters such as spindle speed—3200 rpm, feed rate—0.15 mm/rev., delamination—1.2311, torque—1.483331 N-m and thrust—7.897 N.

Keywords CFRP · GRA · Torque · Thrust

1 Introduction

Composites are most useful advanced materials and widely utilized because of high strength-to-weight ratio, corrosion resistance and adaptability in design. Composites are used in aerospace, automobile, medical and marine industries for reduction of

J. Suthar
Vishwatmak Om Gurudev College of Engineering, Thane, India

S. N. Teli (✉)
Bharati Vidyapeeth College of Engineering, Navi Mumbai, India
e-mail: shivanandteli@yahoo.com

S. Lad
Saraswati College of Engineering, Navi Mumbai, India

V. K. N. Kottur
D. J. Sanghvi College of Engineering, Mumbai, India

© Springer Nature Singapore Pte Ltd. 2020

H. Vasudevan et al. (eds.), *Proceedings of International Conference on Intelligent Manufacturing and Automation*, Lecture Notes in Mechanical Engineering,
https://doi.org/10.1007/978-981-15-4485-9_27

Table 1 Process parameter levels selection by literature review for CFRP

Material	References	Drilling conditions
CFRP	[5, 11–18]	Spindle speed: 800–4000 rpm, Feed rate: 0.10–0.40 mm/rev., Diameter: 4–12 mm, Point angle: 90°–180°, Laminate thickness: 2–10 mm

the weight without affecting strength of the parts [1, 2]. For example, two best commercial Aircraft Boeing 787 and Airbus 380 are used as composite parts for weight reduction [3]. Larger size of composite parts can be made by the assemblies, which require drilling hole on the composite materials. Estimated that 60% of machining parts are rejected due to poor drilling hole quality [4]. Drilling hole quality can be improved by optimizing the process parameters of drilling operation. Tool life can be improved by torque and thrust optimization [5]. By optimizing the parameters like spindle speed, feed rate, helix and point angle of the twist drill and tool material, the defects like delamination can be reduced [6]. Generally, twist drill with a point angle of 120° is used for the reduction in delamination [7–9]. The drilling hole delamination defect can be predicted with the analytical modelling [10]. Table 1 elaborates range of cutting conditions used during drilling on CFRP composite laminates based on the past research.

Mathematical modelling approch used for the predicting delamination (drilling defect) in machining process. Response surface methodology was used for making a model with statistical rules. It showed a good agreement with the experimental data [14, 19] with less than 5% error [20]. Many optimization techniques developed at a global level gave a best solution. Taguchi method is used for the single objective and gave local optima values. Metaheuristic technique is based on the natural phenomena or biological phenomena in the form of mathematics. Like mimic swarm of birds used in PSO, mimic metal annealing process is used in simulated annealing and mimic genetic chromosomes are used in the genetic algorithm. Metaheuristic optimization technique is used for the complex problem for reduction of calculation time [21, 22]. Öktem et al. (2005) used RSM for the modelling and that model was optimized by using genetic algorithm to improve surface finish of machining surface [23]. Palanisamy et al. (2007) used genetic algorithm optimization method to reduce machining time with independent parameter such as spindle speed and feed rate [24]. Spindle speed and feed rate are most critical independent parameter [25]. Optimization of the thrust and torque as dependent parameter is needed to improve tool life [26]. Hence, in this study, independent parameter such as spindle speed and feed rate and dependent parameter such as delamination, torque and thrust for CFRP material were considered. It was not explored before this study.

Table 2 Fibre properties

Fibre properties	Availability
Density (g/cm^3)	1.8
Filament diameter (10^{-6} m)	7
Tensile strength (MPa)	3450
Tensile modulus (GPa)	230
Elongation (%)	1.5

2 Research Methodology

The objective of this study was to optimize independent parameter for delamination, torque and thrust by using grey relation analysis. This research paradigm is positivism. Hypothesis considered was: Independent parameter significantly impacts on dependent parameter and improves drilling quality by optimization of process parameter. The descriptive type of nature is considered for this research (tried for correlation and structured full factorial design used) and quantitative design means data collection in numerical form. Research tool used is grey relation analysis.

3 Experiment Work

3.1 Specimens

The CFRP composite sheet manufactured by hand layup process with fibre orientation of 0°/90° and dimension of 150 × 25 × 3 mm is used in experimentation work. Its content is bidirectional woven, carbon fibre and epoxy. Here, CFRP sheet is made with fibre volume of 50% at 25 °C room temperature having nine laminar layers at dry condition. The CFRP material properties are mentioned in Table 2.

3.2 Drilling Tool

High speed steel twist drill tool with point angle of 120°, diameter 8 mm and 27° helix angle was used in this experimental work.

3.3 Drilling Parameter and Levels

The choice of the drilling parameter and their level based on the literature is mentioned in Table 1 and constrains of the VMC machine. Test run was conducted, which

Fig. 1 Test run

Table 3 Drilling parameters and levels

Level/factor	Spindle speed (rpm)	Feed rate (mm/rev.)
1	800	0.05
2	1600	0.1
3	2400	0.15
4	3200	2.0

is shown in Fig. 1, for checking the feasibility of levels and factors. Spindle speed and feed rate are considered for response variable delamination. The levels and factors for this study are mentioned in Table 3. Data collection methodology is considered based on the full factorial design of experiment. A full factorial design of experiment considered every possible combination for the levels and factors for the making an accurate predicting model.

3.4 Experimental Set-up

In this study, vertical milling center is used for drilling holes in CFRP composite. The workpiece is fixed on bed of the VMC machine by using the fixtures. Kistler dynamometer was used for the torque and thrust measurement. The experimental set-up is shown in Fig. 2.

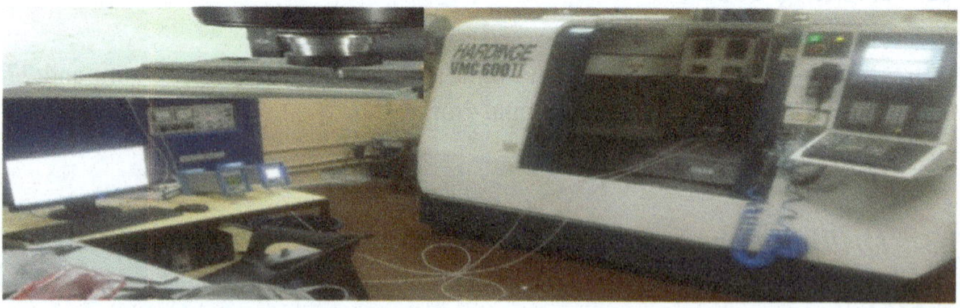

Fig. 2 Experiment set-up

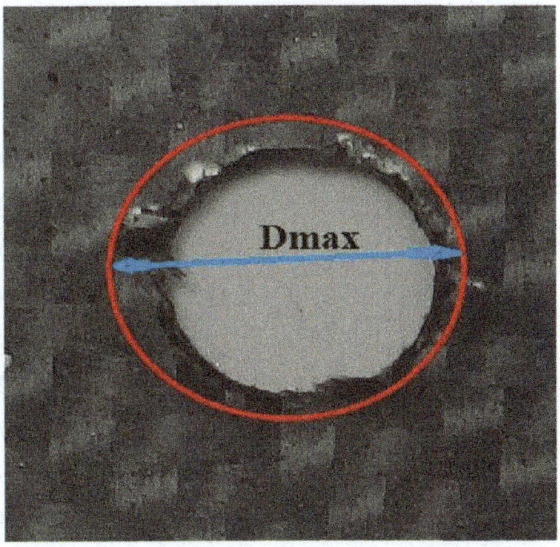

Fig. 3 Photograph illustrating the delamination (drilling defect) around the drilled hole

3.5 *Data Collection by Experiment*

In this study, independent parameters considered were spindle speed and feed rate and dependent parameter considered was delamination. Spindle speed and feed rate were controlled by PLC of VMC machine and collected data as per the full factorial data collection methodology. Delamination was the objective of this study, which was recognized as a most critical drilling defect. Delamination is the ratio of the maximum drilling hole diameter on composite (D_{max}) to the drill tool diameter (D drill) as shown in Fig. 3 [14]. Equation 1 was used for the collection of the delamination [27]. Torque and thrust were measured by the Kistler dynamometer. All the collected data as per the full factorial design methodology were mentioned in Table 4.

$$F_d = (D_{\text{max}})/(D_{\text{drill}}) \tag{1}$$

4 Result Analysis

In multi objective problem, more than one dependent parameter (objective function) which increase complexity in decision making process related to process optimization. In this work, GRA was used as a tool to analyse uncertainty between the independent parameters and optimize the drilling process (shown in Table 5).

By used of grey relation analysis convert multi-objective problem into single objective with single relational grade. The first step of GRA is normalizing there we convert all the data in 0 to 1 range [28]. After that as per the GRA rule we found the

Table 4 Drilling holes data for full factorial design

Ex. No.	Speed (rpm)	Feed (mm/s)	Delamination factor			Torque (N-m)	Thrust (N)
			Entry	Exit	Avg.		
1	800	0.05	1.2031	1.2109	1.2070	3.515542	15.16983
2	800	0.10	1.1797	1.2682	1.2240	4.2823	17.07165
3	800	0.15	1.1817	1.1897	1.1853	4.251746	16.90537
4	800	0.20	1.1769	1.2437	1.2103	3.826533	16.01198
5	1600	0.05	1.1928	1.2799	1.2364	2.948531	11.91906
6	1600	0.10	1.2096	1.3203	1.2650	2.767849	11.92577
7	1600	0.15	1.1719	1.2396	1.2057	2.770394	13.00238
8	1600	0.20	1.1927	1.2813	1.2370	2.588019	11.7699
9	2400	0.05	1.2383	1.2839	1.2611	2.455692	10.62451
10	2400	0.10	1.1979	1.2760	1.2370	1.91739	10.9202
11	2400	0.15	1.2044	1.3151	1.2598	2.163565	12.57398
12	2400	0.20	1.3073	1.3451	1.3262	2.134116	11.8823
13	3200	0.05	1.2969	1.4115	1.3542	2.053138	11.61078
14	3200	0.10	1.1667	1.2945	1.2306	1.915737	12.28054
15	3200	0.15	1.2031	1.2591	1.2311	1.483331	7.897059
16	3200	0.20	1.2383	1.2891	1.2637	1.680845	8.993346

grey relation factor and calculate grey relation grade. GRG show the correlation level between parameters. On the base of the grey relation grade higher to lower values provided ranking. Higher grade of grey relation shows better drilling quality. So, experiment 15 have a highest GRG value 0.883 which is the optimum solution (Bold in Table 5). In this work, according to grey relation analysis, Experiment 15 is the optimum solution. Optimum spindle speed-3200 rpm, feed rate-0.15, delamination-1.2311, torque-1.483331 N m and thrust-7.897 N.

5 Conclusion

In this study, an application of the multi-objective grey relation analysis was used and investigated the effects of drilling parameters (spindle speed and feed rate) on the delamination factor, thrust and torque in dry drilling of CFRP composites. After the experimentation, the following optimum parameters for delamination were observed. Torque and thrust force: spindle speed—3200 rpm, feed rate—0.15, delamination—1.2311, torque—1.483331 N-m and thrust—7.897 N.

Table 5 Grey relation analysis

Grey relation analysis							
Step 1—Normalization			Step 2—Grey relation factor				
Delamination	Torque	Thrust	Delamination	Torque	Thrust	GRG	Rank
0.87152161	0.273943	0.2073	0.12847839	0.726057	0.792708	0.530	13
0.77087034	0	0	0.229129663	1	1	0.451	16
1	0.0109162	0.0181	0	0.989084	0.981876	0.558	9
0.85198342	0.1628339	0.1155	0.148016578	0.837166	0.884499	0.502	15
0.69745411	0.4765215	0.5616	0.302545885	0.523478	0.438385	0.548	10
0.52812315	0.5410746	0.5609	0.47187685	0.458925	0.439116	0.523	14
0.87921847	0.5401653	0.4435	0.120781528	0.459835	0.556463	0.600	5
0.69390172	0.6053232	0.5779	0.306098283	0.394677	0.422127	0.574	7
0.55121374	0.6526003	0.7027	0.448786264	0.3474	0.297283	0.581	6
0.69390172	0.8449218	0.6705	0.306098283	0.155078	0.329512	0.662	3
0.5589106	0.7569698	0.4902	0.441089402	0.24303	0.509769	0.566	8
0.16577857	0.7674912	0.5656	0.834221433	0.232509	0.434378	0.531	12
0	0.7964225	0.5952	1	0.203577	0.404783	0.532	11
0.73179396	0.8455124	0.5222	0.268206039	0.154488	0.477785	0.642	4
0.72883363	**1**	**1**	**0.271166371**	**0**	**0**	**0.883**	**1**
0.53582001	0.9294333	0.8805	0.464179988	0.070567	0.119492	0.734	2

References

1. Adsit NR (1983) In: Compression Testing of homogeneous materials and composites, symposium. ASTM Special Technical Publication, pp 175–186 (1983)
2. Suthar J, Teli SN, Lad S (2017) Review of machining on composite and connected optimisation techniques. Int J Sci Eng Res 8(3):259–261
3. Garnier C, Pastor ML, Eyma F, Lorrain B (2011) The detection of aeronautical defects in situ on composite structures using non destructive testing. Compos Struct 93:1328–1336
4. Abrate S, Walton DA (1992) Machining of composite materials. Part I: traditional methods. Compos. Manuf. 3:75–83
5. Lin SC, Chen IK (1996) Drilling carbon fiber-reinforced composite material at high speed. Wear 194:156–162
6. Turki Y, Habak M, Velasco R, Laurent JN, Vantomme P (2013) An experimental study of drilling parameters effect on composite carbon/epoxy damage. Key Eng Mater 554–557:2038–2046
7. Singh D, Venkateswara Rao P (2007) Optimization of tool geometry and cutting parameters for hard turning. Mater Manuf Process 22:15–21
8. Panchagnula KK, Palaniyandi K (2018) Drilling on fiber reinforced polymer/nanopolymer composite laminates: a review. J Mater Res Technol 7:180–189
9. Davim JP, Reis P, António CC (2004) Experimental study of drilling glass fiber reinforced plastics (GFRP) manufactured by hand lay-up. Compos Sci Technol 64:289–297
10. Tsao CC, Hocheng H (2003) The effect of chisel length and associated pilot hole on delamination when drilling composite materials. Int J Mach Tools Manuf 43:1087–1092
11. Bhatnagar N, Ramakrishnan N, Naik NK, Komanduri R (1995) On the machining of fiber reinforced plastic (FRP) composite laminates. Int J Mach Tools Manuf 35:701–716

12. Rahman M, Ramakrishna S, Prakash JRS, Tan DCG (1999) Machinability study of carbon fiber reinforced composite. J Mater Process Technol 89–90:292–297
13. Piquet R, Ferret B, Lachaud F, Swider P (2000) Experimental analysis of drilling damage in thin carbon/epoxy plate using special drills. Compos Part A Appl Sci Manuf 31:1107–1115
14. Davim JP (2009) Book Review: machining of polymer composites by Jamal Y Sheikh-Ahmad. Int J Mach Mach Mater 6:322
15. Faraz A, Biermann D, Weinert K (2009) Cutting edge rounding: An innovative tool wear criterion in drilling CFRP composite laminates. Int J Mach Tools Manuf 49:1185–1196
16. Shyha IS, Aspinwall DK, Soo SL, Bradley S (2009) Drill geometry and operating effects when cutting small diameter holes in CFRP. Int J Mach Tools Manuf 49:1008–1014
17. Lazar MB, Xirouchakis P (2011) Experimental analysis of drilling fiber reinforced composites. Int J Mach Tools Manuf 51:937–946
18. Krishnamoorthy A, Rajendra Boopathy S, Palanikumar K, Paulo Davim J (2012) Application of grey fuzzy logic for the optimization of drilling parameters for CFRP composites with multiple performance characteristics. Meas J Int Meas Confed 45:1286–1296
19. Gaitonde VN, Karnik SR, Rubio JC, Correia AE, Abrão AM, Davim JP (2008) Analysis of parametric influence on delamination in high-speed drilling of carbon fiber reinforced plastic composites. J Mater Process Technol 203:431–438
20. Suthar J, Teli SN, Lad S (2018) Application of response surface methodology in drilling of carbon fiber application of response surface methodology in drilling of carbon fiber reinforced polymer composite (CFRP). Int J Sci Eng Res 9:347–350
21. Bhowmik S, Jagadis KG (2019) Modeling and optimization of advanced manufacturing processes. Springer International Publishing, Berlin
22. Bhushi U, Suthar J., Teli SN (2020) Performance analysis of metaheuristics optimization techniques for drilling process on CFRP composites. Mat Today Proc
23. Öktem H, Erzurumlu T, Kurtaran H (2005) Application of response surface methodology in the optimization of cutting conditions for surface roughness. J Mater Process Technol 170:11–16
24. Palanisamy P, Rajendran I, Shanmugasundaram S (2007) Optimization of machining parameters using genetic algorithm and experimental validation for end-milling operations. Int J Adv Manuf Technol 32:644–655
25. Callister WD Jr (2000) Materials science and engineering—an introduction, 5th edn. In: Anticorrosion methods, materials, vol 47
26. Kyratsis P, Markopoulos AP, Efkolidis N, Maliagkas V, Kakoulis K (2018) Prediction of thrust force and cutting torque in drilling based on the response surface methodology. Machines 6:24
27. Arul S, Vijayaraghavan L, Malhotra SK, Krishnamurthy R (2006) The effect of vibratory drilling on hole quality in polymeric composites. Int J Mach Tools Manuf 46:252–259
28. Haq AN, Kannan G (2006) An integrated approach for selecting a vendor using grey relational analysis. Int J Inf Technol Decis Mak 5:277–295

Enabling Technologies and Current Research Scenario of Industry 4.0: A Systematic Review

Bhaveshkumar N. Pasi, Subhash K. Mahajan, and Santosh B. Rane

Abstract This research was carried out to review the Industry 4.0 enabling technologies. This study incorporates the recently published research articles on Industry 4.0 enabling technologies and applications of these enabling technologies. To perform this research work, bibliometric analysis was carried out. Further, content analysis was used to identify the various domains within the literature on smart manufacturing. These techniques were used to identify the current trends, various research works and the contribution of researchers in Industry 4.0 domain. It was observed that many research articles on Industry 4.0 are theoretical in nature which did not give a strong benchmark for policymakers. It was found that Germany was the leading country in the publication of research articles on Industry 4.0 followed by USA, UK, and China.

Keywords Smart manufacturing · Industry 4.0 · Bibliometric analysis

1 Introduction

Industry 4.0 concept enables the manufacturer to achieve the maximum operational efficiency of manufacturing plants with the minimum resource utilization [1, 2]. The concept of industry 4.0 was developed in Germany which deals with the development of the smart manufacturing system [3, 4]. Industry 4.0 allows manufacturing with

B. N. Pasi (✉) · S. B. Rane
Department of Mechanical Engineering, Sardar Patel College of Engineering, Andheri, Mumbai, India
e-mail: bnpasi@vishwaniketan.edu.in

S. B. Rane
e-mail: s_rane@spce.ac.in

B. N. Pasi
Department of Mechanical Engineering, ViMEET, Khalapur, India

S. K. Mahajan
Directorate of Technical Education, Mumbai, India
e-mail: skmahajan@dtemaharashtra.gov.in

© Springer Nature Singapore Pte Ltd. 2020
H. Vasudevan et al. (eds.), *Proceedings of International Conference on Intelligent Manufacturing and Automation*, Lecture Notes in Mechanical Engineering,
https://doi.org/10.1007/978-981-15-4485-9_28

information technology [5, 6]. Industry 4.0 integrates the cloud computing, Internet of Things (IoT), and cyber-physical systems (CPS) with an aim to develop a highly efficient production plant [7, 8]. Out of four industrial revolutions, Industry 1.0 deals with the use of water and vapor powers to develop the mechanical production facilities [9, 10]. During the second industrial revolution, electrical energy generation was accelerated which helped to achieve mass production. The third industrial revolution introduced advanced computer technologies that furthered the production automation [11, 12]. Few years ago, technology framework was proposed (which is named as Industry 4.0) to integrate and extend the manufacturing processes [12, 13]. Industry 4.0 concept sounds best at first, but it has multiple challenges such as structural resistance, cyber security, cultural resistance, and training to employees that are to be overcome before implementing Industry 4.0 [14]. Industry 4.0 includes technologies such as rapid prototyping, big data analytics, cloud computing, IoT, and machine learning [15]. It has been found that the potentials of these technologies are not fully utilized. Therefore, it is important to review the current status of research on enabling technologies of Industry 4.0.

2 Literature Review Method

All research papers considered in this research work are in English language. Table 1 shows the inclusion and exclusion criteria of research articles.

In this research article, electronic databases such as SCOPUS, Science Direct, and Web of Sciences were used. Only those research papers are considered which were

Table 1 Inclusion and exclusion criteria of research articles

Criteria	Reason	Explanation
Exclusion criteria	Absent of full-length articles	Research articles in which only abstract and keywords are given but full-text is not given
	Articles which are not closely related	Industry 4.0 is just utilized/referenced for instance
		Industry 4.0 is just referenced in catchphrases
		Industry 4.0 is only used as a tool for future scope without giving strong evidence
Inclusion criteria	Closely related to Industry 4.0	Outcomes of papers are fully devoted to Industry 4.0
	Partially related to Industry 4.0	Industry 4.0 is utilized as an instrument to solve the difficulties that research articles aim to deal with

submitted between March 2013 and March 2019. 164 research papers were included for systematic literature survey. It has been found that 87%, i.e., 143, research articles were found in the SCOPUS reference electronic database. Remaining research articles such as 8%, i.e., 13, research papers were from Science Direct electronic database, and 5%, i.e., 08, research papers were from the Web of Sciences electronic database.

3 Research Article Publications Summary

In order to increase the productivity and utilization of the resources, Industry 4.0 has been emerged. Many researchers have published their research articles and case studies to explore Industry 4.0 concept. After 2014, considerable growth in the publication of research articles on Industry 4.0 has been observed. It is determined that in the coming years, there will be more research on Industry 4.0 Fig. 1 shows Industry 4.0 related articles year-wise.

3.1 Distribution of Publications Across Different Countries

Figure 2 shows country-wise distribution of Industry 4.0 research papers. It has been found that Germany is topping the list by publishing 87 research papers out of 164 research articles on Industry 4.0.

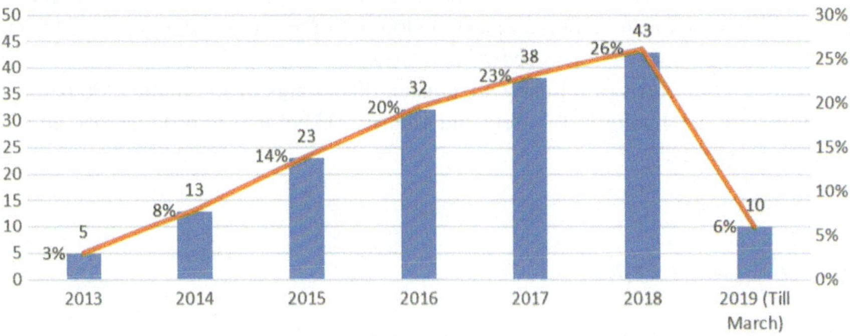

Fig. 1 Industry 4.0 related articles year-wise

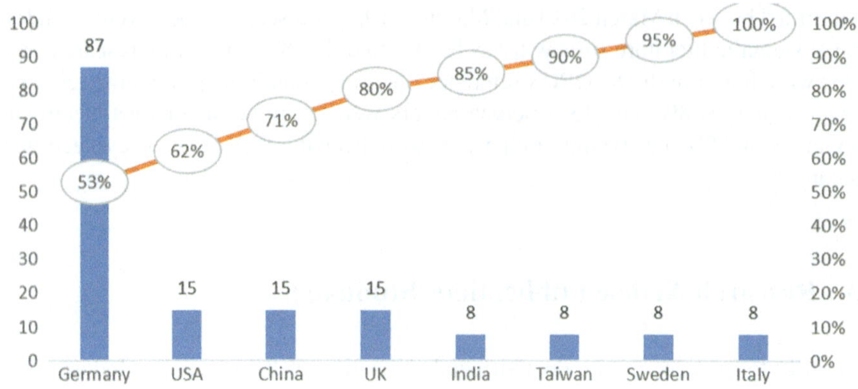

Fig. 2 Country-wise distribution of Industry 4.0 research papers

3.2 Paper Keywords Analysis

In this research article, keyword analysis was carried out to understand the different research areas related to Industry 4.0. In this research article, a total of 87 keywords were considered. Out of 87 keywords, 41 were for enabling technologies cluster, 23 were for smart manufacturing cluster, and 23 were for Industry 4.0 cluster.

Out of 41 keywords of enabling technologies cluster, 12 belongs to big data analytics, 09 belongs to cloud technology, 07 belongs to decision making technology, 05 belongs to machine learning technology, 04 belongs to data mining technology, and 04 belongs to 3D printing technology.

Out of 23 keywords of smart manufacturing cluster, 11 belongs to the automated robot, 05 belongs to machine tools, 04 belongs to the industrial Internet network, and 03 belongs to the wireless sensor networks.

Out of 23 keywords of Industry 4.0 cluster, 08 belongs to automation, 06 belongs to optimization, 04 belongs to safety and security, 03 belongs to productivity, and 02 belongs to quality.

3.3 Research Type

Research outcomes of already published articles are divided into the various categories such as theoretical, framework development, case study solution, literature review, models, and empirical relations development. 30% of research articles on Industry 4.0 are theoretical type, whereas 29% of research articles are of literature review type. 21% of research articles are based on Industry 4.0 case study, whereas 9% of research articles are on framework development. 8% and 3% of research articles are based on Industry 4.0 models and empirical relations development, respectively.

3.4 Journal-Wise Distribution of Research Articles

Out of 164 included research articles, 15 were from energies journal, 16 were from computers and industrial engineering journal, 11 were from process safety and environmental protection, 8 were from journal of manufacturing technology management, 9 were from human resource management international digest, 8 were from journal of intelligent manufacturing, 11 were from sustainability journal, 22 were from international journal of production research, 28 were from computers in industry journal, 22 were from advance manufacturing journal, and 14 were from journal of industrial and production engineering.

3.5 Industry 4.0 Enabling Technologies Research Articles Categorizations

Categorizations of 164 research articles were done in three different categories, i.e., engineering, computer science, and others. It has been found that 56 (34%) research articles were published in the engineering category, 77 (47%) research articles were published in the computer science category, and 31 (19%) research articles were published in the others category.

In computer science research article categorizations, it has been found that out of 77 research articles, 3 (4%) research articles were based on pattern recognition, 11 (14%) research articles were based on information systems, 13 (17%) research articles were based on hardware and architecture, 4 (5%) research articles were based on artificial intelligence, 23 (30%) research articles were based on software, 20 (56%) research articles were based on computer networks and communications, and 3 (4%) research articles were based on computer science applications.

In engineering research articles categorizations, it has been found that out of 56 research articles, 25 (45%) research articles were based on industrial and manufacturing engineering, 10 (17%) research articles were based on control and systems engineering, 4 (8%) research articles were based on aerospace engineering, 15 (26%) research articles were based on electrical and electronics engineering, and 2 (4%) research articles belongs to miscellaneous category.

In other research articles categorizations, it has been found that out of 31 research articles, 4 (13%) research articles were based on business management, 3 (10%) research articles were based on social sciences, 8 (26%) research articles were based on decision sciences, 5 (15%) research papers belongs to energy sector, 4 (13%) research papers belongs to medicine field, 3 (10%) research articles were based on astronomy, and 4 (13%) research articles were based on biochemistry.

4 Research Gaps

After a review of the published research articles, various research gaps were identified in them and presented in this research article. These are as follows:

- Till date, the universally acceptable definition of Industry 4.0 is not defined. Available definition of Industry 4.0 misleads the manufacturer. The available few definitions define Industry 4.0 as a tool which aims at the design and development of a fully automatic manufacturing process. But, there are few definitions which define Industry 4.0 as a tool which aims at the design and development of real-time data-driven manufacturing process.
- Outcomes of many research articles are theoretically based. Unavailability of strong empirical data is the main reason for less adoption of Industry 4.0 in the manufacturing sector. Therefore, there is a need of development of more conceptual methodology.
- Research articles on big data analytics in the manufacturing sector are based on philosophies. Big data technology is considered as new technology in the manufacturing sector.
- Till date, the standard framework model for 3D printing technology in small-scale manufacturing sector is not available. Therefore, the benefits of 3D printing technology are being only experienced by the large-scale manufacturing sector.
- Research articles on the use of virtual reality and application of cyber security in the manufacturing sector are published without rigorous data validation.

5 Discussions

From Fig. 2, it is found that many countries such as China, India, Taiwan, Sweden, Italy, UK, USA, and Germany have published research articles based on Industry 4.0. The contribution of Germany in the research area of Industry 4.0 is very high. Germany has published 87 research articles from January 2013 to March 2019 based on Industry 4.0 research. China is far ahead in Industry 4.0 research across Asia. China has published 15 research articles from January 2013 to March 2019.

Industry performances can be enhanced by integrating Industry 4.0 enabling technologies. These enabling technologies have the ability to optimize the processes of any industry. Figure 3 shows a proposed model with four different levels for the successful implementation of enabling technologies of Industry 4.0.

As shown in Fig. 3, this is the first level of Industry 4.0 implementation strategy. This level includes the digitization of every division of the industry. In order to implement Industry 4.0 enabling technologies effectively, every industry is advised to achieve the digitization of every process in every division.

The second level of Industry 4.0 implementation strategy deals with the integration of software and develop real-time data-driven manufacturing system. Rapid growth

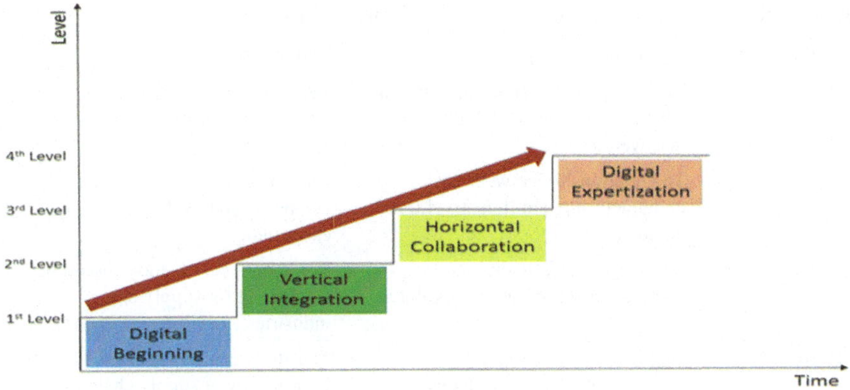

Fig. 3 Industry 4.0 implementation strategy: a proposed model

in Internet-based technology allows the manufacturer to track the manufacturing process in real time. This level allows the manufacturer to use embedded system software.

The third level of Industry 4.0 implementation strategy deals with the integration of customers, suppliers, and vendors with the production line. This level allows the manufacturer to integrate value chain with different stakeholders. This level helps the manufacturer to enhance the quality of work, to reduce the operation costs and to minimize rejection rates.

After effective execution of three levels, any industry can taste the productive after-effects of Industry 4.0 empowering innovations. Fourth, i.e., last, stage of implementation strategy demands an implementer of Industry 4.0 enabling technologies to maintain the environment of digitization in the entire business model and also update processes from time to time.

6 Conclusions

A detailed review of already published 164 research articles on Industry 4.0 (from January 2013 to March 2019) was carried out in this research article. This research article focused on the published articles with respect to type of research used in Industry 4.0, the contribution of various countries and authors in Industry 4.0. According to the substance-based classifications, 28 (17%) of the included research articles are delegated review or survey papers. For the rest, 72 (44%) research articles fall into the discussion classification, 54 (33%) research articles are delegated theoretical solutions, and 10 (6%) research articles have a place in the practical solution class. Few research articles are summarized, and available future research opportunities are discussed in Table 2.

Table 2 Research articles summary and available future research

Authors	Contribution	Future research
Bodrow [4]	The reliance of various empowering advances of Industry 4.0 is discovered	Lack of empirical data is the main drawback of this research article
Buer [5]	The relationship between lean manufacturing and Industry 4.0 is obtained	The outcome of this research article can be strongly validated by showing case studies
Fettermann [6]	Industry 4.0 effects on manufacturing sectors are explored	The same kind of study should be carried out for supply chain industries
Hofmann [9]	Application of Industry 4.0 technologies in supply chain management is suggested	The study can be validated by taking a case study of supply chain management sector
Kong [11]	Proposed a human–cyber–physical framework to support dynamic interaction among production systems	Developed framework can be validated by taking a case study
Li [12]	Industry 4.0 technologies are used for predictive maintenance at work centers	The study can be explored for real-time fault diagnosis and preventive maintenance
Liao [13]	The current status of Industry 4.0 is reviewed	Architecture diagram should be developed for different sectors
Longo [15]	SOPHOS-MS is developed to improve man–machine interactions	The suggested SOPHOS-MS model can be validated by taking a case study of manufacturing line

References

1. Alexopoulos K, Makris S, Xanthakis V, Sipsas K, Chryssolouris G (2016) A concept for context-aware computing in manufacturing: the white goods case. Int J Comput Integr Manuf 29(8):839–849
2. Atzori L, Iera A, Morabito G (2010) The internet of things: a survey. Comput Netw 54(15):2787–2805
3. Brettel M, Friederichsen N, Keller M, Rosenberg M (2014) How virtualization, decentralization and network building change the manufacturing landscape: an industry 4.0 perspective. Int J Mech Aerosp Ind Mechatron Eng 8(1):37–44
4. Bodrow W (2017) Impact of Industry 4.0 in service oriented firm. Adv Manuf 5(4):394–400
5. Buer SV, Strandhagen JO, Chan FTS (2018) The link between Industry 4.0 and lean manufacturing: mapping current research and establishing a research agenda. Int J Prod Res 56(8):2924–2940
6. Fettermann DC, Cavalcante CGS, de Almeida TD, Tortorella GL (2018) How does Industry 4.0 contribute to operations management? J Ind Prod Eng 35(4):255–268
7. Ghobakhloo M (2018) The future of manufacturing industry: a strategic roadmap toward Industry 4.0. J Manuf Technol Manage 29(6):910–936
8. Gonzalez AGC, Alves MVS, Viana GS, Carvalho LK, Basilio JC (2018) Supervisory control-based navigation architecture: a new framework for autonomous robots in industry 4.0 environments. IEEE Trans Ind Inf 14(4):1732–1743

9. Hofmann E, Rüsch M (2017) Industry 4.0 and the current status as well as future prospects on logistics. Comput Ind 89:23–34
10. Kamble SS, Gunasekaran A, Gawankar SA (2018) Sustainable Industry 4.0 framework: a systematic literature review identifying the current trends and future perspectives. Process Saf Environ Prot 117:408–425
11. Kong XTR, Luo H, Huang GQ, Yang X (2018) Industrial wearable system: the human-centric empowering technology in Industry 4.0. J Intell Manuf 29(3):1–17
12. Li Z, Wang Y, Wang KS (2017) Intelligent predictive maintenance for fault diagnosis and prognosis in machine centers: industry 4.0 scenario. Adv Manuf 5(4):377–387
13. Liao Y, Deschamps F, Loures E de FR, Ramos LFP (2017) Past, present and future of Industry 4.0—a systematic literature review and research agenda proposal. Int J Prod Res 55(12):3609–3629
14. Lin KC, Shyu JZ, Ding K (2017) A cross-strait comparison of innovation policy under industry 4.0 and sustainability development transition. Sustainability (Switzerland) 9(5):786
15. Longo F, Nicoletti L, Padovano A (2017) Smart operators in industry 4.0: a human-centered approach to enhance operators' capabilities and competencies within the new smart factory context. Comput Ind Eng 113:144–159

Evaluation of Green Supply Chain Management Practices in Small and Medium Enterprises in Pune Region

Malleshappa T. Bhagawati and P. Venkumar

Abstract . This paper presents the study on GSCM practices in SMEs in Pune region (India). The components for the study (survey) were based on the literature review and also from the feedback received from the academicians and executives from the corporate world. From the survey, it was found that the adoption of GSCM practices in SMEs was still in the early stage. The environmental sustainability awareness was very low among the customers. The legislation and regulatory framework is also lagging in promoting the sustainability of the environment. Data analysis results revealed that vendor–vendee relationship has positive impact on sustainability of product design and the logistics, which further was positively related to economic performance and competitiveness of the organization. This study was conducted with 120 manufacturing industries under SMEs category. AMOS, a statistical software, which stands for Analysis of Moment of Structures is used for structural equation modeling and path analysis, and confirmatory factor analysis was used to measure the overall respondents view on sustainable supply chain management.

Keywords Green supply chain management (GSCM) · Environmental sustainability · SMEs · Structural equation modeling (SEM)

1 Introduction

The supply chain is a chain of activities which starts from customers and end with customers. Through the chain flows all the raw materials and finished products, information and all the transaction processes. Handfield and Nicholas [1] define supply chain and SCM as the Supply Chain encompasses all the activities associated with the flow and transformation of goods from the raw material stage (extraction), through to the end user, as well as the associated information flows. Material and information flow both up and down the supply chain. Supply chain management (SCM) is

M. T. Bhagawati (✉) · P. Venkumar
Department of Mechanical Engineering, Kalasalingam University, Krishnankoil, Virudhunagar, Tamil Nadu, India
e-mail: malleshbhagawati@gmail.com

© Springer Nature Singapore Pte Ltd. 2020

H. Vasudevan et al. (eds.), *Proceedings of International Conference on Intelligent Manufacturing and Automation*, Lecture Notes in Mechanical Engineering,
https://doi.org/10.1007/978-981-15-4485-9_29

the integration of these activities through improved supply chain relationships, to achieve a sustainable competitive advantage. Now, the supply chain includes collaboration from customers, agents, and from suppliers. With government legislations and regulations and customer's awareness, industries have started the integration of environmental factors throughout their firms. Now, the industries have started shifting toward eco-friendly supply chains by the integration of green technologies into their production process activities through product design and development, manufacturing, and distribution processes. All these together led to the incorporation of Extended Production Responsibility (EPR), which further led to the evolution of green supply chain management (GSCM) [2].

The strategy of green supply chain management is to minimize the adverse effect of the firm and their supply chain network on the environment with respect to change of climate, pollution control, and non-renewable resource constraints. Thus, GSCM integrates supply chain management activities for reducing the environmental impact during the whole product life cycle management [3]. The objective of the GSCM is to improve the flow of material by adding value and controlling the material flow, capital, information, and work. Also it includes providing good quality products or services to the customers at the lowest possible cost and environmental impact [4]. Three issues are important to the environmental management and sustainability. They are as follows [5]:

(i) Inclusion of environmental aspect in an integrated supply chain management for the production of goods.
(ii) Integration of innovations of modern technology for the ecologically beneficial outcomes throughout the supply chain activity.
(iii) The larger participation of industrial actors for the environmental management of industrial production, in order to strengthen the capacity building on environment management governance.

Review of extant literature on green supply chain management has shown that there is an important link between supply chain management decisions and green supply chain [6]. For some of the firms, these new greener perspective will be in more transformative, which leads to fresh thoughts, newer markets, with profitability and increased values. And for some others, the focus on environmental aspect will emerge gradually and modestly as their critical element of their corporate strategy [7]. With the passing of time, these companies may find long-term sustainable benefit for being greener but not overnight gains. For most of the large-scale and heavy industries, these gains are very closer to be assured. But very surprisingly, the smaller organizations also found the benefits [8]. The framework of a green supply chain with forward and reverse activities is shown in Fig. 1.

Figure 1 explains the forward as well as reverse activities of green supply chain management. The forward activities include purchasing, engineering (Design), production (Manufacturing), and product distribution to the end-user (Distribution). The backward activities include inspection, sorting, and disassembly for reprocessing, reusing, and redesigning which is done mainly for the demand in primary as well as secondary markets.

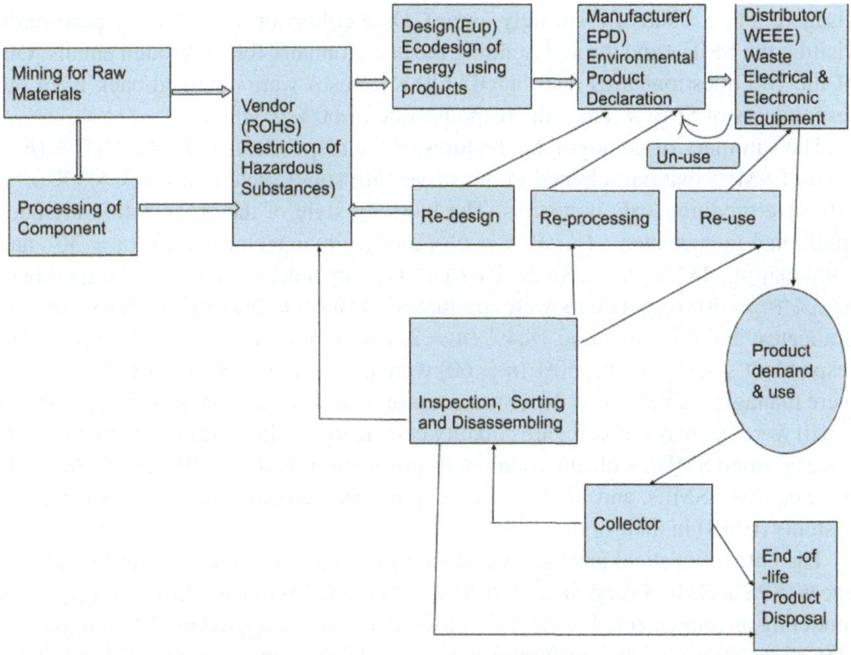

Fig. 1 Framework of a green supply chain with forward and reverse activities with their directives. *Source* Hsiao-Fan Wang and Surendra M Gupta, Green Supply Chain Management, Product Life Cycle Approach

2 Materials and Methods

The sample size of the study is determined by using the formula (Slovin's formula) given below (1). Slovin's formula is used when nothing about the behavior of a population is known at all.

$$n = N/\left(1 + Ne^2\right) \tag{1}$$

where n—Sample size, N—Population size = 500, e—Error of acceptance = 8% Power = 80%, Type I error (level of significance) = 5%.

According to Slovin's formula, the minimum sample size to be studied in the study is 120. Stratified random sampling ($n = 120$) was carried out to collect data from small and medium enterprises (SME's) of Pune region in Maharashtra state (India). Out of these, 110 industries are small scale and ten industries are medium scale. All the factors of green supply chain management such as legislation and regulation (LR), green purchasing (GP), eco design (ED), green logistics (GL), reverse logistics (RL), environmental performance (EP), and supply chain performance (SCP) are given due consideration in the questionnaire. All the variables will be measured by the participant's responses based on the five-point Likert- scales (1 stand for strongly disagree and 5 stands for strongly agree). Data collection was done by personally visiting the SME's and through sending the questionnaire forms through emails. Out of the 200 questionnaires distributed, 120 responses were obtained back from the respondents of SME's. Thus, the response rate of 60% is achieved.

The summary of demographic features of the respondents indicates 91.7% (n = 110) of SMEs that participated in the survey are small-scale industries, 8.3% (n = 10) were medium-scale industries. The working style of the organization includes traditional management, 21.7% ($n = 66$), modern management, 30% ($n = 36$), and combination, 48.3% ($n = 38$). 28.3% ($n = 34$) respondents were diploma qualified, 33.3% ($n = 40$) respondents were graduates, 20% ($n = 24$) respondents were post graduates, and the remaining 18.4% ($n = 22$) were other categories. With regard to respondent's designations, 50% ($n = 60$) were the owners of SMEs, 18.3% ($n = 22$) were managers, 23.3% ($n = 28\%$) were senior managers, and the remaining 8.4% (n = 10) were in charges. As regards quality certifications (ISO-9000), 61.7% ($n = 74$) were certified SMEs, with environmental certifications (ISO-14001), 20.8% ($n = 25$) were certified SMEs, and 17.5% ($n = 21$) implemented environmental management systems (EMS) in their organizations.

The entire statistical analysis was done using Statistical Package for Social Sciences (SPSS, AMOS version 22.0, IBM, USA) for MS Windows. Structural equation modeling as implemented in AMOS is used to test the proposed model on impact of GSCM on environmental performance and supply chain performance (EP and SCP).

2.1 Hypothesis Setting

The hypotheses are formulated based on the objectives of the study:

H1: Medium-scale enterprises are more likely to adopt green purchasing (GP) practices than small-scale enterprise. H2: There is a significant difference between mean ranks of dimensions of GSCM (LR, GP, ED, GL, and RL). H3: Legislation and regulation (LR) is positively related to environmental performance (EP). H4: Green purchasing (GP) practices are positively related to environmental performance (EP). H5: Green logistics (GL) is positively related to environmental performance (EP). H6: Reverse logistics (RL) is positively related to environmental performance (EP). H7: Eco-friendly design (ED) collaboration in SME's is positively related to environmental performance (EP). H8: There is an interrelationship between various dimensions of GSCM (LR, GP, ED, GL, and RL), environmental performance (EP) and supply chain performance (SCP) (Fig. 2).

Structural equation modeling (SEM) is one of the multivariate statistical modeling tools which are widely used to investigate the independent determinants of the outcome variables of interest. Its estimation techniques, modeling capacities, and spectrum of application are expanding rapidly day by day. It is also used to study the relationships among latent constructs indicated by multiple measures. It is also applicable to experimental and non-experimental data, as well as cross-sectional and

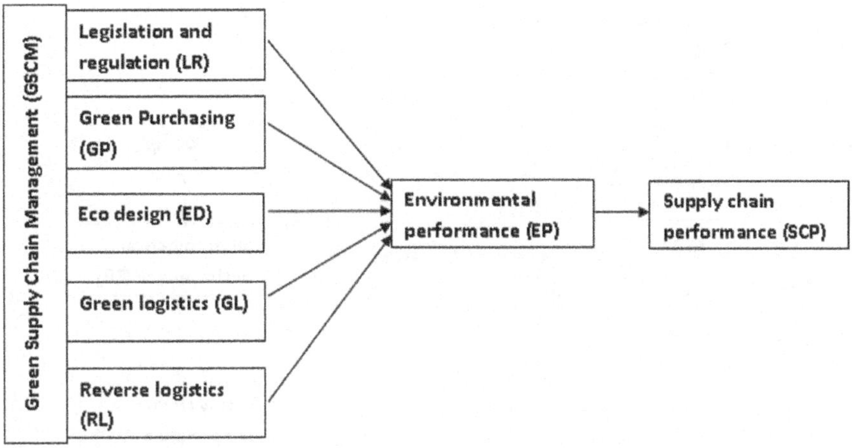

Fig. 2 Conceptual model for components of green supply chain management (GSCM) (LR, GP, ED, GL, and RL), environmental performance and supply chain performance

longitudinal data. In general, SEM analysis goes through the steps of model specification, data collection, model estimation, model evaluation, and possible model modification.

3 Results and Discussions

From Fig. 3 and Table 1, the variables used for structural equation model are given below:

Observed or endogenous variables	Unobserved, exogeneous variables
Legislation and Regulation (LR)	e1: Error term for Legislation and Regulation (LR)
Green Purchasing (GP)	e2: Error term for Green Purchasing (GP)
Eco-friendly design (ED)	e3: Error term for Eco-friendly design(ED)
Green logistics (GL)	e4: Error term for Green Logistics (GL)
Reverse logistics (RL)	e5: Error term for Reverse Logistics (RL)
Environmental Performance (EP)	e6: Error term for Environmental Performance
Supply chain performance (SCP)	e7: Error term for SCP (Supply chain Performance)
	GSCM

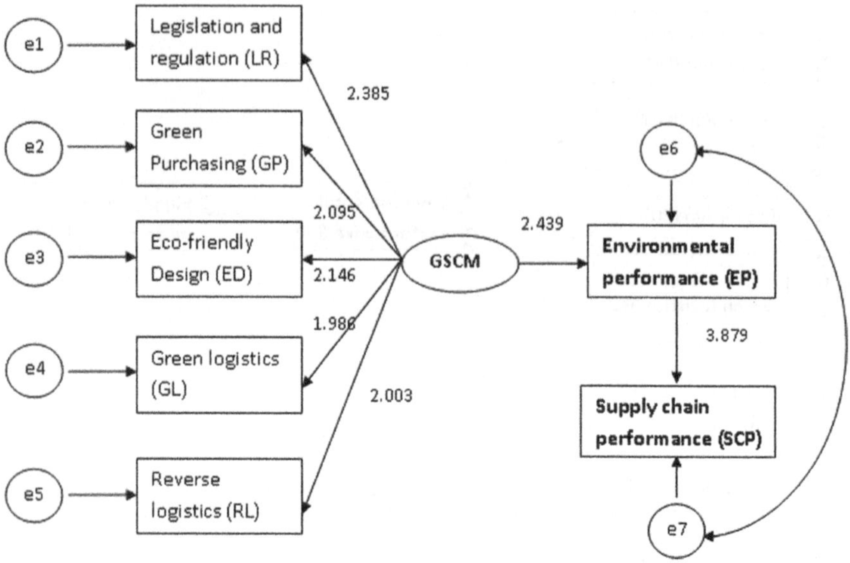

Fig. 3 Structural equation modeling on the impact of GSCM on EP and SCP

Table 1 Distribution of statistical indices through structural equation model (SEM) analysis

Variables	Direction of impact		Un-standardized co-efficient	S.E	Standardized coefficient	*T*-value	*P*-value
Legislation and regulation (LR)	←	GSCM	2.385	0.076	0.814	15.865	0.001***
Green purchasing (GP)	←	GSCM	2.095	0.082	0.716	14.826	0.001***
Eco design (ED)	←	GSCM	2.146	0.077	0.795	15.123	0.001***
Green logistics (GL)	←	GSCM	1.986	0.085	0.703	14.463	0.001***
Reverse logistics (RL)	←	GSCM	2.003	0.089	0.709	14.736	0.001***
Environmental performance (EP)	←	GSCM	2.439	0.088	0.711	15.846	0.001***
Supply chain performance (SCP)	←	EP	3.879	0.074	0.899	18.911	0.001***

Note ***denotes significance *P*-value < 0.001 (Highly significant)

Variable details

Number of variable in the model: 15	Number of observed Variables: 7
Number of unobserved variables: 8	Number of Exogenous variables: 8
Number of Endogenous variables: 7	

In this study, a SEM model was tested and developed by exploring the relationship between GSCM, EP, and SCP. The absolute fit of the dataset for the model test confirms that the model is stable. From the model, it is found that LR, GP, ED, and RL play an important role during the implementation of GSCM with the improvement in EP and SCP.

From Table 1 and Fig. 3, the below interpretation has been made.

For legislations and regulations (LR); the unstandardized coefficient of GSCM is 2.385 which indicates that there will be little or partial effect of LR on GSCM, and the other variables are constant (fixed or adjusted.) The positive sign indicates positive effect. There will be increase of LR score by 2.385 which means that LR increases by 2.385 for each increase in GSCM score. The value of co-efficient is significant at 5% level.

For green purchasing (GP), the un-standardized coefficient of GSCM is 2.095, For eco-friendly design (ED), the un-standardized coefficient of GSCM is 2.146, For green logistics (GL), the un-standardized coefficient of GSCM is 1.986, For reverse logistics(RL), the un-standardized coefficient of GSCM is 2.003, For environmental performance(EP), the un-standardized coefficient of GSCM is 2.439, and finally, for supply chain performance(SCP), the un-standardized coefficient of EP is 3.879.

From the structural equation model (SEM) analysis, it is seen that the unstandardized coefficient of environmental performance (EP) is maximum and indicates some partial effect of EP on SCP. And other variables are constant (fixed or adjusted). The positive sign indicates that it has positive effect and the SCP would increase by 3.879 for each increase in EP score.

From Table 2, it is found that the calculated P value is 0.209, which is greater than 0.05; this indicates a perfectly fit model. Goodness of fit (GFI) value and adjusted goodness of fit (AGFI) value are greater than 0.90; this represents a good fit to the sample data.

Table 2 Structural equation model (SEM) fit summary

Variable	Value	Suggested values
Chi-square	42.913	–
P value	0.209	>0.05
GFI	0.957	>0.90
AGFI	0.933	>0.90
CFI	0.969	>0.90
RMR	0.044	<0.08
RMSEA	0.033	<0.08

The calculated comparative fit index (CFI) value is 0.969, which means that it is perfectly fitted, and that root mean square residual (RMR) and root mean square error of approximation (RMSEA) values are less than 0.08 also indicating the perfect fit [9–13].

Decision based on statistical analysis Null hypothesis (H0) is rejected and alternative hypothesis (H1) is accepted at 5% level of significance (LOS).

4 Conclusion

Among the various dimensions of GSCM, the most influencing dimension is LR, followed by ED, GP, RL, and GL. Green logistics has the minimum influence on GSCM. From this study, one can say that green supply chain management influences environmental performance which in turn has its impact on supply chain performance. The results from this study are helpful for manufacturing organizations for identifying an effective approach toward successful green supply chain management practices in small- and medium-scale enterprises (SMEs).

The GSCM practices are mostly dominated by positively influencing some of the production system performances, particularly with the environmental compliances. The relationship between supply chain performance and the environmental performance in manufacturing sectors of SMEs was being evaluated from this study. The study further concludes that an environmental performance in manufacturing industries (SMEs) in Pune region positively influences the supply chain performances. For future researchers, this study acts a reference source on green supply chain management for SMEs. As the sample size is small ($n = 120$), it is very difficult to draw the optimistic conclusion. A bigger sample size will be helpful for a detailed cross-sectoral comparison.

References

1. Handfield RB, Nicholas EL Jr (1999) Introduction to supply chain management. Prentice-Hall, New Jersey
2. Wang H-F, Gupta SM (2011) Green supply chain management: product life cycle approach. Mc Graw-Hill, New York
3. Sarkis J (2006) Greening the supply chains. Springer London Limited, UK
4. Rao P (2002) Greening the supply chain: a new initiative in South East Asia. Int J Oper Prod Manage 22(6):633–655
5. Berger G, Flynn A, Hines F, Jones R (2001) Ecological modernization as a basis for environmental policy and implications on environmental supply chain management. Innovation 14(1):55–72
6. Seuring S, Muller M (2008) From a literature review to a conceptual framework for sustainable supply chain management. J Cleaner Prod 16(15):1699–1710

7. Gao Y, Li J, Song Y (2009) Performance evaluation of GSCM based on membership conversion algorithm. In Proceeding: IEEE international colloquium on computing, communication, control and management
8. Zhu Q, Sarkis J, Lai K (2007) Initiatives and outcomes of GSCM implementation by Chinese manufacturers. J Environ Manage 85(1):179–189
9. Hervani AA, Helms MM (2005) Performance measurement for green supply chain management. Benchmarking Int J 12(4):330–353
10. Cronbach LJ (1951) Coefficient alpha and the internal structure of tests. Psychometrika 16(3):297–334
11. Nunnally JC (1978) Psychometric theory, 2nd edn. McGraw-Hill, New York
12. Gopal MH (1970) An introduction to research procedures in social sciences. Asia Publishing House, Bombay
13. Kothari CR (2004) Research methodology methods & techniques, 2nd edn. New Age International publisher, New Delhi

Review on Perspectives in Supply Chain Trust Evaluation

Manu Mathew, Justin Sunny, and V. Madhusudanan Pillai

Abstract Trust is an obligatory requirement for each supply chain (SC), as it can greatly improve the overall SC performance. Evaluation of trust among SC actors is not an easy task, but it is essentially needed to take crucial SC decisions. To this point in time, many researchers have attempted to evaluate trust through different perspectives. The purpose of this work is primarily to review the various perspectives of researchers in evaluating SC trust. This paper particularly targets the trust in buyer–supplier relationship. Review reveals that researchers mainly adopt two approaches in evaluating trust, which are survey based and mathematical model based. Benevolence, credibility, ability, integrity, goodwill, and openness are the key trust factors considered by most of the researchers.

Keywords Trust · Supply chain management · Trust factors · Trust evaluation

1 Introduction

Trust is a multi-faceted concept that can be interpreted in many ways. Trust is defined in several ways with regard to various fields including sociology, economics, automation, organizational management, computer and networking, psychology, and political science. Supply chain is a network of organizations to produce and distribute specific products to the end customer.

Trust in a SC can be defined in several ways. Fundamentally, trust is the firm's belief that another company will perform actions that will result in positive outcomes for the firm as well as avoid unexpected actions that would result in negative outcomes for the firm [1].

Trust is an efficient solution to improve cooperative efficiency and it reduces the cooperative uncertainty [2]. Trust is a critical factor in fostering commitment among

M. Mathew · J. Sunny · V. M. Pillai (✉)
Department of Mechanical Engineering, National Institute of Technology Calicut, Kozhikode, Kerala, India
e-mail: vmp@nitc.ac.in

© Springer Nature Singapore Pte Ltd. 2020

H. Vasudevan et al. (eds.), *Proceedings of International Conference on Intelligent Manufacturing and Automation*, Lecture Notes in Mechanical Engineering,
https://doi.org/10.1007/978-981-15-4485-9_30

SC partners [3]. Trust can be considered as a necessary antecedent of information sharing in a SC [4]. In short, trust act as a necessary ingredient in improving SC performance.

Measurement of trust in SC is one of the widely researched areas and still lots of researchers are working on this ground. As there is no universal definition of trust, there is no universal way of measuring it. Researchers are adopting different perspectives for evaluating trust in SCs. The purpose of this work is primarily to review the various perspectives of researchers in evaluating SC trust. To the best of our knowledge, this is the first review paper that discusses various perspectives in trust evaluation after McEvily and Tortoriello [5].

This paper is organized as follows. Section 2 describes the methodology. Review on perspectives in SC trust evaluation is presented in Sect. 3. Section 4 highlights the findings from the review. Section 5 concludes the whole work.

2 Methodology

An intensive literature review was done to study the approaches in SC trust evaluation. This work considers literature published from 2000 to 2019. As the next step, an abstract review was done to filter the relevant materials. This review paper fundamentally narrates about 27 recent research works related to the SC trust evaluation. Literatures are thematically analyzed in the chronological order. Even though the search was not exhaustive, it serves as a comprehensive basis for gaining an overall understanding of the attempts for evaluating trust in different kinds of SCs.

3 Perspectives in Supply Chain Trust Evaluation

Researchers have modeled and evaluated trust in a range of perspectives. Perspectives change with respect to the trust factors considered, methodology followed, scenario considered, etc.

Handfield and Bechtel [6] analyzed the effect of four factors, viz. buyer dependence, contract, site-specific asset, and human-specific asset on trust and interpreted the influence of trust on SC responsiveness. They concluded that trust can improve supplier responsiveness. This survey-based study shows that buyer dependence, human-specific asset, and trust are positively related to responsiveness.

The remarkable developments in the technology and wide use of Internet have already redefined the traditional SC concepts. Several authors have attempted to conceptualize trust in electronic-based SC management or e-supply chain management (e-scm). Xiong and Liu [7] has framed a model for evaluating trustworthiness of a peer in a decentralized information system. It was done by combining three trust factors, viz. amount of satisfaction, number of interaction, and balance factors. In this era of electronic-based business, trust in peer-to-peer decentralized information systems is very important. Wang and Varadharajan [8] considered a decentralized scenario for peer-to-peer interaction and attempted to evaluate the trust of a peer by developing a probability-based model. Trust is evaluated based on the interaction history of a peer with the other peers. Nowadays almost all the SC organizations are net enabled and transactions are done online. In order to evaluate the trust among net-enabled organizations, Lin et al. [9] proposed a trust mechanism for selecting supplier in net-enabled organization. Benevolence, integrity, and ability are considered as trust factors to develop the trust mechanism for selecting the supplier [10].

Batt [11] discussed the distrust developed between growers and market agent in the absence of auction. This work employed a survey-based approach and traced out six trust factors. Study found that factors like relational satisfaction, goal compatibility, and relational investment have positive impact on trust, whereas factors like market agent's power and opportunistic behavior have negative impact. Interestingly, duration of relationship has no impact on trust.

In order to understand the role of trust in the supplier–customer relationship, Sahay [12] has done a survey-based study among the SC managers and consultants from various organizations. From the survey, it was found that factors like ability, calculative, prediction, motive, credibility, integrity, consistency, institutional and behavior have potential influence on trust. Ping and Jing [13] have presented a probabilistic model for evaluating trust among actors in an e-commerce scenario. Numbers of successful trade and total number of interactions were identified as critical factors considered for evaluating partner's trustworthiness. A study has been done to trace the relationship between trust and commitment by a survey-based empirical test [3]. The questionnaire devised in the survey is framed, mainly, by considering two sorts of variables, viz. transaction cost variables and social exchange variables. The model depicts that partner's asset specificity, information sharing, perceived satisfaction, and partner's reputation are positively associated with trust, but the behavioral uncertainty, perceived conflict, and asset specificity of respondent are negatively associated with trust.

Ghosh and Fedorowicz [14] devised the survey-based approach to structure a framework for increasing trust. According to them, trust act as conjunction between the contract and bargaining power. A study found that trust improves SC performance by positively influencing innovativeness [15]. SC performance was measured with parameters such as responsiveness, process improvement, lead time, time to market, and delivery reliability. To measure trust, factors considered were same as the factors considered by [16]. These factors are reputation, willingness, information sharing, likability, similarity, and power.

Laeequddin et al. [17] suggested a multi-perspective and multi-level trust measurement method to measure SC member's trust on a ten-point scale. They have considered characteristic trust, rational trust, and institutional trust in their study. They refer trust as the threshold level of risk bearing capacity of a trustor. Beyond this level, subject of trust will turn into the matter of risk management.

Integrity, reliability, openness, predictability, fairness, benevolence, and honesty were indicated as the key trust factors in the survey-based study [18]. They developed and identified a multi-faceted measure of trust. This trust factors were used as a trust signal and also as a benchmark to know the extent of trust in relationship. The study describes that trustworthiness would develop relationship commitment in a SC.

Reputation and communication can build mutual trust between SC alliance partners [2]. In the mathematical model, they have considered direct transaction record and recommendation of third-party record for trust evaluation. Updating the trust value after each transaction and incorporation of third-party trust value toward trustor has made the model suitable for faultless decision making for transaction. In light of a survey-based study [19], it was found that social interaction, personality traits, competence and reputation, communication, and rule of law are the paramount important factors constituting trust in SC. Structural equation modeling, a statistical approach was used to find the influence of different factors on trust between SC members. Study indicates that all factors have positive effect on trust.

Honesty, credibility, experience, jurisdiction, sincerity, predictability, transparency, goodwill, commitment, respect confidentiality of information exchange, communication skill, shared value, resemblance, sharing working method, and influence in the network are trust factors recognized by [4] to develop trust in a SC. Esmaili et al. [10] developed a mathematical trust model to calculate the amount of trust that is adaptable in SCs. The probabilistic model considers past interaction to know the trust between retailer and supplier. The issue of organizational relationship based on trust was studied by [20]. Benevolence, credibility, openness, calculative, contract, institutional, availability, reliability, and expertness are some of the generalized trust dimension considered for evaluation of trust in this review paper. To give a complete picture of the proposed structure, a rule-based fuzzy ranking model is considered in the paper for trust evaluation.

In a SC, trust can be specified by the extent a supplier relies on the retailer's ordering information. In this perspective, a trust model for a two-tier SC scenario was proposed [21]. According to the findings of this work, instant behavior of SC actors and predetermined factors like reputation, historical transaction, peer recommendation from trustees, emotions, experience, and cognition from trustors have great impact on trust. In order to determine the factors affecting trust, Cho et al. [22] performed a survey-based study. Survey revealed that risk, faith, fear, feeling, valence, power, delegation, control, credit, cooperation, altruism, reciprocation, adoption, social capital, and relational capital are the factors resulting trust in a supply chain. Ozer and Zheng [23] addressed direct information sharing as the major factor in mitigating risk and increasing coordination in the SC. They identified four building block of trust which are personal values and norms, market environment, business infrastructure, and business process design. Xia and Yongjun considered transactional satisfaction, product ability, risk probability of information concealment, reward, and penalty for evaluating trust among SC enterprises under blockchain environment [24]. Considering these factors, a mathematical model was developed. Rehman et al. [25] proposed a trust model capable of determining the trustworthiness of a trustor on a trustee. Irrespective of scenarios, this model facilitates a framework for graphically representing trust in multiple levels. This study fundamentally considers customer trust toward a particular firm. As per the proposed model, trustworthiness is the sum of trust gain and trust deficit based on past interactions between trustor and trustee over a period of time.

Ruel et al. [26] highlights trust as a vital part of collaboration. A questionnaire-based survey has been done as the part of this work to identify the key trust factors. Factors considered are shared values, information sharing, communication, information quality, partner reputation, rule of law, and uncertainty behavior. Results indicate that information quality, communication, and uncertainty behavior have a high influence on trust. Recently, [27] scrutinized the relationship between trust and risk. A mathematical model for evaluating risk is formed based on a literature study. The study proposes that risk and trust are inversely related to each other. Mansouribakvand [28] identified benevolence, openness, and credibility as vital trust factors and modeled SC trust under blockchain environment.

From the literature review, it is evident that different researchers view trust in different perspectives with regards to the scenarios under their consideration. At the same time, several similarities are also there; particularly, in the trust factors considered by the researchers. To analyze the homogeneity among the trust factors, we mapped the key trust factors revealed in the literature with the thirty-seven dimensions of trust identified by [5]. Table 1 shows the mapping.

4 Findings from Literature Review and Discussions

The literature review discloses some imperative facts regarding the trends in SC trust evaluation. From the literature, it is observed that attitude toward SC trust varies from person to person and this can be justified with the obscure nature of trust definition. Out of the 27 articles reported in the literature, 20 deals with survey-based methodology. The ideology in judging trust will be probably different for different SC actors. In such a situation, a collective opinion from the survey would be better to make a valid conclusion. This might have prompted majority of researchers to go with survey-oriented trust evaluation. Surveys are devised either to trace out the relevant factors affecting the trust or to assess the level of influence of already identified trust factors. Some researchers have attempted to formulate regression models using the responses they received from surveys. Six articles reported in the literature, exhibit trust evaluation models framed with probabilistic theories. Remarkably, all these works consider past interaction as an antecedent of trust. Only one literature discusses development of fuzzy logic-based model for evaluating trust.

Mapping of trust factors with 37 trust dimensions unveils few interesting points. Benevolence, credibility, ability, integrity, goodwill, and openness were considered as key trust factors in more than ten articles in the literature. This observation undoubtedly declares the importance of these factors in building the SC trust. Benevolence is defined as the extent of desire to exercise liberality. Credibility is the quality, which shows partners capability to keep promise and rise up to other competence standard. Openness takes away all the restrictions and allows transparency in information. Goodwill between SC members is defined as the intention of not harming

Table 1 Mapping of trust factors in the literature with the trust dimension identified by McEvily and Tortoriello [5]

Author (year)	Benevolence	Credibility	Ability	Integrity	Cognitive	Affective	Openness	Reliability	Fairness
Handfielda and Bechtelb (2001)	✓	✓					✓		
Xiong and Liu (2002)	✓	✓		✓					✓
Batt and Peter (2003)	✓		✓	✓			✓		✓
Sahay (2003)		✓	✓	✓					
Wang and Varadharajan (2004)	✓	✓		✓					✓
Lin et al. (2005)	✓		✓	✓					
Ping and Jing(2007)	✓	✓	✓	✓			✓		
Kwon and Suh (2008)		✓	✓				✓		
Ghosh and Fedorowicz (2008)									
Panayides and Lun (2009)		✓	✓				✓		
Laeequddin et al. (2010)	✓	✓			✓			✓	✓
Jones et al. (2010)	✓			✓			✓	✓	✓
Zhuo and Dai (2010)	✓	✓					✓		
Evilya and Tortoriellob (2011)	✓		✓	✓	✓	✓	✓	✓	✓
Xhuvani and Elbasan (2012)	✓	✓	✓						
Hossain and Ouzrout (2012)	✓	✓		✓			✓		
Esmaili et al. (2014)		✓							
Husseini et al. (2015)	✓	✓					✓	✓	
Han and Dong (2015)	✓	✓			✓				
Cho et al. (2015)		✓	✓	✓			✓	✓	
Esmaeili et al. (2015)		✓			✓				
Ozer and Zheng (2017)		✓							

(continued)

Table 1 (continued)

Author (year)	Benevolence	Credibility	Ability	Integrity	Cognitive	Affective	Openness	Reliability	Fairness
Xia and Yongjun (2017)	✓	✓	✓	✓			✓		✓
Rehman et al. (2017)		✓	✓				✓		
Ruel et al. (2018)	✓	✓							
Piotrcofta (2019)									
Bakvand and Norrman (2019)	✓	✓					✓		

Author (year)	Faith in intention	Predictability	Commitment	Willingness to risk	Availability	Consistency	Discreteness	Opportunity	Receptivity
Handfielda and Bechtelb (2001)				✓					
Xiong and Liu (2002)									
Batt and Peter (2003)						✓		✓	
Sahay (2003)		✓				✓			
Wang and Varadharajan (2004)									
Lin et al. (2005)									
Ping and Jing(2007)									
Kwon and Suh (2008)		✓					✓	✓	
Ghosh and Fedorowicz (2008)				✓					

(continued)

Table 1 (continued)

Author (year)	Faith in intention	Predictability	Commitment	Willingness to risk	Availability	Consistency	Discreteness	Opportunity	Receptivity
Panayides and Lun (2009)		✓							✓
Laeequddin et al. (2010)			✓						
Jones et al. (2010)		✓							
Zhuo and Dai (2010)									
Evilya and Tortoriellob (2011)									
Xhuvani and Elbasan (2012)			✓						
Hossain and Ouzrout (2012)		✓	✓						
Esmaili et al. (2014)									
Husseini et al. (2015)		✓			✓				
Han and Dong (2015)									
Cho et al. (2015)		✓			✓	✓			✓
Esmaeili et al. (2015)		✓							
Ozer and Zheng (2017)									
Xia and Yongjun (2017)									

(continued)

Table 1 (continued)

Author (year)	Faith in intention	Predictability	Commitment	Willingness to risk	Availability	Consistency	Discreteness	Opportunity	Receptivity
Rehman et al. (2017)									
Ruel et al. (2018)		✓							
Piotrcofta (2019)				✓					
Bakvand and Norrman (2019)									

Author (year)	Surveillance	Motives	Institutional	Informal agreement	Influence acceptance	Honesty	Goodwill	Forbearance	Judgment
Handfielda and Bechtelb (2001)			✓		✓				
Xiong and Liu (2002)							✓		
Batt and Peter (2003)					✓		✓		
Sahay (2003)		✓	✓						
Wang and Varadharajan (2004)									
Lin et al. (2005)									
Ping and Jing(2007)									
Kwon and Suh (2008)	✓						✓		
Ghosh and Fedorowicz (2008)			✓		✓				

(continued)

Table 1 (continued)

Author (year)	Surveillance	Motives	Institutional	Informal agreement	Influence acceptance	Honesty	Goodwill	Forbearance	Judgment
Panayides and Lun (2009)		✓					✓		
Laeequddin et al. (2010)			✓		✓	✓	✓		
Jones et al. (2010)						✓			
Zhuo and Dai (2010)							✓		
Evilya and Tortoriellob (2011)									
Xhuvani and Elbasan (2012)		✓	✓				✓		
Hossain and Ouzrout (2012)					✓	✓	✓		✓
Esmaili et al. (2014)							✓		
Husseini et al. (2015)			✓						
Han and Dong (2015)				✓			✓		
Cho et al. (2015)	✓				✓	✓	✓	✓	
Esmaeili et al. (2015)									
Ozer and Zheng (2017)									
Xia and Yongjun (2017)									

(continued)

Table 1 (continued)

Author (year)	Surveillance	Motives	Institutional	Informal agreement	Influence acceptance	Honesty	Goodwill	Forbearance	Judgment
Rehman et al. (2017)									
Ruel et al. (2018)	✓		✓				✓		
Piotrcofta (2019)									
Bakvand and Norrman (2019)									

Author (year)	Dynamism	Coordination	Concern	Communication	Character	Loyalty	Habitualization	Expertness	Avoid taking excessive advantages	Calculative
Handfielda and Bechtelb (2001)										
Xiong and Liu (2002)		✓								
Batt and Peter (2003)		✓	✓	✓						
Sahay (2003)							✓			
Wang and Varadharajan (2004)		✓								
Lin et al. (2005)										
Ping and Jing(2007)										
Kwon and Suh (2008)		✓	✓	✓					✓	✓
Ghosh and Fedorowicz (2008)										
Panayides and Lun (2009)							✓			
Laeequddin et al. (2010)	✓				✓			✓		✓
Jones et al. (2010)										
Zhuo and Dai (2010)				✓						

(continued)

Table 1 (continued)

Author (year)	Dynamism	Coordination	Concern	Communication	Character	Loyalty	Habitualization	Expertness	Avoid taking excessive advantages	Calculative
Evilya and Tortoriellob (2011)						✓				
Xhuvani and Elbasan (2012)		✓		✓						
Hossain and Ouzrout (2012)				✓						
Esmaili et al. (2014)										
Husseini et al. (2015)								✓		✓
Han and Dong (2015)		✓	✓		✓		✓			
Cho et al. (2015)					✓					
Esmaeili et al. (2015)					✓		✓			
Ozer and Zheng (2017)										
Xia and Yongjun (2017)										
Rehman et al. (2017)										
Ruel et al. (2018)				✓						
Piotrcofta (2019)										
Bakvand and Norrman (2019)										

each other's interest. Goodwill pile up when there is a repeated transaction with the same chain member. The expertness and capability of trustee are represented as ability. Integrity implies the probability of a trustee keeping a promise to its partner.

5 Conclusion and Recommendations

In this work, an extensive literature review was done to get a bird's eye view on the different perspectives on SC trust evaluation. Findings from the review point out that trust has different meanings in different SC contexts and hence quantifying trust and identifying the appropriate factors for evaluating trust in SC are very tedious jobs. Researchers view trust in diverse angles. SC trust factors traced by researchers are not exactly same but have some level of homogeneity.

Identification of suitable trust factors is highly important in modeling and evaluating SC trust. While most of the researchers have adopted a survey-based approach in finding these factors, others built trust models, either based on probability theory or with fuzzy logic. Benevolence, credibility, ability, integrity, goodwill, and openness are the trust factors considered in most of the literature.

Latest trends such as incorporation of blockchain or machine learning in decentralized SC network are changing the outlook of trust. Novel trust evaluation models that perfectly suit these kinds of modern SC circumstances are currently limited and thus provide scope for further research.

References

1. Anderson JC, Narus JA (1990) A model of distributor firm and manufacturer firm working partnerships. J Mark 54(1):42–58
2. Zhuo X, Dai F (2010) Dynamic quantitative trust assessment model between partner-enterprises of supply chain alliances. In: ICLEM 2010: logistics for sustained economic development: infrastructure, information, integration, pp 4582–4589
3. Kwon IWG, Suh T (2004) Factors affecting the level of trust and commitment in supply chain relationships. J Supply Chain Manage 40(1):4–14
4. Hossain SA, Ouzrout Y (2012) Trust model simulation for supply chain management. In: 2012 15th International conference on computer and information technology (ICCIT), pp 376–383
5. McEvily B, Tortoriello M (2011) Measuring trust in organisational research: review and recommendations. J Trust Res 1(1):23–63
6. Handfield RB, Bechtel C (2002) The role of trust and relationship structure in improving supply chain responsiveness. Ind Mark Manage 31(4):367–382
7. Xiong L, Liu L (2002) PeerTrust: A trust mechanism for an open peer-to-peer information system. Georgia Institute of Technology
8. Wang Y, Varadharajan V (2004) Interaction trust evaluation in decentralized environments. In: International conference on electronic commerce and web technologies. Springer, Berlin, Heidelberg, pp 44–153
9. Lin FR, Sung YW, Lo YP (2005) Effects of trust mechanisms on supply-chain performance: a multi-agent simulation study. Int J Electron Commer 9(4):9–112

10. Esmaili S, Makui A, Hafezalkotob A (2014) Introducing a mathematical model in supply chain by adding trust flow. J Ind Syst Eng 7(1):80–103
11. Batt PJ (2003) Building trust between growers and market agents. Supply Chain Manage Int J 8(1):65–78
12. Sahay BS (2003) Understanding trust in supply chain relationships. Ind Manage Data Syst 103(8):553–563
13. Ping W, Jing Q (2007) A mathematical trust model in e-commerce. In: 2007 International conference on multimedia and ubiquitous engineering (MUE'07), pp 644–649
14. Ghosh A, Fedorowicz J (2008) The role of trust in supply chain governance. Bus Process Manage J 14(4):453–470
15. Panayides PM, Lun YV (2009) The impact of trust on innovativeness and supply chain performance. Int J Prod Econ 122(1):35–46
16. Doney PM, Cannon JP (1997) An examination of the nature of trust in buyer–seller relationships. J Mark 61(2):35–51
17. Laeequddin M, Sahay BS, Sahay V, Abdul Waheed K (2010) Measuring trust in supply chain partners' relationships. Measuring Bus Excellence 14(3):53–69
18. Jones SL, Fawcett SE, Fawcett AM, Wallin C (2010) Benchmarking trust signals in supply chain alliances: moving toward a robust measure of trust. Benchmarking Int J 17(5):705–727
19. Cerri S (2012) Exploring factor affecting trust and relationship quality in a supply chain context. J Bus Stud Q 4(1):74
20. Husseini ZM, Zarandi MF, Husseini SM (2015) Trust evaluation for buyer-supplier relationship concerning fuzzy approach. In: 2015 Annual conference of the North american fuzzy information processing society (nafips) held jointly with 2015 5th world conference on soft computing (WConSC), pp 1–6
21. Han G, Dong M (2015) Trust-embedded coordination in supply chain information sharing. Int J Prod Res 53(18):5624–5639
22. Cho JH, Chan K, Adali S (2015) A survey on trust modeling. ACM Comput Surv (CSUR) 48(2):1–40
23. Özer Ö, Zheng Y (2017) Establishing trust and trustworthiness for supply chain information sharing. In: Handbook of information exchange in supply chain management. Springer, Cham, pp 287–312
24. Xia J, Yongjun L (2017) Trust evaluation model for supply chain enterprises under blockchain environment. In: 2017 7th International conference on social network, communication and education (SNCE 2017). Atlantis Press
25. Rehman SU, Qingren C, Weiming G (2017) Rise in level of trust and trustworthiness with trust building measures: a mathematical model. J Model Manage 12(3):349–363
26. Ruel S, Shaaban S, Wu J (2018) Factors which influence trust in supply chains. Logistique Manage 26(1):58–69
27. Cofta P (2019) Trust, choice, and self-preservation: a computational approach. Cogn Technol Work, pp 1–17
28. Mansouribakvand G (2019) The impact of blockchain technology on trust in the supply chain. M.S. thesis, Engineering Logistics, Lund University, Sweden, June 2019. Accessed on 28 Sept 2019. [Online]. https://lup.lub.lu.se/student-papers/search/publication/8987756

Building Envelope Optimization and Cost-Effective Approach in HVAC to Support Smart Manufacturing

Shamir Talkar, Amit Choudhari, and Pavan Rayar

Abstract To fulfill the demand for modern products, nowadays industries are growing and expanding their manufacturing units at very faster rate. But due to space constraints, there are some restrictions in expanding longitudinally so the only option is to expand laterally, i.e., increasing the number of floors. In many manufacturing units, there is a need to implement optimized and effective HVAC system. The main power consumers in most buildings are HVAC, apart from lighting systems which accounts nearly 60–65% of the total building load. The first step in energy savings on HVAC systems is to reduce the cooling load. The amount of electricity for air-conditioning systems used depends on the cooling load, i.e., the amount of heat the system must remove. In order to support the purpose of a smart and intelligent manufacturing process refining the measurement of heat loads using different methods, and enhancing the insulation used in the building envelope will result in enormous energy savings in HVAC by reducing the tonnage of heat load by creating a stronger and effective barrier to minimize the heat contribution from the air and results in saving money.

Keywords Demand · Expansion of industries · Space constraints · Number of floors · Optimized and effective HVAC

1 Introduction

In order to understand the cooling load required by the building, major emphasis needs to be given to the building envelope. During the headload calculations, there

S. Talkar (✉)
Arkk Consulting, MEP, Mumbai, India
e-mail: shamirtalkar@gmail.com

A. Choudhari
Worley, Mumbai, India

P. Rayar
Department of Production Engineering, Dwarkadas J. Sanghvi College of Engineering, Mumbai, India

© Springer Nature Singapore Pte Ltd. 2020
H. Vasudevan et al. (eds.), *Proceedings of International Conference on Intelligent Manufacturing and Automation*, Lecture Notes in Mechanical Engineering,
https://doi.org/10.1007/978-981-15-4485-9_31

are different variables like U-factor for wall, glass, roof and some constant factors like outdoor temperature its relative humidity, solar gains, heat dissipation through lights, equipment and humans. There is no control over the constants, except the variables by using glass having innovative film technology, using wall and roof with insulation and air cavity. Therefore, by controlling the U-factor, i.e., by changing the different configuration, the total aggregate heat entering indoors can be restricted [1]. This leads to a reduction in headload tonnage, thereby resulting in savings. By using this methodology in the future, there is a great scope of saving energy and manufacturing various façade wall panels and glass having excellent U-values which will bring about a significant difference in load reduction. After researching and analyzing in the field of HVAC, its ever-increasing demand due to attaining thermal comfort within the premises has led to increased energy demand and higher electricity bills. Also, the conventional methods of generating electricity further lead to environmental impacts. Therefore, a reduction in HVAC load will not only result in lower energy consumption but also reduce the energy bills and have less impact on the environment. In this research paper, there are improved ways of reducing the cooling load by means of improving the building envelope. This research paper shows reduction in the cooling load by comparing three different sets of options by changing the conditions and the U-values.

2 Overview of the Research Paper

The building envelope acts as the door between indoor and external climatic conditions. It potentially controls the building climatic response [2]. The building envelope should be designed to preserve energy significantly. Well-designed building envelope maximizes daylight, natural ventilation (access to fresh air) and views to the exterior and enables to modulate solar heat gain and control/reduce noise. Building envelope components and their configuration largely determine the amount of heat gain or loss and wind that enters inside the building and extent of natural ventilation in the building [3].

The primary components of building envelope which affect the performance of a building are,

(a) walls,
(b) roof,
(c) fenestration (openings with or without glazing),
(d) floor, etc. (Fig. 1).

Heat can be generated in various ways, and they are either directly or indirectly. Some of the sources are as follows:

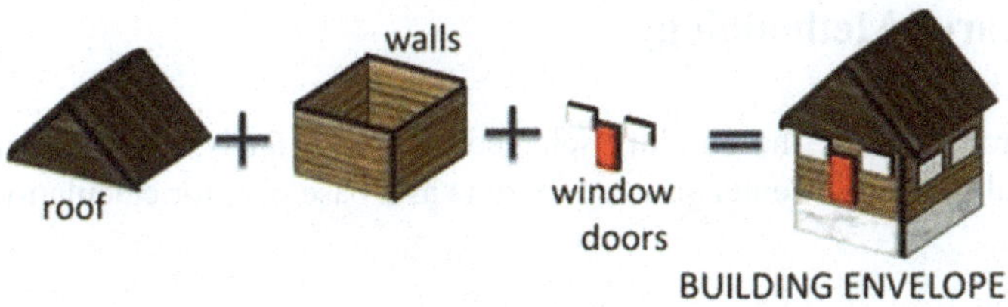

Fig. 1 Basic building envelope incorporating various primary components

2.1 *Solar Heat Gain [4]*

The radiation from the sun enters the interior design spaces in three modes—conduction, convection and radiation.

2.2 *Heat from Human Beings*

Occupants are another major source of heat gain in the interiors of the structure. Humans consume hundreds of calories every day as food, and part of this energy is released as heat to the surroundings during the metabolic processes.

2.3 *Outdoor Air Heat*

The warm outside air is called outdoor air or ambient air. Since it is warmer than the air inside the building spaces, when it enters the spaces it tends to increase the indoor average room temperature.

2.4 *Heat from Electrical and Electronic Appliances*

Indoor spaces are loaded with electrical and electronic appliances such as lightning fixtures, television sets, coffee machines, water heaters, etc. These appliances consume electricity and release a part of it in the form of heat in the air-conditioned spaces.

3 Research Methodology

For research purpose, the building selected to do this analysis is a Ground + 10 Floors Multi-Purpose Center situated in Pune as a base plan for calculations.

3.1 A Conventional Approach to Heat Load Calculations

In this approach, heat load calculation was done based on U-values of standard materials used as building envelope. Architects/civil consultants/engineers use the conventional approach in designing facades. The materials used are as follows:

Summer conditions: [5]
Indoor temperature: 75.2 °F [6]
Outdoor temperature: 101 °F
Glass: double pane regular glass
Properties:
Solar heat gain co-efficient (SHGC)—0.80
U-value—0.55 BTU/HR-FT^2—°F
Wall: (inside) 25 mm plaster + 100 mm brick + 25 mm plaster
Properties:
U-value—0.35 BTU/HR-FT^2— °F
Roof: 100 mm RCC
Properties:
U-value—0.63 BTU/HR-FT^2— °F
Light load: 0.75 W/ft^2.

3.2 B Option 1

In this approach, heat load calculation was done based on improved set of U-values of materials used as building envelope.

The glass and wall configuration used in this approach are as follows:
Summer conditions:
Indoor temperature: 75.2 °F
Outdoor temperature: 101 °F
Glass: double pane regular glass with film
Properties:
Solar heat gain co-efficient (SHGC)—0.28
U-value—0.26 BTU/HR-FT^2— °F
Wall: 25 mm plaster + 200 mm AAC Block + 50 mm insulation + 25 mm plaster
Properties:
U-value—0.06 BTU/HR-FT^2— °F

Fig. 2 Using a fan with an air conditioner leads to savings

Roof: 100 mm RCC + 40 mm insulation
Properties:
U-value—0.09 BTU/HR-FT^2— °F.

3.3 C Option 2

In this approach, heat load calculation was done based on the same set of U-values used in Option 1 as well as increasing the indoor conditions. We have also added ceiling fans in the conditioned space [7]. A perfect combination of saving electricity is making use of the fans as well as the air conditioner. Ceiling fans can create a breeze that makes people in the room feel cooler and more comfortable. With a ceiling fan running, you can raise the thermostat of the air conditioner by 3–4 degrees with no reduction in comfort. Increasing the temperature on the air conditioner can reduce your electricity bills significantly (Fig. 2).

3.4 Headload Comparison

As we can see from Table 1, when the building headload is calculated using conventional approach to walls and glass, the tonnage of refrigeration required to cool the building comes to 240 TR. On the other hand, when these walls and glass configurations are improved the tonnage reduces to 190 TR and when the same U-values are used along with ceiling mounted fans, the tonnage further drops to 170 TR. This

Table 1 The table shows a summary of tonnage required for the entire building premises

Approach	TR	
1	Conventional with normal U-values	240
2	Option-01 with improved U-values	170
3	Option-02 with U-values as per Option-01 and fans	150

show that the tonnage difference between the conventional approach and Option 2 comes to nearly 90 TR (Table 2).

4 Comparison of Capital and Operational Expenditures

The comparison gives us a summary of the initial capital investment required and the operational expenditures for conventional approach as well as Option 1 and 2. Section A shows the capital investment of all 3 options, this included cost for variable refrigerant flow system, wall insulation, roof insulation, glass film, ACC blocks, brick blocks and ceiling fans. Section B shows the operational expenditures for 1 year based on the electrical consumption of all the 3 options. Section C shows the comparison with the annual maintenance charges subjected to every option and 10 years of CAPEX AND OPEX cost difference between the 3 options (Table 3).

5 Results

From Table 4, we understand that,

When we compare conventional system and Option 1, we achieve savings right from the second year. The column named difference shows yearly saving when compared to conventional system, i.e., if the building is designed as per Option 1, we achieve savings shown as in column number 4. Similarly, when we compare conventional system and Option 2, we achieve savings right from the very first year. The column named difference shows yearly saving when compared to conventional system, i.e., if the building is designed as per Option 2, we achieve savings shown as in column number 7.

6 Conclusion

The building envelope is a door to the outside environment and its performance is directly proportional to the HVAC load within the premises. Also, with the rise in urbanization, cities are growing vertically, due to lack of availability of land. This results in tall buildings/towers/skyscrapers. HVAC system deployed in such towers will impact hugely on energy consumption. Smart manufacturing of wall panels with sandwiched air cavity, insulation, even glass panels with low-E coats improved the U-values used for headload calculation leading to reduced tonnage for HVAC system.

From the above results for the envelope of the building, the following parameters were improved;

1. Occupancy comfort

Table 2 Sample heat load of the building premises

Area (Ft^2)	1333.70	Height (FT)	10.50	Vol. (Ft^3)	13,999	Room	Room-1	
	Area/qty	Sun gain		BTU/h		DBT	% RH	GR/LB
Room sensible loads					Outside	101	28	83.2
Glass exp					Room	75.2	55	72
N	289	11	0.80	2545		25.8		11.2
E	155	11	0.80	1363	Occupancy	20		
S	289	11	0.80	2545	CFM/PER	5	100	CFM
W	0	165	0.80	0	CFM/SQFT	0.06	80	
NE	0	11	0.80	0	No. of ACPH	1	233	CFM
SE	0	11	0.80	0	**F.A. CFM**	**233**		
SW	0	113	0.80	0	Correction factor		6	
NW	0	118	0.80	0	**Bypass factor**		0.176	
Walls exp								
N	119	10	0.35	415				
S	119	22	0.35	913	People	20.00		205
W	0	18	0.35	0	Appliances			
NE	0	16	0.35	0	Room latent heat sub-load			
SE	0	24	0.35	0			Safety	5%
SW	0	20	0.35	0	Room latent heat			
NW	0	12	0.35	0	Room total heat			
Roof	0	38	0.63	0	Sensible	233.31	25.8	0.88992
All glass	733	25.8	0.55	10,405	Latent	233.31	11.2	0.56032
Partition	0	20.8	0.40	0	**Grand total heat (BTU/HR)**			
Ceiling	667	20.8	0.44	6103				
Floor	1334	20.8	0.48	13,316	**Sensible heat**		0.92	
O/D air	233	25.8	0.1901	1144	**Factor**			
People	20	1	245	4900	Indicated ADP of		54.59	
HP	0	1	2545	0	Selected ADP of		52	
Lights (W)	1000	1	3.4	3401	Dehumidified rise of		19.12	
APP.	100	1	3.4	340	**CFM/ft²**		**2.00**	
Room sensible heat sub-load				49,009	**SQFT/TR**		**240.35**	
Fan HP %	0.08	Safety	5%	6126	**Dehumidified air CFM**		**2670**	
Room sensible heat				**55,135**	**Tonnage**		**5.55**	

Table 3 Comparison of capital and operational expenditure

S. No.	Description	Conventional system	Option 1	Option 2
	TR required for premises	240	170	150
	Type of system	VRV	VRV	VRV
A	Capital expenses			
	Total AC cost	20,400,000.00	14,450,000.00	12,750,000.00
	Fans			500,000.00
	Roof insulation		550,000.00	550,000.00
	Wall insulation		704,000.00	704,000.00
	Glass film		60,60,032.00	6,060,032.00
	ACC blocks		1,224,248.00	1,224,248.00
	Brick blocks	244,849.00		
	Total cost	20,644,849.00	22,988,280.00	21,788,280.00
B	Operating expenses			
	Power consumption (KW)	264	187	165
	Diversity	0.8	0.8	0.8
	Total load	211.2	149.6	132
	Kw/TR	1.1	1.1	1.1
	Power consumption—TRHRS	493,363.20	349,465.60	308,352.00
	Cost of electricity/KWHR	10.00	10.00	10.00
	Energy charges	4,933,632.00	3,494,656.00	3,083,520.00
	Total electric charges/year	4,933,632.00	3,494,656.00	3,083,520.00
C	Comparison			
	Capital investment	20,644,849.00	22,988,280.00	21,788,280.00
	At the end of first year			
	Electrical charges	4,933,632.00	3,494,656.00	3,083,520.00
	Total recurring costs	4,933,632.00	3,494,656.00	3,083,520.00
	Total accrued cost	25,578,481.00	26,482,936.00	24,871,800.00
	At the end of second year			
	Electrical charges	4,933,632.00	34,94,656.00	3,083,520.00
	AMC charges	450,000.00	318,750.00	281,250.00
	Total recurring costs	5,383,632.00	3,813,406.00	3,364,770.00
	Total accrued cost	30,962,113.00	30,296,342.00	28,236,570.00
	Summary (CAPEX + OPEX)	Conventional system	Option 1	Option 2

(continued)

Table 3 (continued)

S. No.	Description	Conventional system	Option 1	Option 2
	CAPEX	20,644,849.00	22,988,280.00	21,788,280.00
	First year	25,578,481.00	26,482,936.00	24,871,800.00
	Second year	30,962,113.00	30,296,342.00	28,236,570.00
	Third year	34,518,133.00	33,714,549.50	31,615,402.50

Assumption Considering electrical cost for 8 h with 80% diversity

Table 4 Comparison of capital and operational expenditures

Year	Conventional system	Option 1	Difference	% Saving in OPEX	Option 2	Difference	% Saving in OPEX
	(₹)	(₹)	(₹)		(₹)	(₹)	
1	2	3	4	5	6	7	8
CAPEX	20,644,849	2,29,88,280	No returns		2,178,828	No returns	
First year	25,578,481	26,482,936			2,487,180	706,681	2.8%
Second year	30,962,113	30,296,342	665,771	2.2%	2,823,657	2,725,543	8.8%
Third year	34,518,133	33,714,549	803,583	2.3%	3,161,540	2,902,730	8.4%

2. Energy efficiency
3. Reduction in cooling tonnage and energy use
4. Energy cost savings.

This reduction in tonnage due to an optimized envelope reduced the size of the HVAC system, which lead to high energy conservation and lower energy bills.

References

1. Eleftheriadis G, Hamdy M (2018) The impact of insulation and HVAC degradation on overall building energy performance: a case study. Buildings 8(2):23
2. Software: hourly analysis program software metadata architecture
3. Emmerich SJ, McDowell TP, Anis W (2005) Investigation of the impact of commercial building envelope airtightness on HVAC energy use. US Department of Commerce, National Institute of Standards and Technology
4. Bano F, Kamal MA (2016) Examining the role of building envelope for energy efficiency in office buildings in India. Architect Res 6(5):107–115
5. Khurmi RS, Gupta JK (2008) TB of refrigration & airconditioning (ME). S. Chand

6. Homod RZ, Sahari KSM (2013) Energy savings by smart utilization of mechanical and natural ventilation for hybrid residential building model in passive climate. Energy Build 60: 310–329. National Building Code Volume 2 (2017 Edition)
7. Wang P-H, Lin J-Y, Using smart controlled AC and ceiling fan to save energy. Department of Architecture, Kao Yuan University, Kaohsiung, Taiwan

A Review of the Reliability Techniques Used in the Case of Casting Process Optimization

Amit Chaudhari and Hari Vasudevan

Abstract Foundry industries nowadays face problems, such as poor quality of the castings produced as well as wrong practices followed, which finally lead to decrease in productivity. Castings with minimum production cost and no defects have emerged as prime requirements in this vital manufacturing industry. This includes improving productivity and reducing cost by minimizing rejections. To achieve these improvements, various process parameters are necessary to be optimized. Simulation of casting process has become an irreplaceable tool in the production of cost-effective and high performance castings. Most of the casting simulation is based on a purely deterministic approach, in which shop floor iterative trial and error methods are replaced with iterations on computer. Optimization in casting process lacks consideration of reliability of casting production process and casting model. Reliability-based design optimization (RBDO) is an approach for the development of components, which aims to minimize the cost, while constraining the probability of failure. This study provides for a review of the casting process optimization through a reliability approach and concludes that the future studies could develop a comparative analysis of reliability with conventional techniques used in casting process optimization.

Keywords Casting process optimization · Reliability

A. Chaudhari (✉)
Department of Production Engineering, Dwarkadas J. Sanghvi College of Engineering, University of Mumbai, Mumbai, India
e-mail: amit_durlabh@yahoo.com

H. Vasudevan
Dwarkadas J. Sanghvi College of Engineering, University of Mumbai, Mumbai, India
e-mail: principaldjs@gmail.com

© Springer Nature Singapore Pte Ltd. 2020

H. Vasudevan et al. (eds.), *Proceedings of International Conference on Intelligent Manufacturing and Automation*, Lecture Notes in Mechanical Engineering,
https://doi.org/10.1007/978-981-15-4485-9_32

1 Introduction

1.1 Casting Process and Optimization

Quality of any manufactured product depends largely upon the type of manufacturing process used. Casting, as a manufacturing process is widely used to obtain complicated shapes of metal parts, with little or no machining involved, in a very economical way. It is amongst the most common processes of manufacturing. In order to maintain good quality production, cast parts must be defect free (internal as well as external). Quality and strength of the castings depend on various parameters. Reliability of casting depends generally on parameters, such as quality of the mould cavity, parting line, holding time, pouring temperature and the skills of the personnel involved. In casting, conservative design rules are formed due to immeasurable factors, such as mould shift, cold shut, shrinkage, blow holes and porosity inclusion. Unquantifiable factors are the ones with an unknown state of defect. Standardization of production techniques and close control on various parameters can lead to preparing foolproof castings. Due to a wide range of possible factors, reasonable classification of casting defects is considered difficult to achieve.

Simulation of casting process can be worked as a collaborative tool, between the design department and production department to reduce lead times, so as to produce defect free castings and to develop simplified component designs for casting. In the simulation approach, the skillset and knowledge of the simulation personnel determines the effectiveness of the simulation software used to develop the best casting process. There is ample of literature to show that the ability and efficiency of the casting process can be forecast through reliability. It is also known that the redundancies play an essential role in increasing the reliability characteristics of systems [1]. Efficiency of the production can be increased through high reliability of the system [2]. Reliability measures are amongst the primary concerns, while planning, designing and operation of any system or equipment.

1.2 Reliability Approach

Reliability provides the method for assuring that a product or service function, when it is required to do so, by its user. These methods consist of the technique for determining what can go wrong, how can that be prevented from going wrong, and, if something goes wrong, how can it quickly recover and minimize the consequences.

The reliability of an item is the probability that the item will perform a specified function under specified operational and environmental conditions, at and throughout a specified time.

Reliability prediction is the combination of the creation of a proper reliability model, together with estimating the input parameters, like failure rates for a particular failure mode or event and the mean time to repair (MTTR) the system for a particular failure for the model and finally to provide a system-level estimate for output reliability parameters (system availability or a particular function failure frequency). Requirements are specified using reliability parameters. The most common parameter of reliability is mean time to failure (MTTF). Reliability increases as the MTTF increases. Reliability modelling is the process of predicting or understanding the reliability of a component or system prior to its implementation. Two types of analysis, often used to model a system's reliability behaviour are fault tree analysis and reliability block diagram. Reliability must be designed into the system and during the system design, top-level reliability requirements are allocated to subsystems by design engineers, maintainers and reliability engineers working together. Reliability design begins with the development of a (system) model. Reliability model use block diagram and fault trees to provide a graphical means of evaluating the relationship between different parts of the system [3].

2 Literature Review

In the current situation with respect to industry practices, reliability theory has become the most challenging and demanding. During the last 40 years, the reliability theory and the methods of reliability analysis have evolved considerably and have also been acknowledged in a number of publications. These studies have shown a multidisciplinary approach of the reliability. Earlier, the casting process optimization was a focused area for many researchers, wherein they used different traditional and simulation techniques for improvement in the casting process. This section on literature review comprises of two sections, process optimization using reliability and casting process optimization.

(a) **Process Optimization using Reliability**:

For practical optimization studies, reliability-based techniques are getting increasingly popular, due to the uncertainties involved in realizing the design variables and stochastic ties involved in various problem parameters. This section highlights the extant research work done in the process optimization with the help of reliability theory used by various authors.

Gupta and Sharma [4] studied reliability and MTTF evaluation of a two duplex-unit standby system, involving two types of repairs; namely cheaper and costlier. Procurement for a new unit was done immediately, if the costly repair was found in inspection. The unit under repair was replaced by a new unit, if it was delivered before the repair, otherwise, the order for the new unit would be cancelled. The system was analyzed for MTTF, availability, steady-state probability and variance by using the supplementary variable technique (SVT).

Ram [5] discussed the mathematical modelling of a highly reliable complex system, which included three states, i.e. normal, partial failed (degraded state) and complete failed state. The system, "partial failed" was due to the partial failure of internal components or redundancies and "completely failed" was due to the catastrophic failure of the system. Repair rates were general functions of the time spent. All the transition rates were constant, except for one transition, where two types of repair, namely exponential and general were possible and could be tackled with the help of a COPULA approach. By employing the supplementary variable technique, Laplace transformation, various transition state probabilities, availability and cost analysis (predictable profit) were obtained along with the steady-state behaviour of the system. Inversions were also carried out, so as to obtain time dependent probabilities, which find out the availability of the system at any instant.

Park and Yang [6] worked on an automated die-casting process. The authors proposed a linear programming (LP) model for solving the number of products to be cast. They found that the proposed model provided the minimum usage of casting time, i.e. the average time, the efficiency was optimized.

Kumar and Ram [7] worked on the reliability and sensitivity analysis of a coal handling unit of a thermal power plant using a probabilistic approach. The transition state probabilities, availability, reliability, MTTF, sensitivity analysis and cost-effectiveness of the system were evaluated using Laplace transforms and differential equations. The authors concluded that the system failure rates depend on sensitivity of the system, i.e. sensitivity of the system can be reduced by controlling the failures. With this, the reliability model parts or units, which affect the system, could be identified.

Yourui et al. [8] worked on reliability modelling and die-casting process optimization. Firstly, simulation of die-casting process was carried out using finite element analysis (FEA). The multidisciplinary optimization model of die-casting process was generated using the quadratic response surface. Also, the epistemic uncertainty was signified using the evidence theory. Later, a reliability-based multidisciplinary optimization (RBMDO) technique was proposed for die casting. Results confirmed the high computational efficiency with epistemic uncertainty through the RBMDO technique.

Kumar t al. [9] investigated a casting system and used the supplementary technique. System's MTTF was most crucial and sensitive for failure rates of the mould shift. They also found that the failure rate of shrinkage and blow holes were highly sensitive for the reliability of the system.

Hardin et al. [10] integrated the casting simulation with a reliability-based design optimization software tool (RBDO). Uncertainties in input variables and model were considered in this methodology. The outcome of the methodology comprised of the confidence level in design along with the reliable optimum design. Design of riser was optimized by considering the uncertainties in fill level, riser diameter and depth. It was observed that the RBDO provided a different overview for the design than in the case of a traditional approach.

(b) **Casting Process Optimization**:

Due to globalization, short lead time is a challenge for casting set up and foundries. Castings with no defects and minimum production cost have turn out to be the necessity for this indispensable manufacturing industry. Rejection of casting is caused due to defective components. These defects depend on various process parameters, which need to be improved upon, using various methods in optimization. The IT industry in collaboration with manufacturing industry, have come up with various simulation software packages, which could identify the parameters affecting the casting through simulation. This simulation also helps to forecast the defects and provides for the corrective measures to minimize these defects.

Tai and Lin [11] optimized the techniques used to design a runner in a die-casting process. The entire process was mathematically modelled using abductive network technique. This helped them to optimize the runner design in the making stage, by ruling out various discrepancies in the system in the development stage itself.

Tai [12] optimized the control of accuracy of the component which was made using a metal die-casting process. In order to locate the correct gate position of the design mould, a finite element method was used. On the basis of the results obtained from finite element analysis, an abductive network was designed, which could predict important process parameters, such as the injection angle and sectional ratio of the runner and gate and injection positions. Optimizing these process parameters leads to more accurate components.

Rohallah and Davami [13] carried out design optimization of an automatic feeder used in the steel casting process. The design process of the feeder consisted of factors, such as determination of the feeder-neck connection point on the casting surface, initial feeder design and optimization of feeder shape and feeder topology. Process optimization was eventually achieved by introducing an automatic feeding system to the existing process, which helped in increasing the productivity by reducing the process time.

Feng et al. [14] carried out a generalized study on various optimization techniques used for solidification during heat transfer of continuous slab casting. Prior to the study, it was observed that there was a lot of heat loss in this system due to which, this problem was taken into consideration for further research. Numerical simulation tools were used to calculate the optimal cooling rate of the continuous slab casting process, which results in efficient solidification of the slab components.

Dabade and Bhedasgaonkar [15] analyzed various defects in the process of metal casting process and optimized the performance of the system using the design of experiments (DOE). The entire process of metal casting was simulated virtually using a commercial computer-aided engineering (CAE) package MAGMASOFT. The virtual simulation helped to narrow down on defects, such as hot tears and shrinkage porosity. The design of experiments (DOE) model was used to improve the feeding system design and gating locations, which helped them to achieve a reduction in shrinkage porosity by 15% and improved the yield strength by 5%.

Choudhari, Narkhede and Mahajan [16] reduced the defects caused in the casting process by simulating the casting process. The process was simulated using the computer-aided engineering (CAE) package "Auto CAST X", which pointed out major casting defects in the system. The defects, such as shrinkage cavity, porosity and sink were identified in the simulation. A new feeder was designed to reduce the defects, which were detected by "AutoCAST X", which was then performed experimentally. The values attained from the experiment were similar to that simulated by the software, hence validating the process improvement.

Jie et al. (2014) [17] studied on the TiB2/A356 aluminium base composite to be used for investment casting and carried out numerical simulation to optimize the process. Process simulation was performed using a commercial computer-aided engineering (CAE) casting package "Pro Cast". Numerical simulations obtained from "Pro Cast" pointed out the week links in the process, which were then optimized. The results indicated that the problems of shrinkage porosity defects have been resolved by increasing the pouring temperature and casting speed. It was also found that, by adding the insulation materials around the gate and riser, the system could gain more stability.

3 Key Observations and Findings of the Study

Various researchers have applied different techniques for optimization of casting process parameters. They have mainly focused on the design and simulation for optimization of the casting process parameters, using softwares like AUTO CAST X, Z-cast, ESI Pro Cast, MAGMASOFT, etc. Different statistical techniques, such as Taguchi method, ANOVA, DOE, RSM, etc., have also been used for casting process parameter optimization. The extant literature also reveals that the application of reliability optimization methods to casting process design provides for more than just an optimal solution. It provides for an overview of all possible solutions, some of which might be novel and innovative.

It can be found that the reliability-based design optimization and casting simulation can be integrated to go a step forward in the development of optimization methods, by including uncertainties in the process and model variables and determining an optimal solution with known probability of success. Use of reliability in different application areas could prove that it works more efficiently for the process optimization by modelling the system errors. It is also observed that:

- Very few researchers have worked on the reliability in casting areas for casting process optimization.
- Reliability theory has evolved as a discipline with the time.
- Literature suggests that reliability technique is multidisciplinary and is not bound by the specific field.

4 Conclusion

A brief overview of the background of the casting process and its optimization has been done in this study and it was found that the optimization of casting process parameters is a prime need of the industries, suffering from various casting defects. Various conventional techniques are found having used for reducing the defects in the casting process. Various optimization techniques used in the casting process provide multiple possible solutions, some of which may not be countered in conventional techniques. A foundry engineer could check the sensitivity and stability in the process models as well as the actual process against variables and parameters. An overview of reliability was undertaken, wherein the design for reliability, steps in design for reliability, ways to improve reliability by design, etc., were studied.

From the literature, it was found that reliability is a multidisciplinary technique, which can be used as optimization technique to get a reliable solution. It can be easily seen throughout the literature that the optimum casting designs will be unreliable without consideration of the statistical and physical uncertainties in the casting process. Studies on optimization of casting process lack in consideration of uncertainties in the casting process. Use of reliability theory for handling uncertainties in the casting process can provide a reliable solution, instead of the optimal solution.

After reviewing the extant literature, it is concluded and recommended that the future studies could develop the comparative analysis of reliability with conventional techniques used in casting process optimization. The studies may cover implementation of reliability for casting process parameters optimization. The developed methodology can be aimed to use reliability as a multidisciplinary method for use in various application areas. It can also be expected to demonstrate how classical reliability-based concepts can be used in single and multi-objective evolutionary algorithms to enhance their scope in handling uncertainties, a matter which is common in most real-world problem solving tasks.

References

1. Sutaria M, Joshi D, Jagdishwar M, Ravi B (2011) Automatic optimization of casting feeders using feed-paths generated by VEM. In: ASME 2011 international mechanical engineering congress and exposition, Volume 3: design and manufacturing, pp 137–146
2. Ransing RS, Sood MP, Pao WK (2005) Computer implementation of Heuvers' Circle method for thermal optimisation in castings. Int J Cast Metals Res 18:119–128
3. Verma AK, Ajit S, Karanki DR (2010) Reliability and safety engineering, 1st edn. Springer, London. ISBN: 978-1-84996-231-5
4. Gupta PP, Sharma MK (1993) Reliability and MTTF evaluation of a two duplex-unit standby system with two types of repair. Microelectron Reliab 33(3):291–295
5. Ram M (2010) Reliability measures of a three-state complex system: a copula approach. Appl Appl Math Int J 5(10):1483–1492
6. Park YK, Yang JM (2011) Maximizing average efficiency of process time for pressure die casting in real foundries. Int J Adv Manuf Technol 53(9–12):889–897
7. Kumar A, Ram M (2013) Reliability measures improvement and sensitivity analysis of a coal handling unit for thermal power plant. Int J Eng Trans C Aspects 26(9):1059
8. Yourui T, Shuyong D, Xujing Y (2014) Reliability modeling and optimization of die-casting existing epistemic uncertainty. Int J Interact Des Manuf (IJIDeM), pp 1–7. https://doi.org/10.1007/s12008-014-0239-y
9. Kumar A, Varshney A, Ram M (2015) Sensitivity analysis for casting process under stochastic modeling. Int J Ind Eng Comput 6(3):419–432

10. Hardin R, Choi KK, Bechermann C (2015) Reliability-based casting process design optimization. Int J Cast Metals Res 28(3):181–192
11. Tai CC, Lin JC (1998) A runner-optimization design study of a die-casting die. J Mater Process Technol 84:1–12
12. Tai Ching-Chih (2000) The optimization accuracy control of a die-casting product part. J Mater Process Technol 103:173–188
13. Tavakoli R, Davami P (2008) Automatic optimal feeder design in steel casting process. Comput Methods Appl Mech Eng 197:921–932
14. Feng H, Chen L, Xie Z, Ding Z, Sun F (2014) Generalized constructal optimization for solidification heat transfer process of slab continuous casting based on heat loss rate. Energy 66:991–998
15. Dabade Uday A, Bhedasgaonkar Rahul C (2013) Casting defect analysis using Design of Experiments (DoE) and computer aided casting simulation technique. Procedia CIRP 7:616–621
16. Choudhari CM, Narkhede BE, Mahajan SK (2014) Casting design and simulation of cover plate using AutoCAST-X software for defect minimization with experimental validation. In: 3rd International conference on materials processing and characterisation (ICMPC 2014), Procedia Materials Science 6, pp 786–797
17. Jie Z, Dongqi Z, Pengwei W, Gang W, Feng L, Penglong D (2014) numerical simulation research of investment casting for TiB2/A356 aluminum base composite. Rare Metal Mater Eng 43(1):0047–0051

Reinforcement Learning for Inventory Management

Siddharth Singi, Siddharth Gopal, Shashikant Auti, and Rohit Chaurasia

Abstract A comparison between four common reinforcement learning algorithms, namely deep Q network (DQN), double deep Q network (DDQN), prioritized experience reply (DQN + PER) and double DQN + PER; and discussion on the methodology with the limitations and advantages of each algorithm are included in this paper. In order to provide these insights, OpenAI environments that demonstrate the working of these algorithms was used. Mountain car environment was used to generalize our results and prove the consistency of our insights. Insights were derived by evaluating basic parameters like, episode length, minimum rewards, maximum rewards and average rewards. This study discusses strategies for including reinforcement learning in supply chain management by using it for inventory management.

Keywords Reinforcement learning · Inventory management · Deep Q network · Double deep Q network · Prioritized experience replay

1 Introduction

Ever since DeepMind defeated the best Go player in 2013 using reinforcement learning, people have developed a keen interest in this field and have been using it to solve a variety of problems across different industries. Applications for reinforcement learning include: financial investments, robotics, supply chain optimization, etc. Reinforcement learning is generally used where there are practice improves our actions, situations are not similar and there are too many combinations for there to be a definitive right answer. Reinforcement learning can be used for scenarios that involve repetitive actions by humans. The use of reinforcement learning can augment human performance in such fields.

S. Singi (✉) · S. Gopal · S. Auti · R. Chaurasia
Mechanical Engineering Department, Dwarkadas J. Sanghvi College of Engineering, Vile Parle, Mumbai, India
e-mail: siddharth.singi@gmail.com

© Springer Nature Singapore Pte Ltd. 2020

H. Vasudevan et al. (eds.), *Proceedings of International Conference on Intelligent Manufacturing and Automation*, Lecture Notes in Mechanical Engineering,
https://doi.org/10.1007/978-981-15-4485-9_33

2 Background Work

In a reinforcement learning algorithm, an agent learns in an environment through an iterative process. Each time the agent reaches its end goal or completes the maximum number of steps, an episode is completed. The agent takes an action according to a policy and obtains a new state and a reward in return.

Humans inherently know the correct action to choose based on the state they are in. For a reinforcement learning algorithm, it is difficult to choose this action correctly. To teach an agent to choose these actions correctly, i.e., to learn the optimal policy we use unsupervised learning (Fig. 1).

The agent learns in the environment through a feedback mechanism as shown above. It goes through a trial and error process. The agent takes an action according to the policy and obtains a new state and a reward from the environment. The agent does not know the inner workings of the environment and only trains itself from the received new state and reward. This process is repeated for a number of episodes. Our end goal is for the agent to teach itself to navigate the environment to obtain the maximum accumulated reward. This accumulated reward is known as return and is defined as:

$$S_i = \text{ state of the agent at } i\text{th step}$$

$$G_t = R_{t+1} + \gamma R_{t+2} + \cdots = \sum_{k=0}^{k=\infty} R_{t+k+1} \tag{1}$$

A state-action value function is the expected return of while taking an action, 'a' while being in a state, 'S'. It quantifies the goodness of an action based on the state the agent is present in. It is defined as:

$$Q(s, a) = \mathbb{E}(G_t | S_t = s, A_t = a) \tag{2}$$

Fig. 1 Interaction between the agent and environment

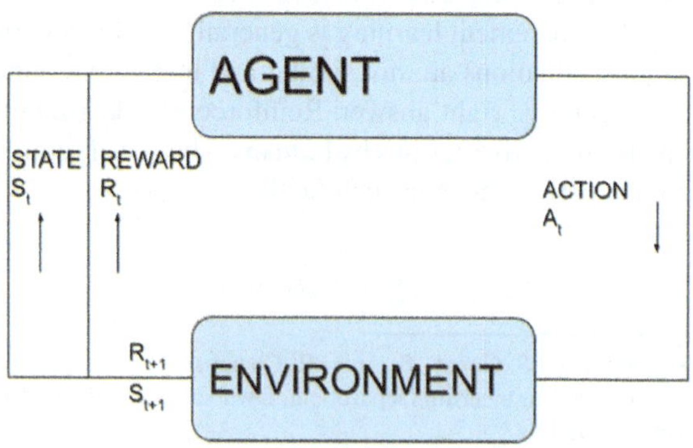

3 Deep Q Network

We use a deep neural network to approximate the value of the state-action value function. We use the states as the inputs for the neural network and it outputs the state-action values for each action. Since we do not have correct state-action values we use one-step TD targets to estimate our state-action values that are used to calculate the error in the loss function of the neural network. The TD target estimates the discounted future return as:

$$\text{TD target} = R(s, a) + \gamma * \max_{a' \forall A} Q(s', a') \tag{3}$$

So the update rule for the Q values becomes:

$$Q(s, a) \leftarrow Q(s, a) + \alpha(R(s, a) + \gamma * \max_{a' \forall A} Q(s', a') - Q(s, a)) \tag{4}$$

3.1 Target Networks

Since TD targets use their own predictions to generate target values for the deep neural nets, it may lead to unstable and undulating updates in the weights of the network. To alleviate this, we use target networks. These are copies of the original neural net and are used to only predict the max values of the next state. The target networks are regularly updated every few episodes. The following equation is used for the target network update:

$$R(s, a) + \gamma * \max_{a' \forall A} Q'\left(s', a'\right) \tag{5}$$

3.2 Experience Replay

The DQN algorithm uses experience replay. Experience replay saves the previous experiences of the agent and feeds it to the neural network in a random order to break the undesirable effects of temporal correlation and avoids forgetting previous memories. Learning while interacting with the environment can be problematic as it can reinforce taking an action over and over again due to the temporal correlation developed. It causes the agent to forget the optimal actions taken during the previous steps due to the effects of the actions taken during the immediate past of the agent.

However, the problem with this is that we are using the predicted values from design network to train design network, which induces a bias in designing system.

This can lead to a lot of instability, thus we use a target network so that the weights for training and the weights for prediction are separate.

4 Double Deep Q Network

When we use our TD targets to train our model, we use target networks. The problem with this, however, is that we are choosing the action which has the maximum of the action values of the next state, however, this action may not be the optimal action of the future state. The expected value of that state may be much lower but since we are using the max operator we overestimate the value of being in that state. The max Q value of the state is then propagated to other states too. This bias of the agent in choosing the max Q value causes the agent to learn to a suboptimal policy.

To avoid this issue, double Q learning decouples the action selection and action value estimation. The behavior network selects the action, and the target network estimates the value of that action. This allows the agent to explore other actions better; hence, the action values of the state become more accurate. Also, by decoupling the action selection and estimation, the agent can converge to an optimal policy faster. Thus, we have used two neural networks in the following format.

$$\text{TD target} = R(s,a) + \gamma * Q'\left(s', \underset{a'\forall A}{\text{argmax}}\, Q\left(s', a'\right)\right) \tag{6}$$

5 Prioritized Experience Replay

When using experience replay the experiences that are called upon to teach the neural net are chosen at random. However, this is not the most efficient method to train the net. Certain experiences that have the maximum TD error should be given greater importance. This is similar to focusing on aspects of learning where you are making the most mistakes in order to learn something quicker. Thus, when inserting a sample into the replay memory instead of the usual tuple of (s, a, r, s, termination) an additional parameter is added to the tuple called priority. The magnitude of the TD error defines the amount of priority to be given to that experience.

$$p_t = |\delta_t| + e \tag{7}$$

A constant is added after to ensure that a small amount of priority is given even to experiences for which the TD error is zero.

$$P(i) = \frac{p_i^a}{\sum_k p_k^a} \tag{8}$$

Fig. 2 Mountain car environment

The process of saving and retrieving experiences from the replay memory is called insertion and sampling. While inserting a fresh memory into the replay memory, it is given the maximum priority value. This ensures that the new memory is used at least once to influence the behavioral network.

Once the priorities of the experiences are known, we select them for sampling with a bias for high-priority sampling. This is done through stochastic prioritization. The probability of an experience being selected for replay is obtained through normalizing the priority value of that experience by the priority values of all the other experiences in the replay buffer. A hyperparameter is introduced in the equation to control the randomness in the selection of the experiences.

6 Comparison Between the Different Algorithms

In order to demonstrate the difference between the algorithms used, the performance of the algorithms is compared by running them on an OpenAI Gym called mountain car. The environment is a one-dimensional scenario, where a car is positioned in a valley between two mountains. The car is not powerful enough to climb the mountain in a single pass. Therefore, it needs to learn to move back and forth to build up the momentum in order to achieve its goal of reaching the mountain top (Fig. 2).

Following are the results obtained while using the four algorithms (Figs. 3, 4, 5 and 6).

7 Using Reinforcement Learning for Inventory Management

In supply chains, one major issue is inventory replenishment. One can use reinforcement learning algorithms to optimize inventory management. Consider the case study of large-scale grocery stores like D-Mart to model and test the reinforcement learning

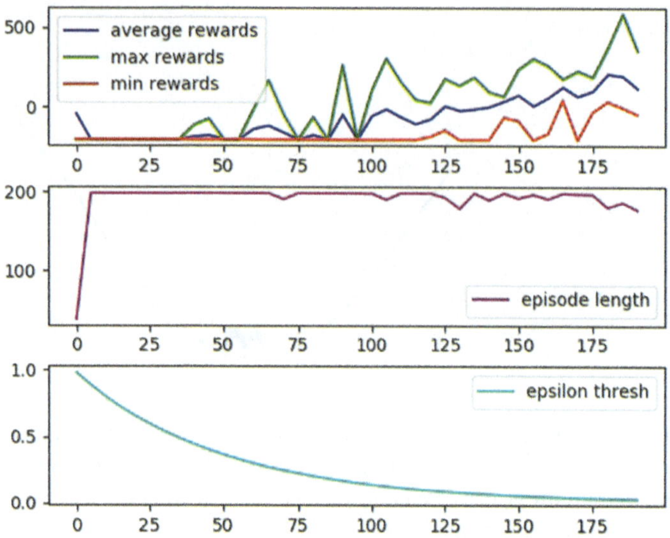

Fig. 3 Deep Q network

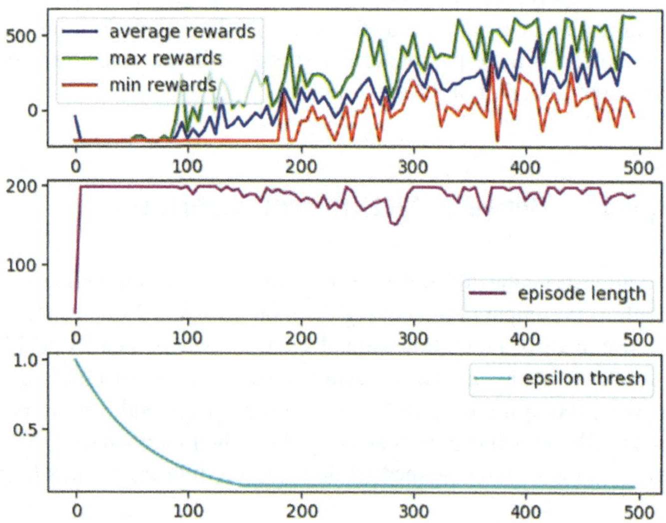

Fig. 4 Double deep Q network

approach. An environment has been built in python to simulate the dynamics of a grocery store. Food stores like this have many factors to consider while reordering their supply, these include: random and changing demand, large number of products, and expiry of food products. This has been incorporated into the environment and deep Q learning is used to find the daily best ordering value of each product.

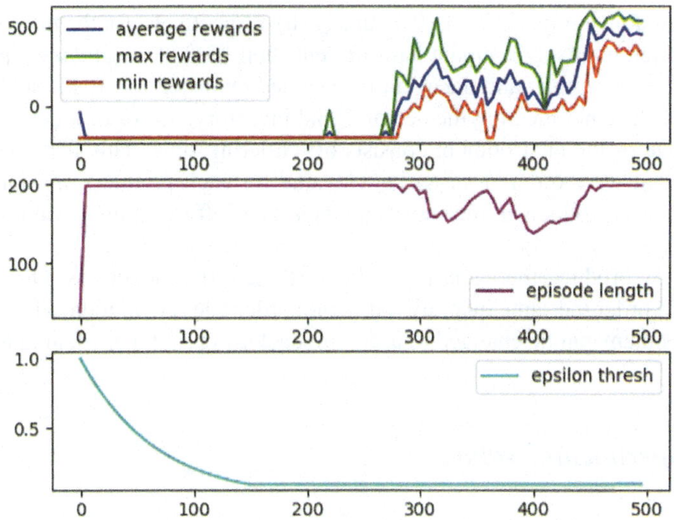

Fig. 5 Deep Q network + prioritized experience replay

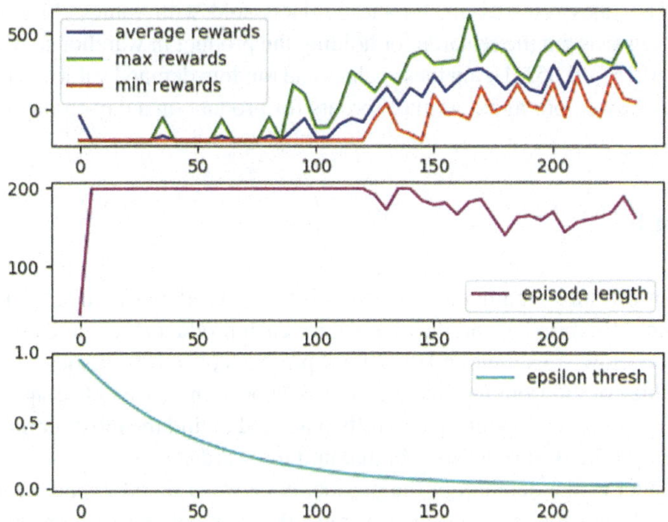

Fig. 6 Double deep Q network + prioritized experience replay

7.1 Environment

There are three products taken into account in the environment. An episode in the environment represents 30 days and for each day the environment chooses values of demand for the product which is not known to the agent. The demand for each product is set for random numbers (0–2), (0–8) and (4–6), respectively, for the three products. Each product in the environment has different shelf lives. A single day represents a step in the environment and the agent is rewarded/punished for: Items sold, expired items, inventory not meeting the demand, and inventory cost of products.

The state of the environment consists of: Ordering state: This is the amount of products that were ordered yesterday (1 value for each product), inventory state: The number of products in inventory and their days left to expire (4 values for each product).

For every product, the agent must choose three different actions: Order 1 more item today, order the same quantity today, and order 1 less item today. The values in the environment can be changed to accommodate different types of situations.

7.2 Experimental Values

Three products were considered, for each having different rewards for expiry, sales, inventory, and opportunity cost. The expired rewards are the rewards the agent receives for expired products whose values are: $\{-0.1, -0.2, -0.1\}$ for each expired product. The sales rewards are for the products sold with values: $\{1.2, 0.9, 1.0\}$. Inventory rewards are the rewards for holding the product in warehouse with values: $\{-0.05, -0.15, -0.05\}$. Using these values and random demand values as mentioned above, we allowed our agent to make orders for product on a day-to-day basis.

8 Results

The sum of inventory maintained by agent for each product ('sum'), the number of items expired ('expired'), and the demand for each product ('demand') was plotted. The results below are shown in Fig. 7 for 3 products over 150 episodes.

The values of environment were carefully chosen in order to test agent in challenging situations. The agent successfully managed to find the most optimum values of ordering products in order to gain maximum rewards.

In 'product 0' the demand of the product is between 0 and 2, and with the high sales reward, and low absolute expiry rewards, it is better to let products expire rather than to not store them at all. This is congruent with our results as the agent has adopted a similar strategy for ordering of products.

As for 'product 1,' our agent has managed to reduce the sum of inventories stored in the warehouse while not letting any products expire, thus reducing the cost of storage.

To demonstrate the capabilities of approach, the agent is placed in tricky situations. If any product is not profitable to store than our agent learns to avoid that product completely. Such a situation has been observed in 'product 2.' This is valuable

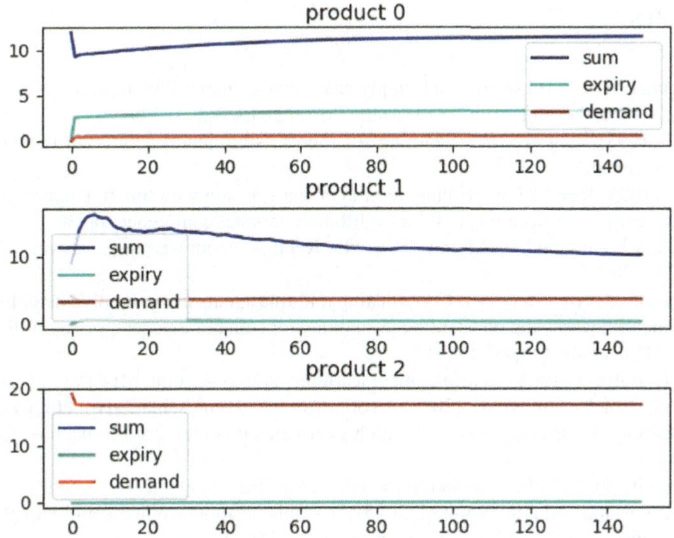

Fig. 7 Inventory orders

information since any human would always tend to expand its range of products but with the use of AI one can understand as to how some products are not worth selling.

Further, agent does not only look at individual products to order but observes the entire state of the warehouse. This means that if a warehouse stores multiple products, than reinforcement learning agent will consider the total inventory costs of the warehouse and maintain a balanced ordering to get the maximum benefits from the available resources.

9 Conclusion

This system has potential of using it for larger corporates like those used by e-commerce Web sites, food storing warehouses, or state run ration centers. A reinforcement learning approach to inventory management can ensure optimized running of warehouses. This would reduce the expiry of products to reduce wastage and reduce costs of warehousing of products. In such large corporates, it is difficult to find the most optimum point of ordering when there are hundreds of products involved and conventional approaches to finding optimum ordering point are not adaptive to changing situations or real-life random demand curves. Reinforcement learning helps in offering key insights to a warehouse manager to efficiently make use of the available warehouse space.

References

1. Christopher M (1992) Logistic and supply chain management. Pitman Publishing, London
2. Watkins CJCH, Dayan P (1992) Q-learning. Mach Learn 8(3–4):279–292
3. Lin L-J (1993) Reinforcement learning for robots using neural networks. Technical report, DTIC Document
4. Baird L (1995) Residual algorithms: reinforcement learning with function approximation. In: Machine learning: proceedings of the twelfth international conference, pp 30–37
5. Kaelbling LP, Littman ML, Moore AW (1996) Reinforcement learning: a survey. J Artif Intell Res 4:237–285
6. Barbuceanu M, Fox MS (1996) Coordinating multiple agents in the supply chain. In: Proceedings of the fifth workshop on enabling technology for collaborative enterprises (WET ICE'96), Stanford University, CA, pp 134–141
7. Sutton R, Barto A (1998) Reinforcement learning: an introduction. MIT Press, Cambridge
8. Diuk C, Cohen A, Littman ML (2008) An object-oriented representation for efficient reinforcement learning. In: Proceedings of the 25th international conference on machine learning, pp 240–247
9. van Hasselt, H (2011) Insights in reinforcement learning. Ph.D. thesis, Utrecht University
10. Sutton RS, Mahmood AR, White M (2015) An emphatic approach to the problem of off-policy temporal-difference learning. arXiv preprint arXiv:1503.04269
11. Van Hasselt H, Guez A, Silver D (2015) Deep reinforcement learning with double Q-learning
12. Schaul T, Quan J, Antonoglou I, Silver D (2016) Prioritized experience replay

Implementation of 5S to Set Up Inventory Control System with HTML Coded Spare Management System

Sandip Mane, Jay Bhuva, and Smit Patel

Abstract This research paper has brought out the importance of inventory management considering the availability of stock in absence of its data or improper accessibility of data that transforms the asset into liability. Process proposed in this paper helped in identifying over inventory and eliminating dead inventory thereby raised level of optimum inventory significantly within two months of process implementation. This process helped to maintain available spares in a systematic manner as well as handle inventory management effectively, which is an important link in supply chain. Inventory models are based on the assumptions. A number of inventory models were analyzed, but none could be implemented because of invalid assumptions. The new inventory model was proposed based on assumptions mentioned in this paper. The proposed model can be implemented effectively in a short span to control the inventory.

Keywords 5S · Optimum inventory · Dead inventory · Over inventory · Spare management system

1 Introduction

Balanced inventory helps organization with numerous benefits such as improved cash flow, high throughput, and reduction in operational expense. In this paper, inventory is further categorized as dead inventory, over inventory, and optimum inventory. Elimination of dead inventory along with effective use of over inventory improves warehouse management.

Mishra et al. [1] studied the inventory model of deteriorating items and contradicted the assumption that items preserve their physical characteristics in storage for long time and shared his thoughts of maintaining inventory of deteriorating items as a challenging and alarming issue. Harris et al. [2] examined the necessity of EOQ

S. Mane (✉) · J. Bhuva · S. Patel
Department of Production Engineering, Dwarkadas J. Sanghvi College of Engineering, Mumbai 400056, India
e-mail: sandip.mane@djsce.ac.in

© Springer Nature Singapore Pte Ltd. 2020
H. Vasudevan et al. (eds.), *Proceedings of International Conference on Intelligent Manufacturing and Automation*, Lecture Notes in Mechanical Engineering,
https://doi.org/10.1007/978-981-15-4485-9_34

model and proposed a formula for determining economic order quantity (EOQ) way back in 1915. Later, Ghaseni et al. [3] stated that assumptions of EOQ model were unrealistic. There are market fluctuations, uncertainty in manufacturing plant, obsolete inventory, availability of disproportional inventory (over inventory and under inventory) in warehouse made it difficult to apply EOQ model in real life. Well-maintained warehouse and knowledge of using appropriate EOQ model can reduce risks associated with usage of EOQ model. Salunkhe et al. [4] investigated the core requirement of inventory management and linked it with models such as 5S, Kanban, and Kaizen. He termed 5S for inventory management as one of the best suited models to enhance efficiency and effectiveness by identifying items in warehouse and maintaining these items in more systematic order and sustenance of this new order. Seyedhoseini et al. [5] emphasized on the fact of decaying inventory and proposed method which would help in using available inventory before it decays. Zhou et al. [6] experimented on time varying demand rate and deterioration of item and designed model of using of two warehouses in minimum cost. Availability of two warehouses may somehow increase holding cost, but it would reduce dead inventory drastically which can be helpful for medium to big manufacturing plant. Arda et al. [7] analyzed the application of multi-supplier strategy which helps in reduction in inventory holding cost and also the shortage cost. Sana et al. [8] focused on imperfect production inventory model and proposed strategy which helped in satisfying demand conditions on time. Baek et al. [9] studied production inventory system and proposed effective outbound process in inventory management considering each customer draw single item from inventory at a time. Hariga et al. [10] studied multi-warehouse concept and proposed model which would provide use of multi-warehouses on the basis of fixed and flexible space leasing contracts.

Various models were studied and analyzed to develop new inventory control system proposed in this paper. 5S is considered to be the best suited aid which would help in effective inventory management. It is used in wide spectrum of industries which includes healthcare, education, and even in government. Its origin is in manufacturing, and inventory management is one of the important segments of manufacturing plant. Standardization is the fourth S of 5S model. More the standardization, less the bottlenecks in system. Inventory can be managed and sustained in much better way when 5S model collaborates with proposed computer controlled inventory management system in this paper. It has not been reported earlier of using such process to manage inventory. Henceforth, it would be discussed that how proposed method helped in eradicating dead inventory and smart use of over inventory helped to reduce large stack of unused inventory. It helped to bring inventory in warehouse at optimum level and to sustain it with simple proposed process. Many models are available to control inventory on basis of assumptions, but most of them are inapplicable because of unrealistic nature. This paper discussed inventory control of spares which was used infrequently, and only 40% of this spare was used frequently, so to maintain such inventory, proposed process is being used.

Proposed model is different from other available inventor models in terms of assumptions and scale. Most of the inventory models available are based upon constant demand rate, total demand is known in advance etc., such models could be excessively used in big production plant, but for department level inventory management, it is incompatible.

The assumptions of inventory model would be as follows:

- Lead time is known
- No holding cost
- Shelf life of inventory is known.

This model facilitates perpetual inventory control system which provides real-time inventory data. Semi-skilled computer operator can easily control inventory management process. Over inventory could be depleted within couple of months of implementation. One of the major advantages is its sustainability with minimal monitoring of inventory. Availability of lead time data gives provision to keep minimal safety stock, and it also helps in costing and for material requirement planning (MRP).

2 Methodology

Dividing methodology in three phases would help to understand process in better way. The first phase started with analysis of available inventory management system and its bottlenecks. Few measures were taken before starting sorting procedure. Spares were kept in cupboards according to their weight. The lighter components were placed in higher shelves and heavier one in lower shelves. This avoided the distortion of shelves and also helped in avoiding injury to person drawing out spare from cupboards. Spares like sensors, connectors, relays, nut and bolts, small batteries, etc., were kept in higher shelves. For heavier spares like pump, motors, gears, etc., lower shelves of cupboards were preferred.

Inventory data was to be stored and maintained in form of Excel sheets. To simplify the maintenance of inventory data, with the help of HTML code, a 'Spare Management System' named form was generated which was attested in Excel sheet containing inventory data. This code helped in provoking form whose stand out feature was particular named UID number. Each spare was given unique six-digit number. This number helped in locating the available spare in warehouse of department. First digit of that UID number signifies the cupboard in which that spare is located. UID number would help in both inbound and outbound process of spare. With the help of this number, new spare would be added in inventory data, and existing spare's data would be updated. In case of outbound process, UID number's sheet would be pasted in cupboards according to shelves. For example, there are ten spares in second shelf of cupboard, and UID number of that ten spares ranges from 100,010 to 1,000,020. So, this ten numbers will get printed on sheet of paper and would be pasted against shelf 2. If a person who is unknown by the name of spare draws spare from shelf, he has to note down the range of UID numbers of spares in that shelf. In the system, he would search for that range of UID number and identify the exact UID number with the help of photograph of that spare to update inventory data.

The situation was such that untidy and disorganized store condition of department made it difficult for supervisor to keep track of inventory properly.

Figure 1 shows the department layout before implementing 5S. Here, cupboards are located at extreme ends of department. Available space was not utilized. Rack was filled with scrap and heavy spares randomly. When ordered inventory was received by department, it was kept in one of those cupboards according to space availability in shelves or else in open nearby rack area.

After analyzing few models, 5S was thought to be implemented. First step of this Japanese concept was sorting, other four stages are set in order, shine, standardize, and sustain. Figure 2 shows the changed layout of department. It helped in availability of more space for cupboards and utilizing department space in much better way. New cupboards were added in store, and storage area was concentrated at extreme left end of department. Rack was removed, and obsolete spare of that rack was disposed. Other spares of rack were shifted in new cupboards to sort it later. In case of inventory

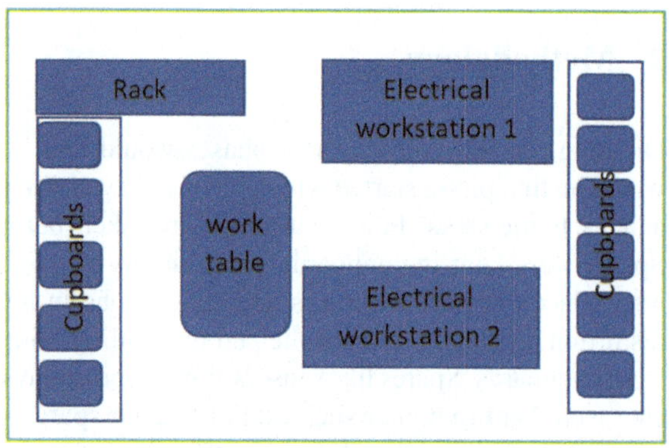

Fig. 1 Before implementing 5S

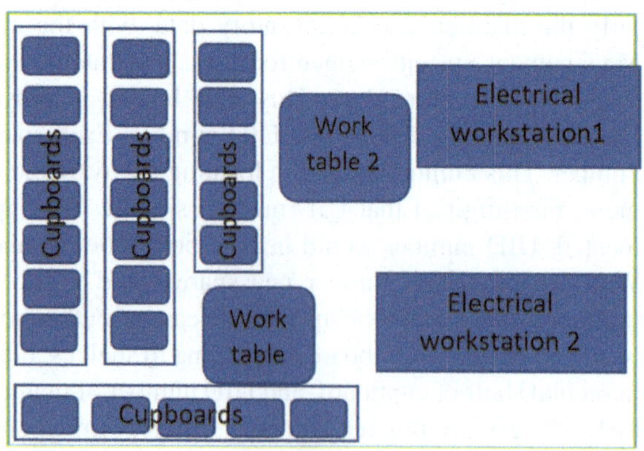

Fig. 2 After 5S implementation

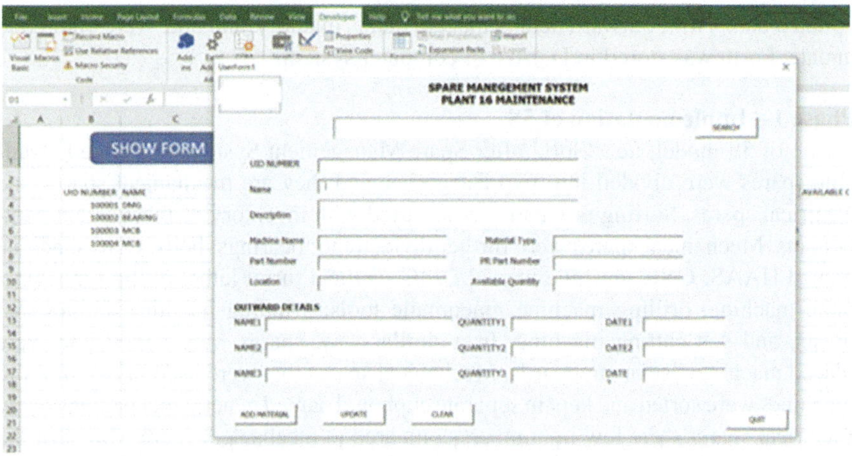

Fig. 3 ‘Spare Management System’ HTML coded form to manage inventory data

management, standardization and sustenance play a major role. Sorting was easier because availability of more cupboards to distribute available spare make it easier to implement 5S in convenient way.

Figure 3 shows ‘Spare Management System’ form attested in Excel sheet of inventory data in upper left corner. Clicking ‘Show Form’ tab pop ups form. Particulars of this form are UID number, Name, Description, Machine name, Material Type, Part Number, PR Part Number, Location, and Available Quantity. Another part of form is named ‘Outward details,’ and it consists of Name, Quantity, and Date. It stored last three entries of outward details at time. There were five tabs named Search, Add Material, Update, Clear, and Quit. ‘Search’ tab is used to view inventory condition.

The six digits UID number should be input of this search box. Whenever new spare which was not there in available inventory data, ‘Add’ tab was used and new UID number was allocated and rest of the data was filled in form. When changing data of available inventory, like adding or subtracting quantity of spare during outbound and inbound process or when location or other available data was changed, it was updated in system through Update tab. Again for update, UID number was used to search that spare, and data was edited according to spare status in warehouse. To clear wrong data filled during adding or updating spare data, Clear tab was used to refill data of that spare again from start. Quit tab helped to disappear form after its data was updated or viewed. In Excel sheet, there will be nine columns as per particulars of form to manage inventory data. In this way, inventory was managed in Excel sheet in proper order, and required data was available within few clicks in spare management system form. This system gave provision to keep track of outward details. When the material was drawn by clicking the ‘Show Form’ tab, the ‘Spare Management System’ form appeared on the screen. Here, reference number of the spares was entered in search box. Data of that spare was available on screen according to last updated data. Now, outward details that are name, quantity, and date were entered to update. Form was stored in D drive of computer system.

Phase 1—Implementation of 5S

Step 1 of 5S model, i.e., 'Sort,' after Spare Management System was coded, available spares were divided into two categories, and they are mechanical spares and electrical spares. Sorting procedure was started with inventory of mechanical components. Mechanical spares were further divided into bearings, belts, CNC machine named HAAS, CNC machine named DMG, vertical turret lathe, milling machine, lathe machine, drilling machine, pneumatic tools, grinding machine, cranes and hoists, and gas cutting machine. In a similar way, spares were sorted and were placed machinewise in cupboards. Mechanical spares that were used in two or more machines were sorted and kept in separate cupboard named general mechanical spare. Electrical spares were few in numbers compared to mechanical spares. Most of the electrical spares were general in nature. Many spares' configuration was such that they can be used in more than one machine. They were divided in numbers proportional to usability of those spares by those machines. Past transactions were referred to do so.

Phase 2—Feeding Inventory data in system

To act upon second step of 5S that is 'set in order,' spares of individual cupboards were further sorted out according to weight, and they were placed in shelves accordingly. Bins were used to avoid disordering of spares in shelf. Inventory data was now started to being noted down.

Inventory calculation started with bearing and belt. Data like bearing number, available quantity, and location was jotted down. Other required data as per spare management system form were noted from ERP software. After bearing, inventory of belts was carried out. All the collected data were added in system through 'Add' tab of the system. Now, inventory data of other machines were noted down. Photographs of each spare were captured to attest it into inventory data in computer system for easy identification for person who is unfamiliar with UID number and with even name of spare. Pictures also helped in avoiding confusion due to identical technical names.

It happened multiple times that spare of other machines were found in cupboard of other machines. In such case, that machine spare was shifted to its respective machine's cupboard, and that spare is entered in inventory data sheet of that machine through add of spare management system. Other thing which happened frequently during set in order procedure was that many times few quantities of X machine's spare were found in Y machine's spare cupboard. In such case, that spare is placed in X machine's spare cupboard and same was updated in system. In similar way, the other 3 S of 5S that is set in order, shine, and standardize were done simultaneously.

Phase 3—Setting up Inventory control process

Figure 4 shows flowchart which was proposed for inbound process of spares, i.e., when spare was received by department:

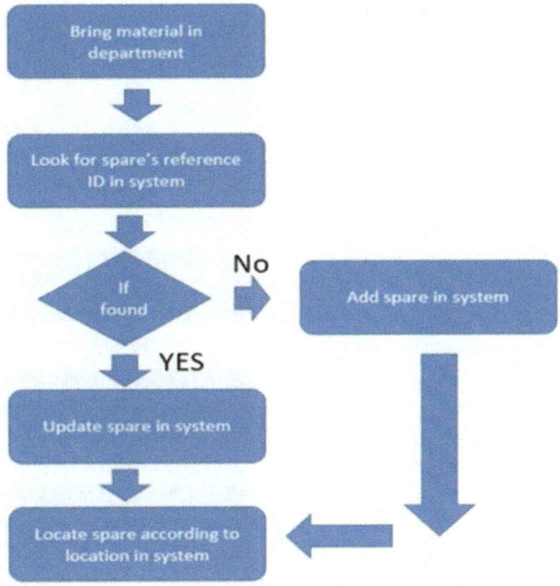

Fig. 4 Inbound process

1. Bringing the spare in department and adding the inventory detail as per requirement in spare management system if particular material is received first time and no data of that spare is available. Else update data according to requirement.
2. Place the spare in cupboard according to location in system.
3. Paste the reference number in cupboard against the spare's shelf, if spare is received first time and no data of that spare was available previously.

Figure 5 shows flowchart which was proposed for outbound process of spares, i.e., when spare was drawn out for use:

1. Required spare was searched through reference number through system or else through name, and in worst case, picture is searched in excel sheet and reference number available in that row is copied to search for spare's location.
2. Spare was drawn from cupboard.
3. Update quantity in system and fill outward details.

After process set up, spares which were critical without which production rate can be hampered were identified. Spares which were more frequently used were also identified in order to keep those spares in optimum quantity in department. Optimum quantity of such spares and critical spares were determined through past transaction data available for this spares in ERP system of company. Accordingly, minimum quantity of this spares were determined. Optimum quantity was 3–5 times of minimum Inventory size of respective spares. Ordering quantity was dependent on criticality of spare, frequent use, cost of spare, and lead time. The spare's whose minimum size was to be maintained were shaded in excel sheet for easy identification

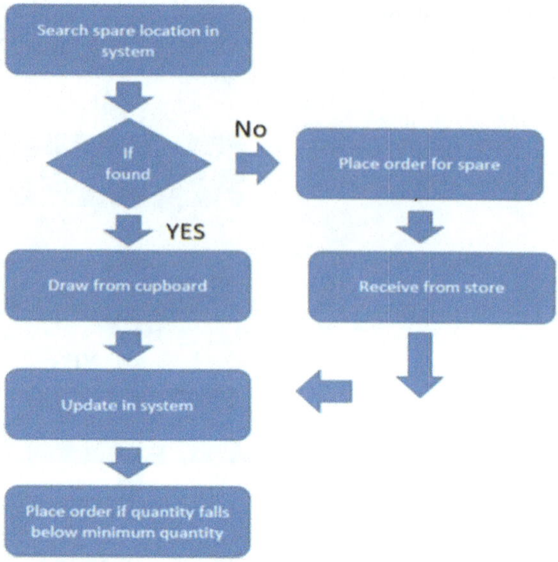

Fig. 5 Outbound process

to keep track during outbound process. Other spares whose minimum size was not mentioned should be available in at least 1–2 unit of quantity.

3 Results and Discussion

There were 180 types of spares which were declared to be dead inventory because of obsolete spares, deterioration, and loss of shelf life. This was disposed which resulted in increase of spaces in store, and at the same time, data reduction was noticed which helped to avail benefits of handling less data. 13% inventory reduction was recorded.

Figure 6 shows the graph of inventory condition before 5S implementation. After

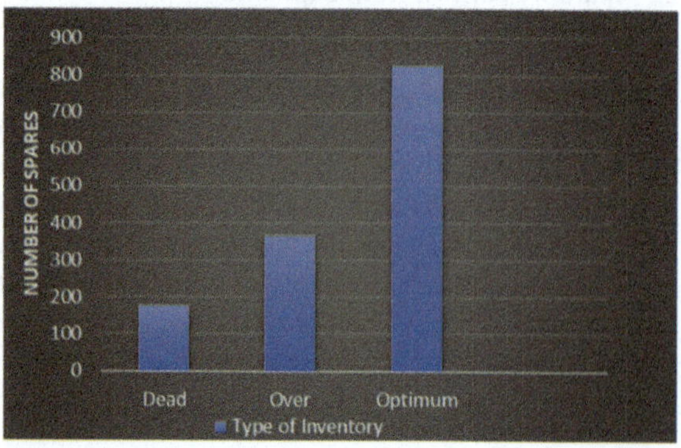

Fig. 6 Effect of dead and over inventory

dead inventory was eliminated, over inventory was targeted. Simple technique was used to tackle over inventory which was to use this inventory spare until it falls below minimum bin size and use of process where refilling up to optimum size helped in tackling over inventory problem. Now, optimum inventory level rose from 59 to 88% within 42 days of implementation. Productivity of workers also improved, and earlier, it used to take around 20–25 min to search for spare. Now, time has been reduced to 3–5 min, and also availability of spare avoids long breakdown hours of critical machines which frequently occur within every 40 days on an average.

Few limitations of this inventory model are as follows:

- More than 2000 spares' inventory control could be complicated.
- Tight monitoring is required sometimes when demand fluctuation is more.
- Technician's discipline is vital to keep track Inbound and outbound process.

With time, staff gets habitual of process, and errors can be reduced considerably within couple of weeks of implementation. Also, carrying out physical inventory twice or thrice within span of one month after implementation can help to adopt inventory model in better way.

4 Conclusion

Dead inventory and over inventory spares were available in the stores for the last 8–12 years costing 164,568 rupees. At 5% of simple interest rate, it costed 49,370 rupees and loss of production at many instances within this period. This can be avoided easily now with the use of new inventory control process. Proposed process helped purchasing department in converting purchase requisition into purchase order easily and quickly, because of availability of PR part number during placing purchase requisitions. This methodology could be proved useful for inventory management of up to 2000 types of spare conveniently. It can be implemented within short span, and inventory data can be improved further without disturbing ongoing process of inventory control. Spares management system can be helpful for managing inventory that are less frequently used and whose availability is important. There are very few models to help such type of inventory management. One computer system and 'Spare Management System' form is sufficient to handle this type of inventory condition conveniently.

References

1. Mishra VK, Singh LS, Kumar R (2013) An inventory model for deteriorating items with time dependent demand and time-varying holding cost under partial backlogging. J Ind Eng Int 9:4
2. Harris FW (1915) Operations and cost. A. W. Shaw Company, Chicago

3. Ghasemi N (2015) Developing EPQ models for non-instantaneous deteriorating items. J Ind Eng Int 11:427–437
4. Salunkhe RT, Kamble GS, Malage P (2007) Inventory control and spare management system through 5S, KANBAN, and Kaizen at ABC industry. J Mech Civ Eng (IOSR-JMCE), pp 43–47. ISSN: 2278-1684
5. Ghare PM, Schrader GF (1963) A model for an exponentially decaying inventory. J Ind Eng 14:238–243
6. Zhou YW (1998) An optimal EOQ model for deteriorating items with two warehouses and time varying demand. Math Appl 10:19–23
7. Arda Y, Hennet J (2006) Inventory control in a multi-supplier system. Int J Prod Econ 104(2):249–259
8. Sana SS (2010) An economic production lot size model in an imperfect production system. Eur J Oper Res 201:158–170
9. Baek J, Moon SK (2014) The M/M/1 queue with a production-inventory system and lost sales. Appl Math Comput 233:534–554
10. Hariga M (2011) Inventory models for multi-warehouse systems under fixed and flexible space leasing contracts. Comput Ind Eng 61:744–751

Automation

Production of Composite Repair Patches for Large Aircrafts Using Advanced Automation Techniques

Amool A. Raina, Boris Manin, and Thomas Gries

Abstract This paper addresses the development of an automated production cell for manufacturing composite patches which are used to repair damages caused on large aircrafts. A step-by-step approach for each process step is provided in detail including damage scanning, sizing and cutting of carbon fibre fabric, kitting and packing of the repair patch.

Keywords Composites · Aircraft repair · Technical textiles · Automation

1 Introduction

In the present day, aircrafts–amongst several other advanced transport vehicles–are no long designed using conventional metal-based materials. Advances in material technology have led to the use of fibre-based materials that exhibit much higher strength and mechanical characteristics as compared to metals. Such fibre-based materials or textiles are used for multidisciplinary applications ranging from traditional apparel used for clothing and upholstery to technical textiles used for the design and manufacturing of structural and non-structural components. Technical textiles, specifically, composites offer unique characteristics owing to their light-weight-to-strength ratios amongst other properties. Several production technologies are used for manufacturing the preforms which are later processed to fabricate the final product. These production technologies include 2D and 3D braiding, multifilament winding, 2D and 3D weaving, tailored non-crimp fabric production, knitting and 3D preforming technologies [1].

A. A. Raina (✉) · B. Manin · T. Gries
Institut für Textiltechnik, RWTH Aachen University, Otto Blumenthal Str. 1, 52074 Aachen, Germany
e-mail: amool.raina@ita.rwth-aachen.de

B. Manin
e-mail: boris.manin@ita.rwth-aachen.de

T. Gries
e-mail: thomas.gries@ita.rwth-aachen.de

© Springer Nature Singapore Pte Ltd. 2020
H. Vasudevan et al. (eds.), *Proceedings of International Conference on Intelligent Manufacturing and Automation*, Lecture Notes in Mechanical Engineering,
https://doi.org/10.1007/978-981-15-4485-9_35

Fibre-reinforced plastics (FRP) have experienced, in recent years, an increasing demand in many industries such as the automotive, aerospace, sports and marine. The development of high strength reinforcements like carbon fibre along with advances in polymer research to produce high performance resins as matrix materials have enabled engineers to meet the challenges posed by the complex designs of modern aircraft. Both large such as Airbus 350 XWB, Boeing 787 and smaller aircrafts such as Lilium, Airbus CityAirbus and Pipistel all use fibre-based composite materials for several key components including fuselage, wings and empennage structures [2, 3]. Due to their increased use, there is an increasing number of damage to components made of FRP. However, the replacement of these parts designed on the basis of high production costs not economical. Damage to aircraft airframe structure causes severe operational interruptions, and the restoration to an airworthy condition has to be demonstrated before commencement of its next flight. Especially, when dealing with composite-based aircrafts, it can also be difficult to assess damage, find and collect relevant information from a wide variety of data sources, while complying with the regulatory record keeping requirements.

It is estimated that on average, a 200 seater aircraft will spend around € 1.52 million per year on maintenance and repair of the aircraft structure [4, 5]. This can be significantly higher depending on the level of damage and complexity of the aircraft being repaired [6].

Hence, two challenges arise from this scenario for the repair of composite-based aircrafts. These two challenges include maintaining airworthy quality of the repaired component and reducing the time required for producing and installing the repair patch onto the aircraft body. This paper addresses these two challenges through the development of an automated production process for manufacturing repair patches for aircraft structures.

1.1 State of the Art

Standards for repair of composite-based aircrafts have been developed by several OEMs and have been ratified and approved by regulatory agencies such as the European Aviation and Safety Agency (EASA) and Federal Aviation Agency (FAA). These procedures are widely documented in repair manuals generated by OEMs as well as in guidelines provided by regulatory agencies [7]. Such standardised repair concepts are also used here as a baseline and benchmark for the design of repair patches for structural components (Fig. 1).

The repair process in general for composite-based aerospace components can be divided into four stages. In the first step, the damage on the component is either visually or digitally detected. As a part of this step, a digital model of the damaged component is created. This enables the maintenance, repair and overhaul (MRO) team to determine the exact strategy for repairing the damaged component. Subsequently, to repair the damaged composites based component, an addition of fibre

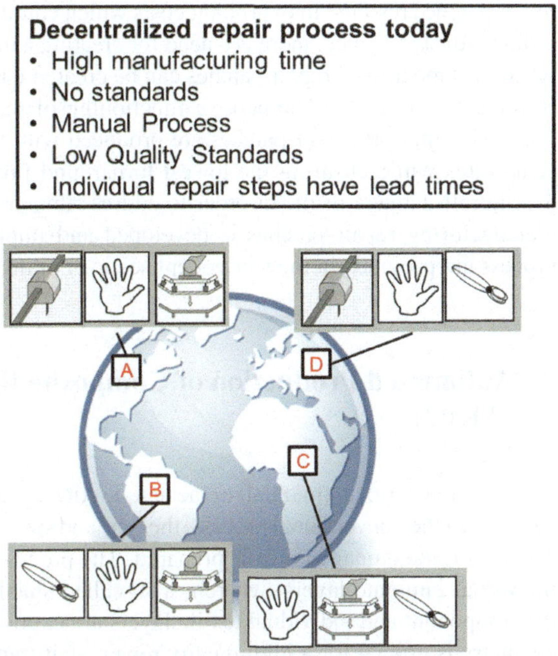

Fig. 1 State of the art of composite repair process for aircrafts

based material to the base structure is required. This added material, when engineered to required dimensions is referred to as a 'patch'. The second step includes the engineering of the repair patch. This includes the selection of materials, sizing of the geometry and determination of number of composite material layers required for repairing the damaged component. In the third step, the damaged area of the component has to be treated and the surface has to be prepared in advance of applicaton of the repair patch. This repair patch is applied to the aircraft body using conventional vacuum-assisted infusion techniques. This patch is then cured as per the guidelines provided in the repair manuals. The final step consists of the quality check and final airworthiness inspection of the repaired component.

The current state of the art, employed by MROs, consists of performing each of the four repair steps (which also includes sub-steps) manually. This manual process is time consuming as well as the quality of the final component cannot be guaranteed. MROs have reported up 30% rework of the repair jobs owing to the non-compliant quality metrics required for airworthiness certification.

1.2 Problem Definition and Need for Automation

Due to the unique nature of the composite part production for aircraft repair, initially, each damage was treated individually. However, it was determined that individualised repair patches resulted in exorbitant costs which could be sustained by the MRO and airline industry. Hence, there is a need for creating a universal repair patch program, wherein standardised repair patches can be created based on well-defined damages. In order to ensure the competitive functioning of the aircraft MRO industry, it is imperative that the repair patches are produced with the highest quality and lowest scrap rates while ensuring the lowest turnaround time for detection and repair of the aircraft damage itself. In order to address this problem, a unique approach for manufacturing repair patches is developed and implemented. The details of this process are provided in the subsequent sections of this paper.

2 Automated Production of Composite Repair Patches (ARP2)

Both the conventional (manual) or the unique ARP2 processes begin with detection of damage. In the conventional process, the type and size of damage is recorded visually. Thereafter, the damaged area is prepared. The process step includes the removal of individual laminate layers, wherein a so-called shank structure is introduced into the component and individual textile layers are exposed (see Fig. 2, left). The shaft geometry is crucial for a high-quality repair, as it transmits the forces of the repair patch and the component. In the current conventional repair processes, the shaft is manually inserted into the component. Therefore, the quality often depends on the human expertise. In the next step, the contours of the exposed textile layers are recorded. Once the contours are known, the individual textile layers can be cut to size. The layer structure of the repair patch usually corresponds to that of the component. In the last step, the repair patch is placed on the damaged area. Optionally, the patch is coated with resin and then cured or in case of prepregs, curing commences after application of the repair patch.

Automated Repair Patch Production (ARP2) Process In the newly developed innovative ARP2 process, individual process steps are fully automated. The damaged area is recorded and digitised using a camera detection system. The repair geometry of the socket is laser milled to the damaged area with a short-pulse laser. This allows the shaft geometry defined in the repair standard to be applied exactly to the component. Thus, the quality no longer relies on or is affected by human intervention. Current laser shank technologies offer a geometric accuracy of 20 μm. At the same time, the fibre orientation of the exposed textile layers can be measured inline in the laser-assisted ablation process. Thus, laminate irregularities caused by the damage or during component manufacture can be detected and taken into account in the shank geometry. In this way, an exact, layer-by-layer removal of the laminate layers is achieved, resulting in high repair quality.

The digital damage image is transmitted to the process chain developed in the ARP2 production scenario. Based on the information of the component's layer structure, the type of damage and its geometry, the material layup structure of the repair

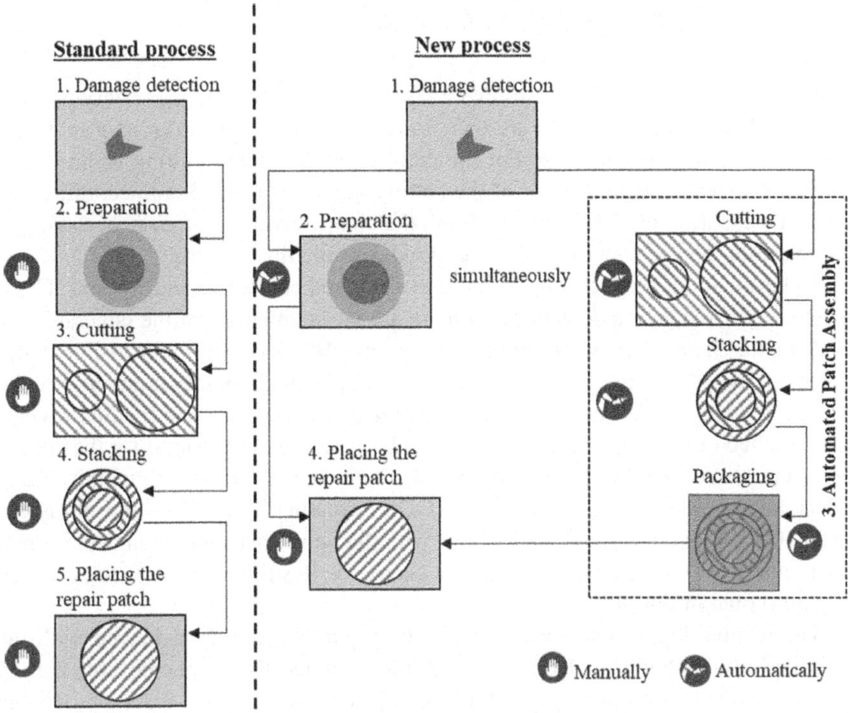

Fig. 2 Overview of conventional and ARP2 repair patch production processes

patch can be determined. The production of the repair patch takes place in the developed automation cell at a different location to the component repair station. In this case, a circular damage spot with a diameter of 25 mm is chosen as a demonstrator. The component consists of eight laminate layers of carbon fibre-based material. The preparation of the repair patch consists of several steps as described below.

2.1 Scanning and Data Evaluation

Scanning and data digitisation is carried out using a nanosecond-pulsed fibre laser system. The average power of this system is 100 W with an operating frequency of 100 kHz. A diameter of 50 μm is readable using this system. The controlling of the scanning system is via a graphic user interface. In the final setup, a hand-held scanning system will be used.

2.2 Sizing of Repair Patch Material

The geometries of the individual textile layers of the repair patch are derived from the recorded damage data and are then co-related to the standard repair geometry size for the particular damage. For the cutting process, three cutting technologies are investigated. The evaluation of the cutting quality is carried out according to a created evaluation model. Figure 3 shows the characterisation of error occurring in cutting tests. The cutting quality is measured over the fibre course.

The evaluation of the cutting technology is ranked between 0 and 3 points. Cases wherein no rovings are detached and there are no undulations in the edge area are evaluated as fault-free and are ranked at 3 points. Blanks with minor visible undulations in the edge area are rated at 2 points. Samples with quality numbers 1 and 0 cannot be used for the repair of structural components, because individual rovings are detached or the cut geometry is deformed. The evaluation criteria for the cutting are summarised in Table 1. According to the defined evaluation criteria, the cutting technologies oscillating drawing knife (ODK), rotating knife (RK) and ultrasonic knife (UK) are examined. Round blanks with diameters between 25 and 117.4 mm are required for the defined demonstrator geometry. Results of the test series carried out are shown in Fig. 4.

The results depict that the highest cutting qualities can be achieved with the rotating knife technology. However, it should be noted that the achievable cutting quality increases with the sample diameter. Furthermore, the test results prove that without auxiliaries for small blanks of less than 64.6 mm, no sufficient quality can be achieved with any of the technologies investigated. The main reason for this is the fibre displacement in the edge area of the sample and the distortion of the sample geometry. If the textile material is fixed with vacuum pressure on a cutting mat, sufficient quality can be achieved with the rotating knife technology even for diameters of 25 mm.

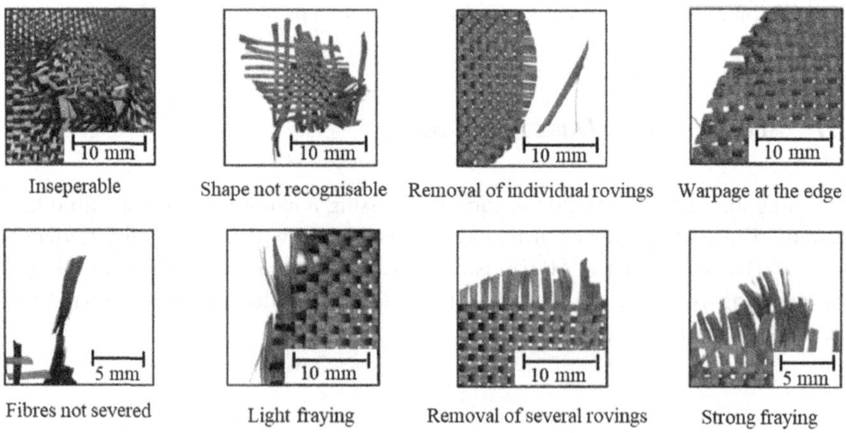

Fig. 3 Characterisation of material defects occurring during the cutting process

Table 1 Font sizes of headings. Table captions should always be positioned *above* the tables

Category	Criteria
0	Several rovings removed from the blank Strong fraying of the rovings and warped geometry
1	Complete severing with distortion at the cutting edge Few rovings removed from the blank
2	Complete severing with no distortion at the cutting edge No rovings removed from the blank
3	Clean severing with no distortion and no fraying of the rovings No rovings removed from the blank

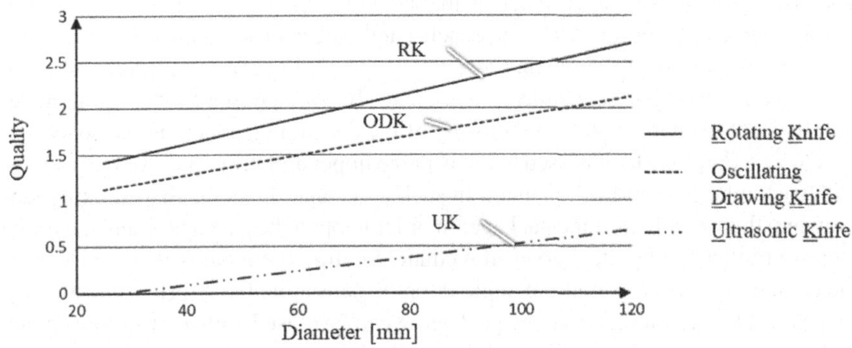

Fig. 4 Achieved quality of individual cutting technologies depending on the sample diameter

2.3 *Handling and Kitting*

Following the cutting process, the individual textile layers are stacked to form a repair patch. The layer structure corresponds to the laminate layers of the component to be repaired. For this stacking process step, a gripping tool is developed. After concept evaluation and elaboration, the gripper system shown in Fig. 5 is implemented. The

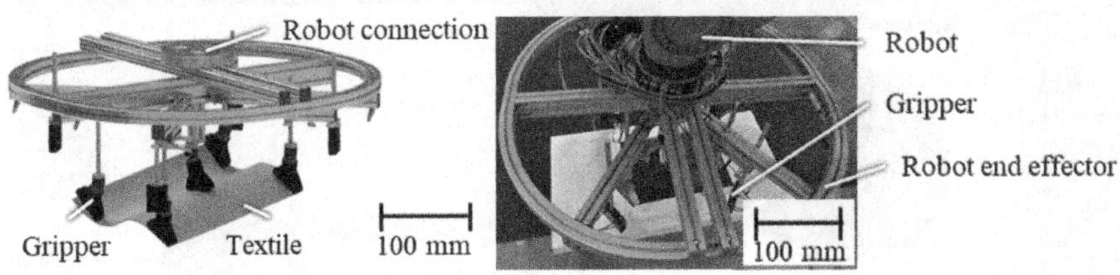

Fig. 5 Handling system developed for transfer of textile (carbon fibre fabrics), stacking and handling of foam structures

system can be adapted to different geometries. By using different gripping techniques (vacuum gripper, needle gripper and Bernoulli gripper), it is possible to grip various types of textiles between 5 mm and 1000 mm diameter. At the same time, the end effector is used to handle the packaging material and interact with the packaging station. The adjustable gripper elements allow gripping of curved geometries.

2.4 Packaging and Transport

The ARP2 process digitally connects the various steps of the repair patch production. For this reason, the data collected from the damage transferred digital to an automation cell, where the repair patch is produced and packed ready for shipment. The goal is to have the repair patch available at the repair location within 24 h after damage detection. For this ARP2 approach, candidate packaging materials are examined with regard to economic and technical aspects. In order to ensure a safe and cost-effective transport, expanded polystyrene (EPS) is chosen as packaging material. In order to enable automated closing of the packaging, a locking mechanism is developed (Fig. 6), which is used to connect the upper and lower parts of the package with each other in a form-fit manner. In such a manner, the hook-shaped fixing tabs snap into the lower half of the package. In order to open the package again, the fixing tabs are broken off by the worker immediately before the repair patch is placed on the damaged area. The optimum angle α_opt is determined in a series of tests. This angle must be adjusted, so that the packaging can be closed with a linear movement, and at the same time, both halves of the packaging do not separate from each other when lifting off.

However, not all requirements can be met by the EPS material alone. As an example, polystyrene packaging does not provide protection against foreign particles and humidity. Hence, the developed packaging is sealed in a vacuum bag in a downstream process step. In this way, the repair patch is isolated from the environment and at the same time protected from mechanical influences. To automate the packaging process, a packaging station is setup, wherein the vacuum bags are opened and the

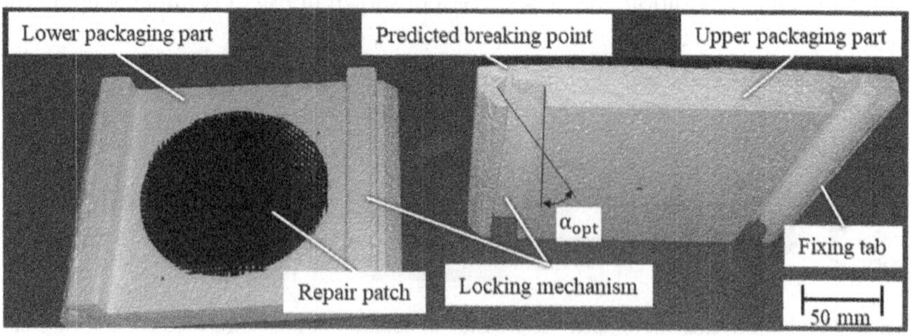

Fig. 6 Two-part styrofoam package with inserted repair patch

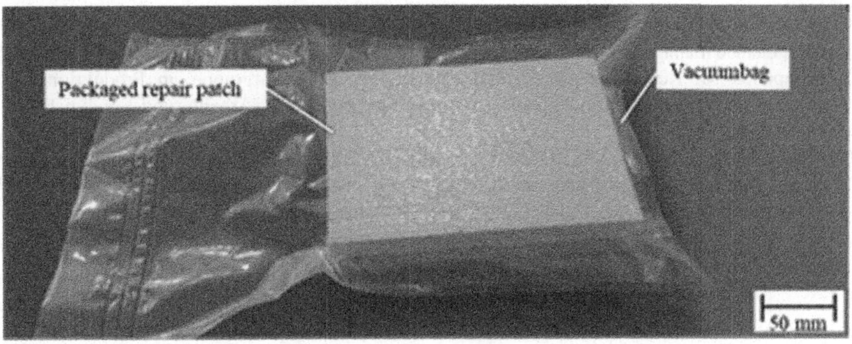

Fig. 7 Final repair patch ready for shipment

EPS unit with the repair patch is inserted. The bag is simultaneously sealed while vacuum is applied to the overall package as seen in Fig. 7.

3 Conclusions

Using the ARP2 process, an automated manufacturing cell for production of repair patches for aircraft is developed. With the rotation knife technology, repair patches of sufficient quality can be produced from a diameter of 70 mm. The precision in stacking the individual textile layers is defined by the industrial robot used and the process dynamics. The developed packaging enables the transport of flat repair patches and protects them from foreign particles and mechanical influences. In further investigations, an economic comparison of the newly developed process with an established process will be carried out. Furthermore, the individual process steps continue to be optimised.

Acknowledgements The authors would like to thank German funding agencies BMBF and DLR for funding the project. We would also like to thank all industrial partners and our project partners from Sabanci University and Turkish Technik, Turkey, for their cooperation.

References

1. Raina A, Huber P, Gries T (2017) Advanced technical textiles for aerospace applications. In: ADMAT conference, Trivandrum, India
2. Raina A, Gries T (2019) Development of a mathematical model for supply chain mapping of composite materials in the aerospace industry. IUP J 2 Supply Chain Manage XVI(2)
3. Dupuy MJ, Wesely D, Jenkins C (2011) Airline fleet maintenance: trade-off analysis of alternate aircraft maintenance approaches, pp 29–34

4. Saltoglu R, Humaira N, Inalhan G (2016) Scheduled maintenance and downtime cost in aircraft maintenance management. World Acad Sci Eng Technol Int J Aerosp Mech Eng 10(3):580–585
5. Kämpchen M, Reimerdes H, Abschlussbericht der Reparaturgruppe, in Produktionstechnik für Bauteile aus nichtmetallischen Faserverbundwerkstoffe, Arbeits- und Ergebnisbericht 1999/2000
6. Ramkumar RL, Bhatia NM, Labor JD, Wilkes JS, Handbook: an engineering compendium on the manufacture and repair of fiber-reinforced composites. Prepared for Department of Transportation FAA Technical Center, Atlantic City International Airport, New Jersey, USA
7. American Airlines, Boeing 737-800 structural repair manual, Rev 25, 10 Nov 2005

Autonomous Real-Time Navigation Based on Dynamic Line and Object Detection

Pavan Rayar, Adarsh Prabhudesai, Samruddhi Pai, and Shaival Parikh

Abstract An Android application has been developed that captures images, detects a line and computes the lateral deviation as well as angular deviation from the path. It also takes into consideration the intensity variation of the surroundings which has been explained in the following paper. We have used the Sobel edge detection algorithm and Hough transform for the angular inference. Colour thresholding and matrix manipulation are done for calculating the lateral deviation. Finally, the counter-active actions and inferences are sent via BLE (low energy Bluetooth). An application was developed for directing the autonomous quadruped robot. The rare-view camera of the mobile was used for taking the input. This Android application can have multiple implementations apart from robotics with slight manipulations, such as the automobile industry, military, industrial automation and blind assistance, and these kinds of researches will boost concept of smart manufacturing.

Keywords Image processing · Filters · Matrix manipulation · Line and object detection · Automation · Robots · Android application

1 Introduction

Computer vision is one of the budding technologies which has an implementation in fields like the automation industry, health sector, financial services, etc. The perception of the real world with an Android application is a very useful tool in today's age. Edge detection is one of the significant fields of computer vision. Out of the multiple edge detection algorithm, Sobel edge detector is used in this application. Hough

P. Rayar (✉)
Department of Production Engineering, Dwarkadas J. Sanghvi College of Engineering, Mumbai, India
e-mail: pavan.rayar@djsce.ac.in

A. Prabhudesai · S. Pai · S. Parikh
Department of Electronics Engineering, Dwarkadas J. Sanghvi College of Engineering, Mumbai, India

© Springer Nature Singapore Pte Ltd. 2020

H. Vasudevan et al. (eds.), *Proceedings of International Conference on Intelligent Manufacturing and Automation*, Lecture Notes in Mechanical Engineering,
https://doi.org/10.1007/978-981-15-4485-9_36

transform is a feature extraction method utilized to define geometrical shapes. Different aspects of visualization are tested in different colour spaces. Colour thresholding is one of the key methods of image segmentation. Only the required information is encoded into pixels to avoid background noise.

2 Literature Survey

Traditionally, line following for autonomous robots is done with infrared sensors. According to a study done by Prakash et al. [1] presented in 'Computer Vision Assisted Line Following Robot', they have used camera for object detection and IR sensor array for line detection and automating the supply chain industries. Object detection is done with the help of colour segmentation [2] for detecting defective shipments. Line following is used for navigation of an automated guided vehicle through various zones [3]. The major drawback of the infrared sensors is the functional inaccuracy when exposed to intensity variations of light. Owing to that in this study, we amalgamate the object detection with line following using an Android device which not only is more reliable and accurate but can handle the intensity variations with better precision [4, 5].

3 Methodology Used

Image acquisition is done with the help of a rear-view camera of the Android device. The video capture rate can be selected according to the dynamicity of the application. For our quadruped robot perception, 10 fps was sufficient. The capture image quality can be decided according to the requirement of the task as well as the computational power of the device. In our application, a 640 × 480 pixel image is captured. The image is parallel processed in two ways for finding out the angular deviation and the lateral deviation Fig. 1.

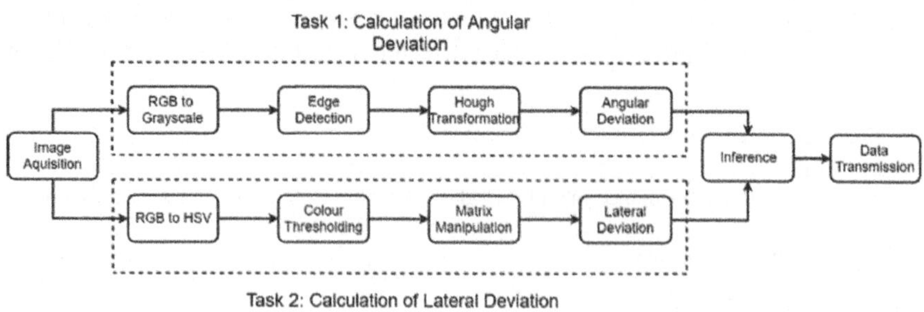

Fig. 1 Flow chart of the complete algorithm

For angular deviation, first, the image is converted into a greyscale image followed by the edge detection. Hough transformation refines the shape further giving out the angular deviation from the axis of the device.

In the second parallel track, RGB is converted into the HSV colour space as it is susceptible to less noise. In this specific application, colour thresholding is used to detect white colour, but any colour can be detected by changing the colour thresholds. Finally, the binary image is interpreted into the lateral deviation via matrix manipulation. Thus, taking these to values into consideration, the position of the robot can be accurately interpreted. The IR sensor of the mobile is utilized to measure the intensity of the surroundings so as to make sure the image is captured in suitable lighting conditions.

Task 1: Angular Deviation

1. RGB To Greyscale Conversion RGB to greyscale conversion is done to encompass the data of three matrices into a single matrix. RGB to greyscale conversion is obtained by computing the weighted sum of the red, green and blue matrices of the RGB image. The weights are adjusted to compensate for the variable wavelengths of the three colours. The equation used to calculate the equivalent greyscale pixels is as follows:

$$X_{ij} = ((0.3 * R_{ij}) + (0.59 * G_{ij}) + (0.11 * B_{ij})) \tag{1}$$

The RGB to greyscale conversion can also be depicted on a 3D plane using a unit cube with one vertex on the origin and three edges coincident with the coordinate axes. The origin corresponds to black, and point (1, 1, 1) corresponds to white. The diagonal joining these two points defines the greyscale threshold (Fig. 2).

Fig. 2 RGB to greyscale correlation in 3D space

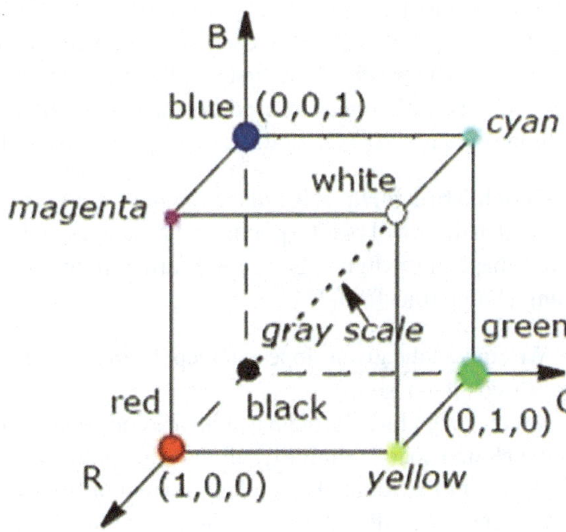

2. Edge Detection In our application, we had to detect lines and objects like a step, ropes, slopes, etc. Sobel edge detection algorithm was used in the first stage of processing, in order to detect these objects.

A discrete approximation of the gradient can be used to detect the significant changes in the greyscale image. Sobel edge detector uses two convolution masks (generally 3×3), one is used for computing gradient along the x-axis and another one for computing gradient along the y-axis [4]. Wherever a sudden change in intensity is detected, a local-maxima is obtained there as an indication of an edge. These gradient values of each pixel are compared with the threshold value, and the one exceeds this value is detected as an edge. After multiple iterations, the threshold value was chosen to be 10 as it gave relatively better results.

Its convolution kernel used is designed in such a way that it smoothens the input image considerably and makes the operator less susceptible to noise. One of the advantages of the Sobel detector is that it does not consider only the single-pixel but takes differential of two rows or columns which enhances the edges making them seem brighter and thicker.

In the Sobel edge detection technique, the centre of the convolution mask coincides with the pixel of interest, and then, it calculates the G_x and G_y (gradient in each direction) for that particular pixel. For digital images, the derivates are approximated by considering the difference of adjacent pixels. The gradient approximation used here is given by,

$$G_x \sim f[i, j+1] - f[i, j] \tag{2}$$

$$G_y \sim f[i, j] - f[i+1, j] \tag{3}$$

where $i =$ negative y-direction, $j =$ positive x-direction.

In this way, the gradient of each pixel is calculated by varying the values of i and j in a loop. The pixels of the first and last rows and columns cannot be calculated if a 3×3 convolution mask is used as while attempting to align the centre of the mask with the concerned pixel, and the mask goes outside the image boundary [2].

3. Hough Transform Edge detection does not give accurately the most prominent line and in it, a lot of pixels (points) are missing, so it becomes difficult to interpret the given shape of the figure. Hence, we further process the image for feature extraction using Hough transform [1].

We convert the given slope-intercept form ($y = mx + c$) into parametric equation ($x = r \cos \phi + r \sin \phi$).

All the given lines through a point are considered, and this yields a sine-like curve in the Hough space. Similarly, all the edge points are considered, and Hough peaks are taken into account. The global maxima of the set are considered to be the most prominent line as the intensity variation would be maximum, and the edge would be the one differentiating the background with the white line on it. The angle and rho value are given as the output of the Hough transform block. The angle of the line is with respect to the axis of the Android device; hence, we get the perceptional angular deviation with respect to the axis.

Task 2: Lateral Deviation

1. RGB To HSV Conversion While RGB model depicts colour as a combination of primary colours, HSV model is used to define an output similar to how the human eye perceives colour. In applications where colour description would be more useful, HSV is preferred over RGB model.

'Hue' stands for the colour, 'saturation' defines the ratio to which the required colour is mixed with white, and 'value' defines the ratio to which the required colour is mixed with black. (grey level).

RGB to HSV conversion formula:

The values of R, G, B are divided by 255 to change their range from 0–255 to 0–1.

$R_{\mathrm{norm}} = R/255$
$G_{\mathrm{norm}} = G/255$
$B_{\mathrm{norm}} = B/255$
$C_{\max} = \max\left(R_{\mathrm{norm}}, G_{\mathrm{norm}}, B_{\mathrm{norm}}\right)$
$C_{\min} = \min\left(R_{\mathrm{norm}}, G_{\mathrm{norm}}, B_{\mathrm{norm}}\right)$
$\Delta = C_{\max} - C_{\min}.$

Hue calculation:

$$\text{Hue } H = \begin{cases} 0^{\circ}, \ \Delta = 0 \\ 60^{\circ} \times \left(\frac{(G_{\mathrm{norm}} - B_{\mathrm{norm}}) \bmod 6}{\Delta}\right), \ C_{\max} = R_{\mathrm{norm}} \\ 60^{\circ} \times \left(\frac{B_{\mathrm{norm}} - R_{\mathrm{norm}}}{\Delta} + 2\right), \ C_{\max} = G_{\mathrm{norm}} \\ 60^{\circ} \times \left(\frac{R_{\mathrm{norm}} - G_{\mathrm{norm}}}{\Delta} + 4\right), \ C_{\max} = B_{\mathrm{norm}} \end{cases}$$

Saturation calculation:

$$\text{Saturation } S = \begin{cases} 0, & C\max = 0 \\ \frac{\Delta}{C\max}, & C\max \neq 0 \end{cases}$$

Value calculation:

$$V = C_{\max}$$

HSV or Hue Saturation Value are used to extract the luminance information from the given image and provide better inference of the surrounding light intensity when compared to RGB model.

2. Colour Thresholding Colour thresholding is done to extract the object of desired colour by masking all the other information of the image. In our application, we had to detect line and various objects. The line was white in colour, and all the other objects were of yellow colour. Thresholding operation was performed on the HSV image where different limits of Hue, Saturation and Value were applied to isolate white and yellow colour from the image. The pixels with all three values within the specified limits were assigned logic 1 and others as logic 0.

If $\text{Hue}_{\text{min}} < H < \text{Hue}_{\text{max}}$, then $H = 1$, else $H = 0$
If $\text{Sat}_{\text{min}} < S < \text{Sat}_{\text{max}}$, then $S = 1$, else $S = 0$
If $\text{Val}_{\text{min}} < V < \text{Val}_{\text{max}}$, then $V = 1$, else $V = 0$

3. Matrix Manipulation For calculating the lateral deviation, we split the processed image matrix into five parts, partitioning column-wise. The matrix would be of 0s and 1s, and the subpart containing a greater number of pixels having value 1 would denote the white area.

$$A_k = \sum_{j=1+(k*\frac{n}{6})}^{(k+1)*n/6} 1 \sum_{i=1}^{m} W_{ij}$$

$$\begin{bmatrix} W(1,1) & \cdots & W(1,m) \\ \vdots & \ddots & \vdots \\ W(m,1) & \cdots & W(m,n) \end{bmatrix}$$

4. Lateral Deviation Calculation The sum of each submatrix is stored into an array named 'Line_detect' and then compared to identify in which part the line lies. The index of the maximum value of the array denotes the position of the line.

Line_detect = [A1 A2 A3 A4 A5]
[max_value, postion] = max (Line_detect)

The max() function outputs the maximum value and the index of the maximum value. The index is the maximum value and is the position of the white line. This position is transmitted via BLE (low energy Bluetooth) to other devices.

4 Applications

This application was primarily built to help a quadruped robot in perceiving its surrounding and according to take actions. The information about the angular as well as lateral deviation from the white line, that the robot had to follow, was transmitted to the on-board controller of the robot. According to the received information, the controller gave information to the actuators that in turn reacted to reduce the deviation. There were different functions designed for climbing the step of crossing the rope.

Whenever a step or rope was detected, the respective functions were called and the actuators responded accordingly. The application as a lag free user interface and multiple functional on-screen buttons. A slider is used to set the threshold value, and the 'Line_Detection' button is used for screen transition from binary output to RGB (as shown in Fig. 3a, b).

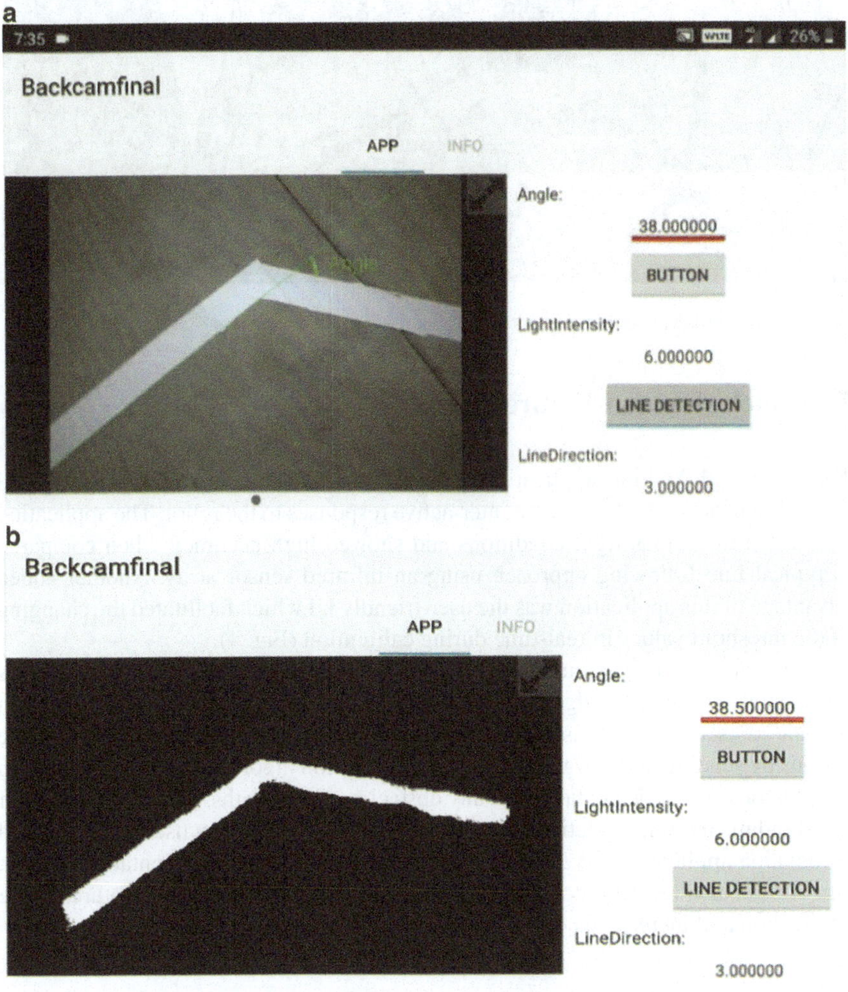

Fig. 3 **a** Input image displayed on the Android GUI. **b** Processed image

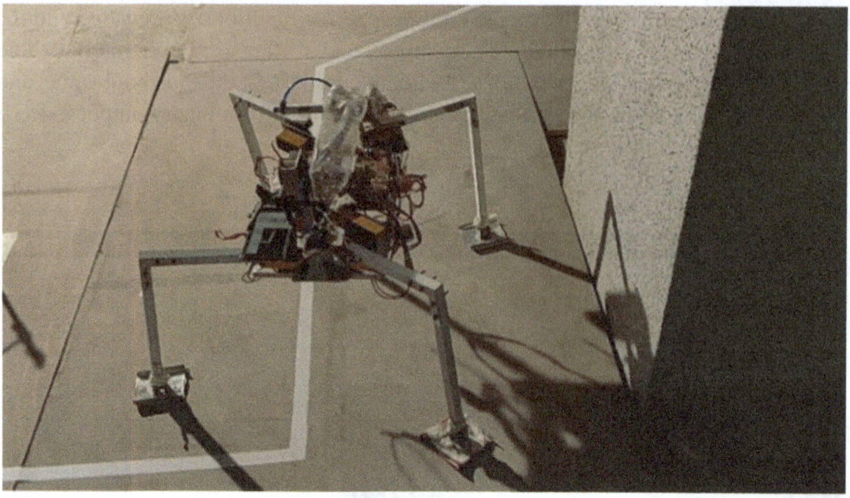

Fig. 4 Algorithm deployed on a quadruped robot

5 Conclusion and Future Scope

The developed Android application was used to guide the autonomous quadruped robot. It sent accurate data and counter-active responses to the robot. The application was tested in various light conditions and showed high accuracy when compared to typical line following approach using an infrared sensor array. Another added advantage of this application was the user-friendly UI which facilitated the changing of the threshold values in real-time during calibration (Fig. 4).

The generalized algorithm developed here can be modified and used for various other applications. This algorithm can be deployed on various platforms and can provide assistance to the user by detecting obstacles and providing counter-active actions by sending audio signals. The object detection algorithm can also be used in autonomous cars to detect lanes and any obstacles in their paths. The dynamic colour thresholding and line detecting part of the algorithm can be very useful in industrial automation applications like parcel segregation based on shape, orientation, colour, etc. We can also incorporate other sensors of the mobile to add more features in the application and increase its functionality and robustness.

References

1. Surya Prakash M, Ajay Vignesh K, Shyamsunthar J (2012) Computer vision assisted line following robot. In: International conference on modeling, optimization and computing (ICMOC-2012)

2. Malik R, Khurshid J, Ahmad SN (2007) Road sign detection and recognition using colour segmentation, shape analysis and template matching. In: 2007 international conference on machine learning and cybernetics, vol 6. IEEE, pp 3556–3560
3. Pakdaman M, Sanaatiyan MM (2009) Design and implementation of line follower robot. In: 2009 second international conference on computer and electrical engineering, vol 2. IEEE, pp 585–590
4. Gupta S, Mazumdar SG (2013) Sobel edge detection algorithm 2(2). ISSN: 2278-733X
5. Gonzalez RC, Woods RE Digital image processing, 2nd edn

Machine Learning as a Smart Manufacturing Tool

Meera B. Kokate, Bhushan T. Patil, and Geetha Subramanian

Abstract In smart manufacturing, machine learning is used in various manufacturing fields at various stages such as for future prediction in the manufacturing system, pattern recognition, fault detection, quality control and monitoring. Machine learning (ML) is used for classification and regression purpose which can be achieved using the past data. Machine learning algorithms and combination of algorithms are widely used in various machining processes. This paper reviews different machine learning algorithms used for specific applications in the product life cycle.

Keywords Machine learning (ML) · Smart manufacturing · Big data · Smart machining · Support vector machine (SVM) · Support vector regression (SVR) · Neural network (NN)

1 Introduction

Smart manufacturing means implementing advanced technologies of Industry 4.0 in manufacturing. Main components of Industry 4.0 are big data, cyber-physical systems (CPS), Internet of things (IOT), cloud computing and machine learning (ML) [1, 2]. Data obtained from various sensors is stored in big data. By using this data, machine learning can help to acquire knowledge and make decisions automatically and intelligently [3, 4].

In machine learning, data is split into training and testing data set. Model is trained using training data (70–80%), and then, performance is evaluated using test data (20–30%). Machine learning handles high-dimensional, multivariate or complex data, structured or unstructured data and also gives relationship between large data

M. B. Kokate (✉) · G. Subramanian
Production Engineering Department, Fr. Conceicao Rodrigues College of Engineering, Bandra, Mumbai, India
e-mail: meera.kokate@gmail.com

B. T. Patil
Mechanical Engineering Department, Fr. Conceicao Rodrigues College of Engineering, Bandra, Mumbai, India

© Springer Nature Singapore Pte Ltd. 2020
H. Vasudevan et al. (eds.), *Proceedings of International Conference on Intelligent Manufacturing and Automation*, Lecture Notes in Mechanical Engineering,
https://doi.org/10.1007/978-981-15-4485-9_37

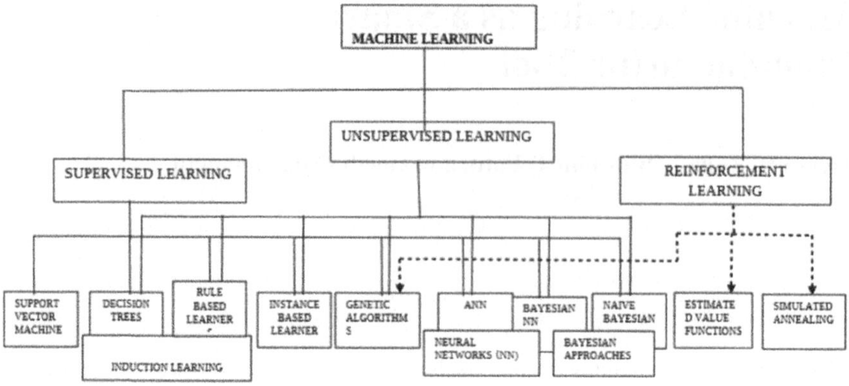

Fig. 1 Structure of ML techniques and algorithms

sets. Also, the major advantage of ML is that it adapts to fast-changing and dynamic manufacturing environment. ML algorithms are divided into three categories supervised learning, unsupervised learning and reinforced learning. The algorithms are performed to attempt two main tasks: classification and regression. Commonly used algorithms are shown in Fig. 1 [5]. Prediction performance of each ML algorithm can be validated using processes like variance R2, mean absolute error MAE, correlation coefficient R, root mean squared deviation RMSD.

Major challenge in using ML is selection of algorithm. Selection of the algorithm depends on whether the data is labeled or unlabeled or the experts available. The data set can be high-dimensional data, or it may contain high degree of irrelevant and redundant data, or some data may be missing. Hence, the biased data and the negative influence should be reduced as much as possible [5].

1.1 Applications of ML in Manufacturing

Nowadays, ML in product life cycle increases the solution effectively in the areas shown in Fig. 2 [6].

ML has a great impact on pattern detection and classification, discovering of faults and calculation of future working conditions and the rest of the useful life.

1.1.1 ML for Material Design and Engineering

ML methods were first introduced in 2006 for the design of materials, which predicts crystal structure, i.e., atomic arrangement in the materials. Valuable information related to properties and other factors of the material is extracted from big data. Influencing factors for design of material are chemical composition, process parameters

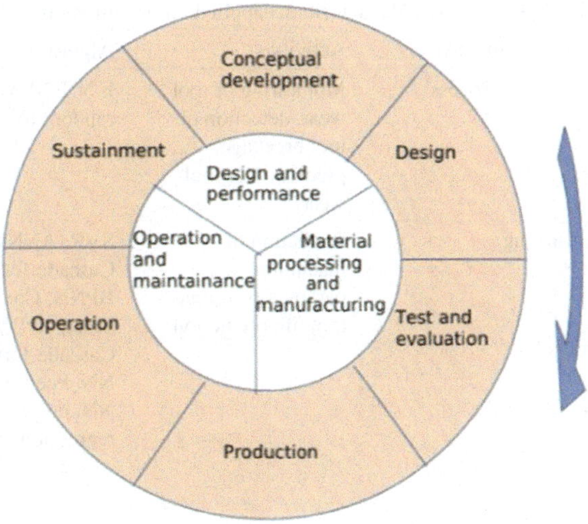

Fig. 2 Machine learning in smart manufacturing

and microstructures. Relationship between these variables can be nonlinear as well. Model for predicting material properties based on nonlinear programming has also been presented. Multiple properties of the material are also taken into consideration. The model is solved using interior-point algorithm [7, 8].

1.1.2 ML in Cutting Process

ML algorithms have been applied to various cutting processes shown in Table 1. Milling and turning are the fields where most of the ML algorithms have been applied, and ANN is most commonly used in cutting processes [9]. ML in machining processes improves productivity rates and product quality levels. It also optimizes design and process parameters and monitors the health of systems and gives cost-efficient production. It is therefore called smart machining [4].

In milling operation, ANN, RF and SVR have been used for tool wear prediction. Data is collected from 315 milling machines. After experimentation and comparison of these three algorithms, results show that RF generates more accurate prediction than ANN and SVR, but the training time for RF is longer than ANN and SVR [22].

1.1.3 ML in Quality Monitoring

Using image's pixels or sensor data about surface roughness, ML can classify good and defective parts into groups. Automated visual inspection using ML is done. Naive Bayes and decision tree classifier are performed to classify good and defective samples. Extraction and classification of histogram features are set. Decision tree classification is superior to Naive Bayes because 9.6% is miss-classification in

Table 1 Different ML algorithms applied on machining processes: (Ref. 10–21)

Machining process	Activity	Algorithms	Input
Milling process	Monitoring of tool wear, detection of tool breakage, prediction of tool wear	k-NN, SVM, SVR, random forest (RF)	Tool images, power consumption data, cutting force, vibrations and acoustic emission
Turning	Prediction of machining parameters, online tool life prediction	SVR, ANNs, Cascade-forward BPNN, Computing based on DNA, Cascade-forward NN, Feed-forward NN, polynomial regression	Radius of tool edge, tool coating status, machining time, feed rate, cutting speed, avg. no. of white pixels from tool image, depth of cut, 6 signal features from cutting force, vibrations, acoustic emission sensor, spindle speed
Grinding	Surface roughness and surface shape peak valley monitoring	Interpolation factor SVR	Acoustic emission, vibration, grinding force
Drilling	Evaluation of quality and geometric profile	Logical analysis of data	Thrust force, torque, cutting force
Boring	Chatter prediction (stable, transition, chatter)	SVM	Feed rate, spindle speed and depth of cut
Abrasive water jet	Surface roughness prediction	Feed-forward BPNN, regression model	Abrasive flow, traverse speed, abrasive grit size, jet pressure, standoff distance
Electric discharge machining (EDM)	Predict optimum process parameter for maximum material removal rate (MRR) and minimum wear ratio, minimum surface roughness	BPNN, GA particle swarm optimization, GPR, NSGA-II, GRA—Gray rational analysis	Pulse current, pulse-ON/OFF time
ECDM, ECM	Optimization of process parameters for maximizing MRR and minimizing radial overcut	Teaching learning-based optimization (TLBO)	Electrolyte concentration and flow rate, applied voltage, inter-electrode gap

Rotating machinery

↓

Data acquisition (Data like vibration, temperature et al from sensors)

↓

Feature extraction (techniques like wavelet transform, statistical analysis)

↓

Fault recognition (pattern recognition methods like k-NN, Naive Bayes)

Fig. 3 Fault diagnosis framework using ML

decision tree and 17.3% in Naive Bayes [23]. Fault diagnosis of rotating machinery using ML can be achieved using different ML algorithms. Steps are shown in Fig. 3 [24].

1.1.4 Life Prediction of Cutting Tool Using ML

A model has been presented to assess the tool condition and to predict its remaining useful life (RUL). It is a reduction of nonlinear features and also uses SVR. Features help to learn nonlinear regression model, to identify and predict the stage of tool wear. Few approaches are developed and implemented to monitor the tool wear level. One of the approaches is prognostics and health management (PHM).

Cutting processes are tedious processes and data received from monitoring process signals and observation values of wear level, are nonlinear. The traditional nonlinear regression method is limited to estimate RUL of tool. Logistic regression model is used for assessment of online performance. Support vector regression model is very much applicable for nonlinear features in diverse machine learning problems [25–27].

1.1.5 ML in Cost Estimation

Estimation of cost involved in manufacturing of engine parts of a large civil jet was done and compared using different statistical methods in the field of data mining and machine learning, namely support vector regression (SVR), generalized additive models (GAMs) and gradient boosted trees (GBT). Comparison of these models is done with usual methods, i.e., ANN and multiple linear regressions. Cost is estimated at the beginning level of the design process. Manufacturing cost for the year 2012 is explained using 254 data points and six covariates [28].

1.1.6 Predictive Maintenance Using ML

In predictive maintenance, maintenance activities are initialized once the working condition of equipment is identified. Predictive tools help to identify the condition of the equipment before its failure. Predictive maintenance can use information of run to failure, so supervised approach is used. In case the information related to past failure experiences is available and if the data is continuous then regression problem can be implemented to predict equipment's remaining life. If the available information is in the form of categorical value, then classification method is applied. It helps to make classification of faulty and non-faulty equipment [29, 30].

Multiple classifiers are used in predictive maintenance. This approach is to avoid unexpected failure. Evaluation of classifier is validated by Monte Carlo cross-validation technique. Mostly used classification techniques are SVM and k-NN. Machine learning predictive maintenance is better for unexpected breaks and equipment to reduce operating cost. Machine learning predictive maintenance chooses dynamic maintenance policy [31].

1.1.7 ML for E-Quality Control

Based on the experimental data, support vector classifier model is made to present e-quality control. The application of sensors, computers, Internet and information technologies and communication technologies is used in the manufacturing sector. This leads to complete automation and complete quality inspection [31, 32]. E-quality is one of the new approaches used to identify performance losses in every stage of production. By this new approach, traditional quality control transforms to sampling techniques by vision sensors and above-mentioned technologies. For this approach, complete loop has been made using networked station.

Experimental data are collected from remote inspection station and analyzed using SVM equation. From the outcome of the model, predictions are made to make decisions regarding acceptance or rejection [31]. Model is made with SVM and kernel. SVM is a systematic approach using statistical learning concept for regression, classification and for density calculation [33].

2 Conclusion

This paper reviewed different ML applications in the manufacturing right from design of material to the marketing of the product. In the design of material, crystal structure is predicted. ML has been used in various machining processes, other operations in the industries such as quality control, inspection, maintenance, parameters optimization and cost estimation. Various types of ML algorithms for particular applications and steps to follow for the execution have also been discussed. Most commonly used ML algorithms are SVM and ANN.

3 Future Scope

ML can be used for calculating mechanical properties like stress, strain and temperature using finite element analysis results obtained from ANSYS.

References

1. Wu D, Jennings C, Terpenny J, Gao RX, Kumara S (2017) A comparative study on machine learning algorithms for smart manufacturing: tool wear prediction using random forests. J Manuf Sci Eng 139
2. Ahuett-Garza H, Kurfess T (2018) A brief discussion on the trends of habilitating technologies for industry 4.0 and smart manufacturing. Manuf Lett 15:60–63
3. Shang C, You F (2019) Data analytics and machine learning for smart process manufacturing: recent advances and perspectives in the big data era. Engineering 1010–1016
4. Kim D-H, Kim TJY, Wang X. Kim M, Quan Y-J, Oh JW, Ahn S-H (2018) Smart machining process using machine learning: a review and perspective on machining industry. Int J Precis Eng MFG Green Technol 5(4):555–568
5. Wuest T, Weimer D, Irgens C, Thoben KD (2016) Machine learning in manufacturing: advantages, challenges, and applications. Prod Manuf Res Open Access J 4:23–45
6. Wang J, Ma Y, Zhang L, Gao RX, Wu D (2018) Deep learning for smart manufacturing: methods and applications. J Manuf Syst 48:144–156
7. Guo S, Yu J, Liu X, Wang C, Jiang Q (2019) A predicting model for properties of steel using the industrial big data based on machine learning. Comput Mater Sci 160:95–104
8. Fischer CC, Tibbetts KJ, Morgan D, Ceder G (2006) Predicting crystal structure by merging data mining with quantum mechanics. Nat Mater 5:641–646
9. du Preez A, Oosthuizen GA (2019) Machine learning in cutting processes as enabler for smart sustainable manufacturing. Procedia Manuf 33:810–817
10. Garcia-Ordas M (2017) Wear characterization of the cutting tool in milling processes using shape and texture descriptors. Ph.D. thesis, Universidad de Leon
11. Cho S, Asfour S, Onar A, Kaundinya N (2005) Tool breakage detection using support vector machine learning in a milling process. Int J Mach Tools Manuf 45(3):241–249
12. Wu D, Jennings C, Terpenny J, Gao RX, Kumara S (2017) A comparative study on machine learning algorithms for smart manufacturing: tool wear prediction using random forests. J Manuf Sci Eng 139(7)
13. Jurkovic Z, Cukor G, Brezocnik M, Brajkovic T (2016) A comparison of machine learning methods for cutting parameters prediction in high speed turning process. J Intell Manuf. https://doi.org/10.1007/s10845-016-1206-1
14. D'Addona DM, Ullah AS, Matarazzo D (2017) Tool-wear prediction and pattern-recognition using artificial neural network and DNA-based computing. J Intell Manuf 28(6):1285–1301
15. Karam S, Centobelli P, D'Addona DM, Teti R (2016) Online prediction of cutting tool life in turning via cognitive decision making. Procedia CIRP 41:927–932
16. Zhang D, Bi G, Sun Z, Guo Y (2015) Online monitoring of precision optics grinding using acoustic emission based on support vector machine. Int J Adv Manuf Technol 80:761–774
17. Shaban Y, Yacout S, Balazinski M, Meshreki M, Attia H (2015) Diagnosis of machining outcomes based on machine learning with logical analysis of data. In: Proceedings of the international conference on industrial engineering and operations management (IEOM), pp. 1–8
18. Saravanamurugan S, Thiyagu S, Sakthivel N, Nair BB (2017) Chatter prediction in boring process using machine learning technique. Int J Manuf Res 12:405–422
19. Caydaş U, Hascalık A (2008) A study on surface roughness in abrasive water jet machining process using artificial neural networks and regression analysis method. J Mater Process Technol 202:574–582

20. Majumder A (2015) Comparative study of three evolutionary algorithms coupled with neural network model for optimization of electric discharge machining process parameters. Proc Inst Mech Eng Part B: J Eng Manuf 229:1504–1516
21. Rao RV, Kalyankar V (2011) Parameters optimization of advanced machining processes using TLBO algorithm. In: International conference on, project, and production management (EPPM), pp 20–21. Science Direct, Singapore
22. Shi D, Gindy NN (2007) Tool wear predictive model based on least squares support vector machines. Mech Syst Sig Proc 21(4):1799–1814
23. Ravikumar S, Ramachandran KI, Sugumaran V (2011) Machine learning approach for automated visual inspection of machine components. Expert Syst Appl 38:3260–3266
24. Liu R, Yang B, Zio E, Chen X (2018) Artificial intelligence for fault diagnosis of rotating machinery: a review. Mech Syst Sig Proc 108:33–47
25. Benkedjouh T, Medjaher K, Zerhouni N, Rechak S (2013) Health assessment and life prediction of cutting tools based on support vector regression. J Intell Manuf 1–19
26. Medjaher K, Tobon-Mejia DA, Zerhouni N (2012) Remaining useful life estimation of critical components with application to bearings. IEEE Trans Reliab 61(2):292–302
27. Ribeiro B (2005) Support vector machines for quality monitoring in a plastic injection molding process. IEEE Trans Syst Man Cybern Part C: Appl Rev 35
28. Loyer JL, Henriques E, Fontul M, Wiseall S (2016) Comparison of machine learning methods applied to the estimation of manufacturing cost of jet engine components. Int J Prod Econ 178:109–119
29. Luo M, Xu Z, Chan HL, Alavi M (2013) Online predictive maintenance approach for semiconductor equipment. In: Proceedings of the 39th IEEE annual conference of the industrial electronics society (IECON) pp 3662–3667
30. Heng A, Tan AC, Mathew J, Montgomery N, Banjevic D, Jardine AK (2009) Intelligent condition-based prediction of machinery reliability. Mech Syst Sig Proc 23:1600–1614
31. Susto GA, Schirru A, Pampuri S (2015) Machine learning for predictive maintenance: a multiple classifier approach. IEEE Trans Ind Inform 2(3)
32. Tseng TL, Aleti KR, Hu Z, Kwon YJ (2016) E-quality control: a support vector machines approach. J Comput Des Eng 3(2):91–10
33. Koksal G, Batmaz I, Testik MC (2011) A review of data mining applications for quality improvement in manufacturing industry. Expert Syst Appl 38:13448–13467

Tool Changer Selection for the Robot to Attach and Detach the End of Arm Tooling

Dhanesh Dhanawade, Nilesh Vijay Sabnis, and Pankaj Gavali

Abstract In today's world, industrial revolution is changing to the industry 4.0. Humanless manufacturing has started to increase the productivity and reduce the cost of the production. Automobile industry is one of the areas, in which such drastic changes are happening day by day. As per the requirement of the customers, a number of variants are produced by the organization to sustain in the market, and such operations are performed by the robots. Fixture and grippers are used for holding and material handling of the parts, respectively. Tool changer is the part, which is used in between the robot and gripper. Signals from robot to gripper and gripper to robot are transferred by the tool changer. Fast and automatic tool changing is done by using it. This paper deals with the study of proper selection of tool changer of robot and how this action enhances the flexibility and productivity.

Keywords Fixture design · Robotic simulation · Automation · Tool changer selection · Gripper · Digital manufacturing · Smart manufacturing · CNC · Automation · Robotics

1 Introduction

Digital manufacturing is one of the new trends taking place in the organizations to increase productivity as per the demand of the customers by offering with minimum price [1].

Robotic simulation is the part of digital manufacturing, which is used for simulating the process, layout and tool. In robotics department the team of simulation engineers works in collaboration with the team of tool design engineers. During simulation, some problems in the fixture or gripper design may be found, which are then immediately resolved, and further process of simulation is carried out [2].

D. Dhanawade (✉) · N. V. Sabnis · P. Gavali
Department of Mechanical Engineering, School of Technology, Sanjay Ghodawat University, Kolhapur, India
e-mail: dhaneshdhanawade123@gmail.com

© Springer Nature Singapore Pte Ltd. 2020
H. Vasudevan et al. (eds.), *Proceedings of International Conference on Intelligent Manufacturing and Automation*, Lecture Notes in Mechanical Engineering,
https://doi.org/10.1007/978-981-15-4485-9_38

While doing simulation, some selection criteria are provided for selecting any component from the library such as robot selection, gun selection and tool changer selection. Tool changer is used for attaching and detaching the tool from the robot and also power supplement such as electricity, signals from the sensors, pneumatic energy and hydraulic energy, etc. [3].

While selecting tool changer, the following points are taken into consideration,

- Weight carrying capacity
- Acceleration by which robot moves weight of the EOAT.

2 Literature Review

K. Youcef-Toumi et al. gave information about CAD/CAM workstation of a robotic sheet metal fixturing system, and the work cell consists of modular (reconfigurable) fixture and six-axis industrial robots, which are integrated together to perform the task [4]. The software mainly consists of four modules given as—graphical appearance of the workplace, analysis of the fixture, design and assembly of the fixture and sequential analysis of the fixture. A database is developed for the entire package. Finite element method is a manual method, which is used to analyze the system, which consists of metal parts, a fixture, gripper and required tools for the operations. While designing an accurate fixture for a given operation of different sheet metal processes fixturing, bending, buckling and yielding of parts are considered as major issues. Stress distribution in the part also affects the optimum selection of the fixture design.

S. J. Blacker et al. illustrated that flexible fixturing in general aims to reduce the time of the production and increase the variants of the product [3]. Additionally, the automation requirements of modern computer-controlled manufacturing can be met by flexible fixture. These include automated workpiece loading and unloading and the automatic configuration and control of the manufacturing process.

Jay lee gave information about the simulation software for the robot. Computer-aided workstations with high performance graphics and proprietary software are dedicated to robotic work cell design and simulation [5].

1. Proper robot and gun selection for the given operation.
2. Operations which are to be performed on the work cell are finalized.
3. Location and optimum path are decided to reduce the cycle time.
4. Program finalization and downloading.

Nicola Pedrocchi illustrated that in multi-robot cell system for spot welding has number of locations for the spot projection. These locations are followed by different robots for different times. For proper working of the robots, stepwise movement is given to the robot by sequence of operation command [6]. Offline programming of the robot is required to be done to increase the actual work setup generation. While considering cell productivity, costs, flexibility and reconfigurability, cell design involves resource selection (such as robots and welding guns) and resource configuration. Motion planning involves allocation of welding points to each resource and calculation of collision-free motion plan for each robot. Currently, cell design and motion planning are sequential and on the other hand manual activities are managed by separate industrial functional units.

3 Case Study

(A) Purpose of Robotic Tool Changer

On the actual work cell, in multi-robot cell, a number of operations were performed simultaneously. Sometimes single robot also has to perform more than one operation, one after another. That time robot has to change the tool number of times as operation changes. Manually, tool can be changed, but it takes time to disassemble and assemble the tool.

But tool changer is one of the components of the robot, which helps the robot to change the tool automatically, which increases the automation of the work cell, speed of the operations and definitely reduces the cycle time of the process. Tool changer has mainly two parts, tool plate and master plate. Tool plate is attached to the end effector, and master plate is attached to the robot. Tool changer also transfers signal, electricity and fluid for hydraulic and pneumatic power as shown in Figs. 1 and 2.

(B) Selecting the Proper Robotic ToolChanger

While selecting tool changer, some points are taken into consideration such as tool changer must have good locking mechanism, to lock and unlock the master plate and tool plate with proper strength. Tool changer must have full-proof mechanism.

When suddenly power supply is stopped, then tool changer should not lose the tool. Tool has to stay coupled to master plate. The payload capacity of the robot and weight of the end effector are compared with the tool changers because of its payload rating. Selecting a tool changer based on payload alone will provide a starting point, but most important factor is moment capacity [7].

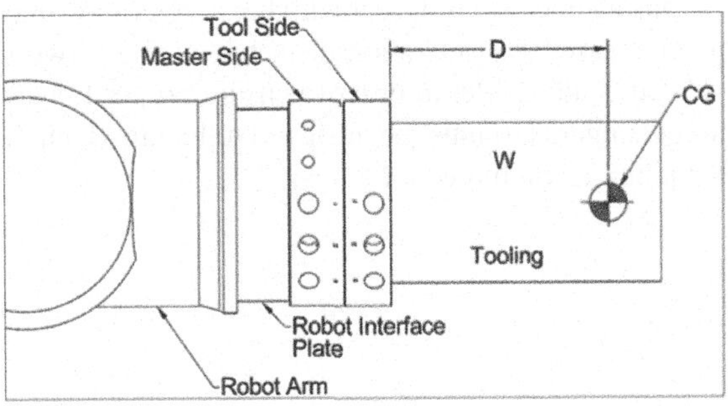

Fig. 1 Tool changer

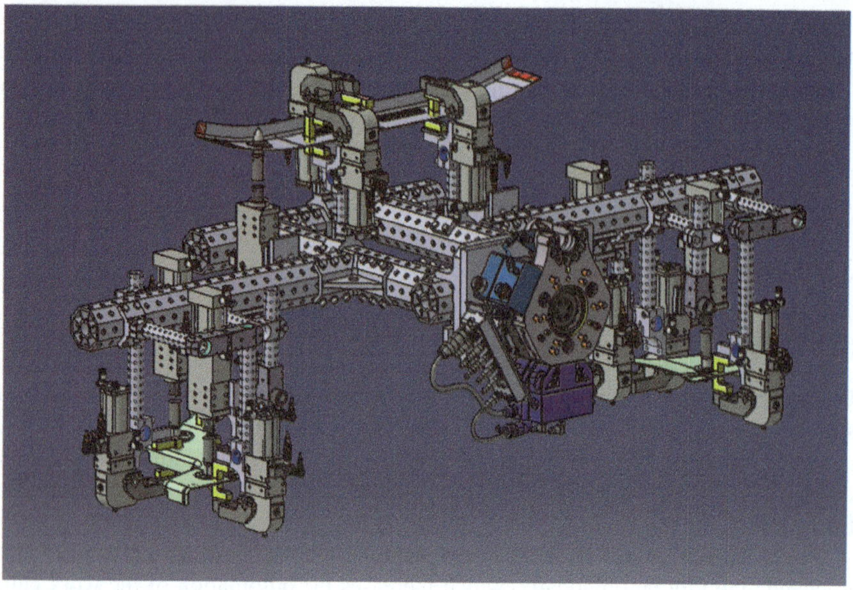

Fig. 2 Tool changer with gripper

(C) Calculating a Static Moment Load

During selection of correct tool changer model, static moment load of the end effector needs to be calculated. Steps below are followed during calculation of static moment:

(a) Get center of gravity (CG) of the gripper to be used.
(b) Calculate distance between mounting face to the center of gravity of the tool plate.
(c) Calculate the weight (mass * acceleration of gravity (acceleration of gravity = 9.8 m/s^2) of the gripper.
(d) Calculate static moment (M) by multiplication of weight and distance.
(e) Compare static moment capacity of the tool changer and calculated static moment.

During operations of robot, higher acceleration may be produced than considered; in such cases, consideration of factor of safety two to three times higher than calculated is advisable. In order to ensure proper longevity and functionality, selected tool changer also must be designed to handle dynamic moments three times higher than their static moment ratings.

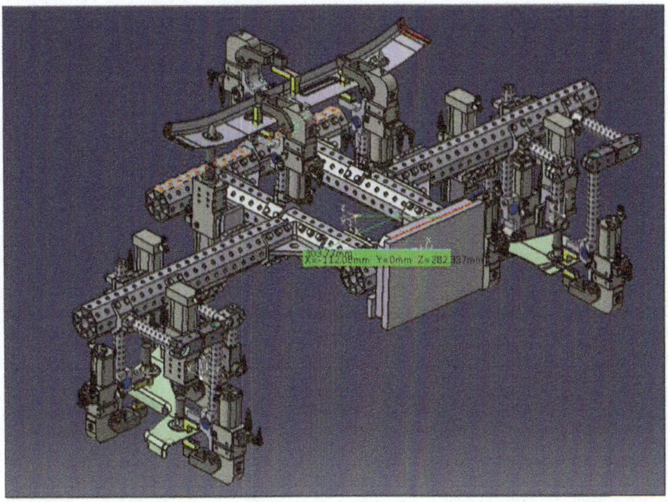

Fig. 3 Bending moment calculation

(D) **Calculations: (Refer** Figs. 3 and 4)

$$
\begin{aligned}
D &= 303.77 \text{ mm} \\
W &= 104.47 * 9.8 \text{ N} \\
&= 1023.8 \text{ N} \\
\text{Moment} &= D * M \\
&= 303.77 * 1023.8 \\
&= 310,999 \text{ Nmm} \\
&= 310.99 \text{ Nm}
\end{aligned}
$$

By considering above calculations, we can say that QC 110 (shown in Fig. 5) is perfect tool changer for given application.

Chart for all different weights are available (Tables 1 and 2) from that we can check suitable tool changer for the given EOAT. Software is also available which gives direct output to the operator about tool changer, but above explained process is the basic method for tool changer selection.

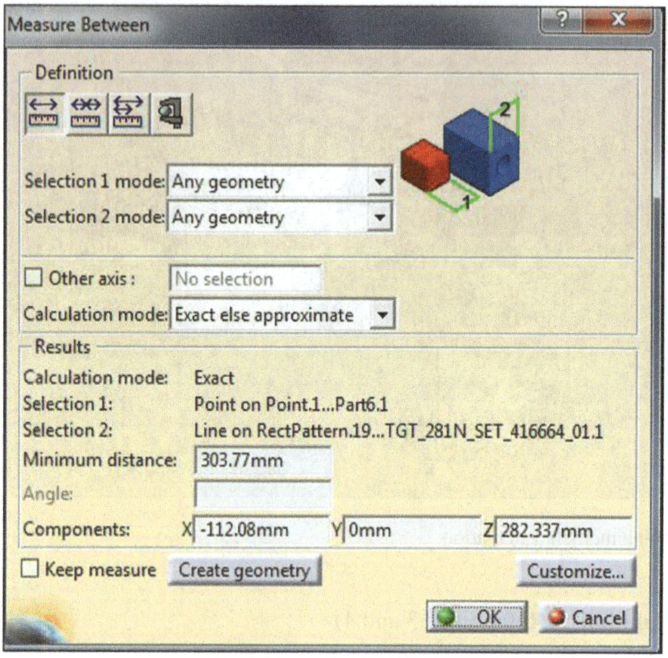

Fig. 4 Observations

Fig. 5 QC 110 tool changer

Table 1 Weight chart 1

QC–62	75 kg	7400 N	197 Nm	294 Nm
QC–71	79 kg	8100 N	395 Nm	395 Nm
QC–76	100 kg	12,000 N	542 Nm	701 Nm
QC–110	150 kg	12,000 N	784 Nm	784 Nm
QC–112	150 kg	12,000 N	780 Nm	780 Nm
QC–160	300 kg	31,000 N	2710 Nm	2260 Nm

Table 2 Weight chart 2

Suggested payload limit	150 kg
Locking force @5.5 bar	12000 N
Moment capacity (*X* & *Y*)	784 Nm
Moment capacity (*Z*)	784 Nm
Positional repeatability	0.0152 mm
Weight when coupled	5.9 kg
Master weight	3.63 kg
Tool weight	2.27 kg
Max. distance between master and tool plate before locking at 6 bar	3.05 mm

(E) **Benefits of Using Robotic Tool Changer**

(a) Time required for change in production line will be reduced.
(b) Tools can be easily changed for purpose of repairing or maintenance, by drastically reducing downtime.
(c) More than one end effector can be used to maximize the flexibility.
(d) Single tool is prepared for number of operations instead of the combination of different grippers with different tools.

4 Conclusion

Money will be saved with robotic tool changer by increasing production capacity and flexibility of the robot. Idle time will be reduced by installing tool changer and increases manufacturing efficiency. Downtime will be reduced by reducing tooling maintenance and replacement by the tool changer. Tool changer provides seamless transition and introduction of new tooling while minimizing equipment idle time. Specific information about the load created by the gripper and its required utilities are required while selecting tool changer model and configuration. Suppliers only consider load calculation, utility module selection and tool changer model selection.

In this case, weight of the gripper was 105 kg, and the distance between center of gravity and master plate was 304 mm. And bending moment of the system was 310 nm; hence, tool changer having model no QC 110 was selected for the given gripper.

References

1. Sabnis NV, Dhanawade DA (2018) A review: state of the art of robotic grippers. Int Res J Eng Technol (IRJET) 5(5):371–375
2. Hirikude SM, Sabnis NV, Kulkarni VV (2017) An interdisciplinary project based learning, a case study. J Eng Educ Transform 31(1). ISSN: 2349-2473, eISSN: 2394-1707
3. Youcef-Toumi K, Bausch JJ, Blacker SJ (1989) Automated setup and reconfiguration for modular fixturing. Robot Comput Integr Manuf 5(4):357–370
4. Youcef-Toumi K, Liu WS, Asada H (1988) Computer aided analysis of reconfigurable fixtures and sheet metal parts for robotic drilling. Robo Comput Integr Manuf 4(3–4):387–393
5. Lee J Computer-aided robot modeling, programming and simulation. Robotic vision systems, Hauppauge, NY, USA
6. Pellegrinelli S, Pedrocchi N, Tosatti LM, Fischer A, Tolio T (2017) Multi-robotspot-welding cells for car-body assembly: design and motion planning. Robot Comput Integr Manuf 44:97–116
7. Gavali MP, Shinde SN (2019) Optimization of machining parameters of CNC turning by using analytical hierarchy process based taguchi method. Int J Trend Res Dev 6(2). ISSN: 2394-9333

Remote Data Acquisition System for Measurement of Ambient Climatic Conditions and SPV Battery Status

Mahesh B. Gorawar, Veeresh G. Balikai, Vinayak H. Khatawate, and P. P. Revankar

Abstract The cost of solar PV (SPV) systems has reduced drastically over the years on account of efficient PV cell manufacturing technologies. Its suitability and economic viability for the installation site have strong correlation to local climatic resources. The access to data on solar insolation, wind speed and other climatic parameters is hence essential for site selection and SPV installation. The remotely accessible data of ambient temperature and solar irradiance at target location was acquired through Arduino-GSM hardware loaded with compatible software. The study showed that temperature and solar irradiance measured through developed system was within 2.23% and 5.83%, respectively, as compared to conventional measurement techniques. The novelty of the device is envisioned in its application to monitor small-scale as well as large-capacity renewable energy systems that have a strong dependence on climatic factors.

Keywords Insolation · Temperature · Solar PV · Remote data acquisition

1 Introduction

The increasing energy demand globally has turned to be a serious concern on account of fossil fuels largely used in thermal power generation. This has lead to exploration of cleaner options to satisfy electric power needs and other domestic consumption for cooking and transportation applications. The renewable energy offers dual benefit of eco-friendliness and resource abundance to meet energy demand of human society. The renewable energy includes nature-based sources like solar, wind and biomass. The SPV occupies foremost place for its high feasible potential for exploitation and hence subjected to rigorous research and field implementation [1].

M. B. Gorawar (✉) · V. G. Balikai · P. P. Revankar
KLE Technological University, Hubli 580031, Karnataka, India
e-mail: mb_gorwar@kletech.ac.in

V. H. Khatawate
Department of Mechanical Engineering, Dwarkadas J. Sanghvi College of Engineering, Vile Parle, Mumbai 400056, India

© Springer Nature Singapore Pte Ltd. 2020

H. Vasudevan et al. (eds.), *Proceedings of International Conference on Intelligent Manufacturing and Automation*, Lecture Notes in Mechanical Engineering,
https://doi.org/10.1007/978-981-15-4485-9_39

The electrical equipment that operates on SPV either needs backup of standby devices or storage to support load during non-generating spells. Batteries constitute the vital balance of system (BOS) for SPV system especially operating on stand-alone mode for storage and help maintain constant load even with SPV panels delivering fluctuating power. The charge controllers prevent battery overcharge and maintain a uniform system voltage. The storage support includes battery monitors to continuously track changes in battery status and initiate necessary action in case of exigencies. The unregulated battery storage normally needs higher storage compared to regulated system owing to protection sought against being overcharged [2].

2 Literature Survey

The research on wireless sensing, Arduino sensing, SPV monitoring and effect of parameters on solar PV is dealt in this section. Galande et al. (2014) developed wireless data monitoring using LPC2148 ARM 7 processor and GPRS SIM900 wireless module for relative humidity, ambient temperature and solar irradiance. The data was stored in SQL database with Keil software which made the system more interactive, user friendly to give rapid and accurate results [3]. Tobnaghi et al. (2014) studied effect of temperature and solar radiation on electrical performance of SPV system installed at Baku (40.4° N, 49.9° E) on silicon-based SPV system at three combinations of ambient temperature and solar insolation like 15 °C, 25 °C and 50 °C and 1000 W/m^2, 800 W/m^2 and 500 W/m^2, respectively [4]. Vianney et al. (2015) reported on SPV-powered Arduino wireless flood sensing to predict natural disaster on basis of model using water level indicator, Arduino UNO and necessary support system. The model exhibited accurate, reliable and secured information on natural disasters for long-term disaster management [5]. Adilah et al. (2015) used GSM technology to monitor panel efficiency of system on basis of Arduino Mega 2560 Microcontroller. The model communicated through alert signals and helped to improve the long-term performance of SPV system [6]. Haider-e-Karar et al. (2015) proposed an effective and efficient GUI, using Arduino interfaced with NI LabVIEW and Web browser. The hardware part consists of Arduino Mega 2560 Microcontroller, ACS712 current sensor, voltage divider circuit for voltage sensing, a charge controller and a 5 V relay for switching to other alternative available energy source [7]. Joshi et al. (2016) designed a monitoring system for 125 Wp SPV systems, using ATmega328P as core processor of Arduino and Wi-Fi module [8]. Moron et al. (2017) reported on dual-axis SPV tracker with Arduino platform to draw a maximum benefit of an incident radiation [9]. Mansouri et al. (2018) reported on solar tracker with two geared dc servomotors, LDR and ATmega328P microcontroller with Arduino. The system successfully tracked sun orientation for wide range of locations [10]. Laseinde et al. (2019) developed cost-effective microcontroller-based maximum power point tracking algorithm for solar multiple-axis tracking systems using Arduino board. It was observed that the designed system increased electrical efficiency of the solar array up to 23.95% [11].

Chowdhury et al. (2019) reported on stand-alone low-cost, high-precision dual-axis closed-loop sun-tracking system using Astronomical Almanac's (AA) algorithm, implemented in an 8-bit microcontroller. The results revealed that incorporation of the sun position algorithm into a solar tracking system helps in outperforming the fixed system and optical tracking system by 13.9% and 2.1%, respectively [12].

3 Details of Experiment on SPV DAQ System

The details provided in Fig. 1 indicate the programmed Arduino configured as main controlling unit with GSM shield used for communication and sensors collected parameters related to SPV system. The data collected was stored using NetBeans software. The parameters acquired were temperature and irradiance, voltage status of battery charged through solar panel.

LM35 temperature sensor measured ambient temperature while LDR was used for solar irradiance. Voltage divider circuit measured battery voltage status. Figure 2 shows hardware part housed in box with provision made to observing data using LCD display and also for the solar panel and battery connections to the hardware unit.

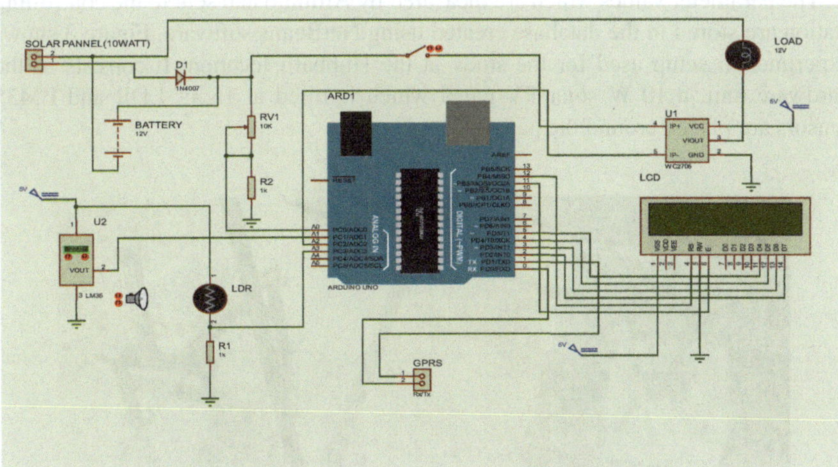

Fig. 1 Circuit diagram of SPV data acquisition system drawn in proteus

Fig. 2 Hardware implementation of the model developed

4 Experimental Investigations on SPV DAQ System

The authorized user has access to data from anywhere using Internet connection by entering URL and username and password. Figure 3 shows the experimental setup used for the study at the Hubballi location during the month of August. It consists of the hardware unit where all the circuits are housed in a box, a 10 W solar PV panel which is tilted at 15.3°, LDR and LM35 sensors are placed around the panel.

The parameter values which are measured by Arduino and sent to the controlling station are stored in the database created using NetBeans software. Figure 3 shows experimental setup used for the study at the Hubballi location. It consists of the hardware unit, a 10 W solar PV panel which is tilted at 15.3°, LDR and LM35 sensors are placed around the panel.

Fig. 3 Experimental setup for the study

5 Results and Discussions

The experiments were conducted, and all data obtained was sent wirelessly using Arduino to database created using NetBeans that was accessed through Internet connection from anywhere in the world and at any time. The experiments were conducted, and all recorded data were sent wireless medium using Arduino to database created using NetBeans and stored in database. The recorded data were accessed remotely through internet.

5.1 Comparison of Instruments

This section is about the comparison of the obtained data of temperature and irradiance from experiment with that of standard instruments, i.e., mercury thermometer and handheld pyranometer. The accuracy of the devices used in the experiment, such as LM35 and LDR sensors used for temperature sensing and solar radiation measurement are assessed using mercury thermometer, and pyranometer respectively. LM35 temperature sensor is compared with the mercury thermometer, and LDR is compared with the handheld pyranometer. From Fig. 4, we can observe that temperature value measured using LM35 sensor showed better accuracy with those of mercury thermometer readings. An accuracy of 1 °C was acceptable as per standards. The measurement of temperature using the developed system and a mercury thermometer indicated a maximum percentage error of 2.23% and a minimum error of 0.86%. From Fig. 5, we can observe that the irradiance value measured using LDR is nearly

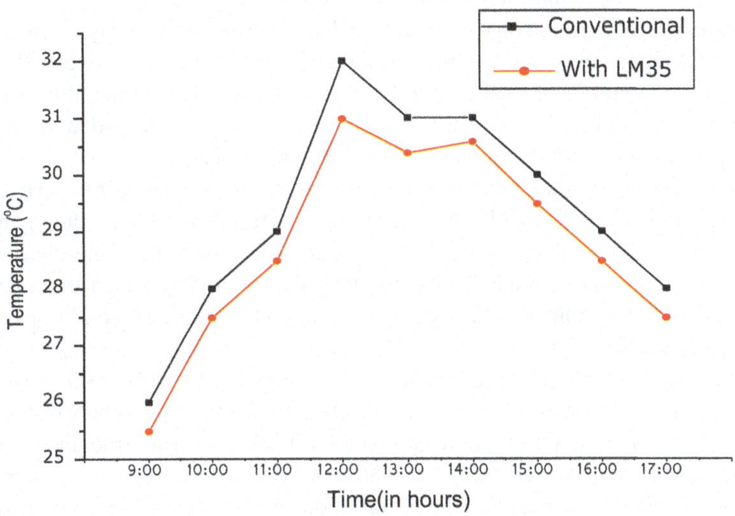

Fig. 4 Comparison of temperature measurement using thermometer and LM35

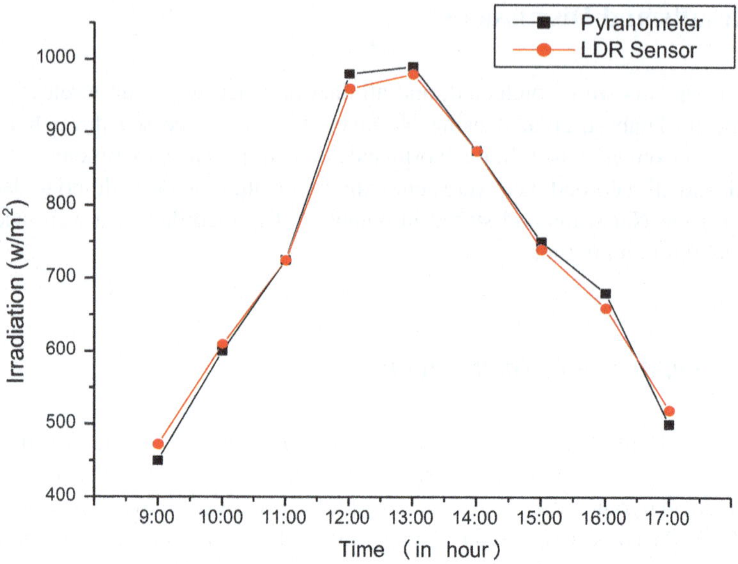

Fig. 5 Comparison of temperature measurement using pyranometer and LDR sensor

accurate at higher insolation level, but at lower insolation level, the difference is high when compared with the handheld pyranometer. The inaccuracy in measurement of solar irradiance is attributed to rapid changes in the intensity of solar energy available at the site on account of climatic variations. The measurement of irradiance using the developed system and a handheld pyranometer indicated a maximum percentage error of 5.83% and a minimum error of 0.99%.

Figure 6 depicts the average temperature data recorded for five days between 9:00 and 17:00 h. It was observed that the variation in temperature was between 25 °C and 32 °C in the month of August for the Hubballi location. The temperature variation indicated that day four had higher temperature while day one recorded lower levels of temperature distribution among the observations made.

The variations in ambient temperature are important with respect to performance characteristics of SPV module. A lower temperature of the SPV module generates higher voltage and results in more energy gain from the panel. The observations of ambient temperature made during the test period have indicated that operating efficiency of the module is influenced by the ambient temperature conditions as well as the solar irradiance.

Figure 7 depicts the average irradiance data recorded between 9:00 and 17:00 h during test period. It was observed that variation in irradiance was between 450 and 1018 W/m^2 in the month of August for Hubballi location. The irradiance variation indicated that day two had higher insolation while day five recorded lower levels of insolation distribution among the observations made. The variations in irradiance with time were between 450 and 1018 W/m^2. Initially, at the start of the day, insolation

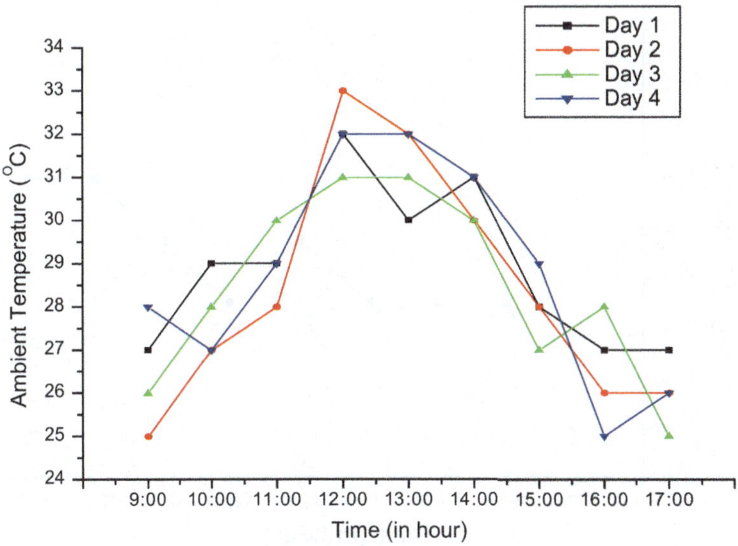

Fig. 6 Variation of temperature with time at the location measured using LM35

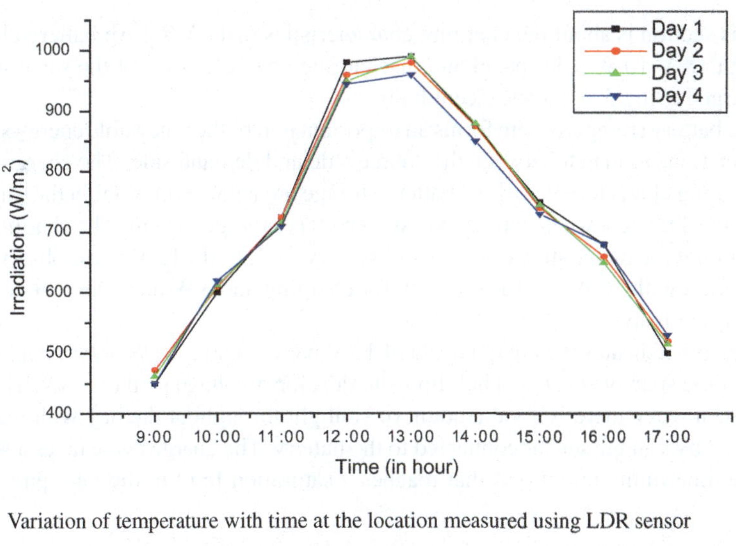

Fig. 7 Variation of temperature with time at the location measured using LDR sensor

levels were lower and increased to a peak value in the duration between 11:00 a.m. and 3:00 p.m. The insolation level was observed to decrease beyond 3:00 p.m. due to the transition of sun downwards the horizon. The variations in insolation levels are important with respect to performance characteristics of SPV module. A higher insolation on the SPV module generates higher load current and results in more energy gain from the panel.

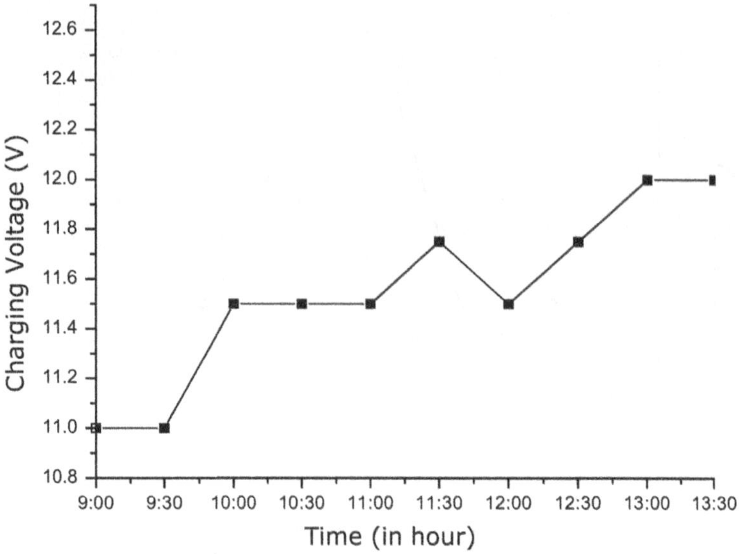

Fig. 8 Variation of battery charging voltage with time

This section is about the charging characteristics of 12 V 7.2 Ah battery charged through 18 V 10 W solar panel and discharging characteristics of the same battery discharged using 5 W incandescent bulb.

The battery storage system forms an important part of the renewable energy system and functions to match between the source side and demand side. The charging and discharging characteristics of the battery storage are hence important in the transient behavior of the renewable energy based on battery storage system. The charging and discharging characteristics of lead–acid battery of 7.2 Ah, 12 V were observed by connecting with 10 W solar PV system for charging and 5 W incandescent lamp for discharging respectively.

Figure 8 indicates the charge cycle of 12 V battery using 10 W solar panel which took a time span of seven and half hours to develop a voltage gain of 2.8 V. The data was taken when there is good amount of sunlight throughout the day with readings recorded by voltage sensor connected to the battery. The charge cycle takes a shorter charge time in the initial part that reaches a saturation limit in the later part of the cycle.

6 Conclusion

The following conclusions were drawn from the study:

- The prototype of a simple wireless solar photovoltaic data acquisition system based on Arduino and GSM communication was developed and successfully tested. The developed system offers the advantage of remote monitoring of SPV-based system.
- The LM35 and LDR-based measuring systems performed satisfactorily for wide range of temperature and irradiance measurement. The errors in measurement of ambient temperature and irradiance lie within a margin of 2.23% and 5.83%, respectively.
- The charging and discharging cycle of the battery storage connected to PV panel and load indicated that charge duration is greater than the discharge duration on account of variations in solar PV panel output. The ratio of charging duration to the discharging duration of the battery was as high as 7.5 owing to direct connection between panel and battery without use of charge controller.

The NetBeans IDE was developed to create a cloud-linked database for long-term storage of climatic information at the location obtained from the developed DAQ system. The memory space of 10 GB available on this platform is adequate for continuous storage and retrieval of the recorded climate data.

References

1. Rai GD (2007) Non-conventional energy sources, 4th edn. Khanna publishers, Delhi, 47–64
2. www.nrel.gov/docs/legosti/old/1448.pdf
3. Swapnali G, Smita K, Patan RD (2014) Solar power wireless monitoring based on embedded system. Int J Innovative Sci Eng Technol 1(4):408–412
4. Tobnaghi DM, Naderi D (2014) The effect of solar radiation and temperature on solar cells performance. Extensive J Appl Sci 3(2):39–43
5. Vunabandi V, Matsunaga R, Markon S, Willy N (2015) Flood sensing framework by arduino and wireless sensor network in rural-rwanda, pp 1–6
6. Adilah AN, Nadzlin TT, Mahadi AM (2015) Development of solar efficiency monitoring system by using GSM technology. In: International conference on space science and communication (IconSpace), Langkawi, pp 362–365
7. Haider-e-Karar Khuwaja AA, Sattar A (2015) Solar power remote monitoring and controlling using arduino, LabVIEW and web browser. In: Power generation system and renewable energy technologies, Islamabad, pp 1–4
8. Joshi S, Jadhav A, Gavate N, Yashwante M (2016) Wi-Fi based parameter monitoring for solar plant. Int J Eng Sci Comput 6(4):16–21
9. Moron C, Ferrandez D, Saiz P, Vega G (2017) New prototype of photovoltaic solar tracker based on arduino. Energies 10(9):1–13
10. Krim MF, Khouni Z (2018) Design of a prototypical dual-axis tracker solar panel controlled by geared dc servomotors. Sci Iranica D 25(6):3542–3558
11. Laseinde T, Ramere D (2019) Low-cost automatic multi-axis solar tracking system for performance improvement in vertical support solar panels using arduino board. Int J Low Carbon Technol 14:76–82
12. Chowdhury ME, Khandakar A, Hossain B, Abouhasera R. (2019) A low-cost closed-loop solar tracking system based on the sun position algorithm. Hindawi J Sens 2019:1–11

Force Sensitive Resistor Based Design and Modeling of Smart Walking Assistance Device by Axial Direction Control for Osteoarthritis

Akshay Vasage, Onkar Padhye, Gajanan Kulkarni, Shivram Kerkar, and Mahesh Kumar

Abstract Walking is a very important activity in human life which reduces many problems related to blood pressure regulation and heart diseases, but osteoarthritis patients suffer from mobility deficiency. The smart walking assistance (SWA) device was developed for mobility assistance. It is necessary to identify the intended direction of the user by the SWA device. While walking in the intended direction, human body reflects certain forearm movement, which can be recognized in terms of force. In this work, a method was proposed to identify the direction for the movement according to the forces exerted by the user's forearms, which are acknowledged by the force-sensitive resistor (FSR) placed on the walker's armrest. Depending on the FSR outputs in the range of 0–1023 scaled by Arduino, an algorithm was implemented for the direction control of the walker-based threshold. The user intended direction was validated by the experiment and monitored using MATLAB.

Keywords Smart walking assistance · Intended direction walking · Osteoarthritis

1 Introduction

Walking is an essential part of human day-to-day activity. Walking significantly impacts on health by lowering the cause of heart diseases but many people all over the world are suffering from mobility deficiencies and are not able to walk due to many reasons such as elderliness and accidental injuries. In elderly people, they face problems such as weakening of bones, lack of power in muscles, pain in joints, arthritis, and fear in mind to fall. According to the World Health Organization (WHO), 9.6% of men and 18.0% of women aged over 60 years have affected by arthritis diseases worldwide. 80% of those with osteoarthritis have constraints in movement, and 25% cannot play out their real day-by-day activities of life [1]. Combination of motion, such as forward and backward motions, left and right motions, oblique

A. Vasage · O. Padhye · G. Kulkarni · S. Kerkar · M. Kumar (✉)
Department of Electrical Engineering, Maharshi Parshuram College of Engineering, Velneshwar, Maharashtra 415729, India
e-mail: maheshkkamad@gmail.com

© Springer Nature Singapore Pte Ltd. 2020

H. Vasudevan et al. (eds.), *Proceedings of International Conference on Intelligent Manufacturing and Automation*, Lecture Notes in Mechanical Engineering,
https://doi.org/10.1007/978-981-15-4485-9_40

motions, rotations, is a part of walking. By enforced this complex motion in walking rehabilitation, early recovery can be expected [2].

There are many methods for direction control of smart walking assistance (SWA) device such as use of touch panel, joystick [3], strain gauges [4], and pressure sensor employed grippers. Working with joystick or touch panel, user has to lift the arm which would be uncomfortable for users having a disability in legs also a good coordination between operators, and functioning modules of the walker become hectic. Investigation in the human–machine interface which decides the human expectation through EEG and EMG has made it conceivable to control a robot without manipulation; however, a gadget for human–machine interface is costly and still unfit for day-to-day use [5] due to complex connections and sensor placement on body parts. Any walking assistance should work in such a manner to coordinate with the intended speed and direction of user. Previous studies reported problems related to the accidents due to walker designs which were considerably high [6, 7]. A commonly reported problem while operating of walker was accidental fall [8]. Researchers had implemented FSR-based omni walker to avoid any specialized operating training previously by use of forearm pressure for walking rehabilitation using omni wheels [9] but the cost of omni wheel is high so an alternative cost-effective walker with low-speed and high torque considering the safety of user is desirable.

In this work, A low-cost prototype for smart walking assistance has been developed with the implementation of a strategy to recognize the user directional deliberate as indicated by force exerted by human forearms on FSRs installed in walker. The strategy for finding the intended direction using FSRs is to qualify the weight from the user's lower arm on installed FSR sensors in the SWA armrest and bearing expectation identified by pressure information from the sensors. Unidirectional wheels and DC geared motors have been used to provide low-speed high torque to the SWA device.

2 System Description

While walking in intended direction body reflect motions, these motions are differentiated based on the variation of pressure distribution produced by the mechanical action of muscles on forearm [10]. The specific pressure distribution corresponding to the motion was sensed by FSRs. In the present prototype of SWA device, four FSRs placed on armrest of SWA are used to recognize forearm pressure by the patient for the intended direction. Further, this sensed output from the sensors fed input the Arduino controller. Depending on the various output of the sensors, it will decide the direction of the walker. The structure of SWA device is shown in Fig. 1.

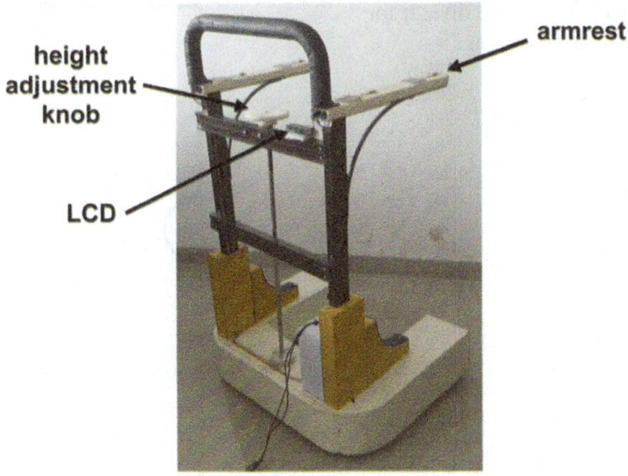

Fig. 1 Smart walking assistance

2.1 Component Description

FSR is one type of sensor, which changes its resistance as the force, pressure, or stress applied to it. Normally, FSR is in open loop that means it has infinite resistance. At normal force, its resistance is about 100 Ω and at the maximum force applied, the resistance is 200 Ω. In this work, FSR was selected with a sensitivity range of 100 g–10 kg [4]. Compared to other force sensors, the advantage of FSRs was their size, low-cost and good shock resistance. In DC gear motor, gear was coupled with DC motor to reduce speed and increase torque. A squared gearbox assembled high torque heavy duty DC gear motor with 50 kg cm^{-1} stall torque, and the maximum speed of 50 rpm was used to drive SWA. Motor was driven using motor driver which serves to govern the motor in some predetermined manner. The wheel was made up of cast iron material with size of 12.7 cm diameter. To obtain proper frictional force between wheels and plane as well as the rough surface, the material which was used for tires is rubber and its frictional coefficient is approximately 0.72 for smooth surface. The block diagram consisting of all the components is shown in Fig. 2.

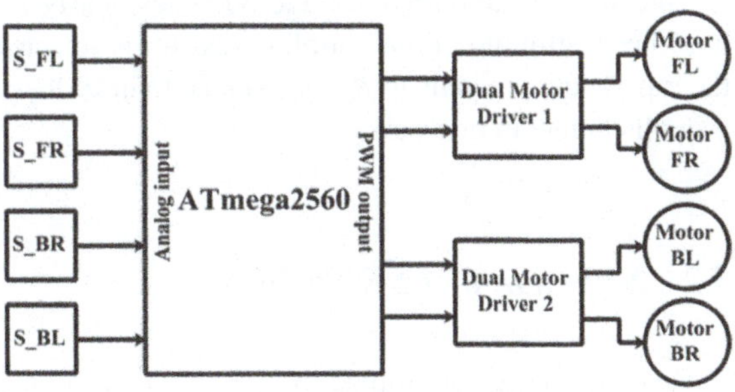

Fig. 2 Block diagram of experimental setup

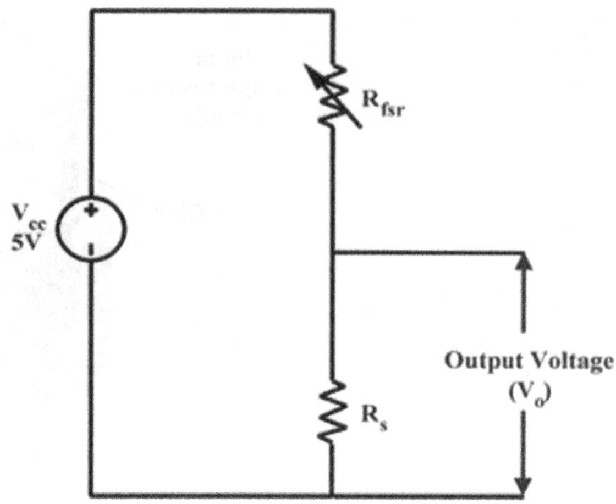

Fig. 3 Voltage divider for effective sensing

3 Positioning and Pressure Measurement

3.1 Selection of Sampling Resistance

Sampling resistance R_s was connected to the input in series with FSR as shown in Fig. 3. It was adjusted so as to get efficient results [4]. To get the desired output, the following cases were be taken into consideration.

$$V_0 = V_{cc} \frac{R_s}{R_s + R_{fsr}} \tag{1}$$

In case 1, as $R_s \gg R_{fsr}$ the voltage across a sampling resistor is nearly equal to input voltage which is applied to the Arduino. Under such condition, applied small pressure on FSR results in saturation of output voltage so further applied pressure cannot be detected as the R_s is dominating over the R_{fsr} resistance.

In case 2, $R_{fsr} \gg R_s$ neglecting R_s as R_s is very small in comparison with R_{fsr} in (1) and we get the output voltage becomes nearly zero. So, the sensitivity of FSR becomes negligible. If the sampling resistor is selected such that the value of R_s is in the range nearly equals to R_{fsr} resistance than FSR will be able to effectively detect a small change in pressure.

3.2 Selection of FSR Position

FSR positions are determined by pressure measurement experiment on one armrest using 3 FSR sensors named as $S1$, $S2$, and $S3$ as shown in Fig. 4. For this experiment, right armrest was selected by considering movement in three directions which are

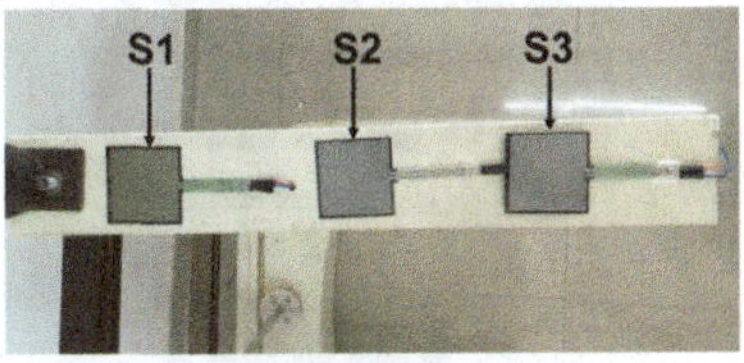

Fig. 4 FSR positions on the armrest

Table 1 Pressure measurements on $S1$, $S2$, and $S3$ positions

Direction	Sensor output (on 1–1023 scale)		
	$S1$	$S2$	$S3$
Forward	600.9	306.8	545.3
	808.9	403.5	14.8
	713.7	64.8	0
	715.5	467.5	323.3
	745.9	336.4	0
Right	381.8	195.6	755.7
	0	379.2	728.4
	225.8	87.7	702.1
	225.8	532.6	614.4
	186.4	182.8	703.8
Backward	82.9	51.8	716.6
	0	0	431.1
	429.5	268.2	764.1
	0	45.3	704.1
	0	75.9	651.9

as follows: forward, right, and backward. SWA was subjected to various forearm pressure with the intended. Pressure on each sensor for suggested direction was measured with the help of MATLAB in terms of the scaled analog outputs of FSR from 0 to 1023 range. Results of pressure on each sensor for three directions are shown in Table 1. Experimental results showed that all three directions variations in pressure for sensor $S1$ and $S3$ are significantly higher as compared to the sensor $S2$ so to sense a change in pressure for different directions, sensor position needs to be selected near the forward end and backward end.

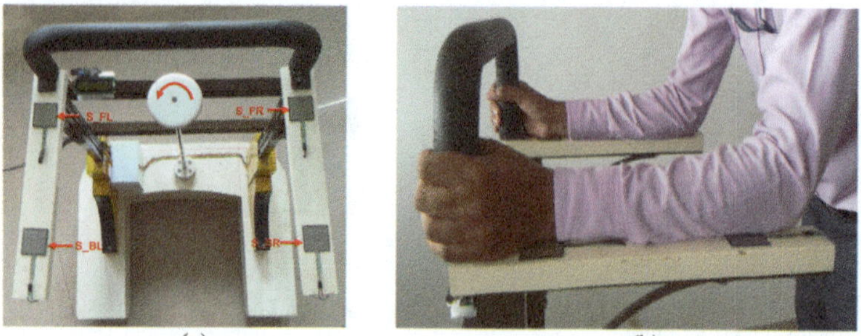

(a) (b)

Fig. 5 Forearm pressure analysis **a** position of sensors, **b** scene of experiment

4 Results and Discussion

4.1 Forearm Pressure Analysis

The respective FSRs assembled on the armrest for pressure measurement as shown in Fig. 5a. For recognition, sensors are named as S-FR, S-FL, S-BR, and S-BL. The user needs to simply hold the grip bar with laying arm on the armrest as shown in Fig. 5b. For a few seconds of time, pressure was measured at a sampling speed of 100 Hz with sampling resistor 1.2 kΩ. Three directions were considered for this experiment on SWA which are as follows: right, left, and forward. Four distinct attempts of pressure results, for all predefined directions on each FSR sensor, are represented graphically in Fig. 6. Outputs of the FSRs were found analogous to pressure for each sensor categories in 3 columns as per the specified three directions right, left, and forward; i.e., each column shows the result for distinct attempt on each FSR for direction mentioned on top of that column. The experimental results point out that for forward movement, the output of FSR sensor S-FR and S-FL, for right direction movement, the output of sensor S-BR, and for left direction movement, the output of sensor S-BL becomes higher than remaining FSRs.

4.2 Threshold Value

All four FSRs have different pressure values for different directions, and particular FSRs have the highest value of pressure for respective direction as shown in Fig. 6. To detect which FSR goes with higher pressure value and finds intended direction, the threshold value must require. When the value on particular FSR goes higher than the threshold, then respective analog input to that FSR goes high. It is essential to select the value of threshold such that it can detect the intended direction accurately. The pressure exerted on each FSR under different moments is varied widely so for each FSR, and a threshold must be different. To select the appropriate threshold

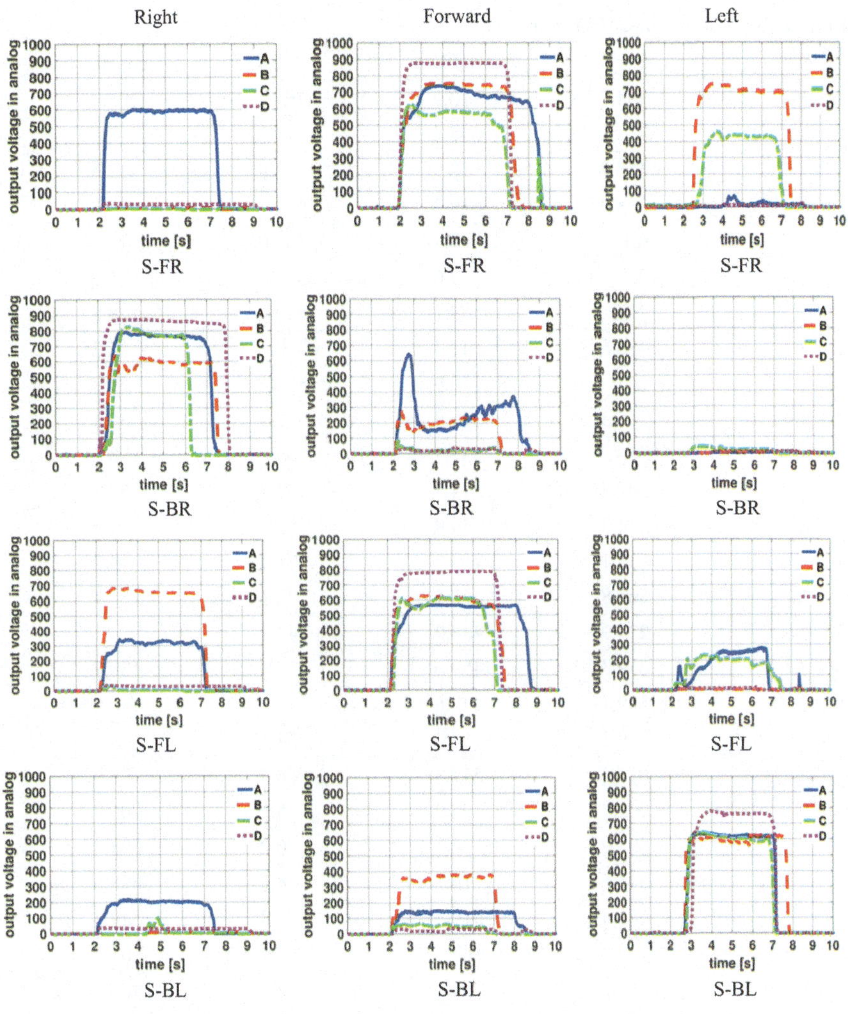

Fig. 6 Experimental results for right, forward, and left intended direction

value, nine distinct forearm pressure combinations have been applied on FSRs with intended move in a predefined direction with the help of the SWA device. The average values of FSR from the experiments were taken.

Results were recorded using interfacing the Arduino embedded platform with MATLAB Simulink which is shown in Table 2. The experimental results showed that calculated mean values for right direction movement on S-BR and for left direction movement on S-BL are greater than 600; whereas for forward direction movement, they are greater than 500 on S-BR as well as S-BL and selected 500 threshold values for forward direction movement and 600 threshold value for the right as well as left direction movement.

Table 2 Selection of threshold value

	S_FR	S_FL	S_BR	S_BL
Right	584.7	310.9	761.8	205.1
	9	648.4	574.6	8.9
	456.7	18.9	632.8	12.5
	338.9	619.2	490.1	5.7
	6.1	7.7	765.6	17.9
	2.5	2.6	656.6	5.7
	23.8	559.1	828.4	28.3
	486.6	47.4	220.1	5.2
	29.9	29.9	854.4	29.8
Mean	215.3	249.3	642.7	35.4
Forward	656.7	532.7	268.8	138.6
	690.3	574.7	203.2	357.2
	644.5	439.9	5.3	5.1
	743.1	715.9	71.1	9.2
	565.8	572.5	27.1	52.7
	227.9	438.9	42.4	3.1
	659.3	459.6	57.3	130.7
	509.1	442.6	7.1	6.8
	859.5	767.4	24.1	6.8
Mean	617.3	549.3	78.4	78.9
Left	20.3	194.9	5.3	706.2
	695.2	6.1	6.2	598.4
	344.6	14.1	8.3	698.6
	461.4	510.5	5.8	549.3
	329.3	187.5	27.4	607.1
	285.3	251.1	1.1	480.4
	638.9	233.3	24.3	626.8
	3.3	49.1	3.1	660.1
	9.3	12.2	12.1	749.7
Mean	309.7	162.1	10.4	630.7

4.3 Axial Rotation Control

The problem related to less space for movement can be overcome by axial rotation control of the SWA. In this, the SWA spins on the spot and rotates on its own axis. This makes the radial distance required to turn very less. For movement, four standard fixed wheels were attached to the DC geared motor and fixed near to the four corners of the walker. These wheels were rotated in clockwise as well as anticlockwise direction, according to the intended direction using twisting force in the system which is generated by the clockwise and anticlockwise rotation of four wheels attached to the base of SWA. For forward movement, all four wheels were run in the forward direction. Similarly, in case of a left turn or right side two wheels run in a forward direction while left side two wheels run in a reverse direction. In case of right turn left side two wheels run in a forward direction while right side two wheels run in a reverse direction.

5 Conclusion

In order to control the walker's direction and to support the user in walking, a method was proposed to specifically recognize a user directional intention that analyzed from pressure exerted by forearm on four force-sensitive resistors on the walker. This relevance between forearm pressure and directions of the walker is decided using programming algorithms on an embedded platform. The direction and movement of the walker are achieved by a fixed standard wheel rather than omnidirectional wheel so as to reduce the cost considerably. However, the pressure and position of the FSR will change according to the different users. Axial direction control using threshold qualifying algorithm was implemented on real-time embedded platform on Arduino MEGA2560. Results of pressures are monitored by interfacing SWA with MATLAB Simulink. Future work will involve the implementation of an adaptive algorithm for user with different weight and height.

References

1. Pal CP, Singh P, Chaturvedi S, Pruthi KK, Vij A (2016) Epidemiology of knee osteoarthritis in India and related factors. Indian J Orthop 50(5):518
2. Jiang Y, Wang S (2010) Directional intention identification for running control of an omni-directional walker. In: International conference on fuzzy systems, IEEE, pp 1–8
3. Hashimoto H, Sasaki A, Ohyama Y, Ishii C (2006) Walker with hand haptic interface for spatial recognition. In: 9th IEEE international workshop on advanced motion control, pp 311–316
4. Yuk G, Chang YY, Park H, Dan BJ, Chang, WS (2013) Smart walker development based on experts' feedback for elderly and disabled. In: 10th International conference on ubiquitous robots and ambient intelligence (URAI), pp 552–553
5. Jiang Y, Wang S, Ishida K, Kobayashi Y, Fujie MG (2012) User directional intention identification for a walking support walker: adaptation to individual differences with fuzzy learning. In: 6th international conference on soft computing and intelligent systems, and 13th international symposium on advanced intelligence systems. IEEE, pp 1207–1210
6. Eakman AM, Havens MD, Ager SJ, Buchanan RL, Fee NJ, Gollick SG, Stevenson KA (2002) Fall prevention in long-term care: an in-house interdisciplinary team approach. Top Geriatr Rehabil 17(3):29–39
7. Craven R, Bruno P (1986) Teach the elderly to prevent falls. J Gerontol Nurs 12(8):27–33
8. Bradley SM, Hernandez CR (2011) Geriatric assistive devices. Am Fam Phys 84(4)

9. Hirata Y, Baba T, Kosuge K (2003) Motion control of omni-directional type walking support system. In: Walking helper, proceedings of IEEE workshop on robot and human interactive communication, vol 2, no A5
10. Li N, Yang D, Jiang L, Liu H, Cai H (2012) Combined use of FSR sensor array and SVM classifier for finger motion recognition based on pressure distribution map. J Bio Eng 9(1):39–47

Autonomous Package Dispatcher Bot Using Video Processing

Manthan Tambe, Sahil Vora, Shaunak Thakar, and Manish M. Parmar

Abstract Basic services, which are provided by hospitals to patients, are food, medicine, blood checks, hospital clothes, etc. To carry out these patient care services, there are other support services that are managed by the hospital. These services are most of the time invisible to users but they do affect the patient care services. The autonomous medical bot described in this paper uses camera and QR codes to dispatch packages in a specified geography. It takes the aid of a line marked on the floor to follow the required path using a camera. Along with delivery, locker security is provided which can only be opened by the one requesting for the service. An attempt to improve the quality of healthcare services provided by hospitals is made by efficiently managing all the logistics tasks.

Keywords Autonomous · Camera · QR codes · Logistics

1 Introduction

Delivery of consumables required by patients is a task which is to be operated 24 × 7. This increasing need for various materials, equipment also increases disposable items which cause complex and chaotic logistic transportation which leads to operational breakdowns, and these operational breakdowns prevent staff of hospitals from working more on patient care services, which directly affects the quality of service as described in [1].

M. Tambe (✉) · S. Vora · S. Thakar · M. M. Parmar
Bhartiya Vidya Bhavan's Sardar Patel Institute of Technology, Bhavan Campus, Munshi Nagar, Andheri West, Mumbai 400058, India
e-mail: manthantambe999@gmail.com

S. Vora
e-mail: sahilvora225@gmail.com

S. Thakar
e-mail: shaunak241997@gmail.com

M. M. Parmar
e-mail: manish_parmar@spit.ac.in

© Springer Nature Singapore Pte Ltd. 2020

H. Vasudevan et al. (eds.), *Proceedings of International Conference on Intelligent Manufacturing and Automation*, Lecture Notes in Mechanical Engineering,
https://doi.org/10.1007/978-981-15-4485-9_41

Traditionally, all logistic transportation and material flow are controlled by staff of hospitals which has some limitations such as limited amount of manpower, human error in timely delivery of supplies also amount of volume or weight each person can carry and frequency of work. This traditional method can be an inefficient way of handling logistics and can also cause more operational breakdown.

One of the methods which can be implemented to cope with this problem is automation of logistics in hospitals [2] which can reduce the operation cost and manpower. Transportation capacities and capabilities can be improved by designing a robot which is capable of performing on-demand delivery autonomously with less delay and more accuracy. Limitation of traditional methods can be overcome using automation. Autonomous bot uses camera for path planning and ultrasonic sensors for accurately avoiding any obstacles in the path to prevent collision which enables a system to work more efficiently. Autonomous systems increase the effectiveness of everyday processes and also enable staff to save a significant amount of time, which can be allocated to more patient-related tasks.

2 Methodology

To drive the autonomous feature of the bot, a path planning algorithm will be used [3]. The algorithm will be based on line following using a camera. A line will be drawn on the floor of the entire arena to be travelled [4]. These lines will be drawn with the help of tapes with their color in contrast with the color of the flooring. The reason for using contrasting colors is to easily distinguish between the floor and the path to be followed.

As shown in Fig. 1, the camera is placed on bottom-center in front of the bot such that the path is clearly visible. The camera will continuously analyze the frames captured by it, and based on the deviation of the bot from its path, it will act accordingly.

At every junction as shown in Fig. 2, a bot will encounter QR code which has metadata about the current location of the bot and the path to be selected in order to reach the destination. These QR codes are nothing but the checkpoints for the bot to navigate through the arena properly.

The QR code will contain information about the current location and the zones/rooms located along the path. This information will be stored using comma-separated values (CSV). For example, if the destination of the bot is a blood testing laboratory in the pathology center and the bot encounters a junction. The QR code gives the information about the zones located on the left, straight and right paths along with its current location. After scanning the QR code, the bot comes to know that the pathology center is located along the straight path. So in order to reach the destination, a straight path will be chosen. If one of the directions is not available at a particular junction, its corresponding value in the QR code will be "not available." Also, this will be verified by no path drawn in that direction.

Fig. 1 Front view of bot

Fig. 2 Junction image

3 Hardware Description

The proposed system consists of two major parts—Android application (Software) and robotic structure (Hardware). Figure 3 shows the block diagram of the system. To implement the package dispatcher bot with on-demand delivery of medicine, food, trash removal [5] wireless communication is required to fulfill this need Raspberry

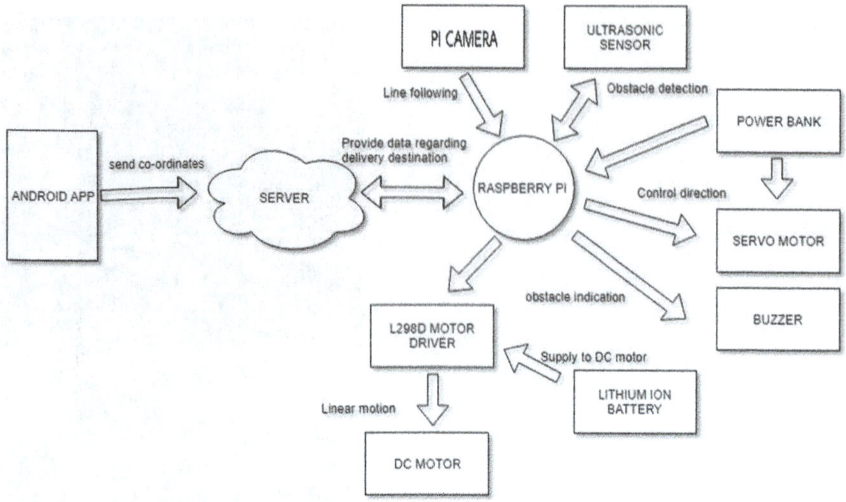

Fig. 3 Block diagram of system

Pi 3B [6] can be used as a central processing unit and to handle data acquisition, communication, and control of robotic system.

Ultrasonic sensors (HCSR-04) [7] and Pi camera can be used as main sensors. These sensors will provide necessary inputs to Raspberry Pi and according to these inputs enable Raspberry Pi to control the entire robotic system. An ultrasonic sensor is to provide intruder detection and also makes sure that collisions of robotic system won't happen. The Pi camera gives live streaming of the path which robot is supposed to follow, and Pi camera [8] shown in Fig. 4 helps in keeping the robot on the specific path using video processing also to provide live stream on android application to make sure the security medical supplies.

To deliver medical supplies to respective rooms according to requirement, robot has to travel across rooms. DC and servo motors are being used in the system in order

Fig. 4 Pi camera

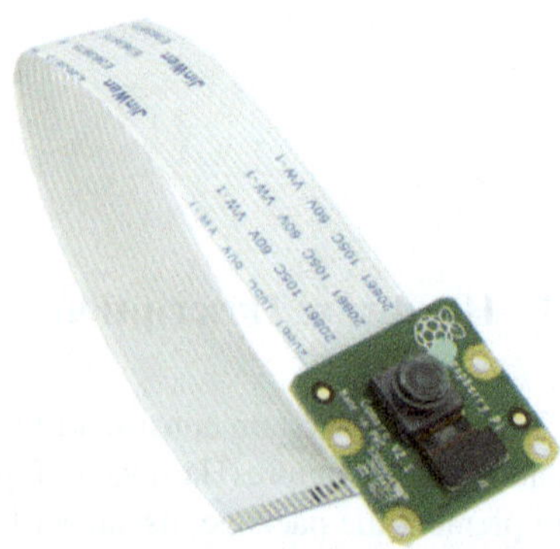

to handle the delivery of supplies. Raspberry Pi 3B works on 5 V, whereas DC motor works at 12 V [9], and hence to provide 12 V to DC motor, external lithium-ion battery is used; also, DC motor driver L298N [10] is used to control DC motor and isolate Raspberry Pi from 12 V external supply. Raspberry Pi provides control signals to a motor driver to control the DC motor. Servo motor takes a supply of 5 V from power bank [11] and control signal from Raspberry Pi. It allows the robotic structure to take left and a right turn at a particular angle, whereas the DC motor grants a robot to move in forward or backward directions. Buzzers can be used in a system to the intimate outside world if they are in robot's path. This robotic system will avoid collisions as well make sure, safe, and accurate delivery of medical supplies without any interruption.

4 Software Description

Three-tier architecture is most popular and used because of its scalability. We can separate user application and physical database in our system which gives support to DBMS characteristics. Software of our system stands primarily on three pillars: server, database, and its Android application. The intercommunication between these entities defines the soul of our system and drives the flow of our system. We have implemented three-tier architecture, i.e., client–server architecture as shown in Fig. 5 [12].

4.1 Server

Server handles the request posted by clients (i.e., application users, bot for fetching tasks). Major tasks performed by the server are mentioned below:

- Accept orders from users and store them in the database to deliver them the service.
- Send the task to be performed by the bot on a first-come first-serve basis.
- Update the user about the status of their ongoing order and show the details of their previous orders.

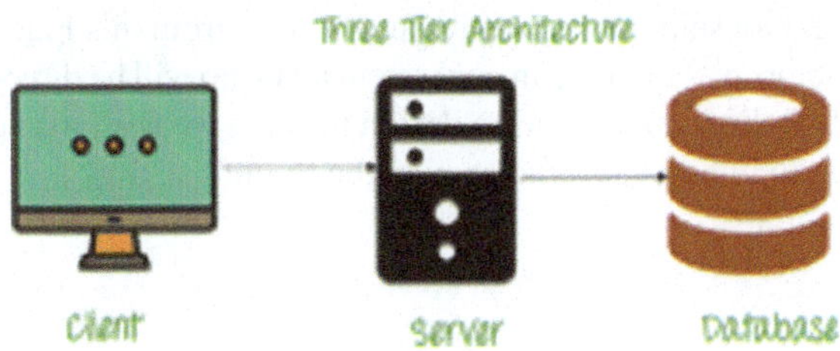

Fig. 5 Three-tier architecture

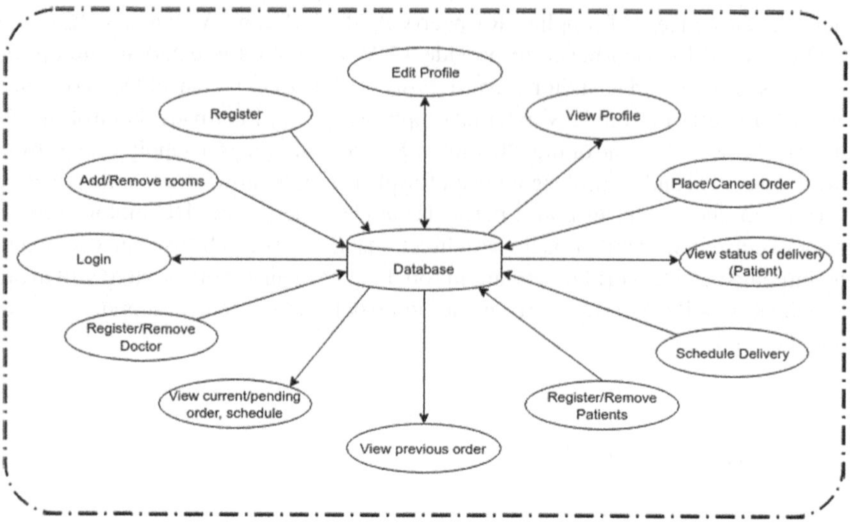

Fig. 6 Information flow diagram

4.2 *Database Management*

Database management is essential to develop a robust system where we can store and access data flawlessly. To achieve R-DBMS model, we used an approach which have designing of following three diagrams:

4.2.1 Information Flow Diagram

Figure 6 illustrates the information flow diagram for our system. Information flow diagram (IFD) helps us to understand the main functions/services to be included in our system, and how that function/service syncs with the database.

4.2.2 E-R Diagram (Entity–Relationship)

In Fig. 7, entity–relationship diagram defines the entities, properties of entity and relationship between different entities. From this Fig. 7, we are able to visualize the actors participating in our system and who will be directly affected by using the same. E-R diagram helps us to design the data we need to capture from the surrounding so as to improve its efficiency in the future.

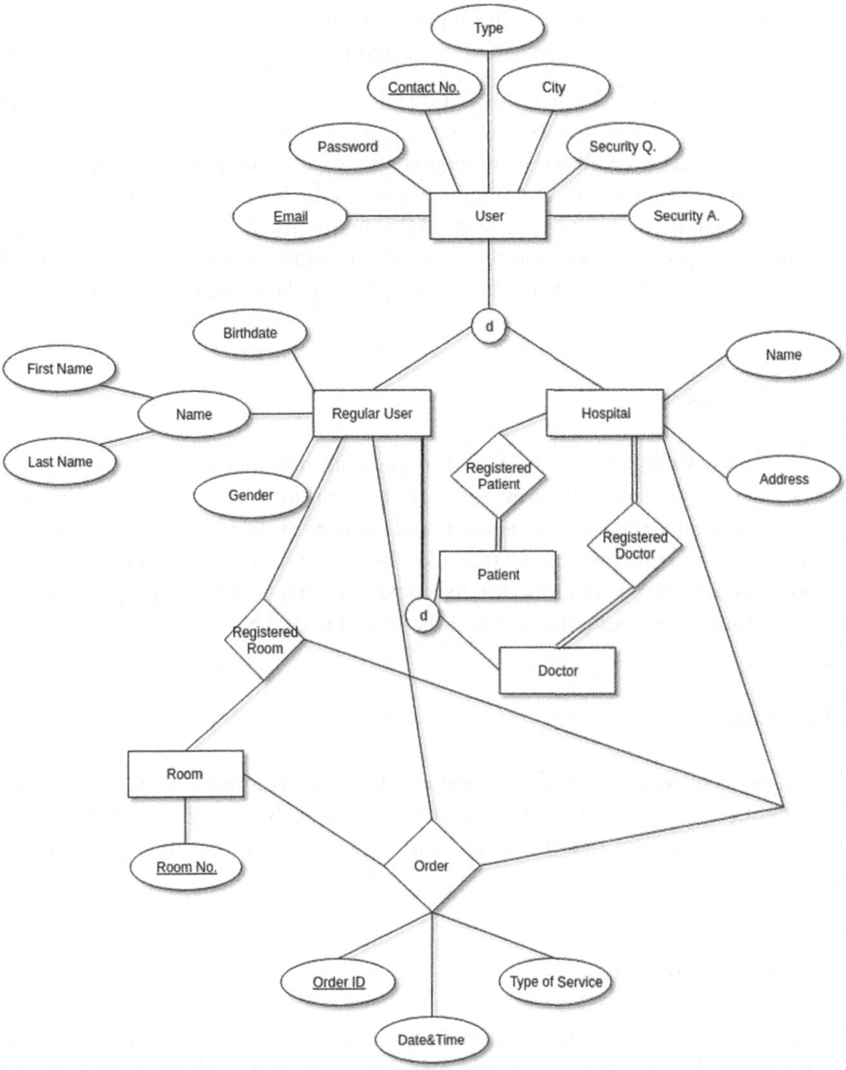

Fig. 7 E-R diagram

4.3 Android Application

Figure 6 shows the information flow diagram of our system. It includes all the features provided by the MedBot application. The main intention behind developing the application is to provide on-demand services to patients, doctors, and hospitals. The users of the MedBot application can be categorized into three based on their type, i.e., patients, doctors, and hospitals. To use all the facilities provided by our system, a user has to first create an account on the application by signing up. Based on the type of the user, the features provided are different.

- Patient is provided with features like placing an order, viewing the status of an order, all the previous orders, and viewing and editing profiles.
- Doctor has access to all the features given to a particular patient but also has an extra feature of adding or removing a patient in the hospital's database.
- As the hospital is the governing body in this hierarchy of users, a number of other features are provided like adding/removing patients and doctors, scheduling all the requests by the patients/doctors and managing the available hospital rooms.

4.3.1 Register

Every user has to register on the MedBot application before using the application. The credentials that are to be provided include email id, mobile number, city, password, security question, security answer, and the type of user, i.e., patient, doctor, or hospital. User-type specific credentials like first name, last name, gender, birth date are asked to patients and doctors, whereas hospitals have to fill in name and address. After creating the account, the user is directed to the dashboard.

4.3.2 Login

Users have to provide email id and password to login. The server then checks if the user is authentic or not. Saving the user login using caching is also implemented in the app so as to ensure that the user is still logged in even after the application is closed.

4.3.3 View/Edit Profile

The user can view or edit his credentials like first name, last name, birth date, and gender using the application. He can also upload a profile photograph from the gallery. All the changes made are saved in the database. But email id and mobile number are unique for each user which are non-editable.

4.3.4 Place Order

The place order feature is provided for both patients and doctors. The patients can order for services such as

- Food
- Clothes

- Medicines
- Test samples
- Reports
- Waste collection.

The patient can order any of the above-mentioned services for himself or a doctor can order the same for a patient during a visit to his/her room. When a particular service is requested, the request goes to the control room when the cart of the bot is filled with the required package. Then according to the registered room number of the patient, the bot sets in motion to deliver the package.

4.3.5 Delivery Status

After placing the order, the user is directed to the order status page. The information is regarding the order such as ID, service type, date, time, and room number. Also, the estimated time of arrival of bot as calculated by the server is displayed to the patient. The user can also cancel a particular order using the cancel order button on the screen.

When the order arrives, i.e., the bot with the required package arrives, the cart can be unlocked by using the application. To ensure that only the person who has ordered is able to unlock the cart, the application sends the username along with the order details so that the server can validate the order. That implies, only the person who has requested for the service will be able to open the cart for retrieving the order.

4.3.6 Schedule Delivery

This feature is only available for the hospitals. Using this, the hospitals can schedule the various orders at a particular instance of time. The hospital coordinator, assigned this task can change the order of deliveries or can assign a priority to a task. Optimization in efficiency of delivery can be achieved through scheduling.

4.3.7 Previous Orders

All the past service requests successfully executed for a patient are displayed in previous orders. The service image, type, order ID, date, and time are displayed for each order. This feature is available for both patients and doctors.

4.3.8 Register/Remove Patients/Doctors

To ensure that a person who is a patient/doctor is the only one using the services, the hospital adds the email id of that user into its database. Only after doing, this step is the user able to request for a service. This feature is very necessary in order to ensure that no fake accounts are created in the database using the MedBot application. At the time of registering a doctor, all the necessary certificates are uploaded for proper registration.

4.3.9 Add/Remove Rooms

When a patient is admitted to the hospital, he is also registered in the MedBot application by hospital's login. This is done instead of the patient specifying the room no. while placing the order. Security is ensured in the process.

5 Conclusion

The proposed approach ensures real-time performance, adaptability, simplified interfacing with human operators, and accurate dispatching of packages. System is capable of delivering medicine supply on-demand which drastically reduces operational breakdowns. Intelligent system allows medical staff to work more on patient care services instead of transportation medical supply which improves the quality of healthcare services and reduces manpower. This system not only improves the accuracy in work but also improves frequency of work.

The target segment can be expanded from only hospitals to any confined area with a specific geography that can be represented using a map. For example, such a system can be used in storage houses (warehouses) for transportation of goods from one place to another. Another possible area of application can be hotels. The room service is carried out manually by the hotel staff. If such a bot is deployed on each floor of the hotel, the room service requests made by the guests can be executed efficiently and with less amount of time and manpower involved.

Future Scope

The system mentioned in this paper is an autonomous package dispatcher bot for the hospital applications. But the target segment can be expanded from only hospitals to any confined area with a specific geography that can be represented using a map. For example, such a system can be used in storage houses (warehouses) for transportation of goods from one place to another. Another possible area of application can be hotels. The room service is carried out manually by the hotel staff. If such a bot is deployed on each floor of the hotel, the room service requests made by the guests can be executed efficiently and with less amount of time and manpower involved.

References

1. Özkil AG, Fan Z, Dawids S, Aanæs H, Kristensen JK, Christensen KH (2009) Service robots for hospitals: a case study of transportation tasks in a hospital. Denmark
2. Fung WK, Leung YY, Chow MK, Liu YH, Xu Y, Chan W, Law TW, Tso SK, Wang CY (2003) Development of a hospital service robot for transporting task. Hong Kong
3. Engin M, Engin D (2012) Path planning of line follower robot
4. Hasan KM, AI-Nahid A, Al Mamun A (2012) Implementation of autonomous line follower robot
5. Krishnamurthy B, Evans J (1992) HelpMate: a robotic courier for hospital use
6. Raspi https://www.terraelectronica.ru/pdf/show?pdf_file=%252Fds%252Fpdf%252FT%252FTechicRP3.pdf
7. Ultrasonic sensor (HC-SR04) https://cdn.sparkfun.com/datasheets/Sensors/Proximity/HCSR04.pdf
8. Pi camera https://cdn.sparkfun.com/datasheets/Dev/RaspberryPi/RPiCamMod2.pdf
9. www.nex-robotics.com/products/motors-and-accessories/100-rpm-side-shaft-37mm-diameter-compact-dc-gear-motor.html
10. Motor Driver L289N https://cdn.sparkfun.com/datasheets/Dev/RaspberryPi/RPiCamMod2.pdf
11. Powerbank https://www.technicahub.com/syska-10000-mah-power-boost-100-blue-specification/
12. Three tier architecture https://www.guru99.com/images/1/091318_0745_DBMSArchite3.png

Improvement in Material Feeding by Introducing Kitting in the Assembly Line

Chinmay Kule, Shantanu B. Patil, and Sandeep Vaity

Abstract The concept of kitting is to ensure the supply of all complete sub-assemblies in the form of kits to the assembly station/line. The main purpose of this study was to analyze the business case and the possibility of the company to implement lean kit assembly concept. Inceptive part of the study was to understand the kitting concept completely and simultaneously study the current material feeding system, where improvement had to be incorporated. The study shows that kitting results in reduction of operator walk time, material search time and material fetch time. It also leads to better planning for material, manpower, and enhancement in productivity through appropriate work allocation. Although the benefits achieved from the implementation of kitting concepts are specific to the company's needs, results make it clear that kitting goes hand in hand with the lean manufacturing process.

Keywords Kit assembly · Centralized feeding · Analytic hierarchy process · Productivity

1 Introduction

ERP/MRP systems typically keep a record of the total quantity of a part type and not how that material is delivered. This inaccuracy and lack of granularity, combined with stockroom personnel mistakes, lead to problems that include: insufficient quantity of components, excess quantity of components, wrong components, incomplete kits, insufficient quantity of component packages (e.g., insufficient quantity of reels for split parts). Thereby, resulting in increased operator idle time, lead times and manpower.

Conventionally, kitting was started by the production control department based on the shop floor orders, as made by the plant's ERP/MRP system. Production control

C. Kule (✉) · S. Vaity
Dwarkadas J. Sanghvi College of Engineering, Mumbai, Maharashtra 400056, India
e-mail: chinmaykule@gmail.com

S. B. Patil
ACG Engineering, Mumbai, India

© Springer Nature Singapore Pte Ltd. 2020

H. Vasudevan et al. (eds.), *Proceedings of International Conference on Intelligent Manufacturing and Automation*, Lecture Notes in Mechanical Engineering,
https://doi.org/10.1007/978-981-15-4485-9_42

had to initially confirm that adequate stock/quantity is available for each part number. Orders are placed for parts in case of any shortages. When executed properly, benefits of kitting include: maximized value added time of operators, easier operator training resulting in reduced training cost, maximized machine utilization—no line stoppage due to part shortages or searching for parts, reduced work in process (WIP), reduced lead times, reduced part damage due to excess handling.

This study of a manufacturing firm in Mumbai will discuss the problem with the current material feeding system in the methodology section. This section will also discuss the various parameters related to the kit formation to be considered before finalizing the kits. Further, the AHP process will be used as a medium to finalize the project parameters. Finally, the results and conclusion section will discuss the current findings and future scope of implementation.

2 Literature Review

The motive of lean is to eliminate unwanted resources and delays from each of the company's activities for maximizing the value added by them. Inventories can be classified into the following categories: (a) Raw material is an inventory that is used for the manufacturing of a product. (b) Work in process (WIP) is an inventory that is currently being used to complete a product. (c) Finished goods are goods where the process of raw material has been completed and those can be either sold to the customer or sent to another company as raw materials. Inventory went stored in excess not only results being wasteful but also hides other relating problems and their possible solutions [1, 2].

With reference to the movement, quantity, and storage of the material used, material feeding principles can be classified, namely into continuous supply and batch supply. In continuous supply, the material is supplied in units suitable enough for handling, which are replaced once empty. In batch supply, the material is sent in batches, either batch of required part numbers or a batch of these part numbers in specific quantities.

AHP is a quantitative methodology that determines the importance of a criterion concerning other multiple criteria by using paired comparison analysis. AHP assigns weights to every criterion according to their importance. AHP is an important decision-making tool in manufacturing industries and can be used to determine which factors are to be considered as the project parameters [3].

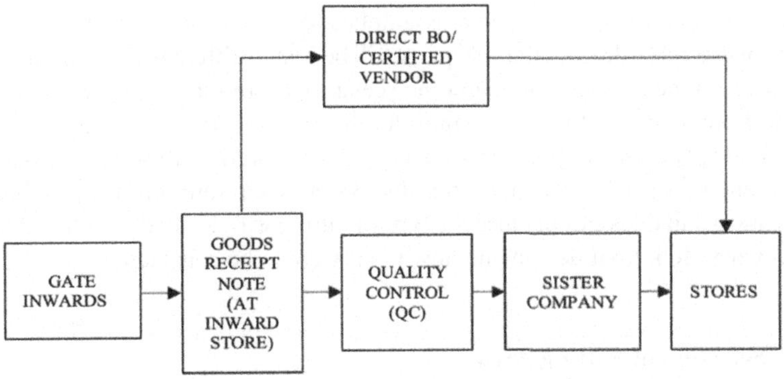

Fig. 1 Material flow in earlier method

3 Methodology

3.1 Earlier Method

Once the sales order is confirmed, a sales order bill of material (SOBOM) is generated in the SAP system. This bill of material (BOM) contains all the parts that are required for the assembly of the machine. The purchase department starts working toward the procurement of these parts. There are about 89 vendors who are supplying these parts. Figure 1 shows the process through which the material flows.

Material received from self-certified vendors is directly transferred from inward store to the warehouse, skipping QC. The mentioned path of material movement is traced in SAP and at every step, and it is posted to the next location in SAP along with the physical transfer. 718 components are required to complete the assembly of the machine. The production manager has a closing deadline, so he instructs to assemble the said machine with available 400 out of 718 components required. Besides, the rest of the machines which were kept aside for the priority of this order also add up to WIP inventory. A few weeks later on arrival of the missing parts the production manager shortcuts few procedures on the waiting machine and sends it for dispatch. Therefore, to improve this process, a study is being conducted which will only analyze the current material feeding system and compare how kit assembly will help the organization eliminate the problems with the current system.

3.2 Method by Using Kitting System

The method of implementing kitting system involves considering the type of kitting, depending upon the assembly and the current manufacturing scenario of the organization. It suits to manufacturing having a parallel flow of sub-assemblies and assemblies consisting of many high valued sub-assemblies [4, 5]. According to Johansson [6], the type of assembly decides the way parts will be sent in different kits to the assembly line. Based on the current manufacturing scenario, kit-to-manufacturing, where different kits are delivered to the shop floor for the assembly of more components, was incorporated [7]. Considering the number of parts required for the machine assembly, sending one part per kit was not a feasible option; therefore, kit-to-manufacturing was the most suitable option. Once the type of kitting was decided, additional factors such as where to kit, kit design and how to kit were to be considered.

3.2.1 Selection of Kitting Area

A centralized material feeding system [8], which involves filling the kits in a particularly assigned area (first floor in this case) and then distributing the kits throughout the assembly line, was implemented. The only way from the stores to the assembly line was through the first-floor elevator being one reason. Another reason was to avoid an increase in the line storage space. In addition to the centralized material feeding system, different color codes were assigned to bins of the different machines to easily identify them in the particularly assigned kitting area. For example, gray color bins for model 1, green color bins for model 2, and blue color bins for model 3, as shown in Fig. 2.

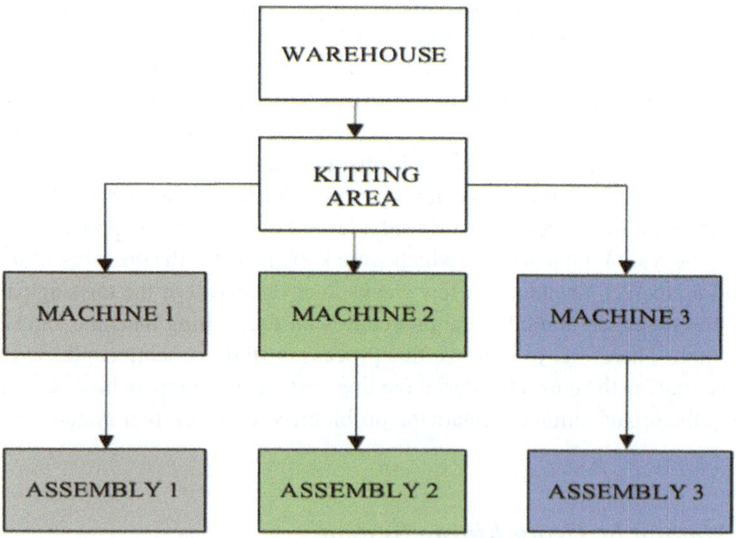

Fig. 2 Centralized material feeding system with the assignment of color codes for easy identification of stationary type bins (kits) of different machines in the kitting area

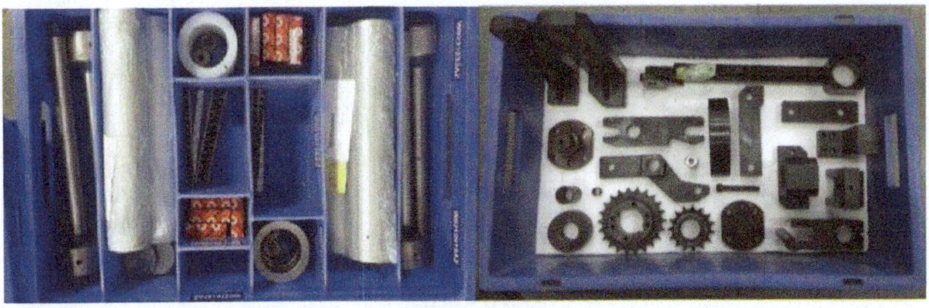

Fig. 3 Left: kit containing separate racks for parts as well as drawings and right: Poka-yoke implemented in the kit to avoid the mistakes while filling of the kits

3.2.2 Design of Kits

The kit may contain all the items required to complete the assembly, may it be drawings, tools or parts relating to that assembly as shown in Fig. 3. Sometimes due to product complexity, assembly parts might include small parts such as nuts, bolts, washers, which might make the installation of the kit a troublesome process. Since these parts were standardized, it was decided to exclude these items from kits, and make them available in separate racks available on shop floor. The purchase of these items was been done as a bulk consumable, and the value of some of these parts was considered to be uneconomical for them to be handled separately for each machine [9].

3.2.3 How to Kit?

The kits were being manually assembled by plant associates assigned to the stores' department. A material requisition (MR) provided by the production team was used by associates from the stores' department to fill kits. A picker to part method, where associate brings the part into the kit (Fig. 4), was used for material withdrawal and kit formation [8] since a system-generated picklist of parts along with the bin numbers was used by the store associates. In addition to the picker to part method, the kitting area was divided into separate kitting zones [8], according to different models of machines available (Fig. 4).

3.2.4 Network Analysis

In addition to the formation of kits considering several factors, network analysis, which involves understanding and segregating the various processes involved in the machine assembly as online and offline processes, was performed, as it would employ additional time-saving and cost-saving. Offline processes can be carried out

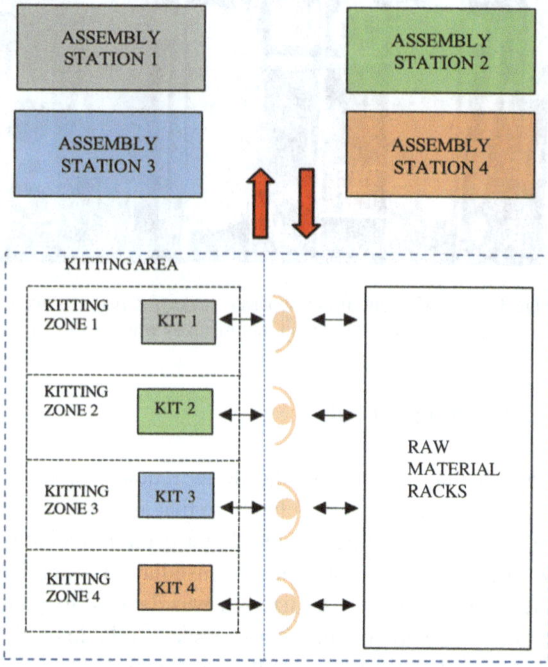

Fig. 4 Stationary kits from different zones in the kitting area

simultaneously with the ongoing assembly process. Online processes are processes that are to be followed in order of the assembly process.

3.2.5 Analytic Hierarchy Process (AHP)

This section will cover the formulation of the AHP matrix using the rankings provided by the various managers. A hierarchy was tabulated consisting of the main criteria and sub-criteria. The main criteria consisted of assembly side inventory, binning time, assembly time/space, part handling, idle time, required kitting space, and assembly side replenishments/day. The sub-criteria were considered at an advance level of decision making. A senior manager from every department was involved in the AHP meeting. Each criterion was discussed, and the ranks provided were incorporated (Table 1) into a relationship matrix using Saaty's scale for the AHP matrix. The rating given to each criterion was based on the scale with 1 (high) to 9 (low). For example, while comparing assembly side inventory and binning time, managers felt binning time was more important, therefore provided a higher rating (1/2 = 0.5). While rating the various criteria, factors such as current requirements, future goals, alignment with manufacturing goals were considered. After the pairwise comparison of every criterion, the normalized weights were obtained using the AHP software and were tabulated in MS Excel. Table 2 exhibits the outcome of the analysis, i.e., idle time, part handling, and assembly time/space are three criteria that are notably very

Table 1 Relationship matrix showing a comparison of the main criteria

Main criteria	Assembly side inventory	Binning time	Assembly time/space	Part handling	Idle time	Required kitting space	Assembly side replenishments/day
Assembly side inventory	1	2	0.5	0.33	0.2	3	4
Binning time	0.5	1	0.5	0.25	0.142	0.5	2
Assembly time/space	2	2	1	0.333	0.2	4	5
Part handling	3	4	3	1	0.333	3	5
Idle time	5	7	5	3	1	5	7
Required kitting space	0.333	2	0.25	0.333	0.2	1	3
Assembly side replenishments/day	0.25	0.5	0.2	0.2	0.142	0.333	1

Table 2 Results of AHP

Criterion	Normalized weights
Idle time	0.413
Part handling	0.212
Assembly time/space	0.133
Assembly side inventory	0.098
Required kitting space	0.064
Binning time	0.05
Assembly side replenishments/day	0.031

important. Therefore, idle time will be considered as the project parameter and the data collection will be based on this project parameter.

4 Results of Kit Implementation

The implementation of the kit assembly was quite successful and it nearly achieved all previously stated goals. In addition to achieving goals, it also brought a few other assemblies related discrepancies in the picture. Table 3 and Fig. 5 provide a statistical comparison between before and after assembly times of various processes. The cells that are highlighted in Table 3 indicate the reduction in the assembly time of that particular operation after implementing the kit for that assembly. Despite the measures taken, certain psychological and technical factors were overlooked and certainly caused hurdles during the implementation stage. The psychological challenges included: noticing the operator idle, the production manager used to force the stores' department to send the kits. There were times when even the technicians tried to send incomplete kits to the assembly station to show goodwill to their reporting line managers. Few of the technical factors include: due to itching missing on the majority of parts a lot of time was wasted in material rechecking. As the lead time consisted of production time as well as waiting time, where the waiting time was largely only due to components unavailability. This gave the illusion that the operation will take more time, and lead to the faulty conclusion that earlier start made will help to achieve the deadline for dispatching machine.

To avoid the occurrence of the challenges faced during the implementation stage again in the future, certain rules were set:

(a) Only complete kits must be sent to the assembly station.
(b) Only one person should be responsible for releasing kits, monitoring materials being filled in kits in the kitting area and monitoring buffers on the shop floor.
(c) Each production engineer will be responsible as a person in charge for the handover of kits of the machine model allocated to him from the warehouse to the assembly line.

Table 3 Results of implementing kitting in assembly

Process	Current assembly time	Kit assembly time	Percentage saving in time (%)
Index, Z-cam and Geneva assembly	8	**6**	**25**
Main shaft assembly	12	**9**	**25**
Mount drive assembly	22	**20**	**9**
Turret ring assembly	6	**3**	**50**
Control panel assembly	2	2	0
Junction box assembly	2	**1**	**50**
Top cover mounting	4	**3**	**25**
Capsule reservoir assembly	2	**1**	**50**
Vacuum valve assembly	3	**2**	**33**
Closing plate assembly	1	1	0
Tamping assembly mounting	12	**10**	**17**
Loader assembly mounting	6	6	0
Sub assembly mounting	10	**6**	**40**
PMM assembly	8	**3**	**62.5**
Second QC	1	1	0
Pendant box mounting	2	**1**	**50**
Powder unit mounting	4	**1**	**75**
Machine wiring	8	**6**	**25**
Capsule/powder trials and QA check sheet	6	6	0
Electrical test chart and machine power on testing	1	1	0
Final QC	4	4	0
F A T	8	8	0
Dispatch	8	8	0

The bold numbers in the "kit assembly time" column indicate the reduction in the assembly time of that particular operation after implementing the kit for that assembly. The bold numbers in the "percentage saving in time" column indicate the percentage of time saved for that particular assembly operation

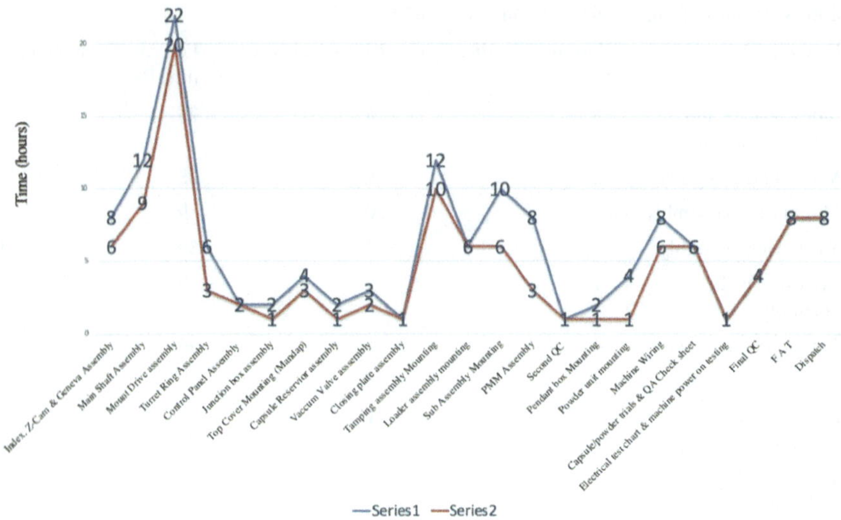

Fig. 5 Series 1: depicts the assembly timings of the current process, and series 2: depicts the assembly timings after implementing kitting

(d) Machines are subject to improvements based on feedback given by the service team or customers for delivering better performance and efficiency. Hence, machines improve with time, changing their parts sometimes if needed too. Hence, there may be a possibility that the kits will also have to be redefined accordingly. This completely depends on the circumstances at that particular time. Thus, the definition of a complete kit should be changed accordingly as a good rule to work smartly [10].

5 Conclusion

The implementation of the kit concept had a positive impact, in terms of functioning and cost involved at several departments in the organization. In terms of functioning, the purchase department, as well as assembly supervisor, knew which parts will be required when and therefore could prioritize the follow up. Shortage lists were also generated kit wise, thus making the process substantially convenient. The concerned person in the purchase department exactly knew as to which part was required when and also had records of lead time for every part. Subsequently, it becomes easier to ensure that each part arrives in the company before it was required on the machine for assembly, therefore, avoiding any delays/stoppages on the assembly line. In terms of cost, as the number of parts handled at a time were less, a lot of time, space and energy was saved. Stocking of parts on the shop floor which are not required immediately was avoided, thereby improving the shop aesthetics and reducing assembly side

WIP inventory. Prioritization of material procurement would avoid delay in assembly, thus helping the production team to adhere to deadlines and commit to lower lead times. Control points that usually occurred at the end of any process, in the case of kit assembly occurred at the end of every individual kit assembled, leading to corrective actions to be taken at the early stages.

References

1. Krajewski J, Malhotra M, Ritzman B (2007) Operation management—processes and value chains, 8th edn. Pearson Education, Upper Saddle River, NJ, USA
2. Karlsson C, Ahlstrom P (1996) Assessing changes towards lean production. Int J Oper Prod Manage
3. Saaty TL (1980) The analytic hierarchy process: planning, priority sett allocations. McGraw Hill, New York
4. Schwind GF (1992) How storage systems keep kits moving. Mater Handl Eng 47(12):43–45
5. Johansson E, Johansson ML (1990) Materials supply systems design in product development projects. Int J Oper Prod Manage 26(4):371–393
6. Johansson MI (1991) Kitting systems for small parts in manual assembly systems. In: Pridham M, O'Brien C (eds) Production research approaching the 21st centaury, pp 225–230
7. Bozer YA, McGinnis LF (1992) Kitting versus line stocking: a conceptual framework and a descriptive model. Int J Prod Econ 28:1–19
8. Brynzer H, Johansson MI (1995) Design and performance of kitting and order picking systems. Int J Prod Econ 41:115–125
9. Johansson B, Johansson ML (1990) High automated kitting system for small parts: a case study from the Uddevalla plant. In: Proceedings of the 23rd international symposium on automotive technology and automation, Vienna, pp 75–82
10. Brynzer H (1995) Evaluation of kitting systems—implication for kitting system design. Licentiate thesis, Chalmers University of Technology, Gothenburg

Advanced Materials

Microstructure and Roughness Analysis of Drum Brakes of Maruti 800

Atul D. Dhale and Swapnil S. Phadnis

Abstract The purpose of this paper was to focus on the basics of drum brake systems, so as to provide an in-depth microstructure analysis of the drum brake and the roughness value of the drum surface. As the roughness affects the braking capacity of linings on drum surface, it also allows the duration of service for the drum. This study has helped to know the amount of wear on drum and to minimum roughness value of drum surface that is required for safe braking.

Keywords Drum brakes · Maruti 800 · Microstructure analysis · Roughness analysis · Tribological analysis

1 Introduction

A drum brake is a brake that presses the set of shoes or pads against a rotating drum called as brake drum; this action causes friction which stops the vehicle. Drum brake consists of wheel cylinder, brake drum, shoe, backing plate and various springs and pins. In normal braking, when the brakes are applied, the brake fluid is pushed from master cylinder and moves to the wheel cylinders, the piston pushes outwards force over the brake linings on the brake drum surface, which creates friction and reduces the speed of the vehicle, and heat is generated that is dissipated through natural convection. Emergency brakes are also drum brakes which are bypass for fluid action; they are connected to hand brakes by means of steel cables. It is fully mechanical brakes, and when regular brakes fail, the cable stretches the lever which directly connected to shoes to stop in emergency.

A. D. Dhale (✉)
Department of Production, Dwarkadas J. Sanghvi College of Engineering, Mumbai, Maharashtra 400056, India
e-mail: atul.dhale@djsce.ac.in

S. S. Phadnis
Department of Mechanical, SSJCE, Dombivli, India
e-mail: swapnilphadnis91@gmail.com

© Springer Nature Singapore Pte Ltd. 2020
H. Vasudevan et al. (eds.), *Proceedings of International Conference on Intelligent Manufacturing and Automation*, Lecture Notes in Mechanical Engineering,
https://doi.org/10.1007/978-981-15-4485-9_43

Grey cast iron is used for the drum brakes as they have good tribological properties and good at heat dissipation. Brake lining is made up of asbestos material such as reinforcements, binder, abrasives, friction modifiers and filler materials [1]. The microstructure reveals the quality of the surface, and the insight the microstructure helps us to check the current state of material at microscopic level. Any changes observed can be noted, accordingly change the etchant, which are used for etching the surface to get a neat microstructure. There are various types of etchants [2]: nital—96–98 mL ethanol, 2–4 mL nitric acid (HNO_3), picral—4 g picric acid ($(NO_2)3C_2H_2OH$), 100 mL ethanol, glyceregia—three-part glycerine, two-part hydrochloric acid (HCl), one-part nitric acid (HNO_3), alkaline sodium, picrate—2 g picric acid ($(NO_2)3C_6H_2OH$), 25 g sodium hydroxide (NaOH), 100 mL distilled water and Klemm I 50 mL sat. Aq. sodium thiosulfate ($Na_2S_2O_3{\cdot}5H_2O$).

Surface roughness is the parameter of the smoothness of the surface; it measures the peaks and valleys of the surfaces, and they are measured in grit which is the number of scratches per inch. For the smooth surface, the number of Ra value is less, and it is the average value of the height of peaks and valleys; higher the value rougher is the surface. Coefficient of friction slightly depends on the roughness of the drum [3].

2 Experimental Work and Discussions

The vehicle selected was Maruti 800 DX and has two different set of brakes. Front brakes are disc brakes, and rear are drum brakes, while drum brakes are selected as they are more important in normal and emergency braking.

Standard dimension of new drum brake was 180 mm (diameter of drum).

The selected vehicle has run 62,900 km, and the wear was checked in the drum brakes of rear wheels as well as thickness of used brake drum. As the brake drums are more costlier, they should be replaced very rarely so our main focus was on brake drum rather than brake linings which are easier to replace Figs. 1 and 2.

Comparison of thickness of brake drum and lining of new and used brake to find the amount of wear is mentioned in Table 1.

Thickness of brake lining was reduced by 1 mm, and brake drum was reduced by 0.4 mm. Wear on brake linings and decrease in thickness of brake lining were observed. There were two ways abrasive wear on lining and drum was observed, as the liner materials was removed or cut by the brake drum and straight lining was seen on drum and material was removed from surface of the drum by liner, so it is expected that to replace the liner after every 5000 km s (Fig. 3).

The sample of $1 \times 1''$ drum brake was used to perform the microstructure analysis and to find out Ra value of the drum brake [4].

Fig. 1 New drum brake for Ra value

Fig. 2 Used drum for microstructure and Ra value

Table 1 Thickness comparison

	Thickness (mm)	
	Drum	Lining
New	3	5
Used	2.96	4

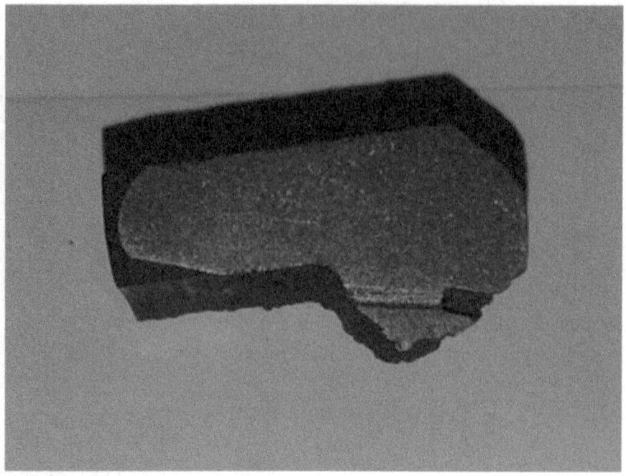

Fig. 3 Etched sample of 1 × 1″ sample of drum brake

2.1 Microstructure Analysis of Drum Brake

The drum under investigation was removed, and new drum was replaced on its behalf. The sample of 1 × 1″ in was submitted to the metallurgical laboratory for microstructure analysis and roughness analysis. Also other sample pieces of old and new brakes were submitted for Ra (roughness) factor analysis. The above-mentioned tests were carried out at ELCA laboratories at Thane. The microstructure analysis was carried as per ASM Handbook. The longitudinal orientation of the sample was considered for the analysis and central location of sample was selected for ease of eaching and 2% nital was used as etchant [nital—96–98 mL ethanol, 2–4 mL nitric acid (HNO_3)].

The microstructure analysis revealed that there was uniform distribution of type A graphite flakes in pearlite matrix, and the whole microstructure was grey cast iron. The image below shows 200× magnified image of the drum brake (Fig. 4).

The graphite flakes in the pearlite matrix lead to unmachinable surface due to presence of graphite, the brake may break during machining, and the microstructure does not change.

2.2 Surface Roughness Test of Drum Brake

Surface roughness test was performed on the drum brake by using instrument Handysurf E-35B [5].

Three readings were taken for the given sample, and values of roughness as below Ra (μm) is 1.866 and 2.033. The maximum and minimum values are from 1.2 to 3.5.

From Table 2, it can be observed that the Ra value was near to lower limit, but was not below lower limit of 1.2, and it does not exceed the limit of 3.5. It was also seen

Fig. 4 Microstructure of brake drum

Table 2 Table of new and old Ra values

Range	Ra value of new drum (μm)	Ra value of used drum (μm) (reading1)	Ra value of used drum (reading 2)	Ra value of used drum (reading 3)
Minimum	1.2	1.9	1.8	1.9
Maximum	3.5	3.0	3.1	3.0

that the drum needs to be replaced when the value decreases below 1.2 or exceeds the limit 3.5.

The roughness of the drum has a slight effect on the coefficient of friction, which for practical purposes may be neglected. The effect of the roughness of brake drums on the wear of linings was probably less in service than on the test machine.

3 Conclusion

The microstructure analysis revealed that the graphite flakes in the pearlite matrix lead to unmachinable surface due to the presence of graphite. The brake may break during machining as the microstructure does not change. Average values of roughness Ra (μm) were 1.866 and 3.033, wherein the increase in Ra value could lead to irregular surface of drum, which leads to replacement of the drum. It can also be concluded that no hot spots are seen on the drum brakes. Excessive braking and rapid cooling will lead to hot spots, damaging of drum and required replacement of drum. It can further be concluded that the above drum brakes are safe for further use.

References

1. Kodgiri VD (2018) Material science and metallurgy for engineers. 42nd edn. Everest Publishing House, Pune
2. Radzi kowska JM Metallography and microstructures of cast iron. Foundry Research Institute, Krakow, Poland
3. Talati J Surface roughness—significance and symbol interpretation in drawings. Hexagon Design Centre, Vadodara
4. Taylor RH, Holt WL (1941) Effect of roughness of cast iron brake drum in wear tests of brake linings. Part J Res Nat Bur Stand 27
5. Rajbongshi SK, Borah A, Choudhury PK (2014) Optimisation of process parameters in turning of grey cast iron with mixed oxide ceramic tool using taguchi's approach. In: 5th international & 26th all India manufacturing technology, design and research conference (AIMTDR 2014), 12–14 Dec, 2014. IIT Guwahati, Assam, India

A Review on Stress Relaxation Cracking in Austenitic Stainless Steel

Indhumathi Dayalan, Prashant Frank Crasta, Sritam Pradhan, and Renu Gupta

Abstract Stress relaxation cracking (SRC) is a high-temperature failure mode occurring in equipment, operational in industries at high service temperature for a long time. SRC has been noticed in thick-walled welded components of austenitic stainless steel, ferritic stainless steel, heat-resistant steel, and nickel-based alloys for many years. Understanding SRC phenomenon is necessary to improve the lifetime of equipments and to save resources. In order to improve the lifetime of the equipments, understanding SRC mechanism and its characteristics is important. In this paper, past experience of SRC in austenitic stainless steel has been reviewed to understand why and how SRC occurs and what are the major causes for cracking. Preventive measures can be taken that could avoid stress relaxation cracking and extend the lifetime of the equipment's working at high temperature.

Keywords Stress relaxation cracking · Austenitic stainless steel · Intergranular cracking · Reheat cracking

1 Introduction

Austenitic stainless steels and nickel (Ni)-based alloys are extensively used in nuclear plants, chemical fertilizers, and petrochemical plants due to their excellent high-temperature properties. These materials provide high strength at elevated temperature. They also possess good corrosion resistance and oxidation resistance at high temperature [1]. Appreciable creep properties at high temperature enable these steels to be a candidate material for equipment's used in boilers and pressure vessels, for example, superheaters, reheater tubes, and steam generators [2]. But austenitic steel and Ni-based alloys are vulnerable to stress relaxation cracking (SRC) that occurs either during service at high temperature or during post-weld heat treatment [3].

I. Dayalan (✉) · P. Frank Crasta
Dwarkadas J. Sanghvi College of Engineering, Mumbai, Maharashtra 400056, India
e-mail: innu.vasu@gmail.com

S. Pradhan · R. Gupta
L&T Heavy Engineering Division, Powai, Mumbai, India

© Springer Nature Singapore Pte Ltd. 2020
H. Vasudevan et al. (eds.), *Proceedings of International Conference on Intelligent Manufacturing and Automation*, Lecture Notes in Mechanical Engineering,
https://doi.org/10.1007/978-981-15-4485-9_44

Stress relaxation cracking is often noticed in weldments of austenitic stainless steels, ferritic stainless steels, and nickel-based alloys. SRC has been noticed for many years in equipments that are operational at high temperature. This phenomenon has been identified either shortly after service or after several hours of prolonged exposure to high temperature. This catastrophic cracking, i.e., SRC, leads to frequent repair work and abrupt plant shutdowns. Moreover, SRC has severe implication on cost-effectiveness and safe work environment.

SRC is referred as low-ductility intergranular fracture in coarse-grained heat-affected zone. The ductility dip is inadequate to hold the plastic deformation linked with stress relaxation. High stresses and temperature during service cause the thick-walled weldments of boilers and pressure vessel to undergo stress relaxation cracking at the heat-affected zone (HAZ). SRC is mostly prevalent in the coarse-grained heat-affected zone (CGHAZ) adjacent to fusion boundary. SRC is featured by coalescence of micro-voids near the grain boundary which resembles creep fracture and plastic deformation in grain boundary regions [4]. Relaxation cracks are generally observed after a prolonged exposure of nearly 10,000–1,00,000 h at high temperature. Stress relaxation cracking is also referred as reheat cracking or strain age cracking or post-weld heat treatment (PWHT) cracking [4]. Strain age cracking and reheat cracking occur in short time during PWHT, whereas stress relaxation cracking occurs after long exposure to high temperature for more than 10,000 h [5]. Stress relaxation cracking and strain age cracking are differentiated by the time to failure.

Reheat cracking has been noticed in stabilized austenitic stainless steel 347H during service or PWHT. PWHT of the weldments is generally done to relieve residual stress formed during welding, regain material properties, and restore corrosion resistance to make the weldments suitable for high-temperature service. Similarly, strain age cracking has been noticed in Ni-based alloys due to formation of precipitates [6]. At high operating temperature, the precipitated carbides strengthen the grain interiors and prevent plastic deformation when the relaxation of residual stress in the HAZ gets converted from elastic strain to creep strain [7]. Repairing these cracks is not a permanent solution since the cracks reappear even after a controlled weld repair. Relaxation cracking depends on the following factors: (1) chemical composition of the base metal, (2) service temperature of the equipment, (3) level of internal stresses formed due to welding and cold deformation, (4) system stresses raised due to geometry of the equipment, and (5) grain size of the material [4]. The reasons for SRC are attributed to weld quality, residual stresses formed due to fabrication process, carbide precipitation, and level of system stress [1].

In the last decade, there has been lot of researches conducted to understand why SRC happened and how to prevent stress relaxation cracking in austenitic stainless steel and other alloy steels. In this paper, we study stress relaxation cracking behavior in equipments used at high temperature to analyze the features of SRC and its mechanism.

2 Stress Relaxation Cracking Mechanism

Stress relaxation cracks are profoundly sited in high carbon austenitic stainless steel 304H, 347H, 321H, and Ni-based alloys like alloy 800H, Inconel 601, and 617 [8]. These steels have been used in equipments such as furnaces, pipes, drums, and towers in petrochemical plants [9]. SRC was first observed in thick-walled welded components from high-temperature Cr-Ni steel with Nb or Ti additions. Similarly, low-alloy and high-strength structural steels containing Cr, Mo, V, and Nb were susceptible to these cracks [10]. Stress relaxation cracking is generally referred to as ductility dip at high operating temperature. In literature, SRC is also referred to as stress-induced corrosion and stress-assisted grain boundary oxidation cracking [8]. Though Ni-based alloys are more resistant to SRC than austenitic stainless steel, they are also susceptible to relaxation cracks.

SRC is featured as a continuously decreasing strain rate called as relaxation through a creep mechanism [4]. The mechanism of SRC is material oriented and considered as the relaxation of residual stresses by plastic deformation that is localized along or near the grain boundary. The primary cause for SRC is lack of ductility [11]. Loss in ductility generally happens when the austenitic stainless steel age hardens between 500 and 700 °C that is the operating temperature of many boilers and pressure vessels [11]. The fine precipitates which are formed during age hardening increase the hardness level, thereby reducing the deformation capacity [11]. Other factors that contribute to SRC are chemical composition of the material where the age hardening elements have a prominent role in cracking [11]. SRC is associated with steels which can be precipitation hardened by certain alloying elements and not in unalloyed steels. Certain elements have both adverse and favorable effects on relaxation cracking, for instance, though carbon and nitrogen have an adverse effect on cracking, they contribute to stress–rupture properties. Hence, to obtain maximum benefits and minimum demerits, C and N contents are limited to 0.03–0.06% and 0.04–0.07%, respectively, for improving creep properties [12]. Coarse-grained and highly stressed materials are more susceptible to SRC [13]. SRC is dependent on the service temperature, amount of residual stresses from deformation process and system stresses available during service. SRC is an alloy-specific failure at high temperature. The susceptible temperature to SRC is dependent on the individual chemical composition of the material and its properties [8]. Various materials with their corresponding susceptible temperature are listed in Table 1.

Table 1 Various materials with different susceptible temperature to SRC Ref. [8]

Material	Susceptible temperature (°C)
AISI 304H and 321H	550–600
AISI 347H	575–650
Alloy 800H	550–650
Alloy 617	550–700

2.1 Identification of Stress Relaxation Cracking

SRC is often observed in CGHAZ and weld metal regions [14]. Cracks are located on the grain boundaries, and these cracks are surrounded with small cavities [11]. These cracks are enclosed with or without metallic filament which are encircled with chromium layer indicating Cr depletion [8]. Cracks are generally found in the areas where the Vickers hardness is higher than 200 Hv10. In general, the hardness values at HAZ and weld metal in as-welded condition are high. The hardness values at HAZ and weld metal further increases during high temperature service and thereby impedes grain deformation [11]. Generally, SRC is found in the susceptible temperature range between 550 and 750 °C. For certain materials, the lower limit is susceptible temperature, whereas few materials have higher limit as their SRC temperature [8].

3 Case Studies

In the following section, we discuss research study on failure analysis of the cracks observed in reactors and pipeline of austenitic stainless steel and working at high temperature. This includes identifying location of crack, features of crack, probable reasons for crack, and remedial measures to avoid cracks.

In the study "Stress relaxation cracking in 304H stainless steel weld of a chemical reactor serviced at 560 °C" [5], the authors conducted a research to identify the causes for cracking in a reactor. The crack was observed in the chemical reactor made of SS30H after 6 years of service. The crack reappeared in two years even after a repair. The reactor was operating at a temperature between 535 and 565 °C. In order to analyze the reason for cracking, a failure analysis at the cracking location that is the weld between reactor and an inlet nozzle pipe was conducted. Since the circumferential crack appeared even after a controlled welding during repair, the chances of poor weld quality leading to cracks were omitted.

The possible causes for cracking were listed as high thermal system load, high thermal stress and material properties of 304H. These causes were individually studied to understand the actual reason behind the cracking. Using AUTOPIPE,[1] software for pipe stress analysis, the system stress was used. The analysis revealed that the stress at the cracking location was apparently less. The thermal stress analysis was studied using ABAQUS software both for designed condition and operating condition. It was revealed that the stress at the cracking location changed from 35.8 to 84.3 MPa during startup and shutdown processes, respectively. This variation in stress in accordance with temperature was assumed to be a contributing factor for cracking. The intrinsic material properties of SA240 304H were reviewed based on ASME boiler and pressure vessel code Sect. 2 part D. As per ASM handbook Chap. 6, stainless steel 304H

[1] https://www.bentley.com/en/products/product-line/pipe-stress-and-vessel-analysis-software/autopipe.

can be chosen for an operating temperature range of 595–815 °C. However, in the current study, the operating temperature 560 °C was in the lower bound range, and still the material was prone to cracking. Thus, SA387 grade 9 or 91 was proposed as a suitable material for an operating temperature less than 600 °C. Since austenitic stainless steel 304H is susceptible to stress relaxation cracking or reheat cracking, either of these two cracking were suspected to be the main cause for cracking in the reactor. Reheat cracks generally occur either during PWHT or in a short period of service. During welding at high temperature, alloy carbides get dissolved in the weld and HAZ regions; these alloy carbides get precipitated in the grain interior during PWHT or short service of the equipment. Precipitation of carbide makes the grain strengthened and weakens grain boundaries. This results in intergranular cracking when the residual stress during welding is relaxed either during PWHT or short service at high temperature, whereas stress relaxation cracking is noticed in weldments that are exposed to high temperature for a long period. Though the intergranular cracking mechanism is similar to reheat cracking, the driving stresses contributing to cracking can be raised due to thermal stress, system stress, or stress concentrating geometry.

In this case study, a thermal stress of approximately 84% of the yield strength is noticed at the cracking location under operating condition. Hence, this could clearly be identified as the reason for stress relaxation cracking that was noticed after a prolonged exposure at high temperature. As a prevention measure, an annealing heat treatment of the weld at 850 °C for 2 hours was proposed. Through annealing heat treatment, the residual stresses built up during welding are substantially reduced. Moreover, coarse carbide precipitation would be obtained in preference to fine carbides which weaken the grain boundaries. To reduce the stress concentration at the cracking location which leads to SRC, the root radius was altered. With the current root radius of 10 mm, a maximum stress of 84.3 MPa is noticed in the cracking location. While changing the root radius to 20 and 30 mm, the maximum stress at the cracking location reduced to 63.8 MPa and 55 MPa, respectively. Thus, changing the stress concentrating areas helps to reduce the stress which contributes to stress relaxation cracking.

In the study "Research on weld cracking of TP321H stainless steel pipeline under elevated temperature" [1], a research on failure analysis of cracking in pipeline is done. The cracks appeared in the welds of pipeline made of TP321H austenitic stainless steel. The pressure pipeline was serviced in a petrochemical plant as a synthetic ammonia equipment. The operating temperature and pressure in the pipeline were 650 °C and 3.4 MPa, respectively. The crack was observed in the pipeline within two years of service. These cracks reappeared repeatedly after repair work. The cracks were located in the T-joint weld junction between the straight pipe and elbows. All of the T-joints in the pipeline layout were observed with cracks. In order to understand the root cause for cracking, stress distribution in the pipeline was analyzed using ABAQUS,[2] a finite element software. The original conditions during service were taken as the parameters for FEA. The results of FEA analysis showed that the

[2]https://www.3ds.com/products-services/simulia/products/abaqus/.

Table 2 Microstructure, hardness, and grain size measurement

Pipe condition	Microstructure	Hardness (HRB)	Grain size (ASTM no)
Old pipe	Austenite+Ferrite	86.6	6
New pipe	Austenite+Ferrite+TiN	86.3	7

stress level at the cracking locations was not high enough to cause cracking. The allowable stress of TP321H is 137.3 MPa at room temperature and 46.7 MPa at high temperature. The maximum stress identified at the cracking location using FEA was 67.2 MPa. Though the stress was higher than the allowable stress of the material, it was still permitted as per the standard. Hence, the slight increase in stress level could be one of the reasons and not the sole reason for cracking.

Chemical compositional analysis was performed on the cracked region and new pipeline. The results of both the old and new pipeline were in accordance with ASME boiler and pressure vessels code Sect. 2 part D specifications. In order to understand the mechanical behavior of the pipeline material at room temperature as well as at high temperature, tensile tests and Charpy test were conducted. Both these tests were conducted at room temperature and operating temperature 650 °C. The tensile tests at room temperature showed a slight increase in tensile values of the old pipe and weld when compared to new pipe. Similarly, the ductility values were lesser in the old pipe and weld when compared to new pipe. The impact toughness values of the old and new pipe obtained through Charpy impact test were normal. But an abnormal impact toughness was noticed on the weld which indicated that the weld had undergone microstructural change when exposed to high temperature for long time. This was confirmed by the fractography of the failed test specimens of impact test. The scanning electron microscope (SEM) analysis revealed a marked brittle fracture failure mode. Microstructure and hardness of the old and new pipe were analyzed. The microstructure hardness and grain size were realized to be normal. The results are represented in Table 2.

The material characterization of the cracked region using SEM revealed that the crack is intergranular in nature and it originated along the fusion zone. To analyze the chemical composition at cracking locations, energy-dispersive spectrometry (EDS) was used. The analysis at three locations, middle of the crack, near to crack, and away from crack, was done. The results showed distinct Cr and C contents at respective locations. The location far away from crack had normal Cr content. The region near the crack showed higher Cr content, whereas at the middle of the crack, low Cr content was noticed. This is due to Cr-rich carbides that formed at grain boundaries resulted in chromium-depleted zone adjacent to the grain boundaries. Sigma phase and chromium precipitation were identified in X-ray diffraction. Since all the mechanical and metallurgical test results were fairly normal and slight changes were permissible by the standards, an analysis of the welding procedure was done. On analyzing the welding procedure that was carried out during pipeline construction, it was found that the welding parameters were not matching with the welding procedure specification (WPS). During welding large areas, the interpass temperature was not controlled, and thus, the subsequent weld temperature could convert few austenite, ferrite, and carbonization in the weld into sigma phase. The formation of sigma phase could make the material more prone to cracking. Hence, a poor weld quality could also lead to reheat cracking. Hence, a good-quality welding with controlled parameters is more important to avoid cracks. From the above case studies of relaxation cracking in austenitic stainless steel welds exposed at high temperature, the following inferences can be drawn; stress relaxation cracks are intergranular cracks found in HAZ of the welded regions when the welds have been exposed to high temperature for prolonged period. At high temperature, the precipitation of the carbides within the grains prevents the plastic deformation at grains. Thus, weakening of the grain boundaries reduces the ductility at high temperature. A high stress level in the equipment formed either due to thermal stress, system stress, or internal stress promotes stress relaxation cracking.

3.1 Preventive Measures for SRC

From the existing literature, the following preventive measures are recommended to avoid relaxation cracking. The preventive measures are proposed based on causes that lead to SRC in the equipments under various operating conditions. The preventive measures are follows:

- Avoid selecting materials for operational in the susceptibility range of SRC. For example, if your operating temperature of the equipment made of SS304H is 580 °C avoid using SS304H for manufacturing the equipment since the susceptibility temperature range of SS304H is 550–600 °C.
- Use controlled welding procedure to obtain good-quality weld. Welding heat input and interpass temperature should be strictly monitored for quality weld [1].
- Do stress relief heat treatment of the welds at specific temperature depending on the base metal. For example, SS304H is recommended for annealing heat treatment at 850 °C for 2 hours. Similarly, a PWHT at 980 °C for 3 h is recommended for Alloy 617. To prove this effect of PWHT on SRC susceptibility, a three-point bend test was used. The heat-treated welded parts surpassed the bending test for 150 hours [15].
- Avoid stress concentrating designs, sharp corners like pipe bends or sudden change in the dimension of the equipment [1].
- Choose welding consumables that possess high strength and creep properties when compared to base metal. Filler metals with titanium give good creep properties and strength to the welds [16].

4 Conclusion

In this paper, the cracking mechanism of SRC in austenitic stainless steel equipment used at elevated temperature was reviewed. It is indicative from the study that stress relaxation cracking cannot be completely avoided, but the cracking can be delayed and henceforth increasing the efficiency and lifetime of the equipment is possible. Stress relaxation cracking needs more detailed study to understand the mechanism of SRC very clearly. Also, a wide study of SRC in different materials like nickel-based alloys and ferritic stainless steels enables the researcher to generalize the mechanism in these different materials. Since there is no standard testing available for stress relaxation cracking, a detailed study on testing for stress relaxation cracking helps to predict and understand the susceptibility of materials to SRC.

References

1. Pan J-H, Fan Z-C, Zong N-S (2016) Research on weld cracking of TP321H stainless steel pipeline under elevated temperature. Int J Press Vessels Pip 148:1–8
2. Ghalambaz M et al (2017) A case study on failure of AISI 347H stabilized stainless steel pipe in a petrochemical plant. Case Stud Eng Fail Anal 9:52–62
3. Li Y et al (2019) Stress-relief cracking mechanism in simulated coarse-grained heat-affected zone of T23 steel. J Mater Process Technol 266:73–81
4. Kant R (2018) Stress relief cracking susceptibility in high temperature alloys
5. Yoon KB, Yu JM, Nguyen TS (2015) Stress relaxation cracking in 304H stainless steel weld of a chemical reactor serviced at 560 °C. Eng Fail Anal 56:288–299
6. Phung-On I (2007) An investigation of reheat cracking in the weld heat affected zone of type 347 stainless steel. Dissertation, The Ohio State University
7. Unnikrishnan R et al (2016) Investigating plastic deformation around a reheat-crack in a 316H austenitic stainless-steel weldment by misorientation mapping. Procedia Struct Integr 2:3501–3507
8. Van Wortel H (2007) Control of relaxation cracking in austenitic high temperature components. In: Corrosion 2007, NACE International
9. Ghalambaz M et al (2017) A case study on failure of AISI 347H stabilized stainless steel pipe in a petrochemical plant. Case Stud Eng Fail Anal 9:∇–∇62
10. Nawrocki JG et al (2003) The mechanism of stress-relief cracking in a ferritic alloy steel. Weld J NY 82(2): 25–S
11. Van Wortel JC (1998) Relaxation cracking in the process industry, an underestimated problem. Plant Mainten Manag Life Perform 637
12. Thomas Jr RD (1984) HAZ cracking in thick sections of austenitic stainless steels-II. Weld J 63(12):355
13. Shoemaker Lewis E et al (2007) Fabricating nickel alloys to avoid stress relaxation cracking. In: Corrosion 2007, NACE International
14. Nawrocki JG et al (2003) The mechanism of stress-relief cracking in a ferritic alloy steel. Weld J NY 82(2):25–S
15. Kuhn B et al An advanced method for evaluation of strain-age/stress-relaxation cracking susceptibility of welded joints
16. Klueh RL, Edmonds DP (1986) Chemical composition effects on the creep of Type 316 and 16-8-2 stainless steel weld metal. Weld J 65(6):156s–162s

Selection of Materials for Manufacturing of Disc Brake Rotor for a Racing Go-Kart Having Single Hydraulic Disc Brake System

Aman Dharmendra Chheda and Ravikant Hattale

Abstract Disc brake or rotor is a common device used in a racing go-kart with single hydraulic brake system for slowing down or stopping the motion of wheel running at a certain speed. In this paper, an extensive study was carried out to develop material selection methods for disc brake rotor. Materials, namely grey cast iron, stainless steel 420, aluminium 7075 T6 and titanium grade 5 alloys, were studied for general performance requirement, thermal and mechanical properties and static structural analysis (FEA). The results obtained were used to identify the promising material among the candidate materials for disc brake rotor.

Keywords Racing go-kart · Disc brake rotor · Grey cast iron · Aluminium · Titanium · Stainless steel · Material selection · Mechanical · Thermal · FEA

1 Introduction

In motorsports industries, to attain better performance, the reduction in vehicle weight is very important irrespective of its component size. In the recent years, the automobile industry has rapidly increased the use of aluminium, carbon fibre and titanium for manufacturing light vehicles. These materials have less weight and higher thermal conductivity as compared to the traditional grey cast irons and are expected to result in weight reduction of up to 50–60% in brake systems. Moreover, these super materials have the potential to perform better under severe service conditions like higher speed and load.

Since disc brake or rotor is one of the most important and stressed components from safety point of view, materials used for brake systems should have reliable wear and frictional properties under severally changing conditions like environment, velocity, load and temperature. There are several factors, which need to be considered

A. D. Chheda (✉) · R. Hattale
Dwarkadas J Sanghvi College of Engineering, Mumbai 400056, India
e-mail: amndchheda@gmail.com

R. Hattale
e-mail: Ravikant.Hattale@djsce.ac.in

© Springer Nature Singapore Pte Ltd. 2020

H. Vasudevan et al. (eds.), *Proceedings of International Conference on Intelligent Manufacturing and Automation*, Lecture Notes in Mechanical Engineering,
https://doi.org/10.1007/978-981-15-4485-9_45

while selecting material for a disc brake rotor. The most important factor is the capability of the brake rotor material to withstand high braking forces with less abrasive wear due to high friction. Another requirement is to withstand the high temperature that is generated due to friction. Weight, manufacturing process ability and cost are also important factors; those are needed to be considered during the design phase. The brake rotor must have enough thermal storage capacity to prevent distortion or cracking from thermal stress until the heat can be dissipated. This is not particularly important in a single stop, but it is crucial in the case of repeated stops from high speed. The materials' selection chart is a very useful document for comparing a large number of materials at the design phase. In this work, certain methods were developed to select the best candidate material for manufacturing of brake rotor for a racing go-kart. All the materials were studied and analysed according to the developed methods and ranked accordingly.

2 Material Selection Methodology

In this work, we have developed certain methods which will help in selecting the best material for disc brake rotor application with optimum combination of the desired properties. The stages involved in the selection of material are shown in Fig. 1.

2.1 General Material Performance Requirement

A schematic view of the brake rotor system is shown in Fig. 2. In this system, a braking force is generated by clamping the brake pads against a rotor contact patch, which is mounted on the axle of go-kart via hub. Due to high mechanical advantage, a smaller lever input force at foot brake pedal with pedal ratio 5:1 is converted to a large brake force at the wheel. This force in turn pushes the rotor against brake pad

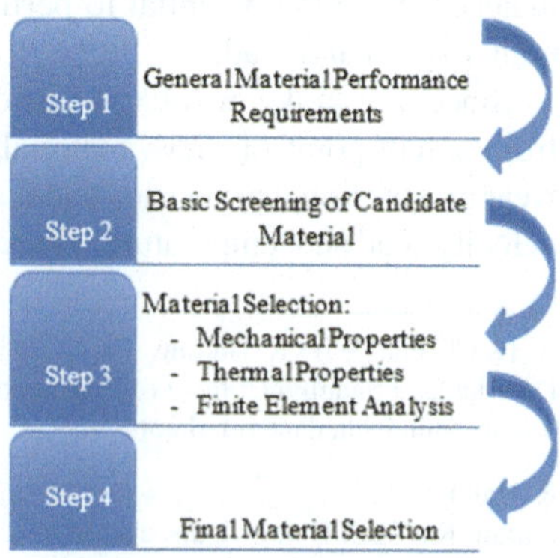

Fig. 1 Stages in material selection

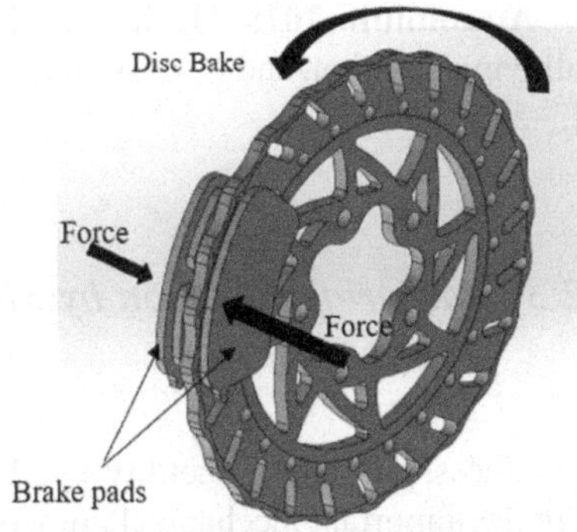

Fig. 2 Schematic representation of disc brake rotor

and hence generates a brake force due to large friction. The more brake power can be achieved with material with high coefficient of friction. The amount of frictional force generated is given by the relation $F_{\text{rotor}} = 2 \cdot Cf_{\text{pad}}\ F_{\text{pad}}$, where Cf_{pad} is coefficient of friction for the pad material and F_{pad} is the force pushing the brake pad.

The candidate material used for disc brake rotor components should have high coefficient of friction, high compressive strength, high wear resistance, lightweight, better thermal properties, and should be economically viable [1, 2].

2.2 *Initial Screening of the Candidate Material*

In the initial screening process, we have chosen following four material based on their properties, cost and manufacturability and availability.

- Grey cast iron
- Titanium grade 5 (*Ti-6Al-4 V, 3.7165, R56400*)
- Martensitic-type AISI stainless steel 420 (*S42000*)
- Aluminium 7075 T6.

Grey Cast Iron: It contains 2% carbon dissolved in Fe matrix. It is most widely used material for disc brake rotor application due to its low cast, easy manufacturing and high temperature stability. But it consumes much fuel due to its high specific gravity. Also it is heavy due to high density which in turn affects the performance of the racing go-kart [3].

Titanium Grade 5: These are lightweight Ti alloys used for disc brake rotor. Commercially, it is known by Ti-6Al-4 V alloy. In comparison with traditional cast iron material, it offers about 37% weight reduction for the disc brake rotor with same dimensions. Along with this material also offers better corrosion and high temperature strength.

Stainless Steel 420: This is martensitic stainless steel material used for disc brake rotor which shows moderate ductility and electrical conductivity among other candidate materials.

Aluminium 7075 T6: It is a 7000-series aluminium alloy with Zn as a main alloying addition, and it is solution heat treated and artificially aged to get T6 temper [3–5].

2.3 *Material Selection by Mechanical and Thermal Properties*

Hardness, brittleness, ductility, toughness, yield strength and ultimate tensile strength are the important mechanical properties, which affect the performance of disc brake rotor. These properties were evaluated for the said materials, and then materials were ranked from 1 to 4, 1 being the best candidate material and 4 being the worst candidate material. Similarly, the ratio of the strength (*usually the tensile or compressive strength*) of an object to its weight, or that of a substance to its density was also measured. It was observed that higher the strength-to-weight ratio points, better is the material for the application of brakes and ranking was done accordingly [3–5].

Thermal properties such as heat capacity and thermal conductivity are also greatly affecting the performance of disc brake rotor since during brake application large amount of heat is generated, which is need to be dissipated properly. In brake application, rotor with less thermal conductivity gives better performance as it conducts less heat and also helps the rotor to keep its temperature as low as possible. In this work, the specific heat capacity and thermal conductivity values for all the alternative materials were obtained based on the observed values. The candidate materials were ranked 1 to 4, 1 being the best and 4 being the worst candidate [6–9].

2.4 *Material Selection by Finite Element Analysis (Static Structural Analysis)*

In this work, first CAD model was developed according to the required geometry for all the candidate materials using SOLIDWORKS 2017 software as shown in Fig. 3, and then, 3 mm tetrahedron mesh was done on the generated model as shown in Fig. 4.

Static structural analysis is basically FEA solver, which is used to check the stress generated, factor of safety, total deformation, creep, etc., in many engineering components that are to be analysed. In this work, static structural analysis for all potential candidate materials were conducted to check factor of safety and total deformation by inputting the parameters as calculated below. All this analysis work was conducted by using ANSYS Workbench 18.1.

Braking calculations, values and consideration [10].

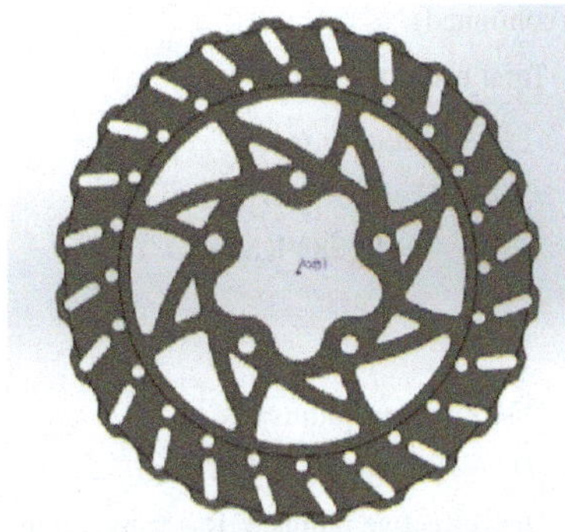

Fig. 3 Geometrical model of disc brake rotor

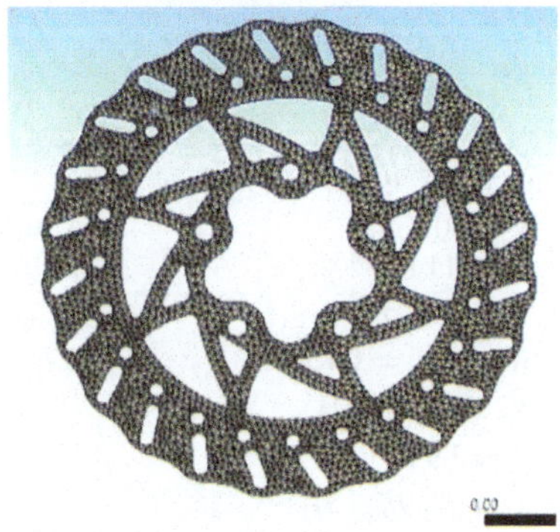

Fig. 4 3mm tetrahedron meshing done on rotor

Considerations	P_{mc} = 9.293 Mpa *For caliper (piston Φ=25.4 mm)* $P_{caliper}$ = P_{mc} [Pascal's law] $A_{caliper} = 0.785 \times (25.4 \times 10^{-5})^2$ $A_{caliper} = 506.707 \times 10^{-6}$ m^2
Max driver force on pedal = 24 kg = 24 × 9.81 = 235.44 N Pedal ratio = 5:1 Mass of the kart with driver = 180 kg μ_{rotor} & pad = 0.5	
Calculation *For master cylinder (piston Φ = 12.7 mm)* $F_{mc} = 235.44 \times 5$ = 1177.2 N	Now $F_{caliper} = P_{mc} \times A_{caliper}$ $= 9.293 \times 10^6 \times 506.707 \times 10^{-6}$ $F_{caliper}$ = 4.7088 KN
Area of master cylinder piston	*Since it is a two-piston caliper used*
$= 0.785 \times 12.7^2$ A_{mc} = 126.677 m^2	$F_{caliper} = 4.7088 \times 2$ $F_{caliper}$ = 9.4177 KN
Pressure in master cylinder	

(continued)

(continued)

Total frictional force	Torque on rotor $= F_{\text{frictional}} \times R_e$
$= F_{\text{caliper}} \times \mu_{\text{rotor}}$ $= 9.4177 \times 10^3 \times 0.5$ $= 4.7088$ KN	$=4.7088 \times 10^3 \times 84.5 \times 10^3$
	Torque on rotor = 397.89 Nm
	Tyre diameter (rear) = 11 in = 0.2794 m
Disc effective radius=	Rear tyre radius = 0.1397 m
(D = 200 mm, d = 138 mm)	Force on rear tyre = 2848.175 N
Disc effective radius = R_e = 84.5 mm	Deceleration (a) = 15.82 m/s^2
Now	

Stopping distance (Sd)		
V = 40 Kmph $= 11.11\ \text{ms}^{-1}$	V = 60 Kmph $= 16.67\ \text{ms}^{-1}$	V = 80 Kmph $= 22.22\ \text{ms}^{-1}$
$V^2 - u^2 = 2aS_d$, where, $V = 0$ Kmph $S_d = \frac{11.11 \times 11.11}{2 \times 15.82}$ $S_d = 3.9$ m	$S_d = \frac{16.67 \times 16.67}{2 \times 15.82}$ $S_d = 8.73$ m	$S_d = \frac{22.227 \times 22.227}{2 \times 15.82}$ $S_d = 15.6$ m
Stopping time (t)		
$U = 11.11\ \text{ms}^{-1}$ $a = -15.82\ \text{ms}^{-2}$ $\therefore v = u + a \times t$	$U = 16.67\ \text{ms}^{-1}$ $a = -15.82\ \text{ms}^{-2}$ $\therefore v = u + a \times t$	$U = 22.22\ \text{ms}^{-1}$ $a = -15.82\ \text{ms}^{-2}$ $\therefore v = u + a \times t$
$t = \frac{-u}{a} = \frac{-11.11}{-15.82}$ $t = 0.7$ s	$t = \frac{-u}{a} = \frac{-16.67}{-15.82}$ $t = 1.054$ s	$t = \frac{-u}{a} = \frac{-22.22}{-15.82}$ $t = 1.405$ s

3 Results and Discussion

The results obtained from the analysis of mechanical properties, such as tensile strength and strength-to-weight ratio both in bending and axial conditions for all the candidate materials, are reported in Table 1. Ranking of the materials is also shown in Table 1. Highest values of tensile strength and strength-to-weight ratio were observed for titanium grade 5 material. Hence, it has ranked 1.

Similarly, the results obtained for the specific heat capacity and thermal conductivity values for all the alternative material are reported in Table 2. Again better thermal properties, i.e. low thermal conductivity, were observed for titanium grade 5 material; hence, it has ranked 1.

Table 1 Mechanical properties

Materials	Tensile strength: ultimate (Mpa)	Tensile strength: yield (Mpa)	Density (kg/m^3)	Ranking
Grey cast iron	260	180	7500	4
Titanium grade 5	1000	910	4400	1
Stainless steel 420	640	380	7700	3
Aluminium 7075T6	560	480	3000	2

Materials	Strength to weight bending	Strength to weight axial	Ranking
Grey cast iron	12 points	13 points	4
Titanium grade 5	50 points	62 points	1
Stainless steel 420	22 points	25 points	3
Aluminium 7075T6	50 points	51 points	2

Table 2 Thermal properties

Materials	Specific heat capacity (J/kg K)	Thermal conductivity (W/mK)	Ranking
Grey cast iron	490	46	3
Titanium grade 5	560	6.8	1
Stainless steel 420	480	27	2
Aluminium 7075T6	870	130	4

The static structural analysis to measure factor of safety and total deformation was conducted for all the candidate materials as shown in Figs. 5, 6, 7, 8, 9, 10, 11 and 12. The measured values for factor of safety and total deformation are also reported in Table 3. The materials were ranked based on factor of safety. The highest factor of safety was recorded for titanium grade 5 material, and hence, it has ranked 1 [11].

The comprehensive result of mechanical properties, thermal properties and finite elements analysis for various materials used for disc brake rotor is given in Table 4. From the analysis of all the material selection stages, the titanium grade 5 material has ranked overall 1 and could be the best material for manufacturing disc brake rotor.

4 Conclusion

The material selection methods for manufacturing disc brake rotor were successfully developed in this study. In order to identify the best material, the mechanical

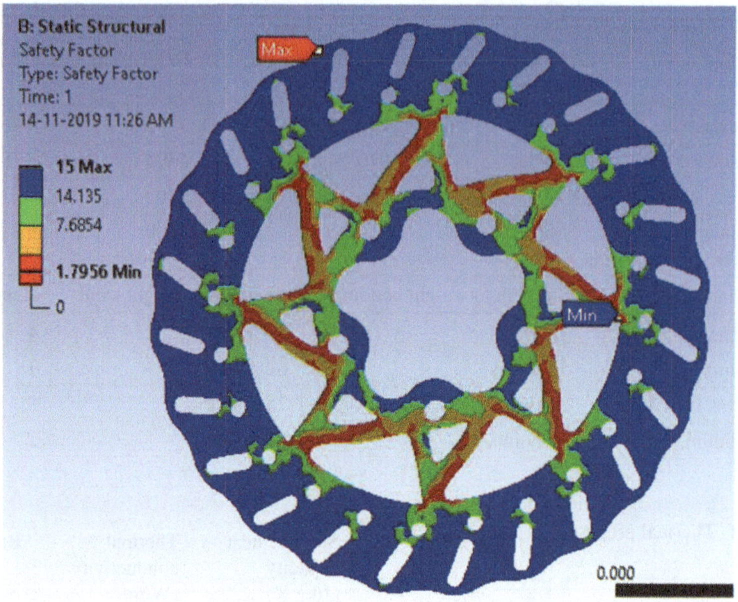

Fig. 5 Factor of safety for grey cast iron

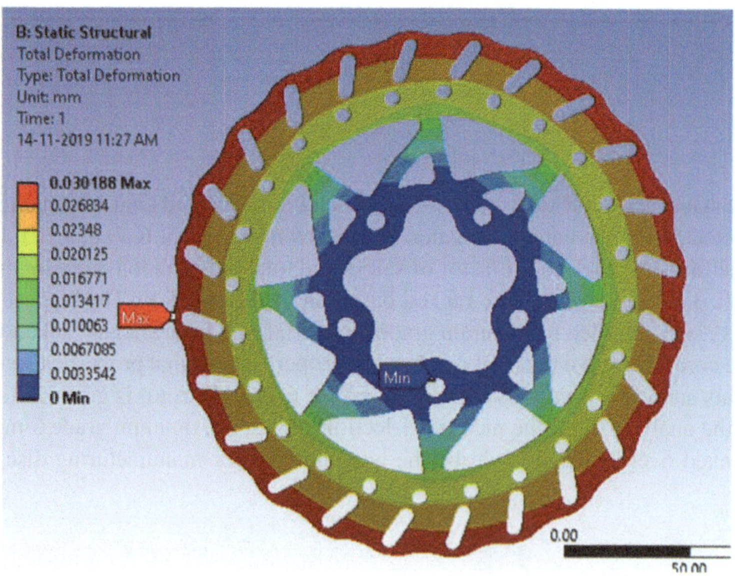

Fig. 6 Total deformation for grey cast iron

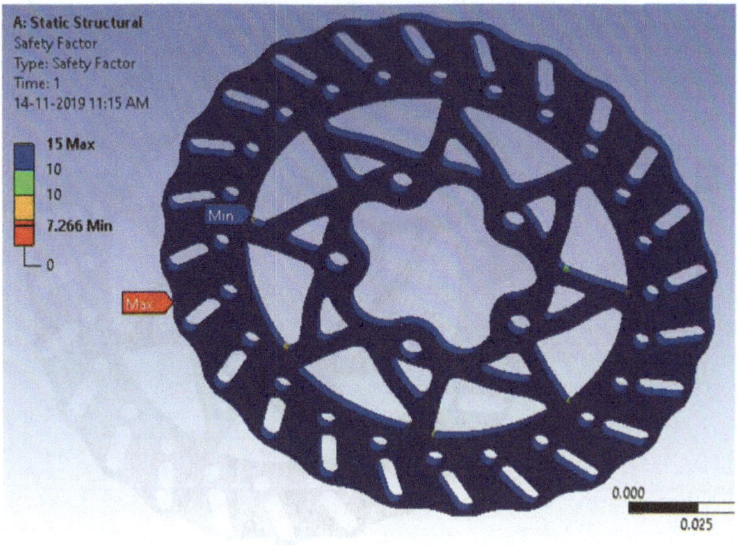

Fig. 7 Factor of safety for titanium grade 5

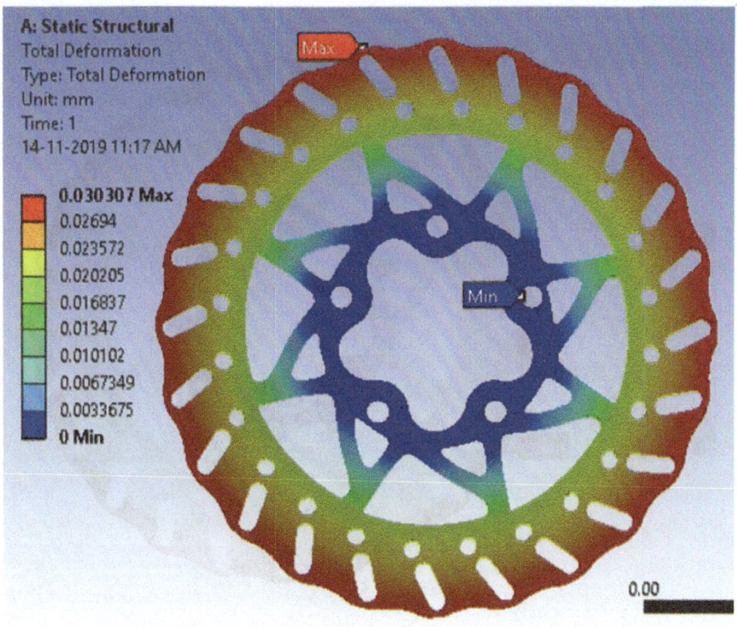

Fig. 8 Total deformation for titanium grade 5

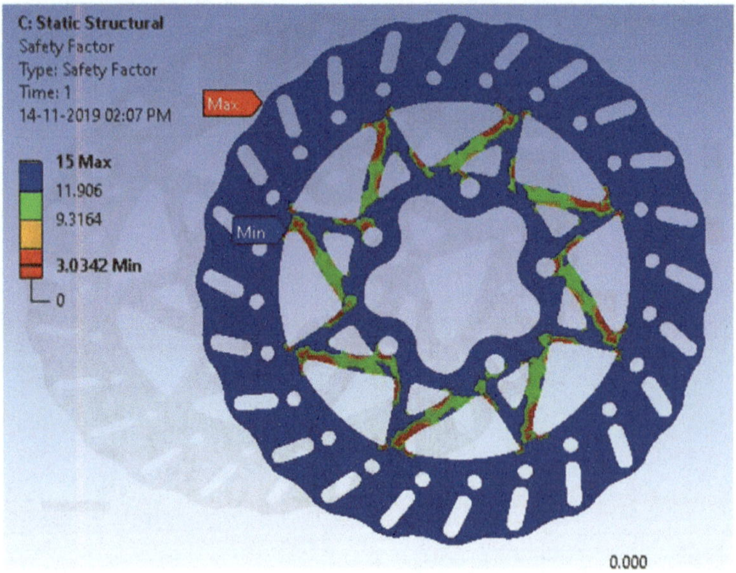

Fig. 9 Factor of safety stainless steel 420

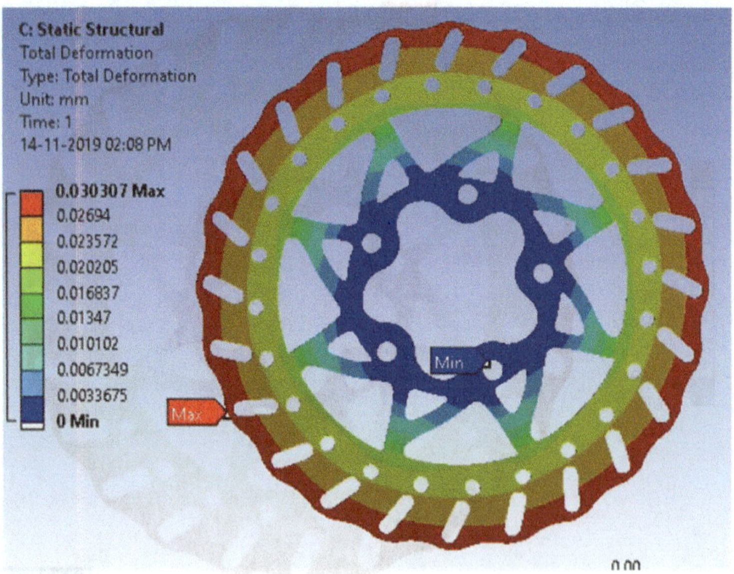

Fig. 10 Total deformation for stainless steel 420

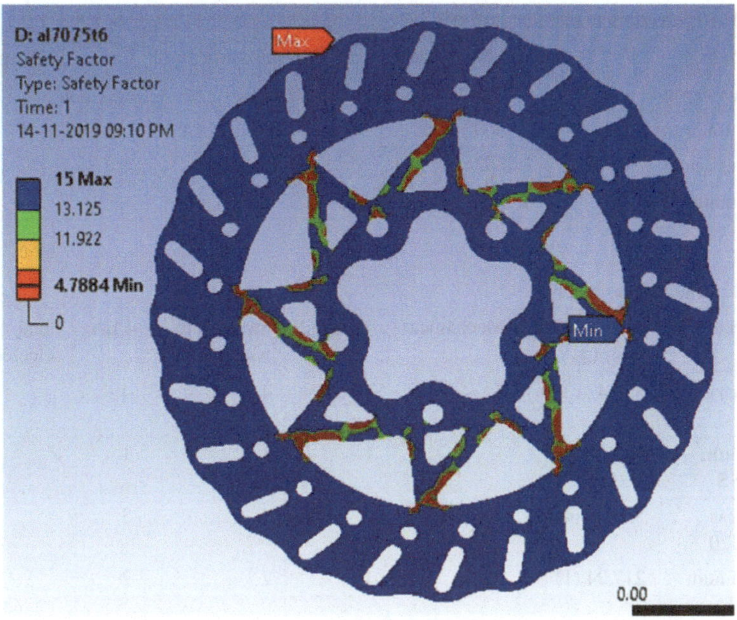

Fig. 11 Factor of safety for aluminium 7075T6

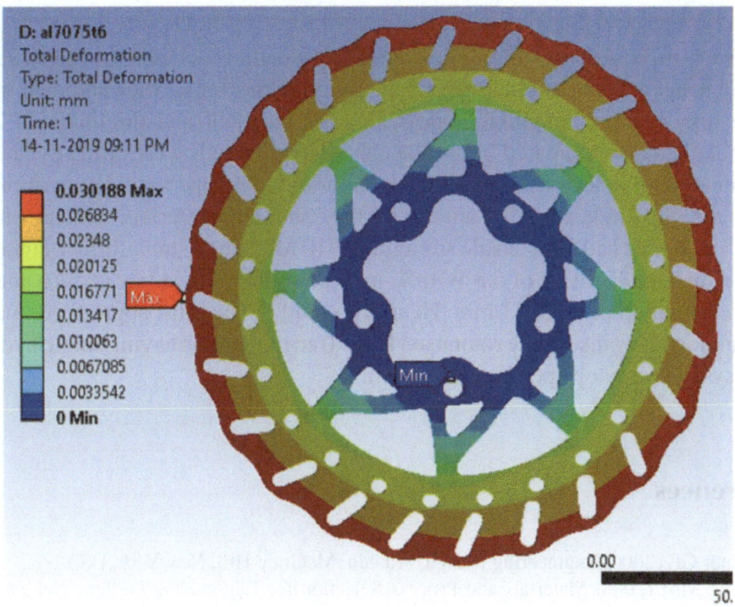

Fig. 12 Total deformation for aluminium 7075T6

Table 3 Result of finite element analysis

Materials	Factor of safety	Deformation (mm)	Ranking
Grey cast iron	1.7956	0.03016	4
Titanium grade 5	7.2660	0.03031	1
Stainless steel 420	3.0342	0.03012	3
Aluminium 7075T6	4.7880	0.03037	2

Table 4 Comprehensive result

Material	Mass of rotor (g)	Mechanical	Thermal	Strength to weight	FEA	Final selected
Grey cast iron	558.42 (3)	4	3	4	4	
Titanium grade 5	**343.49 (2)**	1	1	1	1	✓
Stainless steel 420	597.20 (4)	3	2	3	3	
Aluminium 7075 T6	217.94 (1)	2	4	2	2	

Significance of bold in this table is to show final material selected for manufacturing disc brake rotor. Bold helps in recognizing material easily

properties, such as tensile strength, strength-to-weight ratio, thermal properties like specific heat capacity and thermal conductivity values for all the candidate materials, i.e. grey cast iron, titanium grade 5, stainless steel 420 and aluminium 7075 T6, were obtained and compared. Also, the candidate materials were critically analysed by static structural analysis to check the factor of safety and total deformation. The results obtained indicated that among all the candidate materials studied, titanium grade 5 materials had high tensile strength of 910 Mpa, high strength-to-weight ratio, low thermal conductivity of 6.8 W/mK, maximum factor of safety of 7.26 and least total deformation of 0.03031 mm. Hence, it could be the most appropriate material for manufacturing disc brake rotor used for 150 cc racing kart having single hydraulic brake system with top speed of 110 kmph.

References

1. Dieter GE (2000) Engineering Design, 3rd edn. McGraw-Hill, New York, USA
2. Farag MM (2008) Materials and Process Selection for Engineering Design, 2nd edn. CRC Press, New York, pp 259–280
3. Cueva G, Sinatora A, Guesser WL, Tschiptschin AP (2003) Wear resistance of cast irons used in brake disc. Wear 255:1256–1260
4. Blau PJ, Jolly BC, Jun QU, Peter WH, Blue CA (2007) Tribological investigation of titanium-based materials for brakes. Wear 263:1202–1211

5. Maleque M.A, Dyuti S, Rahman MM (2010) Material selection method in design of automotive brake disc. Proc World Cong Eng III. June 30–July 2, 2010
6. Material Properties Database. https://www.makeitfrom.com/material-properties/Annealed-Grade-5-Titanium (12-11-2019)
7. Material Properties Database. https://www.makeitfrom.com/material-properties/7075-T6-Aluminum (12-11-2019)
8. Material Properties Database. https://www.makeitfrom.com/material-properties/Annealed-420-Stainless-Steel (12-11-2019)
9. Material Properties Database. https://www.makeitfrom.com/material-properties/ASTM-Grade-225-Grey-Cast-Iron (12-11-2019)
10. Puhn F Brake handbook. HPBooks, pp 42–103
11. Kaiwart D, Tembhurne YK (2017) Comparison of structural and thermal analysis of disc brake using various materials. Int J Therm Eng 5:01–10

Comparative Study of Aluminum and Composite Stub Axle Using FEA

Rajnarayan Yadav, Vinayak H. Khatawate, Deval Patel, Sahil Thonse, and Danish Sunsara

Abstract This paper includes the analysis of a front stub axle used in a rear wheel drive BAJA ATV (all-terrain vehicle). An ATV is a vehicle developed to be used on rough terrains and low traction. These types of vehicles are generally known and used for their off-roading ability and maneuverability of the vehicle on rough terrains. The stub axle used in such a vehicle is subjected to different types of loads, depending upon the motion and terrain the vehicle is running on. The analysis of the structure was compared with two different models, on the basis of their strengths and rigidity of the axle during the action of bump force subjected on the tire, change in weight of the axle and cost. This paper deals with the structural analysis of stub axle of BAJA ATV using finite element analysis approach. The objective of this analysis was to study and evaluate the performance of two stub axles made of aluminum and composite material respectively, under severe conditions.

Keywords BAJA · ATV · Stub axle · Structural analysis · FEA

1 Introduction

The Society of Automotive Engineers (SAE) BAJA series is an annual series of competitions, which originated in 1976. The BAJA SAE tasks the students to design, fabricate, and validate a single-seat four-wheel off-road vehicle to take part in a series of events spread over a course of 3 days that test the vehicle for sound engineering practices that have gone into it, the agility of the vehicle in terms of gradeability, speed, acceleration, and maneuverability characteristics, and finally its ability to endure that back-breaking durability test [1, 2]. This durability test checks the structure by dropping the ATV through a 5 ft vertical drop in a pit. This paper specifically deals with the structural analysis of a single component, i.e., the stub axle. The front stub axle is designed and developed based on defined parameters of the suspension,

R. Yadav (✉) · V. H. Khatawate · D. Patel · S. Thonse · D. Sunsara
Department of Mechanical Engineering, Dwarkadas J. Sanghvi College of Engineering, Vile Parle West, Mumbai, Maharashtra 400056, India
e-mail: rajnarayan.yadav@djsce.ac.in

© Springer Nature Singapore Pte Ltd. 2020
H. Vasudevan et al. (eds.), *Proceedings of International Conference on Intelligent Manufacturing and Automation*, Lecture Notes in Mechanical Engineering,
https://doi.org/10.1007/978-981-15-4485-9_46

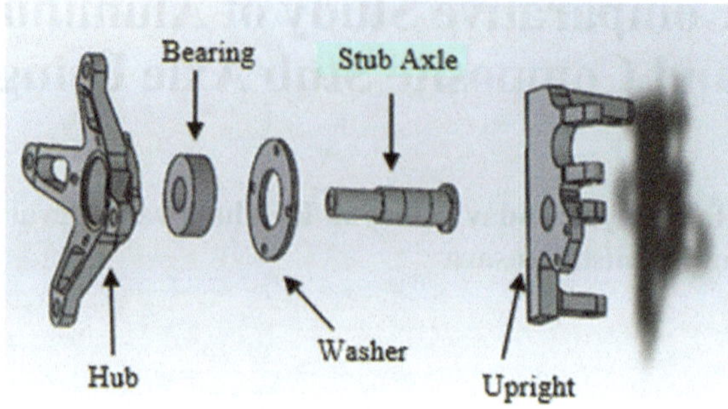

Fig. 1 Assembly of the stub axle

chassis, and brakes that meets the needs for the functional use of the vehicle. The stub axle is one of the major components in the power transmitting system [3]. The stub axle is designed for a BAJA ATV, which is a rear wheel drive. The stub axle designed is used in the front wheel assembly which supports the wheel connected via hub which is supported using deep groove ball bearing on the axle. On the front axle, wheels are mounted, and with the help of steering wheel, the driver can orient the vehicle in different directions [4]. This unit is used to carry the weight of the front part of automobiles as well as for steering and to absorb shocks due to road surface variation [5]. Analysis is made by a finite element analysis software in order to obtain a structure with appropriate value of factor of safety (FOS) and deflection of shaft during the loading condition. The aim of this project is to design a stub axle for front suspension of a rear wheel drive ATV. It should also allow a proper steering control to the wheel so that the driver efficiency is not lost. This component is subjected to discontinuous forces resulting in stress concentration and may result in component failure [6]; hence, the design of the stub axle should consider the forces and moments during 5 ft drop which occurs in an automobile. Figure 1 shows a designed assembly of stub axle.

2 Determination of the Loading Forces

After considering the factors such as calipers mounting point, rim's offset, forces acting on the wheel assembly, bearing support position, available clearance between rotor and the upright and bearing size, a stub axle of 20 mm diameter and length of 70 mm with a desired FOS of 2 is determined. The stub axle is held firmly on knuckle surface by means of tight press fit. The bending force induced on the shaft is observed to be maximum during the 5 ft drop of the vehicle. The magnitude and the direction of the force are calculated to be 4200 N on each front wheel using input parameter velocity: 15 kmp h due to magnitude of force, and center of gravity is considered at same point with respect to weight biasing throughout the motion of the car during the 5 ft drop.

3 Methodology

Following steps are used for the development of stub axle for analysis:

(1) Manual calculations
(2) Modeling and stub axle with materials
(3) Mesh generation
(4) Application of boundary conditions.

3.1 Manual Calculations

Static structural analysis has been performed on the stub axle to find out deformation and factor of safety of the component during 5 ft drop. An ATV of mass 200 kg with driver is considered for analysis.

Input parameters:

5 ft impact force = 4200 N
Rim offset ratio = 3:2
Wheel width = 7 in.

3.1.1 Moment Generated Due to Rim Offset

Cross section of the rim is shown in Fig. 2. In Fig. 3, the distance OA is rim offset displayed.

Ra = 4200 N

Using the free body diagram (Fig. 4), moment = 4200 × 17.78 = 7676 N mm (anti-clockwise)

Force acting on hub = 4200 N, moment on hub = 74,676 N mm.

3.1.2 Force and Moment Transfer on Stub via Hub

The moment and force acting on the bearing cause the bending action of axle. Bearings with inner diameter of 17 and 20 mm were available, and bearing with 20 mm inner diameter is selected considering the constraints of the hub design (Fig. 5).

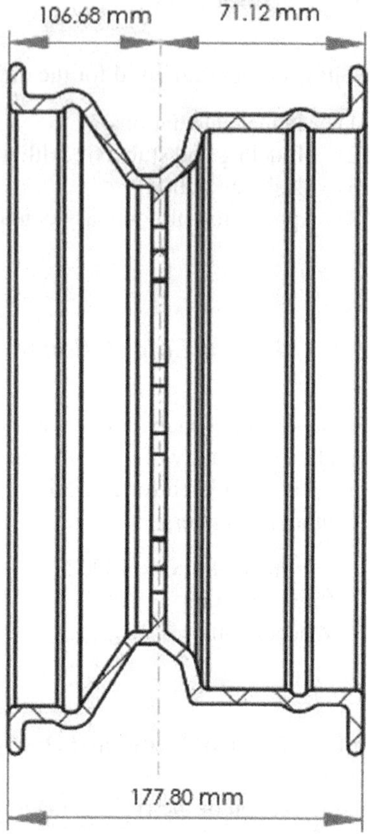

Fig. 2 Cross section of the rim

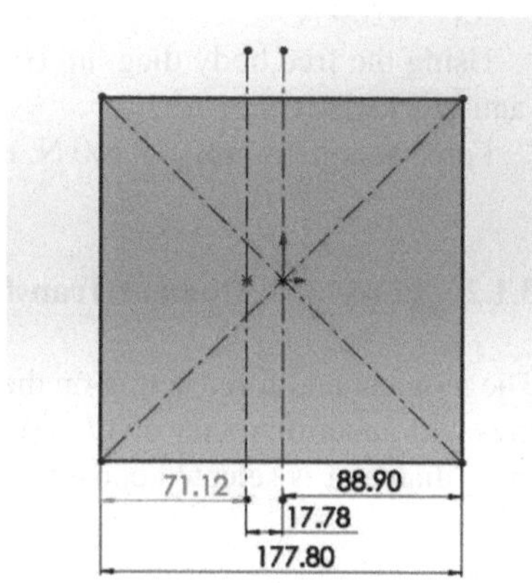

Fig. 3 Rim offset

Fig. 4 Free body diagram of the rim offset

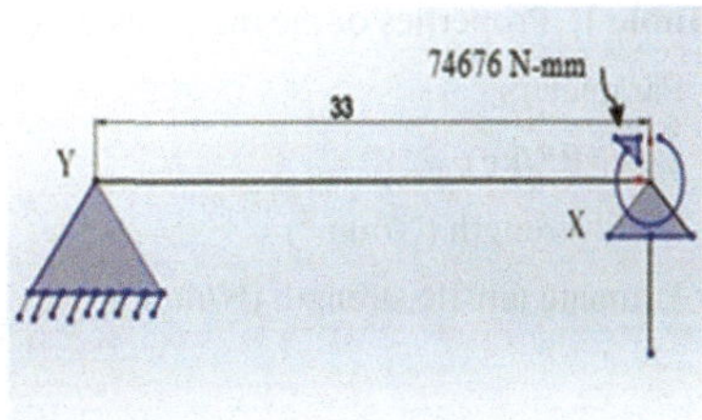

Fig. 5 Bearing support

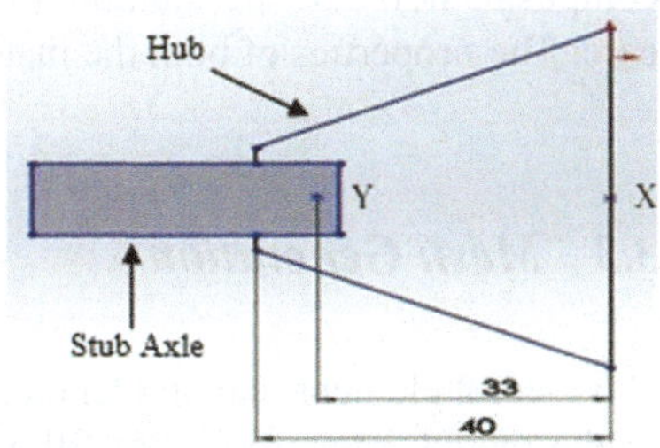

3.2 *Modeling and Stub Axle with Materials*

Geometric modeling of the stub axle is done on Dassault SOLIDWORKS software. Based on the considerations mentioned above, two models of the stub axle were designed shown in below figure. The first model designed is of single homogenous material, i.e., aluminum 7075 T6 (Fig. 6), and the second model designed is a composite of aluminum 6061 T6 outer layer and core of EN 19 steel (Fig. 7).

Fig. 6 Full aluminum stub axle

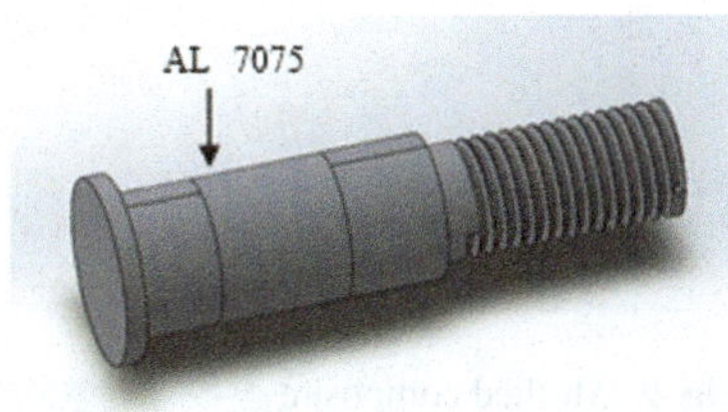

Fig. 7 Composite stub axle

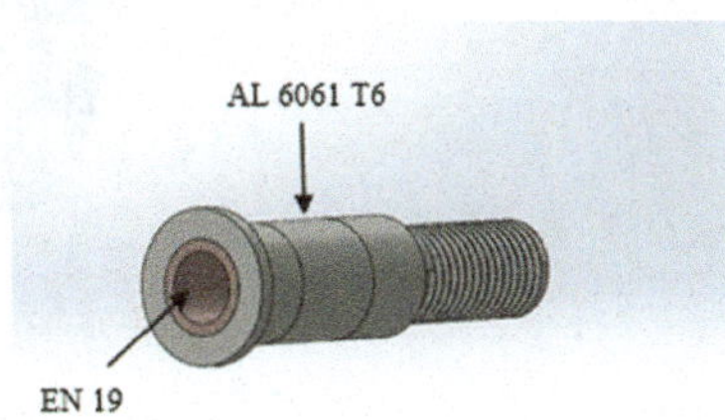

Table 1 Properties of the materials used

Parameters	Al 7075 T6	AL 6061 T6	EN 19
Cost (RS/kg)	750	350	40
Yield strength (N/mm^2)	455	276	755
Ultimate tensile strength (N/mm^2)	525	310	1200

The material used for single stub axle is 7075 T6, whereas the materials used for composite stub axle are aluminum 6061 T6 as outer layer with EN 19 as an inner core. The properties of both the materials are shown in Table 1.

3.3 Mesh Generation

Tetrahedral elements can model most of the solid objects irrespective of complexity of the model. Therefore, three-dimensional tetrahedral elements with eight nodes were used for the following model. Mesh with average size of 2 mm was used. The initial mesh of the single stub axle is shown in Fig. 8, while that of the composite stub axle is shown in Fig. 9.

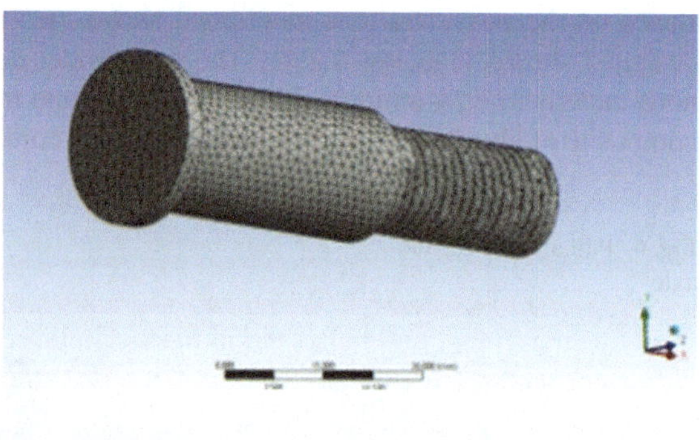

Fig. 8 Meshed single stub axle

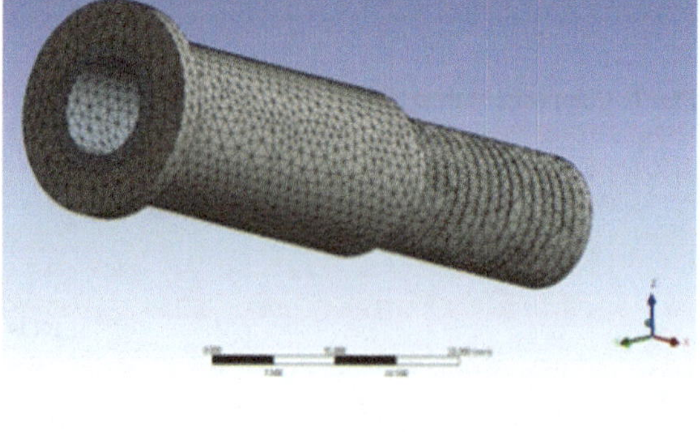

Fig. 9 Meshed composite stub axle

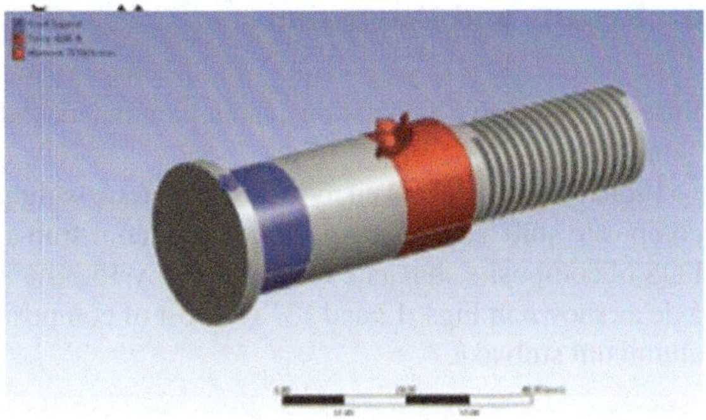

Fig. 10 Application of forces on Al 7075 T6 stub axle

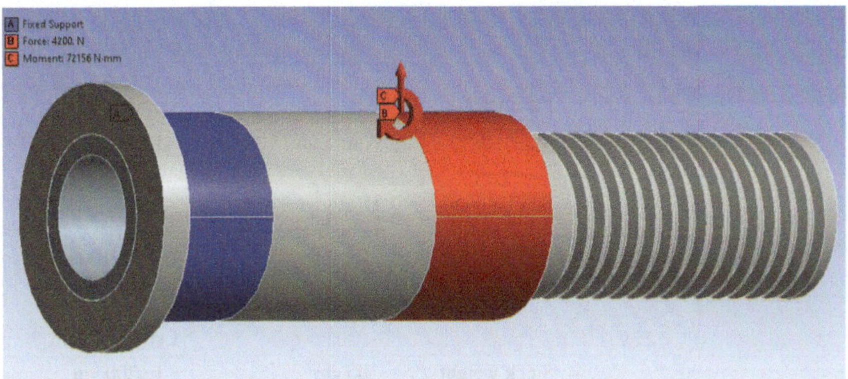

Fig. 11 Application of forces on composite stub axle

3.4 Application of Boundary Condition

The first 12 mm (red region) of the stub axle (Figs. 10 and 11) is where the force of 4200 N and a moment of 74,676 N mm are applied since that part of the axle is completely locked with the knuckle. The last 14 mm (blue region) of the stub axle is where the bearing provides support. Same boundary conditions are fed to both the stub axles in comparison.

4 Results and Discussion

Based on the above calculations, input parameters and results obtained are shown in Table 2.

From result table and analysis, overall stock weight was found to be 5 g more for composite stub axle as compared to aluminum stub axle which was around 8.10%. FOS of composite stub axle is increasing by 15.30% as compared to aluminum stub axle as shown in Figs. 12 and 13. The cost of composite stub axle is 49.86% less ten aluminum stub axle.

Table 2 Result table

Parameters	Aluminum	Composite
Materials	Aluminum 7075 T6	Aluminum 6061 T6 and EN 19
Stock dimension	Díameter 30 mm Length 100 mm	AL: Díameter 30 mm Length 100 mm EN: Díameter 20 mm Length 50 mm
Stock weight	200 gm	AL: 200 gm EN: 130 gm Total: 330 gm
Stock cost	0.200×750 $=$ Rs 150	$0.200 \times 350 +$ 0.130×40 $=$ Rs 75.2
Spindle weight	61.32 gm	66.29 gm
Factor of safety	2.306	2.652
Deformation	0.0651 mm	0.02023 mm

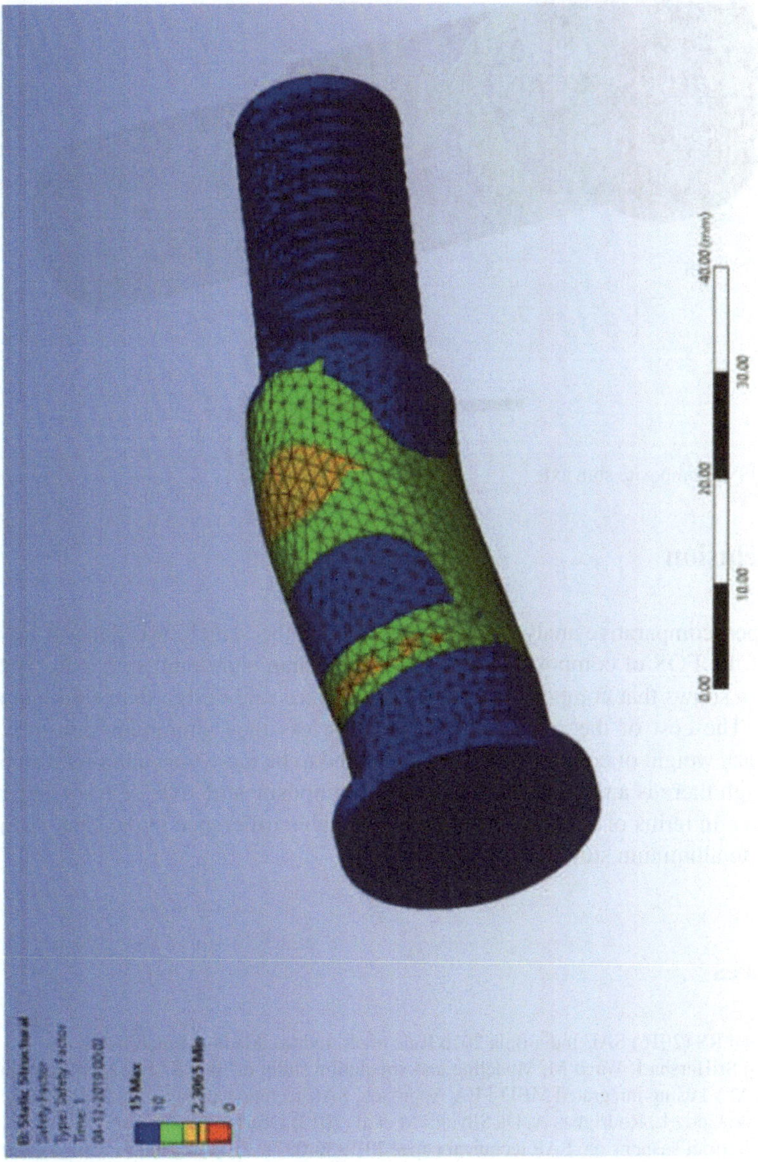

Fig. 12 FOS of AL 7075 T6 stub axle

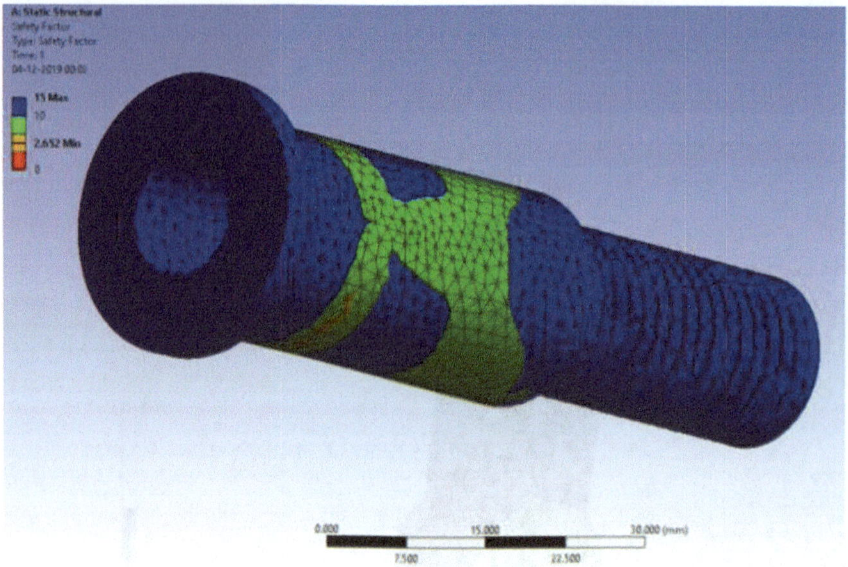

Fig. 13 FOS of composite stub axle

5 Conclusion

In this paper, comparative analysis of the stub axle of the vehicle was done. It was found that the FOS of composite stub axle is more than aluminum stub axle. The comparison shows that composite stub axle gives more safe design than aluminum stub axle. The cost of the composite stub axle is less than aluminum stub axle. Overall stock weight of composite stub axle is found to be more than aluminum stub axle. Though there is an increase in weight, the composite stub axle delivers better performance in terms of FOS, and there is a noticeable difference in the cost when compared to aluminum stub axle.

References

1. Chundawat RS (2016) SAE India Baja 2016 Rule book. Indore, Madhya Pradesh
2. Kakria S, SriHarsha I, Wagh M, Modeling and simulation study of Baja SAE India all-terrain vehicle (ATV) using integrated MBD-FEA Approach, SAE technical paper 2015-26-0219
3. Cervieri A, Gertz L, Rodrigues A, Da Silveira M et al (2012) Design and structural analysis of a stub axle front suspension. SAE technical paper 2012-36-0277
4. Kumar L, Singh C, Dewangan BK, Sen PK, Bohidar SK (2014) Study on the front axle and rear axle attachment to differential system. Int J Innov Res Sci Technol 1(7):268
5. Dey S, P.R.V.V.V. Sri Rama Chandra Murthy D, Baskar P (2014) Structure analysis of front axle beam of a light commercial vehicle. Int J Eng Trends Technol 11(5)
6. Bhandari VB Design of machine elements 3rd edn. McGraw Hill, New York

A Review on Carbon Nanotubes as Novel Drug Carriers in Cancer Therapy

Dhyey M. Rajani, Frank Crasta, and Vijaya Kumar N. Kottur

Abstract Cancer causes one of the pre-eminent health problems all over the world currently. Chemotherapy is used as a conventional treatment, which uses one or more anti-cancer drugs as part of a standardized procedure. The idea of selective treatment to cure cancer is established by carbon nanotube (CNT). In this review paper, the advances in the application of carbon nanotubes as target carriers and drug delivery system for cancer therapies have been studied. CNTs due to their physicochemical and selective targeting abilities and shape act as drug delivery systems. Internal drug loading encompasses decoration utilizes capillarity, encapsulating anti-cancer drugs like Cisplatin. The external loading takes place by linkers. The biocompatible drug targeting mechanisms like active and passive lead to targeted delivery. Hence, eliminating any damage to healthy tissues resulting in negligible side effects. The functionalization of CNTs being crucial in penetrating and increasing hydrophilicity entails non-covalent and covalent methods. This also helps them to penetrate through immunity barrier and ensure a targeted release.

Keywords Carbon nanotubes (CNTs) · Chemotherapy · Drug delivery systems · Cisplatin · Biocompatible

1 Introduction

Cancer is an abnormal growth of cells which tend to proliferate in an uncontrolled way and, in some cases metastasize (spread) leading to blood coagulation and decreasing immunal resistance of human body, leading to rapid mutations in cell behaviour and

D. M. Rajani (✉) · F. Crasta · V. K. N. Kottur
Department of Mechanical Engineering, Dwarkadas J. Sanghvi College of Engineering, Mumbai, Maharashtra 400056, India
e-mail: rajanidhyey@gmail.com

F. Crasta
e-mail: prashant.crasta@djsce.ac.in

V. K. N. Kottur
e-mail: vijayakumar.kottur@djsce.ac.in

© Springer Nature Singapore Pte Ltd. 2020

H. Vasudevan et al. (eds.), *Proceedings of International Conference on Intelligent Manufacturing and Automation*, Lecture Notes in Mechanical Engineering,
https://doi.org/10.1007/978-981-15-4485-9_47

DNA sequences. When the cancer cells coagulate and create a gobbet, they comprise to form a tumour. A tumour is cancerous when it seeds and starts developing in vicinal tissues. The tumour cells can disperse away and start coagulating the blood and lymphatic vessels. Cells may become cancerous due to the existence of tumour micro-environments (TMEs) leading to accumulation of defects, or mutations, in their DNA. Tumour micro-environments (TME) present a unique challenge in tumour therapy due to their complex structures and multiple components, which serve as the basis for tumour development. The complex region of TME consists of immune cells, collagen (amino acids bounded in triple helix) type structures encapsulated within the fibrous tissue structures and twisted (tortuous) blood vessels or lymphatic vessels, in which the traditional therapy is left useless [1]. In cases employing nanomaterials (here CNT), TME plays a pivotal role in synchronizing the nanochemical distribution of the medicaments. Therefore for the interaction of nanochemotherapeutics with the affected cells in tumour milieu, it is essential that nanochemotherapeutics are gathered in the tumour through the vascular connections. Subsequently, this interaction with target cells should be advanced by specific extravasation from tumour vascular chains.

Certain transmutable genetic defects like BRCA1 and BRCA2 mutations can increase the risk of cancer. BRCA1 and 2 produce suppressor proteins, when either of these genes is mutated, such that its protein synthesis does not function correctly, DNA damage may not be repaired. Therefore, cells tend to develop new genetic alterations that can lead to cancer. Most of the time, cells are able to detect and repair DNA damage [2]. If a cell is severely damaged and cannot repair itself, it usually undergoes so-called programmed cell death or apoptosis [3].

In recent few years have seen astounding discoveries in cancer drug research and development leading to more of a therapeutic and clinical approach, but there has been a lack in the drug delivery and targeted supply of these drugs to the accurate centres of malignancy, i.e. the epicentres of mutations, without affecting the healthy adjacent tissues [4]. The inadequacies in the ability to administer therapeutic agents with high selectivity and minimum side effects largely account for the discrepancies. Many numbers of tests have been performed for strategic and selective drug loading and delivery, for example, liposome-induced encapsulation [5]. Systemic toxicity may develop at the same time due to the lack of selectivity of the drugs for cancer tissues and cells, which often leads to the failure of chemotherapies [6, 4]. Hence, considerable efforts are being directed to such a drug delivery system that selectively targets the cancerous tissue with minimal damage to adjoining tissue [7]. The development of advanced drug delivery system does not only involve biomarking of neoplastic regions but also has targeted and extremely selective supply and release system to the neoplastic regions hence rendering them enervated, also rendering the free-floating micro-focuses ineffective. The requirements for new drug delivery systems have led to a marked increase in pharmacological effects and considerable weakening of toxicological effects of the neoplastic cells [8].

In CNTs, the penetration and direct release of the payload are substantial. CNTs have an added advantage of tubular shape, which can be capped at the two ends for

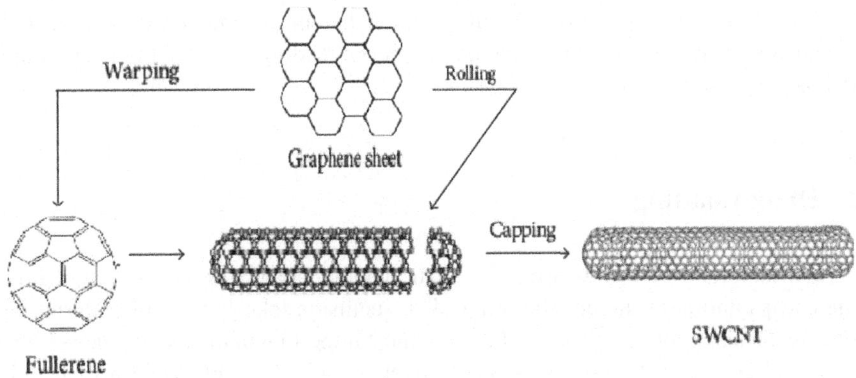

Fig. 1 Capped structure of SWCNT

drug holding, which can serve as an effective way to transcend the cellular restrictions and enter directly into the infected cytoplasmic cavity, allowing the drug to be jettisoned [9, 10]. Basic construction of CNT involves rolling of graphene sheet into single-walled carbon nanotube (SWCNT) or nesting of a number of single-walled nanotubes to form a multi-walled carbon nanotube (MWCNT). The secondary process of warping Graphene sheet into Fullerenes and then capping the nanotube to ultimately form an encapsulating carbon nanotube is as shown in Fig. 1.

2 Importance of CNT in Drug Delivery

CNTs are allotropic conformation of carbon. The nanostructure is of cylindrical nature. The structure of CNTs can be seen as cylindrical rolling up of nanothin graphene sheets consisting of sp^2 hybridized carbon atoms bonded to each other by covalent bonds [11]. There are two types of CNT, namely single-walled carbon nanotube (SWCNT) and multi-walled carbon nanotube (MWCNT). MWCNT consists of a more tubular and dense roll-up of graphene layers, i.e. more number of graphene layers roll up with an annular space of 0.34 nm in average [12]. In recent years, a wide range of different nanoscale therapeutic drug delivery mechanisms and loading systems have been evaluated, wherein, single-walled carbon nanotubes (SWCNTs) have attracted considerable interest, as they offer potential advantages of more stringent and target-oriented drug delivery system than other nanoparticle systems. SWCNTs have been loaded with antibody elements and low molecular weight targeting compounds, for nanotube integration into cells. Their advantages include their ability to carry a high cargo loading, intrinsic stability and structural elasticity, which can increase circulation time and hence the bioavailability of the drug molecules. An added advantage is, SWCNTs have been shown to enter mammalian cells and due to their reliable properties as of needle-shaped penetration mechanism [10]. SWCNT-based materials have already been found as potential delivery vehicles for intracellular movement of immunity establishment nucleic acids and potential candidates for further functionalization and hence leading to a stealthy penetration and biocompatibility [9].

3 Drug Loading

Carbon nanotubes are end-capped by many nanomechanical processes the facilitate encapsulation of therapeutic drugs. This inclusion takes place either during or after synthesis of formatted nanotubes. The most important of the two methods is the post-synthesis method which is marked by a remarkably 70–100% yield, also including different criteria of sensitivity, melting point, etc. [13]. Now the post-synthesis involves melting of end caps to load the drug particles, and the loading is carried out by decoration. Decoration is establishing the bonding between the nanotube's wall with functional groups, and here, the only problem faced is that the carbon atoms in a chain being extremely inert do not allow any kind of bond formation [14]. Also, another drawback is being the small diameter of CNT for capillarity action, for filling CNT. In order to overcome this oxidation of carbon takes place in order to obtain desired efficacy of bonding, this bonding can be done inside or outside the nanotube wall [15]. The capillarity action can be changed by altering surface tension of the solution. Due to the tubular structure, capillarity is possible depending upon factors like the tube diameter and surface tension of the nanomaterial. The chemicals used if they have higher surface tension, it can be reduced by forming a suitable composite with the carbon nanotube which under the influence of other chemicals can be reduced to pristine chemicals without changing the wall properties [15, 16]. After filling the CNTs, the ends can be capped again by passing high electric currents to CNTs as a result encapsulating the drug for uninterrupted delivery [16].

4 Attachment of Drugs (External)

The external attachment of drugs takes place by attaching the drug molecules by amide or disulphide bonds. These attachments serve the function of covalent bonds between the drug and the wall of nanotube, known as linker. The nanotubes injected either subcutaneously or intravenously are administered to follow different routes to the effective site of relevance by means of blood circulation or lymphatic circulation [17]. The folic acid content when increases beyond a certain limit the cancer cells is active in folic acid content. This folic acid affinity can be used in nanotubes to guide them to the cancerous cells, once the sulphide links come into this cancerous environment there is a decrease in reduction potential leading to the release of the drug to the pre-planned region. This marks the importance of linkers [18].

5 Drug Targeting

The traditional cancer treatments involve plenty of after-effects; this is due to the toxic effects of anti-cancer drugs on vicinal healthy cells. In order to avoid this, biocompatible therapeutic drugs along with efficient targeting of nanotubes are pertinent [19]. The ability of CNTs to target the tumour-affected area establishes them as carriers to deliver therapeutic medicament. There are two major types of targeting involved, namely passive and active targeting.

1. Passive targeting: this is mainly dependent on nanoparticle sizing and their proliferation. A tumour cell, when it grows and spreads, requires oxygen and proteins, and for this, it links to a number of blood vessels, by angiogenesis [19]. Drugs that are being delivered intravenously distribute evenly throughout the body. Now, the cancer cells due to their irregulated particle screening have a high intake of a particular type of particles. This tumour cells act as parasites on the blood supply for their nourishment, this property can be exploited and nanotubes can be used as foreign particles in the blood stream to get deposited into tumour cells; this is due to the abnormal lymphatic drainage and positioning of tumour cells [20–22]. All of these factors lead to abnormal molecular and fluid transport dynamics. This mechanism is known as "Enhanced Permeability and Retention (EPR) effect" in solid tumours. The EPR effect is usually employed to describe nanoparticle and liposome delivery to cancer tissue. One of many examples is the work regarding thermal ablation with gold nanoparticles.
2. Active targeting: another path selective process of drug target delivery is active targeting, involving the assimilation of ligands and immune globulins (antigens). These are target-oriented molecules with coordination complexes with proteins. This incorporation of ligands is extremely peculiar and discrete to the types of cells throughout the body. These ligands completely react with target sites, being electron donors, hence safeguard nanotubes from enzyme influence and desolation. The coordination ligands increase the uptake selectivity of nanotubes to which they are bounded. Efficient approaches in identifying the cancer cell receptor are also considered like the "in vivo phage screening" which involves F3 galactosidase to have a strong binding ability with nucleolin [23, 24], which is a major nucleolar protein found on tumour cell surfaces. This mechanism also involves a drawback, i.e. ligands may lead nanotubes to become marginally toxic because of non-specific binding, and also the positive charged ligand particles may decrease drug delivery and reduces the efficiency once inside of cells [25, 26]. Active targeting is pertinent in overcoming multi-drug aversion in tumour cells [27].

6 Functionalization of CNTs

Carbon nanotubes when in the pristine state are highly hydrophobic in nature, i.e. they are highly non-polar and the atoms making the molecule do not establish a static electric field with respect to each other. This is where functionalization is used. The functionalization of nanotubes is defined as a process to enhance the surface characteristics hence making them biocompatible. The meaning establishes that the functionalization of carbon nanotubes involves the linking of organic or inorganic functional groups to their respective tubular structure [26]. Through the functionalization of carbon nanotubes, it is possible to regulate the physicochemical functions, reduce cytotoxicity, gives an opportunity to append molecules of drugs or proteins for delivery system build-up [27]. Majorly, the different methods of functionalization of CNTs include:

1. Non-covalent Functionalization: non-covalent bonding is a better method used in functionalization. This method enhances the biocompatibility along with validity in biological functions. This can be carried out by the creation of molecular dispersed liquid colloidal type structures with amphiphilic molecules studded and dispersed uniformly according to the topological distribution on the walls of CNTs. Another efficient method is piling up of pyrene molecules on the surface area of the CNT as a result establishing π–π bonds with CNT surface. This type of non-covalent π–π attachments preserves the sanctity of aromatic base structure and hence the electronic characteristics. Due to weak π–π bonding, this method is not suitable for the target release of drugs [28].
2. Covalent Functionalization: the attachment of biocompatible functional groups to the surface is a more stringent target delivery method, wherein a stronger arena is needed [29]. CNT functionalization process involves oxidation of the walls of CNT using concentrated acids of sulphuric acid (H_2SO_4) or nitric acid (HNO_3). The solution is then heated and agitated by means of soniferous methods of sound agitation. This process results in side wall of CNT forming covalent bonding with carboxylic acid, which makes CNT water soluble, as shown in Fig. 2. The carboxylic group is having a high negative charge, which leads to an increased amount of hydrophilicity. [29, 30]. The CNT can also be coated with polyethylene glycol as a part of maintaining bio-metastability [31]. This

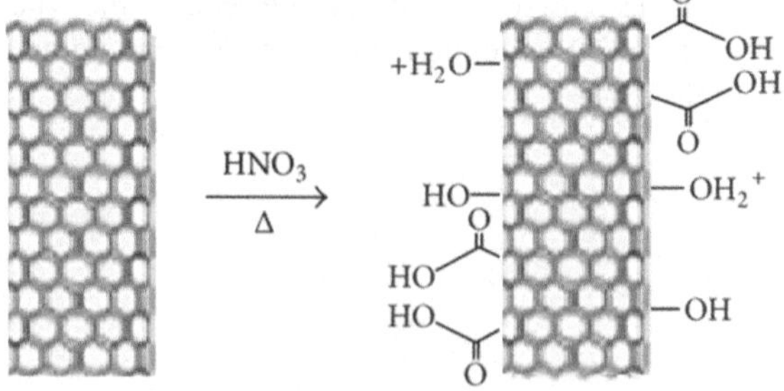

Fig. 2 Covalent functionalization (oxidation) reaction

covalent functionalization leads to a stronger bond formation between the drug and CNT. The only drawback is being the wear out of walls of CNT resulting in transmuting certain features of CNTs [28].

7 Conclusion

This paper reviews a major application of carbon nanotubes (CNTs) as novel drug carriers and selective anticancer treatment vehicles which can be used for drug targeting and delivery. The paper also explains about the prime method of loading and encapsulating the drug within the carbon nanotube (CNT). An effective and substantial substitution of chemotherapy, which is the most traditional and one-handed way to deal and treat the cancer-causing elements, is nanotechnology. Since nanoparticles (nanotubes, in this case) are highly efficient carriers in high pH levels and highly selective in targeting the tumour micro-environment (TME), which goes unabated in case of direct chemotherapy. Before CNTs are loaded with drugs, they need to undergo in vivo and in vitro clinical testing and preliminary checking for their toxicity to healthy tissues. Also, the hydrophobic nature needs to be overcome in order to enhance biocompatibility while penetrating inside the cell and overcoming the immunity barriers, by functionalization and manipulation of surface chemistry using either covalent or non-covalent functionalization. The drug targeting plays a major role in this technology to undergo selective analysis and positioning on nanocarrier to the target cells. There are two types of targeting, namely passive and active, such that passive involves the concept of ERP and active circumscribing around the mechanism of targeting by ligands. The drawbacks which need involved are regarding the toxicology effect of the CNTs without any attenuation in pharmacological efficacies, lack of reliable attachments of drugs externally and risk of releasing the drug before the target site.

References

1. Li H, Fan X, Houghton J (2007) J Cell Biochem 101:805
2. Godet I, Gilkes DM (2017) BRCA1 and BRCA2 mutations and treatment strategies for breast cancer. Integr Cancer Sci Ther PMCID: PMC5505673
3. Chen J, Chen S, Zhao X, Kuznetsova LV, Wong SS, Ojima I (2008) Functionalized single-walled carbon nanotubes as rationally designed vehicles for tumor-targeted drug delivery. J Am Chem Soc 130(49):16778–16785
4. Alderton PM, Gross J, Green MD (1992) Comparative study of doxorubicin, mitoxantrone, and epirubicin in combination with ICRF-187 (ADR-529) in a chronic cardiotoxicity animal model. Cancer Res 52(1):194201
5. Chonn A, Cullis PR (1995) Recent advances in lipo-somal drug-delivery systems. Curr Opin Biotechnol 6(6):698–708
6. Batra R, Davies JN, Wheatley D (2009) Extensive arterial and venous thrombo-embolism with chemotherapy for testicular cancer: a case report. Cases J. 2:9082

7. Misra R, Acharya S, Sahoo SK (2010) Cancer nanotechnology: application of nanotechnology in cancer therapy. Drug Discov. Today 15(19-20):842–850
8. Iijima S (1991) Helical microtubules of graphitic carbon. Nature 354(6348):56–58
9. Mu QX, Broughton DL, Yan B (2009) Endosomal leakage and nuclear translocation of multiwalled carbon nanotubes: developing a model for cell uptake. Nano Lett 9(12):4370–4375
10. Kostarelos K, Lacerda L, Pastorin G, Wu W, Wieckowski S, Luangsivilay J (2007) Cellular uptake of functionalized carbon nanotubes is independent of functional group and cell type. Nat Nanotechnol 2(2):108–113
11. Tasis D, Tagmatarchis N, Bianco A, Prato M (2006) Chemistry of carbon nanotubes. Chem Rev 106(3):1105–1136
12. Singh R, Pantarotto D (2005) Binding and condensation of plasmid DNA onto functionalized carbon nanotubes: toward the construction of nanotube-based gene delivery vectors. J Am Chem Soc 127(12):4388–4396
13. Monthioux M (2002) Filling single-wall carbon nanotubes. Carbon 40(10):1809–1823
14. Gao H, Kong Y, Cui D, Ozkan CS (2003) Spontaneous insertion of DNA oligonucleotides into carbon nanotubes. Nano Lett 3(4):471–473
15. Qiang FU, Gisela W, Su DS Selective filling of carbon nanotubes with metals by selective washing. New Carbon Mater 23(1):7–20
16. Dejonge N, Doytcheva M, Allioux M, Kaiser M, Mentink S, Teo K, Lacerda R, Milne W (2005) Cap closing of thin carbon nanotubes. Adv Mater 17(4):451–455
17. Liu X, Tao H, Yang K, Zhang S, Lee S-T, Liu Z (2011) Optimization of surface chemistry on single-walled carbon nanotubes for in vivo photothermal ablation of tumors. Biomaterials 32(1):144–151
18. Ferrari M (2005) Cancer nanotechnology: opportunities and challenges. Nat Rev Cancer 3:161–171
19. Wang X, Wang Y (2009) Advances of cancer therapy by nanotechnology. Cancer Res. Treat 41(1):1–11
20. Matsumura Y, Maeda H (1986) A new concept for macromolecular therapeutics in cancer chemotherapy: mechanism of tumoritropic accumulation of proteins and the antitumor agent smancs. Cancer Res 46:6387–6392. PMID: 2946403
21. Duncan R, Sat Y-N (1998) Tumour targeting by enhanced permeability and retention (EPR) effect. Ann Oncol 9:1–39
22. Vasey PA, Kaye SB, Morrison R (1999) Phase I clinical and pharmacokinetic study of PK1 [N-(2-hydroxypropyl) methacrylamide copolymer doxorubicin]: first member of a new class of chemotherapeutic agents-drug-polymer conjugates. Clin Cancer Res 5(1):83–94. PMID: 9918206
23. Taylor A, Turnbull B (2011) The potential use of carbon Nanotubes for cancer treatment. Nature 321(2):1–116
24. Christian S, Pilch J, Akerman ME, Porkka K, Laakkonen P, Ruoslahti E (2003) Nucleolin expressed at the cell surface is a marker of endothelial cells in angiogenic blood vessels. J Cell Bio 163(4):871–878
25. Wu H, Zhua L, Torchilin V (2013) pH-sensitive poly(histidine)-PEG/DSPE-PEG co-polymer micelles for cytosolic drug delivery. Biomaterials 34:1213–1222
26. Cajota S, Van Butselea S, Paillardb A, Passiranib C, Garcionb E, Benoitb J, Varshneyc S, Jérômea C (2012) Smart nanocarriers for pH-triggered targeting and release of hydrophobic drugs. Acta Biomater 8:4215–4223
27. Foldvari M, Bagonluri M (2008) Carbon nanotubes as functional excipients for nanomedicines: II. Drug delivery and biocompatibility issues. Nanomedicine. 3:183–200
28. Liu Z, Sun X, Nakayama-Ratchford N, Dai H (2007) Supramolecular chemistry on water-soluble carbon nanotubes for drug loading and delivery. ACS Nano 1:50–56
29. Klumpp C, Kostarelos K, Prato M, Bianco A (2006) Functionalized carbon nanotubes as emerging nanovectors for the delivery of therapeutics. Biochim Biophys Acta 1758:404–412

30. Bekyarova E, Ni Y, Malarkey EB (2005) Applications of carbon nanotubes in biotechnology and biomedicine. J Biomed Nanotechnol 1:3–17
31. Kam NW, O'Connell M, Wisdom JA, Dai H (2005) Carbon nanotubes as multifunctional biological transporters and near-infrared agents for selective cancer cell destruction. Proc Natl Acad Sci USA 11600–11605

A Review on Squeeze Casting of Aluminium-Based Alloys and Its Composites

Dhiraj Nigade, Dhananjay Shukla, and Ravikant Hattale

Abstract Aluminium-based alloys have a better strength-to-weight ratio and hence are widely used in aerospace and automotive industries. Squeeze casting is one of the most suited options for manufacturing of aluminium-based alloys and its composites. This paper provides a review on squeeze casting and its advancements, like ultrasonic squeeze casting. Optimization of various process variables and their influence on output parameters carried out by various authors are also discussed.

Keywords Squeeze casting · Microstructures · MMC · Aluminium-based alloys · Ultrasonic squeeze casting

1 Introduction

Casting is a near-net shape manufacturing process, most suitable where complex and intricate shaped metal parts are to be produced. In this method, liquid material is usually poured into a mould, which contains a hollow cavity of the desired shape, and then allowed to solidify. Shrinkage is one of the undesirable phenomenon highly observed during casting can be eliminated by squeeze casting.

Squeeze casting (SC) is a comprehensive term used for production technique where high pressure is applied for solidification [1]. This high integrity technique involves preheated die set and uniform pressure applied through the hydraulic press in the cavity till solidification [2]. The uniform pressure results in the elimination of shrinkage [3]. Squeeze casting is categorized into two main categories, viz. direct SC and indirect SC. If the pressure is applied on the entire surface of the liquid

D. Nigade (✉) · D. Shukla · R. Hattale
Department of Mechanical Engineering, Dwarkadas J. Sanghvi College of Engineering, Vile Parle, Mumbai, Maharashtra 400056 , India
e-mail: Nigade.Dhiraj@djsce.ac.in

D. Shukla
e-mail: Dhanajay.Shulka@djsce.ac.in

R. Hattale
e-mail: Ravikant.Hattale@djsce.ac.in

© Springer Nature Singapore Pte Ltd. 2020
H. Vasudevan et al. (eds.), *Proceedings of International Conference on Intelligent Manufacturing and Automation*, Lecture Notes in Mechanical Engineering,
https://doi.org/10.1007/978-981-15-4485-9_48

metal during solidification by a punch, then the process is called as direct squeeze casting and it produces castings of full density. If the metal is injected into the die cavity by a small diameter piston, then the process is indirect squeeze casting [4]. Standard process parameters such as squeeze pressure, die preheating temperature, melt temperature, cooling rate, size of reinforcement particle and compression time are highly influenced the performance measures like microstructure and mechanical properties of cast product [5, 6].

Non-ferrous metal parts made by aluminium, magnesium and copper-based alloys with special treatments are primarily used for this process [7]. Metal matrix composites (MMCs) are widely used in the automotive and aerospace industries. Squeeze casting is an efficient method for manufacturing of MMC components on large scale apart from this part manufactured possesses better overall mechanical and tribological properties. Squeeze casting process and its parameters with respect to different aluminium alloys are covered in the subsequent discussion of this study.

2 LM Alloys

Chemical composition LM 13 alloy contains 82–85% Al, 10–13% Si, 0.8–1.5 Mg, 0.1% Zn, 1.0% Fe, 0.7–1.5% Cu, 1.5% Ni, 0.1% Pb, 0.1% Sn, 0.1% Ti and 0.15% Mn. Squeeze pressure of 100 MPa decreases the grain size and secondary dendritic arm spacing (SDAS). Grain size does not get significantly influenced by melt temperature and die preheating temperature [8]. Density of squeeze cast components increases with an increase in temperature. Macrostructure and hardness of the alloy improved by decreasing melt temperature and die preheating temperature [9]. LM13 alloys are mostly used in automotive and aerospace industry where dry sliding wear is most predominant factor. Wear rate increases with increasing load up to 50 N with sliding velocity of 1–5 m/s [10]. Squeeze pressure of 140 N/mm^2 and die preheating temperature of 250 °C produce good surface finish of LM6 aluminium alloy [11]. For evaluating the optimized squeeze casting parameters of LM24 aluminium alloy, overall objective function is formulated using weighted sum approach. Genetic algorithm and weighted sum approach method give optimized value of SC process parameter for the said alloy as under: squeeze pressure (105.901 N/mm^2), die preheat temperature (35.10 °C) and pressure duration (15.124 s). Ultimate tensile strength and hardness values are 272.258 N/mm^2 and 98.013 BHN [12]. Universal testing machine can be used to apply the pressure for squeeze casting of LM20 alloy. Taguchi analysis has shown that squeeze pressure applied by UTM gives a better surface finish and improves casting density [13].

3 Al–Si–Mg Alloys

The squeeze pressure in the range of 30–100 MPa does not have a significant influence on tensile properties of Al–7Si–0.7Mg aluminium alloy. Oxide inclusion in the alloy will lead to significant elongation of the Si particles [14]. Squeeze casting along with heat treatment increases the strength of the investigated Al–6Si–0.3Mg alloys. Hardness of Al–6Si–0.3Mg alloy increases with increasing squeeze pressure [15]. The best temperature for the squeeze casting of aluminium alloy with 13.5% silicon is either 690 or 660 °C, and better properties are observed at the bottom section of the casting [16]. Mechanical properties and microstructure of squeeze cast Al–Zn–Mg–Cu alloys are strongly influenced by casting temperature. Grain size and dendritic arm spacing increase with increasing casting temperature [17]. Connecting rod fabricated by semi-solid squeeze casting process with Al alloy (9% Si, 0.25% Mg, 0.5% Mn) exhibits best mechanical properties and microstructure at pouring temperature of 575 °C [18].

4 Al-Based Hybrid Composites

Al-based metal matrix composites with high volume fraction of SiC particles possess high tensile strength, and it can be enhanced by specific matrix alloying and heat treatment. Hybrid preforms containing SiC whiskers and nanoscale SiC particles can be made by wet blending. Cu as well as Zn + Mg addition enhances the strength of composite. Elastic modulus and strength of Al MMCs increased by the addition of SiC whiskers and SiC nanoparticles [19, 20]. The optimum parameters for manufacturing the preform are as preheating temperature 773–793 K temperature of the melt 1073 K, infiltration pressure 2–3 MPa and solidification pressure 50–60 MPa [20]. SiC particles distribution in A356 alloy composites was found to be highly dependent on secondary dendritic arm spacing (SDAS). Clustering of SiC particles is larger at smaller arm spacing [21]. Stir and squeeze casting methods are used to produce AA 7075 based single and hybrid reinforced nanocomposites with different preheating temperatures. The composites are prepared with reinforcement of 2% nano-alumina particles by wt. and 4% silicon carbide particles by wt. Density, hardness and ultimate tensile strength of single and hybrid reinforced nanocomposites are found superior to base alloy through mechanical characterization [22]. A413–B4C composite material squeeze casted to analyse the effect of process variables on mechanical properties. Mathematical model developed using artificial neural network (ANN) shows that die preheating temperature has less effect on mechanical properties [23].

5 Other Alloys

Optimum process parameters for AC2A aluminium alloy non-symmetrical component estimated with the help of mathematical model are as: squeeze pressure 100 MPa, die preheating temperature, melt temperature, compression holding time for the alloy are 100 MPa, 200, 725 °C, 45 s, respectively [24]. Squeeze cast aluminium alloy AA2024 shows an increase in hardness from 125 BHN to 152 BHN at constant pouring temperature of 700 °C and squeeze pressure of 140 MPa. Dendritic arm spacing of AA2024 decreases when squeeze pressure increases [25]. Cylindrical samples of an aluminium alloy (EN-AB46000) are squeeze cast. Nominal composition (wt%) of EN-AB46000 8.0–11.0 Si, 0.6–1.1 Fe, 2.0–4.0 Cu, 0.55 Mn, 0.15–0.55 Mg, 0.15 Cr, 0.55 Ni, 1.2 Zn, 0.35 Pb, 0.25 Sn, 0.2 Ti and remainder is Al. Cooling rate and mechanical properties (yield strength and Vickers hardness) are correlated using predictive model based on Hall–Petch effect [26]. Mechanical properties and microstructure of squeeze cast Al–Zn–Mg–Cu alloys are strongly influenced by casting temperature. Grain size and dendritic arm spacing increase with increasing casting temperature [27]. Mathematical models formulated for squeeze casting, die casting and continuous casting using teaching learning-based algorithm show that comparatively die casting is superior to squeeze casting [28].

6 Ultrasonic-Assisted Squeeze Casting

Ultrasonic treatment is widely used for refining microstructure and removing the gases from the melt [29]. This advantage of ultrasonic technology is utilized in squeeze casting for producing defect-free castings. Here in ultrasonic-assisted squeeze casting, ultrasonic probe and disc springs are connected to punch, which induces ultrasonic vibration in punch during casting. These vibrations disperse the solid nucleus formed and break the dendrites to form fine equiaxed grains [30]. Indirect ultrasonic vibrations used for squeeze casting produce casting with uniform distribution of silicon particles, which results best mechanical properties [31].

7 Conclusion

The publications studied on squeeze casting explain that the process is being successfully developed through optimization of variables involved in it. The study also reveals that advanced materials like hybrid aluminium alloy composites can be treated successfully. For aluminium alloys and its composites, squeeze pressure influences largely on output parameters such as microstructure, mechanical properties and wear resistance. Decreasing die preheating temperature and melt temperature improves hardness and microstructure. An advanced integrated technique thus devised known as ultrasonic-assisted squeeze casting is incorporated to produce defect-free casting and forms fine grain structure. This technique is now widely accepted and put into practice.

References

1. Ghomashchi MR, Vikhrov A (2000) Squeeze casting: An overview. J Mater Process Technol 101(1):1–9. https://doi.org/10.1016/S0924-0136(99)00291-5
2. Vijian P, Arunachalam VP (2005) Experimental study of squeeze casting of gunmetal. J Mater Process Technol 170(1–2):32–36. https://doi.org/10.1016/j.jmatprotec.2005.03.033
3. Yong MS, Clegg AJ (2005) Process optimisation for a squeeze cast magnesium alloy metal matrix composite. J Mater Process Technol 168(2):262–269. https://doi.org/10.1016/j.jmatprotec.2005.01.012
4. Yue TM, Chadwick GA (1996) Squeeze casting of light alloys and their composites. J Mater Process Technol 58:302–307
5. Dhanashekar M, Senthil Kumar VS (2014) Squeeze casting of aluminium metal matrix composites—an overview. Procedia Engineering 97:412–420. https://doi.org/10.1016/j.proeng.2014.12.265
6. Radha R, Sreekanth D (2019) Mechanical and corrosion behaviour of hydroxyapatite reinforced Mg–Sn alloy composite by squeeze casting for biomedical applications. J Magnesium Alloys. https://doi.org/10.1016/j.jma.2019.05.010
7. Vijayaram TR, Sulaiman S, Hamouda AMS, Ahmad MHM (2006) Fabrication of fiber reinforced metal matrix composites by squeeze casting technology. J Mater Process Technol 178(1–3):34–38. https://doi.org/10.1016/j.jmatprotec.2005.09.026
8. Maleki A, Shafyei A, Niroumand B (2009) Effects of squeeze casting parameters on the microstructure of LM13 alloy. J Mater Process Technol 209(8):3790–3797. https://doi.org/10.1016/j.jmatprotec.2008.08.035
9. Maleki A, Niroumand B, Shafyei A (2006) Effects of squeeze casting parameters on density, macrostructure and hardness of LM13 alloy. Mater Sci Eng A 428(1–2):135–140. https://doi.org/10.1016/j.msea.2006.04.099
10. Chelladurai S, Arthanari R, Thangaraj A, Sekar H (2017, November) Dry sliding wear characterization of squeeze cast LM13/FeCu composite using response surface methodology. Overseas Foundry 14(6). https://doi.org/10.1007/s41230-017-7101-3
11. Vijian P, Arunachalam VP (2006) Optimization of squeeze cast parameters of LM6 aluminium alloy for surface roughness using Taguchi method. J Mater Process Technol 180(1–3):161–166. https://doi.org/10.1016/j.jmatprotec.2006.05.016
12. Vijian P, Arunachalam VP (2007) Modelling and multi objective optimization of LM24 aluminium alloy squeeze cast process parameters using genetic algorithm. J Mater Process Technol 186(1–3):82–86. https://doi.org/10.1016/j.jmatprotec.2006.12.019
13. Patel GCM, Krishna P, Parappagoudar MB (2014) Optimization of squeeze cast process parameters using Taguchi and grey relational analysis. Procedia Technol 14:157–164. https://doi.org/10.1016/j.protcy.2014.08.021
14. Chen ZW, Thorpe WR (1995) Microstructure and mechanical properties of indirect squeeze cast Al–7Si–0.7Mg alloys containing 0.1–0.6 wt% Fe. Cast Metals 8(3):179–191. https://doi.org/10.1080/09534962.1995.11819206
15. Abou El-Khair MT (2005) Microstructure characterization and tensile properties of squeeze-cast AlSiMg alloys. Mater Lett 59(8–9):894–900. https://doi.org/10.1016/j.matlet.2004.11.041
16. Yang LJ (2003) The effect of casting temperature on the properties of squeeze cast aluminium and zinc alloys. J Mater Process Technol 140(1–3):391–396. https://doi.org/10.1016/S0924-0136(03)00763-5

17. Balasubramanian I, Maheswaran R, Manikandan V, Patil N, Raja MA, Singari RM (2018) Mechanical characterization and machining of squeeze Cast AZ91D/SiC magnesium based metal matrix composites. Procedia Manuf 20:97–105. https://doi.org/10.1016/j.promfg.2018.02.014
18. Dao V, Zhao S, Lin W, Zhang C (2012) Effect of process parameters on microstructure and mechanical properties in $AlSi_9Mg$ connecting-rod fabricated by semi-solid squeeze casting. Mater Sci Eng A 558:95–102. https://doi.org/10.1016/j.msea.2012.07.084
19. Zhang XN, Geng L, Wang GS (2006) Fabrication of Al-based hybrid composites reinforced with SiC whiskers and SiC nanoparticles by squeeze casting. J Mater Process Technol 176(1–3):146–151. https://doi.org/10.1016/j.jmatprotec.2006.03.125
20. Beffort O, Long S, Cayron C, Kuebler J, Buffat PA (2007) Alloying effects on microstructure and mechanical properties of high volume fraction SiC-particle reinforced Al-MMCs made by squeeze casting infiltration. Compos Sci Technol 67(3–4):737–745. https://doi.org/10.1016/j.compscitech.2006.04.005
21. Çetin A, Kalkanli A (2008) Effect of solidification rate on spatial distribution of SiC particles in A356 alloy composites. J Mater Process Technol 205(1–3):1–8. https://doi.org/10.1016/j.jmatprotec.2007.11.065
22. Kannan C, Ramanujam R (2017) Comparative study on the mechanical and microstructural characterisation of AA 7075 nano and hybrid nanocomposites produced by stir and squeeze casting. J Adv Res 8(4):309–319. https://doi.org/10.1016/j.jare.2017.02.005
23. Soundararajan R, Ramesh A, Sivasankaran S, Vignesh M (2017) Modeling and analysis of mechanical properties of aluminium alloy (A413) reinforced with boron carbide (B4C) processed through squeeze casting process using artificial neural network model and statistical technique. Mater Today Proc 4(2):2008–2030. https://doi.org/10.1016/j.matpr.2017.02.047
24. Senthil P, Amirthagadeswaran KS (2012) Optimization of squeeze casting parameters for non symmetrical AC2A aluminium alloy castings through Taguchi method. J Mech Sci Technol 26(4):1141–1147. https://doi.org/10.1007/s12206-012-0215-z
25. Jahangiri A, Marashi SPH, Mohammadaliha M, Ashofte V (2017) The effect of pressure and pouring temperature on the porosity, microstructure, hardness and yield stress of AA2024 aluminum alloy during the squeeze casting process. J Mater Process Technol 245:1–6. https://doi.org/10.1016/j.jmatprotec.2017.02.005
26. Boschetto A, Costanza G, Quadrini F, Tata ME (2007) Cooling rate inference in aluminum alloy squeeze casting. Mater Lett 61(14–15):2969–2972. https://doi.org/10.1016/j.matlet.2006.10.048
27. Fan CH, Chen ZH, He WQ, Chen JH, Chen D (2010) Effects of the casting temperature on microstructure and mechanical properties of the squeeze-cast Al–Zn–Mg–Cu alloy. J Alloy Compd 504(2):42–45. https://doi.org/10.1016/j.jallcom.2010.06.012
28. Rao RV, Kalyankar VD, Waghmare G (2014) Parameters optimization of selected casting processes using teaching-learning-based optimization algorithm. Appl Math Model 38(23):5592–5608. https://doi.org/10.1016/j.apm.2014.04.036
29. Eskin GI, Eskin DG (2015) Ultrasonic treatment of light alloy melts, 2nd edn. CRC Press, New York
30. Chen G, Yang M, Jin Y, Zhang H, Han F, Chen Q, Zhao Z (2019, October) Ultrasonic assisted squeeze casting of a wrought aluminum alloy. J Mater Process Technol 266:19–25. https://doi.org/10.1016/j.jmatprotec.2018.10.032
31. Lü S, Wu S, Dai W, Lin C, An P (2012) The indirect ultrasonic vibration process for rheo-squeeze casting of A356 aluminum alloy. J Mater Process Technol 212(6):1281–1287. https://doi.org/10.1016/j.jmatprotec.2012.01.018

A Review on Carbon Fibre Reinforced Polymer Composites and the Methods of Their Manufacture, Disposal and Reclamation

Aman M. Chulawala, Frank Crasta, and Vijaya Kumar N. Kottur

Abstract Fibre reinforced composites can be tailored to meet the demands of the industry for various applications. Carbon fibre reinforced polymer (CFRP) holds the advantages of being lightweight, possessing high tensile strength, superior corrosion resistance and anti-fatigue properties. Their unique ability to possess customisable directional strength is a major factor, which promotes their use. Through bonding CFRP sheets, the ultimate load and post-elastic stiffness of specimens can be significantly increased. However, the carbon footprint and direct cost are the biggest obstacles in the way of adopting CFRP in industrial application. In this paper, the prevalent techniques of manufacture and disposal of CFRP have been studied. New techniques like inculcating thermoplastic matrix-based CFRP in additive manufacturing and manufacture of CFRP sheets by electro-activated deposition resin holding increase the scope of use of the composite. Despite its tremendous capabilities, CFRP and its products pose an issue at the end of their life cycle due to the difficult and cumbersome methods of disposal. Also, hampering their value in the industry is their high purchasable cost and the lack of a method of manufacture which provides high productivity and formability.

Keywords Carbon fibre composites · Reinforcement · CFRP reclamation · CFRP manufacturing · Additive manufacturing · Electro-activated deposition resin holding

A. M. Chulawala (✉) · F. Crasta · V. K. N. Kottur
Department of Mechanical Engineering, Dwarkadas J. Sanghvi College of Engineering, Vile Parle, Mumbai, Maharashtra 400056 , India
e-mail: aman.chulawala@gmail.com

F. Crasta
e-mail: prashant.crasta@djsce.ac.in

V. K. N. Kottur
e-mail: vijayakumar.kottur@djsce.ac.in

© Springer Nature Singapore Pte Ltd. 2020

H. Vasudevan et al. (eds.), *Proceedings of International Conference on Intelligent Manufacturing and Automation*, Lecture Notes in Mechanical Engineering,
https://doi.org/10.1007/978-981-15-4485-9_49

1 Introduction

The tale of reinforced materials is a long one, and the discovery is attributed to a French gardener, Joseph Monier, who was the first to create reinforced concrete in 1849. Reinforced concrete is one of the primary ingredients which results in the tall structures with which we are familiar with in today's world [1]. However, the practice of adding fibres to strengthen materials dates back to ancient times. Horsehair and straw were among the first natural fibres to be used for strengthening bricks [1]. Romauldi and Batson's paper on fibre reinforced concrete in the year 1963 was the first of its kind to explore the mechanism of fibre reinforcement and discuss their properties in conjunction with those of the prevalent materials of their time [2]. A century's worth of research and experiments on different materials has contributed to the composite structure employed by the industry sector today.

The use of structural materials is limited to a certain extent by deficiencies like brittleness, poor tensile strength and poor resistance to impact strength, fatigue, low ductility and low durability, which vary with the material under consideration [3]. The brittleness, for example, of plain concrete, can be compensated by the addition of structural members as reinforcements, such as steel tubes added to concrete resulting in RCC [3]. However, this does not improve the basic property of the base material. It is mainly a method of using two materials for the required performance [3].

The use of composite technology in aircraft applications has increased exponentially, with glass fibre and carbon fibre-based composites occupying the top spots [4]. By the 1950s, fibre reinforced polymer (FRP) boat hulls and FRP car bodies were developed, wherein glass fibres were the major reinforcement used. The advantage of glass was its non-conductivity resulting in its use as an insulator to prevent galvanic corrosion in metals [4]. However, under certain conditions of exposure, glass fibres proved to be sensitive to alkaline environments and moisture attack. In addition, creep affected glass fibres more than any other types of fibres [4]. The higher performance carbon fibre reinforced polymers (CFRPs) were developed in 1963 for specialised applications and are now being standardised for use in industry.

2 Carbon Fibre Composites

As the name suggests, carbon fibre reinforced polymer composites (hence abbreviated as CFRP) comprise two distinct materials, the carbon fibre sheets which acts as the reinforcement and a matrix such as epoxy resin to bind the sheets together and allow for the equal distribution of load to be carried by the composite [5]. Binding under the epoxy resin is what gives the frail sheets of carbon fibres their strength and toughness. CFRP shows directional properties [6]. The properties depend on the layout of the carbon fibre and the proportion of carbon fibre relative to the polymer [7]. The fracture toughness of these CFRP composites is dependent on their interlaminar strength. Thus, the fracture can occur by: (1) debonding between the carbon fibre and polymer matrix, (2) fibre pull-out or (3) delamination between the CFRP sheets. [8]

$$E_c = E_m V_m + E_f V_f \tag{1}$$

$$\frac{1}{E_c} = \frac{V_m}{E_m} + \frac{V_f}{E_f} \tag{2}$$

Here,

E_c Modulus of composite
E_m Modulus of matrix phase
E_f Modulus of carbon fibres
V_m Volume fraction of matrix
V_f Volume fraction of carbon fibre

Equation (1) gives the value of the modulus of elasticity of the composite along the direction of the applied load. Equation (2) gives the modulus of the composite with the fibre is oriented transverse to the applied load. Figure 1 shows the different types of weaves used in fabric manufacture.

Traditional strengthening techniques of steel beams involve an expensive process of welding/bolting steel plates to the tension flange of the section [9]. These plates are heavy, impractical and face many difficulties in site installation [9]. Thus, fibre reinforced polymer (FRP) laminates/strips which are characterised by their high strength-to-weight ratio offer an attractive strengthening alternative [9]. Lately, FRP laminates/strips were widely used to retrofit reinforced and pre-stressed concrete structures [9].

Fig. 1 Twill and Bidrectional woven carbon fabrics

3 Manufacturing Methods in Industry

The first step is the actual production of carbon fibre sheets which act as the reinforcing material. They are produced by the pyrolysis of an organic precursor fibre at temperatures as high as 980 °C in an inert atmosphere [10]. This precursor fibre is usually polyacrylonitrile (PAN), which are converted to strands of pure carbon. After an appropriate surface coating, the fibres are ready to be woven into sheets [10]. These sheets are now used with resins to form CFRP laminates.

Manufacture of CFRP can take place in the following ways:

1. Hand lay-up: this method involves the manual placement of the carbon fibre sheets in the desired orientation in alternative layers of sheets and bonding agents. The commonly used bonding agent here is epoxy resin, which is applied evenly over the layers of carbon fibre sheets. Due to the absence of any vacuum consolidation, the required thickness of the composite sheet is not always achieved.
2. Resin infusion by vacuum bagging: resin infusion is a process where the voids in a vacuumed stack of fibre sheets are filled with a liquid resin. When the resin solidifies, the solid resin matrix binds the assembly of materials into a unified rigid composite. The vacuum bagging process provides high-quality defect-free, non-porous composites when through-thickness pathways and pre-preg formats are used [11].
3. Oven cured pre-preg: pre-preg refers "pre-impregnated" composite fibres where a thermoset polymer matrix material, such as epoxy, or a thermoplastic resin is already present [12]. This prevents the need of an external bonding matrix in the manufacturing phase. However, the sheets start to cure immediately and must be stored under 20 °C, decreasing the life of the sheets [12].
4. Additive manufacturing: an additive manufacturing process creates parts by adding material layer by layer while referencing a computer-aided design model. This is also known as 3D printing [13]. Polymer additive manufacturing (AM) has been gaining momentum with time and is only limited to the properties of the material used for the process. Carbon fibre offers low density, a low coefficient of thermal expansion and high thermal conductivity, making it an ideal material for additive manufacturing [13]. Thermosetting epoxy matrices are used in most CFRPs where a high strength and stiffness to weight ratio is necessary. On the other hand, thermoplastic matrices do not form strong bonds readily due to high melt viscosity and economic constraints. However, their key advantage lies in their melt processability, which allows for rapid manufacture of parts by AM [13]. Despite the ease of the procedure, additive manufacturing fails to comply with the strength criteria which the CFRP parts are expected to be put to [14].
5. Electro-activated deposition resin holding: this is an effective method for the manufacture of CFRP products employing long carbon fibres [14]. Here, the carbon fibre fabric preform with the required directional stability is immersed in an electro-activated deposition solution polymer with epoxy groups. Energisation reposts in the precipitation of the resin on the surface of the fibres, where they are

gradually impregnated between the fibre strands. The impregnated fibre is flexible and easy to mould [14]. After thermal curing, a moulded CFRP part with an epoxy matrix made of thermosetting epoxy can be achieved. This method makes the use of autoclaves for high temperature and pressure conditions obsolete. Thus, energy and time can be saved. The only limitation is the restriction of the thickness of the CFRP member possible by this method [14].

4 CFRP as Reinforcement Materials

Steel tubes are now used heavily in the industry for their cheap and aesthetic values. However, these members degrade with time due to natural and loading stresses, fatigue and creep [15]. Therefore, there is an urgent requirement in the market for a suitable reinforcement material which can elongate the life cycle of the part under consideration. The increasing research efforts concerning the application of FRP to structures yielded very positive results. High stiffness fibres, such as carbon fibres, can easily improve the properties of steel structures such as strength and stiffness, while also significantly improves the fatigue resistance of the subject structure [15]. However, the limited use of these reinforcements in the industries is largely attributed to one factor, the bonding strength between the structure and the carbon composite [15].

CFRPs when used as face plane must show a good adhesive characteristic [16]. The main methods of failures of sandwich panel composites are debunking and delimitation, in other words, the composite's strength is measured by the strength of its weakest bond, the resin which joins steel and CFRP [16]. These adhesive resins must be capable of transferring forces efficiently between these varying load-carrying members. Other requisites of these bonds are good strength at elevated working temperature and good chemical inertness post-curing [16].

One such method of increasing the inter-laminar strength of the composites is the addition of graphene oxide (GO) to the fibre surface [17]. Coating the surface with about 5% GO has shown to increase the inter-laminar strength significantly. An external medium like hexamethylene diisocyanate (HDI) terpolymer used as a coupling agent can also significantly improve the interfacial properties [17]. However, the process is cumbersome and long due to extensive chemical reactions and waiting periods required.

5 Disposal and Reclamation of CFRP

The production of virgin carbon fibre is a process demanding 50–150 kWh/kg. This high cost of carbon fibre dictates its high market price [18]. Nonetheless, demand for carbon fibre products in the market is growing by 10% annually, with the market requirement of CFRP crossing 100,000 tons per year [18]. The Airbus is comprised of 20% CFRP by weight, while the Boeing B787 variants are expected to cross the 50% mark for CFRP content. Shipbuilding industry is another major employer of the CFRP composite sections [19].

High production rates warrant a significant rise in the waste produced by the respective industries. So far, CFRP waste is disposed by landfills, which results in a pile-up of the waste to hazardous levels until proper recycling methods are made available. Generally, carbon fibre recycling involves mechanical and chemical processes, thermal or microwave decomposition methods. A suitable method for this carbon fibre disposal is participation in the carbon fibre reuse programme, where these fibres can be reclaimed from end-of-life composites at an energy cost of around 2–9 kWh/kg, which results in a price lower than that of virgin carbon fibres [19]. Carbon fibre reclamation can be achieved by different methods like pyrolysis (thermal decomposition at elevated temperature in inert environment), solvolysis (slow heating of the composite in a chemical bath, usually of acetone and water) and a fluidised bed process, each of which comes with their own share of advantages and limitations [19]. These are relatively new methods, and active research is being carried out to improve their effectiveness and improve availability.

6 Conclusions

Research in the field of developing new manufacturing and disposal techniques for CFRP parts, like the method of additive manufacturing, inculcating thermoplastic CFRP is currently necessary. A carbon fibre composite used as a reinforcement is as strong as the adhesive bond which permits the transfer of load effectively between the layers of the composite. New methods like graphene oxide (GO) addiction to the surfaces are being tested to increase the inter-laminar strength. Despite its advantages, the non-economic purchasability drawback severely limits its use. Thus, methods are being developed to reclaim the carbon fibre discarded as waste at the end of the product life cycle, such that the reclaimed fibres possess the maximum properties of the pre-discarded carbon fibre material.

References

1. Jain D, Kothari A (2012) Hair fibre reinforced concrete. Res J Recent Sci 1(ISC-2011):128–133. ISSN 2277 - 2502
2. Perez-Pena M, Mobasher B (1994) Mechanical properties of fibre reinforced lightweight concrete composites. Cem Concr Res 24(6):1121–1132. https://doi.org/10.1016/0008-8846(94)90036-1
3. Grija S, Shanthini D, Abinaya S (2016 November–December) A review on fibre reinforced concrete. Int J Civ Eng Technol (IJCIET) 7(6):386–392 (Article ID: IJCIET_07_06_042)

4. Seica MV, Packer JA (2007) FRP materials for the rehabilitation of tubular steel structures, for underwater applications. Compos Struct 80(3):440–450. https://doi.org/10.1016/j.compstruct.2006.05.029
5. Kopeliovich D (2012) Carbon fiber reinforced polymer composites. Archived from the original on 14 May 2012. substech.com
6. Corum JM, Battiste RL, Liu KC, Ruggles MB (2000 February) Basic properties of reference cross-ply carbon-fiber composite. ORNL/TM-2000/29, Pub57518, Oak Ridge National Laboratory
7. Courtney T (2000) Mechanical behaviour of materials. Waveland Press, Inc., United States of America, pp 247–249. ISBN 1-57766-425-6
8. Chawla K (2013) Composite MATERIALS. Springer, United States of America. ISBN 978-0-387-74364-6
9. Faikha VP, Nandhu R, Shilpa S, Sulthan N, Adila Abdulla K (2019 May) Flexural strengthening of steel beams using CFRP sheets. Int Res J Eng Technol (IRJET) 6(5)
10. Yu Chou T, Tsai H-Y, Chuen Yip M (2018) Preparation of CFRP with modified MWCNT to improve the mechanical properties and torsional fatigue of epoxy/polybenzoxazine copolymer. Compos Part A: Appl Sci Manuf. https://doi.org/10.1016/j.compositesa.2018.11.026
11. Muralidhara B, KumareshBabu SP, Suresha B (2019) Utilising vacuum bagging process to prepare carbon fibre/epoxy composites with improved mechanical properties. Mater Today Proc. https://doi.org/10.1016/j.matpr.2019.09.051
12. Giorgini L, Mazzocchetti L, Giangiacomo Minakb, Dolcini E (2012) Investigation of a carbon fiber/epoxy prepreg curing behaviour for thick composite materials production: an industrial case-study. In: AIP confernce proceedings of 6th international conference on times of polymers (TOP) and composites. 1459:190–192 (2012). https://doi.org/10.1063/1.4738439
13. Van de Werken N, Tekinalp H, Khanbolouki P, Ozcan S, Williams A, Tehrani M (2019) Additively manufactured carbon fiber-reinforced composites: state of the art and perspective. Addi Manuf 100962. https://doi.org/10.1016/j.addma.2019.100962
14. Katagiri K, Honda S, Minami S, Tomizawa Y, Kimu D, Yamaguchi S, Sasaki K (2018) CFRP manufacturing method by using electro-activated deposition and the effect of reinforcement with carbon fibre circumferentially around the hole. Compos Struct. https://doi.org/10.1016/j.compstruct.2018.09.084
15. Danilov AI (2016) Some aspects of CFRP steel structures reinforcement in civil engineering. Procedia Eng 153:124–130. https://doi.org/10.1016/j.proeng.2016.08.091
16. Abramovich H, Kalnins K, Wieder A (2017) Test results on the stability and vibrations of composite shells. Stab Vibr Thin Walled Compos Struc 619–691. https://doi.org/10.1016/b978-0-08-100410-4.00013-2
17. Sun T, Li M, Liang SZM, ChenY, Zou H (2019) Multi-scale structure construction of carbon fibre surface by electrophoretic deposition and electropolymerization to enhance the interfacial strength of epoxy resin composites. Appl Surf Sci 143929. https://doi.org/10.1016/j.apsusc.2019.143929
18. Van de Werken N, Reese MS, RedaTaha MM, Tehrani M (2019) Investigating the effects of fiber surface treatment and alignment on mechanical properties of recycled carbon fiber composites. Compos Part A: Appl Sci Manuf. https://doi.org/10.1016/j.compositesa.2019.01.012
19. Jeong J-S, Kim K-W, An K-H, Kim B-J (2019) Fast recovery process of carbon fibres from waste carbon fibres-reinforced thermoset plastics. J Environ Manage 247:816–821. https://doi.org/10.1016/j.jenvman.2019.07.002

A Review: In Vitro Investigation of Dental Composite Materials and Tooth Enamel by Using Pin-on-Disc Tribometer

Abhijeet Suryawanshi and Niranjana Behera

Abstract Present study mainly deals with the review of in vitro investigation for assessing the wear action of different dental materials. With in vitro testing, the researcher gets more control over the experimental variables leading to more accurate results. Whereas, other methods have limited contribution towards the tribological study of dental materials. Therefore, in vitro method was developed for studying the wear mechanisms. The wear testing devices like pin-on-disc or artificial mouth were used in the study of in vitro. By selecting different test parameters like load, speed, temperature and use of artificial saliva, an artificial oral environment can be created on a pin-on-disc tribometer. The filler particle percentage is an essential parameter for studying the tribological and mechanical characteristics of dental composite materials.

Keywords In vitro · Pin-on-disc tribometer · Wear · Artificial saliva · Test parameters

1 Introduction

Human teeth play a vital role in our daily life by helping in the process of oral communication, digestion and also add grace to the face. However, tooth damage such as cavities and tooth tissue loss will gradually occur due to ageing, poor oral hygiene and traumas. To overcome these damages, artificial dental materials were invented. Nowadays, ceramics and composite materials, metals and alloys are extensively used for restoring dental structure and their transplants. Tribology deals with the study of three parameters, i.e. friction, wear and lubrication of interacting surfaces in relative motion [1]. The resistive force which occurs when two solid surfaces slide against

A. Suryawanshi · N. Behera (✉)
School of Mechanical Engineering, VIT University, Vellore, Tamilnadu, India
e-mail: niranjanabehera@vit.ac.in

A. Suryawanshi
e-mail: suryawanshi.abhijeet1@gmail.com

© Springer Nature Singapore Pte Ltd. 2020

H. Vasudevan et al. (eds.), *Proceedings of International Conference on Intelligent Manufacturing and Automation*, Lecture Notes in Mechanical Engineering,
https://doi.org/10.1007/978-981-15-4485-9_50

each other is known as friction. Wear is the process in which particles of solid surfaces gradually deteriorate due to the mechanical and chemical action of surfaces in contact [2].

The factors affecting tooth wear are muscular forces, nature of saliva, eating habits of patient and the type of the material used for refurbishment. Among all these factors, material selection is under the control of the dentist. Therefore, to reduce the tooth wear, the prime focus of all researchers has been the development of dental materials with better wear control properties [3]. There are four different types of wear that may occur in dental materials. These are attrition, abfraction, abrasion and erosion [4].

2 Testing Methods

2.1 *In Vivo Observation and Measurements*

For years, in vivo method is being utilized in medical research for noting down the parameters related to damage of dental tissue. As the tooth structure and its movement are complicated, tooth wear has multiple causes due to its interaction with the physical and chemical processes. Hence, to actually test all the facets of the tooth wear and properties of artificial dental material, in vivo method is useful. However, due to some disadvantages of this method, its contribution to the development of new dental material is limited. Firstly, individual wear processes such as abrasion, attrition and erosion cannot be isolated and studied by in vivo methods. Although testing conditions can be unified by taking some actions, it is still evident that interpretation of results is influenced by subjectivity. Secondly, important variables including dietary intake, chewing force or environmental factors may influence the tribological behaviour and hence limits the wear mechanisms involved in the tribology of dental materials [5]. Volunteer compliance influences the wear process involved for research work of in vivo; as a result, this method is expensive and time-consuming. In general, either indirect methods or clinical categorization systems are used for wear measurement in dentistry [6].

The categorization index given by Smith and Knight (1984) distinctly calculates the wear over the surface of tooth corresponding to bare dentine surface area [7]. Cvar and Ryge [43] have given a standard scoring system for the United States Public Health Service, by classifying the wear of restorative material as 'Charlie', 'Bravo' and 'Alpha'. An 'Alpha' indicates no wear, 'Bravo' indicates visible wear and 'Charlie' indicates extreme wear condition where new restoration is required [8]. Penny Fleur Bardsley (2008) states that there is no such index that satisfies the requirements of the teams working on dental material research [9, 10]. The chairside method for testing the wear parameters can be easily implemented without any separate set-up but is a less sensitive and time-consuming process. There are various wear measurement techniques implemented using sample models where the samples of artificial dental materials are compared with some threshold levels. This method works quickly and is cost-effective as well but tests only the outer edges of the restorative materials. The most suitable method to measure wear is to compare the 3D image sequence of the material under test. As this method requires high-cost equipment to measure the wear, some of the clinical studies have been conducted using 3D scanning technology [2, 11].

2.2 *In Vitro Laboratory Simulation*

There are two groups of In vitro testing: (1) dental-wear-testing instruments and (2) toothbrush testing. This assessment is mainly done to find out the similarity between clinical wear and in vitro laboratory wear and clinical. As the clinical trials are not cost-effective and take more time from preparing the material according to the formula to final product approval, in vitro wear testing is a good approach. As per the findings by McCabe and Smith (1981), the similarity between clinical observations and in vitro tests is most of the time-poor [12]. Therefore, as there is a change in the nature of the test, how different levels of wear are produced due to the nature of the wear may also change. Harrison (1976) said that most of the assessment methods conducted using abrasion-testing machines fail to recreate the clinical environment [13]. Due to the variety of experimental procedures and measuring equipment, the outcomes of these findings cannot be compared directly. So, to avoid these conditions, individual ranking of material under the test of the respective studies is considered for the conclusion. Considering the current in vitro data, predicting the actual nature of artificial dental materials is a critical task. Therefore, to obtain realistic results, the in vitro experiments should be able to build the actual oral environment [14, 15].

2.3 *In Situ Testing*

In the in situ testing method of wear testing, the material under test is kept in an apparatus held inside the mouth and then taken out for ex vivo testing. It helps these materials to get a feel of the actual oral environment. Hence, in situ testing can be seen as partially combining the advantages of the in vitro and clinical conditions [16, 17].

3 In Vitro Investigations of Dental Composite and Tooth Enamel Using Pin-on-Disc Tribometer

Since their inception in this profession, artificial dental materials are being widely used due to their versatile nature. Considering the extended usage of artificial dental composites across a variety of applications, their characteristics and performance is a prime factor [18]. In the pin-on-disc test, calculated wear is applied to the specimens under observation. The specimen under test is placed on a circulating disc with a given sliding force between the pin or ball and the sample surface for creating the wear conditions. The wear rate of the material is estimated by calculating volume lost. The wear rate can be used to calculate wear resistance as the action performed on all samples is same [19].

Masouras et al. (2008) determined the bulk modulus, Young's modulus, Poisson's ratio and shear modulus of restorative material and correlated them with filler content. It was observed that the elastic module had noticeable variations and there was more similarity index between elastic modulus and filler content [20]. The nano-sized filler particle improves the mechanical properties over the micro-sized particles. The improved mechanical particles by nanoparticles containing composite are due to better filler/polymer interaction [21, 22].

A comparative study of artificial dental composite material with medium and high filler content was carried out by Nagarajan et al. (2004). Both two-body contact wear method and in vitro assessment of medium filled dental materials are not affected by slight changes in chemical formulation, variations in size of filler particles and their distribution. When the wear behaviour of the medium filled composite material is compared with a highly filled composite consisting of alumina fillers along with silica, it was seen that the two-body wear performance of these composites is minimized due to chemical action between the particles and filler loading instead of the changes in the size of particles and their distribution [23]. Callaghan et al. (2006) investigated the relationship between the wear behaviour of glass fibre-superimposed dental composites with changing fibre volume fractions and lengths [24]. Also, the wear rate and surface hardness of dental composite and enamel change remarkably with a different pH value of saliva solution [25, 26]. Enamel wear was most sensitive for varying pH, while composite material was least influenced by acid. The composite material was most wear-resistant than traditional glass cement [27].

4 Specimen Preparation

Generally, the shape of the pinned specimen is spherical or cylindrical. Diameter of a typical spherical or cylindrical pin samples ranges from 2 to 10 mm. Diameters of the disc samples are in the range of 30–100 mm with a thickness of 2–10 mm range. It is usually recommended a ground arithmetic average surface roughness of 0.8 μm (32 μin.) or less [28]. Mitov et al. stated that various finishing methods affect the tribological behaviour of enamel and dental composite materials. Also, there was no major linear correlation observed between pretesting surface roughness and abrasive wear [29].

5 Test Parameters—Load, Speed, Temperature

While selecting suitable dental material for different oral conditions, it is required that the clinicians should have a sound knowledge of tribological and mechanical properties of the same. This would help the manufactures of dental material for enhancing the quality of their products [30, 31 and 32]. Hu et al. (2002) proved that the testing conditions have a high impact on deciding the wear mechanisms. For example, two-body and three-body wear mechanisms affected the rankings of wear behaviour differently [33, 34].

The variation of friction with applied load and rotating speed in the presence of artificial saliva was studied by Li et al. (2011). As normal load increases, the coefficient of friction for nano-metric ZrO_2 ceramics rises. Also, the friction coefficient is low for high rotation of speed [35, 36]. Micro-hardness, modulus of elasticity modulus, flexural resistance and wear were greatly affected by temperature [30]. Also, replication of different temperature levels in the oral environment using thermo-cycling process suggested that the conventional dental composite was more sensitive to change in temperature as compared to nano-composite and nano-hybrid composite [31].

6 Significance of Artificial Saliva

In most of the studies conducted so far, the tribological behaviour of human enamel and dental composite were tested in dry and wet conditions (artificial saliva) [37, 38]. The fundamental requirement of an artificial oral condition is to have artificial saliva which behaves with the material under test in the same way as that of natural saliva like the actual oral environment required for in vitro test. The properties of artificial saliva should be similar to natural saliva for the real oral environment. Various recipes of artificial saliva have been introduced from 1931 but no one had explained the stability of tooth material, i.e. hydroxyapatite (HAP). Leung and Darvell (1997) studied the stability of HAP of various artificial saliva recipes and have compared and investigated how the ionic components individually affected the stability of HAP [39].

The tribological properties of the human tooth and dental composite material were not significantly affected by saliva viscosity. Also, above a certain threshold viscosity value, there were no noticeable variations in the wear outcomes of the three- and two-body mechanisms [40]. Mayworm et al. (2008) compared the wear hardness and resistance of two different dental nano-hybrid composites when stored in the presence of artificial saliva. It was observed that the micro-hardness decreases and wear resistance increases for both materials after storage in artificial saliva [41]. Again, wear behaviour varies with the use of different types of artificial saliva solutions [42].

7 Conclusion

Numerous in vitro mechanisms have been evolved in the last thirty years with each scientist formulating his methodology and wear testing device. Of these, few experimental set-ups and mechanisms match the qualifying standards and validation that currently exist in different areas of medical product assessment. An important aspect of future in vitro testing is that wear assessment mechanisms are required to recreate typical oral environment and biomechanics; otherwise, the mismatch between the outcomes of in vivo and in vitro studies is bound to happen. In pin-on-disc tribometer, by setting different test parameters, the real oral environment can be simulated with some necessary modifications.

References

1. Bonse J, Kirner S, Griepentrog M, Spaltmann D, Kruger J (2018) Femtosecond laser texturing of surfaces for tribological applications. Materials
2. Zhou ZR, Zheng J (2008) Tribology of dental materials: a review. J Phys D Appl Phys 41
3. Lee A, He H, Lyons K, Swain MV (2012) Tooth wear and wear investigations in dentistry. J Oral Rehabil 39:217–225
4. Sarode GS, Sarode SC (2013) Abfraction: a review. J Oral Maxillof Pathol 17:222–227
5. Hahnel S, Schultz S, Trempler C, Ach B, Handel G, Rosentritt M (2001) Two-body wear of dental restorative materials. J Mechan Behav Biomed Mater 4:237–244
6. Mair LH, Stolarski TA, Vowles RW, Lloyd CH (1996 January-March) Wear: mechanisms, manifestations and measurement. Report of a workshop. J Dentist 24(1–2):141–148
7. Smith, B, Knight, J (1984) An index for measuring the wear of teeth. Br Dent J 156:435–438
8. Bayne, SC, Schmalz, G (2005) Reprinting the classic article on USPHS evaluation methods for measuring the clinical research performance of restorative materials. Clin Oral Invest 9:209–214
9. Bardsley, PF (2008) The evolution of tooth wear indices. Clin Oral Invest 12:15–19
10. Lopez-frias FJ, Castellanos-Cosano L, Martin-Gonzalez J, Llamas-Carreras JM, Segura-Egea JJ (2012) Clinical measurement of tooth wear: tooth wear indices. J Clin Exp Dent 4(1):e48–53
11. Hao Z, Yin H, Wang L, Meng Y (2014) Wear behavior of seven artificial resin teeth assessed with three-dimensional measurements. J Prosthet Dent 112(6):1507–1512
12. McCabe J, Smith B (1981) A method for measuring the wear of restorative materials in vitro. Br Dent J 151:123–126
13. Harrison A (1984) Wear of dental materials, Part-I modes of wear. Dental Advertiser 39:8–11
14. Mohd. Zainal Abidin Mohd. Sulong, Ramlah Abdul Aziz (1990) Wear of materials used in dentistry: a review of the literature. J Prosthet Dentist 63(3):342–349
15. Ruben JL, Roeters FJM, Montagner AF, Huysmans MCDNJM (2014) A multifunctional device to simulate oral ageing: the "Rub & Roll". J Mechan Behav Biomed Mater 30:75–82

16. Zhou Z-R, Yu H-Y, Zheng J, Qian L.-M, Yan, Y (2013) Dental biotribology, 1st edn. Springer, New York
17. Zhou ZR, Zheng J (2006) Oral tribology. Proc Inst Mech Eng Part J J Eng Tribol 220(8):739–754
18. Ferracane JL (2010) Resin composite—state of the art. Dental Mater (2010)
19. Maru MM, Tanaka DK (2007) Consideration of stribeck diagram parameters in the investigation on wear and friction behavior in lubricated sliding. J Braz Soc Mech Sci Eng 55–62
20. Konstantinos Masouras, Nick Silikas, David C. Watts (2008) Correalation of filler content and elastic properties of resin-composites. J Dental Mater 24(7):932–939
21. Foroutan F, Javadpour J, Khavandi A, Atai M, Rezaie HR (2011) Mechanical properties of dental composite materials reinforced with micro and nano-size Al_2O_3 filler particles. Iranian J Mater Sci Eng 8(2)
22. Karthick R, Sirisha P, Ravi Sankar M (2014) Mechanical and tribological properties of PMMA-sea Shell based biocomposite for dental application. Procedia Mater Sci 6:1989–2000
23. Nagarajan VS, Jahanmir S, Van Thompson P (2004) In vitro contact wear of dental composites. Dental Mater 20:63–71
24. Callaghan DJ, Vaziri A, Nayeb-Hashemi H (2006) Effect of fiber volume fraction and length on the wear characteristics of glass fiber-reinforced dental composites. Dental Mater 22:84–93
25. Sajewicz E (2007) Tribological behaviour of human enamel in red wine and apple juice environments. Wear 262:308–315
26. Aliping-Mckenzie M, Linden RWA, Nicholson JW (2004) The effect of Coca-Cola and fruit juices on the surface hardness of glass–ionomers and 'composers'. J Oral Rehabilitation 31:1046–1052
27. Shabanian M, Richards LC (2002) In vitro wear rates of materials under different loads and varying pH. J Prosthet Dentist 87(6)
28. Standard test method for wear testing with pin on disk apparatus. ASTM Int (2004)
29. Mitov G, Heintze SD, Walz S, Woll K, Muecklich F, Pospiech P (2012) Wear behavior of dental Y-TZP ceramic against natural enamel after different finishing procedures. Dental Mater 28:909–918
30. Ramalho A, Braga de Carvalho MD, Antunes PV (2013) Effects of temperature on mechanical and tribological properties of dental restorative composite materials. Tribol Int 63:186–195
31. Ayatollahi MR, Yahya MY, Karimzadeh A, Nikkhooyifar M, Ayob A (2015) Effects of temperature change and beverage on mechanical and tribological properties of dental restorative composites. Mater Sci Eng C 54:69–75
32. Condon JR, Ferracane JL (1997) Factors effecting dental composite wear in vitro. J Biomed Mater Res (Appl Biomater) 38:303–313
33. Zheng J, Zhou ZR (2007) Friction and wear behavior of human teeth under various wear conditions. Tribol Int 40:278–284
34. Hu X, Shortall AC, Marquis PM (2002) Wear of three dental composites under different testing conditions. J Oral Rehabilitation 29:756–764
35. Li C, Liu Z, Liu G, Ding Y (2011) Experimental investigations of mechanical characteristics and tribological mechanisms of nanometric zirconia dental ceramics. Open Mater Sci J 5:178–183
36. Li CH, Qi LY, Li LG, Ding YC (2011) Theoretical analysis and experimental investigations of the frictional wear characteristics of nanometric zirconia ceramics. Adv Mater Res 211–212:26–30
37. Zheng L, Li Y, Zheng J, Wen M, Zhang YF, Qian LM, Zhou ZR (2013) A comparative study on the sliding wear behaviors of human tooth enamel, Cu–Zn alloy and Al_2O_3 ceramic. J Wear 301:308–315
38. Li H, Zhou ZR (2002) Wear behaviour of human teeth in dry and artificial saliva conditions. Wear 249(10–11):980–984
39. Leung VWH, Darvell BW (1997) Artificial saliva for in vitro studies of dental materials. J Dentist 25(6):475–484
40. Sajewicz E (2009) Effect of saliva viscosity on tribological behaviour of tooth enamel. Tribol Int 42:327–332

41. Mayworm CD, Camargo SS, Bastian FL (2008) Influence of artificial saliva on abrasive wear and micro hardness of dental composites filled with nanoparticles. J Dentist 36:703–710
42. Turssi CP, Faraoni JJ, de Menezes M, Serra MC (2006) Analysis of potential lubricants for in vitro wear testing. Dental Mater 22:77–83
43. Cvar JF, Ryge G (1971). Criteria for the clinical evaluation of dental restorative materials. San Francisco: U.S. Dept. of Health, Education, and Welfare, Public Health Service, National Institutes of Health, Bureau of Health Manpower Education, Division of Dental Health, Dental Health Center.

Design

Development of Spring Life Test Apparatus and Life-Cycle Assessment of Extension Springs

Greegory Mathew, Santosh B. Rane, Yogesh Patil, and Sanjay V. Mohan

Abstract Life-cycle test of moulded case circuit breakers (MCCB) on mechanical endurance test apparatus exhibited springs as the weak links. Therefore, improvement in the reliability of the circuit breaker mechanism was possible only with an improvement in the life of the springs. To evaluate spring life, a life-cycle test apparatus was developed which subjected springs to actual load conditions as experienced in MCCB. This paper describes the development and working of the test apparatus. The paper also explains how reliability of springs can be estimated using life data provided by the apparatus.

Keywords Reliability · Fatigue failure · Hook failure · Weibull plot

1 Introduction

Circuit breakers are complex engineering systems used for protection of distribution feeder systems, transformers, etc. Mechanical springs are the heart of a circuit breaker operating mechanism. They provide the driving force necessary for the instant opening of the contacts of a circuit breaker and also play a vital role by ensuring adequate contact pressure between the fixed and the moving contacts. Therefore, failure of the extension springs must be avoided. Their reliability is crucial for the safety and the overall reliability of the breaker.

The best engineering practice is to build reliability into a product at the earliest in the design and development phase [1]. One of the ways to determine the inherent reliability is to test under controlled environments based on the product usage and development requirements [2]. Fatigue tests are increasingly being performed to

G. Mathew (✉)
Dwarkadas J. Sanghvi College of Engineering, Vile Parle, Mumbai, Maharashtra 400056, India
e-mail: greg_12@ymail.com

S. B. Rane
Sardar Patel College of Engineering, Andheri, Mumbai, India

Y. Patil · S. V. Mohan
L&T EAIC, Mumbai, India

© Springer Nature Singapore Pte Ltd. 2020

H. Vasudevan et al. (eds.), *Proceedings of International Conference on Intelligent Manufacturing and Automation*, Lecture Notes in Mechanical Engineering,
https://doi.org/10.1007/978-981-15-4485-9_51

simulate and investigate real-world failures of springs in various applications [3–10]. The circuit breaker selected for the study had a designed mechanical life of 30,000 cycles. But mechanical endurance tests on the circuit breakers terminated before 30,000 cycles because of premature failures of the extension springs of the breaker. A need was therefore felt to estimate and improve the life of springs to improve the life of the circuit breaker. Reliability assessment of a mechanical spring is required to evaluate different spring designs for various spring deflection applications [11]. In the conventional process of evaluating the life of helical extension springs, the circuit breaker with extension springs was mounted on a mechanical endurance test apparatus. A pneumatic actuator then operates the circuit breaker (switches on and switches off) at the standard frequency cycle. The test cycle at which the endurance test terminates is noted, and the circuit breaker is analysed to identify the root cause for the premature termination of the test. During the investigation, if the spring is found to be failed, then the value of the cycle at which the test terminated denotes the life of the springs. The conventional process of testing spring life is time-consuming and expensive. Hence, a need was felt to establish a life-cycle test apparatus to eliminate the time constraints (i.e. it should test multiple springs at the same time) and to provide adequate and reliable data for plotting reliability of the operating life of springs.

2 The Life-Cycle Test Apparatus

The life-cycle test apparatus consists of two stacks of mounting space for springs, i.e. the lower stack between the bottom plate and the mid-plate(nos 6 and 3, respectively) and the upper stack between the top plate and the mid-plate (nos 2 and 3, respectively). The top and the bottom plates are stationary while a double-acting pneumatic cylinder actuates the mid-plate (provides vertically up/down motion to the mid-plate) in order to flex the springs between their 1st and 2nd operating limit.

Grooved spring rods (no. 5) are used to support the springs. Grooves on the spring rod support the spring hooks and prevent their tilting when the springs are flexed between their two operating lengths. Support blocks (no. 7) hold the spring rods as indicated in the diagram. To prevent the spring rods from sliding away during operation (because of vibrations), stopper blocks (no. 10) are provided which restrict the translational movement of the rods. Support block screws (no. 12) and support block nuts (no. 13) fix the support blocks and stopper blocks to the respective top plate, mid-plate and base plate. The apparatus is bolted to the test bench through holes provided in the base plate. Guide pillars (no. 4) welded to the base of the base plate guides the longitudinal traverse of the mid-plate. The distance adjusting bolts (no. 1) are used to maintain the required gap between the top plate and the base plate. Roller guide bushes (no. 8) fixed to the mid-plate reduce the friction between the mid-plate and guide pillar. Stopper pins (no. 9) which prevent the travel of the mid-plate beyond the set limits are fixed to the mid-plate using stopper nuts (11).

Springs are mounted on spring rods between the base plate and mid-plate and between the mid-plate and top plate. The distance between the base and top plates is calculated and then adjusted using distance adjusting bolts such that when the mid-plate traverses upwards (towards the top plate) on being actuated by a pneumatic cylinder, the springs in the lower stack will be flexed to the 2nd operating length, while the springs in the upper stack will be relieved to the 1st operating length. The upward motion of the mid-plate beyond the calculated limit is prevented by stopper pins which stop against the top plate and restrict the further upwards motion of the mid-plate. The advantage of the apparatus is that the relaxing force exerted by the springs in the upper stack (when the springs in the upper stack relax from its 2nd operating limit to its 1st operating limit during the upwards movement of the mid-plate) will aid the pneumatic force in flexing (stretching) the springs in the lower stack to their 2nd operating limit (they were initially at 1st operating limit). The process is reversed during the downward moment of the mid-plate. Figures 1 and 2 indicate various components of the apparatus.

Figure 3 shows the 3D CAD model of the apparatus.

Springs with operating lengths and diameters can be easily tested for life on the apparatus. Before testing, the length between the top and base plates should be readjusted such that when the mid-plate traverses upwards, the springs in the lower stack will be flexed to the 2nd operating length, while the springs in the upper stack will be relieved to the 1st operating length, and vice versa. The length can be adjusted by rotating the distance adjusting bolts, clockwise or anticlockwise, depending on whether the distance needs to be increased or decreased. Spring rods need to be changed to suit the hook diameter of the selected spring. Stopper pins need to be redesigned to maintain the 1st operating length of the new springs in both the stacks.

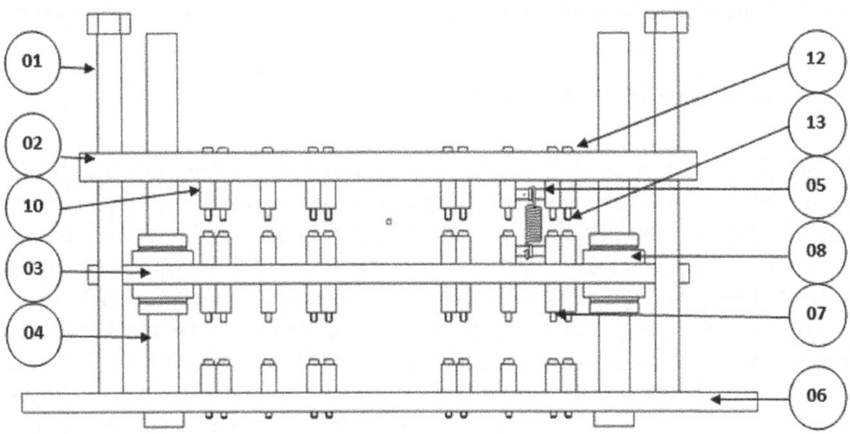

Fig. 1 New experimental apparatus

Fig. 2 Actual apparatus

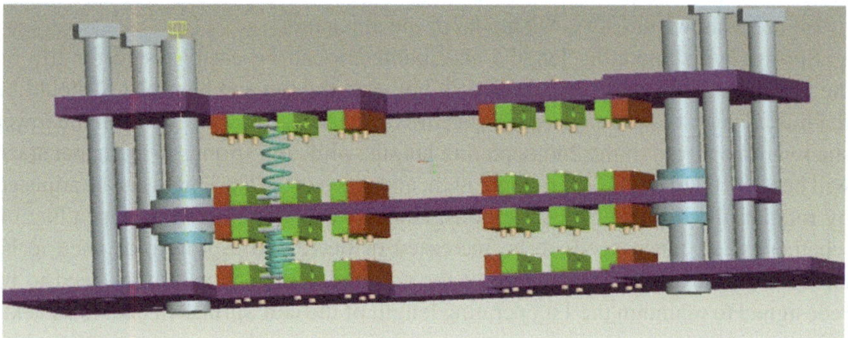

Fig. 3 3D model of the apparatus

3 Testing and Improvement

For the purpose of the study, 8 circuit breaker springs(of initial design), accepted by the quality control department of the company, were selected. The springs mounted on the apparatus were flexed between its 1st and 2nd operating limits(38.3 mm and 48.5 mm, respectively) with a standard test frequency of 240 cycles/hr. Times to failures of these springs as provided by the test apparatus are tabulated in Table 2.

Examination of the failed spring specimens clearly indicated failure at the hooks. The hooks were redesigned to reduce the stress. The physical properties of the springs with improved hooks are tabulated in Table 1 under the column 'Design 2'. Figures 4a and 4b show the condition of the spring (from initial design category) after testing.

To further improve the life of the springs, springs of initial design parameters but with a different material (changed from grade 3 to grade 4) were manufactured and tested. Their physical characteristics are tabulated in Table 1 under the column 'Design 3'.

Table 1 Spring properties through different phases of improvement

S No.	Parameter	Initial	Design 2	Design 3
1	Wire dia in mm	1.7	1.7	1.7
2	Mean dia in mm	9.7	9.7	9.7
3	Active no. of turns	10.5	10.5	10.5
4	Free length in mm	30.2	30.2	30.2
5	Material as per IS 4454: (Part I): 2001: patented and cold drawn unalloyed steel wire	Grade III	Grade III	Grade IV
6	Permissible shear stress Kgf/mm^2	92.49	92.49	98.89
7	Permissible bending stress Kgf/mm^2	176.49	176.49	189.16

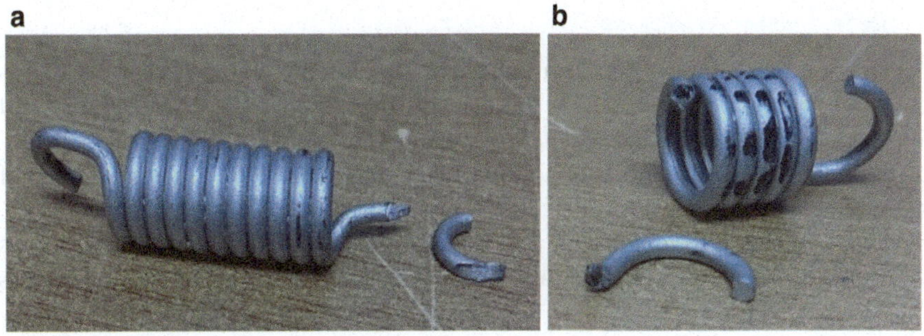

Fig. 4 **a** Failed spring (initial design). **b** Failed spring (initial design)

4 Observations

The life-cycle test for each design phase was run for a maximum duration of 30,000 cycles, and the springs surviving the test were suspended. The interval censored life data of the springs obtained from the life-cycle test apparatus are tabulated in Table 2. The 'initial design' column tabulates the observations on the life-cycle test apparatus for the springs with the initial design. 'Design 2' and 'Design 3' columns tabulate the observations on the life-cycle test apparatus for the springs from the first and the second design modification phases, respectively. The 'from cycle' and 'to cycle' columns define the interval in which a spring failure was reported. For example, three spring failures were observed between 28,000 and 29,250 cycles for springs from Design 3 phase.

5 Results

Life data distribution plots were used to estimate the reliability of springs. The relationship of a distribution plot with failure data differs from plot to plot and

Table 2 Time to failure for springs

Failure No.	Initial design		Design 2		Design 3	
	From cycle	To cycle	From cycle	To cycle	From cycle	To cycle
1	20,800	21,770	18,000	18,770	25,050	26,000
1	24,005	25,065	18,770	19,500	28,000	29,250
1	26,400	27,240	18,770	19,500	28,000	29,250
1	26,400	27,240	19,500	20,000	28,000	29,250
1	26,400	27,240	19,500	20,000	29,250	29,680
1	28,110	29,820	19,500	20,000	29,250	29,680
1	28,110	29,820	20,500	21,000	29,250	29,680
1	28,110	29,820	22,500	24,000	30,000	Suspended

depends on the nature of the data collected. To estimate reliability correctly, it is essential to select the distribution plot that fits the data closely. To select the most appropriate distribution, failure data were analysed using Minitab, and the goodness-of-fit plots for various distribution plots with failure data were compared. Some of these plots are shown in Fig. 5.

As it is difficult to identify the distribution plot which fits the data best, values of Anderson–Darling coefficient and correlation coefficient calculated for various

Fig. 5 Goodness of fit for failure data (Weibull, lognormal, exponential and loglogistic distribution plots)

Table 3 Estimated correlation coefficient for springs with initial design

S. No.	Distribution	Anderson–Darling coefficient	Correlation coefficient
1.	Lognormal	10.892	0.940
2.	3-parameter Weibull	10.875	0.969
3.	3-parameter lognormal	10.892	0.950
4.	3-parameter loglogistic	10.890	0.957

plots were compared with each other. The values of theses coefficients for some distributions are summarized in Table 3.

The distribution with correlation coefficient close to 1 and Anderson–Darling value close to zero will give the most appropriate results. By analyzing the values summarized in Table 3 for initial spring design, it can be concluded that 3-parameter Weibull plot fits the life/failure data of the springs closely.

The reliability graphs for the spring through its various design phases are plotted in Fig. 6. The reliability plot of springs in the initial design phase shows a gradual decrease in reliability which indicates that the fatigue properties of the springs are inconsistent (even though they are from the same lot). If the properties of the springs were identical, the failures would have been clustered around a single time interval and there would be a sudden decrease in the slope of the graph. Defects in the material or inconsistencies in the manufacturing techniques lead to above-said inconsistencies in fatigue properties of springs.

Compared to initial spring designs, 'Design 2' phase shows a decrease in reliability of springs. Analysis of the springs revealed that the material quality was inconsistent with that described in design which led to a decrease in reliability (even though

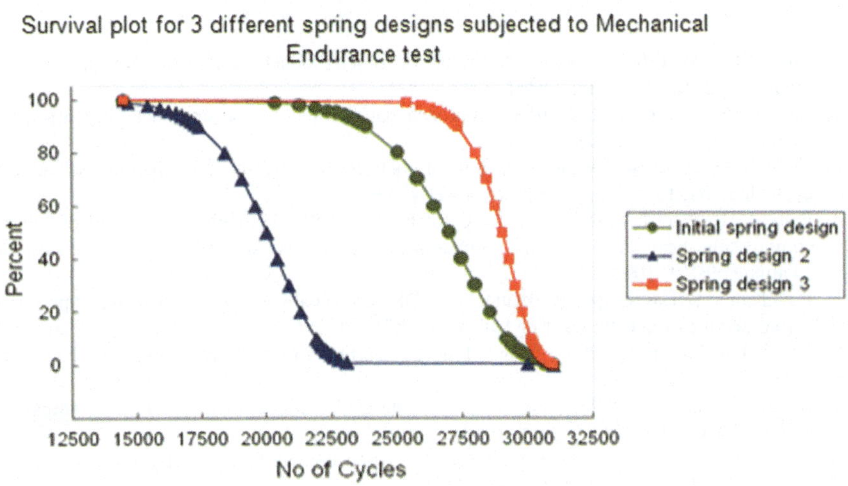

Fig. 6 Survival plot for three different spring designs

theoretical design calculations suggest an improvement in design). Tool marks at the hook bends induced during the spring production process resulted in increased stress and led to decreased spring life. Springs of Design 3 phases were manufactured from grade IV material which has better mechanical strength. Critical zone in the plot of Design 3 phase (between 28,000 and 29,680 cycles) shows 75% springs failures and hence suggested that these springs had consistent fatigue properties. The reliability plot also indicates an increase in reliability of the springs as compared to springs from initial design and Design 2 phases.

6 Conclusion

The fatigue life-cycle test apparatus provides for a quicker and an economical process to evaluate the fatigue life and reliability of springs. The modular construction of the test apparatus makes it versatile enough to permit easy assembly/disassembly of the springs and test helical extension springs of varied lengths (20 mm to 190 mm). Improvements can be incorporated into the springs by changing design parameters such as coil diameter and wire diameter. The improved springs can be tested on the apparatus to estimate the improved life/reliability.

Acknowledgements We express our special appreciation and thanks to Larsen and Toubro, EAIC, Mumbai, for encouraging this research and for providing the necessary resources, support and the motivation for carrying out this study.

References

1. Cui Y, Khan AW (2007) A case study—reliability based product development and improvement model. Qual Reliab Eng Int 24:361–373
2. Zanoff J, Ekwaro-Osire S (2009) An Approach that can quickly assess product reliability. Qual Reliab Eng Int 26:571–578
3. Chen J, Xia Z (2014) Fatigue behaviour of coke drum materials under thermal-mechanical cyclic loading. Theor Appl Mech Lett 4(4):041006
4. Pyttel B, Brunner I, Kaiser B, Berger C, Mahendran M (2013) Fatigue behaviour of helical compression springs at a very high number of cycles—investigation of various influences. Int J Fatigue 60:101–109
5. Ravi Kumar B, Das Swapan K, Bhattacharya DK (2003) Fatigue failure of helical compression spring in coke oven batteries. Eng Fail Anal 10(3):291–296
6. Schulz BA, Moran AL (2001) Fatigue failure of an F/A-18 fin retainer spring. Eng Fail Anal 8(2):167–172
7. Berger C, Kaiser B (2006) Results of very high cycle fatigue tests on helical compression springs. Int J Fatigue 28(11):1658–1663
8. Rivera R, Chiminelli A, Gómez C, Núñez JL (2010) Fatigue failure analysis of a spring for elevator doors. Eng Fail Anal 17(4):731–738
9. Nohut S, Schneider GA (2009) Failure probability of ceramic coil springs. J Eur Ceram Soc 29(6):1013–1019

10. Das SK, Mukhopadhyay NK, Ravi Kumar B, Bhattacharya DK (2006) Failure analysis of a passenger car coil spring. Eng Fail Anal 14:158–163
11. Cheng JX, Yan WC (2004) Mechanical spring reliability assessments based on FEA generated fatigue stress and monte carlo simulated stress/ strength distributions. In: International compressor engineering conference (Paper no: 1693)

Design and Manufacturing of Test Rig for Pyrolysis of Waste Tyres of Two-Wheeler Vehicles (ELVs)

S. M. Auti and W. S. Rathod

Abstract The automobile industry works as a pillar in the manufacturing sector in a developing country like India. Due to a huge number of vehicles on the verge of the end-of-life, guidelines for the management of it came in existence with a law, AIS132, a couple of years back. This research has focused on environment-friendly methods for tyre recycling in the automobile industry. In pyrolysis process, the tyre is heated to 500–600 °C in the absence of oxygen, which leads to thermal cracking of waste tyres, and the output of this process was oil after condensing of the evolved gases. The properties which were considered to compare with diesel properties were cetane number, flashpoint, pour point, viscosity at 40 °C, calorific value, carbon content, sulphur content and density. Results of the study show that tyre pyrolysis liquid generated during the process has the properties like that of diesel. Hence, the generated liquid can be blended with diesel and this increases the commercial value of waste tyres and further helps in alleviating the ELV problem.

Keywords End of vehicle (ELV) · Automotive shredder residue (ASR) · Waste tyre pyrolysis · Tyre pyrolysis oil (TPO)

1 Introduction

1.1 Overview of Automobile Sector

The automobile industry is considered as the backbone of the Indian manufacturing sector. In the year 2010, the numbers of vehicles were around 11 crores, which

S. M. Auti (✉)
Research Scholar, Department of Mechanical Engineering, Veermata Jijabai Technological Institute (V.J.T.I.), Matunga, Mumbai 400019, Maharashtra, India
e-mail: shashikant.auti@gmail.com

W. S. Rathod
Assistant Professor, Department of Mechanical Engineering, Veermata Jijabai Technological Institute (V.J.T.I.), Mumbai 400019, Maharashtra, India

© Springer Nature Singapore Pte Ltd. 2020

H. Vasudevan et al. (eds.), *Proceedings of International Conference on Intelligent Manufacturing and Automation*, Lecture Notes in Mechanical Engineering,
https://doi.org/10.1007/978-981-15-4485-9_52

included commercial vehicles, passenger vehicles, two wheelers (2 W) and three wheelers (3 W).

It is observed that vehicle ownership increased drastically in India. In the year 2014–2015, 23,366,246 vehicles were produced. According to statistics out of total vehicles produced in this year, two wheelers have the highest percentage of 80% according to number, but due to smaller size, it is lower in weight percentage which is around 40% [1].

1.2 Defining End-of-Life Vehicle (ELVs)

An end-of-life "ELV" is defined as a vehicle which will not serve its purpose fully after multiple repairs and servicing of various parts, which leads to the scrapping of such vehicle due to its deteriorated condition.

The definition provided by European Union Directive, 2000 "end-of-life" means a vehicle which is a waste. Based on the unsafe driving condition and higher emission than designated emission standards, an owner of vehicle will term a vehicle as ELV. A vehicle can be termed as ELV based on various other parameters like vehicle age, less comfort in driving, accidental damage, non-repairable parts and request of owner [2].

Due to increase in spending power of population, there is an increase in no. of vehicles; it was then expected that around 90 lakh vehicles will be termed as ELV in the year 2015, out of which more than 80% will be two-wheeler vehicles. By 2025 this number is expected to grow up to 3 times, i.e. more than 2 crores and again two wheelers will be a major part of it [3].

1.3 Challenge in Recycling ELVs

There are several challenges to be dealt regarding increasing the effectiveness and self-sustainability of ELV recycling in India as some of the wastes are economically valuable materials. This matter is of larger concerns due to:

1. Growing demand for vehicles
2. Anticipated huge growth of ELVs.

2 Existing ELVs

Overview: In India, till 2015, it was an informal sector which was handling ELVs; in 2015, Indian government introduced a new law for handling of ELVs, and in 2016, CPCB issued guidelines for ELVS handling [4].

Opportunities and Strengths in ELV's Recycling

(I) Creation of employment
In a developing country like India, unemployment is a big challenge before government; ELVs recycling will create huge employment due to its huge volume. So, conversion of the semiformal sector of recycling to the formal sector will generate employment as it is sustainable business due to its profit percentage. Also, due to huge emission from the old vehicle, its need of time to have stringent rules regarding the use of old vehicles leads to more generation of ELVs.

(II) Material available at low cost

- End product of recycling gives a new source of material for manufacturing industry. So, efficient way of recycling will give way to a source of material at low cost [5].
- Most parts of ELVs are in a usable condition so it can be used again instead of new spare part in vehicle repair, such parts will affect the performance of vehicles which leads to environmental damage [6] Formal sector of recycling will eliminate this problem,
- Safety of human life involved in the non-formal sector is less; formal recycling will increase the safety of them and will eliminate exploitation of labours involved in non-formal sector recycling process [7].

Automotive Industry Standards (AIS 129) for ELVs
AIS 129 help in converting non-formal sector of recycling to the formal sector, and these laws discuss the guidelines for ELVs recycling [3].

- Primarily, it discusses responsibilities of last owner, definition related to ELVs, authorised centre for recycling, certification involved in this process, technical requirements of collection and dismantling centres.
- This standard also sets guidelines for manufacture, extended responsibility concept comes in the picture and this guideline specifically mentions material which should not be used in the vehicle which makes recycling difficult.
- These standards will be a guideline for effective recycling of ELVs; this norm also restricts non-formal sector functioning of recycling.

3 Generation of Tyre Pyrolysis of Tyre Pyrolysis Oil

3.1 Collection of Waste Tyre

As shown in Fig. 1, this phase is comparatively easy as waste tyre is easily available at garages and service centre where replacement of old tyre takes place. For mass

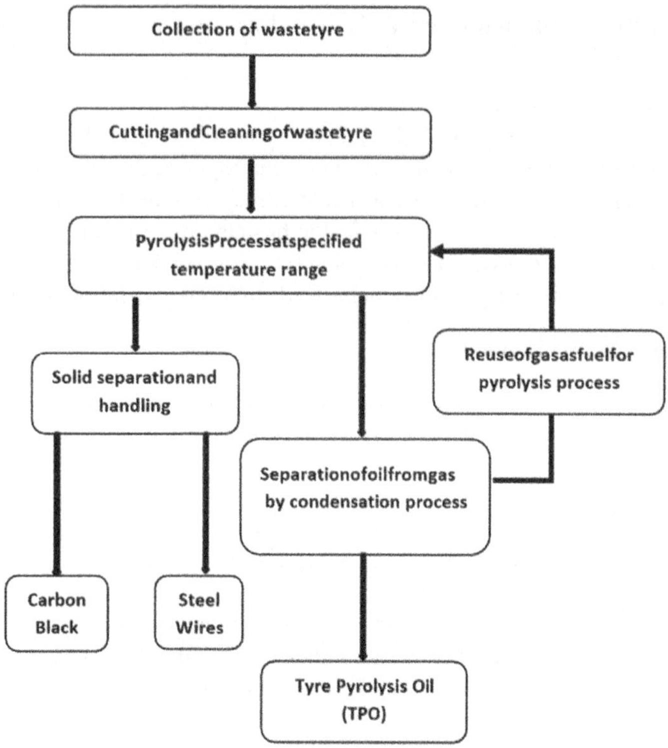

Fig. 1 Methodology followed during generation of oil from waste tyre

collection, cost of the waste tyre will be very less as it does not have functional quality. After collecting waste tyre, segregation according to its size is followed; the content of tyre vary according to its type and manufacturer; for experimental purpose, sample tyres were collected from a local garage.

3.2 *Cutting and Cleaning of Waste Tyre*

The used tyre normally contains elastomer, carbon black, steel, zinc oxide and sulphur. Waste tyre collected may contain dust particle and oil traces which may affect heat transfer rate during pyrolysis process so cleaning of this tyre is required, and it was done with the help of water followed by high-pressure compressed air. Then, these tyres were cut into small pieces to accommodate more quantity inside the reactor; this will also improve heat transfer rate during pyrolysis process, and small particle will lead to high heat transfer rate to reduce pyrolysis time but at the same time, endothermic reaction will increase due to small particle size which increases energy consumption.

3.3 *Tyre Pyrolysis Plant*

Test set-up consists of structure, reactor, gas burner, stirrer and condenser. Figure 2 shows CAD model of pyrolysis plant and actual test set-up.

- Structure

Prepared from angles and the frame is sturdy enough to hold the reactor and condenser on it. The size of the frame is 630 mm × 320 mm × 180 mm and is made up of angles made from MS. The frame is fully bolted to easy assembly and modifications. The mounting for the burner and reactor is a separate assembly which is bolted to the frame.

- Reactor

It is a large vessel-like structure which is used for holding the shredded tyre pieces into it. Reactor for this project has a capacity of 10 litres and is made up of stainless steel.

The dimensions of reactor are as follows:

ID = 220 mm; OD = 224 mm; base thickness = 7 mm; and internal height = 280 mm

Reactor/pressure vessel has three outlets on the top. The central outlet contains stirrer, out of the remaining outlet one is used for collecting the gases produced from the pyrolysis process, and the other outlet is used to mount a temperature gauge which was used for monitoring the temperature in the reactor.

- Gas Burner and LPG

For this process of producing tyre pyrolysis oil, heat is very important; so for heating the reactor, we have used standard LPG gas. The gas burner will supply the required quantity of gas and the burning will take place; the burner is made of cast iron.

Fig. 2 CAD model of tyre pyrolysis plant and test rig for pyrolysis of waste tyre

- Stirrer

Once the heating is started as soon as the temperature rises to a point where the gas production starts, the tyre present in the pressure vessel starts to settle at the bottom of the reactor. Continuous stirring ensures that the base of the pressure vessel is not covered completely because if the entire base is covered then the efficiency of heating will reduce. Stirrer ensures an overall constant rate of heating. Apart from this, because of continuous stirring, it is possible to obtain gas at a constant pace which ensures that the process is continuous.

- Condenser

Once the gases start generating due to thermal decomposition of tyres, to condense this condenser is used. Condenser made up of copper tube which has a size of 3/8″ The tube is a soft copper tube which is coiled in a rectangular shape. This is normal steel box structures; an outlet is provided at the bottom. Water having a low temperature enters from the top and leaves through a drain valve provided at the bottom of the condenser. The gases which entered the condenser form the reactor will get condensed in the form of liquid, and non-condensable gases were taken out with a pipe and burned in a free area.

- Temperature and Pressure Sensor

A K-type temperature is used for measuring the internal temperature of the reactor. The range of temperature sensor is from −50 to −1250 °C. The pressure sensor is Bourdon tube pressure gauge which measures pressure up to 7 Bar. These sensors are needed to monitor reactor temperature and pressure for the safety of the reactor.

3.4 Working of Pyrolysis Plant

It is the process in which shredded tyres are heated up to 500 °C in reactor in the absence or limited presence of oxygen, which led to thermal decomposition of the tyre to give various gaseous, liquid and solid outputs. The whole set-up was assembled. The leak test was done by converting water into steam and condensing it back to the liquid state. After a successful leak test, the actual experiment was conducted.

We started with 1 kg of shredded tyres. The tyre started melting around after 1.5 h of heating and fully melted around 4 h. The vapour generation of tyre also started after that and we got TPO after 6 h of heating.

Table 1 Engine specifications

Engine	Make Kirloskar, Model TV1, Type 1 cylinder, 4 stroke Diesel, water cooled, power 5.2 kW at 1500 rpm, stroke 110 mm, bore 87.5 mm. 661 cc, CR 17.5
Dynamometer	Type eddy current, water cooled
Software	"Enginesoft" Engine performance analysis software
Piezo sensor	Range 5000 PSI, with low noise cable
Crank angle sensor	Resolution 1 Deg, speed 5500 RPM with TDC pulse
Data acquisition device	NI USB-6210, 16-bit, 250 kS/s
Temperature sensor	Type RTD, PT100 and Thermocouple, Type K
Rotameter	Engine cooling 40–400 LPH; calorimeter 25–250 LPH
Load sensor	Load cell, type strain gauge, range 0–50 kg
Fuel flow transmitter	DP transmitter, range 0–500 mm WC
Load indicator	Digital, range 0–50 kg, supply 230 VAC
Airflow transmitter	Pressure transmitter, range 250 mm WC

4 Performance Testing on Kirloskar Diesel Engine (Model TV1)

Oil generated from waste tyre blended with diesel as it has fuel properties similar to diesel, and this blend is tested on the diesel engine to analyse performance characteristics. Tyre pyrolysis oil blended in proportion as 25:75 with diesel by volume. Major parts of the set-up are single-cylinder four-stroke engine, and dynamometer used is eddy current type.

Sensors were used for measurements of combustion pressure and crank angle. For water flow measurement, rotameter was used. With the help of this set-up, various engine performance parameters can be analysed which are brake power, indicated power, frictional power, brake mean effective pressure, indicated mean effective pressure, brake thermal efficiency, mechanical efficiency, volumetric efficiency, specific fuel consumption, A/F ratio and heat balance (Table 1).

5 Results and Discussion

Tyre pyrolysis oil generated during the experiment was tested for its various fuel properties [8]; calorific value of this oil is seen near to the calorific value of diesel. Below table shows various properties of generated tyre pyrolysis oil (Tables 2 and 3).

Two test fuels have been taken for the performance test, which are 100% diesel fuel and 75% diesel with 25% tyre pyrolysis oil (TPO 25). Below table shows the performance results for blended fuel.

Table 2 Properties of tyre pyrolysis oil

Sr. No.	Test	Unit	TPO	Diesel	Method
1	Density at room temperature	g/ml	0.926	0.830	ASTM D 4052
2	Flash point, by PMCC	°C	33	64	IS 1448 (Part 21)
3	Fire point, by COC	°C	89.5	71	IS 1448 (Part 69)
4	Kinematic viscosity at 40 °C	cSt	5.94	2.700	ASTM D 445
5	Gross calorific value	MJ/Kg	42.50	39	D6751
6	Pour point	°C	−24	−40	ASTM D 97
8	Sulphur content	%	0.92	0.02	IS 1448 (Part 33)
9	Cetane value	–	47	49	D613

Table 3 Performance results for diesel and TPO: diesel blend

Fuel	Brake power (kW)	Brake specific fuel consumption, kg/kWhr
100% diesel	0.45	0.86
	0.55	0.81
	0.65	0.80
75% diesel + 25% DTPO	0.45	0.88
	0.55	0.85
	0.65	0.84

At 0.45 kW, the BSFC is 0.86 for diesel, whereas for blending of 25% TPO with diesel, it is 0.88. At 0.55 kW, the BSFC is 0.81 for diesel, and for blend of 25% TPO with diesel, it is 0.85. At 0.65 kW, the BSFC is 0.80 for diesel, for blends of 25% TPO with diesel, it is 0.84. The BSFC of TPO-diesel blends is higher compared to diesel fuel. Increase in BSFC occurred due to low calorific value of TPO. This behaviour was because the engine consumes more fuel with TPO-diesel fuel blends than diesel Fuel, to get the same output as of diesel fuel.

6 Conclusion

Pyrolysis process was carried out in temperature range 500–600 °C. Gases evolved in this process were condensed to get oil, which has properties similar to that of diesel. Properties considered were cetane number, flash point, fire point, pour point, viscosity at 40 °C, calorific value, carbon content, sulphur content and density. Results of the study show that tyre pyrolysis liquid generated during the process have the properties like that of diesel. These properties could be improved by various chemical processes. Odour of this oil is very bad and hence chemical treatment is required to reduce its odour.

This TPO can be blended with diesel to run on diesel engine without any design change, but efficiency is comparatively low and high emission limits its direct use. It could be used after improving its fuel properties by using processes like distillation.

References

1. Khodier A, Williams K, Dallison N (2017) Challenges around automotive shredder residue production and disposal. Waste Manag 73:566–573
2. Directive 2000/53/EC of the European Parliament and of the Council on end-of life vehicles, 18 September 2000
3. Automotive Industry Standard AIS 129 on End-of-Life Vehicles by Ministry of Road Transport and Highways, Government of India (2015)
4. Guidelines for Environmentally sound management of end of life vehicles (ELVs) By Ministry of Environment Forests and Climate Change Central Pollution Control Board Government of India, New Delhi (2015 August)
5. Chen K-c, Huang S-h, Lian I-w (2010) The development and prospects of the end-of-life vehicle recycling system in Taiwan. Waste Manag 30:1661–1669
6. Sakai S-i et al (2013) An international comparative study of end-of-life vehicle (ELV) recycling systems. J Mater Cycles Waste Manag 16:1–20. Springerlink.com https://doi.org/10.1007/s10163-013-0173-2 16 Aug 2013
7. Nayak KR, Auti S (2019) Reviewing the problem of ELVs in India and checking possibilities of pyrolysis as a solution. In: Vasudevan H, Kottur V, Raina A (eds) Proceedings of international conference on intelligent manufacturing and automation. Lecture notes in mechanical engineering. Springer, Singapore
8. Rushdi AI, BaZeyad AY, Al-Awadi AS, Al-Mutlaq KF (2013) Chemical characteristics of oil-like products from hydrous pyrolysis of scrap tires at temperatures 150–400. Fuel 578–584

Development of Mathematical Model for Reduction of Process Time for Peddle-Driven Sewing Machine

Swapna Ghatole, Yashpal, Mahesh Bundele, and J. P. Modak

Abstract By using response surface curve (RSM) method of regression, a model is developed. In the present investigation, a fairly large data of workstations is collected. Based on collected data in terms of independent parameters and dependant parameters, the mathematical co-relationships of effects as function of causes are established. Since the data based on which these relationships are established is a field data, the deduced models are conceptualized as field data-based models. Pedal sewing operation is a man–machine system, comprising of (1) an operator, (2) pedal sewing machine, (3) workstation, and (4) environmental conditions. This operation takes place at a workstation usually known as single person owned premises meant for stitching clothes of the clients of the shop owner. The field study indicated that there is a considerable musculo-skeletal disorder on the part of the operator. Hence, it was decided to execute an assignment in which substantially large field data of this operation is collected. Based on this data, mathematical co-relationships are established between various causes and effects of this activity. The paper reports on: (1) Planning of collection of data. (2) Execution toward collection of data. (3) Presentation of collected data. (4) Processing of the data for formulation of cause–effect relationships. Upon formulation of model, the reliability of model and its optimization is executed. The data collected comprises of 30 sewing machine operators working on stand-alone pedal-driven sewing machines. The collection of data was done on 36 causes/input parameters/independent variables and 1 effects/output parameters/dependant variables. The causes are clubbed and converted into 5 independent dimensionless terms known as Pi terms such as Π_1, Π_2, Π_3, Π_4, and Π_5. The effects/output/dependant parameter is nomenclated as Z1.

S. Ghatole (✉) · Yashpal
Research Scholar, Poornima University, Jaipur, India
e-mail: swapnaghatole72@gmail.com

M. Bundele
Poornima College of Engineering, Jaipur, Rajasthan, India
e-mail: maheshbundele@poornima.org

J. P. Modak
Priyadarshini College of Engineering, Nagpur, Maharashtra, India
e-mail: jpmodak@gmail.com

© Springer Nature Singapore Pte Ltd. 2020

H. Vasudevan et al. (eds.), *Proceedings of International Conference on Intelligent Manufacturing and Automation*, Lecture Notes in Mechanical Engineering,
https://doi.org/10.1007/978-981-15-4485-9_53

Keywords Sewing machine · Mathematical model · Causes · Effects · RSM concept

1 Introduction

This approach is the actual working observed and noted down of a particular type of man–machine system taking place in the society at various locations. The complete data of the thirty man–machine working system is gathered. Data comprises of all independent parameters (causes) of activity, all dependent parameters (effects) of activity, and all extraneous variables of the activity. Based on this data, the analytical relationships are formulated. These analytical relationships are considered here as field database modeling. Several workstations of pedal sewing machine operation are identified with the variation of causes of the system over a certain specific range. The various responses of the system are identified for every workstation. Based on this gathered data, the analytical co-relationships between causes and effects are formulated. These analytical relationships are termed in this investigation as field database modeling for sewing machine operation. The optimization of these models is logically expected to indicate magnitudes of causes/independent variables/inputs to be set during the operation of pedal-driven sewing machine so as to realize the targeted performance of the operation.

2 Materials and Methods

Independent and dependant parameters are identified and nomenclated [1, 2, 3, 4, 5, 6, 7]. They are shown in Table 2. In this work, dependant quantities are grouped into dimensionally homogeneous groups. As per these groups, independent and dependant Π terms are defined. Rewriting all the quantities into dimensionless formula, total five independent Pi terms and one dependant Π term are defined. These are as follows (Table 1).

Independent and dependant quantities
See Table 2.

Table 1 Pi terms

S. No.	Pi terms
1	Π_1 related to anthropometric data of operator
2	Π_2 related to personal data of operator
3	Π_3 related to environmental conditions of workplace
4	Π_4 related to specification of workstation
5	Π_5 related to specifications of subsystems/parts of pedal sewing machine and their condition
6	Z1 related to process time

Table 2 Nomenclature of independent and dependant quantities

S. No.	Name of variable	Units	Symbol	Dimensional parameters
1	Operators height	cm	A1	*L*
2	Operators weight	kg	A2	*M*
3	Sitting of height	cm	A3	*L*
4	Sitting eye height	cm	A4	*L*
5	Elbow sitting height	cm	A5	*L*
6	Knee height	cm	A6	*L*
7	Buttock knee height	cm	A7	*L*
8	Popliteal height.	cm	A8	*L*
9	Elbow-to-elbow span	cm	A9	*L*
10	Hip breadth sitting	cm	A10	L
11	Inclination of head	(α)	A11	0
12	Inclination of eyeball (β)		A12	0
13	Inclination of back with respect to table (θ)		A13	0
14	Age of operator (years–months)		P1	*T*
15	Experience of operator (years–months)		P2	*T*
16	Skill of operator		P3	*T*
17	Health of operator		P4	Rating scale (1–0)
18	Enthusiasm of operator		P5	Rating scale (1–10)
19	Habits of operator		P6	Rating scale (1–10)
20	Ambient temperature at workstation		Tw1	00
21	Noise of workplace		Nw1	Decibels
22	Noise of workplace while working		Nw2	Decibels
23	Illumination of workplace		Iw1	Lumens
24	Height of stool (cm)		H1	*L*
25	Width of stool		W	*L*
26	Surface area of stool		SA	*L2*
27	Ht. of working table	cm	H2	*L*
28	Surface area of working table	cm^2	H3	*L2*
29	Back support length	cm	H5	*L*
30	Area of paddle	cm^2	H6	*L*
31	Arm reach (elbow to working table) (workstation)	(cm)	H7	*L*
32	Speed of operator (take in operator)		S0	$T-1$

(continued)

Table 2 (continued)

S. No.	Name of variable	Units	Symbol	Dimensional parameters
33	Length of leg (take in operator)	cm	LL	L
34	Angle of seat (if any) (workstation)	degree	D11	0
35	Size of treadle (area)	cm	Td	L^2
36	Length of push rod	cm	Psrd	L
37	Axle size	cm	Ax	L
38	Diameter of leather belt	cm	Bd	L
39	Pulley diameter. On m/c head	cm	Pmc	L
40	Diameter of big pully V-belt drive pulley	cm	Fwh	L

3 Mathematical Modeling

Here, the methodology for model definition known as response surface procedure (RSM) [13, 14] is utilized, which relates the connections among the needy variable and the free factors in a second-request condition. The relapse examination was executed to appraise the reaction work as a second-order equation:

$$\mathrm{Yk} = Ao + Ai\,\pi\, ini = 1 + Aii\,\pi\, ni = 12 + Aij\,\pi\, inj = i \\ + 1\pi\, jn = 1j = 1 + \mathrm{e} \quad (1)$$

where Yk = π6 is the anticipated reaction, k = process time, $A0$, Ai, Aii, and Aij are steady coefficients evaluated from relapse and e is an irregular blunder. They speak to the direct, quadratic, and intelligent impacts of πi, $\pi i2$, and $\pi i\pi j$ on reaction variable π6. Upgrading the reaction variable Yk, it is accepted that the free factors are persistent and controllable by the examination individual with immaterial error. The quality of fit of the second-order polynomial was given by the coefficient of determination R^2. Usually, a best-fit polynomial is confirmed. The parameters of the polynomials are estimated by the method of least squares. A statistical software package MATLAB2018 is used for regression analysis of the data obtained and to estimate the coefficient of the regression equation. The investigation aims at deciding the most optimum conditions at which the response variables will have desired values for the reduction of processing time with maximization of work. In this approach, the phenomenon is plotted in three-dimensional rectangular coordinate form observing right-hand screw rule. The effect or the dependent variable is assumed to be a random variable. In the sewing machine process, the independent Pi terms Π_1, Π_2, Π_3, Π_4, and Π_5 have been grouped into two groups. It is necessary to find a suitable combination of Pi terms X (product of $\pi1$, $\pi2$, $\pi3$ Pi term) and Y (product of $\pi4$, $\pi5$). The observed response $Z1$ is a function of the X and Y.

For $Z1 = f\ (\Pi_1, \Pi_2, \Pi_3, \Pi_4, \Pi_5)$ the various combinations of independent Pi terms are considered. These combinations are considered by using multiplication and/ or divisions of independent Pi terms (Ref: Table 3), which can be expressed as

$$\mathrm{Yk} = f(\mathrm{X}, \mathrm{Y}) + \mathrm{e} \tag{2}$$

The $Z1$ is the nomenclature for process time. It is the term of

$Z1$	(Tshl/Shl) * Stl where
Tshl	Time required per length of shirt
Shl	Stitching length of shirt, and
Stl	Stitching length per day

Accordingly, the independent Pi terms Π_1, Π_2, Π_3, Π_4, and Π_5 have been grouped in two groups, for example, on OX-axis, log of the combination of $((\Pi_1 * \Pi_2)/\Pi_3)$ is plotted and on OY-axis, log of combination of (Π_4/Π_5) is plotted and the log of Z1 is plotted on OZ-axis. This results into three-dimensional mapping of activity under study. One is required to get the simplest possible such three-dimensional mapping. For this, the approach would be to consider various combinations of independent quantities in log form on OX-axis and OY-axis. The log of dependant quantity is plotted on OZ-axis. This will yield several three-dimensional maps of the activity under study. The simplest shape is finally chosen and the corresponding algebraic representation of correlation of dependant quantity as a function of independent quantities is considered for three-dimensional form of the model. The approach for choosing the simplest possible form of three-dimensional mapping is as under. Decide algebraic representation of three-dimensional mapping for various combinations of independent Pi terms and dependant Pi term. From among these algebraic representations, the one which nearest has a value of $R2$ to 1 is finally chosen as model. The numerical details of the above approach of getting three-dimensional form of the model are presented in Table 3.

The various combinations formed are above the table.

4 Methodology

Graphs are plotted for Log $Z1$ versus all 30 combinations of Π_1, Π_2, Π_3, Π_4, and Π_5 in Ms Excel and MATLAB. The combination of independent Pi terms is plotted on abscissa and corresponding variation of Z1 on ordinate.

Table 3 Combinations of independent Pi terms

S. No.	Combinations of independent Pi terms as OX-axis	Combinations of independent Pi terms as OY-axis	S. No.	Combinations of independent Pi terms as OX-axis	Combinations of independent Pi terms as OY-axis
1	$(\Pi_1 * \Pi_2)/\Pi_3$	Π_4/Π_5	16	$(\Pi_2 * \Pi_4)/\Pi_5$	Π_1/Π_3
2	$(\Pi_1 * \Pi_3)/\Pi_2$	Π_4/Π_5	17	$(\Pi_4 * \Pi_5)/\Pi_2$	Π_1/Π_3
3	$(\Pi_2 * \Pi_3)/\Pi_1$	Π_4/Π_5	18	$(\Pi_2 * \Pi_5)/\Pi_4$	Π_1/Π_3
4	$(\Pi_1 * \Pi_4)/\Pi_2$	Π_3/Π_5	19	$(\Pi_2 * \Pi_3)/\Pi_5$	Π_1/Π_4
5	$(\Pi_1 * \Pi_5)/\Pi_2$	Π_3/Π_4	20	$(\Pi_3 * \Pi_5)/\Pi_2$	Π_1/Π_4
6	$(\Pi_2 * \Pi_4)/\Pi_1$	Π_3/Π_5	21	$(\Pi_2 * \Pi_5)/\Pi_3$	Π_1/Π_4
7	$(\Pi_2*\Pi_1)/\Pi_4$	Π_3/Π_5	22	$(\Pi_2 * \Pi_3)/\Pi_4$	Π_1/Π_5
8	$(\Pi_2 * \Pi_5)/\Pi_1$	Π_3/Π_4	23	$(\Pi_2 * \Pi_4)/\Pi_3$	Π_1/Π_5
9	$(\Pi_1 * \Pi_2)/\Pi_5$	Π_3/Π_4	24	$(\Pi_3 * \Pi_4)/\Pi_2$	Π_1/Π_5
10	$(\Pi_3 * \Pi_4)/\Pi_5$	Π_1/Π_2	25	$(\Pi_1 * \Pi_3)/\Pi_5$	Π_2/Π_4
11	$(\Pi_3 * \Pi_5)/\Pi_4$	Π_1/Π_2	26	$(\Pi_1 * \Pi_5)/\Pi_3$	Π_2/Π_4
12	$(\Pi_4 * \Pi_5)/\Pi_3$	П1/П2	27	$(\Pi_3 * \Pi_5)/\Pi_1$	Π_2/Π_4
13	$(\Pi_1 * \Pi_4)/\Pi_5$	Π_2/Π_3	28	$(\Pi_1 * \Pi_3)/\Pi_4$	Π_2/Π_5
14	$(\Pi_4 * \Pi_5/\Pi_1$	П2/П3	29	$(\Pi_1 * \Pi_4)/\Pi_3$	Π_2/Π_5
15	$(\Pi_1 * \Pi_5)/\Pi_4$	Π_2/Π_3	30	$(\Pi_3 * \Pi_4)/\Pi_1$	Π_2/Π_5

As suggested by various studies [8, 9, 10, 11, 12, 13, 14], R^2 value nearing to 1 can be suggested as very perfect fit data as given by the theoretical logic of response surface modeling. R^2 is a statistical value, giving information about the goodness of fit of a model. In regression, the R^2 is the value for coefficient of determination in a statistical measure, showing how well the regression line approximates the real data points. An R^2 of 1 indicates that the regression line perfectly fits the data. In general, we cán model the expected value of Y as an nth degree polynomial, yielding the general polynomial regression model [13, 15]

$$y = \beta 0 + \beta 1x + \beta 2 \times 2 + \beta 3 \times 3 + \cdots + \beta n \times n + \varepsilon$$

5 Model Formation

$(\Pi_1 * \Pi_4) / \Pi_2$ (X)	Z1	Π_3 / Π_5 (Y)
−0.57371	5	−2.39283
−0.47822	5.128709	−2.33637
−0.42903	5.079181	−2.30686
−0.19178	5.077557	−2.07395
−0.19088	5.304801	−2.04554
−0.1645	5.350442	−2.04003
−0.12261	4.845098	−2.02743
−0.11192	5.130334	−2.01215
−0.06096	5.350442	−2.00645
−0.05047	5.253532	−1.97154
−0.035	5.120574	−1.94083
−0.02719	4.886491	−1.9286
−0.02719	5.304684	−1.91424
−0.02086	5.077441	−1.82194
−0.01742	4.995635	−1.76779
−0.01702	4.995635	−1.76572
−0.00574	5.077441	−1.7594
0.005383	5.253648	−1.7189
0.012537	5.217484	−1.69451
0.016061	4.995635	−1.65056
0.030117	5.174351	−1.63653
0.030911	5.019449	−1.63076
0.040314	5.041393	−1.57314
0.045844	5.350442	−1.54943
0.06501	5.217484	−1.54304
0.070324	4.944483	−1.51843
0.086782	4.995635	−1.51019
0.113412	5.171726	−1.35405
0.133943	5.217484	−1.33961
0.378974	5.041393	−1.30093

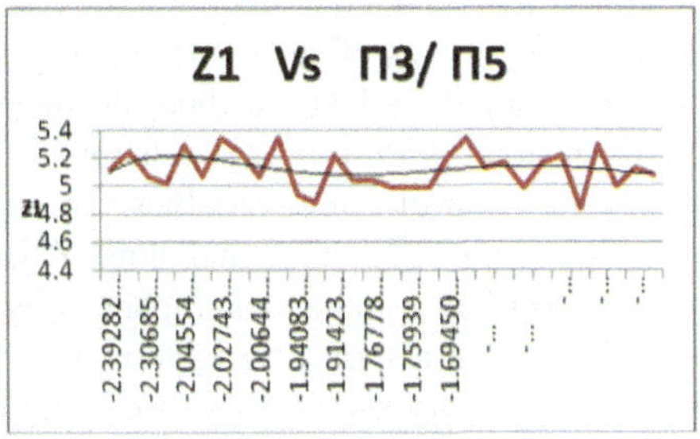

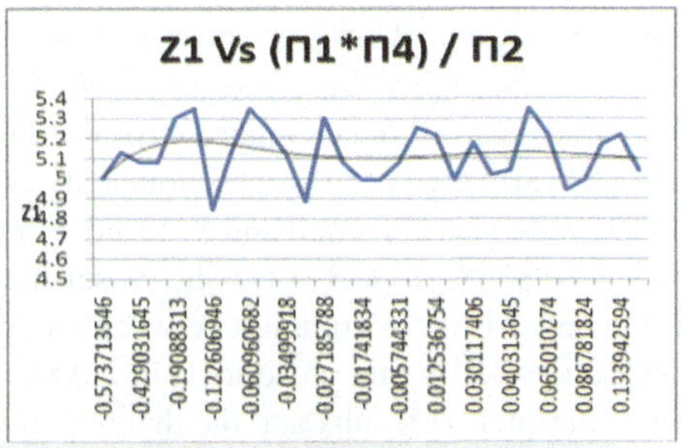

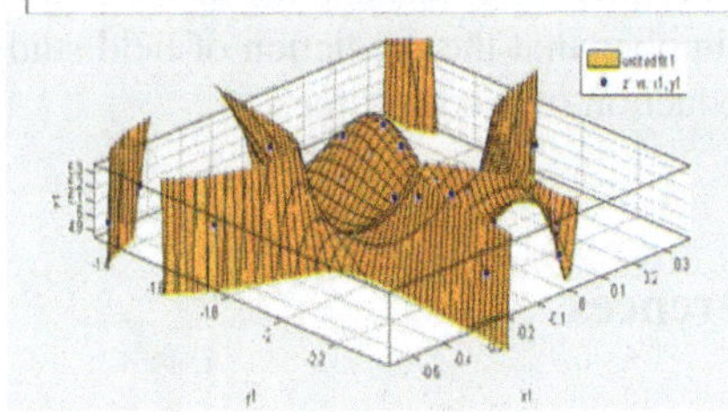

To examine whether the data is well fitted in the model or not by using the above methodology for model of $Z1$, the graph with R^2 value nearing to 1 is selected from all 30 combinations. The R^2 value for Z1 is 0.8668, and the other values of SSE is 0.075. Adjusted R^2 is 0.04 and RMSE value is 0.091 where x is normalized by mean −0.04983 and std 0.1859 and y is normalized by mean −1.804 and std 0.2883

The combination of graph 4 is selected for the model.

$$\begin{aligned} Z1 = f\{f(x,y)\} &= 4.902 + 0.03569x - 0.2643y + 0.2493x^2 \\ &+ 1.775x * y + 0.4338y^2 - 0.02723x^3 + 0.8961x^2 * y \\ &- 0.5514 * x * y^2 - 0.06586 * y^3 - 0.473x^4 \\ &- 3.07 * x^3 * y + 0.1488x^2 * y^2 - 0.455x * y^3 - 0.1046y^4 \\ &+ 0.175x^5 + 0.4755 * x^4 * y + 1.651x^3 * y^2 - 0.7349x^2 * y^3 \\ &- 0.1174x * y^4 + 0.01659y^5 \end{aligned}$$

6 Conclusion

The present work gave an insight into the use of dimensional analysis. The investigation can be applied to decrease the quantity of cause variables used to optimize the process time as effect variable using response surface methodology. Utilizing the dimensional analysis method, 36 input/cause variables have been lessened to 05 dimensionless Pi terms. This helps in constructing a response surface approximation function of a few variables. These 05 Pi terms are further grouped in two variable groups of X and Y and along Z the output variable Pi6 is set as variable to response surface method. The higher-order response surface model for 05 Pi terms and 01 effect or response variable is developed from the gathered data field study. One model equation was developed using RSM in MATLAB R2018a software. To test whether the data is well fitted in the model or not, the values of SSE, R, R^2, adjusted R^2, and RMSE are observed. In general, the more appropriate regression model is the higher the values of R^2 (R^2 is correlation coefficient) and the smaller the values of root mean square error (RMSE). From the developed models, calculated RMSE value of the regression analysis on cycle time is l within 1% limits, i.e., 0.091, which is very small and R^2 value for response variable(process time) is around $87\% = 0.8668$, respectively. The closer the value of R^2 (correlation coefficient) to 1, the better is the correlation between the experimental and predicted values. Here, the value of R^2 being close to 1, i.e., 0.8668 indicated a close agreement between the experimental results and the theoretical values predicted by the model equation. This implies that the prediction of field study data is quite satisfactory for helping the reduction of process time.

References

1. Modak JP et al (2011) An approach to simulation of a complex field activity by a mathematical model. Ind Eng J II(20):11–16
2. Chopade IK (2003) Mathematical simulation modeling of working of an enterprise, March 2003, VNIT, Nagpur, Ph. D. Thesis of I. K. Chopade under supervision of Dr. J.P. Modak, VNIT Nagpur
3. Bansod SV (2005) Evaluation of existing seats on the basis of ergonomic criterion of seat design, July 2005, Ph.D. Thesis of S. V. Bansod under supervision of Dr. J.P. Modak, S. G. Amaravati University, Amaravati
4. S. G. Patil, Assessment of Present Productivity and suggesting Improvement in the Working of Manufacturing Operations in a Garment Manufacturing Unit||, October 2005, Ph. D. Thesis of S. G. Patil under supervision of Dr. J.P. Modak, S. G. Amaravati University, Amaravati
5. Awate AU (2007) Ergonomics study of animal drawn weeders for raw crop planting in Western Vidarbha, September 2007, Ph. D. Thesis of A. U. Awate under supervision of Dr. J.P. Modak, S. G. Amaravati University, Amaravati
6. Bihade OS (2011) Ergonomics evaluation of construction activity, July 2011, Ph. D. Thesis, R.T.M. Nagpur University under the supervision of Dr. J.P. Modak and Dr. S. S. Kulkarni
7. Mishra SP (2011) Mathematical simulation of man-machine-system of some construction activities, October 2011, Ph. D. Thesis, Sant Gadge Baba University, Amaravati, under the supervision of Dr. D. K. Parbat and Dr. J.P. Modak

8. Shank H Jr (1967) Theories of engineering experimentation‖. McGraw Hill Inc., New York
9. Modak JP (2015 May) Overview of mathematical modeling. Lecture No. 1, Poornima University, Jaipur, India, 1, 2 May 2015
10. Murrell KFH (1956) Nature of ergonomics. Ergonomics (Man In His Working Environment), Chapman and Hall, London, New York
11. Eastman Kodak Co. Ltd (1983) Chapter V Environment. Ergonomic design for people at Work, Van Nustrand Reinhold, New York
12. Mahakalkar SG, Tatwawadi VH, Giri JP, Modak JP (2015) Corrugated box production process optimization using dimensional analysis and response surface methodology. Int J Eng Sci Adv Technol [IJESAT] 3(3):96–105. ISSN:2250-3676
13. Waghmare SN, Undirwade SK, Sonde VM, Singh MP, Sakhale CN (2014) Formulation and comparison of Experimental based Mathematical Model with Artificial Neural Network Simulation and RSM (Response Surface Methodology) Model for Optimal Performanceof Sliver Cutting Operation of Bamboo. Procedia Materials Science 6:877–891.
14. Awari GK, Sakhale CN, Modak JP, Kadu RS (2014) Formulation of Mathematical Model for the Investigation of tool wears in Boring Machining Operation on Cast Iron using Carbide and CBN tools. Procedia Materials Science 6:1710–1724.

Design and Development of an Anti-rolling Mechanism for Hand-Driven Tricycles

Vishal Nadar, E. Narayanan, Greegory Mathew, and Rameshbabu Udayar

Abstract Most disabled individuals depend on hand-driven tricycles for their commute between their house of residence and office. They often need to drive over flyovers or slopes during commute. Road conditions or other traffic situation may sometimes force the disabled driver to halt the tricycle on the slope. Starting the tricycle from this stop condition is more difficult as the tricycle has a natural tendency to roll down along the slope. This paper discusses the design and development of an anti-rolling mechanism to prevent the self-roll of the tricycle in the reverse direction. The mechanism is based on the principle of working of a pawl and ratchet. Solidworks software was used to model the mechanism and to evaluate the induced stresses and strains. Calculation for limiting values of deceleration to prevent self-roll is presented for different slope inclinations.

Keywords Tricycle · Pawl and ratchet · Anti-rolling

1 Introduction

Physical disability generally impairs an individual's ability to move around. Such individuals find it difficult to travel to office independently or to visit nearby stores to meet their general needs. Various researchers have developed diverse models of hand-driven tricycles, electrically operated tricycles [1], hybrid tricycles [2], solar-powered tricycles [3, 4], etc., to address these mobility issues of the disabled individuals. Electric, solar and hybrid tricycles have managed to considerably reduce the human effort required to propel the tricycle; however, they are unaffordable. Hand-driven models are comparatively economical. Therefore, most disabled individuals continue to rely on the hand-driven tricycles despite the great effort required to propel them.

V. Nadar (✉) · E. Narayanan · G. Mathew
Dwarkadas J. Sanghvi College of Engineering, Vile Parle, Mumbai, Maharashtra 400056 , India
e-mail: vishalnadarayya@gmail.com

R. Udayar
R. Vishal Fabrics Pvt. Ltd., Boisar, India

© Springer Nature Singapore Pte Ltd. 2020
H. Vasudevan et al. (eds.), *Proceedings of International Conference on Intelligent Manufacturing and Automation*, Lecture Notes in Mechanical Engineering,
https://doi.org/10.1007/978-981-15-4485-9_54

Most disabled individuals travel to their places of work by hand-driven tricycles. In cities like Mumbai, they may regularly need to drive over flyovers or other slopes along their travel route. Climbing slopes are always a difficult task as extra effort has to be applied to work against the gravitational force. Sometimes, road conditions like potholes, diversions, etc., or traffic situations may make it necessary to halt the tricycle on a slope. Starting the tricycle from this stop condition is more difficult as the tricycle has a natural tendency to roll down along the slope. Hence, there is a need for a mechanism that will prevent the self-roll of the tricycle in the reverse direction while climbing a flyover or a slope. The anti-roll mechanism when engaged will ensure that the tricycle will only move forward and never backwards. In worst case, the tricycle will remain at its position if the human effort applied is insufficient.

2 Literature Review

Pawl and ratchet mechanisms and their modifications have been used successfully by many researches and product developers to transmit power and motion in one direction. These mechanisms prevent the motion and the power transfer in the reverse direction. Pawl and ratchet-based anti-roll back mechanisms have been developed for vehicles moving on hilly roads [5, 6]. Pawl and ratchet-based automatic braking systems with integrated sensors to automatically prevent the rolling back of cars have been developed [7]. Electronically operated pawl ratchet mechanism, automobile reverse locking differential mechanism and pawl and ratchet-based self-locking wheel mechanism have also been developed to prevent the rollback of vehicles [8–10].

After reviewing the literature and followed by extensive deliberations, a ratchet and pawl-based mechanism were developed to prevent the self-roll of the tricycle in the reverse direction on slopes.

3 The Mechanism and Its Working

In this work, ratchet and pawl mechanism are used to arrest the backward rotation of the tricycle wheels on a slope. Figure 1 shows the 3D model of the tricycle with anti-roll mechanism. The components of the mechanism are listed in Table 1 and indicated in Fig. 2.

The ratchet, a toothed wheel similar to a chain sprocket, is welded to the wheel hub of the rear wheel. This prevents relative rotational motion between the ratchet and the wheel. The ‘v’ pawl is pivoted on the tricycle frame using the pawl pin. The stopper pin, which engages with the ratchet teeth, is fixed to one arm of the ‘v’ pawl using a hexagonal head bolt. The gravitational force on the other arm of the pawl keeps the stopper pin in contact with the ratchet teeth. The lifting lever is pivoted on the tricycle frame using lever pin. The pawl and the lever pins are fixed to the

Fig. 1 3D model with anti-roll mechanism

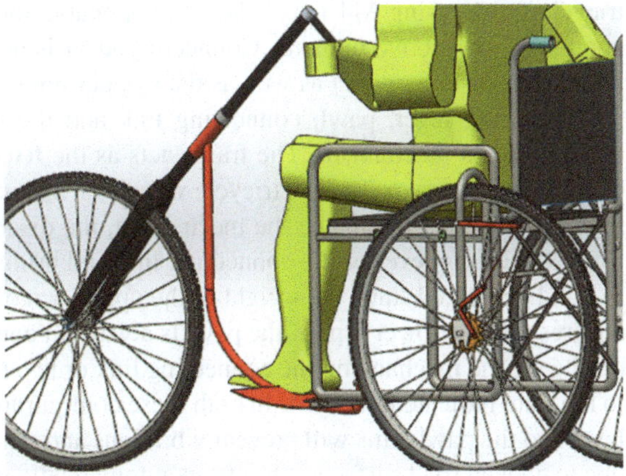

Table 1 Mechanism part list

Part No.	Component name	Material	Quantity
1	Ratchet	Mild steel	01
2	V Pawl	Mild steel	01
3	Stopper pin	Stainless steel	01
4	Connecting link	Mild steel	01
5	Lifting lever	Mild steel	01
6	Lever pin	Stainless steel	01
7	Pawl pin	Stainless steel	01
8	Nuts (M6)	Stainless steel	03

Fig. 2 Anti-roll mechanism components

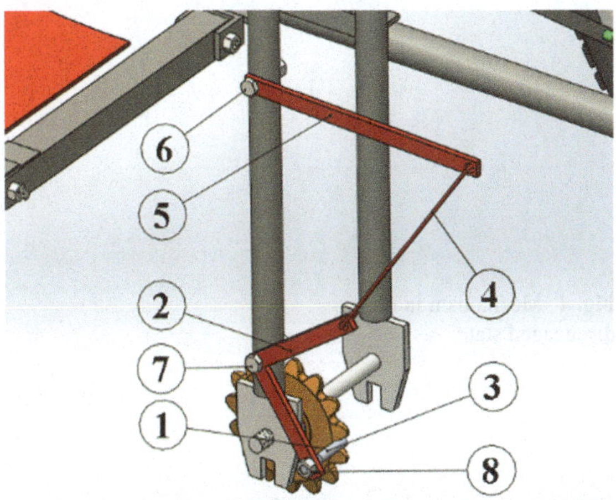

tricycle frame using M4 nuts. Theses pins enable the pawl and the lever to rotate about their respective pin axis. Connecting lever is used to connect the lifting lever and the pawl. The ratchet as well as the stopper pin is case hardened to 50 HRC.

The lifting lever, pawl, connecting link and the tricycle frame form a simple four bar-chain mechanism. The frame acts as the fixed link for the mechanism. To prevent the reverse roll of the tricycle wheels on a slope, the mechanism needs to be in engaged state. To engage the mechanism, the disabled individual has to lift and rotate the lifting lever or the connecting link and bring it to the position indicated in Fig. 3. In engaged state, the weight of the upper arm of the pawl and the lifting lever ensures that the lower arm of the pawl is always in contact with the ratchet teeth. To disengage the mechanism, the connecting link or lever has to be rotated and brought to the state indicated in Fig. 4. In both these configurations, the weights of the lifting lever and the pawl arms will prevent vibrations and other external factors (like forces due to uneven road surfaces, etc.) from automatically switching over the mechanism to the other state.

The pawl stopper pin has a line contact with the ratchet teeth and is allowed to slide over the teeth surface. The stopper pin and the ratchet thus form a higher pair. In engaged state, when the wheel begins to rotate backwards (clockwise), the pawl (with its stopper pin) gets locked between the ratchet teeth. The pawl, ratchet and the frame then form a three-link locked chain and arrest the wheel rotation. This condition is depicted in Fig. 5. However, if the tricycle is moving up a slope, the ratchet is free to

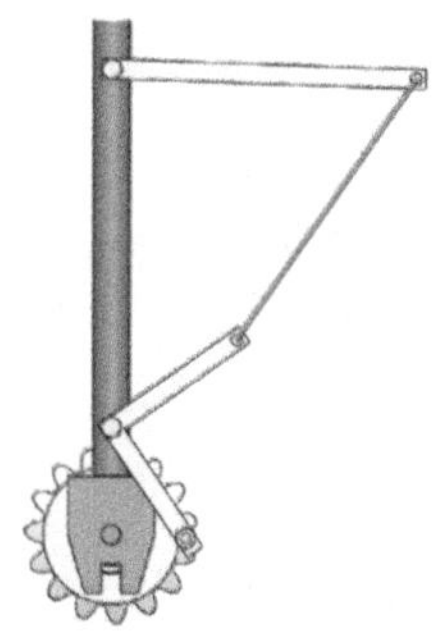

Fig. 3 Mechanism in engaged state

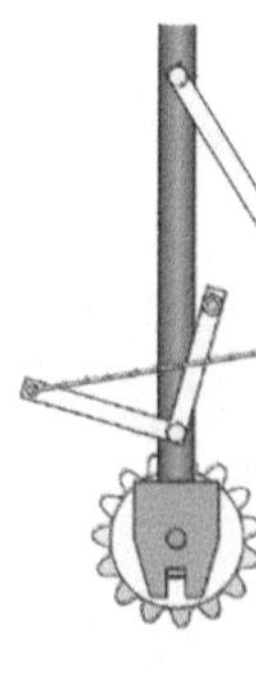

Fig. 4 Mechanism in disengaged state

Fig. 5 Ratchet in locked state

Fig. 6 Ratchet in unlocked state

rotate in the anti-clockwise direction. The stopper pin slides over the tooth surface of the ratchet and allows the wheel to rotate. This condition is depicted in Fig. 6.

4 Calculations

For calculations, the disabled driver was modelled using a mannequin. Figure 7 shows the developed model of a tricycle with a mannequin. The mass of the tricycle is 18.6 kg while the mass of the mannequin is 72 kg. The location of the combined centre of gravity (c.g.) of the tricycle and the mannequin was determined using Solidworks. The coordinates of the c.g. with respect to the origin O are given in Table 2.

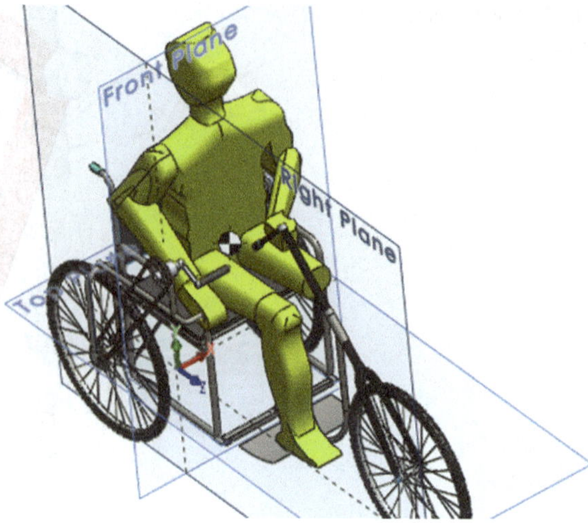

Fig. 7 Location of c.g.

Table 2 Coordinates of the combined c.g.

Axis	X	Y	Z
Distance in mm	261.00	380.15	18.25

The anti-roll mechanism must be engaged only when the tricycle is stationary. If the driver wishes to engage the mechanism while driving up a slope, he is required to halt the tricycle before engaging the mechanism. However, there may arise a situation where the tricycle has begun to self-roll backwards on the slope. The disabled driver may panic and directly engage the anti-roll mechanism before applying brakes. This will lock up the rear wheel. The sudden deceleration may cause the tricycle to roll longitudinally about the rear wheel axis. This situation must be avoided as it will affect the driver safety.

The free body diagram (F.B.D.) when the tricycle is about to roll is shown in Fig. 8. When the tricycle roll is just imminent, the front wheels will lose contact with the ground. The reaction by the ground on the front wheel will therefore be zero. The total weight at the c.g. has been resolved into parallel and perpendicular components along the slope. The abbreviations used in the FBD and in Eq. 1 are listed in Table 3.

The formula for the limiting value of deceleration was derived by applying the three conditions for equilibrium. The derived equation is

$$a \leq g\left(\frac{x\cos\theta}{y} - \sin\theta\right) \tag{1}$$

The limiting value of deceleration for different slope inclinations is presented in Table 4. The values for the limiting gradient have been taken from Table 2.7 of manual of specifications and standards [11].

Fig. 8 Free body diagram

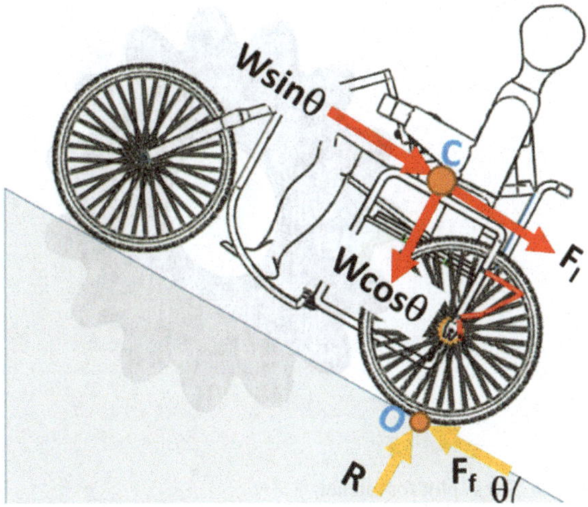

Table 3 Abbreviations used in the FBD

Notation	Abbreviation
θ	Angle of the slope inclination in degrees
a	Deceleration when the mechanism is suddenly engaged in m/s^2
g	Acceleration due to gravity in m/s^2
m	Total mass of the model (tricycle and the mannequin) in kg
W	Total weight of the model (tricycle and the mannequin) in N
C	Centre of gravity for the model
O	Contact point about which the model will roll
R	Normal reaction at the rear wheel in *N*
F_I	Inertia force due to deceleration (i.e. m × a) in N
F_f	Frictional force between the tyre and the ground in N
x	Distance of the c.g. along *X*-axis (from point O) in mm
y	Distance of the c.g. along *Y*-axis (from point O) in mm

Table 4 Calculations for limiting value of deceleration

Nature of terrain	Limiting gradient (%)	Inclination θ in degrees	Distance x mm	Distance y mm	Deceleration a m/s^2
Plain and rolling	5.0	2.86	249.07	741.47	2.80
Mountainous	6.0	3.43	240.41	739.80	2.59
Steep	7.0	4	238.99	739.76	2.48

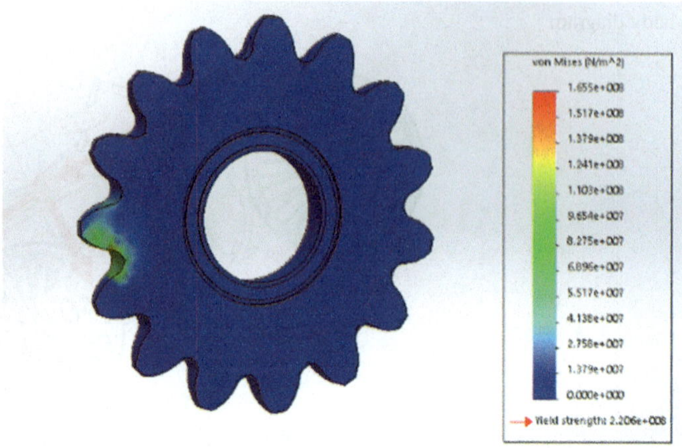

Fig. 9 Stress plot for ratchet

5 Finite Element Analysis

According to The Americans with Disabilities Act of 1990 (ADA), people on wheel chair will find it difficult to manage slopes greater than 4.76°. However, for the purpose of FEA analysis, we have assumed the slope angle to be 10°. The force trying to push the tricycle downwards will then be 886.9 N. If the anti-roll mechanism is engaged, the torque on the mechanism ratchet will be 46.82 Nm. Finite element analysis of the ratchet wheel and the v pawl of the mechanism were performed using Solidworks Simulation.

Figures 9 and 10 indicate the stresses and the deformation in the mechanism ratchet. The maximum stress induced is 165.5 N/mm^2 and the maximum deformation is 0.036 mm. Figures 11 and 12 indicate the stresses and the deformation in the mechanism pawl. The maximum stress induced is 101 N/mm^2 and the maximum deformation is 0.13 mm. Since the maximum-induced stresses in the ratchet and the pawl are less than the yield stress of the selected material (220.6 N/mm^2), both these components are safe from failure.

6 Conclusion

The newly designed anti-roll mechanism prevents the self-roll of the tricycle in the reverse direction while driving up a slope. Stress analysis of the mechanism components rules out any chances of failure. All stresses are within limit with sufficient factor of safety. The limiting values of deceleration to prevent the longitudinal roll of the tricycle along the rear wheel axis are 2.80, 2.59 and 2.48 m/s^2 for slope inclinations of 2.86°, 3.43° and 4°, respectively.

Fig. 10 Displacement plot for ratchet

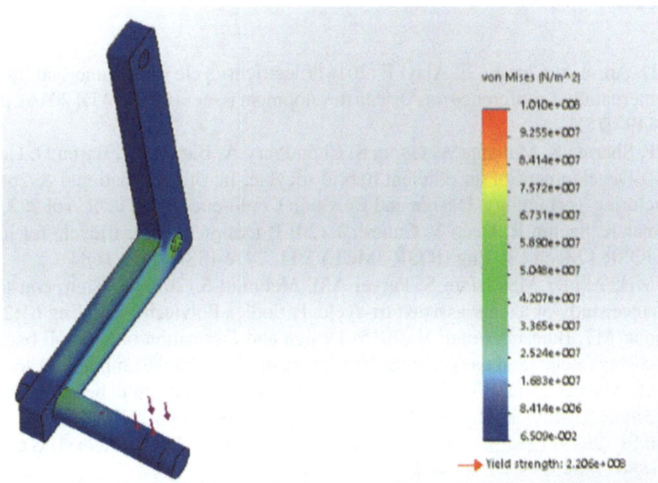

Fig. 11 Stress Plot for 'v' pawl

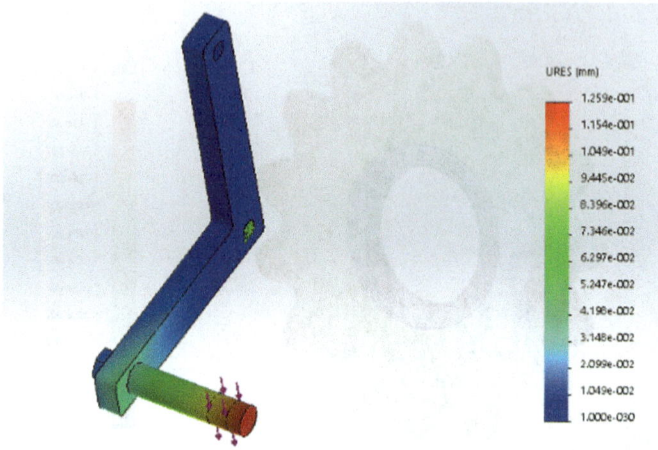

Fig. 12 Displacement Plot for 'v' pawl

References

1. Arinze D, Adisa A, Christie E, Ajayi F (2016) Electric tricycle for commercial transportation. In: 3rd international conference on African development issues (CU-ICADI 2016), pp 376–382. ISSN 2449-075X
2. Dutta PP, Sharma S, Mahanta A, Gupta S, Choudhury A, Barman K, Barua D, Gogoi R, Das A (2014) Development of an efficient hybrid tricycle. In: 5th international & 26th All India Manufacturing Technology, Design and Research Conference, AIMTDR, vol 572, pp 1–7
3. Ravikumar K, Sachin R, Deep V, Ganesh T (2013) Design of solar tricycle for handicapped person. IOSR J Mech Civ Eng (IOSR-JMCE) 5:11–24. e-ISSN 2278-1684
4. Masud MH, Akhter MS, Islam S, Parvej AM, Mahmud S (2017) Design, construction and performance study of a solar assisted tri-cycle. Periodica Polytech Mech Eng 61:234–241
5. Arunkumar MT, Balasubramani V (2015) Design and fabrication of anti-roll back system in vehicles using ratchet and pawl mechanism. Int J Emerg Technol Comput Sci Electron 12:1–4
6. Akash AS, Mahess SP (2017) Analysis of Pawl Ratchet mechanism in heavy vehicles. Int Res J Eng Technol 4:2469–2471. p-ISSN: 2395-0072, p-ISSN: 2395-0072
7. Vishvajit SP (2017) Design and development of intelligent braking system (IBS). Int J Sci Res Dev. 5:1880–1884. ISSN 2321-0613
8. Abhishek SY, Akshay W, Jain C, Kaushik K (2017) Design and fabrication of electronic anti roll back and anti roll front system. Int J Innov Res Sci Eng Technol. 6. ISSN (Online) 2319-8753. ISSN (Print) 2347-6710
9. Mrunmay R, Chetan G, Rahul M, Ashutosh M (2016) Automobile reverse locking differential mechanism. Int Res J Eng Technol 3:1–3
10. Benssin KB, Abraham B, James C, Mathew J, James C (2017) Design and fabrication of self-locking wheel mechanism for manual transmission 4 wheeled automobiles subjected to positive gradients. Int J Innov Res Sci Technol 3:1–5
11. Two-laning of Highways through Public Private Partnership (2007) manual of specifications and standards, Planning Commission Government of India, The Indian Roads Congress

Investigating Red X Parameter for Short Shot-Type Defect in Plastic Injection Moulds Using Shainin's Design of Experiments

Rajendra Khavekar, Hari Vasudevan, and Dharam Ranka

Abstract The study has investigated key parameter(s), which causes short shot-type defect in the case of a plastic injection moulding process using Shainin DoE methodology. An average rejection rate of around 11% was recorded over a period of three months, due to the presence of short shot-type defect for a bulb holder component (E27). Shainin's DoE methodology of Red X, based on progressive elimination search principle, was adopted to identify key parameter (s), which caused such defect among the variables selected for the study. Selective tools from Shainin's DoE methodology were adopted and a particular variation reduction roadmap was prepared to investigate the process. The influential factor identified was the injection time, a solid Red X, i.e. a dominant variation causing variable.

Keywords Product/process search · Variable search · Shainin approach

1 Introduction

Injection moulding is an important polymer processing operation in plastic industries. In this process, polymer is injected into a mould cavity and is allowed to solidify to the shape of the mould required. Optimizing the parameters of the injection moulding process is critically important to enhance the productivity of the process. Rejection rates will be larger, when design and process variables run at high variations from the required tolerance. Shainin as a design of experiments (DoE) tool put forth a statistical technique to control the variation and the philosophy involved is, 'Talk to the parts; they are smarter than the engineers'. Shainin DoE methodology is dominant, because it does not affect the on-line production and the data acquired from the production

R. Khavekar (✉) · H. Vasudevan · D. Ranka
Dwarkadas J. Sanghvi College of Engineering, Vile Parle, Mumbai, Maharashtra 400056 , India
e-mail: khrajendra@rediffmail.com

H. Vasudevan
e-mail: principaldjs@gmail.com

D. Ranka
e-mail: dharam.ranka@djsce.ac.in

© Springer Nature Singapore Pte Ltd. 2020

H. Vasudevan et al. (eds.), *Proceedings of International Conference on Intelligent Manufacturing and Automation*, Lecture Notes in Mechanical Engineering,
https://doi.org/10.1007/978-981-15-4485-9_55

is analysed to spot the suspected source of variable (SSV) and variation for the respective SSV is reduced by optimizing the process parameters [1].

2 Literature Review

Khavekar et al. [2] used Shainin DoE method in their study to find SSV in aluminium casting process and found 86% reduction in the rework of the process. Khavekar et al. [3] compared two DoE methods and concluded that the Shainin Method is very easy to deploy on the shop floor. Khavekar et al. [4] deployed Shainin DoE in NVH testing in automotive industry to find unknown variable responsible for the vaibration. Kiatcharoenpol and Vichiraprasert [5] had used variable search method to find significant parameters affecting the quality of plastic products in their case study. According to them, less numbers of experiments are required in Shainin method and it can be applied for enhancing the quality of manufacturing process. Chitali and Rajiv [6] used Shainin method in their study to eliminate the oil leakage defects in V series diesel engines. Jagdheesson et al. [7] used Shainin's roduct search tool as a clue generating tool and B versus C as a validation tool in their case study to reduce the peak failure load of hot-staked joints in starter motor armature.

3 Problem Statement

An average rejection rate of around 11% was recorded over a period of three months in an injection moulding firm, due to the presence of short shot-type defect for a bulb holder component. Short shot-type defect could largely occur due to lower barrel temperature (zone1 and zone2), insufficient injection pressure, lesser injection time, inadequate degassing and flow of material. With the guidance of the experts in the firm, the parameters identified for analysis in the study were listed as in Table 1, as they could be regulated and controlled.

Short shot-type defect was consistently observed on the component mould. Pareto analysis was done to identify the defect-wise rejection as shown in Fig. 1 and it confirmed that short shot was the major defect observed. Green Y (response factor) selected was short shot-type defect for the component considered, which was of attribute nature.

Table 1 List of selected parameters for the study

Label	SSV	Level
A	Injection time	4 s
B	Injection pressure	48 bar
C	Zone 1 temperature	265 °C
D	Zone 2 temperature	260 °C

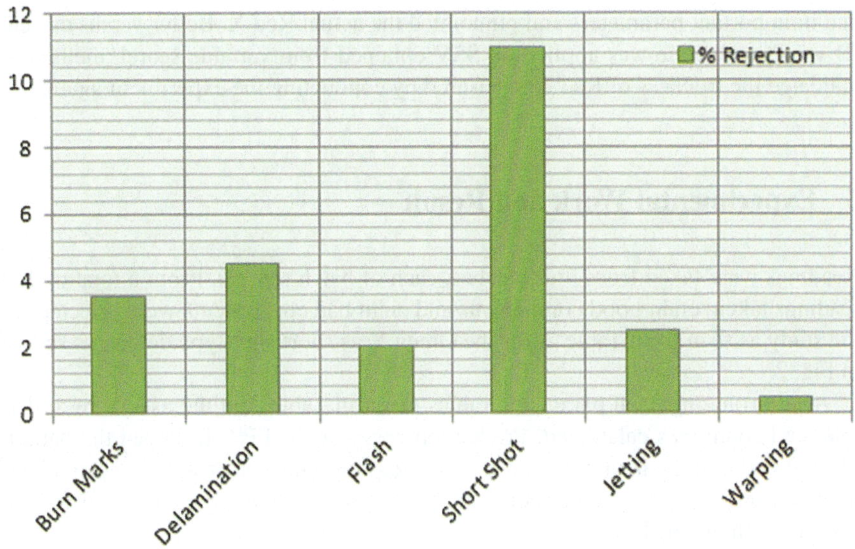

Fig. 1 Major defect by Pareto charting

4 Research Methodology

The first step was defining the problem by historic data. Clue generation tools assisted in identifying the influential SSV and preparing its list. Product/process search (PPS) was the tool selected for clue generation. Variable search (VS) was the formal DoE tool selected for pinpointing the Red X, Pink X and so on. Result validation was achieved by B versus C technique. Product/process search (PPS) separated the important process parameters from the unimportant ones. It followed the pattern of total end count calculation. From trials performed, a batch was selected, which contains 8 good parts & 8 bad parts. The combination of eight good parts and eight bad parts called worst of worst (WOW) and best of best batch (BOB). To identify the influential SSV, the data of each SSV was arranged in the ascending order and the total count (addition of top count and bottom count) with respect to Green Y was calculated. Counting of all batches was made. If the total count (TC) was greater than or equal to 6, then that SSV was the reason for the problem. If the total count was less than 6, then that SSV was not the reason for the problem.

Variable search technique of formal DoE selects the parameters from PPS, which are highly influential, i.e. significant factors with two levels; best (+) setting and marginal best (−) setting. Running the experiment with these settings and calculating the D/d ratio for each SSV were to pinpoint the critical SSV. D is the difference between the median values of the best and the marginal Green Y and d is the average of the two differences (or ranges) within the all best Green Y and the all marginal Green Y. If D/d ratio was greater than 1.25 for SSVs, it indicated that the right process parameters have been captured for analysis. The tool separated the important and unimportant parameters and pinpointed the actual Red X. Better versus current (B vs. C) technique was applied on SSV obtained from variable search method; it validated the trueness of Red X, measured by conducting the experiment again.

5 Experimental Work and Result

Readings were taken from four batches, each of batch size 50. From a total of 200 readings taken, eight good components and eight bad components were selected for the study as seen from Table 2, for the Green Y as short shot-type defect, as shown in Fig. 2.

After arranging each parameter values given in Table 2 in the ascending order, total end count was calculated. Product/process search (PPS) followed the pattern of total end count calculation and it separated important process parameters from unimportant ones. Total end count calculation, whose total count was $\geq$6 for the SSV, is as shown in Table 3.

PPS funnels down the number of SSV from 4 to 2 parameters. These two parameters were further analysed by variable search (VS) tool. With the help of the firm experts, the influential SSVs were set at two levels; best (+) setting and marginal best (−) setting as seen in Table 4, for starting the VS technique. In the VS method,

Table 2 Eight Best of best (BOB) and eight worst of worst (WOW) readings

S. No.	Reading No.	A	B	C	D	Response
1	121	6	50	270	262	B
2	152	6	50	271	261	B
3	69	4	48	268	260	B
4	46	4	48	272	258	B
5	123	6	50	270	262	B
6	49	4	48	271	259	B
7	192	6	50	269	259	B
8	188	6	50	268	258	B
9	11	4	48	266	260	W
10	55	4	48	269	259	W
11	44	4	48	269	261	W
12	72	4	48	268	262	W
13	25	4	48	272	258	W
14	62	4	48	271	259	W
15	29	4	48	270	259	W
16	59	4	48	269	260	W

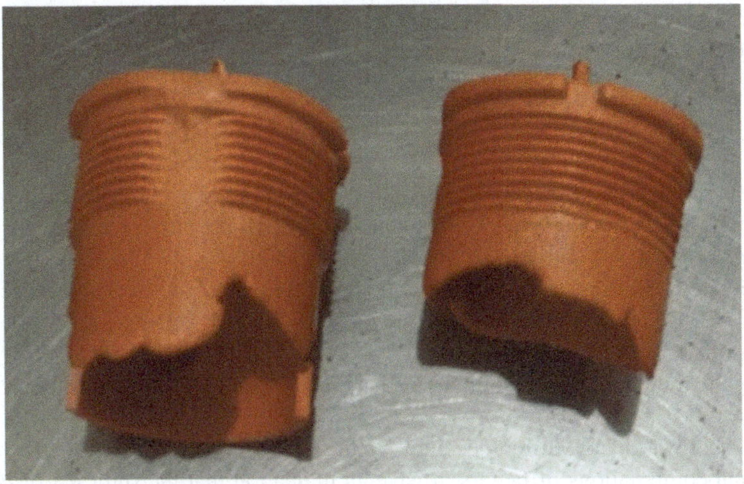

Fig. 2 Component with short shot-type defect

Table 3 Calculation of total count (TC)

A	Response	B	Response	C	Response	D	Response
4	B	48	B	266	W	258	B
4	B	48	B	268	B	258	B
4	B	48	B	268	B	258	W
4	W	48	W	268	W	259	B
4	W	48	W	269	B	259	B
4	W	48	W	269	W	259	W
4	W	48	W	269	W	259	W
4	W	48	W	269	W	259	W
4	W	48	W	270	B	260	B
4	W	48	W	270	B	260	W
4	W	48	W	270	W	260	W
6	B	50	B	271	B	261	B
6	B	50	B	271	B	261	W
6	B	50	B	271	W	262	B
6	B	50	B	272	B	262	B
6	B	50	B	272	W	262	W
TC = 8		**TC = 8**		**TC = 3**		**TC = 2**	

Table 4 Influential process parameter obtained from PPS technique for Green Y with best (+) and marginal best (−) setting

Label	+ setting	− setting
A	6	4
B	52	48

Table 5 First run for VS technique

	All at +	All at −
	Green Y (results)	
Run 1	08	48

initially, running two experiments, first all factors at their best levels and second all factors at their marginal levels were taken as seen in Table 5.

For run 1, all at + settings:

Weighted defect score = number of defective units * defect-type Likert scale = 2 * 4 = 08

For run 1, all at − settings:

Weighted defect score = number of defective units * defect-type Likert scale = 6 * 8 = 48

As there was a large difference between the Green Y's of the all-best and the all-marginal combinations of factors, it gave an indication that the right list of factors was captured. Two more experiments were run with the same setting. This meant that, altogether; there were now three all-best and three all-marginal best levels as seen in Table 6.

For run 2, all at + settings:

Weighted defect score = number of defective units * defect-type Likert scale = 5 * 2 = 10

For run 2, all at − settings:

Weighted defect score = number of defective units * defect-type Likert scale = 7 * 8 = 56

For run 3, all at + settings:

Weighted defect score = number of defective units * defect-type Likert scale = 4 * 1 = 4

For run 3, all at − settings:

Weighted defect score = number of defective units * defect-type Likert scale = 8 * 6 = 48

As all three of the all-best Green Y's were better than all three of the all-marginal Green Y's, with no overlap, the first test of significance was cleared. Now, calculating the D/d ratio for second test of significance: $D = 40$ and $d = 7$, $D/d = 5.71$, which was greater than 1.25:1 indicated that the right process parameters have been captured for analysis. Both the tests of significance were passed, and it was concluded that right factors have been captured, even though the Red X, Pink X, etc., have not been

Table 6 All runs VS results

	All at +	All at −
	Green Y	
Run 1 (initial)	08	48
Run 2	10	56
Run 3	04	48

Table 7 Running each parameter at (+) and (−)

Test	Combination	Results	Median	Decision Limits	Conclusion
1	A−R+	42	8	−02.73 to 18.73	Complete reversal, Red X
2	A+R−	12	48	37.26 to 58.73	
3	B−R+	32	8	−02.73 to 18.73	Partial reversal, B important with another factor
4	B+R−	24	48	37.26 to 58.73	
5	C−R+	16	8	−02.73 to 18.73	C not important
6	C+R−	56	48	37.26 to 58.73	
7	D−R+	14	8	−02.73 to 18.73	D not important
8	D+R−	48	48	37.26 to 58.73	

pinpointed. Now, running a pair of tests for each parameter at (+) best setting and remaining all at (−) marginal best setting and vice versa was done as given in Table 7. Also, calculating the high side and low side of decision limits using the formula: The decision limits were: median ± $(2.776 * d)/1.81$.

As parameter C and D showed results inside the low side and high side of decision limits, factors C and D, along with all of its associate interaction effects, were considered unimportant and it could be eliminated from further study. As there was a complete reversal, i.e. A−R+ became the original all-best level and A+R− became the original all-marginal level, A was the only Red X. Parameter, A, i.e. injection time, was the solid Red X. Now, as both pairs of tests for factor B showed results outside the low side and high side of decision limits, respectively, but not a complete reversal, it could not be eliminated along with its associated interaction effects. Hence, there was definitely the presence of Pink X. On discussion with the firm experts in the firm, 5% risk allowance, i.e. confidence level of 95%, running six more trials, three samples of B and three samples of C were selected to validate the results using B versus C technique. The results of the three B tests, and three C tests, i.e. six pack tests are as given in Table 8. As seen from Table 8, testing was done in random order sequence (run order), the three B's outranked the three C's with 95% confidence (5% risk) and the tool validated the selected parameters.

Table 8 B versus C results

Run order		Rank order	
B or C	Results	B or C	Results
C	52	B	08
B	10	B	10
B	12	B	12
C	40	C	36
C	36	C	40
B	08	C	52

Table 9 Spotted SSV

Label	Label description	Remark
A	Injection time	Red X
B	Injection pressure	Pink X

Table 10 Optimized settings for parameters selected

Label	Label description	Optimized setting	Allowable variation
A	Injection time	6 s	–
B	Injection pressure	52 bar	–
C	Zone 1 temperature	265 °C	± 2 °C
D	Zone 2 temperature	260 °C	± 2 °C

6 Conclusion

Without disturbing the on-line production, Shainin's approach could be implemented easily in the study. Shainin envisages for process improvement using convergent strategies by reducing the variation causing variable(s) with progressive elimination search technique. Narrowing down the critical factor leads to uncomplicated experimental set-up and runs for changing factor levels. Validating and optimizing the process performance by Shainin tool-bag during full production run were of ease as the root cause was known.

Product/process search technique is funnelled down the factors to 2 using Tukey test of end count calculation. These two factors were further taken for study in variable search and the tool has successfully identified the Red X and Pink X and can be seen in Table 9.

The optimized setting obtained after applying Shainin's progressive elimination principle with the guidance of the firm experts can be seen in Table 10.

Shainin DoE is very easy to implement on the shop floor without changing the set-up of the manufacturing process.

References

1. Bhote KR, Bhote AK (2000) World class quality: using design of experiments to make it happen (2nd ed). American Management Association. New York
2. Khavekar R, Vasudevan H, Ranka D (2016) Suspecting dominant variable that causes porosity defect in a cast made up of aluminium alloy (Lm2) using Shainin's DoE approach. Int J Adv Mech Civ Eng 3(5):124–130

3. Khavekar R, Vasudevan H, Modi B (2017) A comparative analysis of Taguchi methodology and Shainin system DoE in the optimization of injection molding process parameters. Mater Sci Eng 225(1):012183–012192
4. Khavekar R, Vasudevan H, Desai H (2017) Application of Shainin DoE tool to explore unknown variables causing ghost noise in 5th gear cycle of transaxles during NVH testing. Adv Intell Syst Res 137:271–276
5. Kiatcharoenpol T, Vichiraprasert T (2017) Implementation of Shainin's DOE: a case of plastic injection molding process. In: Proceedings IEEE IEEM, pp 760–764
6. Chitali G, Rajiv B (2016) Elimination of oil leakage defect in V series diesel engines using the six sigma technique. In: Proceedings of 6th international & 27th All India Manufacturing Technology, design and research conference (AIMTDR-2016), pp 2151–2155
7. Jagdheesson AJ, Karunamoorthy L, Arunkumar N (2015) Investigation and improvement in peak failure load of hot staked joints in Starter Motor Armatures using complementing problem-solving tools. J Mech Sci Technol 29(4):1731–1736

Structural Analysis of the Upright of a FSAE Race Car

Vinayak H. Khatawate, Jinesh Sheth, and Prakriti Tulasyan

Abstract Upright of a FSAE car is normally subjected to various mechanical loads as it handles tire forces as well as the reaction forces from wishbones. This paper aims to analyze the structural integrity of the rear upright for a FSAE vehicle. The upright has been modeled using 3D modeling software SOLIDWORKS and its finite element analysis (FEA) was done using ANSYS. The methodology consisted of generating good quality mesh, considering various mesh quality criteria, applying forces and moments for different load scenarios and further analyzing their effects.

Keywords FSAE · Upright · Statics structural · FEA · Mesh quality

1 Introduction

The FSAE competition challenges teams of engineering students to conceptualize, design, manufacture, and fabricate formula-style race cars. Upright is one of the most complicated and critical design of the suspension system which locates the upper ball joint and lower ball joint of the suspension system and holds the wheel hub with the help of bearings for rotary motion of the vehicle [1]. For the purpose of application on a high-performance FSAE race car, it should be lightweight to keep up good power to weight ratio. Also, it must have optimum stiffness to ensure low system compliance and support design geometries.

The upright transfers the longitudinal, lateral, and vertical forces from the tire to the wishbone in the form of tensile and compressive forces. Various load cases like acceleration, braking, cornering, impact loads, etc., are analyzed and the best compromise between weight and strength is achieved [2].

The basic design process consists of creating or selecting configurations, materials, shapes, and dimension of the upright. The main aim of this paper is to assess the strength of upright used in FSAE race car by using finite element analysis.

V. H. Khatawate · J. Sheth (✉) · P. Tulasyan
Department of Mechanical Engineering, Dwarkadas J. Sanghvi College of Engineering, Vile Parle, Mumbai, Maharashtra 400056 , India
e-mail: jineshsheth66@gmail.com

© Springer Nature Singapore Pte Ltd. 2020

H. Vasudevan et al. (eds.), *Proceedings of International Conference on Intelligent Manufacturing and Automation*, Lecture Notes in Mechanical Engineering,
https://doi.org/10.1007/978-981-15-4485-9_56

2 Literature Review

Strength and durability are utmost important for upright, hub, wheel centers, and rims of the wheel assembly of a race car. FEA provides an exact virtual testing type analysis results of such components before manufacturing them.

Static analysis is used to find the displacements, stresses, and strains of the part in an approximate condition where the applied load does not vary with respect to time. The upright is designed considering the elastic limit of the material as the stress exceeding the elastic limit will result into permanent deformation of the material causing failure of the part [3].

Equations are solved at cell/nodal locations of FEA meshing. For high-solution gradients and fine geometric shell, refined/smaller cells are required. The solution accuracy and stability deteriorate as mesh cells deviate from ideal shape. Good quality mesh means that mesh quality criteria are within correct range, mesh is valid for studied physics, solution is grid independent, and important geometric details are well captured.

This paper thoroughly focuses on the mesh metrics such as element quality, aspect ratio, Jacobian ratio to check the mesh quality of the part to get the solution close to the exact solution.

Forces and moments acting on the tire when the car is in motion are calculated using basic engineering mechanics and vehicle dynamics formulae in MS Excel.

3 Upright Modelling

3.1 CAD Model

3D model of the upright is made using SOLIDWORKS and is shown in Fig. 1. The concepts and considerations before the design of the upright included the size of the wheel bearing used, position of the upper and lower ball joints of the wishbone and positioning of other components of the wheel assembly.

3.2 Material Selection

For selecting the material for the upright, parameters like high tensile strength, more mass density, light weight, more fatigue strength, machinability, and availability are considered. EN24 steel is the best material for the stiffness but it does not favor the weight criteria. Hence, two Aluminum alloys Al 7075 T6 and Al 6061 T6 are shortlisted as they are lighter than EN24 steel. Table 1 tabulates material properties of aluminum alloys.

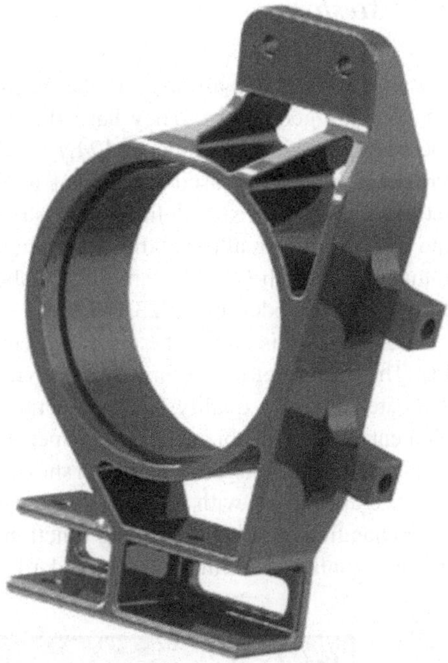

Fig. 1 CAD model of the upright

Table 1 Material properties of aluminum alloys

Properties	Aluminum 7075 T6	Aluminum 6061 T6
Density	2.81 g/cc	2.70 g/cc
Ultimate tensile strength	572 MPa	310 MPa
Tensile yield strength	503 MPa	276 MPa
Fatigue strength	159 MPa	96.5 MPa
Young's modulus	71.7 GPa	68.9 GPa
Machinability	70%	50%

To further aid the selection, as upright undergoes heavy stresses and high fatigue cycles; the tensile yield strength and the fatigue strength of the material have to be sufficient enough to withstand the loads with certain factor of safety. The properties of the aluminum alloys, Al 7075 T6 and Al 6061 T6 are compared in Table 1 which shows every property of Al 7075 T6 better than the properties of Al 6061 T6 with better strength-to-weight ratio [3]. Hence, Aluminum 7075 T6 alloy is finalized.

3.3 Mesh

Tetrahedral elements are used in the meshing process because they can fit in better complex geometries. They have the highest quality factor value among other mesh shapes with the value of 124.7 with others having the value less than 100. Proximity/curvature-based refinement is chosen to resolve the feature where the tetrahedral element size defined on a surface is larger than needed. It subdivides the element into smaller subdivisions automatically. After the refinement, the minimum edge length is 0.023 mm, the number of elements generated is 804,153 and the number of nodes are 1,223,149.

Mesh quality is one of the most important things to consider during the analysis [4]. The mesh metric option helped to view mesh metric information and thereby evaluate the mesh quality. After generating the mesh, the mesh metrics such as element quality, Jacobian ratio, and aspect ratio are analyzed and shown in Figs. 2, 3, and 4. Using the histogram bar graph shown in Fig. 6, areas with low mesh quality are rectified and solved with local mesh refinement. It is solved by checking the values after changing the shape checking function. It is used to increase the general global element quality of the mesh. The standard mechanical shape checking is changed to

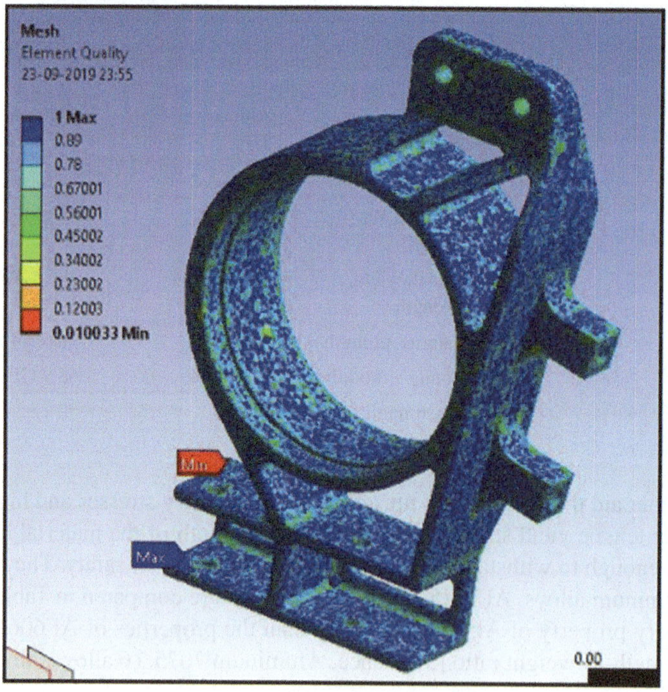

Fig. 2 Element quality of the mesh

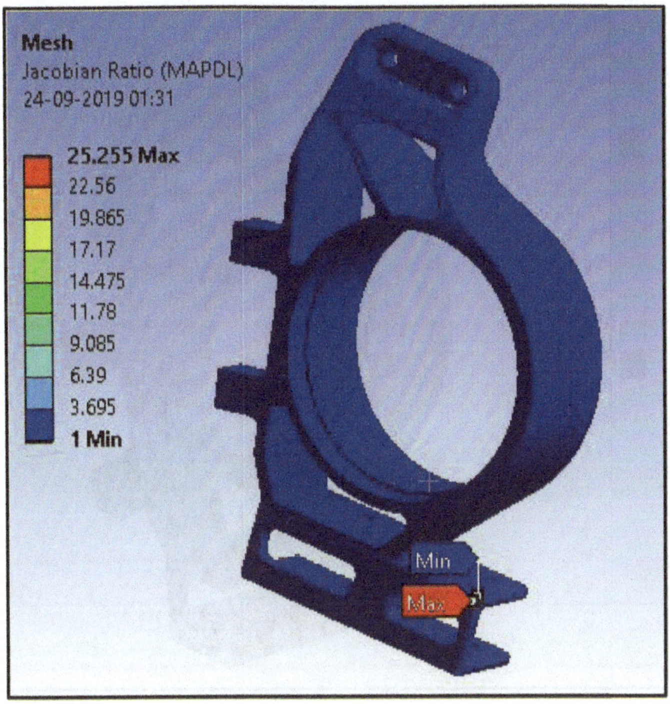

Fig. 3 Jacobian ratio of the mesh

aggressive mechanical and non-linear mechanical thereby concluding the non-linear mechanical gave the best results (Fig. 5).

The mesh quality is finalized by referring Table 2. The final meshed diagram is shown in Fig. 5.

4 Static Structural Analysis

For applying the loads, the magnitude and direction of the forces and moments are calculated using the parameters of the race car.

The extreme load scenario of deceleration and cornering is considered. A braking torque of 163,155 N.mm is added. The longitudinal force (F_x), lateral force (F_y), vertical force (F_z), overturning moment (M_x), rolling moment (M_y), and yaw moment (M_z) are calculated using engineering mechanics and vehicle dynamics formulae [5] as shown below:

$$\text{Load transfer} = \frac{\text{Total Weight} \times \text{longitudinal deceleration} \times \text{CG Height}}{\text{Trackwidth or Wheelbase}}$$

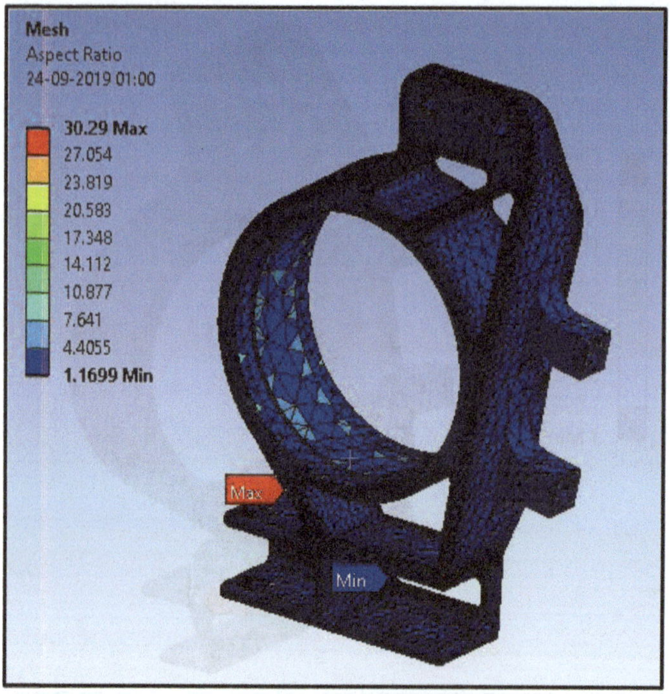

Fig. 4 Aspect ratio of the mesh

Vertical force (F_z)	=	Static load ± Load transfer
Longitudinal force (F_x)	=	Braking friction × vertical force
Lateral force (F_y)	=	Lateral acceleration × vertical force
Braking moment	=	Clamping force × wheel radius.

Table 3 tabulates different load scenarios. The forces and moments transfer from tire contact patch to the upright via tire contact patch, rim, rim bolts, wheel center, lug bolts, wheel hub, wheel bearing and upright. The forces and moments are applied on the remote point (tire contact patch) and its effect is shown at the wheel bearing surfaces on the upright by fixing the 6 degrees of freedom at the upper ball joint, lower ball joint, and toe rod outboard point [6].

The application of forces at remote point, braking moment on caliper mount and moments at remote point are shown in Figs. 7, 8, and 9 respectively.

As shown in Fig. 10, the maximum total deformation is 0.13525 mm. The total equivalent von-Mises stress is 389.64 Mpa with a minimum factor of safety of 1.3859 as shown in Figs. 11 and 12, respectively.

Fig. 5 Final mesh

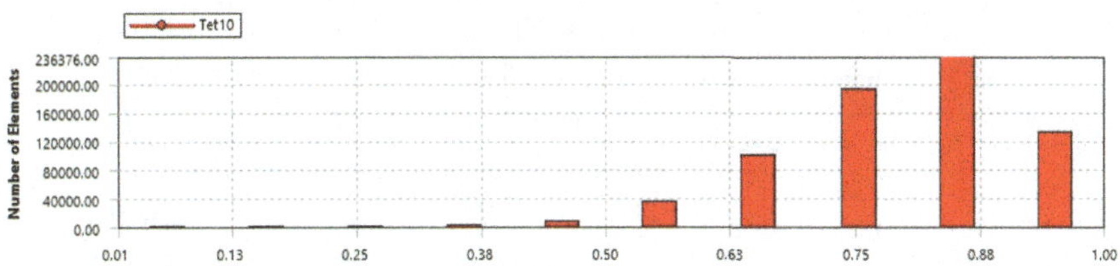

Fig. 6 Histogram of the element quality

Table 2 Mesh metric criteria

Criteria	Bad	Good	Excellent
Element Quality	0.001-0.10	0.20-0.69	0.95-1.00
Aspect Ratio	>10	5-10	0-5
Warping	>0.4	0.2-0.4	0-0.2
Jacobian ratio	>30	10-30	1
Skewness	>0.75	0.25-0.75	0-0.25

Table 3 Different load scenarios

Linear acceleration with cornering			
Units of force N		Units of moment Nm	
Front left tire		*Front right tire*	
F_x	0.000	Fx	0.000
F_y	1452.923	Fy	−149.365
F_z	854.661	Fz	87.360
M_x	2.657	Mx	0.273
M_y	−18.255	My	1.876
M_z	31.034	Mz	−3.190
Rear left tire		*Rear right tire*	
F_x	1785.173	F_x	500.058
F_y	2528.995	F_y	708.416
Fz	1487.644	F_z	416.715
M_x	92.843	M_x	−26.000
M_y	−37.012	M_y	−10.367
M_z	−48.491	M_z	48.834
Linear braking with cornering			
Units of force N		Units of moment Nm	
Front left tire		*Front right tire*	
F_x	3589.461	F_x	1374.533
F_y	2596.632	F_y	994.343
F_z	1527.430	F_z	584.907
M_x	4.750	M_x	−1.819
M_y	32.625	M_y	12.493
M_z	66.627	M_z	16.964
Rear left tire		*Rear right tire*	
Fx	1914.955	F_x	−654.504
F_y	1385.286	F_y	−473.471
F_z	814.874	F_z	278.512
M_x	50.856	M_x	17.381
M_y	20.274	M_y	−6.929
M_z	153.978	M_z	29.067

5 Results

The structural analysis results of the upright of an FSAE race car according to the loading conditions mentioned above are given below:

1. Maximum total deformation—0.13525 mm
2. Total equivalent von-Mises stress—389.64 Mpa

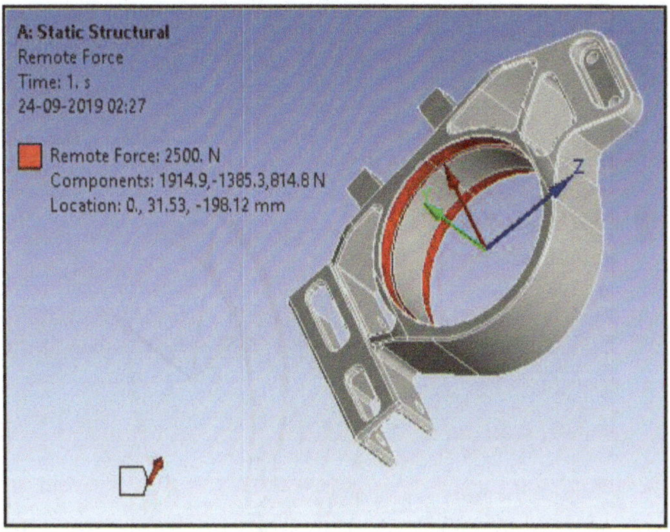

Fig. 7 Application of forces at remote point

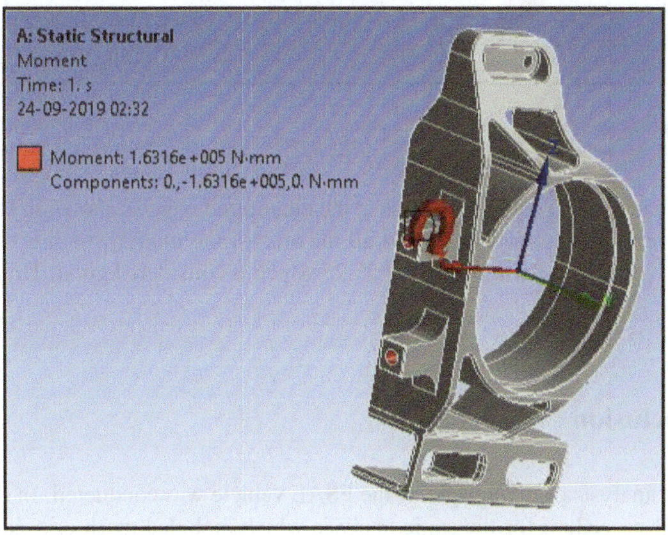

Fig. 8 Application of braking moment

3. Factor of Safety—1.3859.

The maximum equivalent Von-Mises stress of 389.64 MPa does not exceed the tensile yield strength of 503 Mpa for Aluminum 7075 T6. In the mesh metrics, the average element quality is 0.75 which is considered good from Fig. 2. The aspect ratio is below 5 at 90% of the model which is considered excellent. This is shown in

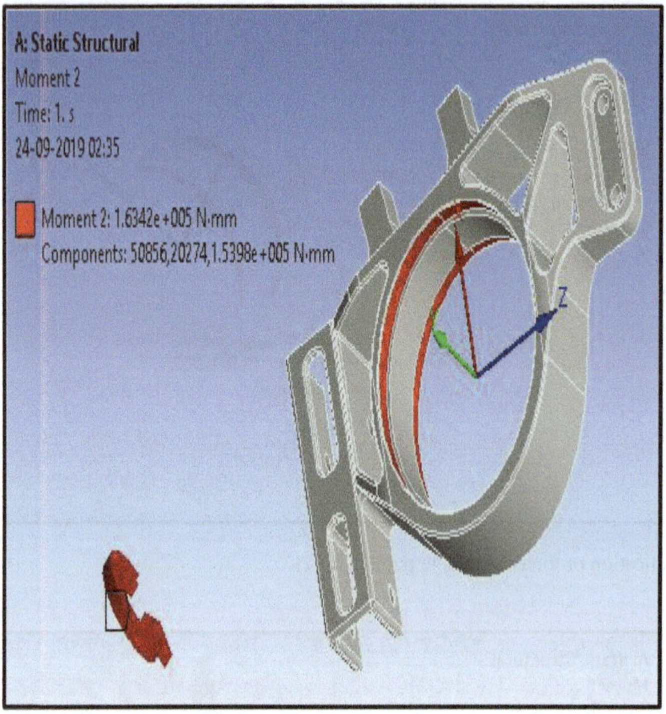

Fig. 9 Application of moments at remote point

Fig. 4. A bad aspect ratio can be seen at the chamfered edge as shown in Fig. 4. The Jacobian ratio is close to 1 at almost all the area of the model which is considered excellent with the maximum value of 25.25 which is considered good. This is shown in Fig. 3.

6 Conclusion

The static analysis of the upright of the FSAE vehicle was conducted. In this paper, the overall procedure for the static analysis which included mesh criteria and load analysis was systematically studied and implemented. From the results of FEA, the maximum equivalent von-Mises stress was found to be less than the tensile yield strength of the material. Hence, it was concluded that the upright had not undergone plastic deformation at any load scenario. Considering the results of the study conducted, it can be concluded that major improvements in terms of static life enhancement, performance prediction, vehicle reliability, weight reduction and other design parameters could be achieved by carrying out the static structural analysis of the components.

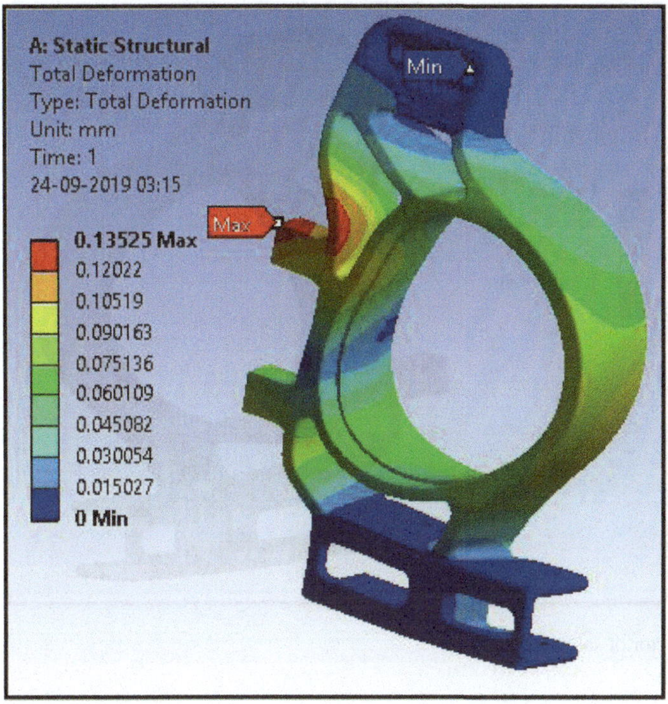

Fig. 10 Total deformation

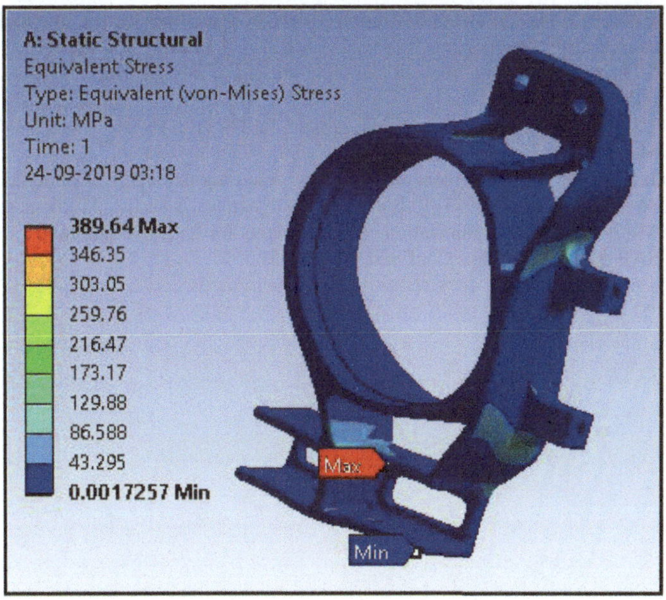

Fig. 11 Equivalent von-Mises stress

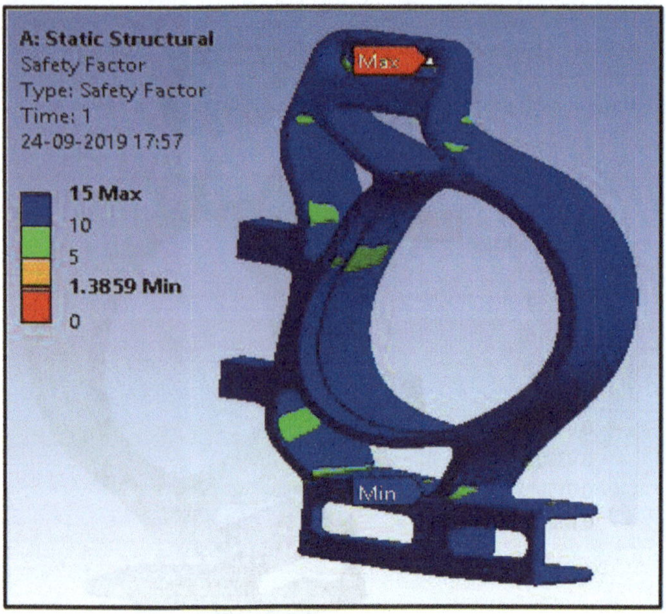

Fig. 12 Factor of safety

References

1. Jen Wong A (2012) Design and optimization of upright assemblies for formula SAE Race car. Int J Mech Indus Eng (IJMIE) 2(1). ISSN 2231 -6477
2. Ranjan R, Dhakar A (2016) Force calculation in upright of a FSAE Race car. Int J Mech Eng Technol (IJMET) 7(2):168–176
3. Garg A (2017) Fatigue analysis of an FSAE upright, Int J Sci Res (IJSR) 6(9):1983–1988. ISSN 2319-7064
4. Bhavikatti SS (2005) Finite element analysis. New Age International (P) Limited Publishers
5. Milliken WF, Milliken DL (1995) Race car vehicle dynamics. SAE Inc. Milliken, pp 678–684
6. Azmeer M (2017) Design optimization of rear uprights for UniMAP automotive racing team formula SAE racing car. J Phys Conf Seri 908:012051

Design and Analysis of Components of a Rotary Car Parking System

Rajnarayan Yadav, Sanjay Kumar, Salil Gavankar, and Suraj Amin

Abstract Parking infrastructure has been unable to keep with the rapid rise in the vehicle sales numbers. This often leads to an increased strain on current parking resources and further causes inconvenience to the general public. Research efforts were channeled into the development of newer technology capable of providing and handling parking spaces to ensure that the current demand is satisfied. This success of the implementation of a parking system depends on various factors viz. effective utilization of area, power requirements, ease of access, and most importantly, cost. One such technique, which ticks all the right boxes is the 'Rotary Car Parking System.' The 'Rotary Car Parking System' features a mechanism that is like the concept of a Ferris wheel, commonly seen in fun-fairs. A modular design, low power consumption, ease of use, and installation are some of its salient features. The parking pallets will be sized, such that they could accommodate vehicles of all dimensions available in India.

Keywords Rotary parking system · Survey · Pallet · Chain-pellet link · Attachment · Frame · Finite element analysis

1 Introduction

The advent of new age technologies coupled with the growing desire for comfort has led to the evolution of vehicles from being just a mode of transportation between A and B to fully fledged cocoons of luxury, comfort, thrill, and ultimately, convenience. This evolution has taken place at a rapid pace in which even the mammoth expansion of city limits could not cope with. In a city like Mumbai with its huge population and paucity of space, the need for parking has increased by leaps and bounds. Citizens travel in personal vehicles to various locations in the city citing the reasons of convenience and comfort instead of utilizing public transportation. Drivers generally need

R. Yadav (✉) · S. Kumar · S. Gavankar · S. Amin
Dwarkadas J. Sanghvi College of Engineering, Mumbai Metropolitan Region (MMR), Maharashtra 400056, India
e-mail: rajnarayan.yadav@djsce.ac.in

© Springer Nature Singapore Pte Ltd. 2020

H. Vasudevan et al. (eds.), *Proceedings of International Conference on Intelligent Manufacturing and Automation*, Lecture Notes in Mechanical Engineering,
https://doi.org/10.1007/978-981-15-4485-9_57

to spend a significant amount of time circling the blocks around their destination searching and waiting for available parking spaces. The demand for parking spaces is generally much greater than the supply due to the lack of a well-planned policy for parking facilities. One such rule states that tenants be provided with parking space proportional to the area of their residence. This has proven to be very difficult to implement in new projects due to the high housing demands and ever increasing sizes of homes. There is a desperate need for a modern technology which can curb the parking problems faced by civilians. This is where our project, the 'Rotary Car Parking System' tries to make its mark. It is a mechanism inspired by the Ferris wheel found at fun-fairs. It consists of multiple platforms or 'pallets' on which the cars are loaded. Each of these pallets is attached using a special attachment to a pair of conveyor chains. The conveyor chains form the basis of the elliptical motion that makes our system much more compact than a circular system. Not only that, the conveyor chains are strong and built for a direct load application just as the system demands. They function as both power transmitting and load bearing elements (Fig. 1).

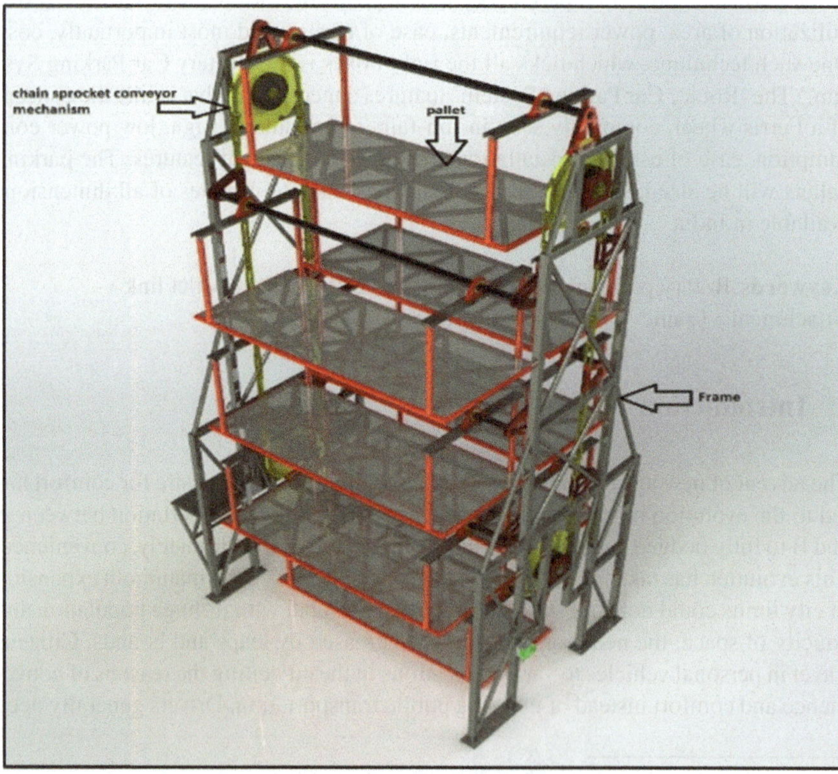

Fig. 1 CAD model of rotary car parking system

2 Literature Review

Due to significant improvements in material science, material selection can be the make or break decision [1]. Materials apart from mild steel must be researched into and analyzed for performance/price ratio before the decision is made. The selection of standard Original Equipment Manufacturer(OEM) components must be optimized by taking performance, cost, and usage criticality into consideration. The bearing housing design is an untouched design issue and in-depth designing and analysis should be conducted. Bearing selection process also plays an important role and should be given its due focus [2].

The platform design is very basic and does not include any modularity as a whole. Designing mass consumer-oriented systems with an overly specific use case is frowned upon today [3]. The inclusion of modularity in the platform design is an absolute must. Modularity may be included via a one size that fits all approaches or real-time updated designs as per requirement [4]. Additionally, platform mountings and supports such as rods, hangers, and bars may be designed to include modularity as well so as to reduce complications [5]. The frame is another component that is not discussed in detail. Weight and size reduction are the top priorities keeping strength and rigidity considerations in check. The coupling methods must be reviewed as well. A direct chain drive requires a large-sized sprocket as well as a high torque prime mover. Incorporating the use of gears, two-stage chain reductions, etc., should be considered during the design phase [6]. The design of the chain drive should expand to include attachments to make designing easier to achieve. The selection parameters should be discussed in depth considering cost, performance, and ease of operation. The chains should be designed to include both power transmission as well as supporting the platforms. Indexing apart from the use of chain drive does not find any application. A proper indexing mechanism must be incorporated into the system if possible to ensure proper arrival and departure of the platforms [7]. Chain misalignment is an issue that has been neglected. The chain must also be provided with guide rails to ensure misalignment does not occur.

The selection of the drive system is of utmost importance as the entire operation of the system depends on the same. Instead of using a standard motor and a standard coupling, the usage of geared motors must be looked into [8]. The drive system must also have inbuilt control system to ensure easy and safe operation of the system. Drive system integration with the conveying mechanism is the most important performance parameter [9].

Computer-aided engineering (CAE) analysis of the designed components must contain in-depth detail regarding the nature of the actual load, its magnitude, and justify the analysis conducted instead of just including images of stresses and deformations in the system. Fatigue analysis of the said components must also be conducted if possible for maximum safety [10].

3 Survey

A survey to establish the need for the 'Rotary Car Parking System' in the Mumbai Mumbai Metropolitan Region (MMR) was created using Google forms. The survey was released to the public and responses were accepted for two weeks. The survey garnered a total of 133 responses from all across the Mumbai MMR region. We believe the survey had a good spread and a variety of permutations and combinations in the responses to justify the inherent randomness every survey must possess.

The survey had helped us gain valuable insight into the mind-set of the potential consumers for this system. We aim to proceed with the design of our project considering the relevant data from the survey to achieve the best consumer satisfaction.

3.1 Summary of Survey

The summary of the data collected from the survey is as follows:

1. More than 90% of the respondents were from the 18–35 demographic.
2. More than 80% of the respondents drove a car.
3. Sedans and Hatchbacks were the most common vehicles.
4. There was almost an equal split between societies with dedicated parking and societies without dedicated parking. They both were the most common responses.
5. Over 60% people complained of parking space issues in their residential complex.
6. Almost 70% people were ready to invest for the installation of the system.
7. 30% of the people were convinced that the Rotary Car Parking System could solve their parking problems while another 50% were of the view that it might be able to solve the parking problem
8. Almost half of the people believe that such a system will fit in their residential complex without any issue. The other half was split almost equally into three parts viz. responses with height as the constraint, responses with length and width as the constraints, and lastly, responses with all three constraints.
9. At least half the people had cost as their main priority. The other two responses, i.e., ease of operations and ease of accessibility were of equal popularity among the respondents.

3.2 Breakdown of Survey

The survey data obtained was thoroughly analyzed to check for trends, relationships, and correlation. The respondents were questioned on the following parameters:

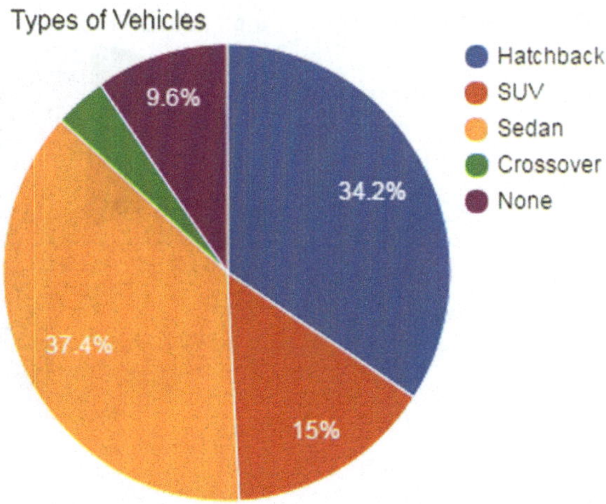

Fig. 2 Types of vehicle

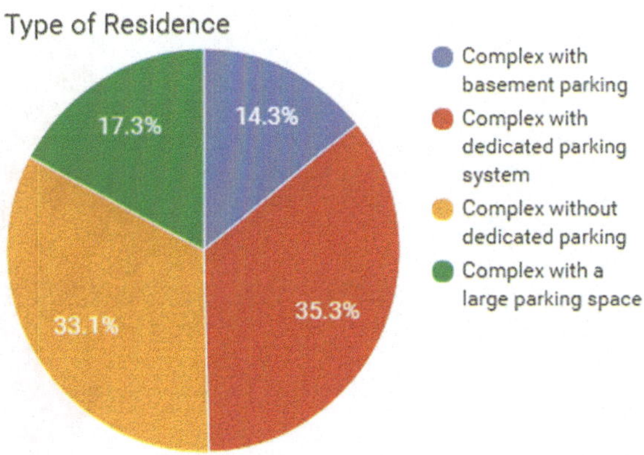

Fig. 3 Types of residence

1. Age group 2. Driving capability 3. Type of vehicle 4. Type of residence 5. Parking space issue 6. Readiness to invest 7. Resolution of parking issues 8. Installation problems 9. Operating parameter preferences

The analysis based on the data led to many inferences that have helped to support the decision-making aspects of the project. We aim to ensure our project meets all the desired criteria according to public perception and need (Figs. 2, 3, 4, 5, 6, 7 and 8).

4 Design

The design phase of the project began after sufficient and detailed analysis of the requirements of the consumers and the constraints of the industry. The research conducted also influenced the design process and decision greatly. We aim to strike a

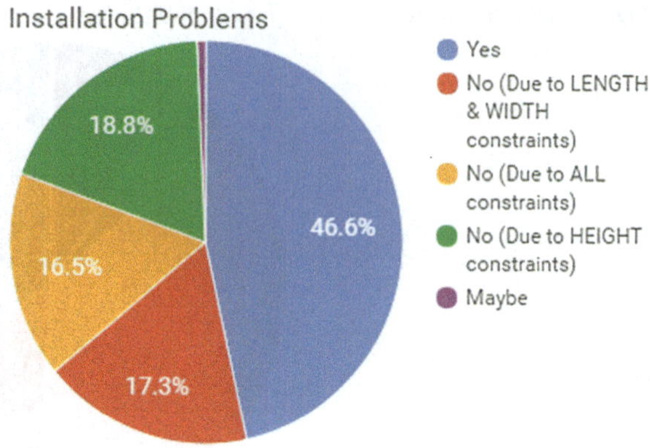

Fig. 4 Installation problems

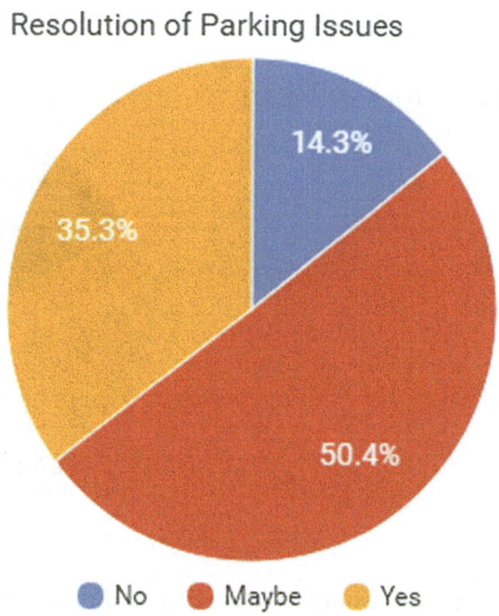

Fig. 5 Resolution of Parking Issues

compromise between these two factors within the available limits to ensure a perfectly acceptable cost/performance ratio. The design phase of the project includes but is not limited to the following sub-assemblies:

1. Pallet
2. Drive system (selection based)
3. Conveying mechanism(selection based)
4. Frame.

The design of each of these components is vital to the operation of the system and hence, great attention to detail needs to be present. After continuous research and brainstorming and by following a diligent methodology, we have progressed in

Fig. 6 Parking space issue

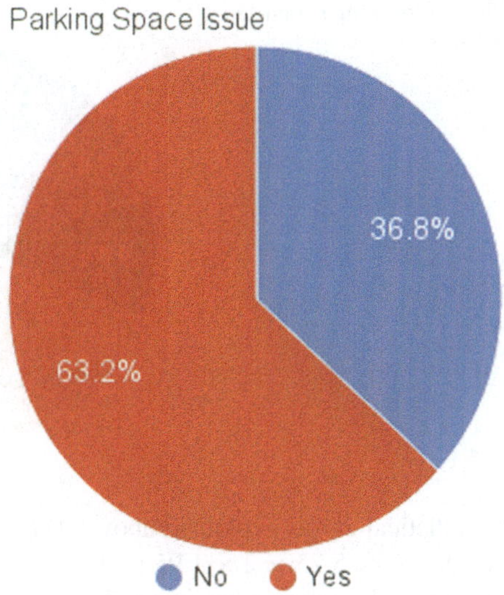

Fig. 7 Readiness to Invest

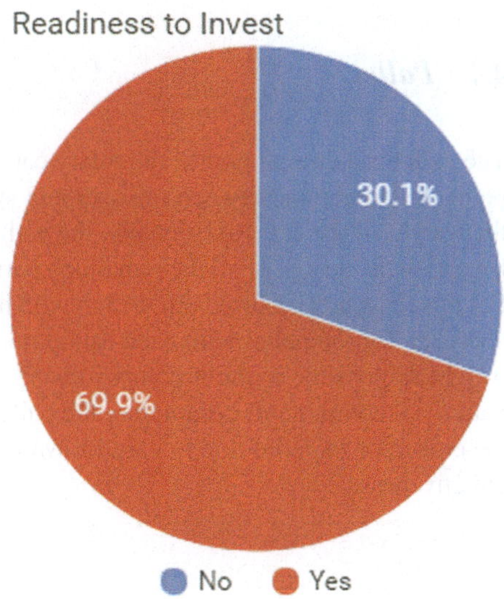

the design of the following components in the system, namely the drive system, the pallet, and the conveying mechanism.

To make our system compatible with all kinds of consumer vehicles, a hypothetical car of dimensions and weight sourced from an amalgamation of the biggest cars available in the country at present was used as a reference in deciding the outer planar dimensions of the pallet. To reduce the difficulties faced by drivers while parking the car in an enclosed space, we ensured that enough clearance was provided on all four sides of the pallet for even the largest car available which we decided to base on a

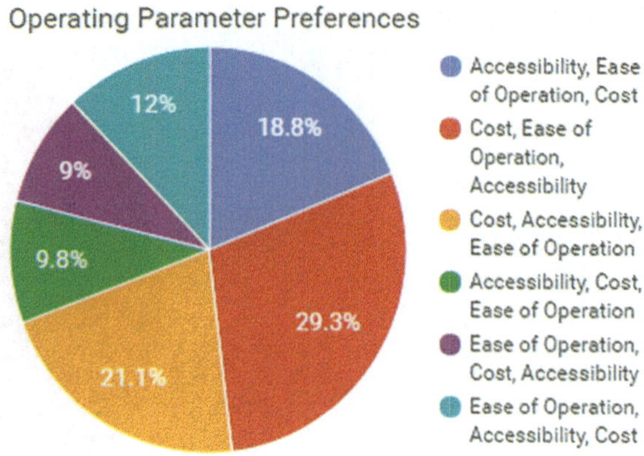

Fig. 8 Operating parameter preferences

hypothetical car as explained above. The dimensions of this hypothetical car were decided to be 5.25 × 2 × 1.95 m (L × B × H) and its mass as 2500 kg. The design of each of the sub systems is detailed in the following texts.

4.1 Pallet

The pallet is a critical part of the rotary car parking system, not just because it is the point of interaction between the user and system but also because it has to support the weight of the car. Additionally, the weight of the pallet itself will determine the forces for the design of all the remaining parts and the power consumed by the motor. For this reason, we started our design with the pallet. We decided to take an iterative approach for the pallet with each future iterations having an aim to reduce as much weight as possible without compromising on the integrity and functionality of the pallet. All components comprising the pallet have been designed with Mild Steel as the choice of material due to ease of availability, machining, weld ability, and cost effectiveness.

4.2 Drive System

The drive system is the heartbeat of the entire system which is entirely on selection from available catalogues. The whole operation of the system depends on the efficiency and smooth operation of the drive system. The drive system consists of many different components that integrate to create a seamless power output. The main components of the drive system are the prime mover, control system, and any reduction units if required.

4.3 Conveying Mechanism

The basic purpose of the rotary car parking system is to convey vehicles in pallets that keep rotating. Thus, the selection of the conveying mechanism is one of the most important design choices to be made. It should be able to convey a fully loaded system at the required speed without any hassles. Our choice of the conveying mechanism is that of a conveyor chain. A conveyor chain is a special kind of chain that is designed for both power transmission and load carrying applications. A vehicle conveyor is one of the prime applications of a conveyor chain.

4.4 Frame

The frame is one of the most important components in the design of the system. It is the component which will ensure proper system integration amongst the various sub-assemblies in the system, namely the sprockets, motor, pallets, etc. The purpose of the frame is to support the weight of all sub-assemblies during static and dynamic conditions as well as withstand and transmit vibrations that emerge during system operation to the ground. It must also provide excellent rigidity to the system as it the basis of assembling the system. Any errors in the frame will cause problems during assembly. The frame is designed keeping in mind the relative position of all the components used in the system (Figs. 9 and 10).

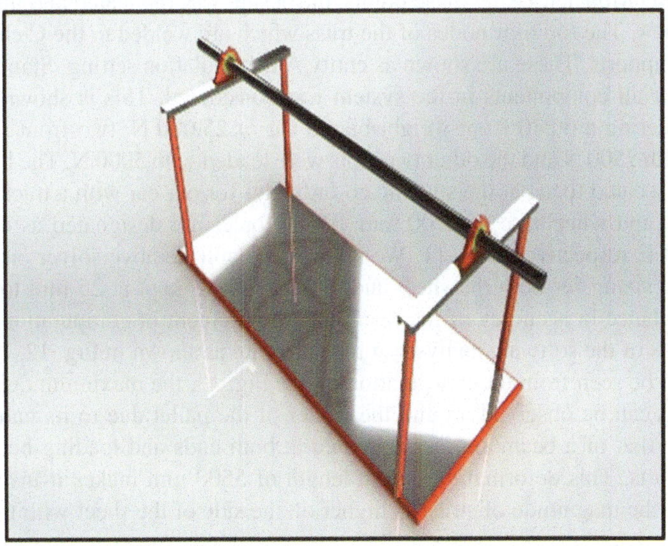

Fig. 9 Pallet

Fig. 10 Frame with motor mount

5 Finite Element Analysis

5.1 Pallet

The pallet was initially modelled on SolidWorks and the nodes were imported to ANSYS 3D Modeler. Using these points, the design was replicated and imported to Model View. The top four nodes of the truss which are welded to the C channel are fixed as supports. These are shown as entity A in simulation setting. Standard earth gravity for all components in the system was considered. This is shown as entity B. Considering a 60:40 front weight-biased car of 25,000 N, two front slots were loaded with 7500 N and the other two slots were loaded with 5000 N. The loads were remotely located to act at the specific co-ordinates for our car with a track width of 1600 mm and wheelbase of 3200 mm. These forces are designated as entities C, D, E, and F, respectively Fig. 11. We used the default iterative solver on ANSYS. The mesh sizing for both the sheet and the beams was kept at 25 mm to ensure a perfect balance in accuracy of the result and requirement of computational power. The results of the software analysis of the pallet are as shown in Fig. 12.

As can be seen from the deformation image Fig. 13, the maximum deformation of 21 mm can be observed around the center of the pallet due to its nature being similar to that of a beam simple supported at both ends and loading between the two supports. This deformation over a length of 5500 mm makes it insignificant. However, the magnitude of stress is higher on the side of the sheet with the greater

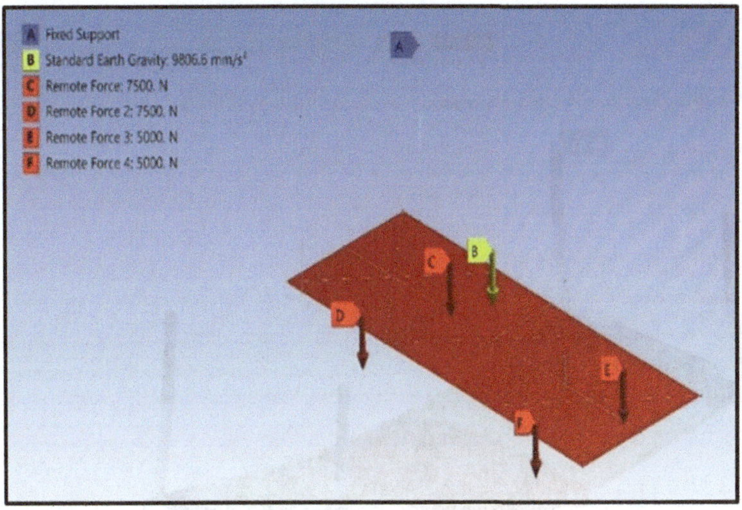

Fig. 11 Pallet analysis: Points of load application

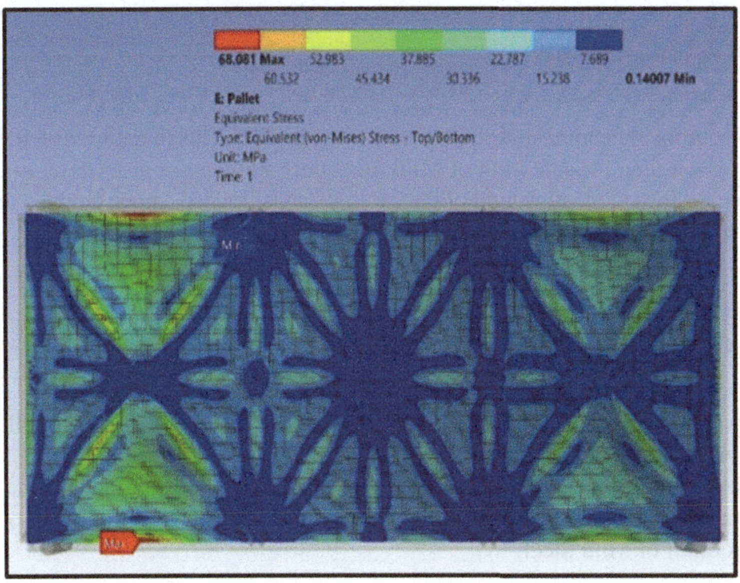

Fig. 12 Sheet stress

loading as the vehicle is a front-biased vehicle, i.e., higher weight of the front half of the car. The maximum magnitude of stress, however, is 68 MPa, well within the safe stress limit for Mild Steel. The critical analysis is the stress on the beams that form the truss structure. The stress is maximum at the center nodes of the truss structure

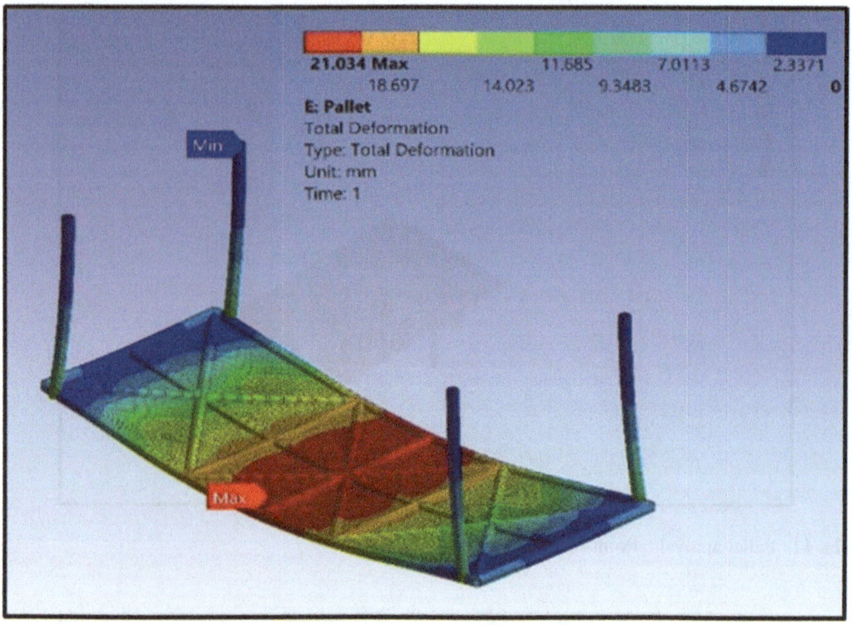

Fig. 13 Pellet (frame) with deformation

due to combined bending and shear stresses acting on it. This magnitude of 108 MPa is also lower than the safe limit of Mild Steel thus allowing us an FOS of around 2. This means the safe load weight of the car that can be loaded can be doubled and the design would still be safe (Fig. 14).

5.2 *Chain-Pellet Link Attachment*

In conveying mechanism, chain will be selected and based on that chain-pellet link attachment can be designed and analyzed as shown in Fig. 15.

The value of the deformation is only 0.15 mm and is insignificant. The more important parameter is the maximum stress induced which at 75 MPa is also under the safe limit of Mild Steel.

5.3 *Frame*

The deformation results show a maximum deformation of 4.6 mm near the bases of the outer supports. All the four bottom supports are fixed and load tries to get the two ends of the frame together (Fig. 16).

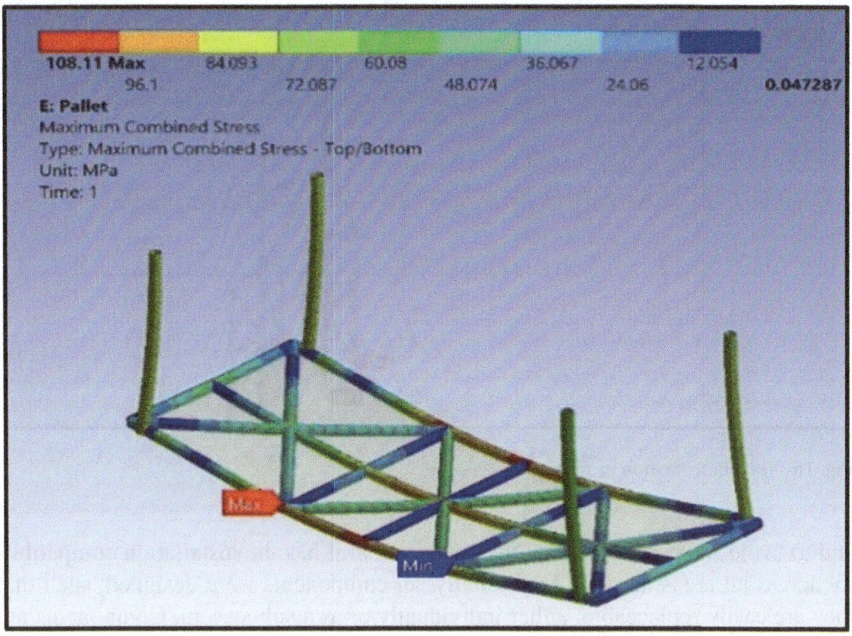

Fig. 14 Tube stresses (without sheet)

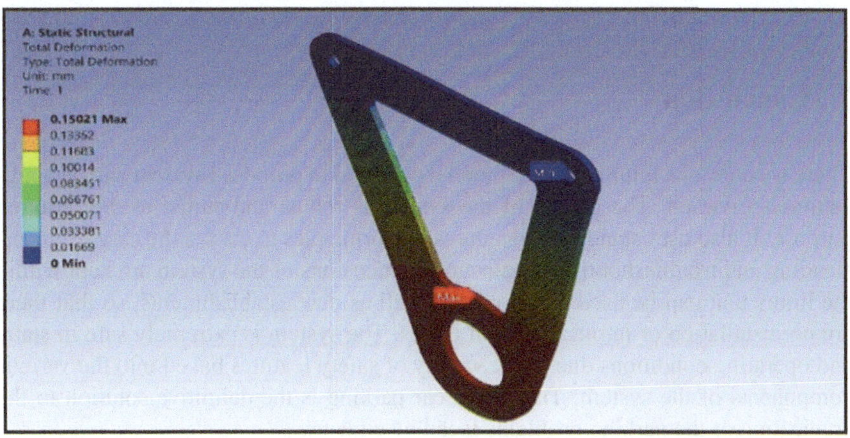

Fig. 15 Total deformation chain-pellet link attachment

6 Results and Discussion

The main objective of fitting the entire system within the width of two parking spaces was successfully achieved considering all the system clearances for easy assembly

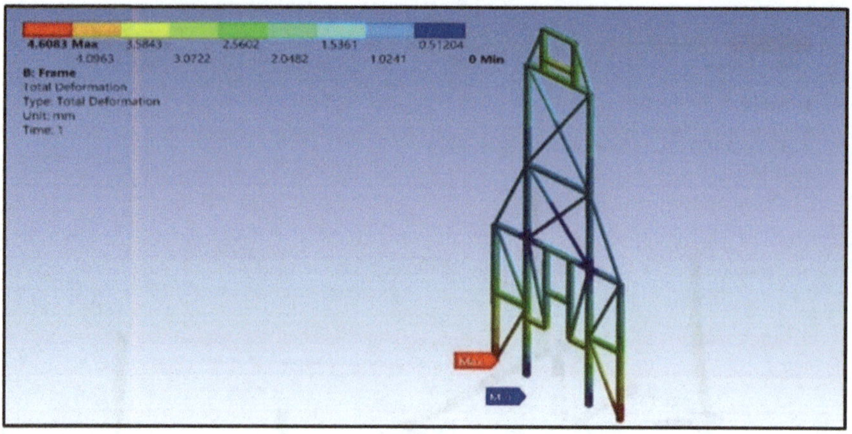

Fig. 16 Total deformation of frame

and to avoid interference. Also, this ensures that it has the installation compatibility across all areas needed. Additionally, all components were designed, such that they are easily replaceable, either individually or as a sub-system. Components are also designed for long uniform life and uniform wear and tear so that maintenance requirement, when required, is only a total quality maintenance and not routine maintenance.

7 Conclusion

There is enormous future and commercial potential in proposed system and its applications are varied. The design of the system is robust and suited to the required purpose. It also uses standard procedures and principles to ensure the ease of understanding and troubleshooting. The overall dimensions of the system are kept within the limits that can be used by current as well as new establishments, so that there are no installation or implementation issues. The system is extremely safe in static and operating conditions due to the variety of safety features baked into the various components of the system. The rotary car parking is the definitive solution to the eradication of the parking problems faced by citizens.

References

1. United States Patent Publication on 'Vertical rotary car parking system'. Dal Young Park, Seoul (KR) (2004, August)
2. Prasad PPC (2011, April) Smart Car Parking, Department of Mechanical Engineering, Gokaraju Rangaraju Institute of Engineering and Technology (Approved by AICTE, New Delhi and Affiliated to JNTU, Hyderabad)

3. Nasir N, Rashid H, Abdullah AH, Jusoh MAM (2011) Rotary car park (pallet design) computer aided design analysis study. Faculty of Mechanical Engineering, Universiti Teknologi MARA Malaysia
4. Biradar RS (2012, May) Conceptual design of automatic smart parking systems for shopping centers
5. Sarayu S, Rajendra SS, Bongale VV (2013, September 1) Design and fabrication of prototype of automated smart car parking system using programmable logical controllers (plc). Int J Sci Eng Technol 2(9):857–860. ISSN 2277-1581
6. Prashanthkumar TJ, Vitala HR, Praveen MP (2014, July) Concept design and proto build of roto parker for two wheeler. Int J Innov Sci Eng Technol (IJISET) 1(5)
7. Kolekar RJ, Gawade SS (2014) Design and development of lift for an automatic car parking system. Mechanical Engineering Department, PG Student, R.I.T. Islampur (M.S.), India., Mechanical Engineering Department, Faculty, R.I.T. Islampur (M.S.), India
8. Narone SG, Chabukswar SS, Hirapure VRB, Solapure VR (2015, April) Vertical car parking—a prototype. Int J Emerg Technol Adv Eng 5(4). ISSN 2250-2459 (ISO 9001:2008 Certified)
9. Bhujade KD, Shahu AS, Saiyyad AA (2016, March–April) Modelling of vertical car parking system. Int J Emerg Trends Eng Basic Sci (IJEEBS) 3(2):59–64. ISSN (Online) 2349-6967
10. Pashte P. Narkhede V, Nade S, More S, Maske YL (2016) Design and analysis of rotary automated car parking system. Int J Sci Res Dev (IJSRD) 4(4). ISSN (online): 2321-0613

Designing a Cowl Template with DFSS Methodology

Sandip Mane, Smit Patel, and Jay Bhuva

Abstract This paper illustrates the application of modern ideas along with the integration of new techniques to achieve optimum process cycle efficiency. The paper uses design for six sigma (DFSS) methodology in designing a metal template for tack welding of brackets on the cowl. The cowl is a part of the fork truck located near the driver's position. The design uses six sigma methodology, considering five steps known as DMADV, which resembles define, measure, analyze, design and verify. Lean and six sigma tools, 3D modeling software, finite element analysis, jigs and fixtures theories were utilized to determine and eliminate bottlenecks with the help of the proposed methodology. The sample readings were recorded and analyzed. A detailed comparison showed a surge of 27% in the process cycle efficiency resulting in savings of manpower cost and reduction in time required for the overall process.

Keywords Lean manufacturing · Six sigma · DMADV · Process cycle efficiency

1 Introduction

The cowl is a sub-assembly part of driver's module in a fork truck onto which all the brackets for electric wirings are fitted, and the driver's module contains the front and the overhead parts like cowl, driver's seat, steering, overhead guard, etc. The cowl resides as a front end of the fork truck below the steering assembly and above the engine transmission assembly.

Figure 1 shows the cowl for which the template has been designed is the part of platform trucks driver's module. Originally, with no template, the operator faced difficulty in tack welding the small brackets and wire mountings onto the cowl. A new metal template was designed and fabricated which ensured in improving the

S. Mane (✉)
Department of Production Engineering, Dwarkadas J. Sanghvi College of Engineering, Vile Parle, Mumbai 400056, India
e-mail: sandip.mane@djsce.ac.in

S. Patel · J. Bhuva
Dwarkadas J. Sanghvi College of Engineering Mumbai, Vile Parle, Mumbai 400056, India

© Springer Nature Singapore Pte Ltd. 2020

H. Vasudevan et al. (eds.), *Proceedings of International Conference on Intelligent Manufacturing and Automation*, Lecture Notes in Mechanical Engineering,
https://doi.org/10.1007/978-981-15-4485-9_58

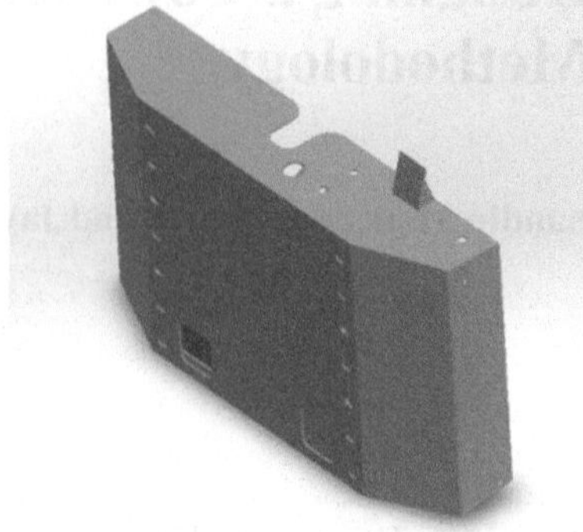

Fig. 1 Platform truck's cowl 1

process cycle time for fabrication, i.e., tack welding from 21 to 48%, leading to a major reduction in the overall cost for tack welding, up to 64%.

1.1 Lean and Six Sigma Techniques and Tools

The impression of lean management was conventionally applicable to various process in manufacturing. Womack et al. [1] described in detailed how diverse the techniques effect in the automobile domain and how six sigma methodology helps in reduction of waste and enhances the quality, thereby reducing in the cost. Ultimately, the prime focus was to move towards value addition process and implementing new techniques such as lean manufacturing in six sigma. Kumbhar et al. [2] explored the importance of jigs and fixture and stated their vital role in designing and fabricating the various components in manufacturing stages, as from which this paper concludes into the use of location and clamping principles in designing the cowl template. This paper studied on the use of modular size pins for location purpose in intricate areas. Luan et al. [3] studied the various tools such as 3D modeling software for designing and for the better understanding of the complexity of the parts. Ulrith et al. [4] reported that the lean manufacturing is not only constrained to minimize the cost or cycle time, but it can be utilized at a much broader scale to maximize the value of the product. Lean in current scenario can be approached with an ability to create value for the customer as well as eradicate the non-value adding activities. Such involvement of combinative ideas and concepts not only enhances the productivity but also improves the quality of the product at a significant mark, thereby fulfilling customer's requirements. Browning [5] studied and explored the numerous product design techniques and flowcharts, which can be implemented in various design task, allowing better understanding the functionality of the end product. A similar design charts are utilized in the paper while designing the cowl template. Naslund et al. [6] stated that the application lean manufacturing and six sigma tools for the organization proved to be a challenging competitor in the market with decrease in cycle time, increase in cycle efficiency and the reduction in the manufacturing cost by the elimination of non-value adding activities. Pepper and Spedding et al. [7] observed and analyzed the integration of lean and six sigma techniques and the complexity faced during the initial stages of implementation. Lean manufacturing and six sigma concepts have been always implemented in isolation. The paper studies the use of these models, which helped in reducing the inter-departmental complexities and increase the communication throughout the firm. Haggerty et al. [8] addressed some typical lean engineering tools that can be used to stream line and improve the design process. These include design for manufacturing and assembly (DFMA), solid model-based design, variability reduction and dimensional management and product simulation with finite element analysis (FEA). Stampfer et al. [9] studied the challenges faced in manufacturing boxed-shaped parts and his inputs made it possible to automize the setup for fixture system. Cecil et al [10] studied the design of automated fixture and suggested improved clamping design for fixtures. DFMA procedure has been applied to reduce the complexity and design the cowl template, thereby cutting down the overall cost incurred in the process. A variety of similar 3D modeling software can be utilized for improved visualization and understanding of the cowl template while working with the programs like computer aided manufacturing (CAM) during the manufacturing stage. These 3D models prove to be an easy source for better understanding. Based on the literature review, the paper narrow downs to design with six Sigma and lean tools and technique to increase the process cycle efficiency by the elimination of loss and reducing the non-value adding activities and increasing the quality of the product with ensuring operator's safety.

2 Proposed Methodology

The DFSS and the DMADV methodology address the practical life use of lean and six sigma tools and techniques to design a metal template for the parts to be tack welded on the cowl. In order to apply design for six sigma to satisfy the requirement, a flowchart has been drafted defining the necessary steps. The DMADV tools are utilized to identify the problem. The DMADV acronyms stand for define, measure, analyze, design and verify. These set of tools are applied along with DFSS approach to counter the problems in the design. Figure 2 presents the steps in DMADV method-

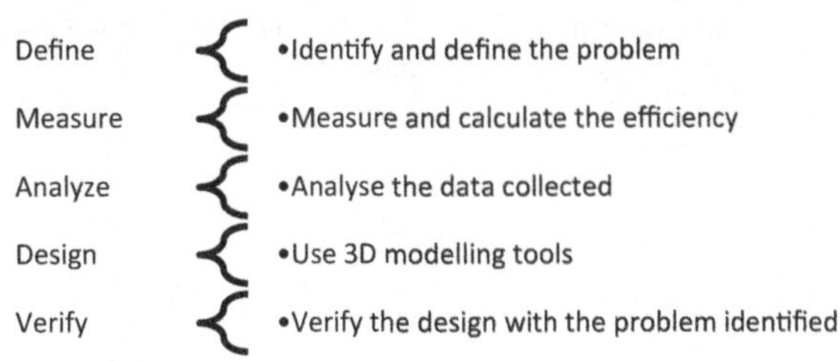

Fig. 2 DMADV methodology

ology, and the flowchart summarizes the major steps in the design process. Each step is explored in detailed along with the reference to the flowchart.

2.1 *Identifying the Problem in the Welding Process*

The introduction to the new design of the cowl led to the operator's facing difficulty in tack welding the parts to the cowl resulting in unnecessary time consumption. Manual welding has reduced the efficiency due to which there was an urge to switch the process towards robotic welding. Before the robotic welding on the cowl is performed, the small brackets and plates with slot for wire mountings and fittings are to be tack weld to the respective positions for which the tack welding of the part is performed. The process included marking and locating the parts to their desired place simultaneously inspecting with the design drawings. Such repetitive processes involved two operators and a large amount of non-value adding activities. Necessary action were to been taken to solve the problem and improve efficiency.

2.2 *Measuring the Time Consumed by the Welding Process*

Time study along with time analysis of the tack welding of parts on the cowl is observed and summarized. The time for three quantities of the parts to be tack welded on the cowl is observed and recorded in the table. Table 1 shows tack welding time for three consecutive days, and the time is segregated into setup time + marking time and the actual tack welding time. From Table 1, it has been observed that the setup time consists of parts being brought from the store and tools being gathered to perform the tack welding, and the marking time includes setting up the parts on the required places throughout the cowl eventually tack welding the parts for robotic welding. Table 1 shows the recorded time for the tack welding process. It has been observed that a large amount of non-value adding activities are been performed resulting in loss. To counter this problem, necessary actions were to be taken.

Table 1 Recorded time for the tack welding process

Day	Number of cowl tack welded (quantity)	Setup time + marking time (minutes)	Tack welding time (minutes)	Total time (minutes)
1	1	182	39	221
2	1	168	47	215
3	1	172	46	218

2.3 Analyzing the Process with the Help of Lean and Six Sigma Tools

In this step the process cycle efficiency (PCE), a six sigma tool has been utilized to arrange and evaluate the process efficiency. The metric compares value-added process time to the actual process time. From the product's viewpoint, value-added time are the stages in the process that add form, function and value to the final product or process. Non-value added time are steps that do not add form, function or value to the end product or process. To perform this process, the activities are divided into value-added time, non-value adding time and loss time. The process calculation has been performed below.

$$\text{Process Cycle Efficiency} = (\text{VALUEADDEDTIME/TOTAL PROCESS TIME}) \times 100\%.$$

Based on the readings observed in the above table, the overall process is averaged to be 216 min which are divided into 170 min of marking on the parts and locating the parts on the cowl plate and 46 min of actual tack welding. To calculate the efficiency, the above formula was utilized.

$$\text{Process cycle efficiency} = \frac{\text{Tack welding time}}{\text{Total process time}} \times 100$$

$$\text{PCE} = \frac{46}{2216} \times 100 = 21.29$$

This shows that only 21% time spent on process is the actual value-added time. One of the main criteria for designing the template was to improve the process cycle efficiency.

2.4 Designing with the Help of 3D Modeling Software

The flowchart has been drafted defining the important steps towards designing the template. Figure 3 shows the flowchart for the design process. The flowchart shows the major steps which are taken to counter the problem and fulfill the requirements. The design process has been planned carefully and executed systematically. Specifically, the engineering design process must integrate the many aspects of designing in such a way that the whole process becomes logical and comprehensible. In accordance with the previous statement, the design flowchart has been drafted considering basic and necessary key points.

Figure 3, a flowchart can be implemented to any new or similar design as it is or with minimum alteration to customize the particular application. The lean designing tool solid works modeling software was utilized to design the metal template.

Fig. 3 Design step flowchart

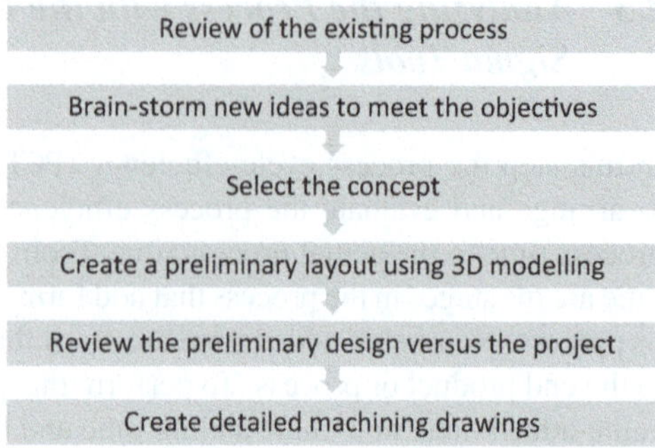

The use of 3D modeling software allowed better visualization of the design before manufacturing.

The laser-cut metal plate is machined around the edges to ensure that the surfaces are smooth from the safety point of view. Later, additional parts like support plates of appropriate size and shapes along with ribs to grip the parts perpendicular to the bottom plates are fully welded with proper gauging to maintain parallelism. Figure 4 shows the main design steps included, while designing and fabricating the cowl template which resembles appropriate location of the template with the bottom cowl to ensure the location, the cowl is suspended and referenced with the top left corner reducing the operator's effort. Ribs and plates are attached to align and support the plate perpendicular to the bottom surface. Another location is provided with a central pin that holds the two plates with holes maintaining their concentricity. Slots are provided to insert pins which would lock the template's position with the

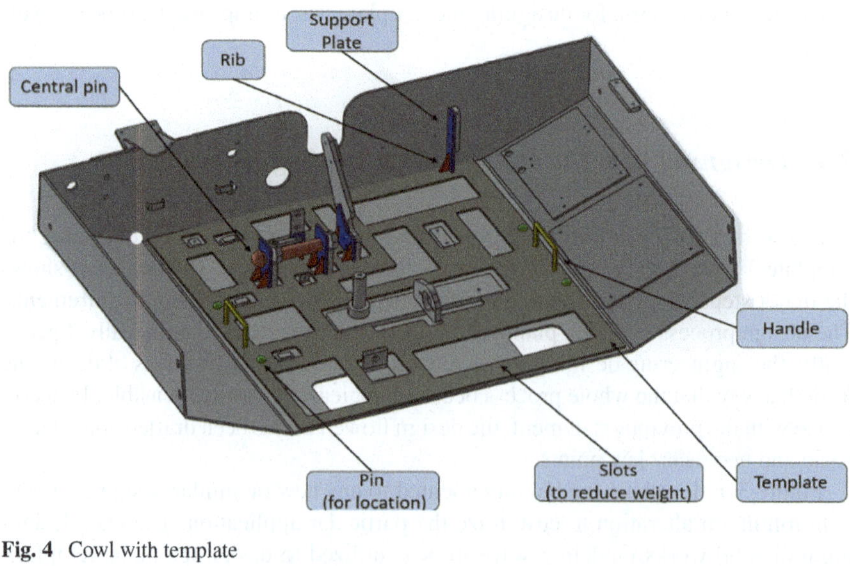

Fig. 4 Cowl with template

Table 2 Recording of the new timings

Day	Number of cowl tack weld (quantity)	Setup time + marking time (minutes)	Tack welding time (minutes)	Total time (minutes)
1	1	38	39	77
2	1	45	47	92
3	1	43	46	89

cowl. Additional slots/cavities are provided wherever possible to reduce the weight of the metal plate for the easy removal process. Handles are provided for ease in the handling process.

2.5 *Verifying the Process with the Design Objectives*

In order to verify whether the innovative design fulfills the design objectives of reducing the overall tack welding time and increases the efficiency, a sample cowl which was tack weld for 3 consecutive days were observed and tabulated (Table 2).

Table 2 shows a drastic reduction in the non-value adding activities, with an average of 42 min of setup time and 38 min of tack welding time with an overall of 86 min for the process. Since process cycle efficiency is defined as the value-adding time divided by the total process time, this resulted in an increase of 48% in the overall process cycle efficiency.

$$\text{Process cycle efficiency} = \frac{\text{Tack welding time}}{\text{Total process time}} \times 100$$

$$\text{PCE} = \frac{42}{84} \times 100 = 50\%.$$

The process cycle efficiency is 50% with minimum manpower consumption.

3 Result and Discussion

The process cycle efficiency calculated above shows time reduction in the overall tack welding process, considering the man-hour cost (Rs.48 per hour) and an increase in the process efficiency. Table 2 shows the time taken for tack welding the parts on the cowl. It shows a reduction in time consumption, which directly affects the man-hour cost. The man-hour cost saved is estimated as follows with an average of Rs.48 per hour and the average time after the utilization of template which summarizes to 86 min from Table 2, the cost incurred previously without a template is summed up to Rs.172.8 for 3.6 hours, and current cost estimated sums up to Rs.60.48.

Table 3 Wages calculation

Method	Wages per hour (in Rs.)	Time required for one quantity (in hour)	Total cost (in Rs.)
Old method	42	3.36	141.12
New method	42	1.4	58.8

From Table 3, a major reduction in time is being observed, which directly impacts the manpower cost saved. It is concluded that there is a 64% reduction in the manpower cost of the total cost incurred for the tack welding process, and it leads to an increase in the process cycle efficiency by 27%.

4 Conclusion

With the results accounted for in the study, the paper concludes the use of DFSS methodology along with six sigma and lean manufacturing tools and techniques for designing and improving the process of tack welding. From the outcomes stated, an increase of 27% in the process cycle efficiency was recorded along with 65% saving in the manpower cost. The designed template results in an overall improvement of the weld quality and accuracy of the work. The reduced time can be utilized in more value-adding activities which add to the betterment of the firm. The DMADV design flowchart was drafted to suit the necessary requirements. This flowchart could be altered as per the demand and can be utilized in similar design tasks. The designed cowl template for tack welding not only increased the process cycle efficiency but also reduced operator's efforts by providing safer and operator friendly process.

References

1. Womack JP, Jones DT, Roos D (1990) The machine that changed the world. Free Press, New York. ISBN-13: 978-0-7432-9979-4
2. Kumbhar JV, Pandit HC (2017) A review article on jigs and fixture. Int J Sci Res (IJSR). ISSN (Online): 2319-7064
3. Luan X-D (2008) Research and development of 3D modelling. IJCSNS Int J Comput Sci Netw Secur 8(1)
4. Ulrich K, Eppinger S (2011) Product design and development. McGraw-Hill Education
5. Browning TR (2000) Value-based product development: refocusing lean. In: Proceedings of the IEEE engineering management society, IEEE, Albuquerque, NM, USA
6. Naslund D (2008) Lean, six sigma and lean sigma: Fadsor real process improvement methods. Bus Process Manag J 14(3):269–287
7. Pepper MPJ, Spedding TA (2010) The evolution of lean six sigma. Int J Qual Reliab Manag 27(2):138–155
8. Haggerty A, Lewis D, Mc Manus H, Murman E, Weigel A (2012) Lean Engineering Basics V7.6, MIT Open Course Ware

9. Stampfer M (2008) Automated setup and fixture planning system for box-shaped Parts. Int J Adv Manuf Technol 45:540–552. https://doi.org/10.1007/s00170-009-1983-1
10. Cecil J (2008) A clamping design approach for automated fixture design. Int J Adv Manuf Technol 18:784–789

Reliability Estimation of Molded Case Circuit Breaker in Development Phase

Greegory Mathew and Santosh B. Rane

Abstract Life test of molded case circuit breakers (MCCB) on mechanical endurance test setup showed degradation in the functional deliverables of a MCCB, such as reset force and contact pressure. Degradation in functional deliverables affects the satisfactory working of the circuit breaker and decreases its reliability. Functional deliverables of a MCCB in development phase can be measured using two systems of measurement. A gage repeatability and reproducibility study using 2 operators, 5 parts and 3 trials was performed to select the best measurement system. MCCBs were then tested on a life cycle test setup, and the degradation in the deliverables was measured using the selected measurement system. These measured values were analyzed, and the circuit breaker reliability was then predicted using Minitab software.

Keywords Repeatability · Reproducibility · Degradation analysis · Reliability

1 Introduction

Molded case circuit breakers (MCCBs) are complex engineering systems used for protection of distribution feeder systems, transformers, etc. They are used in all low voltage (less than 1000 V) applications, such as, in residential electrical distribution panel and in main power feed panels used in large buildings such as offices, hospitals and shopping centers. A circuit breaker mechanism is required to perform six functions, namely on, off, reset, trip, over travel and contact reset. The force required to switch on, switch off, reset, trip the MCCB and the functions of over travel and contact reset is critical deliverables of the MCCB and should not degrade in quality. So, reliability of the circuit breaker can be estimated by measuring the ability of the circuit breaker to perform the above functions adequately. However, if the measuring instrument and the method are incapable of making correct, repeatable

G. Mathew (✉)
Dwarkadas J. Sanghvi College of Engineering, Vile Parle, Mumbai, Maharashtra 400056, India
e-mail: greg_12@ymail.com

S. B. Rane
Sardar Patel College of Engineering, Andheri, Mumbai, India

© Springer Nature Singapore Pte Ltd. 2020
H. Vasudevan et al. (eds.), *Proceedings of International Conference on Intelligent Manufacturing and Automation*, Lecture Notes in Mechanical Engineering,
https://doi.org/10.1007/978-981-15-4485-9_59

and reproducible measurements, the observed data can have significant inconsistencies and will involve error. A measurement system analysis (MSA) of the available measurement systems is thus necessary to prevent incorrect measurement.

The performance of a measurement system can be evaluated by finding the variations generated in measurements using a study of repeatability and reproducibility (R&R) [1]. Various researchers have successfully demonstrated and applied repeatability and reproducibility-based studies [2–8]. For the purpose of the study, only repeatability and reproducibility of the off force measurements are explored. The selected measurement system is then used to measure degradation in the MCCB deliverables. This paper explains how reliability of MCCBs in development phase can be estimated by studying the degradation in the functional deliverables. The flowchart for the research is presented in Fig. 1.

Two system of force measurement are available.

Measurement system type 1 (MS1). A trained operator measures the off forces with a calibrated gauge. To measure the force required to switch off the breaker:

1. The breaker is switched on if not in on position.
2. The force measuring gauge is attached to the operating fork of the circuit breaker.
3. The operator then pushes the fork with the gauge till the fork flips over, i.e., the breaker switches off.
4. The reading indicated by the gauge corresponds to the force required to switch off the breaker.
5. Care is taken to keep the gauge horizontal while measuring as inclination of the gauge while measuring will yield incorrect data.

Measurement system type 2 (MS2). A hole is drilled in the fork of the MCCB. A weighing pan is suspended through wire rods of sufficient strength. A trained

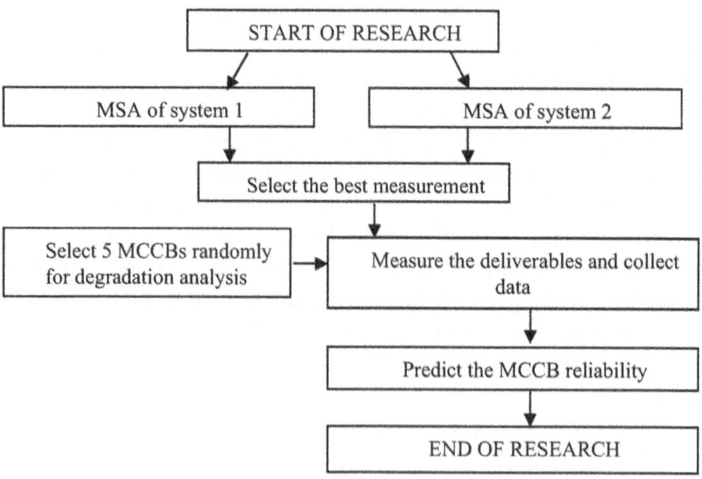

Fig. 1 Flowchart of research

operator then measures the off forces by adding calibrated weights to the pan. To measure the force required to switch off the breaker:

1. The circuit breaker whose force is to be measured is mounted, on the test bench.
2. To measure off force, the circuit breaker is first switched on.
3. The operator than adds weight in small increments to the weighing pan till the circuit breaker is switched over to the off state by the weights.
4. The required off force is now equal to the total weight of the pan.

2 Measurement System Analysis

To select the best measurement system for force measurement, five circuit breaker units (parts), assembled as per standard code of assembly practice, were selected in a random sequence. Two operators were randomly selected to measure the required forces. Each operator was made to repeat the measurement, thrice on each circuit breaker. Both operators measured the forces in a random order to avoid interaction between operator and part. The forces were initially measured using MS1. The process was repeated for MS2. The measurement values obtained from the study for both MS1 and MS2 are presented in Table 1. Though the forces were measured in random sequence, for ease of readability, the force values measured are re-organized and presented in tabular form.

2.1 Discussion on MS1

From Fig. 2, X bar chart for off force measurements shows that the values for unit 4 lie above the UCL while that for unit 5 lie below the LCL. The value for unit 1 as measured by operator A lies below the LCL. As points are above or below the statistical control limits, these results indicate that part-to-part variation is much greater than the variation in the equipment or appraiser. From R chart for off force, the measured data points for both operators lie within the control limits. It can be concluded that the operators measure the off force consistently. From the operator part interaction chart, the interaction remains parallel only between units 3 and 4. As the interaction lines are not parallel throughout, it can be concluded that an operator's ability to measure a part depends on the part being measured during off force measurements. From the off force measurement by operator graph, the line between the average measurements of all units by operator A and that for operator B is inclined to the horizontal axis. It conveys that on an average, operators A and B are measuring the units differently.

Table 1 Force measurement observations

Breaker No.	Off force(kg)—MS1						Off force(kg)—MS2					
	Operator A (trials)			Operator B (trials)			Operator A (trials)			Operator B (trials)		
	T1	T2	T3	T1	T2	T3	T1	T2	T3	T1	T2	T3
1	11.4	11.2	10.6	11.4	11.2	11.4	11.2	11.4	11.2	11.5	11.1	11.2
2	12	12.4	11.6	12.2	12	11.6	11.8	12.2	12.1	12.2	12.3	12.4
3	11.6	11.4	11.4	11.8	11.4	11.6	11.4	11.5	11.3	11.3	11.5	11.6
4	12.8	12.8	12.6	13	12.8	12.8	12.8	12.9	12.9	12.7	12.7	12.9
5	11.2	10.8	10.8	11	10.8	11	10.8	10.8	10.8	10.7	10.9	10.8

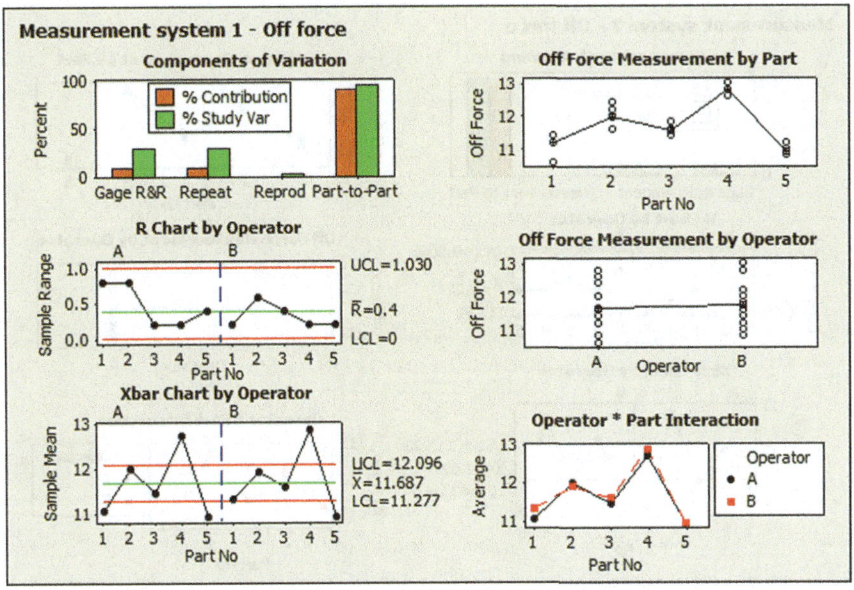

Fig. 2 Measurement system 1—off force

2.2 Discussion on MS2

From Fig. 3, X bar chart shows that the values for unit 1 lie above the UCL while that for units 2, 3 and 5 lie below the LCL for both the operators. These results indicate that part-to-part variation is much greater than the variation in the equipment or appraiser. From R chart, the measured data points for both operators lie within the control limits. It can be concluded that the operators measure the off force consistently. The operator part interaction remains parallel only between units 4 and 5. It can be concluded that an operator's ability to measure a part depends on the part being measured. From the off force measurement by operator graph, the line between the average measurements of all units by operator A and that for operator B is inclined to the horizontal axis. It conveys that on an average, operators A and B are measuring the units differently.

2.3 Selection of the Best System

Results in Table 2 show that the variation in the off force measurement system due to gage repeatability and reproducibility as percent of total variation for MS1 is 29.71%. As per AIAG guidelines, variation below 10% is a good system of measurement. Variation between 10 and 30% of the total variation is a cause of concern and should be corrected. Variation above 30% may defeat the entire purpose of measurement and hence needs immediate correction. Percentage of component variation shows

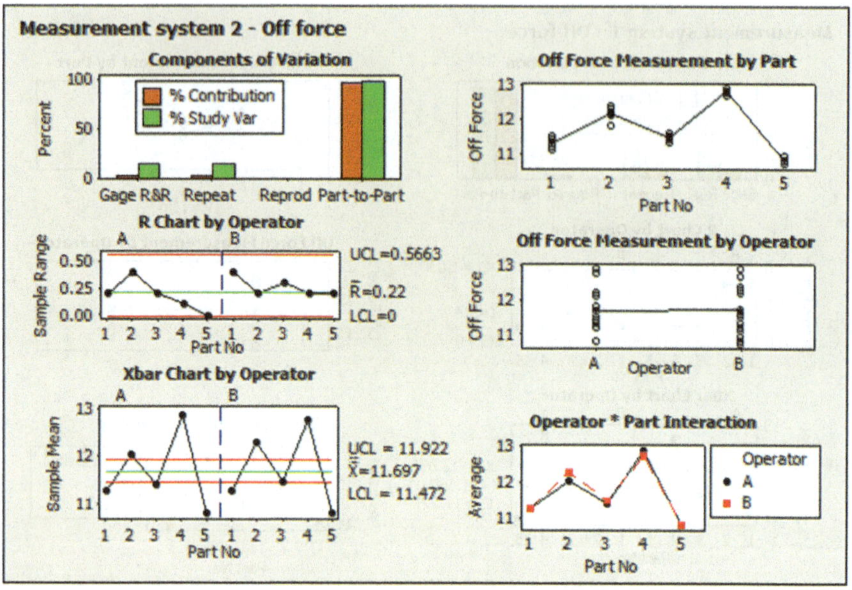

Fig. 3 Measurement system 2—off force

Table 2 Summarized results of MS1 and MS2

Source	Off force—MS1		Off force—MS2	
	SV	% SV	SV	% SV
Repeatability	1.395	29.50	0.767	15.54
Reproducibility	0.167	3.53	0.013	0.26
Total gage R&R	1.405	29.71	0.768	15.54
Part variation	4.516	95.48	4.879	98.79
Total variation	4.730	100.00	4.940	100.00

that gage R &R variation is 15.54% for MS2. It can be concluded that MS1 cannot be used to measure off force. Circuit breaker forces were therefore measured using MS2.

3 Degradation Observation

Degradation in certain deliverables such as reset force, on force and off is beneficial to a limit as it reduces the effort required by the operator to reset, switch on and switch off the circuit breaker, respectively, but degradation in deliverables such as contact pressure reduces the withstand capacity of the circuit breaker against overload current. Increase in contact pressure over a certain threshold is also not advised as it would affect the ability of the mechanism to trip the circuit breaker in case of a high fault current. Decrease in trip force below a design limit will nuisance trip the circuit breaker whenever it is switched on or off and thus affect the functionality of the circuit breaker. Increase in trip force above a design limit will make it impossible for the under-voltage release, shunt release and the thermo magnetic release to rip the circuit breaker whenever the need arises. Increase in reset force, on force and off force above the design limit will make it impossible for the electrically operated mechanism to operate the circuit breaker. In manual operation, the operator will require a lot of muscular energy to operate the circuit breaker.

A total of five circuit breakers were randomly selected for the study and tested on life endurance test setup. The measured values for MCCB 1 are presented in Table 3. The upper limit and the lower limit for the on force are 7.5 and 5.5 kg, respectively. From the values tabulated in Table 3, it is clear the circuit breaker 1 has no functional degradation in on force up to 50,000 cycles. The limits for the force required to switch off the MCCB are 16.5 and 12 kg, respectively. From Table 3, it is evident that the force required to switch off the circuit breaker reaches the upper limit after 32,100 cycles. The limits for the force required to trip the MCCB are 750 and 450 g, respectively. The force required to trip the MCCB is always within the acceptable limits till 50,000 cycles for MCCB 1. The limits for the force required to reset the breaker are 23.5 kg and 18 kg, respectively. However, MCCB 1 could not be reset after 23,500 cycles. The functions of contact reset and over travel could be successfully observed up to 41,100 cycles and 23,500 cycles, respectively, for MCCB 1. Of the six functional deliverables of MCCB1, the earliest cycles to failure were for reset force and over travel at 17,500 cycles. Hence, the life of MCCB 1 is 17,500 cycles.

Table 3 Observed degradation in reliability deliverables for circuit breaker 1

S. No.	Cycle	On force (kg)	Off force (kg)	Reset force (kg)	Trip force (gm)	Contact reset	Over travel
1	0	7.0	14.7	20.9	600	Yes	Yes
2	1800	6.9	14.6	20.7	590	Yes	Yes
3	4300	6.9	15.5	21.0	590	Yes	Yes
4	8000	7.0	15.5	23.5	580	Yes	Yes
5	12,300	6.8	15.7	23.5	580	Yes	Yes
6	17,500	6.6	16.2	23	580	Yes	Yes
7	23,500	6.5	16.3	27.5	580	Yes	No
8	26,900	6.5	16.1	33.0	570	Yes	No
9	32,100	6.5	16.3	–	570	Yes	No
10	36,900	6.3	16.5	–	570	Yes	No
11	41,100	6.3	16.5	–	570	Yes	No
12	50,000	6.1	16.5	–	560	No	No

Table 4 Observed functional degradation for the five circuit breakers

MCCB No.	On force (kg)	Off force (kg)	Reset force (kg)	Trip force (g)	Contact reset	Over travel
1	50,000	32,100	17,500	50,000	32,100	17,500
2	50,000	35,000	20,000	50,000	38,500	19,500
3	48,000	31,000	18,000	50,000	29,500	16,500
4	50,000	33,200	15,000	50,000	41,000	20,300
5	49,500	33,700	22,000	50,000	33,000	17,300

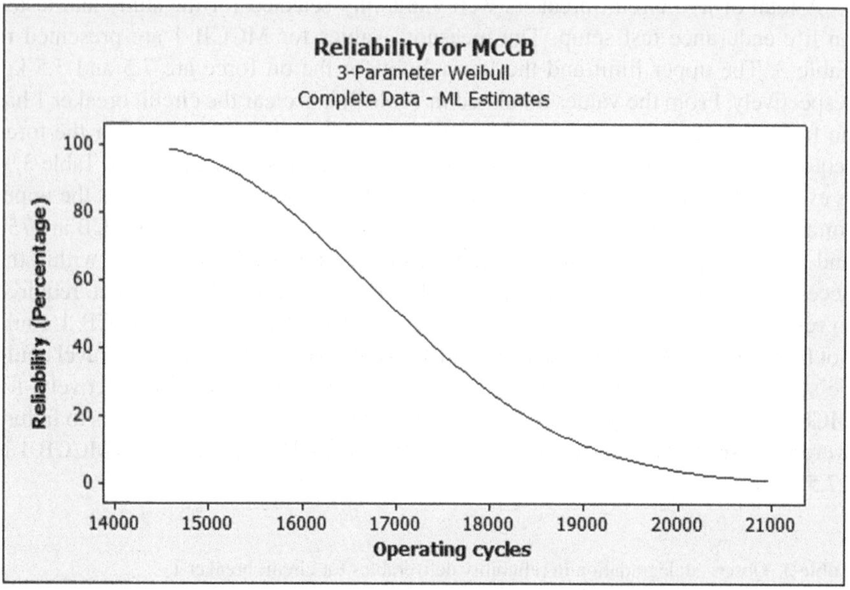

Fig. 4 Survival plot for the MCCBs

The observed cycles to failure for the deliverables of the five MCCBs are presented in Table 4. From the values in Table 4, the life of MCCB 2, 3, 4 and 5 can be estimated to be 19,500, 16,500, 15,000 and 17,300 cycles, respectively.

4 Reliability Estimation

The life data of the MCCBs was analyzed in Minitab software. Three-parameter Weibull plot was used to estimate the various distribution parameters and predict the reliability of the MCCBs. The shape factor, the scale factor and the threshold factor were 2.0656, 3263.08 and 14,265.7, respectively. Reliability of the MCCB at 10,000 cycles was calculated to be 100%. Reliability at 15,000, 20000 and 30,000 cycles is calculated as 95.5%, 4.06% and 0%, respectively. Figure 4 shows the survival plot for the circuit breaker at various operating cycles.

5 Conclusion

Reliability of a MCCB can be estimated by measuring the degradation in the functional deliverables. At the design and development stage of a MCCB, measurement systems to measure the deliverables are not well established. A gage repeatability and reproducibility study, using 2 operators, 5 parts and 3 trials established that the measuring system using pan and calibrated weights was a better one. MCCBs were then tested for life data on life cycle test setup. The degradation in the MCCB deliverables was measured using the system with pan and calibrated weights. This data was then analyzed in Minitab software to estimate the reliability of the MCCBs. From reliability analysis, it was concluded that the MCCB would survive minimum 10,000 cycles without any loss of performance. Since the MCCB is in development phase, the causes for the degradation in the MCCB deliverables can now be identified. Improvements can be made to the MCCB mechanism to improve the reliability.

References

1. Kuo CC, Huang PJ (2012) Repeatability and reproducibility study of thin film optical measurement system. Optik 124(18):3489–3493
2. Peruchi RS, Balestrassi PP, Paiva AP, Ferreira JR, Carmelossi MD (2013) A new multivariate gage R&R method for correlated characteristics. Int J Prod Econ 144(1):301–315
3. Cottenden AM, Rothwell JG, Leander H, Grau C, Brooks RJ (2001) An investigation of the repeatability and reproducibility of ISO 11948-1 (the Rothwell method) for measuring the absorption capacity of incontinence pads. Med Eng Phys 24:159–163
4. Abenhaim GN, Desrochers A, Tahan AS (2013) An investigation of the repeatability of nonrigid parts measurements: a case Study of an aluminium panel. CIRP Conf Comput Aided Tolerancing 10:105–111
5. Lenis YA, Agudelo A, Pérez JF (2012) Analysis of statistical repeatability of a fixed bed downdraft biomass gasification facility. Appl Therm Eng 51(1–2):1006
6. Monnet T, Thouzé A, Pain MT, Begon M (2012) Assessment of reproducibility of thigh marker ranking during walking and landing tasks. Med Eng Phys 34(8):1200–1208
7. Al-Refaie A, Bata N (2010) Evaluating measurement and process capabilities by GR&R with four quality measures. Measurements 43(6):842–851
8. Voor MJ, Yang S, Burden RL, Waddell SW (2007) In vivo micro-CT scanning of a rabbit distal femur: Repeatability and reproducibility. J Biomech 41(1):186–193

Design and Development of Easy Access Crisper/Shelf in a Refrigerator

Addanki Sambasiva Rao, Vinayak H. Khatawate, and Sumit Mane

Abstract The category of premium refrigerators available in market is different from regular household refrigerators, with the former being bigger in size. In such refrigerators, the depth can go up to 70–80 cm, and the items placed at the back side or in corners are ergonomically less accessible. The taller items need to be placed at the back side of the refrigerator to easily access the short items. Also, due to the limitations of the support structure, the full extension of the crisper is not achievable. Currently, the crisper is pulled out only in a linear manner. Hence, it is advantageous to make the crisper rotate to make it accessible from more number of sides. In this paper, a mechanism to achieve the rotational motion of the crisper was developed. Finite element analysis (FEA) was carried out to check the permissible deflection of the crisper and is validated using experimentation.

Keywords Shelf · Crisper · Accessibility · Mechanism · Refrigerator · FEA

1 Introduction

Refrigerators are apparatuses configured to store food under low temperature conditions. Such a refrigerator includes a main body provided with a storage compartment, and a movable door connected to the main body to open and close the storage compartment. The storage compartment may be divided into a refrigerator compartment and a freezer compartment.

A. S. Rao (✉) · S. Mane
Department of Mechanical Engineering, Veermata Jijabai Technological Institute (V.J.T.I.), Matunga, Mumbai 400019, India
e-mail: asrao@me.vjti.ac.in

V. H. Khatawate
Department of Mechanical Engineering, Dwarkadas J. Sanghvi College of Engineering, Vile Parle, Mumbai 400056, India

© Springer Nature Singapore Pte Ltd. 2020
H. Vasudevan et al. (eds.), *Proceedings of International Conference on Intelligent Manufacturing and Automation*, Lecture Notes in Mechanical Engineering,
https://doi.org/10.1007/978-981-15-4485-9_60

1.1 Crisper

Crispers are nothing but sealed drawers that help to keep fruits and vegetables. Vegetables require higher humid conditions, while fruits require lower humid conditions, which is why many refrigerators have two separate (marked) drawers called crispers.

1.2 Shelf

Shelf is a horizontal plate attached to the inner side of the cabinet in order to facilitate the placing of items in the refrigerator compartment.

2 Literature Review

Hoshbin et al. designed a rotary type fruit and vegetable box for a refrigerator, wherein a box body is an object containing drawer formed by a quadrilateral bottom surface and four walls surrounding it [1]. Yuan Kaijun et al. designed a rotary shelf which swings together with a door of the refrigerator for being externally drawn out when the door is opened [2]. Lazy Susan, a fixed small rotary compartment, is already present in many refrigerators (Fig. 1). A 1954 refrigerator of certain brand consisted of two rotary shelves. It resulted in less storage area available on shelves (Fig. 2).

2.1 Outcome of Literature

After studying the above literature, the following points are extracted:

Fig. 1 Lazy Susan

Fig. 2 GE refrigerator

1. A quadrilateral box is designed with upward protruding cylindrical cavity for assembling the rotating shaft. This reduces useful volume of the crisper, and also, the crisper is rotated at its fixed axis [3].
2. In one model, there are four chambers in which refrigerator body is divided and an actuating unit in the power chamber has an upper and a lower transmission shafts extending out along the centre line of a circular chamber. This system is not reliable and too complex [4].
3. A 1948 general electric model was launched, but it came with semi-circular refrigerator cabinet, thus wasting too much volume of the refrigerator.

2.2 Problem Statement

In a refrigerator, it is generally less convenient to access the items placed at the back side of the shelf/crisper. The taller items need to be placed at the rear/back side of the shelf/crisper, otherwise they would obstruct while reaching to comparatively shorter products which are placed at the back. Finally, the full extension of the crisper is not achievable because of which there are some constraints while accessing the items at the back side.

3 Concept Formation

In order to facilitate the rotation of the crisper, a smooth curvature has to be provided at the outside of the crisper (Fig. 3).

In this case, the crisper is to be pulled out by 150 mm, i.e. half the length of the crisper, and then, it is turned out about the pivot point which is located at the centre of two crispers. The area lost in this case is 2.4% of the original available area which is acceptable. The above generated concept gives us the following benefits:

Fig. 3 Shape of crisper

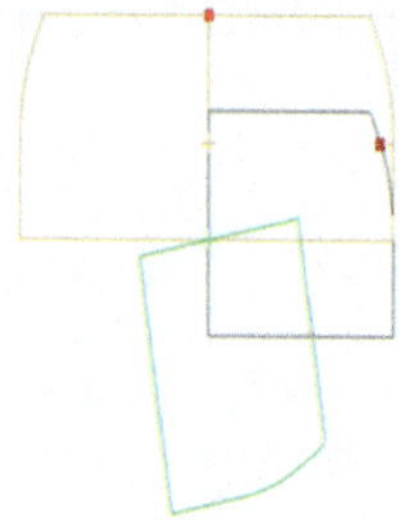

Fig. 4 Guideways with rollers

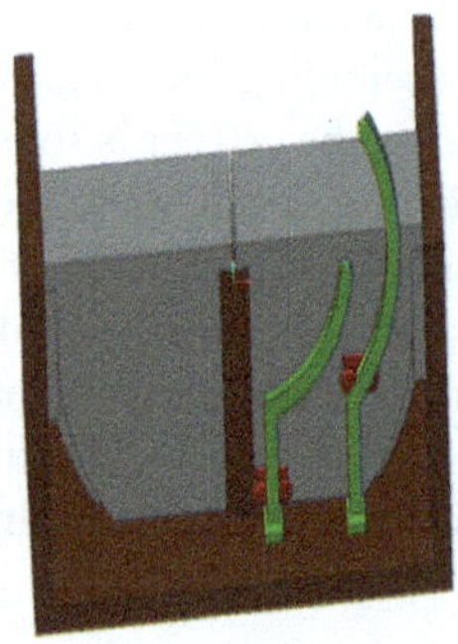

1. As the crisper is first pulled out and then rotated as its position is shown in green, the area at the backside bottom corner is easily accessible.
2. The rear side of the crisper becomes accessible, hence increasing the more number of accessible sides.
3. The full extension of the crisper is achievable, opposed to the current partial (85–90%) extension.

Two concepts are generated here for the achievement of the rotary and linear motion of the crisper. These concepts are presented below in brief and are compared to each other using Figs. 4 and 5 as reference, to see the relative merits and demerits (Table 1).

4 Concept Selection

For finalizing the concept, a tool called Pugh matrix [6] was used in which various considerations are given points and the best concept is identified (Table 2).

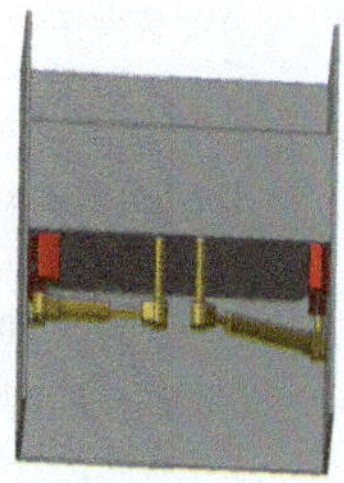

Fig. 5 Mechanism with extraction links

Table 1 Comparison of concepts

Guideways with rollers (Fig. 4)	Mechanism with extraction links (Fig. 5)
In this concept, the guideways are given the curvature in such a way that they give the desired motion sequence of the crisper	In this concept, the linear motion is provided with the help of guideways, and the precisely cut curved slots at the bottom of the crisper facilitate the rotation
There are two guideways for each crisper with each guideway provided with a set of two rollers which are connected to the crisper via a roller block	The endcaps on the glides will match the profile of that of the curved slots
When the crisper is pulled, the roller will roll along the surface of the track, and the crisper will come out and rotate	There is a provision of the extracting links at the either side of the bottom of the crisper, which supports the overhanging portion of the crisper when it is at fully extracted position
Due to the complete length of the tracks being the cantilever, the bending stress at the support will be excessive with large deflection at the end of the tracks which would result in failure of the system [5]	When the crisper is being closed, the links retract back, obtaining their initial position
A sharp change in the direction of the tracks from linear to curvature will be difficult to manage considering the required continuous and smooth motion of the rollers [5]	This mechanism gives the desired motion output in the most effective way

5 Concept Development

To develop the concept further, various components and ideas were generated. It includes the parts, joints, working/function of various parts as follows.

Table 2 Pugh matrix

S. No.	Key criteria	Importance	Guideways with rollers (Fig. 4)	Mechanism with extraction links (Fig. 5)
1	Design safety	5	4	4
2	User experience of crisper opening	5	2	4
3	User experience of crisper closing	5	2	4
4	Manufacturing/assembly complexity	4	−4	−3
5	Cost	3	−3	−3
Weighted sum of positives			40	60
Weighted sum of negatives			−25	−21
Total			15	39

5.1 Glides

The glides were of adjustable type, wherein the length of the glide extension can be controlled.

5.2 Endcaps

To maintain the constant contact of the glides with the crisper, endcaps were needed. To make sure that endcaps would always remain in contact with the crisper even in turning motion, they were given specific curved profile corresponding to the distance from the pivot point.

5.3 Crisper with Slots at Bottom

At the bottom of the crisper, curvilinear slots will be cut to accommodate to endcaps in them. The radius of the slots will be determined by the distance of the starting point of the slot from the pivot point of rotation.

5.4 *Extending Links*

5.4.1 Fixing Member

One end of the link will be a structural member which will be fixed to the support structure and will transmit all the loads and forces to it. This member will also have a provision for a rotary joint for the next part.

5.4.2 Turning Joint

This joint will connect the fixed member with the rotating link. This joint and the fixed member will have snap fit for the ease of the assembly and operation.

5.4.3 Cylinder (Outer Link)

It will be joined to the fixed member through rotary joint. It will act as the outer member for the sliding pair with next member.

5.4.4 Piston (Inner Link)

It will form a sliding pair with the cylinder. The rotation of the crisper will be facilitated by this joint, i.e. it will act as a telescopic link.

6 CAD Design

The final CAD design which is made from the idea is shown in Figs. 6 and 7. As could be seen in Fig. 6, curved wall for crisper is created at the rear-right side of the crisper. Figure 8 shows the front view of the crisper subsystem which clearly shows

Fig. 6 Final CAD design

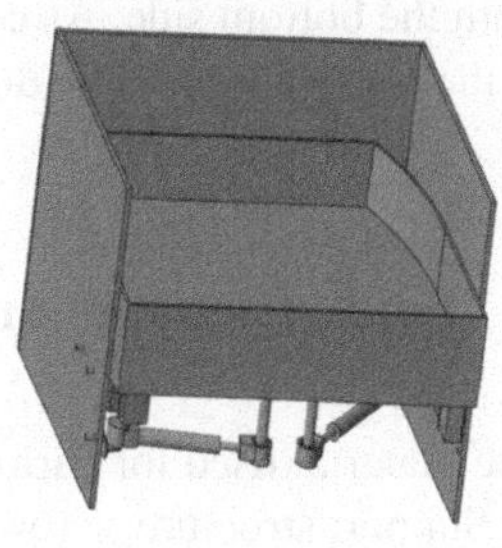

Fig. 7 3D crisper view

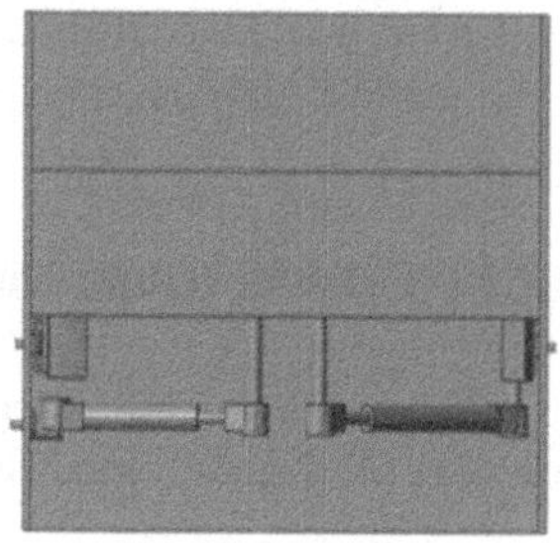

Fig. 8 Extended crisper view

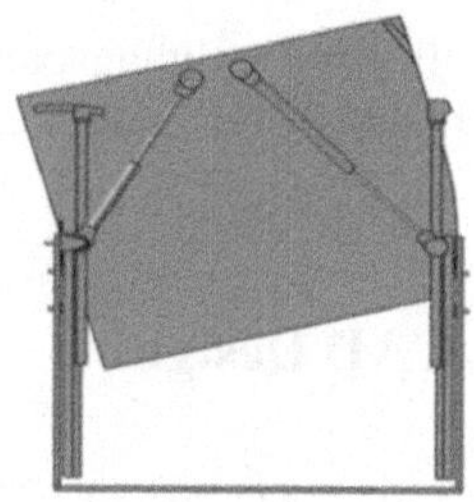

Fig. 9 Extended crisper bottom view

the position of the extending links just below the crisper bottom surface. Figure 8 shows the final extended position of the crisper, and Fig. 9 shows the same position from the bottom side. As could be seen from Figs. 8 and 9, the maximum extension of the crisper is possible due to the extension of the links as clearly seen in Fig. 9.

7 Finite Element Analysis

The material used for various components is as follows.

Support structure, screws, glides, bracket, cylinder, piston rod—Mild Steel (MS)

Crisper—Acrylic and endcaps, screw caps, connectors—ABS plastic

Thus, allowable working stress calculated for each material is as follows:

MS—Yield point stress/FOS = 200/3 = 66.6 MPa

Acrylic—Yield point stress/FOS = 70/3 = 23.3 MPa

ABS—Yield point stress/FOS = 40/3 = 13.3 MPa.

Figures 10 and 11 show the boundary conditions applied and deformation results, respectively.

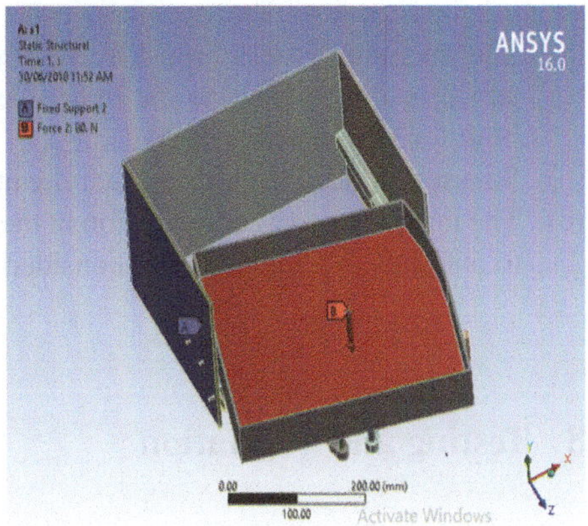

Fig. 10 Boundary conditions

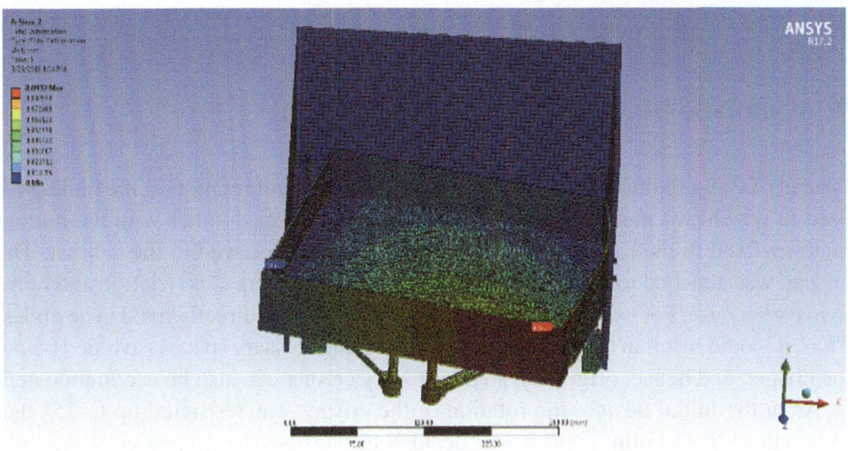

Fig. 11 Total deformation

Table 3 Comparison of results

	Simulation (mm)	Actual (mm)	Permissible (mm)
Position 1 (closed)	0.022	0.07	3
Position 2	0.093	0.21	3

7.1 Results and Conclusion of Simulation

(a) The maximum stress (von-Mises stress) induced at the position 2 is 31 MPa which is also induced at the right side screws, is almost half of the allowable limit for steel which s 66 MPa.
(b) Stresses induced in ABS and acrylic are up to 5 MPa which is also safe.
(c) The maximum total deformation in the position 2 is 0.09 mm which occurs due to overhanging which is safe considering allowable limit is 3 mm.

8 Testing and Validation

The prototype was built and then tested for the maximum deflection under specified lab conditions (Table 3).

9 Prototype Modification and Final Design

Instead of using the side by side structure glides, new configuration of the glides was used in which two members were assembled on top of each other with the bottom member fixed to the L-shaped bracket which was then screwed to the cabinet. The endcap was attached to the top moving member so that there is no friction and noise as in earlier case. For extending links, the brackets are to be directly fixed to he glides. Thus, it would result in the saving of the space as unnecessary space is avoided below the crisper, and hence, original system of pantry crisper can also be accommodated.

As in the initial design, the rotation of the crisper was restricted up to 25° due to length of the RH link, and it was decided to increase the extension of the link. As more than twice the extension was quite difficult to achieve, it was decided to increase the initial length of the link itself, for this purpose, the fixing point of the link to the glide was moved towards the back side of the glide so as to increase the length as shown. The small connector which was used at RH link was also eliminated and was replaced by the additional bracket. Modified design and prototype of crisper are shown in Figs. 12 and 13, respectively.

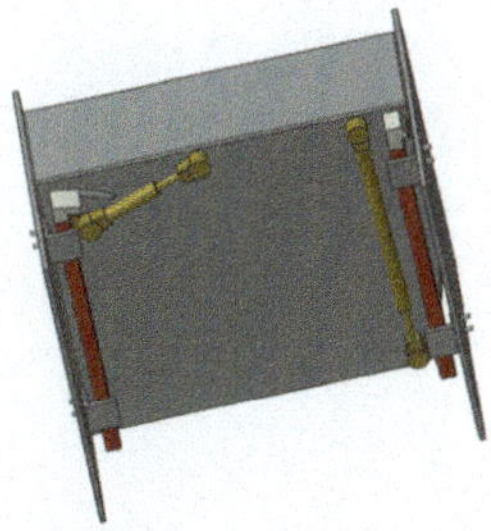

Fig. 12 Modified design

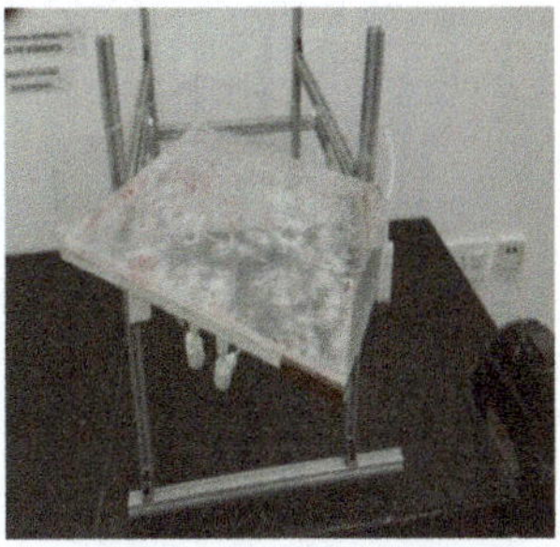

Fig. 13 Prototype

10 Conclusion

In this work, different concepts were conceived for the rotation of the crisper to increase the accessibility. For the selection, Pugh matrix was used in which different considerations were taken. After giving weightage to each criterion, the best concept was selected. The selected concept was then developed in detail, such as generating the mechanism, preparing the CAD model and performing FEA. FEA was run on the CAD design to determine its safety, and results were then validated with the actual test results. The final FEA tests were run on new design, and it passed the tests.

References

1. Hoshbin, Qian Xubo (2010) Rotary type fruit and vegetable box for refrigerator, US Patent CN201680665U
2. Yuan Kaijun, Qian Xubo, Hoshbin (2010) Rotary fruit and vegetable box of refrigerator, US Patent CN101806526A
3. Swindell TR, Sisson CW, Ocharzak DJ (1997) Adjustable refrigerator shelf, US Patent US5913584
4. Zhang Liyu, Zhang Youyu (2002) Cylindrical internal-rotary-shelf refrigerator, US Patent CN2548091Y
5. Beauregard MR, Dermott RM, Mikulak RJ (1996) The basics of FMEA, CRC Press, 2nd edn
6. Pugh S (1991) Total design—integrated methods for successful product engineering. Prentice Hall, Harlow

Design, Analysis and Optimization of a Single-Pass Straight Pipe Resonator for an Exhaust System of a Single Cylinder Engine

Omkar Samant, Gulammoin Kasmani, Jay Saple, Jayraj Ranade, and Vinit Katira

Abstract The purpose of this research was to design a method to attenuate exhaust sound waves within a specified frequency range, in an exhaust system of an internal combustion engine for a formula-type car by resonance. A resonating chamber was designed along with the muffler that targets the frequency range of the exhaust sound. The engine data obtained through dynamometer was used as an input to Ricardo WaveBuild software which provided the data and was further analyzed by a fast Fourier transform model. The resonating chamber design was finalized after iterations of the same which were tested virtually for their transmission loss. The parameters such as location and orientation of exit pipe, length of perforation (internal structure) start and the distance between holes were observed to have a significant effect on transmission loss.

Keywords Resonator · Sound · Frequency · Reflective · FFT analyzer · Muffler · Exhaust · Noise · Ricardo · Wave

1 Introduction

Designing of an exhaust system requires understanding about its effect on output of an engine and noise attenuation. Thus, a sound knowledge in the fields of internal combustion engines, fluid mechanics and acoustics is essential. At the same time, the rules imposed by FSAE competition needs to be taken into consideration. Sonal [1] observed that the engine exhaust noise originates from the exhaust tailpipe openings and is transmitted through the cabin walls, firewall and nose gear bay. Various parameters such as transmission loss, back pressure to the engine due to the restriction of the flow, insertion loss required and available space must be taken into consideration before designing an exhaust system. Since the prime motive was to design a car for the Formula SAE events, the rules prohibit use of absorptive or hybrid mufflers. Therefore, this research is limited to the design and analysis of reactive muffler and

O. Samant (✉) · G. Kasmani · J. Saple · J. Ranade · V. Katira
Dwarkadas J. Sanghvi College of Engineering, Mumbai, India
e-mail: omisamant@gmail.com

© Springer Nature Singapore Pte Ltd. 2020

H. Vasudevan et al. (eds.), *Proceedings of International Conference on Intelligent Manufacturing and Automation*, Lecture Notes in Mechanical Engineering,
https://doi.org/10.1007/978-981-15-4485-9_61

resonator. The volume flow rate, internal diameter of the pipe and other parameters were calculated on the basis of the approach specified by Rahul [2]. Simply by putting a reflective obstacle at a distance of $n/4$ like Shubham et al. [3]. We considered making tunable resonator, but there was a considerable risk of leaking. A tunable resonator is the one whose length can be varied by using a piston that can be set at different positions. The expansion chambers and perforated pipes were designed to relieve back pressure, and the muffler is tested on SimScale GmbH. Multiple design iterations were carried out, and the optimum design was chosen which was then tested on Ricardo WaveBuild software to analyze the acoustics of the exhaust system. The maximum attenuation occurs when $L = n/4$ as reported by Shubham et al. [3]. Testing procedures that Sonal [1] carried out tests by measuring sound pressure level at four locations fixed by making a rig [resonator], measured the noise at different settings of the resonator. Doige and Thomas [4] measured the noise at two locations across the muffler elements. It was decided to perform the noise test before and after installation of the resonator in order to observe the noise attenuation. The frequency log is obtained with and without the resonator to compare it with the original data.

2 Design

The resonator acts like a buffer for the muffler in exhaust system. It is not the actual muffler; hence, it should be designed keeping that in mind. The resonator is a chamber which has a pipe running throughout its length. The pipe running through the resonator has perforations so as to allow air to expand in this chamber considerably and then move on. This expansion helps in noise minimization if the length and diameter of the resonator are kept such that the waves of the desired frequencies are dampened. Since a large magnitude of sound reduction, i.e., 30 dB, is required, a more intricate and effective resonator had to be designed. After some iterations of distance of start of perforation from inlet end, perforation size, etc., the final design is shown in Fig. 1.

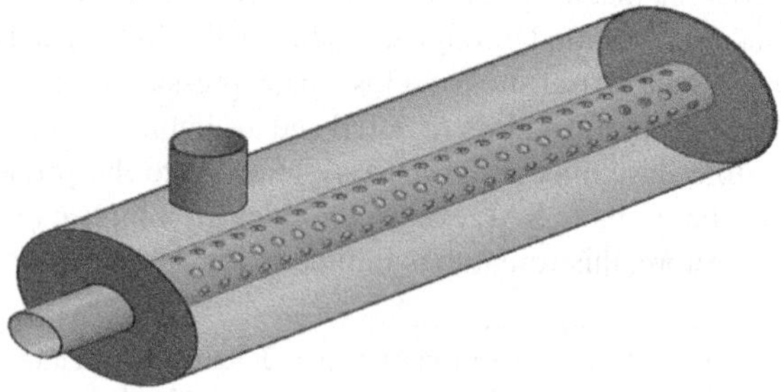

Fig. 1 Resonator design after performing multiple iterations

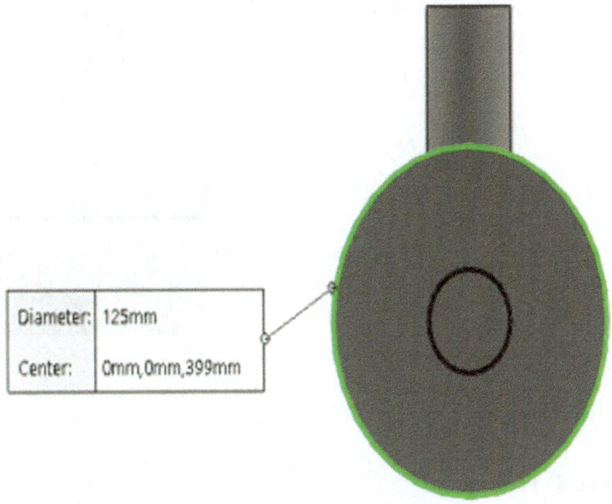

Fig. 2 Resonator can diameter

3 Theoretical Analysis and Calculations

The simplest way to produce a wave of opposite phase is to put a reflective obstacle at a distance of $n\lambda/4$, where $n = 1, 3, \ldots$ other odd integers. A value of $n = 1$ has been used in the design. And the frequency range is taken from 180 to 500 Hz.

The wavelength λ corresponding to frequency 180 Hz $= V_s/f = 1.83$ m and the wavelength λ corresponding to frequency 500 Hz $= V_s/f = 0.66$ m.

Maximum length of resonator cylinder $L_{\text{max}} = \lambda/4 = 0.458$ m.

Minimum length of resonator cylinder $L_{\text{min}} = \lambda/4 = 0.165$ m.

Length of tuning of resonator is given by $L_{rt} = L_{\text{rmax}} - L_{\text{rmin}} = 0.29$ m.

The Resonator Pipe ID was kept 35 mm and OD 37 mm which was similar to the exhaust runner. Each perforated hole had a diameter of 7 mm and distance between two holes was kept 14 mm with eight holes making a circular pattern on the circumference of the pipe. The perforation starts at 44 mm from inlet end all the way till the end. Outlet pipe was 52 mm from one end of the resonator as shown in Figs. 2, 3, 4 and 5.

4 Simulations

There are two types of simulations carried out, and they are as follows:

1. Ricardo wave software simulations to determine the transmission losses in different designs of resonators.
2. CFD analysis on SimScale software to validate the flow of gases inside the resonator and to examine the pressure and velocity variations inside the resonator.

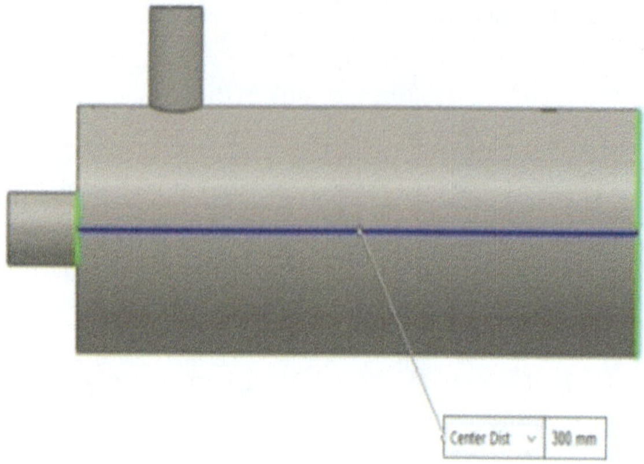

Fig. 3 Resonator can length

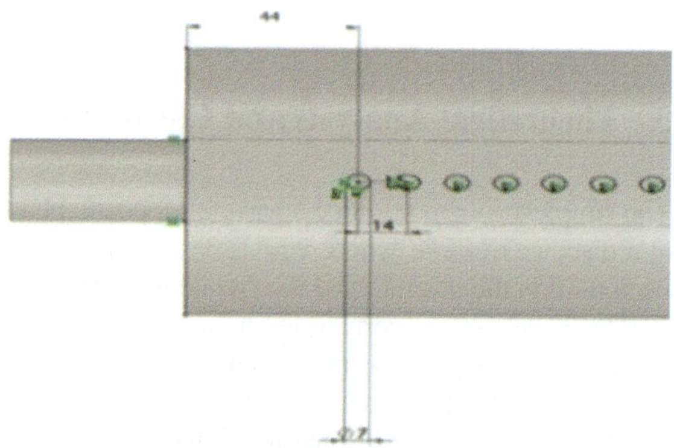

Fig. 4 Perforations dimensions

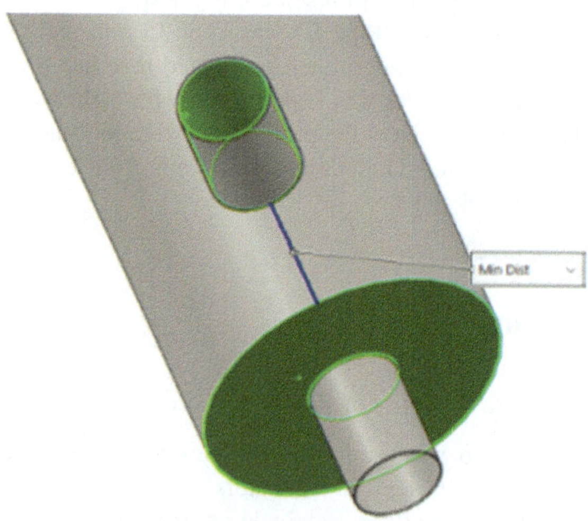

Fig. 5 Distance from inlet

4.1 *Ricardo Simulation*

4.1.1 Engine Plot on Ricardo WaveBuild

A virtual engine dynamometer helps to compute the dB level for the exhaust system. The engine model was created on Ricardo wave along with the complete intake and exhaust system. The WaveBuild model was used to find out the acoustic data required for designing an exhaust system. A pressure sensor was placed just near the exhaust port which was used to collect real-time pressure data in the exhaust runner as shown in Fig. 6. This data was used to extract an audio file from Ricardo Wave-Post.

The extracted audio file created using the pressure sensor was analyzed in FFT analyzer (DSSF3), and SPL versus frequency graphs were obtained as shown in Fig. 7. The analyzer was set to dB(*c*) and fast weighing. The sampling rate was set the same as the audio file.

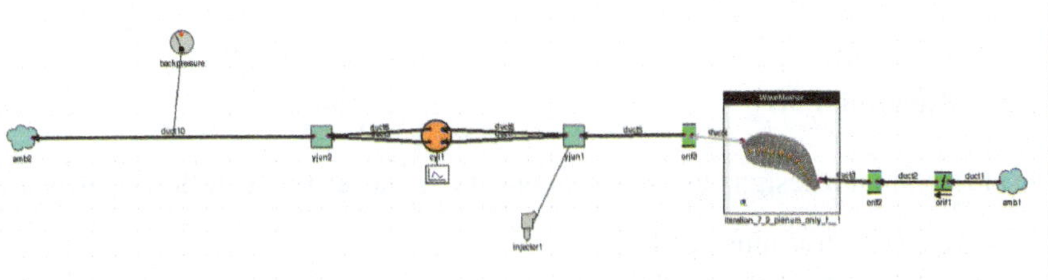

Fig. 6 Engine plot on WaveBuild

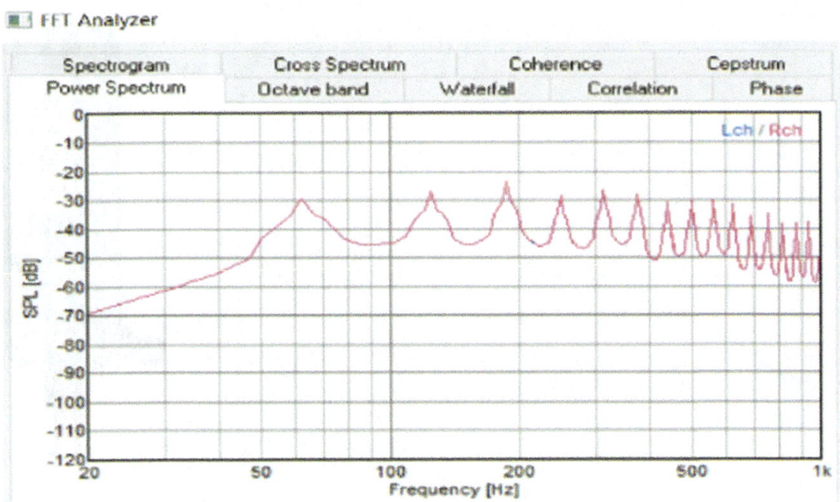

Fig. 7 SPL versus frequency graph

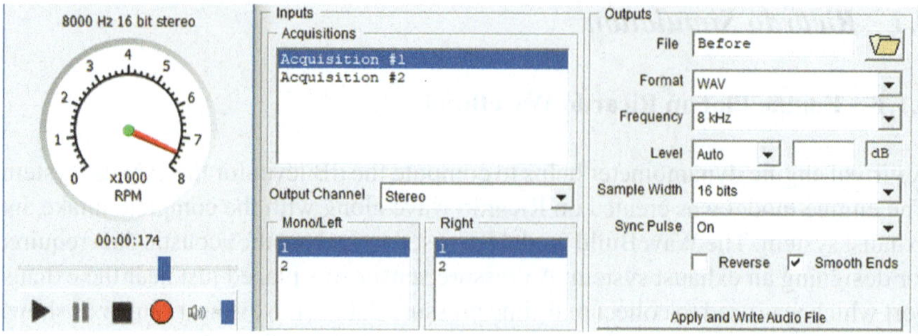

Fig. 8 Input to FFT analyzer

It was observed from the above plots that the peaks in the range of 180 and 500 Hz are to be attenuated, so resonator design was iterated to get the desired transmission loss (Fig. 8).

4.1.2 WaveBuild 3D

The most optimum design represented in Fig. 9 was chosen for the following reasons:

1. Nearly 20 dB transmission loss
2. No nodes in the range of operation
3. Routing and packaging are better.

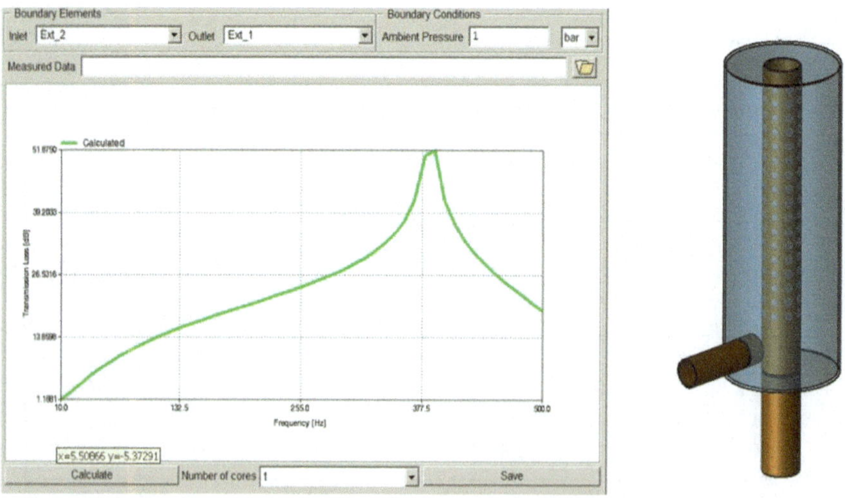

Fig. 9 Iteration 1

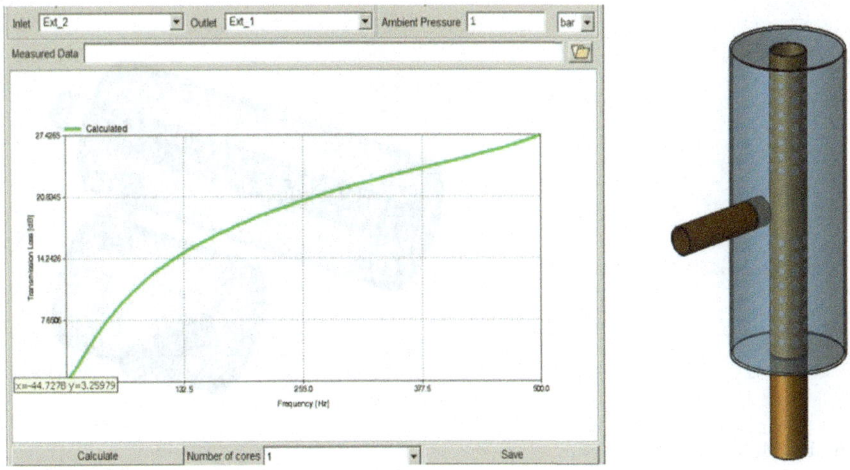

Fig. 10 Iteration 2

The resonating chamber design was finalized after iterations of the same which were tested virtually for their transmission loss. The parameters such as location and orientation of exit pipe, length of perforation (internal structure) start and the distance between holes were observed to have a significant effect on transmission loss (Fig. 10).

4.2 CFD Analysis on SimScale

A bounding box was also specified to generate an accurate mesh. After meshing, the analysis type was chosen. Incompressible type of analysis was done, and k-omega, SST turbulence model was chosen. The pressure at inlet of resonator and mass flow rate at outlet of muffler and wall were the three boundary conditions used for analysis as shown in Figs. 11, 12, 13 and 14. The simulated results of pressure drop and velocity variations are shown in Fig. 15.

The design represented was finalized due to the following reasons:

1. This design gives the highest pressure drop between inlet and outlet.
2. This design gives almost constant velocity between inlet and outlet.

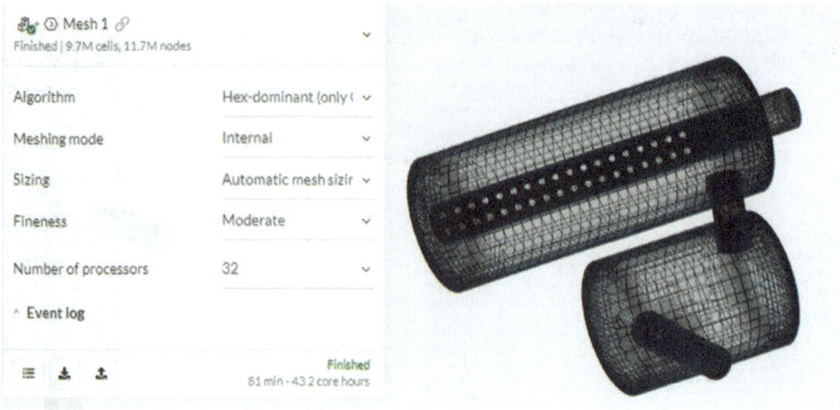

Fig. 11 Mesh size

Fig. 12 Analysis type

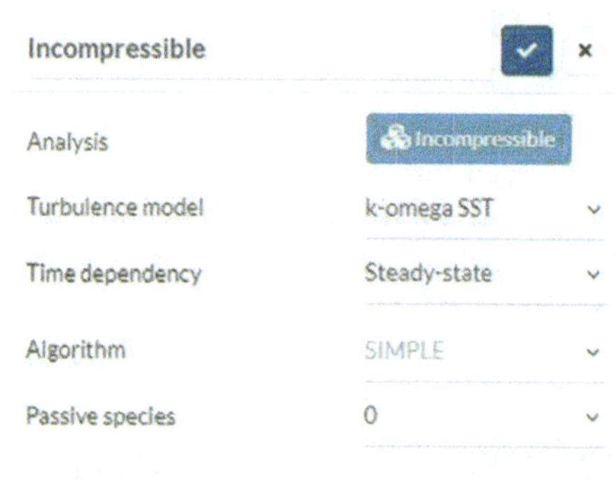

Fig. 13 Velocity condition

Fig. 14 Inlet pressure

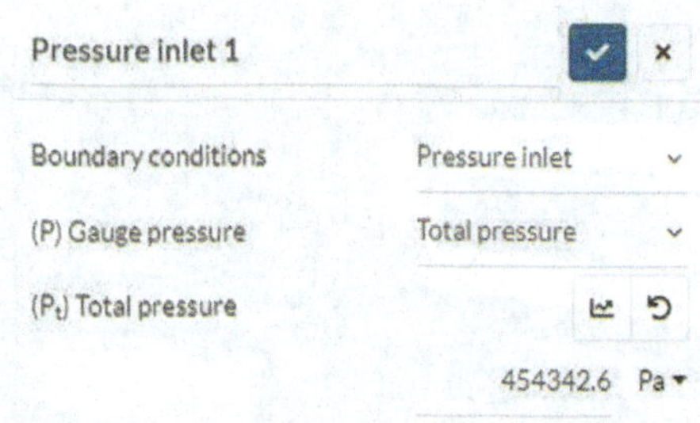

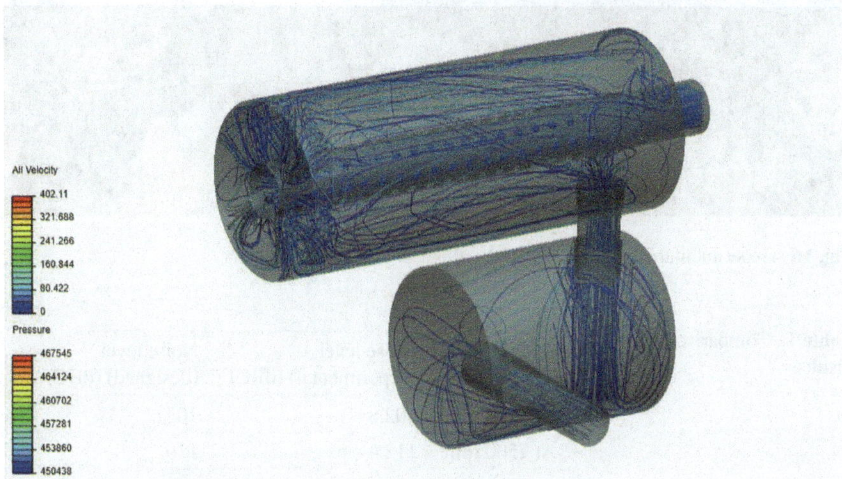

Fig. 15 Analyzed results

5 Experimental Analysis

5.1 Test Procedure

1. The test was carried out with the engine mounted on the chassis of car as shown in Fig. 16. The gaseous emissions from the exhaust of the engine include hydrocarbons, carbon monoxide and oxides of nitrogen.
2. During a prescribed sequence of warmed up engine operating conditions, the amounts of the above gases in the exhaust shall be examined continuously. The prescribed sequence of operations consists of a number of speed and power modes which span the typical operating range of petrol engines.
3. During each mode, the concentration of each pollutant, exhaust flow and power output shall be determined and the measured values weighted and used to calculate the noise level of the resonator.

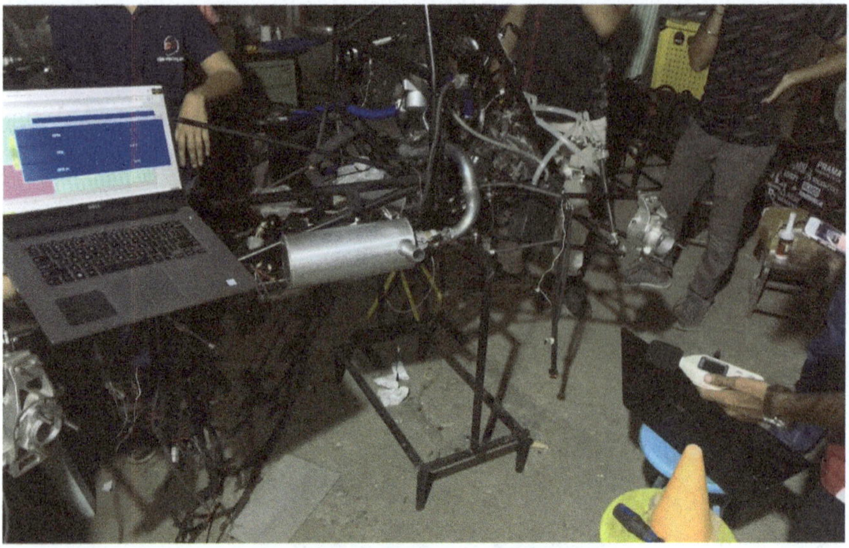

Fig. 16 Experimental Test Setup

Table 1 Comparison of results

Condition	Noise level (experimental) (dBC)	Noise level (designed) (dBC)
At 3100 rpm	102.8	103
At 7500 rpm	112.4	110

5.2 *Test Results*

The experimental results were compared with the simulated results at idling rpm and at 7500 rpm. The obtained experimental results are shown in Table 1.

6 Conclusion

The exhaust resonator designs were iterated to obtain the most optimum design on the basis of maximum pressure drop and sound minimization. The selected design was simulated using Ricardo WaveBuild and SimScale software packages. The selected design was fabricated, and its performance parameters were obtained. The exhaust sound attenuation results obtained from the experimental tests were compared with simulated results and were observed to be in a good agreement. The difference between the two experimental and simulated values could be attributed to the uncontrollable parameters like the temperature of surrounding, wind speed, localized pressure differences, calibration of decibel meter, etc.

References

1. Jebanski R. Absorption mufflers in exhaust systems
2. Nazirkar CD, Meshram SR, Namdas AD, Navagire SU, Devarshi SS. Design and optimization of exhaust muffler and design validation
3. Pal S, Golan TS, Kumar V, Jain V, Ramdas N, Sharma OP. Design of a muffler & effect of resonator length for 3 cylinder SI engine
4. Panicker VB, Munjal ML. Aeroacoustics analysis of straight through mufflers with simple and extended tube expansion chambers
5. Potente D. General design principles for an automotive muffler
6. Zakhmi S. Study of noise generated by single cylinder two stroke petrol engine
7. Parrott TL. An improved method for design of expansion-chamber mufflers with application to an operational helicopter
8. Gupta AK, Tiwari A. Performance of transmission loss on hybrid muffler by using rock wool and glass fiber as absorbing materials
9. 2019 Formula SAE Rules. SAE International

Design and Development of Cost-Effective Solar Water Heating System

Sarvesh Kulkarni, Vijaya Kumar N. Kottur, and Prasad Shirodkar

Abstract One of the major problems faced by the thermosiphon solar water heating system is mixing of hot and cold water. The mixing is often caused due to the disturbance in the stratification layers. The incoming cold water jet creates turbulence that causes the mixing of hot and cold water. The current work is based on the development of effective control system which controls the flow of incoming cold water. A solenoid valve is actuated according to the temperature and water level inside the tank. With 10-bit microcontroller, the programming was done with the help of mikroC PRO software, and the circuit diagram was developed with the help of Proteus Design Suite software. The compact device which detects level was developed that controls the flow of incoming cold water. A thermistor was incorporated in the system separately to monitor the temperature inside the tank.

Keywords Thermosiphon solar water heating system · Control system · Level sensor

1 Introduction

Among the available solar water heating systems, the simplest and cheapest is thermosiphon solar water heating system which works on the principle of natural convective heat transfer. In the thermosiphon system, the hot water is settled at top of the tank due to low density, whereas cold water is settled at the bottom of the tank due to high density. However, the system is facing the major problem of mixing during the replacement of hot water with cold water in the storage tank. Due to this, the customer/user is unable to get hot water for a longer period. The mixing degrades

S. Kulkarni (✉) · V. K. N. Kottur · P. Shirodkar
Mechanical Engineering, Dwarkadas J. Sanghavi College of Engineering, Mumbai, India
e-mail: sarvesh.94.kulkarni@gmail.com

V. K. N. Kottur
e-mail: Vijaykumar.kottur@djsce.ac.in

P. Shirodkar
e-mail: Prasad.shirodkar@djsce.ac.in

© Springer Nature Singapore Pte Ltd. 2020
H. Vasudevan et al. (eds.), *Proceedings of International Conference on Intelligent Manufacturing and Automation*, Lecture Notes in Mechanical Engineering,
https://doi.org/10.1007/978-981-15-4485-9_62

the performance of the system. This causes an additional use of auxiliary electrical source. The cost of electricity found out to be Rs. 16/day. Also, the amount of solar fraction of the system found out to be 40% [1].

One of the methods to stop the mixing problem is the development of the control system. Many attempts have been done already to increase the performance of the system. Ahmed et al. [2] have developed a control system in which they have kept the temperature of the tank in a specified range, using two Arduino-based devices for monitoring the temperature and to control the speed of the motor based on the values of temperature. They have used this system for solar PV thermal collector. In the attempt, Tahat et al. [3] have developed a remote monitoring control system. The installed device shows the level of heated water inside the tank, temperature of heated water and amount of water retrieved from the tank. The data can be obtained globally with the help of the cellular network. Buniyamin et al. [4] have developed a low-cost solar water heater. In this, they have used a PLC for controlling the sensors and actuators. The purpose of this is to provide constant hot water. Tasnin and Choudhury [5] developed a solar water heater controller in which they provide water at a required temperature. When the temperature is heated above the desired temperature, controller drops down the value by mixing cold water into it. In Indian market, a private company is providing a non-mixing technology along with their water heating system.

The current work focuses on maximum work utilization of the system by avoiding mixing of hot and cold water. The next focus is on the development of the control system which is simple in design, assembly and maintenance. Finally, the main aim is to provide the 100% hot water at desired temperature. We are using a thermistor as a temperature sensor for the measurement of temperature and resistive water level sensor for the measurement of water level inside the tank. The microcontroller used is 10-bit Pic 16f877a having inbuilt ADC convertor.

2 Methodology

The following step-by-step approach is followed for the development of control system.

2.1 *Design of Control System for Non-mixing Solar Water Heating System*

The installation of the sensor device outside the tank makes assembly simpler, and the maintenance will become easier. The schematic diagram of the arrangement of a solar water heating system along with the sensor device is shown in the following Fig. 1. In the figure, cold water is taken out of the storage tank. A solenoid valve

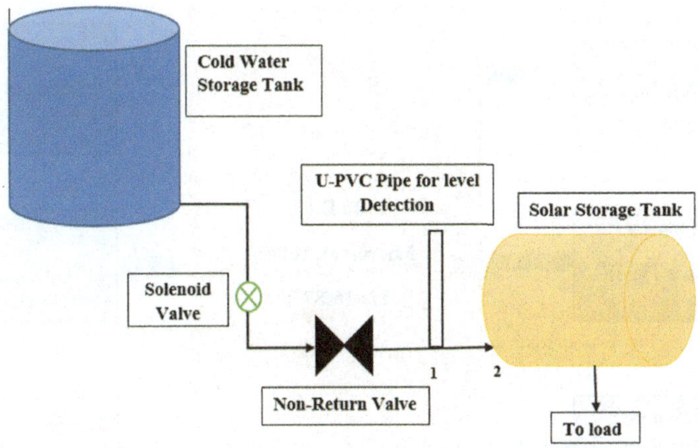

Fig. 1 Schematic arrangement of the proposed system

will get actuated when 75% of the hot water removed from the solar storage tank. A U-PVC pipe will be connected in between solar storage tank and non-return valve. A non-return valve is used for the safety purpose to prevent the reverse flow of water from the solar storage tank to the solenoid valve. The hot water will be taken out from the bottom of the tank called "drain" to the load.

A U-PVC pipe will acts as a sensor device in the proposed study. Whatever, the water level inside the tank the same water level will be there inside the pipe.

The Bernoulli's equation is based on the conservation of energy inside the fluid. The total energy at any point inside the fluid remains constant. The energy includes pressure energy, kinetic energy and datum energy.

$$\frac{P_1}{\rho g} + \frac{V_1^2}{2g} + z_1 = \frac{P_2}{\rho g} + \frac{V_2^2}{2g} + z_2 \tag{1}$$

In Fig. 1, at point "1" and point "2", the flow is constant. Therefore, velocity and pressure head remain same. As a result, the datum head will also remain same.

2.1.1 Development of Block Diagram of Control System

The block diagram of the control system is as shown in Fig. 2. The block diagram will indicate what are the inputs required and what the possible outputs of the control system are. The 230 V AC power supply is given to the microcontroller. After processing the data, it will be indicated on display unit. The solenoid valve will actuate according to the command given to it. The level sensor is of resistive type. The cost of the sensor is very low. The principle behind this is voltage divider rule. The voltage

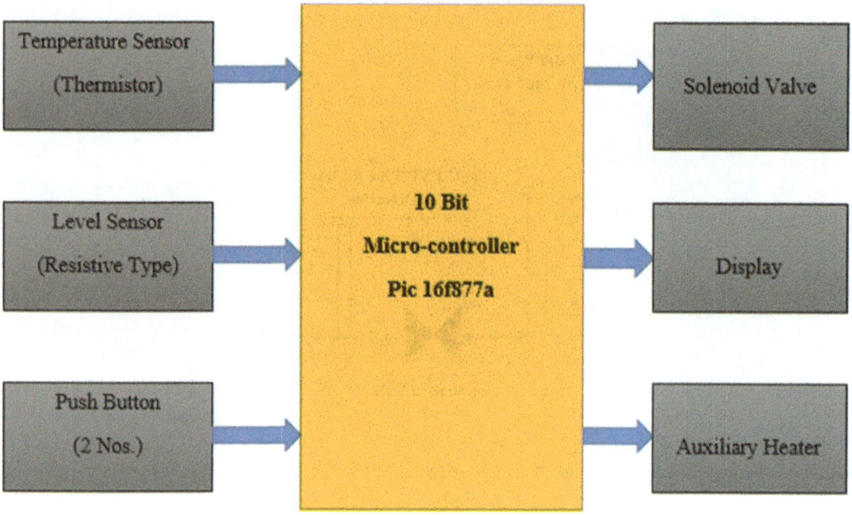

Fig. 2 Block diagram of the control system

divider rule is applicable to the resistors connected in series. The voltage across any of the resistor is given by the following equation.

$$V_{\text{out}} = V_{\text{in}} * \left(\frac{R_1}{R_{\text{total}}}\right) \tag{2}$$

A 10 kΩ NTC type thermistor is used as a temperature sensor. The thermistor is having very high sensitivity and fast response time [6]. Its resistance value decreases when there is an increase in temperature.

A solenoid valve is a combination of the electromechanically operated valve. It allows the flow of water only when an electrical signal is given to it. When there is no requirement of hot water, a push button is provided to turn on the solenoid valve so that the tank gets filled with cold water. Another push button is provided to turn on the auxiliary heater. The auxiliary heater is located at the middle of the tank. A thermostat is provided to turn off the power supply when set temperature achieved. A proper care has taken to avoid the dry run of the heater.

2.2 Development of Circuit Diagram of the Control System

The primary function of the circuit diagram is to convert 230 V, 50 Hz AC supply into 5 V DC supply. For this transformer, rectifier, filter, the regulator is used. A step-down transformer will convert 230 V AC voltage into 12 V AC. Then, it gets rectified with the help of rectifier. The rectifier will convert AC voltage into DC voltage. The value of the will be $12\sqrt{2}$. Then the same signal gets filtered by the filter in which unwanted noise gets eliminated. The regulator will maintain the required voltage in the circuit. For this, 7805 regulator is used. This will provide the 5 V DC to flow through it. An input from level sensor and thermistor is given to it. LCD is connected to show the water level inside the tank and outlet water temperature.

The circuit diagram of the given system is as shown in Fig. 3.

2.3 Design of Sensor Device

In the design part, as we have seen in Fig. 1, a U-PVC pipe has been used as a sensor device. The diameter of the pipe is 15 mm. The height of the pipe is equal to the diameter of the solar storage tank.

$$\text{Height of pipe} = \text{Diameter of storage tank}$$

For detection of the different levels like 0, 25, 50, 75 and 100%, a total of five screws were connected. The resistors were connected in between screws. A total of four resistors having the value of 10 kΩ, ½ were connected in series. To avoid the leakage, proper insulation was provided. The actual model of a sensor pipe is as shown in Fig. 4.

3 Analysis of Storage Tank

The storage tank is used to store the sensible heat which is gained from the collector. The analysis of the storage tank is done with the help of hourly variation of temperature inside the storage tank. The temperature of water drops down only when load is high on system. The heat lost from side face and also from the longitudinal surface has been considered. Also, the heat lost due to the support structure has been also considered. However, the radiation heat loss has been neglected during the analysis.

After installation of the system, the output temperature of water has been monitored hourly. In the regular system, the output hot water has been taken out from the top of the tank, whereas in current research work, it has been taken out from downside of the tank. The pipe layout of the tank is as shown in Fig. 1.

The hourly temperature variation inside the tank is given by the following equation [1].

$$\frac{Q_u - Q_{\text{load}} - (\text{UA})_t (T_l - T_a)}{Q_u - Q_{\text{load}} - (\text{UA})_t (T_{\text{li}} - T_a)} = \mathrm{e}^{\left[-\frac{(\text{UA})_t * t}{(\rho v c_p)_e}\right]} \tag{3}$$

where

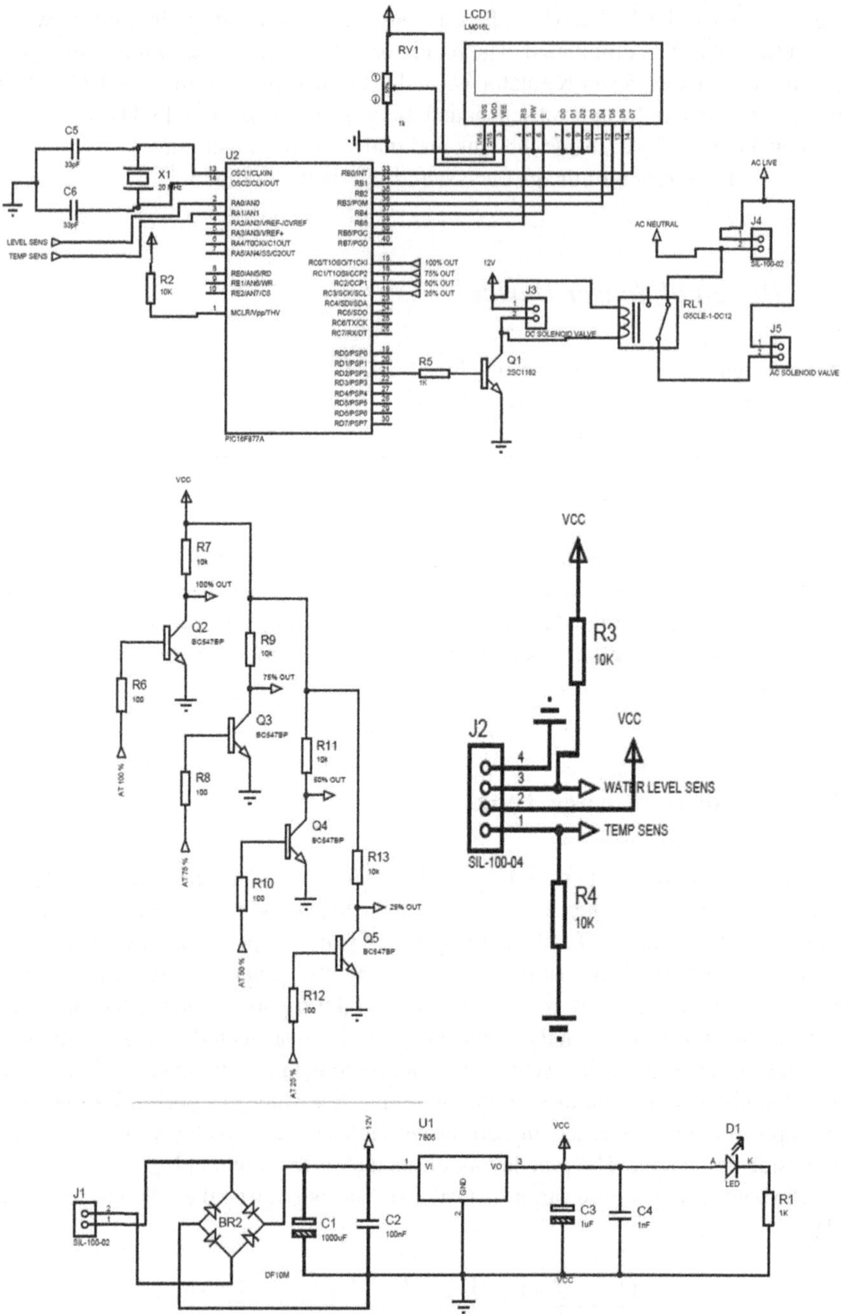

Fig. 3 Circuit diagram of control system

Fig. 4 Actual model of sensor pipe

Q_u = heat gain by the system in w
Q_{load} = extraction of energy to load in w
$(\text{UA})_t$ = product of overall heat transfer coefficient and area of the tank
T_{li} = temperature of liquid at tank $t = 0$ in K
T_a = ambient temperature in K
t = time in s.
$(\rho v c_p)_e$ = effective heat capacity of tank.

The heat gain (Q_u) by the system can be found out with the help of the following equation

$$\eta_i = \frac{Q_u}{A_c * I_T} \tag{4}$$

where

η_i = instantaneous efficiency of collector
I_T = intensity of solar radiation on tilted surface in (KJ/m^2 h)
A_c = collector area in m^2.

The load on the system has been calculated with the help of the following equation.

$$Q_{\text{load}} = m^*_{\text{load}} * c_p * (T_l - T_i) \tag{5}$$

where

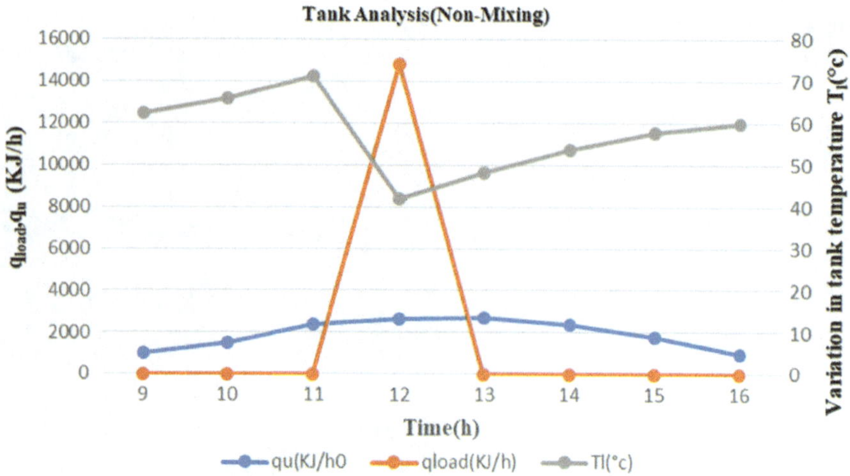

Fig. 5 Tank analysis of the system

m^*_{load} = quantity of water removed from the tank per hour
c_p = specific heat of water = 4.195 kJ/KgK
T_l = temperature of liquid inside the tank
T_i = temperature of makeup liquid.

Figure 5 shows the hourly variation of temperature at the end of hour. For the given load profile, the user starts drawing hot water from the system. As there is no cold water coming inside the tank, the load on the system becomes zero. This prevents the temperature drop of water inside the storage tank. The water receives heat energy from the sunlight which results in increase in temperature of water.

When a 75% of water is drawn, a solenoid valve allows cold water to come inside the tank. This causes a huge load on the system. The water starts circulating through the collector. A main purpose of the current research work is to decrease the load on the system during the requirement of water. From Fig. 5, the temperature of water reaches up to 71 °c. The temperature drops down to 36 °c when cold water comes inside the tank. The load profile (m^*_{load}), heat gain by the system (Q_u) and the instantaneous collector efficiency (η_i) are summarized in Table 1.

4 Cost Analysis

For the proposed control system, the incoming cold water will go inside the tank around 11 am. The tank gets filled with the cold water, and solenoid wall gets turned off automatically. The user will get water at constant temperature till solenoid valve turns ON. The system has provided the hot water at constant temperature. At 11 AM because of incoming cold water, the load on the system is high and temperature drops

Table 1 Variation of tank temperature (per hour)

Time (h)	q_u (KJ/h)	T_a (°C)	q_{load} (KJ/h)	T_l	η_i (%)	m^*_{load} (Kg/h)
9	997.02	30.2	0	62.2	31.6	60
10	1535.6	31.2	0	65.6	39	10
11	2363.5	31.9	0	71.1	42.4	10
12	2655.1	32.3	14,806.6	41.8	43.6	0
13	2695.0	32.4	0	48.1	43.5	0
14	2363.5	32.4	0	53.5	41.7	0
15	1774.3	32.2	0	57.6	37.5	10
16	989.36	32	0	59.8	29.1	10

down to 36 °C. The water gets circulated through the collector, receives heat, and it gets stored inside the tank. This process continues until there is enough sunlight available. The auxiliary heater has to be turned on at 2 o'clock because of load requirement of the user.

The amount of auxiliary heat required to reach the temperature of water can be calculated as follows:

$$Q_{\text{aux}} = m * C_P * \Delta T \tag{6}$$

$$Q_{\text{aux}} = 100 * 4.195 * (60 - 53.5) = 2726.75\,\text{KJ}$$

Therefore, the cost of electricity required can be calculated as,

$$\begin{aligned} \text{Cost} &= \text{unit of energy required} * \text{cost of electricity(Rs./kWh)} \\ &= (2726.75/3600) * 8 \\ &= \text{Rs.}\,6.095/- \end{aligned}$$

The solar fraction is the amount of solar energy used for the heating. The solar fraction can be calculated as follows,

$$F = \frac{Q_1 - Q_{\text{aux}}}{Q_1} \tag{7}$$

$$F = \frac{14806.6 - 2726.75}{14806.6} = 0.81$$

5 Conclusion

The domestic solar heating systems currently available are based on thermosiphon principle. The main problem encountered with thermosiphon-based solar water heating system is distraction of the stratification layers. Through this research work, an attempt has been made to significantly reduce the distraction of stratification layers through the use of a control system. The control system approach was found to be effective. The Haier company has developed a control system which incorporates a sensor device, which is installed inside the storage tank. The constructional features of the sensor device are not user friendly, making it rather difficult to install and maintain.

The sensor device envisaged currently is simple in design and construction, permitting ease of maintenance and installation. It permits installation at the outer surface of the storage system and does not get soaked in water, preventing contamination from sediments like mud present in the water.

The energy cost incurred while using the existing systems works out to be Rs.10/- per day. As against this, the improved version of the water heating system costs only Rs.6 per day. Because of cold water getting mixed up with hot water, the efficiency of these systems is rather poor and permits only a bare solar faction of 40%. As against this, the developed system permits a solar fraction of 80%, resulting in effective energy utilization. An efficient and affordable solar water heating system was explored through this research work.

References

1. Sukhatme S, Nayak J (2013) Thermal energy storage. In: Solar energy. McGraw Hill Education (India) Private Limited, pp 263–264
2. Ahmed WA, Aggour M, Bennani F (2017) Automating a solar water heating system. J Energy Syst 1(2):56–64
3. Tahat A, Abukhalaf M, Elmuhesen O (2011) Solar energy water heater remote monitoring and control system. In: International conference on electronic devices, systems and applications
4. Buniyamin N, Salah KNY, Mohamed A, Mohamad Z (2011) Low-cost solar water heater. In: International conference on electronic devices, systems and applications
5. Tasnin W, Choudhury PK (2015) Design and development of an automatic solar water heater controller
6. National Instruments (Online). http://www.ni.com/en-in/innovations/white-papers/06/overview-of-temperature-sensors.html#section–1635560423. Accessed 19 Apr 2019

Development of Storage System by Designing a Magazine for Forged Rings

Meet Karelia, Mehul Prajapati, and Vinayak Salian

Abstract A forged groove ring is one of the vital parts, widely used item in aerospace engineering. It is an essential step to store these rings carefully and keep them handy, when they are required. This paper helps the manufacturing shop to store these rings efficiently and keep it handy, whenever they are needed. By designing a magazine for storage system, stresses were calculated by selecting proper material and its availability. This provides machine shop safety as well as helps in efficiently handling the rings.

Keywords Forged groove rings · Shell segment

1 Introduction

The forged groove rings are used for connecting two shell segments in order to manufacture rocket launching vehicle. They have to be fixed with other in a proper full proof way that is convenient to assemble and dismantle at any given point of time. It is often found that it is difficult to roll a large shell of the entire length of the rocket. Hence, the connected shell segments with rings not only increase the convenience factor but also help in the strength and flexibility of the launch vehicle. When ring arrives into the shop, they are not stacked properly and do not have any specific space for storage. It was decided to develop a storage cum retrieval system to access these rings in the shop floor.

M. Karelia (✉) · M. Prajapati
Department of Production Engineering, Dwarkadas J. Sanghvi College of Engineering, Vile Parle, Mumbai, India
e-mail: kareliameet23@gmail.com

V. Salian
Defence & Aerospace Department, Larsen & Toubro Pvt Ltd, Powai Works, Mumbai, India

© Springer Nature Singapore Pte Ltd. 2020
H. Vasudevan et al. (eds.), *Proceedings of International Conference on Intelligent Manufacturing and Automation*, Lecture Notes in Mechanical Engineering,
https://doi.org/10.1007/978-981-15-4485-9_63

Fig. 1 Old storage system

2 Methodology

a. **Old method**

In Fig. 1 shown, the rings are left open with just wooden blocks separating them, and they are stacked one above each other. These rings are placed in an open unassigned area completely, where crane is used to pick the rings up and place them one over another. In old method, if the bottom rings are to be taken out from the magazine, the rings above are disturbed, an ineffective way to remove the rings which takes 10–15 min to handle.

b. **Modified method**

In this new and proposed method shown in Fig. 2, a new design has been created for storing and handling these forged rings. As mentioned, the previous method was inefficient and time consuming, whereas this method eliminates the handling issues and moreover reduces the risk of any accident that can be caused while stacking these rings [1]. The design which had been selected for rings is in horizontal position, and it can stack up to 10–11 rings.

3 Design

In this new and proposed method in Fig. 3, a new design has been created for storing and handling these forged rings; as mentioned above, the previous method was inefficient and time consuming, whereas this method eliminates the handling issues and moreover reduces the risk of any accident that can be caused while stacking these rings. The design which had been selected for rings was in horizontal position, and

Fig. 2 Modified storage system

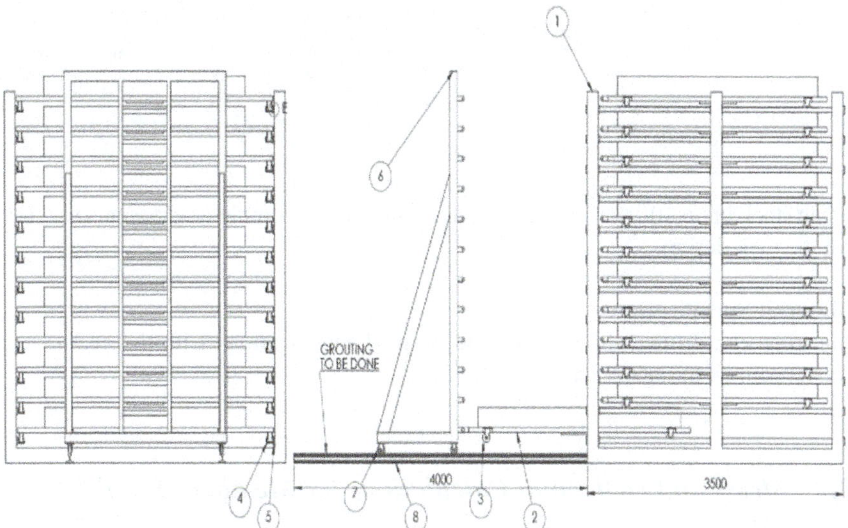

Fig. 3 Storage system design

Table 1 Design description

S. No.	Description
1	Storage rack
2	Trolley
3	Caster wheel
4	ISMC 100
5	Spacer plate
6	Moving trolley
7	Track wheel
8	Rail beam

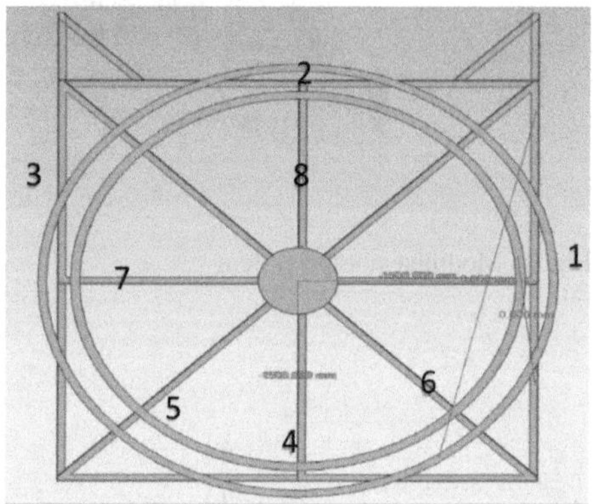

Fig. 4 Top view of 2 rings with frame

it can stack up to 10–11 rings [2]. Table 1 enlists the assembly components required for the rack design.

Material Specification

Inferring Fig. 3, the following are the detailed specifications with the beam cross section shown in Figs. 4 and 5) (Table 2).

3.1 Stress and Deflection Calculations for Members 1 and 3 [3]

$$\text{Max Weight acting on the frame considering both rings} = 400 \times 2 = 800\,\text{Kg} = 8000\,\text{N}$$

Considering load factor, Total Load, $P = 16{,}000$ N

Load at centre of each beam (considering No. of beams = 4) = $\frac{16{,}000}{4} = 4000\,\text{N}$

Total load at centre (weight of rings + self-weight) (Fig. 6);

$$W = 4000 + \frac{(3200 \times 92)}{1000} = 4294.4\,\text{N}$$

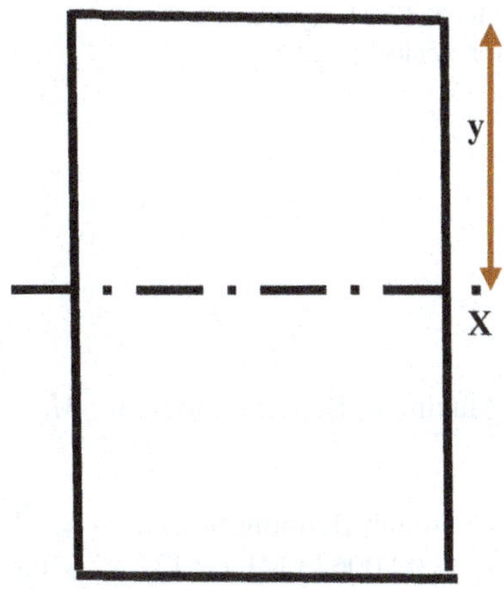

Fig. 5 Rectangular cross section of beam

Table 2 Material specification

Material	ISMC 100 Carbon steel
Self-weight	w = 92 N/m
y	50 mm
Moment of inertia for rectangular section	IXXX = 186.7×10^4 mm^4
Yield stress	σ_y = 247 MPa
Factor of safety	FOS = 1.8
Allowable stress	$\sigma_{allowable}$ = 137.22 MPa
Beam span along edge (2,4, 7 & 8)	L = 3000 mm
Beam span along edge (1 & 3)	L = 3200 mm
Beam span along diagonal (5 & 6)	L = 4242 mm
Young's modulus	E = 210 GPa
Weight of groove	400 kg
Load factor	2

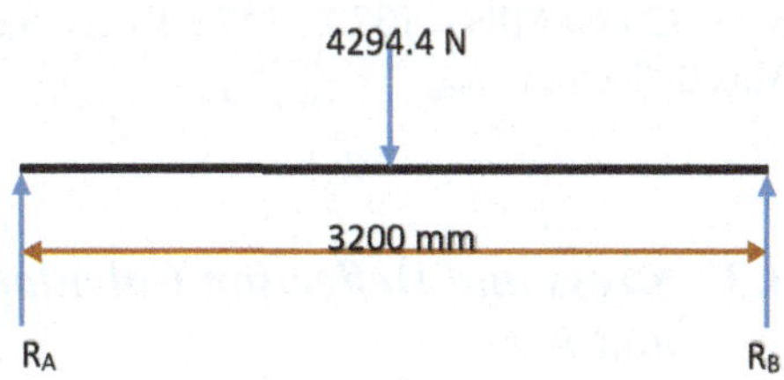

Fig. 6 Free body diagram for member 1 and 3

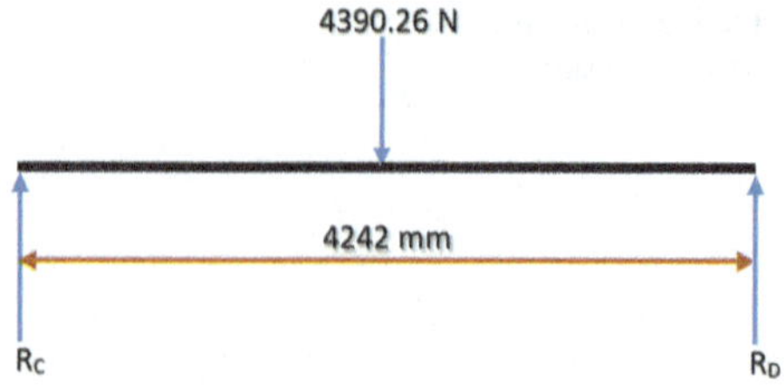

Fig. 7 Free body diagram for member 5 and 6

$$\text{Maximum Bending moment, } M_{max} = \frac{W \times L}{4} = \frac{4294.4 \times 3200}{4} = 3{,}435{,}520\,\text{N mm}$$

Maximum Bending Stress, $\sigma_b = \frac{M \times y}{I} = \frac{3{,}435{,}520 \times 50}{186.7 \times 10^4} = 92.00643\,\text{MPa}$
$\sigma_b = 92.00643$ MPa < 137.22 MPa. Hence Safe
Max deflection, $\delta_{max} = \frac{W \times L^3}{48 \times E \times I} = \frac{4294.4 \times 3200^3}{48 \times 2.1 \times 10^5 \times 186.7 \times 10^4} = 7.48\,\text{mm}$

3.2 Stress and Deflection Calculations for Members Diagonal Members 5 and 6

Max Weight acting on the frame considering both rings together = 4000 N
Let, Load factor = 2 Therefore, Total Load, $P = 8000$ N
Load at centre(due to rings = 2 rings) = $\frac{8000}{2} = 4000\,\text{N}$
Total load at centre(rings + self − weight); (Fig. 7)

$$W = 4000 + \frac{(4242 \times 92)}{1000} = 4390.264\,\text{N}$$

$$\text{Maximum Bending moment, } M_{max} = \frac{W \times L}{4} = \frac{4390.264 \times 4242}{4} = 4{,}655{,}874.972\,\text{N mm}$$

Maximum Bending Stress, $\sigma_b = \frac{M \times y}{I} = \frac{4{,}655{,}874.972 \times 50}{186.7 \times 10^4} = 124.69\,\text{MPa}$
$\sigma_b = 124.69$ MPa < 137.22 MPa. Hence Safe
Max deflection, $\delta_{max} = \frac{W \times L^3}{48 \times E \times I} = \frac{4390.264 \times 4242^3}{48 \times 2.1 \times 10^5 \times 186.7 \times 10^4} = 17.81\,\text{mm}$

3.3 Stress and Deflection Calculations for Members 2, 4, 7 and 8

Max Weight acting on the frame considering both rings together = 4000 N
Let, Load factor = 2 Therefore, Total Load, $P = 8000$ N
Load at centre(due to rings = 2 rings) = $\frac{8000}{2} = 4000\,\text{N}$

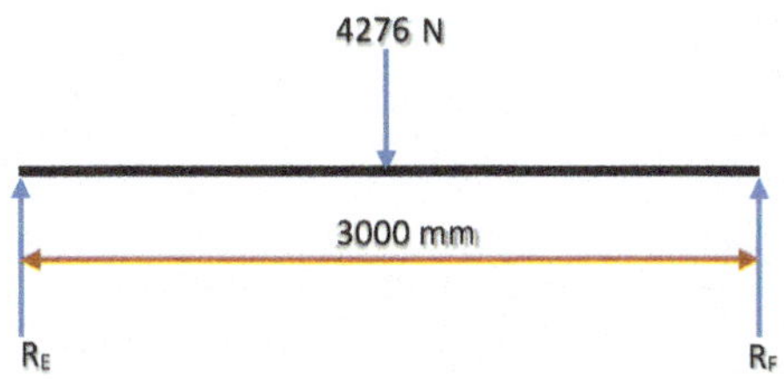

Fig. 8 Free body diagram for member 2, 4, 7 and 8

Total load at centre(rings + self − weight); (Fig. 8)

$$W = 4000 + \frac{(3000 \times 92)}{1000} = 4276\,\text{N}$$

Maximum Bending moment, $M_{\max} = \frac{W \times L}{4} = \frac{4276 \times 3000}{4} = 3{,}207{,}000\ \text{N mm}$
Maximum Bending Stress, $\sigma_b = \frac{M \times y}{I} = \frac{3{,}207{,}000 \times 50}{186.7 \times 10^4} = 85.88\ \text{MPa}$
$\sigma_b = 85.88\ \text{MPa} < 137.22\ \text{MPa}$. Hence Safe
Max deflection, $\delta_{\max} = \frac{W \times L^3}{48 \times E \times I} = \frac{4276 \times 3000^3}{48 \times 2.1 \times 10^5 \times 186.7 \times 10^4} = 6.135\,\text{mm}$

4 Result

Table 3 shows manpower cost where total manpower required for job is 3. Each worker's wage is Rs.48/hr, and time required to complete job was 90 shifts. Total manpower cost was Rs.1152 per shift, so the cost of product/shift and number of shift give total manpower cost which is Rs.103,680.

In Table 4, it shows comparison between material costs for C channel beams. Based on the criteria of whether to make the storage system from new material or from scrap material, the cost analysis for 3000 kg of material was required for beam. New material costs Rs.45/kg whereas for scrap material was Rs.28 /kg. So

Table 3 Manpower cost

Manpower required	Cost/h	No. of hrs. worked	Total cost (in Rs.)	No. of shifts manpower operated	Cost/shift	Total cost (in Rs.)
3	48	8	1152	90	1152	103,680

Table 4 Cost comparison of material categories

C channel beams	Material used (in kgs)	Cost/kg (in Rs.)	Total cost (in Rs.)
New material	3000	45	135,000
Scrap material	3000	28	84,000

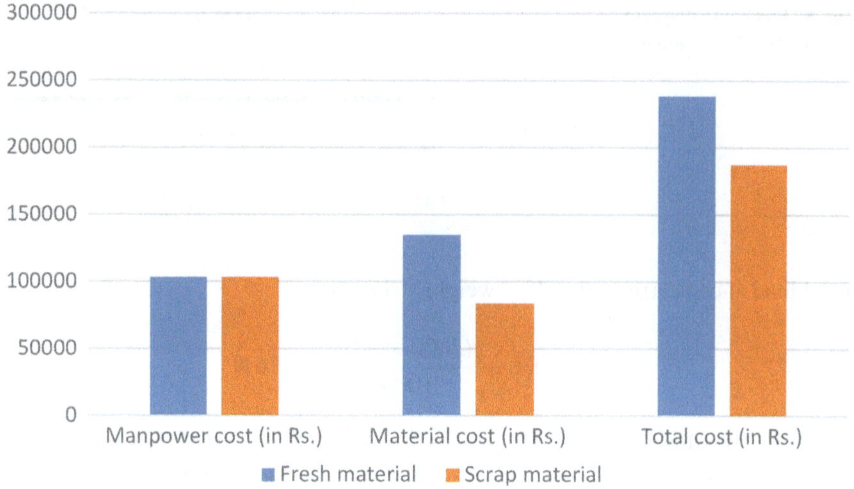

Fig. 9 Benefits of using scrap material

total cost for new material would have been Rs.135,000. The scrap material which was actually used cost Rs.84,000.

Figure 9 concludes Tables 1 and 2, which adds total cost, i.e. manpower cost and material cost for both fresh material as well as scrap material. When the cost is compared, it is found that fresh material expenditure is Rs.238,680, whereas scrap material usage expenditure is Rs.51,000 less than fresh material usage. Hence, scrap material usage helps in saving 21.36% cost.

5 Conclusion

Handling material in general is very expensive, and hence, with designing a storage system, materials can be efficiently and quickly moved with less cost per unit time.

Every time during material movement there is handling cost, and hence, effective use of storage system will reduce the same.

Health and safety of workers depend on the type of materials handling and storage systems employed, the equipment operated and the level of training among operators. A faulty system of poorly trained personnel working in the storage system can lead to serious accidents, and hence, a good storage system can eliminate all of these wastes. With storage system design, the storekeeper can easily access the stock of rings, and hence, inventory can be managed properly.

References

1. Kim KS, Eom JK (1997) An expert system for selection of material handling and storage systems. Int J Ind Eng Theory Appl Pract 4(2):81–89
2. De Koster RBM, Johnson AL, Roy D. Warehouse design and management
3. Gudhate S, Gudhate SS, Kulkarni PP, Mali PK (2016) Design and analysis of material handling system. Int J Innov Res Sci Technol 2(12). ISSN (online): 2349–6010

Design of Shredder Machine for ELV Tyres

S. M. Auti, Jinesh Sheth, Prakriti Tulasyan, Asmita Gaikwad, and Purnima Bagwe

Abstract Waste tyres generated in India make up around 7% of the total waste tyres of the world. With the advancements in the automotive industry, the tyre industry is rapidly increasing through projected growth of 6–8%. With this development, the amount of tyre's waste increases and the risk to the environment increases. Recycling of tyres helps to reduce the negative effects. Tyre shredding is the first step towards recycling as well as reducing the space needed by waste tyres. This study has mainly focused on detailed shredder machine design and analysis. The components of the system were modelled using 3D modelling software, and ANSYS was used for its finite element analysis (FEA). In this analysis, a couple of criteria for cost-effective shredding of scrap tyres were addressed. More emphasis was on cutting blade design. Analysis revealed that design was safe taking into account safety and strength criteria.

Keywords Shredder · Gearbox · Tyre shreds · End of life (ELV)

1 Introduction

When an automobile life is prolonged through repeated repair and reconditioning, the vehicles ultimately become unusable and have to be scrapped. At this stage before scrapping, the vehicles are termed as "end-of-life vehicles" (ELVs). With such a huge number of ELVs in India, waste tyres generated from them are huge in volume [1]. Waste tyre production is present even during an automobile's lifetime and not just at the end of life. Every year about 1.5 billion tyres are manufactured globally, which eventually turns into waste tyres. A shredder is a mechanical device for cutting the tyre into small pieces. Nowadays, recycled tyre material is used in various applications,

S. M. Auti (✉)
Research Scholar, Department of Mechanical Engineering, Veermata Jijabai Technological Institute (V.J.T.I.), Mumbai, Maharashtra 400019, India
e-mail: shashikant.auti@gmail.com

J. Sheth · P. Tulasyan · A. Gaikwad · P. Bagwe
Undergraduate Student, Department of Mechanical Engineering, Dwarkadas J. Sanghvi College of Engineering, Mumbai, India

© Springer Nature Singapore Pte Ltd. 2020

H. Vasudevan et al. (eds.), *Proceedings of International Conference on Intelligent Manufacturing and Automation*, Lecture Notes in Mechanical Engineering,
https://doi.org/10.1007/978-981-15-4485-9_64

from garden pavers to playground floors. It also includes by-products such as gravel substitute, crumb rubber, landfill medium, filters for wastewater treatment, garden mulch, steel wires and fuel derived from tyre. The project includes design of the tyre shredder and analysis of the same with concerns regarding portability, strength, durability and ergonomic factors.

The main aim of this research is to design and analyse a prototype of scrap tyre shredder with the goal of achieving high productivity cost-effectiveness and safe operation. The secondary goals considered for this prototype are motor selection to give the required amount of torque and gearbox design to obtain the required turning velocity and shear force.

2 Literature Survey

Rodgers et al. evolution of different types of low-speed shredders was addressed in their study. Type 1 shear shredder combines two counter-rotating shafts, which do not inter-mesh with each other. Instead, they cut against a stationary anvil bar equipped with replaceable and adjustable knives. Type 2 shear shredders can be considered as a tool for simple size reduction or shredding before further processing. Type 3 design must take into consideration all downstream process specifications (a) the shredder output must be reasonably consistent in shape. (b) The raw material is bulky and not easy in the interface between the cutters [2].

M. Sakthivel et al., in their research work titled Design and Analysis of twin shaft shredder using PRO-E and HYPERWORKS SOFTWARE, recognized the selection criteria for the choice of crushing/shredding machine. The first factor is choosing the type of material for the reduction. The second factor is the size of the material for shredding. The third factor is the productive capacity. The fourth and the final factor is the final product particle size. This research helps for deciding the required factors or parameters to be used in designing [3].

Malina Vatskicheva and Irena Grigorova, in their research work titled Study of Two-Shaft Shredder for Crushing of Concrete, Rubber, Plastic and Wood, proposed a shredder for concrete railway sleepers crushing. According to the research, two-shaft shredder is the optimal combination of structural and economical parameters. It also approached to allow detail assessment of the machine elements, under various load conditions. All the power calculations are taken through this and estimated various parameters like number of blades, their position, shaft rpm, number of cogs, cutting area of blade, material volume to be cut, etc. [4].

Albert J. Shih and Ryan C. McCall presented the kinematics of the relative motion of the adjacent disk cutters in their research work. It was discussed that the wear of tool blades made of AISI D2 and CW tool steels is comparatively less [5].

Sekar L. R. and Vinoth Kumar, in their research work, discussed the methodology behind deciding number of cutting edges. Design of shredder blade is being upgraded. In the upgraded blade, it is studied about how reversing action of the blade may also damage the cutting edges. This research focused on criteria for material selection and different hardness increasing processes for cutting blade [6].

3 Shredder Design

A counter-rotating twin-shaft model was designed for the shredder machine. The machine has a good strength and is compact with an overall dimension of 620 mm × 290 mm × 220 mm. The components of the machine include cutting blades, hex and circular shafts, spacers, side support knives, frame, custom gearbox, bearings, coupler and a motor. It is a secondary shredder (Type 3) machine which requires pre-cutting of the vehicle tyre in a primary shredder.

The material selection is done on the basis of strength required and cost-effectiveness [4]. EN24 steel is selected for the majorly loaded components of the machine. It is a standard component of steel with high tensile and shear strength, good wear resistant properties, high Brinell hardness number for safety in pitting failure in gears and easily available in the market with standard costs.

The power requirement of the shredder is based on the productive capacity of the shredder [4]. A quarter of standard car tyre primary shreds were assumed to be reduced to fine and uniform-sized particles per minute, giving the productive capacity of 180 kg/h. The required speed, torque and power from the calculations were 73 rpm, 267 Nm and 3.067 kW, respectively. A very low-speed, high-torque motor would be required for driving this shredder which was practically not cost effective. Hence, a reduction mechanism was beneficial for a certain motor to drive a very low-speed, high-torque shredder.

According to the available motors, the most convenient motor was 3.7 kW, 960 rpm which results in the requirement of the reduction ratio of 13.15. Hence, a two-stage gearbox was used with input speed (N1) of 960 rpm and output speed (N4) of 73 rpm. The reduction ratio was kept into ascending order in order to minimize the load and smooth functioning. The value becomes 3.60 and 3.65 for Stage 1 and Stage 2, respectively. Spur gear pair is used in both the stages for high transmission efficiency (98%) and economical design. Design criteria were based on bending and wear, since sliding velocity is very less. The dimensions of the gearbox are shown in Table 1. The two hex shafts run at different speeds. This differential rpm gives different relative tip velocities between the blades on adjacent rotors. This feature provides better shearing action [3]. CAD model of the gearbox is shown in Fig. 3.

A hex shaft is used for placing the cutting blades in a particular phase difference between the consecutive blades for continuous shredding action and to eliminate the inter-rotation between the cutting blade and shaft. Shafts are designed and analysed under point load, bending moment and torsion. Figures 1 and 2 show the vertical and horizontal bending moment diagram, respectively.

Thus, the equivalent bending moment 616.574 Nm along with the torsional moment 267.525 Nm resulted in a minimum shaft diameter of 30 mm. CAD model of the hex shaft is shown in Fig. 4.

Table 1 Gear dimensions

	1st stage		2nd stage		
	G1	G2	G3	G4	G5
PCD (mm)	36	130	54	198	168
Addendum (mm)	2	2	3	3	3
Dedendum (mm)	2.5	2.5	3.75	3.75	3.75
Addendum diameter (mm)	40	134	60	204	174
Dedendum diameter (mm)	29.75	123.75	44.625	188.63	158.63
Speed (rpm)	960	265.85	265.85	73	89

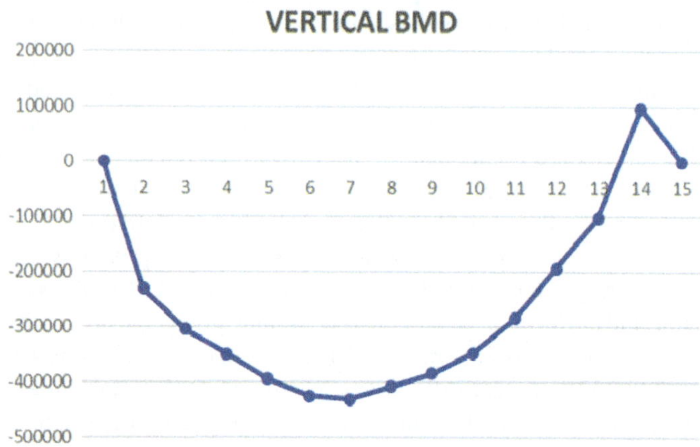

Fig. 1 Vertical BMD

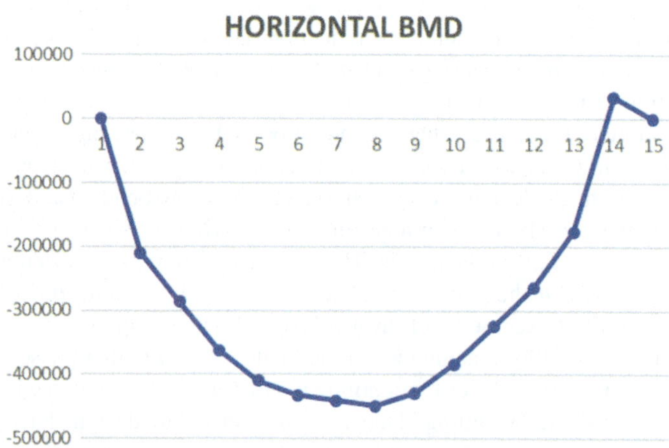

Fig. 2 Horizontal BMD

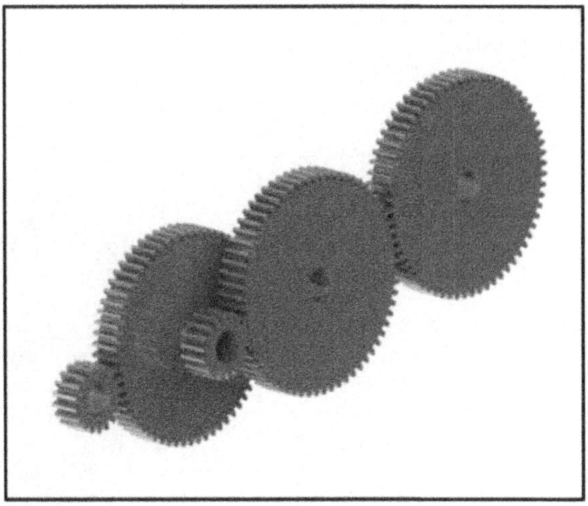

Fig. 3 CAD model of the gearbox

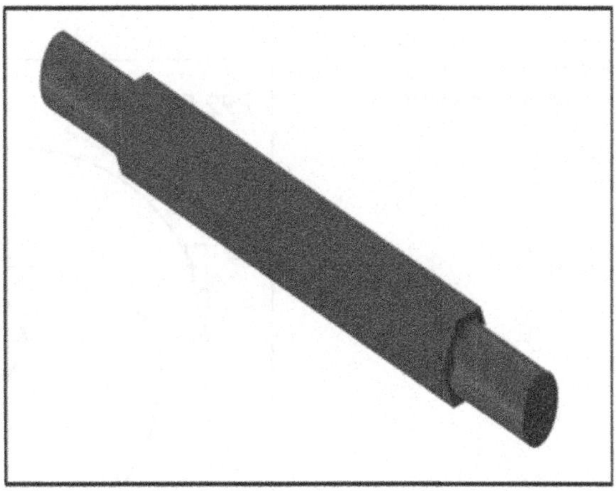

Fig. 4 CAD model of the shaft

The conventional design of cutting blade is modified for obtaining optimum cutting action with available torque. The rake angle of 6° and top/back clearance angle of 12° are used in the design of cutting blade. The wear and tear of cutting edges is reduced with this developed design [6]. The blades are in phase of 30° with each other with three cutting edges on each blade. CAD model of the cutting blade is shown in Fig. 5, and the representation of the above parameters in terms of 2D sketch is shown in Fig. 6.

Motor shaft is connected to the gear shaft using a muff coupler. Keys and keyways are used to lock the rotation between the gear and shaft. To take the radial load of shaft under highly fluctuating conditions, bearings are selected.

Three-phase induction motor has a high starting torque, good speed regulation and reasonable overload capacity. An induction motor is a highly efficient machine

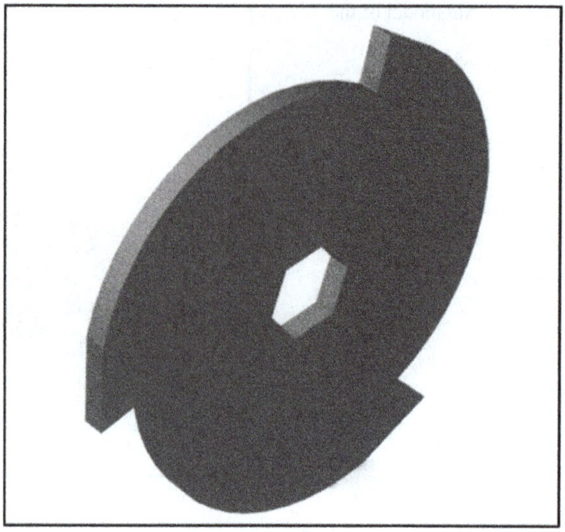

Fig. 5 CAD model of the cutting blade

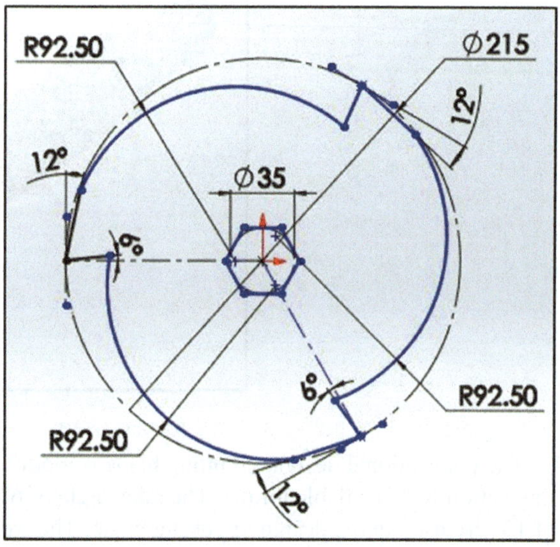

Fig. 6 2D sketch of the cutting blade geometry

with full load efficiency varying from 85 to 97%. It is good at maintaining their speed even under load. It does that by drawing more current in order to maintain the speed. Hence, considering the allowable power requirements and less cost requirement of motor, three-phase induction motor was selected for the shredder.

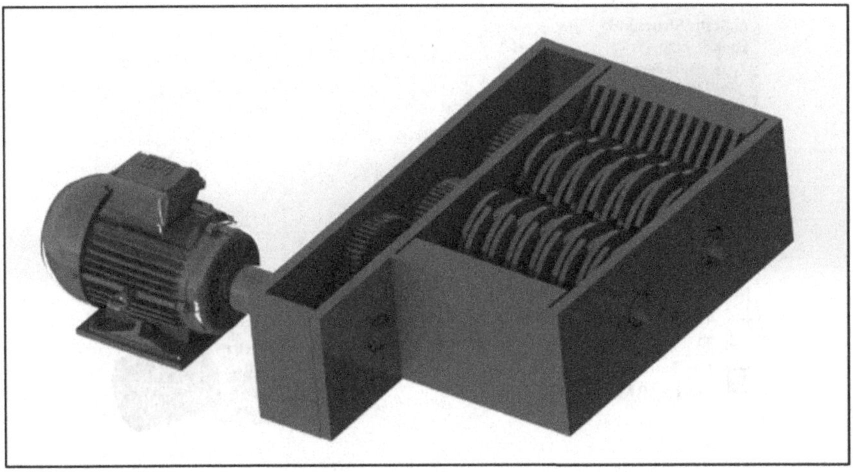

Fig. 7 CAD model of the tyre shredder machine

4 CAD Model of the Assembly

The final assembly of the shredder machine for ELV tyres is shown in Fig. 7 [3].

5 Results and Discussion

The shredder machine is structurally analysed by doing FEA on various parts of the machine. The equivalent von Mises stress for shaft is 136.64 N/mm^2, factor of safety for shaft is 1.8296 and maximum total deformation for shaft is 0.016757 mm. This analysis is shown in Figs. 8 and 9.

The equivalent von Mises stress for blade is 88.375 N/mm^2, factor of safety for blade is 2.8289 and maximum total deformation for blade is 0.01542 mm. This analysis is shown in Figs. 10 and 11.

The equivalent von Mises stress for blade and shaft assembly is 436.82 N/mm^2, factor of safety is 1.9733 and maximum total deformation is 0.14967 mm. This analysis is shown in Figs. 12 and 13.

The maximum total deformation and factor of safety for gear assembly are 0.017733 mm, 2.0026, respectively and are shown in Figs. 14 and 15, respectively.

Design of this shredder was focused on the particle size. Particle sizes produced by the existing shredders were large and non-uniform of sizes 50–150 mm. This shredder produces fine and uniform particle size of 5 mm. Hence, 90–97% reduction in particle size is achieved. The low speed, high torque generating gearbox reduces the requirement of higher power motor to 5hp, aiming at a small-scale productivity of 180 kg/h with high-quality shreds. The size of cutting chamber is 537 × 175 mm

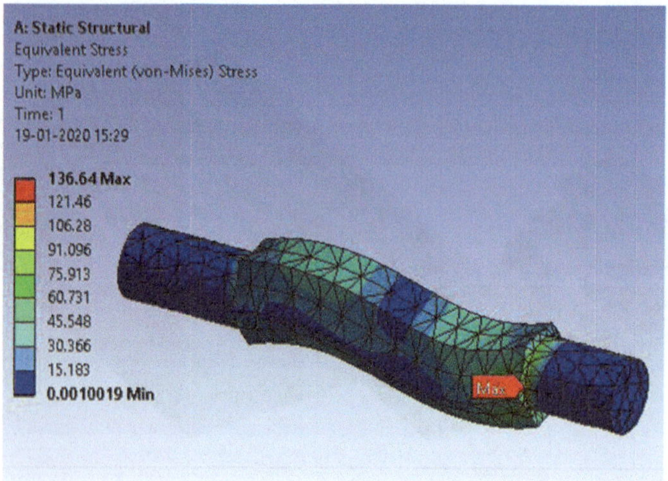

Fig. 8 Equivalent von Mises stress for shaft

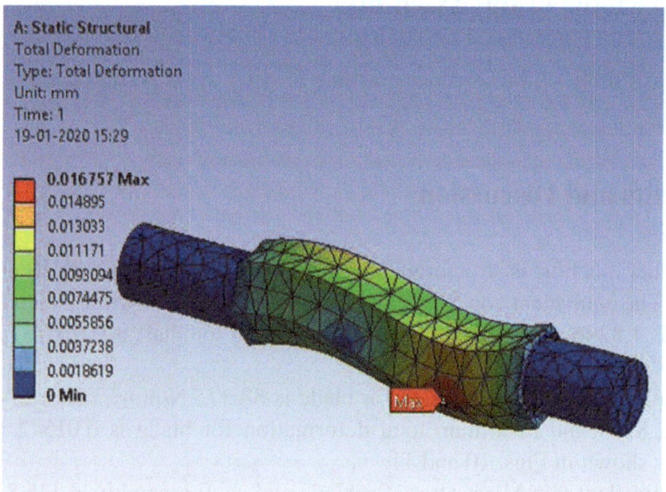

Fig. 9 Total deformation for shaft

which is very compact as compared to the conventional shredders having the cutting chamber sized 1200 × 800 mm.

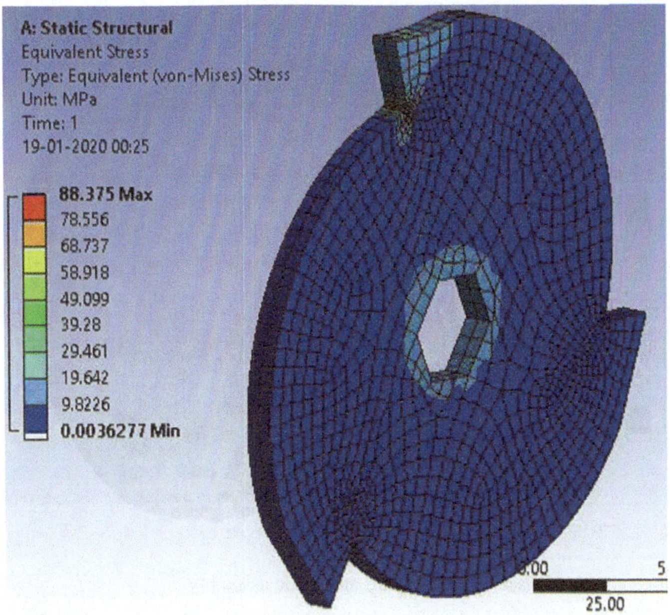

Fig. 10 Equivalent von Mises stress for blade

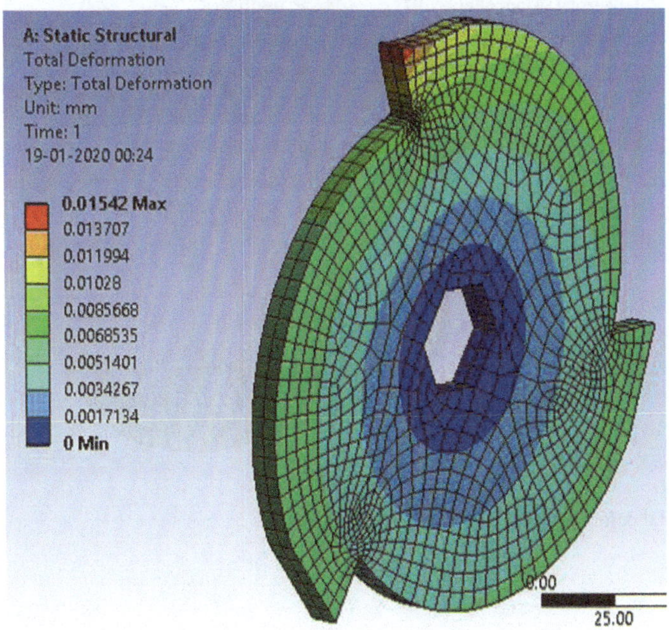

Fig. 11 Total deformation for blade

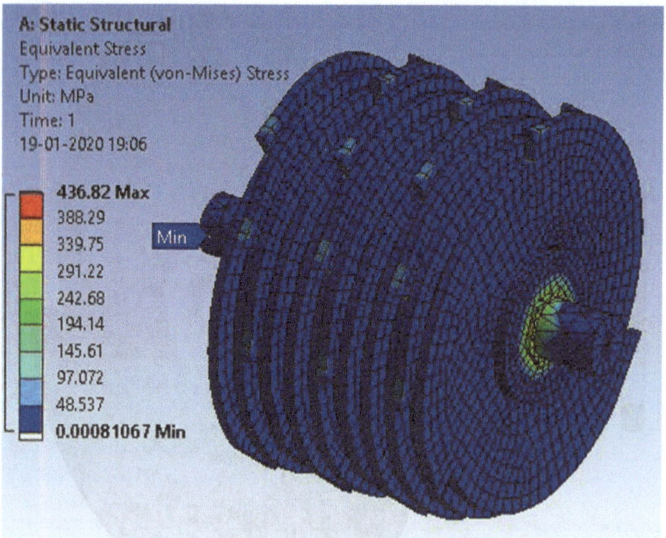

Fig. 12 Equivalent von Mises stress for shaft and blade assembly

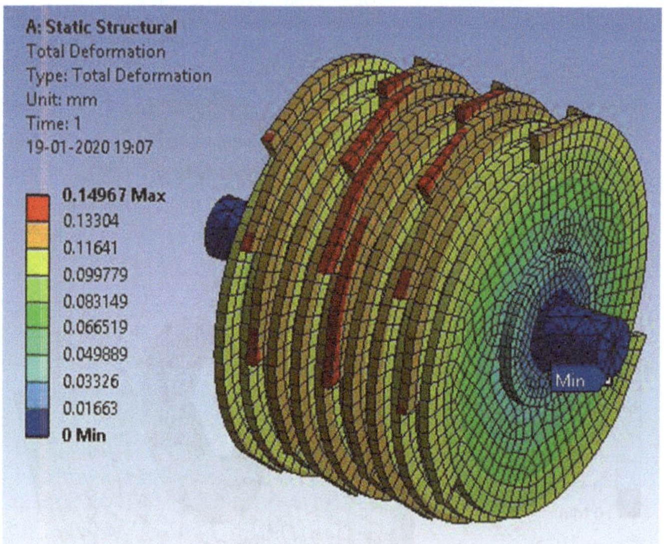

Fig. 13 Total deformation for shaft and blade assembly

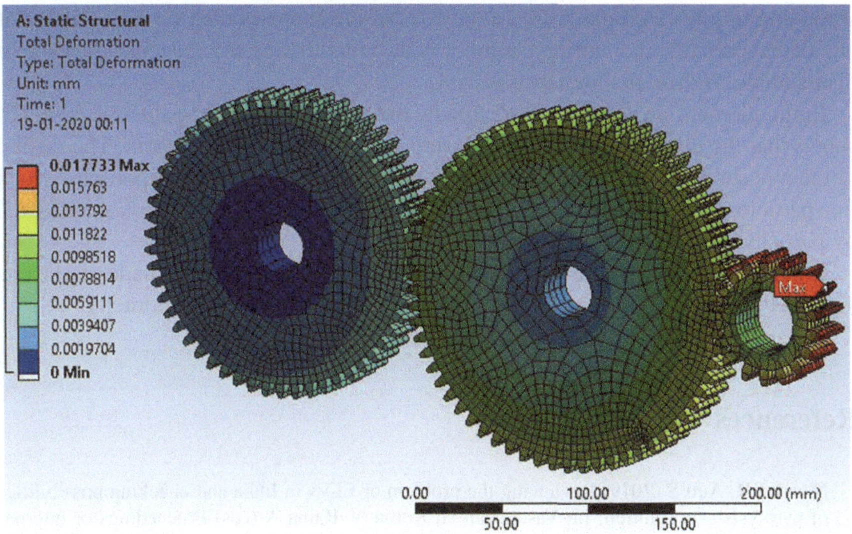

Fig. 14 Total deformation for gear assembly

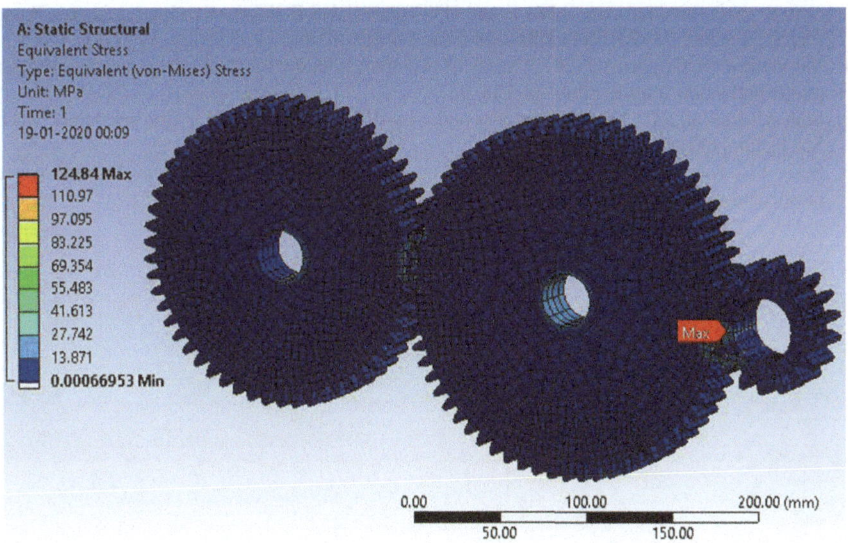

Fig. 15 Equivalent von Mises stress for gear assembly

6 Conclusion

In general, conventional tyre shredders are a primary shredder that shreds the whole tyre into large pieces. This tyre shredder is a secondary shredder that reduces primary shredding in fine-sized particles. In the case of blade sharpening or replacement of any damaged parts, the cutting chamber is designed taking into account the ease of disassemble. Hence, maintenance is easy.

Finite element analysis of shaft, blade and gearbox, performed using ANSYS, shows that the design was safe considering safety and strength criteria. The design of the tyre shredder method is perfect for achieving fine shredding of particles. Such fine particles were used for the process of pyrolysis where process yields are affected by particle size.

A similar configuration can be used to shred other pieces of ELV made of plastics such as PVC and ABS by making minor design changes in blade geometry.

References

1. Nayak KR, Auti S (2019) Reviewing the problem of ELVs in India and checking possibilities of pyrolysis as a solution. In: Vasudevan H, Kottur V, Raina A (eds) Proceedings of international conference on intelligent manufacturing and automation. Lecture notes in mechanical engineering. Springer, Singapore
2. Hill RM (1986) Three types of low speed shredder design, Dallas, TX, pp 265–274
3. Sakthivel M, Rajeshkannan G, Naveenkumar M, Muralimanokar M (2017) Design and analysis of twin shaft shredder using PRO-E and HYPERWORKS SOFTWARE. Int J Adv Res Basic Eng Sci Technol (IJARBEST) 3(Special Issue 24):836–858
4. Vatskicheva M, Grigorova I (2017) Study of two-shaft shredder for crushing of concrete, rubber, plastic and wood. J Int Sci Publ 11:238–253
5. Shih AJ, McCall RC (2004) Kinematics and wear of tool blades for scrap tire shredding. Mach Sci Technol 8(2):193–210
6. Sekar LR, Kumar V (2018) Utilization of upgraded shredder blade and recycling the waste plastic and rubber tyre. Int J Sci Res (IJSR) 7(9):160–165

Design and Development of a Foldable Hand-Driven Tricycle

Vishal Nadar, E. Narayanan, Greegory Mathew, and Pascol Fernandes

Abstract Hand-driven tricycles are a primary mode of transportation for most disabled individuals. These tricycles provide a good riding experience and have solved the mobility issues of disabled individuals to a great extent. However, these tricycles cannot be folded. They can neither be carried along in trains and buses, nor be stored in small residential and office spaces. This research paper focusses on the design and development of a foldable tricycle. The paper also explains the procedure to fold the tricycle, the FEA analysis of the frame and the calculation for the time taken to fold and unfold the tricylce.

Keywords Foldable · Compact · Tricycle · Wheel chair

1 Introduction

According to definitions, a disability or functional impairment is an impairment that may be cognitive, developmental, intellectual, mental, physical, sensory or some combination of these. Disability substantially affects a person's life activities as they find it difficult to travel or move around. Various researchers have conceived and developed different configurations of hand-driven tricycles, wheelchairs, retrofitted vehicles, etc., to enable persons with disabilities. Hassan [1] designed a motorised tricycle to suit wheelchair occupants of healthy upper torso with pelvic to foot restraint. Mohekar et al. [2] designed and fabricated a motorised retrofitted tricycle that can allow the disabled person to wheel up or down his wheelchair onto or down the tricycle. Vaidya et al. [3] developed a single slider-drive mechanism to increase the tricycle speed and reduce the human effort required. Arinze et al. [4] designed an electric solar-powered tricycle for use as a commercial means of transportation. Nigam and Sharma [5] fabricated a hybrid electro-mechanical tricycle with regenerative braking

V. Nadar (✉) · E. Narayanan · G. Mathew
Dwarkadas J. Sanghvi College of Engineering, Mumbai, India
e-mail: vishalnadarayya@gmail.com

P. Fernandes
Fellowship of the Physically Handicapped (FPH), Mumbai, India

© Springer Nature Singapore Pte Ltd. 2020

H. Vasudevan et al. (eds.), *Proceedings of International Conference on Intelligent Manufacturing and Automation*, Lecture Notes in Mechanical Engineering,
https://doi.org/10.1007/978-981-15-4485-9_65

system. However, these tricycles cannot be folded. A high-priority requirement for a tricycle in the urban environments is not only their riding performance but also their ability to be folded into compact units for storage and transportation. Foldable tricycles can be carried along in trains, buses, or in any other mode of public transportation. Their compact size (in folded state) will also enable them to be stored in small residential and office spaces. A foldable tricycle must be similar to a regular tricycle in all aspects, but its size must be adjustable to enable it to be carried to different locations such as the inside of a building or through narrow passages. One of the most important parameters for foldable tricycles will be therefore the extent to which their size can be minimised by folding.

2 Tricycle Design

A survey was carried out at All India Institute of Physical Medicine and Rehabilitation (AIIPMR), Mumbai, and also at Fellowship of the Physically Handicapped (FPH), Mumbai, to identify and discuss various areas of improvement for a tricycle. The inputs received from the disabled individuals and the staff at AIIPMR are listed below:

- Efforts required to fold the tricycle should be minimum.
- In folded state, it should be compact and should occupy the least space.
- It should be cost-effective.
- Functional controls of the tricycle (like brake, driving pedal, etc.) should be within the reach of the person driving it.
- The tricycle should incorporate a roof to protect the disabled driver from the sun's rays and rains.

Considering all these inputs received, a foldable tricycle was designed for disabled individuals. However, the design does not incorporate the provisioning for attaching a roof. Important dimensions like width, total length, distance between wheel base and shoulder support, distance between wheel base and hand rest, and positions of functional controls, such as handle, brakes, pedal, were finalised after considering ADA standards. Additionally, sufficient care has been taken to ensure that the centre of gravity of the foldable tricycle is as closest to the ground as possible. Figures 1 and 2 show the side view and the rear view of the existing tricycle. All-important dimensions are indicated in these 2D CAD models.

Table 1 presents these data for the normal tricycle as well as the newly designed foldable tricycle.

Figure 3 shows the 3D CAD model of the foldable tricycle. Figure 4 shows the 3D CAD model of the foldable tricycle in a completely folded state.

The sequential steps to fold the tricycle are elaborated below.

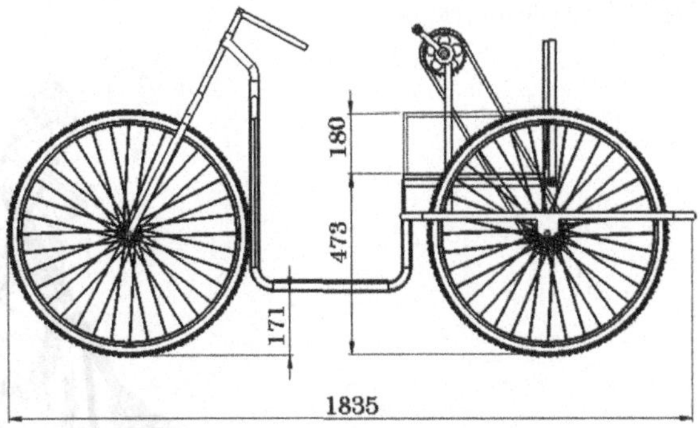

Fig. 1 Side view of existing tricycle

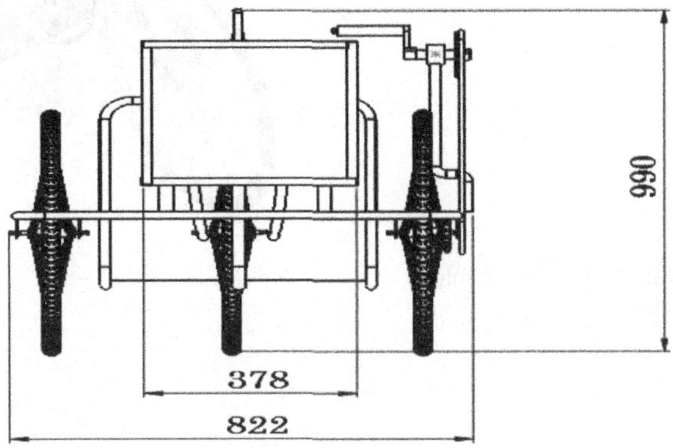

Fig. 2 Rear view of the existing tricycle

Table 1 Important dimensions of the tricycle

Parameter/size	Existing tricycle	Foldable tricycle	
		Non-folded condition	Folded condition
Overall length (mm)	1835	1800	830
Overall width (mm)	822	810	643
Overall height (mm)	990	990	1164
Front wheel diameter	24 × 1½″	24 × 1½″	24 × 1½″
Rear wheel diameter	24 × 1½″	24 × 1½″	24 × 1½″
Fork length	22″	22″	22″
Seat height from floor (mm)	473	560	560
Distance between seat and foot rest (mm)	302	375	375
Seat width (mm)	378	504	504
Height of armrest from seat (mm)	180	180	180
Clearance between footrest and floor (mm)	171	155	15

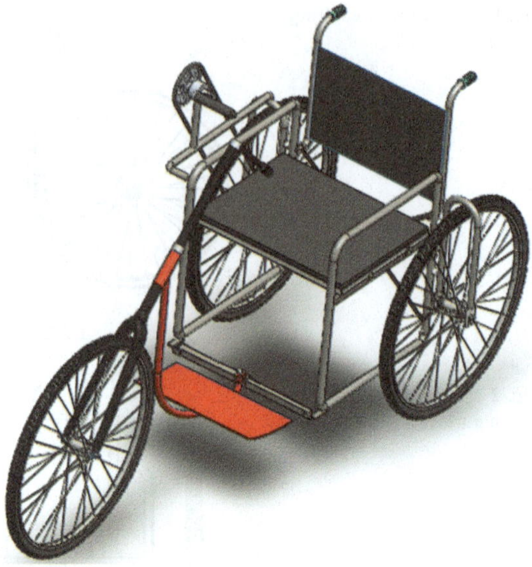

Fig. 3 Foldable tricycle

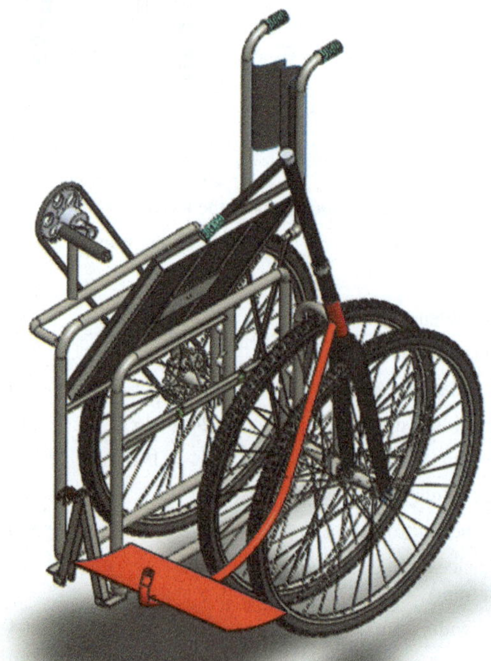

Fig. 4 Tricycle in folded state

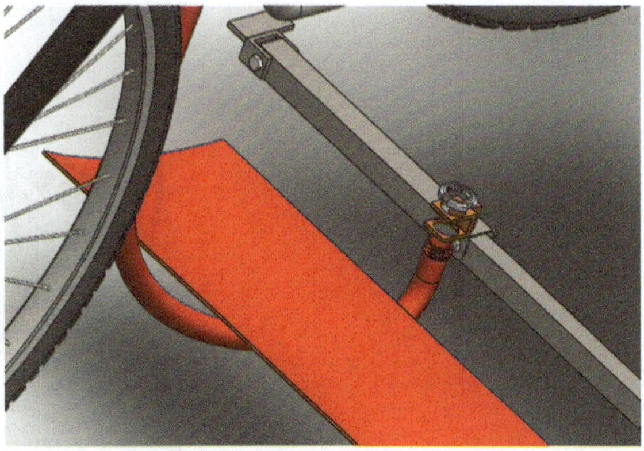

Fig. 5 Unscrewing the lock nut

- **Step 1**. The front wheel assembly is detached from the main frame assembly by unscrewing the lock nut from the threaded end of the curved down tube as shown in Fig. 5
- **Step 2**. The tricycle seat is unlocked from the tricycle frame by rotating the seat clamps. Figure 6 shows the seat clamps in unlocked and locked condition.
- **Step 3**. The seat is hinged to the tricycle frame at the other end. The unclamped seat can be lifted (by rotating about the hinged end) as shown in Fig. 7.
- **Step 4**. The main frame has two scissor link mechanisms; one each on the front and the rear side. The main frame can be folded by lifting the centre pivot of the front scissor link mechanism and moving the wheels towards each other. This is shown in Fig. 8.

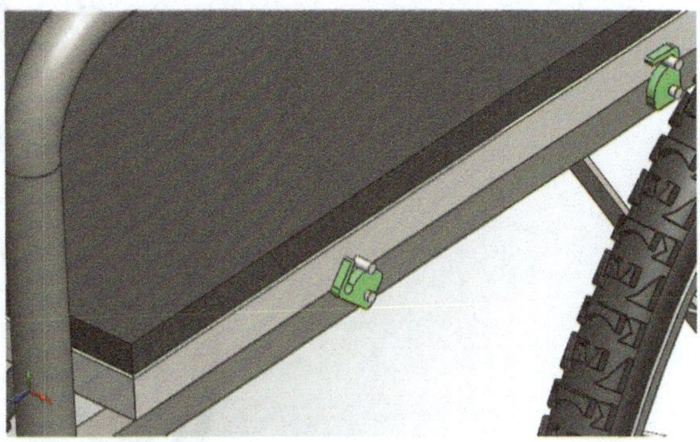

Fig. 6 Unlocking the seat

Fig. 7 Lifting the seat

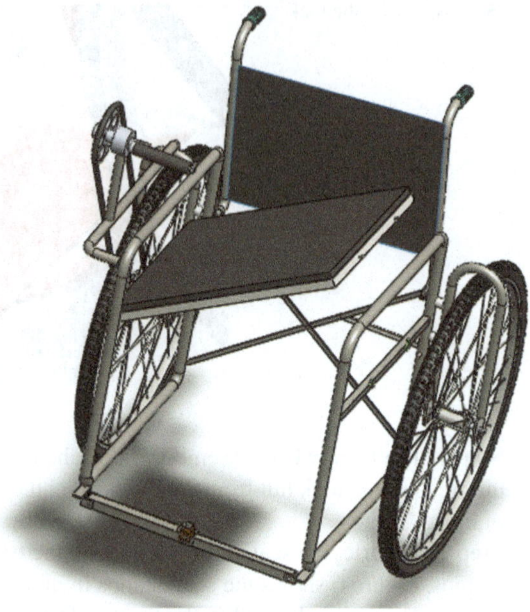

Fig. 8 Folding the frame

3 Frame Analysis

It is a well-established fact that the effort required to drive a tricycle is proportional to its total weight. To reduce the overall weight, the frame should be fabricated from a material that offers high-strength-to-weight ratio. Considering the high tensile strength and low weight requirements, the frame was fabricated from wrought stainless steel tubes. The material properties for wrought stainless steel are listed in Table 2 while details of the frame tube are listed in Table 3.

Finite element analysis of the frame was performed using SolidWorks simulation. Tetrahedral elements were used for the meshing process. The minimum edge length was 0.6 mm while the number of generated elements and the number of nodes were 502,632 and 937,385, respectively. The weight of the frame was 20 kg. The weight of the disabled individual was assumed to be 80 kg. The frame was analysed considering three possible scenarios.

- **Situation 1**. The tricycle is on a level road.
- **Situation 2**. The tricycle is moving up a slope inclined at 10° to the horizontal.
- **Situation 3**. The tricycle is moving down a slope inclined at 10° to the horizontal.

When the tricycle is on a level road, the maximum induced stress in the tricycle frame is 223.1 N/mm^2 (Fig. 9). The maximum deformation is about 1.6 mm (Fig. 10). According to The Americans with Disabilities Act of 1990 (ADA), people on wheel chair will find it difficult to manage slopes greater than 4.76°. Based on similar considerations, we concluded that the maximum recommended slope angle for tricycles should be 5°. However, for the purpose of analysis, we have assumed the slope angle to be 10°. Figures 11 and 12 indicate the stresses and the deformation in the tricycle frame, when the tricycle is inclined upwards at 10° to the horizontal. The maximum stress induced is 206 N/mm^2, and the maximum deformation is 1.547 mm. When the tricycle is inclined downwards at 10°, the maximum stress is 383.6 N/mm^2 (Fig. 13),

Table 2 Material specification

Property	Value
Density	8000 kg/m^3
Poison's ratio	0.26
Tensile strength	517.02 MPa
Yield strength	206.80 MPa
Elastic modulus	200 GPa

Table 3 Details of the tricycle frame tube

Parameter	Value
Pipe material	Wrought stainless steel
Thickness	16 B&S Gauge
Outer tube diameter	3/8 in.

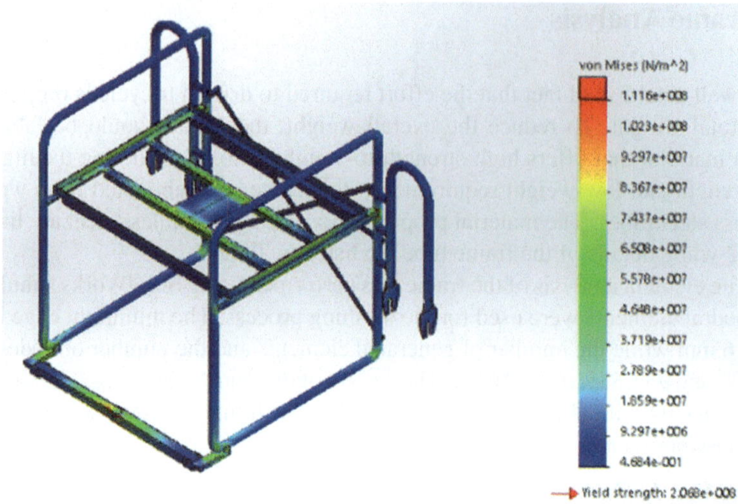

Fig. 9 Von Mises stress plot—Situation 1

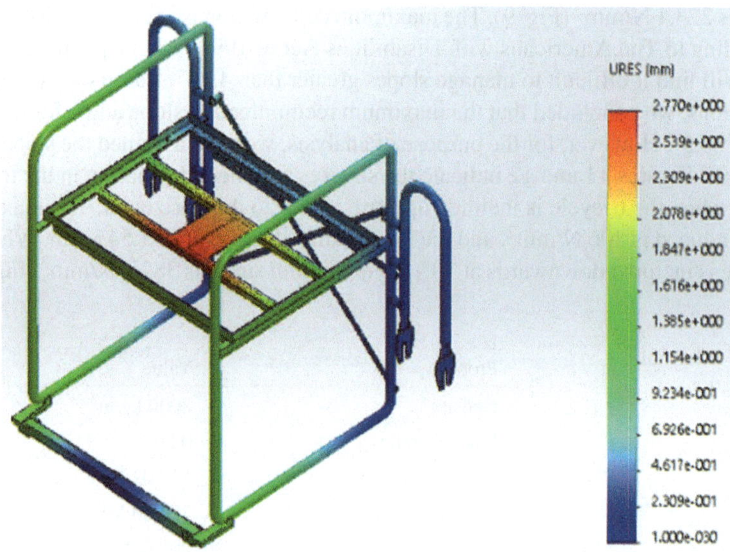

Fig. 10 Displacement plot—Situation 1

and the maximum deformation is 2.832 mm (Fig. 14). Since the maximum induced stress is less than the yield stress of the tricycle frame material (800 N/mm^2) in all three situations, the frame is safe from failure.

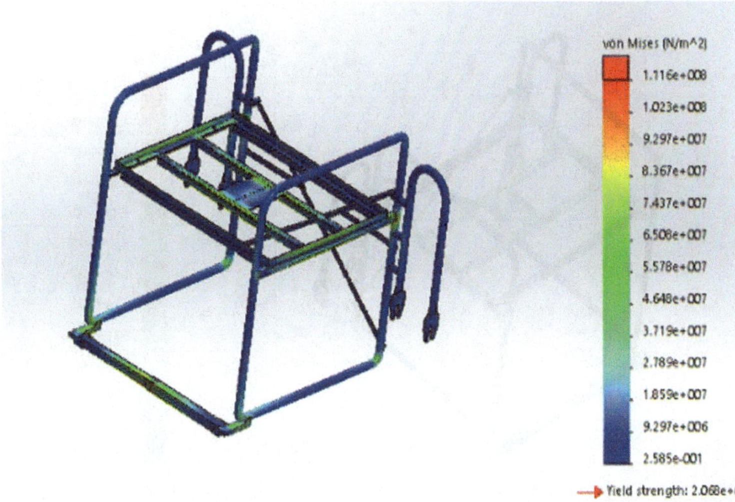

Fig. 11 Von Mises stress plot—Situation 2

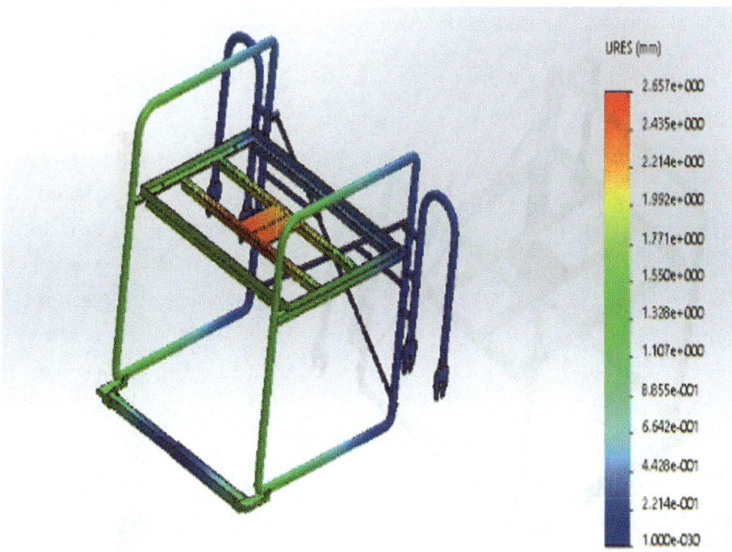

Fig. 12 Displacement plot—Situation 2

4 Calculations for Time and Space Requirement

To measure the time required to fold and unfold the tricycle, trial experiments were conducted at FPH. 10 volunteers belonging to different age groups were selected for

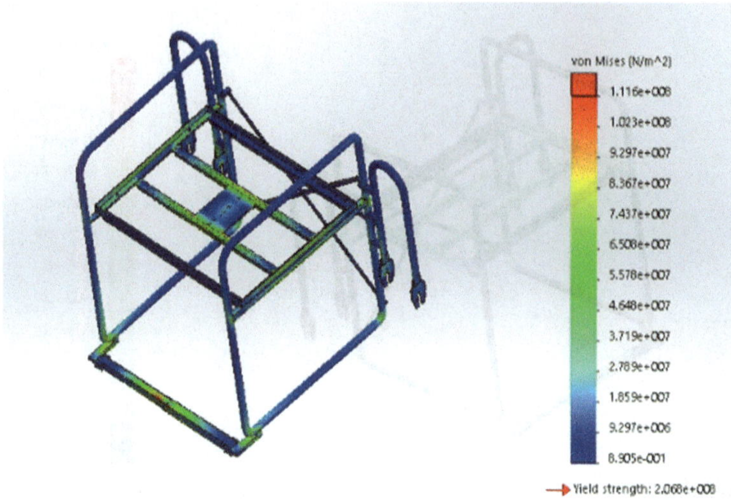

Fig. 13 Von Mises stress plot—Situation 3

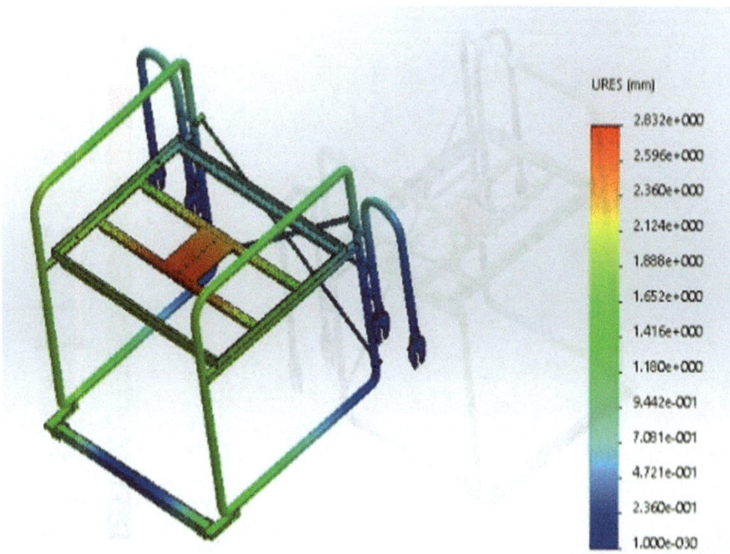

Fig. 14 Displacement plot—Situation 3

the experiment. The folding and the unfolding of the tricycle were demonstrated to the volunteers to familiarise them with the tricycle assembly. Each volunteer then practised these steps independently. The volunteers were then made to fold and unfold the tricycle, and the time required by each volunteer was measured. Each volunteer

Table 4 Trial experiments to measure time taken

Volunteer No.	Age (in years)	Height (in cm)	Weight (in kg)	Gender	Average time (in seconds) taken to	
					Fold the tricycle	Unfold the tricycle
01	39	153	59	Male	1 min 59 s	2 min 47 s
02	23	161	64	Male	1 min 42 s	2 min 27 s
02	29	167	70	Male	1 min 55 s	2 min 31 s
04	21	167	68	Male	1 min 36 s	2 min 18 s
05	47	171	70	Female	2 min 52 s	3 min 17 s
06	30	176	65	Female	1 min 43 s	2 min 22 s
07	49	157	63	Male	2 min 49 s	3 min 21 s
08	54	161	74	Male	3 min 09 s	3 min 49 s
09	45	160	56	Female	2 min 57 s	3 min 35 s
10	51	162	56	Female	3 min 11 s	3 min 53 s
Average time					2 min 23 s	3 min 02 s

folded and unfolded the tricycle thrice and the average measurement for each of them is tabulated in Table 4.

The length (l), width (w) and height (h) of the normal tricycle are 1835 mm, 822 mm and 990 mm, respectively. The total volume of space required by the tricycle (V_1) is therefore

$$\begin{aligned} V_1 &= l \times b \times h \\ &= 1835 \times 822 \times 990 \\ &= 1.493\,\text{m}^3 \end{aligned}$$

The length (l_f), width (w_f) and height (h_f) of the foldable tricycle in folded state are 830 mm, 643 mm and 1164 mm, respectively. The total volume of space required by the tricycle in folded state (V_f) is

$$\begin{aligned} V_f &= l_f \times b_f \times h_f \\ &= 830 \times 643 \times 1164 \\ &= 0.621\,\text{m}^3 \end{aligned}$$

Reduction in space is required to store the tricycle when not in use

$$\begin{aligned} &= \frac{V_1 - V_f}{V_1} \times 100 \\ &= 58.4\% \end{aligned}$$

5 Conclusion

The newly designed foldable tricycle will reduce the mobility issues of the disabled individuals. The tricycle can be folded and carried along in trains or other modes of public transport. The tricycle can be folded and stored conveniently when not in use. The volume of space required for storage is 58.4% lesser than that required by a normal tricycle. The average time required to fold and unfold the tricycle are 2 min 23 s and 3 min 2 s, respectively.

Acknowledgements This project was successful only because of the support extended by the staff at FPH, Mumbai. We specially thank their entire workshop team for providing their resources and helping us in fabricating the foldable tricycle.

References

1. Hassan AB (2012) Design and fabrication of a motorized prototype tricycle for the disable persons. IOSR J Eng 2:1071–1074. ISSN: 2250-3021
2. Mohekar AA, Kendre SV, Shah TN (2015) Design of an innovative retrofitted tricycle for a disabled person. Int J Adv Res Sci Eng 4(7):275–288. ISSN: 2319-8354
3. Vaidya VB (2016) Design and fabrication of wheelchair cum tricycle for physically challenged and elder people. Int J Sci Res Dev 4(4):804–807. ISSN: 2321-0613
4. Arinze D, Adelakun A, Etukudor C, Ajayi F (2016) Electric tricycle for commercial transportation. In: 3rd international conference on african development issues (CU-ICADI 2016), pp 376–382. ISSN:2449-075X
5. Nigam A, Sharma S (2016) Fabrication of hybrid electro mechanical tricycle with regenerative braking system. IOSR J Mech Civ Eng (IOSR-JMCE) 13(5):36–39. e-ISSN: 2278-1684, p-ISSN: 2320-334X

Assessment of Local Stresses and Strains Using NSSC Rules

Vinayak H. Khatawate, M. A. Dharap, Atul Godse, Veeresh G. Balikai, and A. S. Rao

Abstract Practically, many machine components and structures contain irregularities in spite of careful and detailed design. Such stress concentration site drastically diminishes the properties of the material. The service cracks get initiated in the vicinity of stress raiser sites. For the reliable design of machine components, the knowledge of stress and strain near the stress raising site is essential. In particular, for high-strength and high-performance components, better insights of the behavior of these components are required for the reliable design. In this paper, the criteria commonly used to estimate the local stresses and strains, viz Neuber's rule and equivalent strain energy density (ESED) methods, are presented. An Excel-based program for the stress-strain response was developed which used a constitutive equations together with Neuber and the ESED method.

Keywords Neuber's rule · ESED · Inelastic · NSSC (Notch stress-strain conversion) rule · Local stress · Local strain

1 Introduction

Engineering structures and components when subjected to loading fail in the region of high stress concentration; therefore, it is necessary to know the inelastic stresses and strains. To assess the crack initiation and fatigue lives, notch stress/strain analysis

V. H. Khatawate (✉)
Department of Mechanical Engineering, Dwarkadas J. Sanghvi College of Engineering, Mumbai 400056, India
e-mail: vinayakhk@gmail.com

M. A. Dharap · A. S. Rao
Department of Mechanical Engineering, Veermata Jijabai Technological Institute (V.J.T.I.), Mumbai 400019, India

A. Godse
ANSYCAD Solutions, Navi Mumbai 400406, India

V. G. Balikai
School of Mechanical Engineering, KLE Technological University, Hubli, India

© Springer Nature Singapore Pte Ltd. 2020

H. Vasudevan et al. (eds.), *Proceedings of International Conference on Intelligent Manufacturing and Automation*, Lecture Notes in Mechanical Engineering,
https://doi.org/10.1007/978-981-15-4485-9_66

and low-cycle fatigue concepts are often used. For such analysis, it is useful and essential to estimate the local inelastic notch stresses and strains [1–3]. Different methods, viz experimental methods [4, 5], inelastic FEA, and the NSSC rules, are available to estimate notch root stresses and strains. Experimental method is tedious and consumes lot of time. Among these, Neuber method [6] and equivalent strain energy method [7] are most popular and commonly used methods by designers. The objective of this paper is to review and assess the local stresses and strains in the vicinity of notches using NSSC rules, viz Neuber's rule and ESED method. Both the methods are programmed using Microsoft Excel to find the local plastic stresses and strains.

2 Johnson–Cook Elastic Plastic Material Model

Johnson–Cook elastic plastic material model [8] is used to define the material nonlinearity. In this paper, Aluminum 6063T7 material is used, even though in the present methodology any material can be used.

At normal strain rate and room temperature, the Johnson–Cook model is given by

$$\sigma = \left(a + b \in_p^n\right) \tag{1}$$

a is yield stress at low strains. b and n are strain-hardening constants.

Plastic strain, $\in_p = \in - \in_e$ and $\in_e = \frac{\sigma}{E}$

σ and $\in$ are true stresses and strains. E is Young's modulus of the material.

For Aluminum 6063T7, Johnson–Cook parameters are

$a = 90.26$ MPa, $b = 223.13$ MPa, $n = 0.374$, and ultimate stress (σ_{m}) = 175 MPa.

The stress–strain curve is generated using Eq. (1) as shown in Fig. 1.

3 Notch Stress–Strain Conversion Rules

3.1 Neuber's Rule

The relation between local stress ($\sigma_{\max}$), local strain ($\in_{\max}$), theoretical or linear stress concentration factor (k_t), and nominal stress (σ_n) is given by Neuber's rule

$$\sigma_{\max} \in_{\max} = k_t^2 \frac{\sigma_n^2}{E} \tag{2}$$

The actual or local stresses and strains may be obtained by solving Eqs. (1) and (2) as shown in Fig. 2.

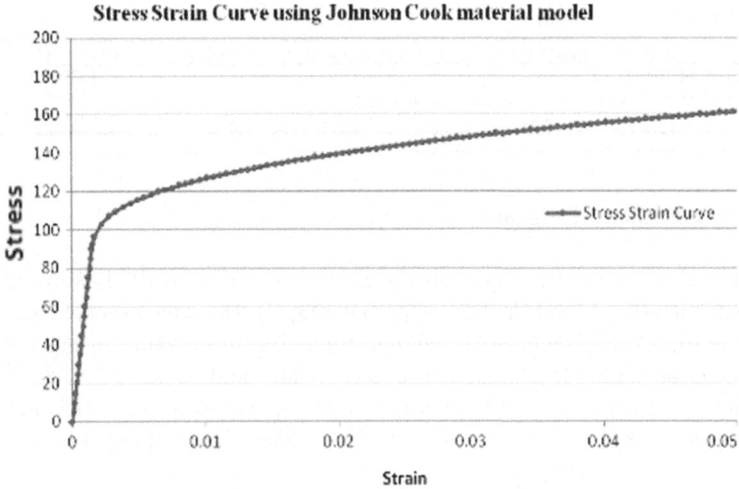

Fig. 1 Stress-strain response using Eq. (1) [9, 10]

Fig. 2 Assessment of local stresses and strains using Neuber's rule

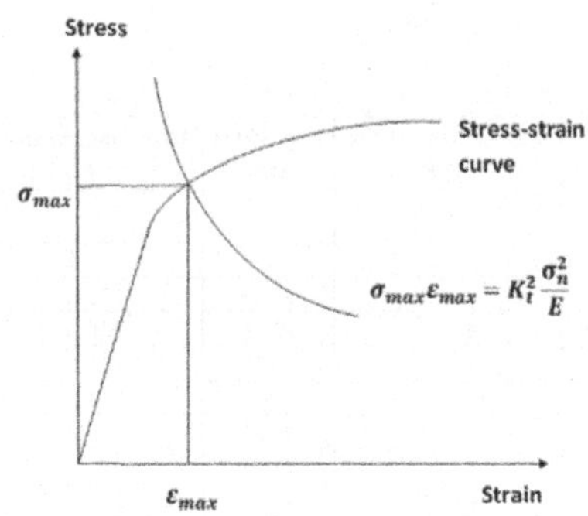

3.2 *Neuber's Rule to Calculate Local Stresses in Plastic Region*

An Excel-based program is prepared to calculate the local stresses for given nominal stress (σ_n). Johnson–Cook material model for Aluminum 6063T7 and Neuber's hyperbola is plotted in Excel sheet. For given values of nominal stress (σ_n), local stress ($\sigma_{\max}$) can be calculated graphically using the Excel-based program.

3.2.1 Input to the Program

In the worksheet, only yellow colored cells as given in Table 1 are to be given as input to find local stress and strain.

Table 1 Input to the program

Only Enter the yellow Colored Data to Find Local Stresses and Strains at Notch by using Neuber's Rule			
Modulus of Elasticity	E	60400	MPa
Theoretical Stress Concentration factor based on Geometry	k_t	2.42	-
Nominal Stress at which Local Stress and Strains to be calculated	σ_n	65	MPa

3.2.2 Excel-Based Program

To find local stress and strain by Neuber's rule, two equations need to be solved, stress-strain equation (Eq. 1) and Neuber's hyperbola (Eq. 2). The screen shot of Excel sheet shown in Fig. 3 represents stress and corresponding strain values using Johnson–Cook material model for Aluminum 6063T7 with yield stress (a) = 90.26 MPa, hardening parameter (b) = 223.13 MPa, hardening exponent (n) = 0.374618, and Young's modulus (E) = 60,400 MPa. The Excel sheet shown in Fig. 4 presents data

Johnson Cook Material Model Data

$$\sigma = \left(a + b\varepsilon_p^n\right)$$

Modulus of Elasticity (MPa)	Yield Stress (MPa)	Hardening Modulus (MPa)	Hardening Exponent	Elastic strain	Plastic Strain	Strain	Stress (MPa)
E	a	b	n	ε_e	ε_p	ε	σ
60400	90.26	223.13	0.374618	0	0	0	0
60400	90.26	223.13	0.374618	8.27815E-05	0	8.27815E-05	5
60400	90.26	223.13	0.374618	0.000165563	0	0.000165563	10
60400	90.26	223.13	0.374618	0.000248344	0	0.000248344	15
60400	90.26	223.13	0.374618	0.000331126	0	0.000331126	20
60400	90.26	223.13	0.374618	0.000413907	0	0.000413907	25
60400	90.26	223.13	0.374618	0.000496689	0	0.000496689	30
60400	90.26	223.13	0.374618	0.00057947	0	0.00057947	35
60400	90.26	223.13	0.374618	0.000662252	0	0.000662252	40
60400	90.26	223.13	0.374618	0.000745033	0	0.000745033	45
60400	90.26	223.13	0.374618	0.000827815	0	0.000827815	50
60400	90.26	223.13	0.374618	0.000910596	0	0.000910596	55
60400	90.26	223.13	0.374618	0.000993377	0	0.000993377	60
60400	90.26	223.13	0.374618	0.001076159	0	0.001076159	65
60400	90.26	223.13	0.374618	0.00115894	0	0.00115894	70
60400	90.26	223.13	0.374618	0.001241722	0	0.001241722	75
60400	90.26	223.13	0.374618	0.001324503	0	0.001324503	80
60400	90.26	223.13	0.374618	0.001407285	0	0.001407285	85
60400	90.26	223.13	0.374618	0.001490066	0	0.001490066	90
60400	90.26	223.13	0.374618	0.001494371	0	0.001494371	90.26
60400	90.26	223.13	0.374618	0.001532535	0.000005	0.001537535	92.56513683
60400	90.26	223.13	0.374618	0.001584794	0.00005	0.001634794	95.72153487
60400	90.26	223.13	0.374618	0.001602202	0.00008	0.001682202	96.77300978
60400	90.26	223.13	0.374618	0.001708608	0.0005	0.002208608	103.1999533
60400	90.26	223.13	0.374618	0.001772129	0.001	0.002772129	107.0366006
60400	90.26	223.13	0.374618	0.001817692	0.0015	0.003317692	109.7885915
60400	90.26	223.13	0.374618	0.001854483	0.002	0.003854483	112.0107993
60400	90.26	223.13	0.374618	0.001885881	0.0025	0.004385881	113.9071854

Fig. 3 Screen shot image of stress and strain values for Al 6063T7

Only Enter the Coloured Data to Find Local Stresses and Strains at Notch by using Neuber's Rule

Modulus of Elasticity	E	60400	N/mm2
Theoretical Stress Concentration factor based on Geometry	Kt	2.42	--
Nominal Stress at which Local Stress and Strains to be calculated	σ_n	65	N/mm2

Local Stress Strain Equation based on Neuber's Rule

$$\varepsilon_{\max} \cdot \sigma_{\max} = K_t^2 \cdot \frac{\sigma_n^2}{E}$$

Young's Modulus	Linear Stress Concentration factor	Nominal Stress at which Local Stress,Strains to be calculated	Local Stress by formula	Local Stress for plotting graph	Local Strain	Constant
E	K_t	ε	σ_{max}	σ_{max}	ε_{max}	$\sigma_{max} \cdot \varepsilon_{max}$
60400	2.42	65	0	0	0	0
60400	2.42	65	819.3142384	819.3142384	0.0005	0.409657119
60400	2.42	65	409.6571192	409.6571192	0.001	0.409657119
60400	2.42	65	273.1047461	273.1047461	0.0015	0.409657119
60400	2.42	65	204.8285596	204.8285596	0.002	0.409657119
60400	2.42	65	163.8628477	163.8628477	0.0025	0.409657119
60400	2.42	65	136.5523731	136.5523731	0.003	0.409657119
60400	2.42	65	117.0448912	117.0448912	0.0035	0.409657119
60400	2.42	65	102.4142798	102.4142798	0.004	0.409657119
60400	2.42	65	91.03491538	91.03491538	0.0045	0.409657119
60400	2.42	65	81.93142384	81.93142384	0.005	0.409657119
60400	2.42	65	74.48311258	74.48311258	0.0055	0.409657119
60400	2.42	65	68.27618653	68.27618653	0.006	0.409657119
60400	2.42	65	63.02417219	63.02417219	0.0065	0.409657119

Fig. 4 Screen shot image of stress-strain values for Neuber's hyperbola

for Neuber's hyperbola. Figure 4 also shows screen shot image of Excel sheet to generate stress–strain curve for Neuber's hyperbola.

3.2.3 Output of the Program

The data in the two sheets (Figs. 3 and 4) is used to plot a graph of stress-strain curve and Neuber's hyperbola. As explained in Sect. 3.1, the intersection of these two curves gives local stress and strain at the notch and is shown in Fig. 5.

3.3 Equivalent Strain Energy Density (ESED) Method

One more popular method to estimate local stress is ESED approach given by Molski and Glinka [7]. The strain energy per unit volume is called as strain energy density, and mathematically, it is expressed as $W = \int_0^{\in} \sigma(\in) \, . d \in$, and graphically, it is considered as area under stress-strain curve (Fig. 6). From stress-strain constitutive relation, within an elastic limit $\sigma = E \, . \in$. From the above equation, $w = \frac{\sigma^2}{2E}$.

Let, w_S = Strain energy density due to nominal stress and
w_σ = Strain energy density due to notch stress

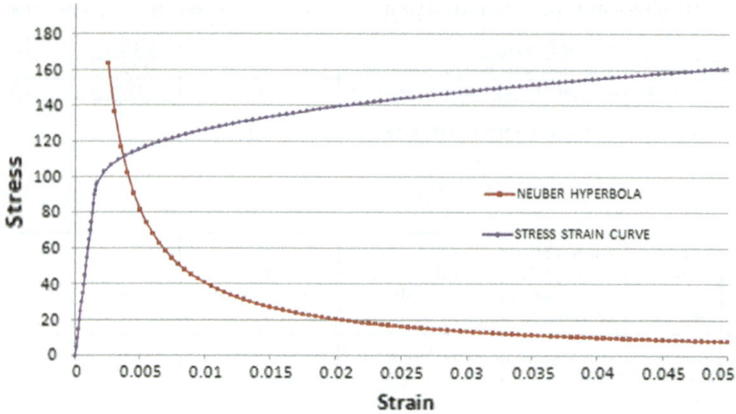

Fig. 5 Determination of notch stress using Neuber

Fig. 6 Energy density due to nominal and notch stress [7]

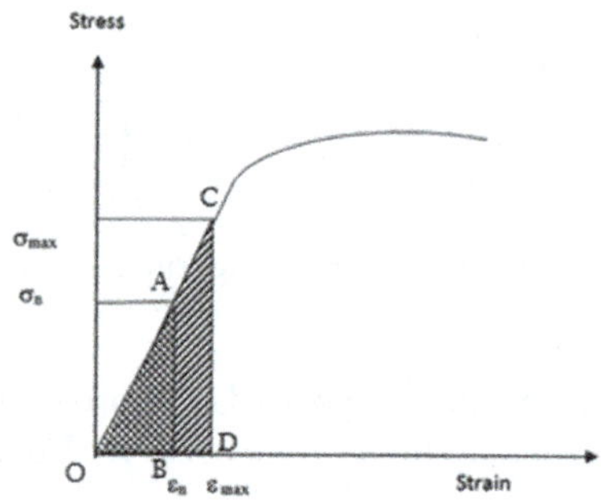

$$w_s = \frac{\sigma_n^2}{2E} \tag{3}$$

$$w_\sigma = \frac{\sigma_{\max}^2}{2E} \tag{4}$$

$$k_t = \frac{\sigma_{\max}}{\sigma_n} \tag{5}$$

From Eqs. (3), (4), and (5), the relation between k_t, w_s, and w_σ is

$$w_\sigma = k_t^2 w_s \tag{6}$$

Graphical interpretation of w_σ and w_s is shown in Fig. 6. As per Glinka's approach [7], k_t, w_s, and w_σ, can be correlated as $k_t^2 = \frac{W_\sigma}{W_s}$.

For finding maximum stress $\sigma_{\max}$ in the plastic region, ESED method [7] states that "it is reasonable to assume that the energy distribution does not change significantly if

Fig. 7 Comparison between Glinka's and Neuber's method [7]

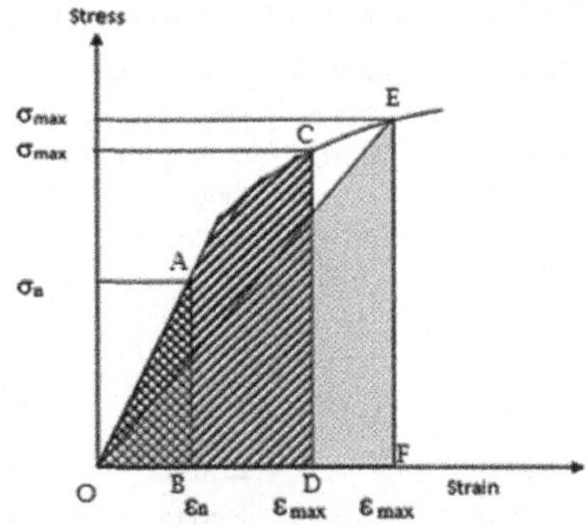

localized plasticity is surrounded by predominantly elastic material," i.e., the energy ratio given in Eq. (6) should be almost the same for elastic region and local yielding due to the effect of relatively large volume of the elastically strained material. The volume of the material that is strained elastically is more near the material that has undergone plastic deformation. Therefore, the assumption made in the above statement is reasonable. The relation can also be used if the material undergoes plastic deformation at the notch root. Equation (6) is used to find local stress in plastic region.

The comparison between ESED method and Neuber's method is shown in Fig. 7. The strain energy w_σ is represented by the triangle as shown in Fig. 7 by Neuber's rule which can be expressed as follows.

From Eqs. (2) and (6)

$$\frac{1}{2}\sigma_{\max} \in_{\max} = w_\sigma \tag{7}$$

Left hand side of Eq. (7) is area of a triangle under stress-strain curve as shown in Fig. 7, and RHS of the Eqn. is strain energy density at notch root. According to ESED method [7], the same energy w_σ is represented by area as shown in Fig. 7. From Fig. 7, it can be seen that the stresses and strains calculated on the basis of Neuber's rule are always more than or equal to those obtained using ESED method, i.e., Neuber's rule overestimates the results.

3.3.1 ESED (Molski and Glinka) Method to Calculate Local Stresses in Plastic Region

To find local stress and strains in plastic region knowing E, σ_n and k_t, first $W_s = \frac{\sigma_n^2}{2E}$ is determined, then $W_\sigma = k_t^2 W_s$, and finally, $\sigma_{\max}$ is located on stress-strain curve such that area under the curve is W_σ. Excel sheet is prepared to calculate W_s and W_σ for

Table 2 Data for ESED (Molski and Glinka's) method

Modulus of elasticity E (MPa)	Nominal stress σ_n (MPa)	Theoretical stress concentration factor k_t	W_s	W_σ	Local stress from area under curve using hypergraph 2D σ_{max} (MPa)	Local strain from area under curve using hypergraph 2D ε_{max}
60,400	0	2.3042	0	0	0	0
60,400	10	2.3042	0.000827815	0.004395147	23	0.00038
60,400	20	2.3042	0.003311258	0.017580588	46	0.00076
60,400	30	2.3042	0.007450331	0.039556323	69	0.00114
60,400	40	2.3042	0.013245033	0.070322353	92.11	0.00153
60,400	45	2.3042	0.016763245	0.089001728	97.6	0.00173
60,400	50	2.3042	0.020695364	0.109878676	100.5	0.00193
60,400	55	2.3042	0.025041391	0.132953198	102.8	0.00216
60,400	60	2.3042	0.029801325	0.158225294	104.7	0.00240
60,400	65	2.3042	0.034975166	0.185694963	106.4	0.00266
60,400	70	2.3042	0.040562914	0.215362206	107.9	0.00294
60,400	75	2.3042	0.04656457	0.247227022	109.4	0.00323
60,400	80	2.3042	0.052980132	0.281289411	110.8	0.00355
60,400	85	2.3042	0.059809603	0.317549375	112.1	0.00388
60,400	90	2.3042	0.06705298	0.356006911	113.3	0.00421

known values of E, σ_n, and k_t as shown in Table 2. Locating σ_{max} for given σ_n is done using hypergraph [11]. Stress-strain curve is imported to hypergraph. Hypergraph has facility to calculate area under curve between two limits. First point is set to (0, 0), and second point (σ_{max}, ε_{max}) is manipulated using curser in order to get area under curve equal to W_σ as shown in Fig. 8. The procedure is repeated for all values of σ_n and (σ_{max}, ε_{max}) values are tabulated in Table 2.

4 Conclusion

From the reported work on nonlinear analysis of assessment of local stresses and strains, it was observed that the past researchers used ESED method proposed by Glinka and Neuber for comparing their methodology. Therefore in this paper, the procedure to evaluate local stresses and strains using NSSC rules (Neuber and ESED) is explained. An Excel-based program was developed to assess inelastic stresses and strains by using Neuer's rule. The systematic approach was also explained to assess inelastic stresses and strains by area under stress-strain curve using hypergraph-2D

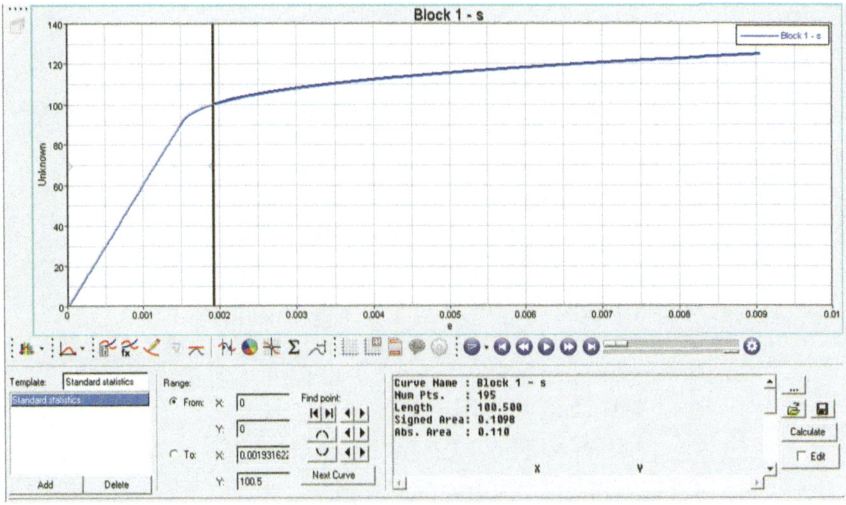

Fig. 8 Calculation of local stress and strains by area under stress-strain curve using ESED method (Tool used: Hypergraph-2D)

for ESED method. The method used in this paper would help design engineer to assess notch stresses and strains by NSSC rules without tedious numerical calculations.

References

1. Pluvinage G (1998) Fatigue and fracture emanating from notch, the use of the notch stress intensity factor. J Nucl Eng Des 185:173–184
2. Stephens RI, Fatemi A, Stephens RR, Fuchs HO (2000) Metal fatigue in engineering, 2nd edn. Wiley, New York
3. Penny RK, Marriott DL (1995) Design for creep, 2nd edn. Chapman and Hall, London
4. Leis BN, Gowda CVB, Topper TH (1973) Some studies of the influence of localized and gross plasticity on the monotonic and cyclic concentration factors. J Test Eval 1:341–348
5. Zeng Z, Fatemi A (2001) Elastic-plastic stress and strain behavior at notch roots under monotonic and cyclic loading. J Strain Anal 36:287–300
6. Neuber H (1961) Theory of stress concentration for shear-strained prismatic bodies with arbitrary nonlinear stress-strain law. ASME J Appl Mech 28:544–550
7. Molski K, Glinka G (1981) A method of elastic-plastic stress and strain calculation at a notch root. Mater Sci Eng 50:93–100
8. Johnson GR, Cook WH (1983) A constitutive mode and data for metals subjected to Large strains, high strain rates and high temperatures. In: Seventh international symposium on Ballistics, The Hague, Netherlands, Apr
9. Khatawate VH, Dharap MA, Moorthy RIK (2019) Analysis of stress and strain concentration factors on notched plate beyond yield point of the material. Aust J Mech Eng 17:76–89
10. Khatawate VH, Dharap MA, Moorthy RIK (2017) Estimation of the effective notch root stresses and strains for plate with opposite U-shaped notches by nonlinear analysis. Int J Des Eng 6:326–350
11. Altair Hyperworks Radioss Theory manual

Numerical Comparison of Tube Bank Pressure Drop of an SHTX Using Elliptical and Flat Face Header with Different Nozzle Positions

Kartik Ajugia and Mihir Sanghvi

Abstract Among distinguishing types of a heat exchanger, this report analyzes the pressure drop in the tube bank in a shell and tube heat exchanger. Pressure drop is one of the most important parameters that describe the efficiency of a shell and tube heat exchanger and its acceptance in any engineering application. Studies show that if the pressure drop across the tube side is more, the heat transfer rate decreases; hence, it affects the performance of the heat exchanger, and a greater pumping power is required, thereby increasing the cost. This paper numerically compares pressure drop in the tube bank in single-pass shell and tube heat exchanger using: (1) Elliptical face header (2) flat face header at different nozzle position, i.e., conventional nozzle position and inline nozzle position. The numerical comparison is done with the help of ANSYS 14.5, considering the flow to be steady and isothermal.

Keywords Shell and tube heat exchanger · Numerical comparison · Tube bank pressure drop · Elliptical face header · Flat face header

1 Introduction

A piece of equipment that is used for efficient heat transfer between two mediums in thermal contact with each other at different temperatures is known as a heat exchanger. In complex engineering applications where there are energy generation and energy transformation, heat exchanger is an essential component.

As there are different mechanical designs for front and rear head and also shell structure of a shell and tube heat exchanger, there is a system of designation for each type of heat exchanger based on their front and rear heads and shell design as described by TEMA. Shell and tube heat exchangers are available with various

K. Ajugia (✉) · M. Sanghvi
Department of Mechanical Engineering, Dwarkadas J. Sanghvi College of Engineering, Mumbai, India
e-mail: Kartik.ajugia@djsce.ac.in

M. Sanghvi
e-mail: m.sanghvi1998@gmail.com

© Springer Nature Singapore Pte Ltd. 2020

H. Vasudevan et al. (eds.), *Proceedings of International Conference on Intelligent Manufacturing and Automation*, Lecture Notes in Mechanical Engineering,
https://doi.org/10.1007/978-981-15-4485-9_67

geometrical specifications, but for the current analysis, TEMA BEM and TEMA AEN types of heat exchangers are used for SHTX with an elliptical head and flat head, respectively. The different components on the tube side of SHTX consist of the inlet and outlet nozzle, inlet and outlet header and the outlet nozzle. The pressure drop is an important factor in determining any heat exchanger's efficiency [1]. The maximum pressure drop (ΔP_{T}) in the tube side of a single-pass shell and tube heat exchanger involves a drop in pressure in the horizontal tubes (ΔP_{TT}), a pressure drop at the inlets of the tube, outlets and reversals (ΔP_{TE}), and the pressure drop in the nozzles (ΔP_{TN}) [1].

The uniform flow distribution of a shell and tube heat exchanger in the pipe bundle is an assumption in the traditional layout of the heat exchanger as stated by Bejan and Kraus [2]. In his research, Traub found that increased turbulence rates at low pitch-to-diameter ratios for tube banks with a few lines of flat circular tubes lead to an improvement in heat transfer. Passive methods, on the other hand, seek to modify the movement by altering the arrangement's own geometry [3].

According to the study by Achenbach, he observed that rough surfaces on inline cross-flow system tubes result in the pressure drop being reduced while simultaneously increasing heat transfer, at least within the limited range of Reynold's number, which can be calculated using roughness parameter [4].

2 Physical Modelling

Three-dimensional physical models were created for elliptical and flat face header with conventional and inline nozzle position using ANSYS ICEM 14.5. While designing the heat exchanger, tube thickness was neglected as pressure drop is independent of tube thickness [5]. Figures 1 and 2 show the shell and tube heat exchanger having Elliptical header with conventional and inline nozzle position, respectively. The design of SHTX using flat head with either nozzle positions was same as SHTX with elliptical head as geometrical specifications were kept the same.

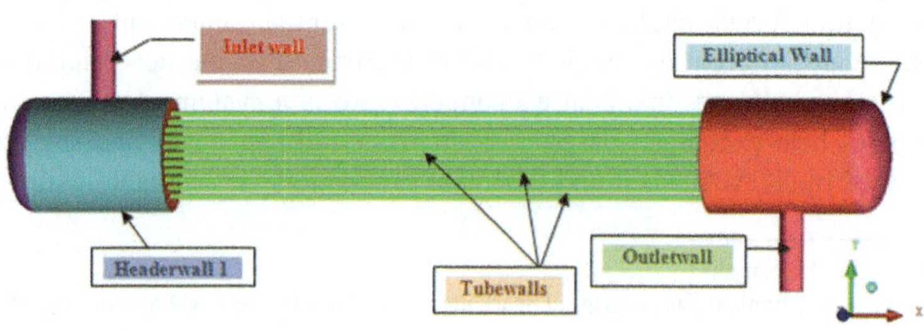

Fig. 1 3D model of SHTX with an elliptical head and conventional nozzle position

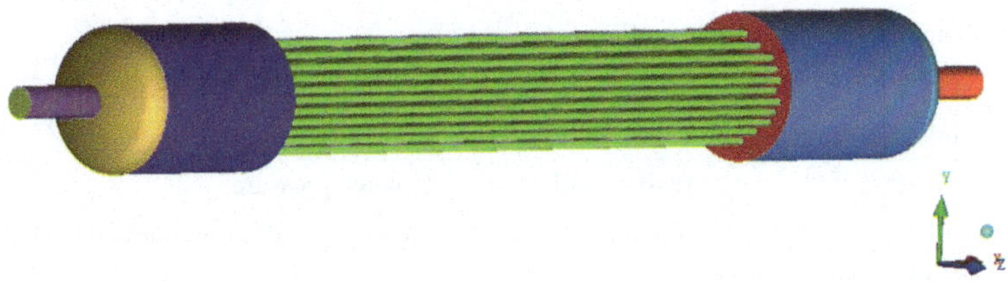

Fig. 2 3D model of SHTX with an elliptical head and inline nozzle position

Table 1 Number of nodes and minimum mesh quality

Case No.	Type of geometry	Number of nodes		Minimum mesh quality
1	Elliptical header with conventional nozzle position	Entire geometry mesh	634,357	0.22
2	Flat face header with conventional nozzle position		633,611	0.26
3	Elliptical header with inline nozzle		636,580	0.21
4	Flat face header with inline nozzle		632,593	0.3

3 Meshing

1. ANSYS ICEM 14.5 was used for creating the mesh for both elliptical face header and flat face header.
2. The surface mesh was developed using the completed global mesh setup, surface mesh design, and curve mesh setup. Tetra/mixed robust octree method was the approach used for creating a volume mesh setup.
3. Furthermore, 3D meshing was done using the prism mesh by creating the two layers near the curve surfaces for the elliptical face header to accurately capture the boundary layer flow.

The number of nodes and the minimum quality of the mesh is tabulated in Table 1.

4 Solver Setup

In order to obtain the numerical solution, continuity equation and Navier–Stokes equation in three dimensions were used as the governing equations by the solver. The boundary conditions used for the solver setup are as given in Table 2 [5].

Table 2 Components' boundary conditions

S. No.	Name of the zone	Boundary condition	Boundary characteristics
1	Inlet	Velocity inlet	3.31 (m/s)
2	Outlet	Pressure outlets	0 gauge pressure
3	Inlet wall	Wall	No-slip, surface roughness 0.5, stationary
4	Outlet wall		
5	Header walls		
6	Tube walls		

Second-order upwind scheme was used to define the spatial discretization, and "SIMPLE" (Semi-Implicit Method for Pressure-Linked Equations) algorithm was used as the method to obtain the solution.

5 Results and Analysis

5.1 Analytical Pressure Drop Across the Tube

The input velocity at the nozzle was selected as 3.31 m/s. Tube-side pressure drop (ΔP_{total}) is the sum of pressure losses due to sudden expansion and contraction of the tubes (ΔP_{r}) accounted by four velocity head per pass and pressure losses due to friction inside the tubes (ΔP_{t}) [1, 5, 6]. Therefore, tube-side total pressure drop can be calculated by Eq. 3.

$$\Delta P_{\text{t}} = 4fLN_{\text{p}}\rho U_m^2/(2d_i) \tag{1}$$

$$\Delta P_{\text{r}} = 4N_{\text{p}}\rho U_m^2/2 \tag{2}$$

$$\text{Thus,}\quad \Delta P_{\text{total}} = \left(4fLN_{\text{p}}/d_i + 4N_{\text{p}}\right)\rho U_m^2/2 \tag{3}$$

where

L—Tube length (1520 mm)
N_{p}—Number of passes (single pass)
d_i—Tube inner diameter (15.75 mm)
f—Friction factor, ρ—Fluid density (998.2 kg/m^3)
U_{m}—Mean velocity of fluid inside the tubes.

5.2 *Numerical Pressure Drop Across the Tube Bundles*

A number of *Z*-coordinates are developed, i.e., vertical segments in the *X*–*Y* plane going through the pipe arrays through the middle with the central one being the Z–0, in order to find the numerical pressure drop [5, 7]. Figure 3 demonstrates the geometric mode pipe array.

An *X*-coordinate was also produced from the start of the tube at a distance of 5 mm. A 4 mm gap was left between the start of the tube and the *X*-coordinate to catch the pressure which changes from the header because of the fluid inside the tube [5, 7].

Figure 4 shows the *Z*-0 coordinate of total pressure contours for flat head that captures the number of tubes—5 in the centermost array. An X coordinate was also produced from the start of the tube at a distance of 5 mm. A 4 mm gap was left between the start of the tube and the X coordinate to catch the pressure changes from the header because of the fluid inside the tube [5, 7]. The pressure contour thus obtained using the X Coordinate are as shown in (Fig. 5). Similar representation was done considering elliptical header.

Fig. 3 Tube bundle array

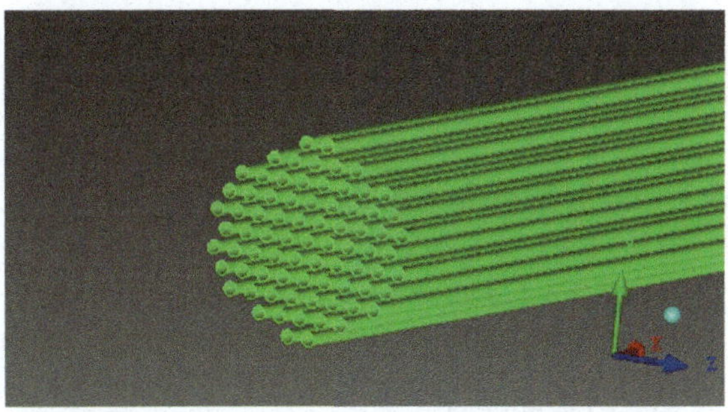

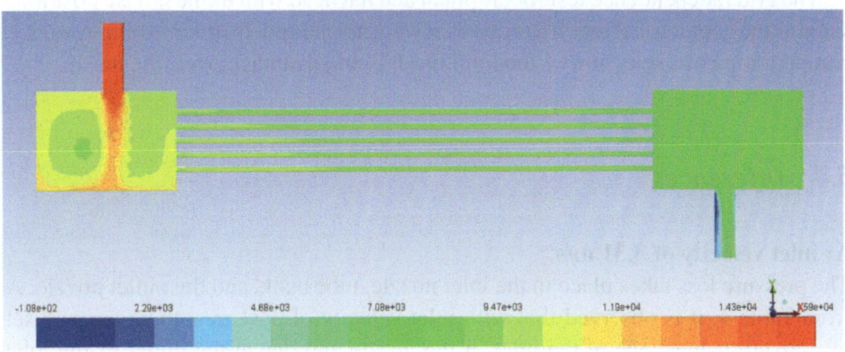

Fig. 4 *Z*-0 coordinate (flat header)

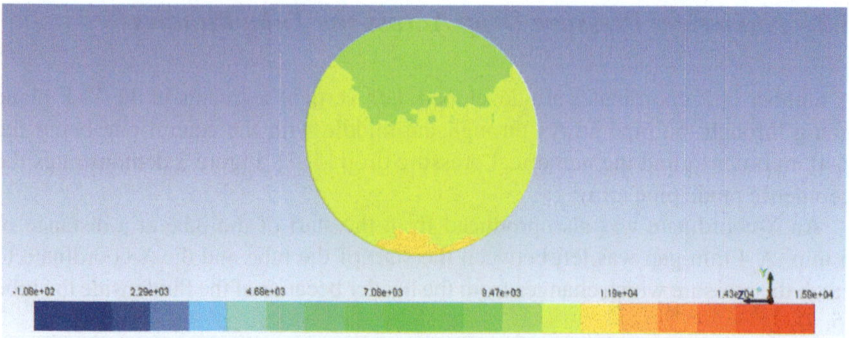

Fig. 5 *X*-coordinate at 4 mm away from the tube bundle

Table 3 Grid independence test for elliptical head and flat head with conventional nozzle position

S. No.	Mesh type	Number of cells		Average pressure drop across tube Side (Pa)	
		Elliptical	Flat	Elliptical	Flat
1	Coarse	1,986,668	1,984,538	1918	2187
2	Medium	2,519,375	2,515,432	1943	2136
3	Fine	3,709,885	3,706,225	1912	1997

5.3 Grid Sensitivity Test

The geometric model was meshed in order to obtain a coarse, medium, and fine mesh for three different dimensions. The numbers of cells and the average pressure drop thus obtained for elliptical head and flat head with conventional nozzle position is mentioned in Table 3.

The grid independence test for elliptical and flat head with inline nozzle position is same as the geometrical specifications that were not altered. In order to obtain reduced computation cost, selection of medium mesh is an advantage over fine mesh.

5.4 Inferences

At inlet velocity of 3.31 m/s:
The pressure loss takes place in the inlet nozzle, tube bank, and the outlet nozzle. As from Table 4, it is observed that with inlet velocity of 3.31 m/s, there are two sets of pressure values, one at the inlet of the nozzle and one at the outlet of the tubes with elliptical header and flat header having conventional nozzle position. Also, at the same inlet velocity, there were five sets of pressure values at the inlet in the form of concentric rings and one at the outlet of the tubes having elliptical header and

Table 4 Comparison of tube bank pressure drop at 3.31 m/s with theoretical and numerical values

S. No.	Geometrical specification	Pressure at the inlet of tube bank (no of tubes)	Average pressure (Pa)		ΔP		% Diff.
			Inlet	Outlet	Theor.	Num.	
1	Elliptical conventional	10282-(31); 11090-(46)	10,764	8666	2030	2098	3.5
2	Flat conventional	10271-(25), 11070-(52)	10,810	8674	2030	2136	5.2
3	Elliptical inline	10430-(39), 11112-(25), 11794-(6), 12476-(6), 13158-(1)	10,592	8384	2030	2568	26.54
4	Flat inline	10599-(1), 11585-(41), 12571-(19), 13557-(9), 14553-(7)	12,316	9613	2030	2703	33.14

flat header with inline nozzle positions. Figures 6 and 7 represent *Z* and *X* pressure contours for elliptical header with conventional nozzle position, and Figs. 8 and 9 represent *Z* and *X* pressure contours for elliptical header with inline nozzle positions. Similar pressure contours were obtained for flat header with both conventional and inline nozzle position.

The *X*-coordinates for elliptical and flat headers with conventional nozzle position indicated that the inlet header was divided into two zones of pressure in the vicinity of the tube bank with the lower pressure zone being on the upper side and vice versa.

The *X*-coordinates for elliptical and flat headers with inline nozzle position indicated that the inlet header was divided into five zones of pressure in the vicinity of the tube bank in the form of a concentric distorted ring. The numerical tube bank pressure drop for the elliptical and flat headers with conventional nozzle position just

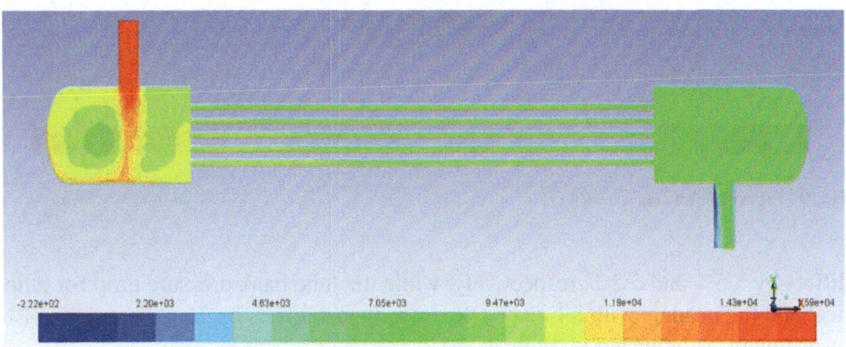

Fig. 6 Pressure contours $Z = 0$

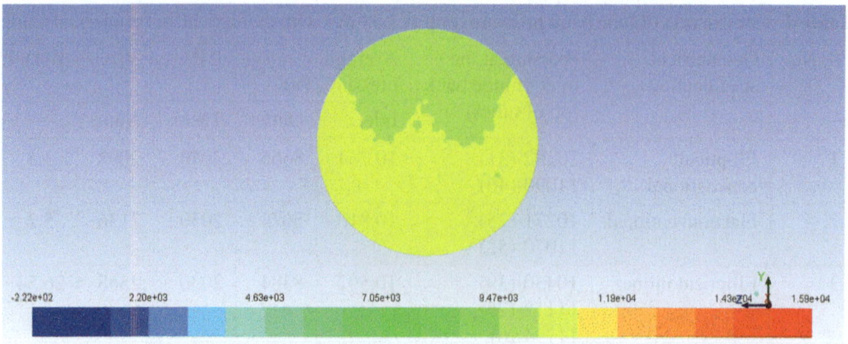

Fig. 7 Pressure contours $X = 387$

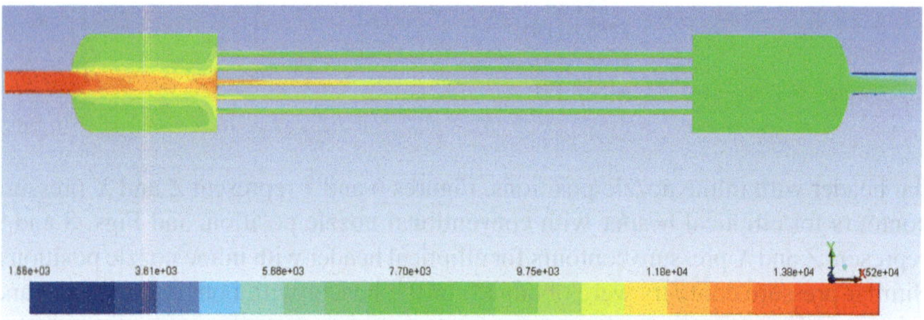

Fig. 8 Pressure contours $Z = 0$

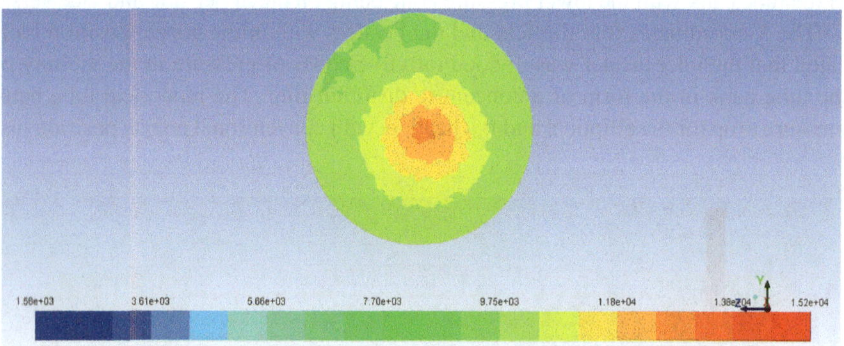

Fig. 9 Pressure contour $X = 387$

differs by 3.5% and 5.2%, respectively, while the tube bank pressure drop for elliptical and flat headers with inline nozzle position just differs by 26.54% and 33.14%, respectively, as indicated in Table 4.

5.5 *Parametric Study*

At the inlet nozzle, simulations were performed for numerous values of input velocities of 5 and 7 m/s along with 3.31 m/s, and the relationship between the theoretical and numerical pressure drops was observed. Table 5 indicates the parametric study (Fig. 10).

As the input velocity increased or the Reynold's number increased, the theoretical as well as the numerical pressure drop was increased as indicated in Table 5.

Table 5 Overall pressure drop analysis

S. No.	Geometrical specification	Pressure drop with different velocity (Pa)		
		3.31 m/s	5 m/s	7 m/s
	Theoretical	2030	4439	8422
1	Elliptical conventional	2098	4444	8843
	% Difference	3.38	0.1	5
2	Flat conventional	2136	4683	7916
	% Difference	5.2	5.38	−6
3	Elliptical inline	2568	5624	11,117.04
	% Difference	26.54	26.7	32.24
4	Flat inline	2703	6569	12,268
	% Difference	33.14	48	45.67

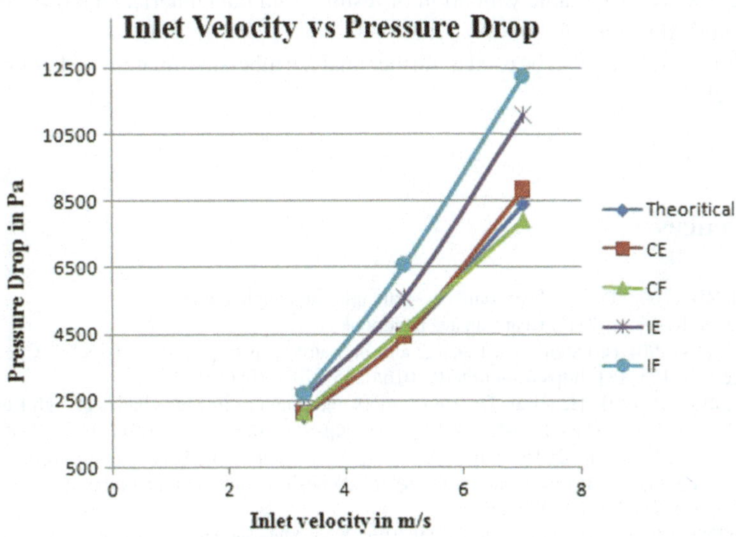

Fig. 10 Overall pressure drop analysis

The percentage deviation with respect to the theoretical pressure drop is least in case of elliptical header with conventional position and the maximum in case of flat header with inline nozzle position with increasing input velocity.

The pressure drop for the elliptical and flat header with conventional nozzle position is in close vicinity with the theoretical pressure drop while the pressure drop for the elliptical and flat header with inline nozzle position is far more than the theoretical pressure drop at all input velocities. The largest pressure drop takes place in the SHTX with flat inline nozzle at all velocities among all cases.

6 Conclusion

This study has concluded that:

- The pressure drop with elliptical header is less than flat header with either nozzle positions. The largest pressure drop takes place in the SHTX with flat inline nozzle position and lowest is in the case with elliptical header with conventional nozzle positions at all velocities among all cases.
- This empirical analysis indicated a pressure drop in a shell and tube heat exchanger across each tube, whereas the theoretical expressions would have just given us an average drop in pressure across the tube bundle.
- Through this numerical analysis of determining the pressure drop across the tube bank, it was possible to successfully identify the two different pressure zones, i.e., the low-pressure drop zone approximated in the upper half while the high-pressure drop zone in the lower half of the inlet header [5, 7].
- There was considerable validation of results from the numerical analysis and the theoretical results.
- This research can also be used to further analyze tube-side-related work for similar aspects.

References

1. Serth RW (2007) Process heat transfer—principles and applications
2. Bejan A, Kraus A (2003) Heat transfer handbook
3. Traub D (1990) Turbulent heat transfer and pressure drop in plain tube bundles. Chem Eng Process 28:173–181. https://doi.org/10.1016/0255-2701(90)80018-Z
4. Achenbach E (1991) Heat transfer from smooth and rough in-line tube banks at high Reynolds number. Int J Heat Mass Transfer 34:199–207. https://doi.org/10.1016/0017-9310(91)90186-I
5. Ajugia K, Bhavsar K (2015) Numerical comparison of the tube bank pressure drop using a conventional nozzle position and an inline nozzle position of a shell and tube heat exchanger. Int J Eng Res Gen Sci 3:487–495
6. Kern DQ (1965) Heat Transfer_Kern.Pdf, 1965th ed. McGraw Hill Book Company

7. Shah TP, Ajugia K, Shah A, Rawool R, Saini S (2018) Numerical analysis of the tube bank pressure drop of a shell and tube heat exchanger using flat face header. Int J Curr Res 10:73693–73697. ISSN: 0975-833X

Modal and Static Analysis of Luggage Rack Systems Used in Mumbai Suburban Railway Trains

Jash Patel, Neel Sanghvi, Ujwal Sutaria, Deven Shetty, Vaibhav Shah, and Vijaya Kumar N. Kottur

Abstract In this study, the authors have made an attempt to perform a numerical investigation of a luggage rack system, which is used in Mumbai suburban railway trains. The first part of the investigation focuses on modal analysis to calculate the natural frequencies of the system, which then is studied in comparison with the phenomenon of hunting oscillation observed in railway trains. The second part of the exercise focuses on the stress-strain behavior of the system, total deformation, stress, and strain under a given load. It is observed that the given system performs safely for the given practical load, and also for the excitation frequencies observed in trains.

Keywords Hunting oscillation · Klingel's formula · Natural frequency · Resonance

1 Introduction

In this paper, the authors have attempted to perform a modal analysis and also an analytical stress-strain simulation for a given luggage rack system, both using ANSYS Workbench.

The Mumbai suburban railway system is one of the busiest mass transit systems globally and is used by an estimated 7.5 million passengers daily. In order to ensure maximum space and convenience for passengers inside the trains, luggage racks are provided in all coaches along the top portion of the walls. Thus, the commuters can use these to store their bags and other luggage while they proceed with their journey, often standing throughout.

A very important aspect of engineering design for transport applications is the natural frequency characteristics of the component being designed. The reason for this is that if the natural frequency of the object under consideration becomes equal

J. Patel (✉) · N. Sanghvi · U. Sutaria · D. Shetty · V. Shah · V. K. N. Kottur
Department of Mechanical Engineering, Dwarkadas J. Sanghvi College of Engineering, Mumbai, India
e-mail: jash.ravariya98@gmail.com

© Springer Nature Singapore Pte Ltd. 2020

H. Vasudevan et al. (eds.), *Proceedings of International Conference on Intelligent Manufacturing and Automation*, Lecture Notes in Mechanical Engineering,
https://doi.org/10.1007/978-981-15-4485-9_68

to that of the vehicle (car, motorcycle, or train in this case) for the given velocity, then the component enters into a state of resonance. In this condition, the component exhibits vibrations with a large amplitude which is high enough to severely damage the component. In the state of resonance, the system is able to store and transfer energy between two or more different storage modes, such as potential energy and kinetic energy in a simple pendulum. Physical systems often have multiple resonant frequencies [1]. This energy transfer manifests itself as a very high amplitude motion of the system, which is often harmful to the system itself. For example, resonance was responsible for the collapse of the Tacoma Narrows Bridge located in Washington, USA, in 1940 [2].

The above aspect of design is particularly important for mass transit systems because of the gigantic number of people who use these systems every day. A compromise in this area of design process will pose a serious threat to the safety of passengers. Making an attempt to apply the knowledge of resonance to railway systems, we focus our attention to the phenomenon of hunting oscillation, which is a serious issue in railway systems. We can define hunting as the oscillations due to the velocity dependency of the "lateral and yaw coupled" stability [3]. These oscillations occur with a frequency which is termed as the excitation frequency of the train system. Hunting often can have serious consequences, as after the initial onset of hunting, the wheels have a tendency to slip along the rail and the wheel flanges collide with the rail, which can cause structural damage to both the rail and the wheel. It has been observed that hunting occurs only when the velocity of the train exceeds the critical velocity, which can be defined for each body.

Research by Klingel [4, 5] provides a method to calculate the wavelength (λ) of hunting oscillation.

$$\lambda = 2\pi\sqrt{\frac{r * d}{2 * k}} \tag{1}$$

In this formula, r is the rolling radius of the wheel, d is the track gauge, and k is the tread taper (the slope of tread in the horizontal direction perpendicular to the track).

Using the velocity–wavelength–frequency relationship, we can derive a frequency corresponding to the wavelength obtained above. This frequency can be defined as the excitation frequency for the given railway system. A sound logical analysis leads us to conclude that if the natural frequency of the luggage rack equals to that of the excitation frequency, then the luggage rack system will enter into a state of resonance, vibrating with a very high amplitude.

Further, we define a few background terms that will be used repeatedly in this paper. We define "velocity" (v), which in this paper implies the maximum possible velocity of the train. The "frequency" (f) can be defined as the excitation frequency of the train's hunting motion. Further, the "wavelength" (λ) can be defined as the wavelength of the hunting motion. The three quantities are related as:

$$V = f * \text{wavelength} \tag{2}$$

2 Analytical Calculations

The rail head design chosen shown in Fig. 1 confirms to the IRS-52 design developed by the Jindal Steel and Power Limited [6]. The IRS model is used for the western railways arm of the Mumbai Suburban Railway Network.

The following wheel dimensions shown in Fig. 2 [7] have been considered for the calculation of wavelength (distance travelled by a wave), and a 2D front view is made using SolidWorks for the same.

Rolling radius of wheel when it is running straight, $r = 545.65$ mm

Track gauge, $d = 1676$ mm

Tread taper, $k = 2.86° = 0.0499$ rad

Maximum speed of train = 100 kmph (as reported by the Mumbai western railways)

Wavelength of the hunting motion of wheel/axle assembly (λ)

$$\lambda = 2\pi\sqrt{\frac{r * d}{2k}} \quad \text{[By Klingel's Formula]}$$

$$= 2\pi\sqrt{\frac{545.65 * 1676}{2 * 0.0499}}$$

$$\lambda = 19{,}019.919\,\text{mm}$$

$$\lambda = 19.019\,\text{m}$$

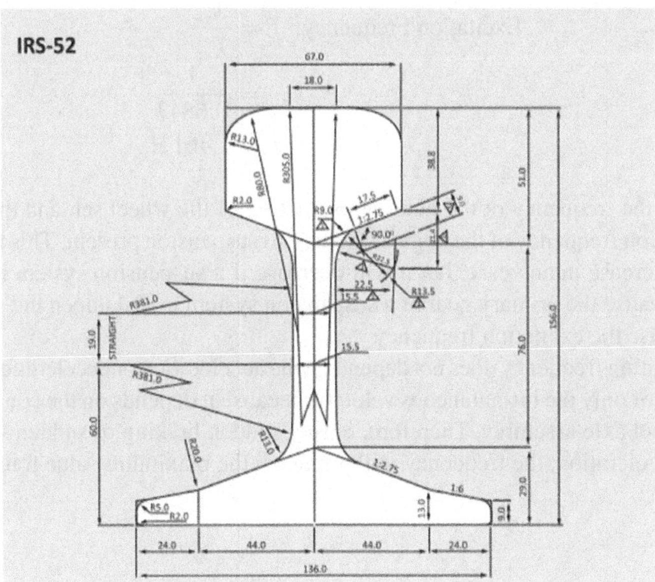

Fig. 1 Dimensions of rail head as per IRS-52

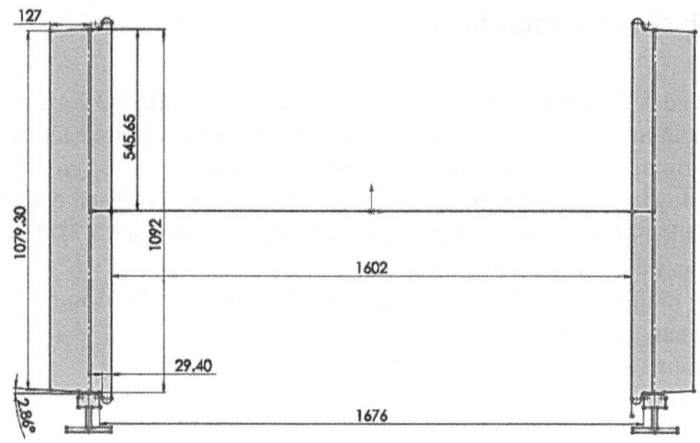

Fig. 2 Dimensions of wheel set

$$\text{Speed} = \frac{\lambda}{\text{Time period}\,(T)} \tag{3}$$

$$T = \frac{\lambda}{\text{Speed}} = \frac{19.019}{\left(\frac{100*1000}{3600}\right)}$$

$$T = 0.6847\,\text{s}$$

$$\text{Excitation Frequency},\ f = \frac{1}{T} = \frac{1}{0.6847}$$

$$f = 1.461\,\text{Hz} \tag{4}$$

This is the frequency of the hunting oscillation of the wheel set, and this will be the excitation frequency of the bogie if there is no suspension present. This frequency will not increase in any case, but it will decrease if a suspension system is present. This is because the primary goal of a suspension system is to dampen the vibrations and decrease the excitation frequency.

The hunting frequency does not depend on the acceleration or deceleration, but it is a function of only the instantaneous velocity, because it depends on the conical shape of the wheel axle assembly. Therefore, during sudden braking or sudden starting or sudden acceleration, the frequency will not reach the maximum value found above.

Fig. 3 Meshed CAD model

3 Finite Element Model

The 3D CAD model of luggage rack system was made using SolidWorks 2017 [8], and its finite element analysis (static, structural, and modal) was done using ANSYS Workbench 2018 [9]. To prepare the CAD model, design considerations were the actual dimensions of the luggage rack system available in Medha EMU bogies traversing in western arm of the Mumbai Railway Suburban Network. Stainless steel was used for the analysis as per the actual rack material used. Nylon bushings are present at all the four corners of the luggage rack system. Tetrahedral three-dimensional elements with eight nodes are used for creating the mesh of the model. Fixed supports given on bolting points, one-dimensional force in negative *Y*-direction (1800 N), and compression only support for the nylon bushings which are the boundary conditions applied for the analysis. It has been observed that the luggage rack has a net capacity of six standard luggage bags. We assume a maximum weight of 30 kg for each bag. Thus, we arrive at a net load of 180 kg or approximately 1800 N on the luggage rack. The required evaluations total deformation (no vibrations), total deformations (with vibrations), equivalent elastic strain, and equivalent stress (von-Mises) are found out after the results were processed. Figure 3 shows the meshed CAD model of luggage rack system, and Fig. 4 shows the input conditions for the analysis.

4 Results and Discussion

In this study, a comprehensive modal analysis has been done for a luggage rack system which is used in Mumbai suburban railway trains. A static analysis has also been done to analyze the stress-strain characteristics for the same. Talking about

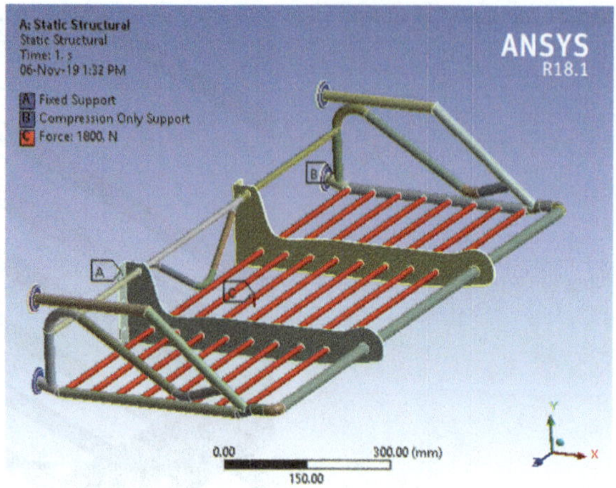

Fig. 4 Input condition for analysis

the deformation of the system, a maximum deformation of 0.9 mm is observed, while the maximum stress recorded is 308.91 MPa. The maximum strain observed is 0.002, which signal toward a good safety characteristics. These observations are summarized in Figs. 5, 6, and 7. Regarding the modal analysis to find the natural frequencies of the system, the natural frequencies for the first three modes are 67.335 Hz, 93.992 Hz, and 139.86 Hz, respectively. Figures 8, 9, and 10 show the total deformation under resonance condition for first three modes of vibration of luggage rack, that is, when the excitation frequency becomes equal to the natural frequency for that particular mode.

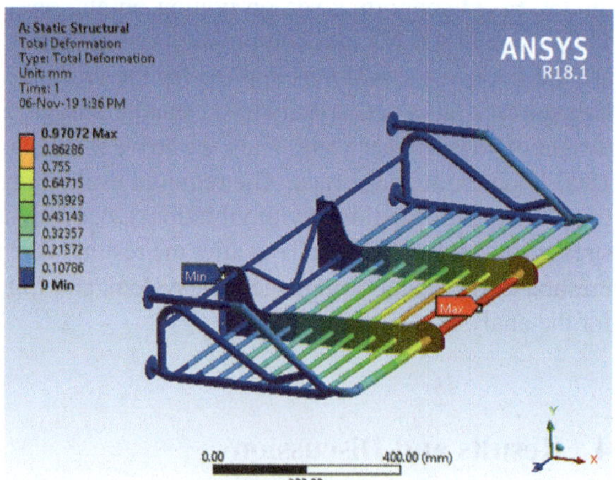

Fig. 5 Deformation results

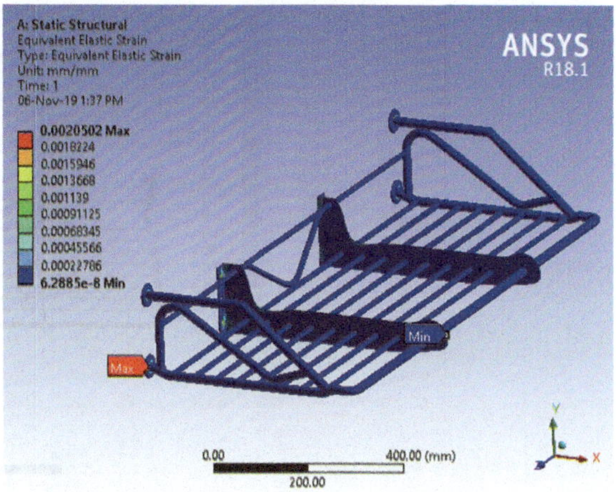

Fig. 6 Strain results

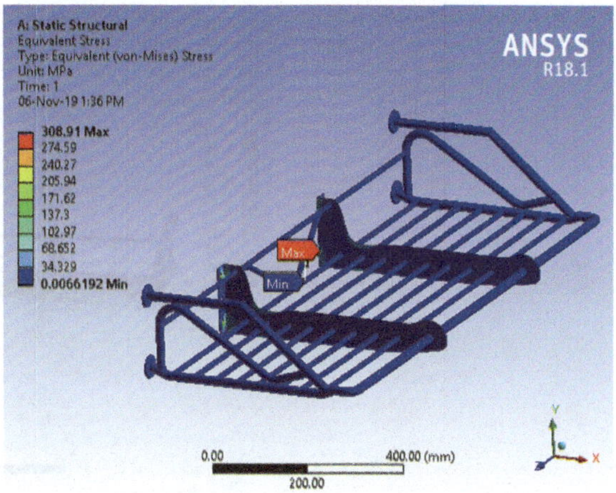

Fig. 7 Stress results

5 Conclusion

In this paper, the authors have performed numerical investigations regarding the natural frequencies and stress-strain behavior of the luggage rack system used in Mumbai suburban railway trains. It was found that the hazardous condition of resonance due to the excitation frequency of hunting oscillation is not met for the worst case scenario of oscillations as the excitation frequency of the bogie (1.461 Hz), and

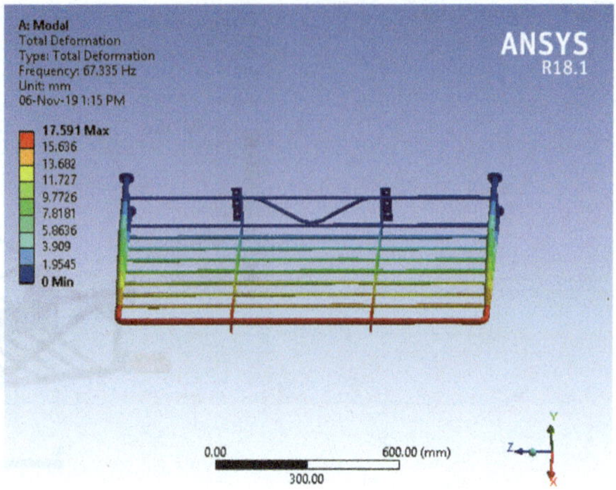

Fig. 8 First mode of vibration

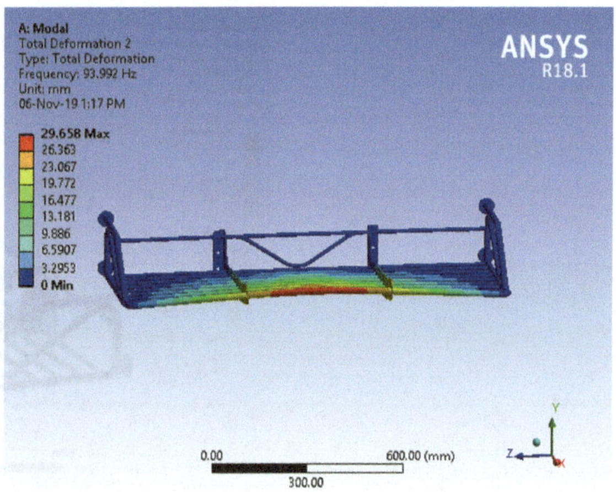

Fig. 9 Second mode of vibration

the natural frequency of the luggage rack (67.335, 93.992, and 139.86 Hz) has a huge difference. Regarding the stress-strain behavior, the system is reported to be safe for practical purposes, considering the total deformation observed.

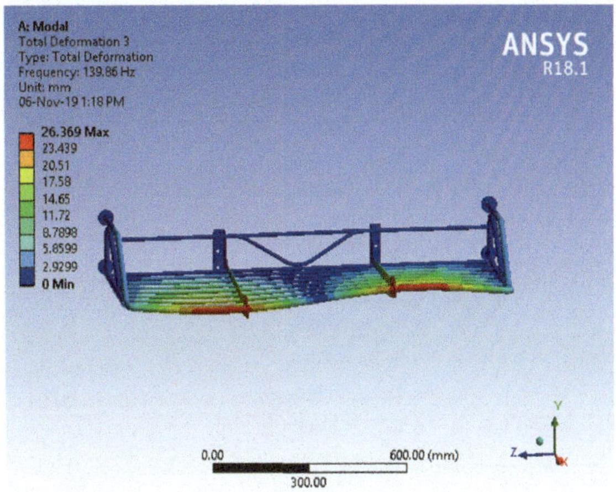

Fig. 10 Third mode of vibration

References

1. Spicci L. Design of a mechanical resonant station to free jammed micro-mechanisms. Every wave SRL, Via Martiri delle Foibe 2/1, 35010, Vigonza (PD), Italy
2. Irvine T (2009) The Tacoma Narrows Bridge failure—revision A, Apr 7
3. Uyulan C, Gokasan M, Bogosyan S (2017) Dynamic investigation of the hunting motion of a railway Bogie in a curved track via bifurcation analysis. Math Probl Eng 2017, Article ID 8276245, 15 p. https://doi.org/10.1155/2017/8276245
4. Mazilu T (2009) An analysis of Bogie hunting instability. UPB Sci Bull Ser D 71(2). ISSN 1454-2358
5. Estvald C, Markine V, Shetsov I. Shape optimization of a railway wheel profile
6. Indian Railways Schedule of Dimensions 1676 mm Gauge (BG) Revised, 2004
7. Jindal Steel and Power Ltd. Product brochure for Rail heads
8. Solidworks 2017 (2017) Introducing Solidworks, Dassault Systems
9. Autodesk ANSYS (2018) ANSYS Theory Reference, 11th edn. SAS IP. Inc.

Design and Analysis of Onion Harvester

Dhairya D. Mehta, Omkar Atale, Tanvi Hodage, S. M. Auti, and Rohit Chaurasia

Abstract The harvesting of any plant is equally important as the seeding and taking care of it. It is of paramount importance to extract the onion bulb from the soil without any damage to the onion. If the bulb is damaged, it would be rejected leading to increased waste and decreased farm productivity. This study was conducted to bring out the reliable solution for harvesting onions with minimal or no damage to the onion bulb, effectively amplifying farming efficiency. By loosening the soil, the harvesting machine digs the onion bulb out and pushes it onto the conveyor belt, which ultimately transits the onion from front to rear. The size of the harvester was decided with respect to the agro-technical features of the crop. Software used for analysis was ANSYS. Strength analysis revealed that strength point of view design is safe.

Keywords Draft force · Conveying mechanism · Cultivation · Windrowing system

1 Introduction

While India ranks second in agricultural production worldwide and 44th in technology. A boost to the agricultural sector with improved technology will improve farm productivity and reduce manpower involved.

Harvester consists of digging, conveying and windrowing as shown in Fig. 1. The digging and conveying system consists of a curved teeth-shaped plate to remove the dirt. The curved plate is so built that it has slots that cause the soil to pass through them due to gravity. The teeth in front of the plate are bolted and secured giving the power to reach into the soil and pull out the onion bulb. The conveyor system used

D. D. Mehta (✉) · O. Atale · T. Hodage
Undergraduate Student, Department of Mechanical Engineering, Dwarkadas J. Sanghvi College of Engineering, Vile Parle (W), Mumbai 400056, Maharashtra, India
e-mail: ddmehta96@gmail.com

S. M. Auti · R. Chaurasia
Assistant Professor, Department of Mechanical Engineering, Dwarkadas J. Sanghvi College of Engineering, Vile Parle (W), Mumbai 400056, Maharashtra, India

© Springer Nature Singapore Pte Ltd. 2020
H. Vasudevan et al. (eds.), *Proceedings of International Conference on Intelligent Manufacturing and Automation*, Lecture Notes in Mechanical Engineering,
https://doi.org/10.1007/978-981-15-4485-9_69

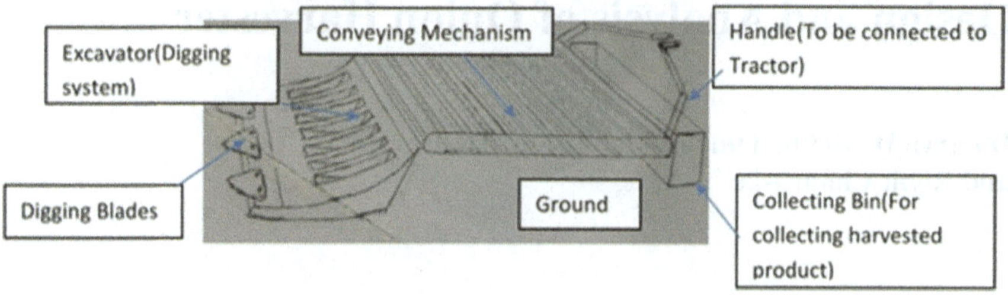

Fig. 1 Onion harvesting system elements

allows the onion to travel from the front end to the back end that was coupled to the engine.

2 Digging Mechanism

The main role of the digging mechanism was for penetration into the soil, lift the soil along with the onions, and transfer it to the conveyor unit. The design of the digging blades as shown in Fig. 2 was made to accommodate for the draft, as well as to facilitate proper penetration and digging action. The average properties of onion depth are in the harvesting range of 7.5–10 cm. Hence, to enable digging of onions without damage, the depth of operation was considered to be the maximum of 10 cm [1, 2].

Fig. 2 Digging blade design

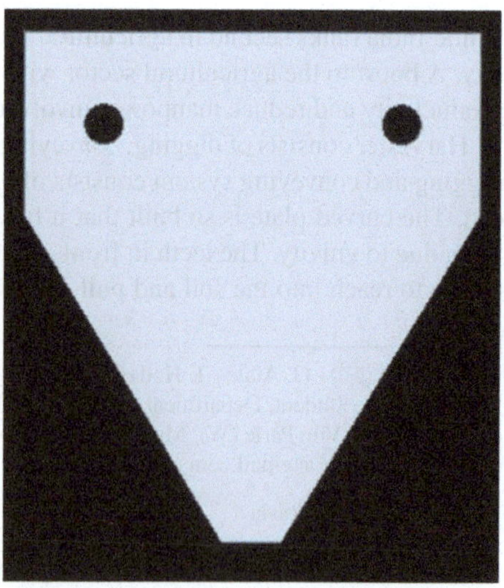

Table 1 Digging blade design parameters [1]

Specification	Value
Blade geometry (mm)	76 × 83 × 3
Draft force (N)	3256.8
Shear blade rake angle	20°
Material	EN42

The equation to determine the draft force on the digging system was formulated using various properties of soil, the speed of traversal, and the material of the digging blade.

$$\text{Draft} = F \times i \times \left[A + B \times v + C \times v^2\right] \times W \times d. \tag{1}$$

where

F = Soil texture parameter. It is selected from ASABE Standards:-
$F = 1$ for different implements like plow, cultivator, etc. [3]
i = Soil factor. It is selected based on soil type ($i = 1$—Fine soil: $i = 2$—Medium soil: $i = 3$—Coarse soil)
A, B, C = Machine parameters as per ASAE
v = Velocity of travel (kmph)
W = Working width (m)
d = depth of cut (in cm)

Values obtained though theoretical calculation were as follows

$F = 1$
$i = 2$
$A = 216$
$B = 11.2$
$C = 0$
$v = 0.1$
$W = 0.75$
$d = 10$

And hence, the calculated draft force was 3256.8 N (Table 1).

3 Digging System Analysis

Excavator design consists of three blades, which were mounted on it and the whole digging system was analyzed for von Mises stress, safety factor, and range of deflection of blades by using analysis software INVENTOR.

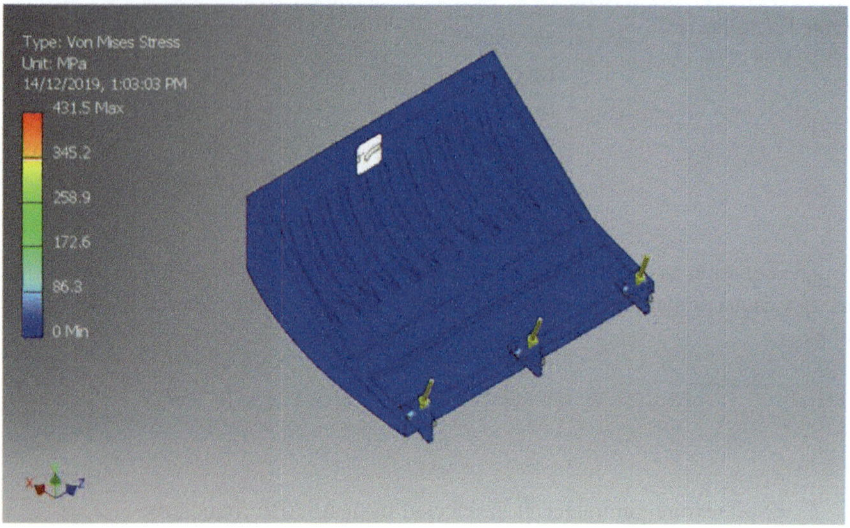

Fig. 3 Von Mises stress on digging system

Backside of excavator was fixed as it is attached to some machine like tractor. Digging blades were given normal force of 3500 N which can act on it while digging. Material selected for excavator and blade was mild steel based on strength criteria.

As seen in Fig. 3, it is observed that on excavator body, stress is in range of 0–86.3 MPa; whereas on blades, it is up to 172.6 MPa. So it can be considered safe considering von Mises stress.

As seen in Fig. 4, maximum displacement of blades was 1.208 mm which was less than permissible displacement.

As seen in Fig. 5, factor of safety value was minimum at digging blades center and maximum at the excavator. So a chance of failure at blade center was there. In order to avoid this, blade thickness was increased and sharpness decreased.

4 Pick-Up Belt/Conveyor Mechanism

This was used to pick up the material from digging system and then transfer only the onion bulbs to the windrowing system [1].

Parameters considered in designing are:

Length of conveyor was kept equal to the distance between digging system and windrowing system.

Width was kept same as the digging system or little less than it giving space for its mountings. We selected chain driven roller conveyors which has space between the rollers so that soil, stones, etc. fall on the ground while onions move ahead.

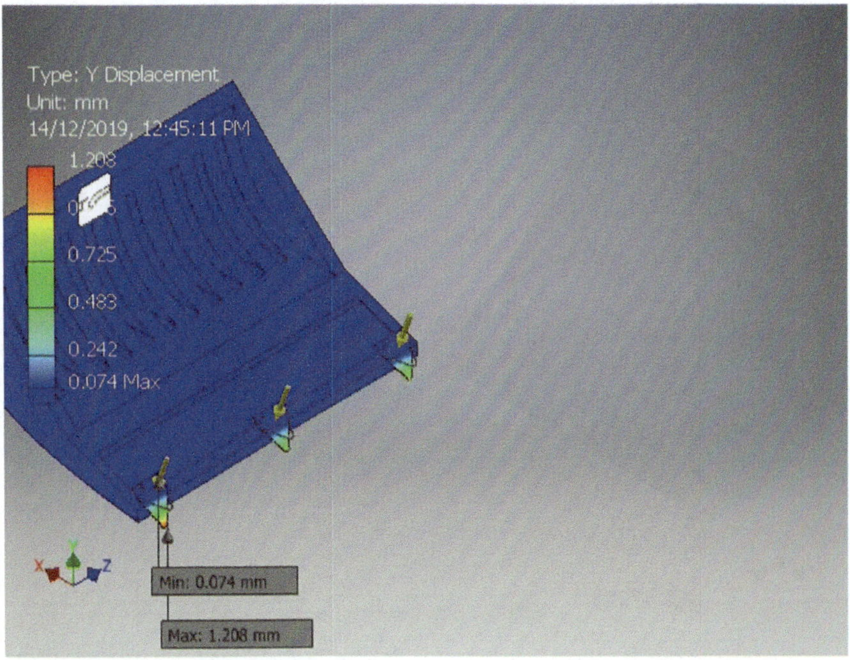

Fig. 4 *Y* Direction displacement of blades

Spacing between the cleaning elements was taken as 2 cm considering average diameter of onion bulb as 70 mm.

Conveyor was designed as shown in Fig. 7 by considering cultivation pattern shown in Fig. 6, vertical lines in the figure (Front view of field) indicates position of onion bulbs. Intersection of horizontal and vertical lines (Top view of field) indicates position of onion bulbs.

Material carried over conveyor:

Taking working depth = 10 cm for digger, working width = 750 mm; Considering speed of harvester while digging equal to 0.1 m/s, volume flow rate of soil with onions = $0.1 \times 0.75 \times 0.1 = 7.5 \times 10^{-3}\ \text{m}^3/\text{s}$.

As harvester speed is 0.1 m/s so onions digged will be 10 per second.

Considering average weight of onion bulb as 80 g then, we get 0.8 kg/s onions picked up by conveyor. Density of onion bulb is 1250 kg/m^3 which was calculated by Archimede's principle.

So volume flow rate of onions is $(0.8/1250) = 6.4 \times 10^{-4}\ \text{m}^3/\text{s}$, i.e., Volume flow rate of onions. And it would be around 1.12 kg/s taking leaves of onion into account.

Volume flow rate of soil is then obtained as:

Volume flow rate of soil is $(0.0075\text{–}0.00064) = 6.86 \times 10^{-3}\ \text{m}^3/\text{s}$.

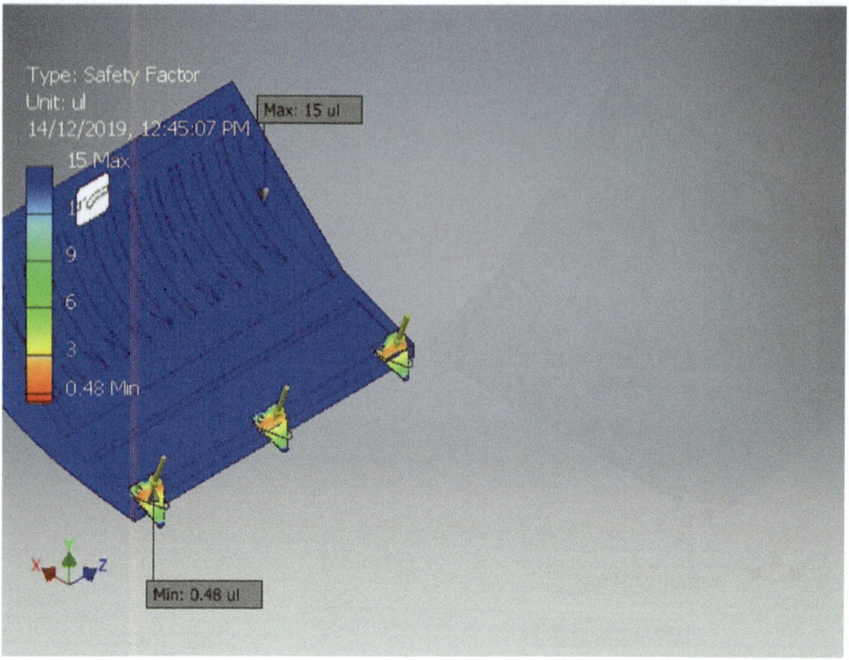

Fig. 5 Safety factor analysis

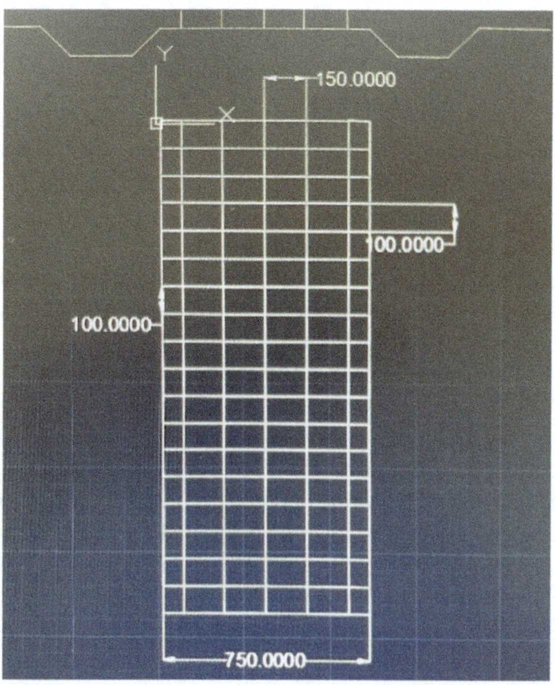

Fig. 6 Cultivation pattern of onions

Fig. 7 Cad model of conveyor

5 Windrowing System

A windrow is a linear pile of crops. In this case, the onions, with the help of the windrowing system, were allowed to fall at the rear end of the harvester in a single straight line. This was achieved by using simple inclined plates on the rear end of the onion harvester.

6 Conclusion

Assembly to develop an onion harvester could be accomplished by designing the digging, conveying and windrowing systems. The design of the digging and the conveyor system helps to automate harvesting process in the agricultural sector. Finite element analysis performed using INVENTOR shows that the design was safe considering safety and strength criteria. The self-propelled onion harvester would ensure high efficiency, by harvesting more onions in less times as compared to traditional manual methods, in low cost, as well as reduce manpower involved. Further developments in the onion harvester are also possible in blade design.

References

1. Budhale KC, Patil AG, Shirole VS, Patil SS, Desai RS, Salav SB (2019) Design & development of digging & conveyor system for self-propelled onion harvester
2. Hong S, Lee K, Cho Y, Park W (2014) Development of Welsh onion harvester for tractor
3. Awadhwal NK, Takenaga T (1988) Groundnut digger for hard soils

Design, Analysis, Prototyping and Testing of Aerofoils for High-Lift at Low Reynolds Number

Pavan Rayar

Abstract Increased payloads, reduced aircraft noise, shortened takeoff and landing distances and lowered stall speeds can all be achieved from the direct effects of improved high-lift aerofoil aerodynamics. It is, therefore, not surprising that the high-lift aerofoil design has been and will remain a topic of interest. Along with the interest, the opportunity is to make it more cost-effective. One possible way to achieve the objective is to build a cheaper yet simpler high-lift system. This presents a challenge to the high-lift aerodynamicist: to design a less complex flap that maintains high-lift and meanwhile lowers the flow separation. Furthermore, by designing the high-lift system to achieve longer attachment of flow onto the flap, aircraft noise will be reduced. This research article presents the design, analysis and prototyping and testing of aerofoil which was carried out after performing a literature survey on the performance of many aerofoils at low Reynolds number. This research was achieved using Vortex generators for high lifts. The prototype was tested in the wind tunnel.

Keywords Lift to drag ratio · CFD · Wind tunnel

1 Introduction

The aerofoil is the wing, rotor blade or sail shape as seen in cross-section. Such an aerodynamic body is meant to produce an aerodynamic force if it is moved on a fluid. The component which is perpendicular to the direction of motion is called the lift and the component parallel to the direction of motion is called the drag. Subsonic flight aerofoils have a characteristic shape with a rounded leading edge, followed by a sharp trailing edge, often with asymmetric camber.

Though the earliest serious work on the development of aerofoil sections began in the late 1800s, the development of the modern aerofoils for wind turbines began

P. Rayar (✉)
Department of Production Engineering, Dwarkadas J. Sanghvi College of Engineering, Mumbai, India
e-mail: pavan.rayar@djsce.ac.in

© Springer Nature Singapore Pte Ltd. 2020

H. Vasudevan et al. (eds.), *Proceedings of International Conference on Intelligent Manufacturing and Automation*, Lecture Notes in Mechanical Engineering,
https://doi.org/10.1007/978-981-15-4485-9_70

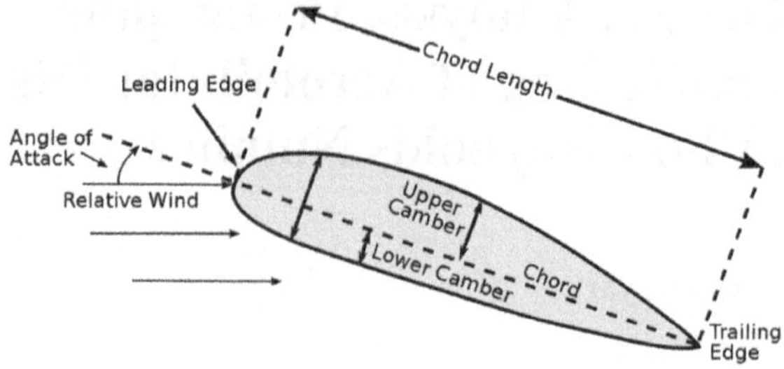

Fig. 1 The nomenclature of aerofoil

in the 1980s. The requirements for such aerofoil differ from standard aviation aerofoils, because of structural reasons and extensive aerodynamic off-design operation conditions. During high wind speeds, wind turbine aerofoils usually operate under fully separated flow when the stall is used for power regulation (Fig. 1).

For simplicity, aerodynamicists follow a convention that allows for the summation of all the very complex aerofoil components in one quantity, the lift coefficient (CL). It determines how well the aerofoil works as a source of lifts. In addition to the lift, there is always a force that directly opposes the wing's movement through air, and is called drag. This component of force is conveniently described by the coefficient of drag (CD). Drag Force is also influenced by the shape of aerofoil and angle of attack

$$\mathrm{FL} = \frac{1}{2}\rho v^2 \mathrm{CLA} \tag{1}$$

$$F = \frac{1}{2}\rho v^2 \mathrm{CDA} \tag{2}$$

Here,
FL = Lift Force, FD = Lift Force, ρ = Density of fluid, v = velocity of free stream, CL = Lift Coefficient, A = Wing Area.

2 Literature Survey

Miley [1] cataloged the results of test conducted on NACA 66(2)-415 aerofoil. For high-lift at low Reynolds number. This test was conducted in 1982 for Rockwell international corporation energy system groups which were a part of the United States Department of energy for the wind energy department under the federal wind energy program. The aerofoil was tested in a close circuit wind tunnel at 40 m/s and Reynolds number of about 7×10^5. An investigation was conducted in the Langley low-turbulence pressure tunnel to evaluate the two-dimensional low-speed aerodynamic characteristics of a 17% thick aerofoil designed for general aviation applications. The findings are compared to the standard older NACA 65 series aerofoil section. A comparison of experimental data and projections is also provided, based on a theoretical framework for measuring the viscous flow about the aerofoil range from $-10°$ to $24°$. Reynolds numbers, based on the aerofoil chord, were varied from about 2.0×10^6 to 20.0×10^6. The results of the investigation indicate that maximum section lift coefficients increased rapidly at Reynolds numbers from about 2.0×10^6 to 6.0×10^6 attained values greater than 2.0 for the plain aerofoil and that for an aerofoil with 20% chord split greater than 3.0. Stall characteristics were seen generally gradual and of the trailing edge type either with or without the split-flap [2]. The NACA 4 and 5 digit sections investigated consisted of the NACA 0012, and the NACA 44 and 230 series sections of 12 and 15% conditions and also included the determination of the effectiveness of the different aerofoils at various Reynolds number when equipped with split flaps [3]. Roughness on the leading edge materially reduced the effect of Reynolds number on maximum lift coefficient, but Mach number effects [4]. The test results revealed considerable maximum lift effects on the three-element landing configuration for Reynolds number variations and significant Mach number effects on the four-element aerofoil [14]. The research paper [18] describes a separate eddy simulation (DES) for a single flow over aerofoil of NACA 00012. Differences were plotted between 2 and 3 dimensional contours, as well as contours of vorticity and entropy. As compared to 2D and 3D URANS, the 2D DES showed an increase in lift and drag predictions. The 3D URANS, DES lift and drag, however, are similar to the experimental results.

3 Statement of Problem and Methodology

It is recommended that the Cambered aerofoil Shape and certain other devices be used to improve lift at the same working conditions by reviewing all of the above Papers and carrying out research work carried out to date on the design of a high-lift aerofoil giving more lift at low Reynolds number. The easiest and best way to improve lift at the same working condition is to have Vortex generators working on the principle of delayed boundary layer separation.

Methodology applied to address the statement of the problem was as follows:

1. To design the best suitable aerofoil for high-lift at lower speeds the NACA 2415 aerofoil will be analyzed first at various angle of attacks to study its performance and draw conclusions regarding the same, thus bettering our understanding of the phenomenon of achieving higher lifts at low Reynolds number or low speeds.
2. To find the separation and reattachment points onto the aerofoil the Ansys 14.0 will be used.
3. The different types of Vortex generators are added onto the upper surface of aerofoil and analyzed for the same speed at various angle of attacks.
4. After getting satisfactory results in the analysis the same model will be 3D printed or Prototyped to validate the results obtained in analysis with the experiments value being conducted on Wind Tunnel.

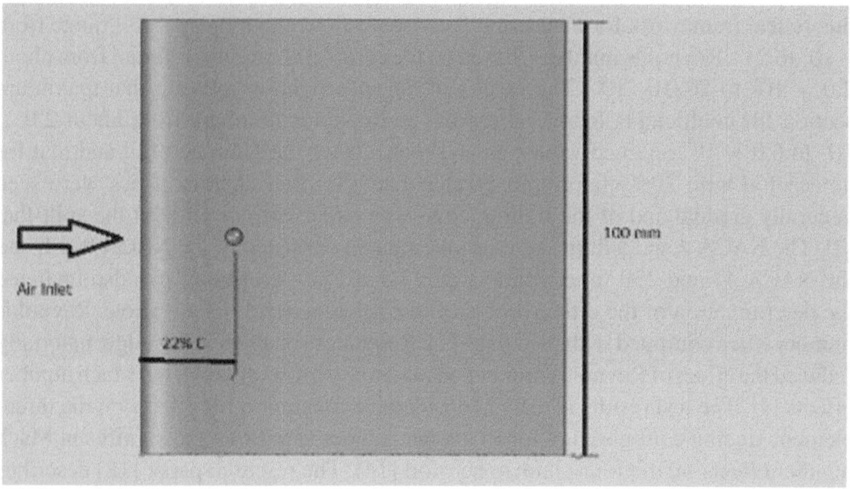

Fig. 2 Top view of NACA 2415 aerofoil with a Single Dimple at a distance of 22% of chord from leading edge

4 Design and Analysis

The aerofoil selected is NACA 2415, which is a low Reynolds number and is perfectly suited for the available wind tunnel set-up. The NACA 2415 aerofoil has a maximum thickness of 15%, Camber 2% located 40% back from the leading edge. Using mathematical relationships, (m p and t) these values can be used to measure the coordinates for a whole aerofoil and then the same thing is CAD modelled in SolidWorks 10.0.

The CAD model of NACa 2415 is analyzed on Ansys to find the separation point then the aerofoil is modified by adding the Vortex generators on the above side (called DIMPLE MODEL) and again analyzed and compared with the earlier model (Figs. 2 and 3).

This section reflects the results obtained from the numerical evaluation on both the Plain and Dimpled Models for Velocity, Pressure, coefficient of lift and coefficient of drag Figs. 4, 5, 6, 7, 8, 9, 10, 11, 12, 13, 14, 15, 16, 17, 18, 19, 20, 21, 22, 23, 24 and 25.

5 Prototyping and Experimentation

Both the models (plain model and Dimpled model) were prototyped on a 3D printer with SLA technology (Stereolithography) and have been analyzed for Cl and Cd on Wind Tunnel. The following observations were made for both the models (Fig. 26).

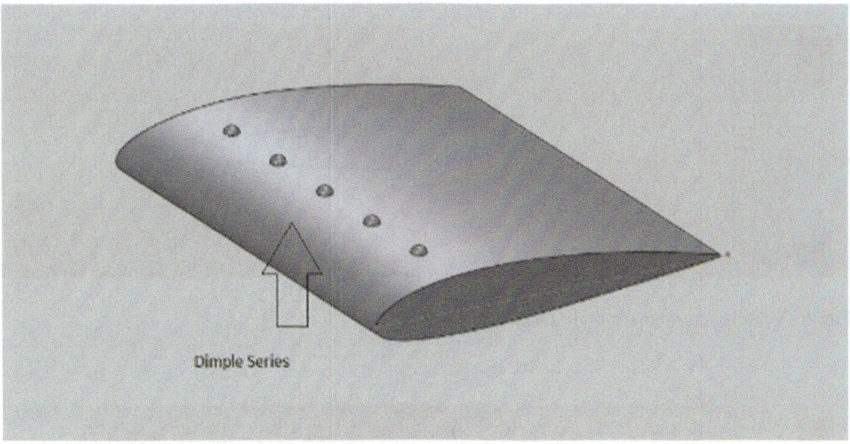

Fig. 3 Isometric view of NACA 2415 aerofoil with Dimple series

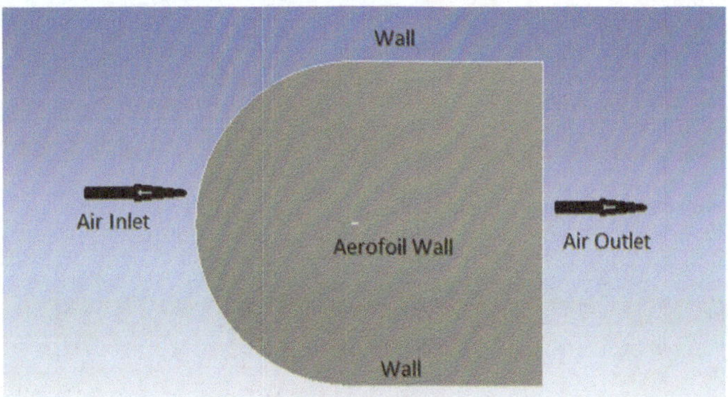

Fig. 4 The Ansys environment (boundary conditions)

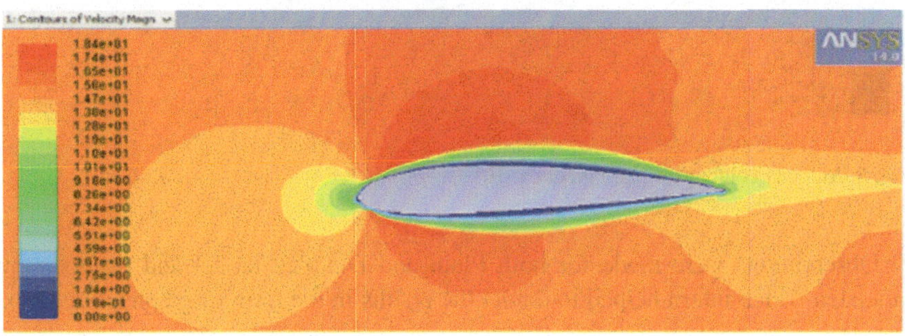

Fig. 5 Velocity contour for plain model at 0°

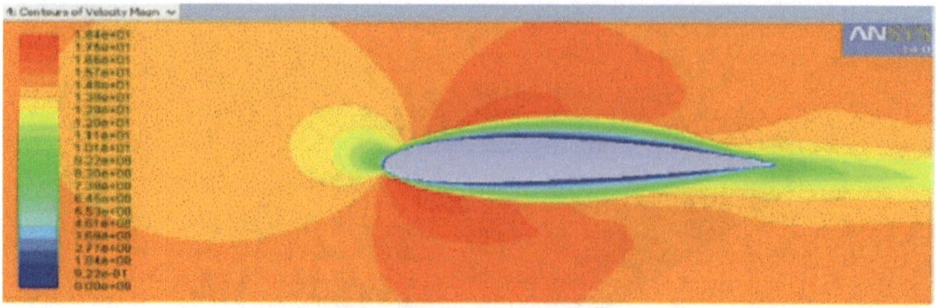

Fig. 6 Velocity contour for plain model at −5°

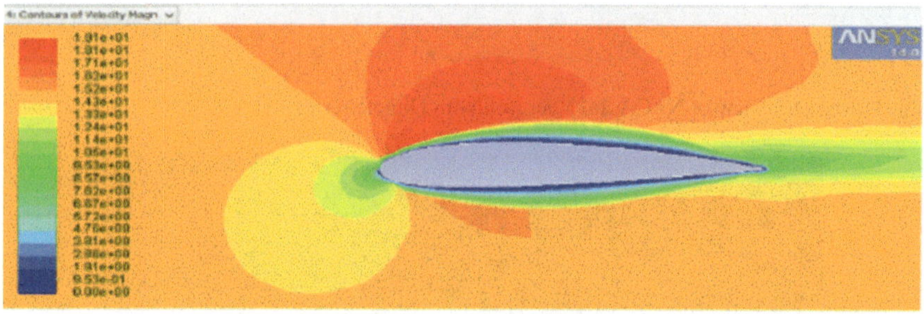

Fig. 7 Velocity contour for plain model at 5°

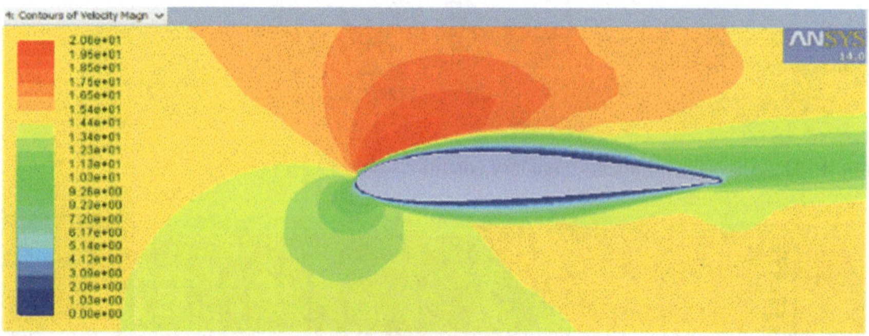

Fig. 8 Velocity contour for plain model at 10°

Observations were made for both Plain and Dimpled model and results were tabulated which intern helped to depict the results in the form of graphs in the next section.

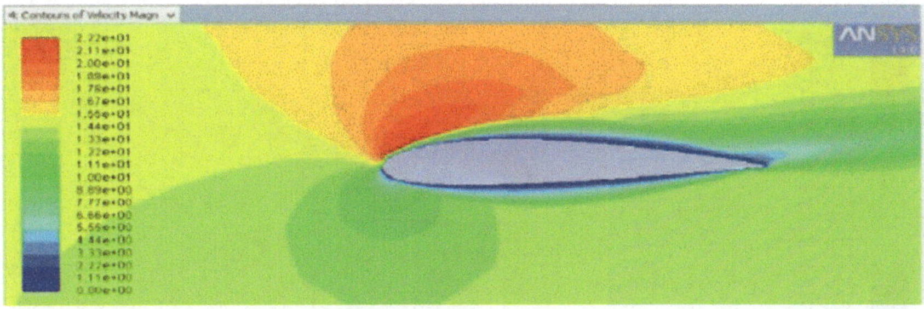

Fig. 9 Velocity contour for plain model at 15°

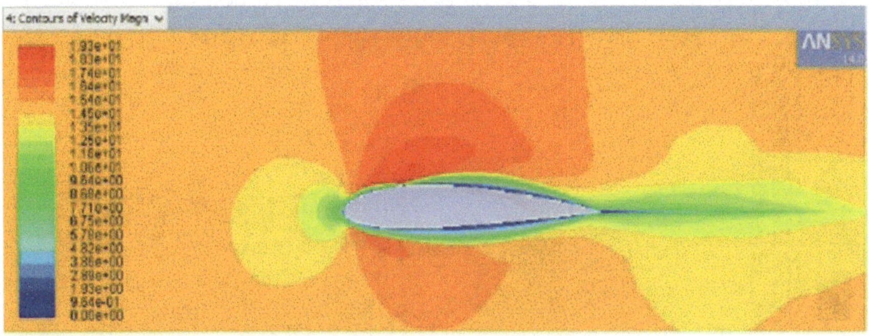

Fig. 10 Velocity contour for Dimpled model at 0°

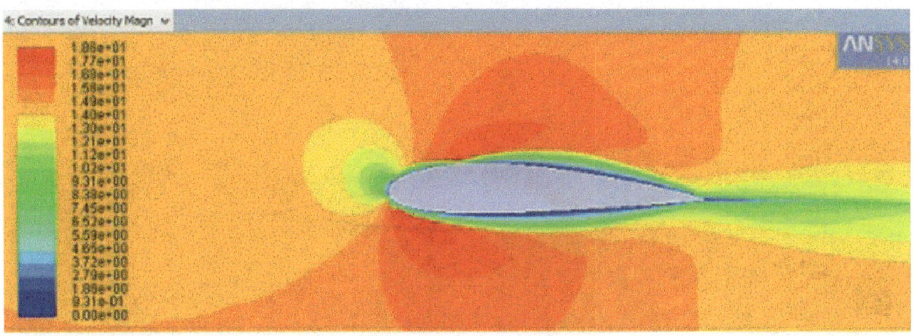

Fig. 11 Velocity contour for Dimpled model at −5°

6 Results

The following graphs were plotted for coefficient of lift and coefficient of drag for both the models against different angle of attack (AOA) and compared to find which model worked more efficiently as given under.

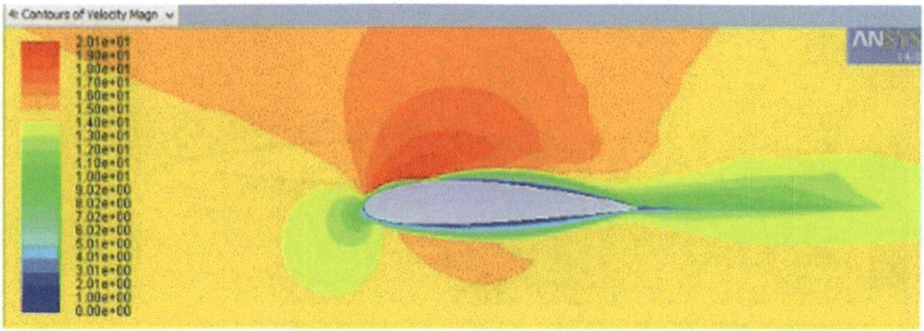

Fig. 12 Velocity contour for Dimpled model at 5°

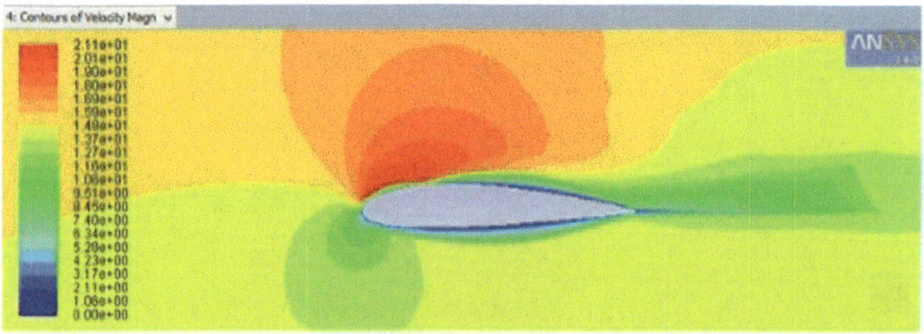

Fig. 13 Velocity contour for Dimpled model at 10°

Fig. 14 Velocity contour for Dimpled model at 15°

Figure 27 shows the variations of Cl against different angle of attack for Dimpled model (Fig. 28).

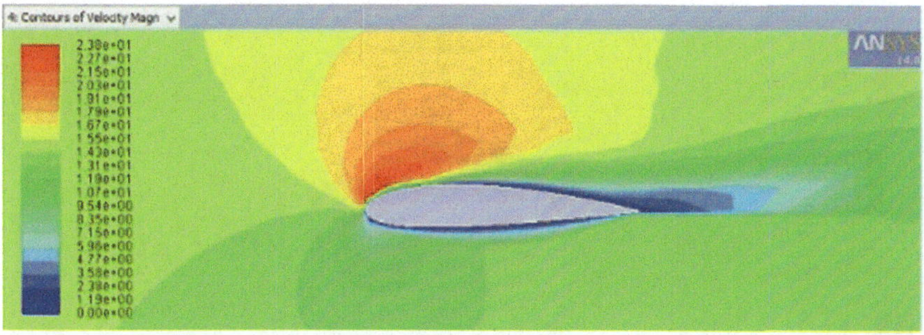

Fig. 15 Velocity contour for Dimpled model at 18°

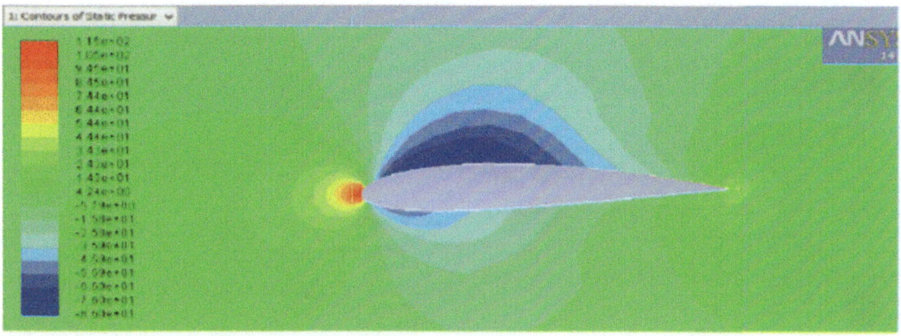

Fig. 16 Pressure contour for plain model at 0°

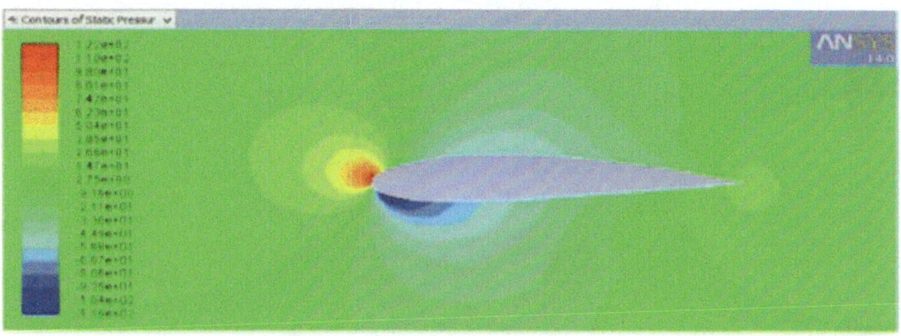

Fig. 17 Pressure contour for plain model at −5°

7 Conclusion and Future Scope

By the comparison held for both the Plain and Dimpled models for NACA 2415 under the same working conditions, it was observed that the Dimpled model has performed well and has achieved a 11% high coefficient of lift and 9% less coefficient of drag as compared to the plain model.

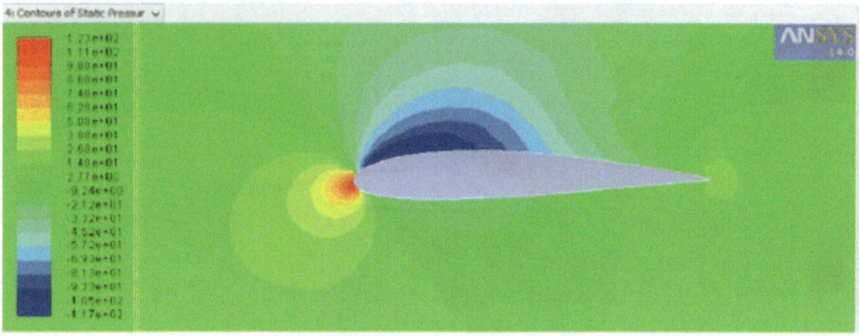

Fig. 18 Pressure contour for plain model at 5°

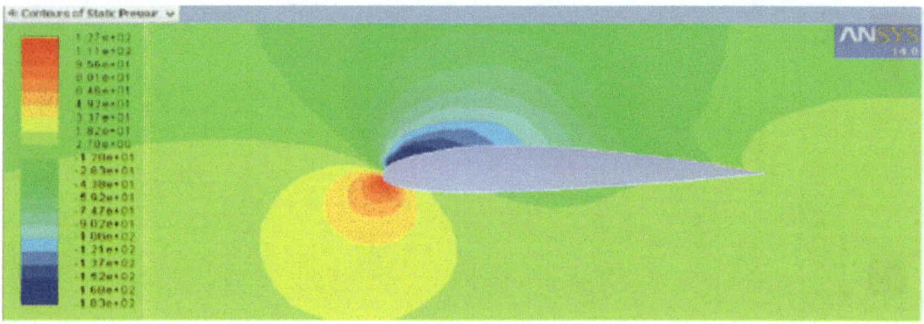

Fig. 19 Pressure contour for plain model at 10°

Fig. 20 Pressure contour for plain model at 15°

The experimental values were depicted in Graphical form. The Dimpled model performed well in the same environment given to the plain model by producing 11% more coefficient of lift compared to plain model and 09% less coefficient of drag compared to plain model. From 4° to 17° the CL increased continuously and the ratio

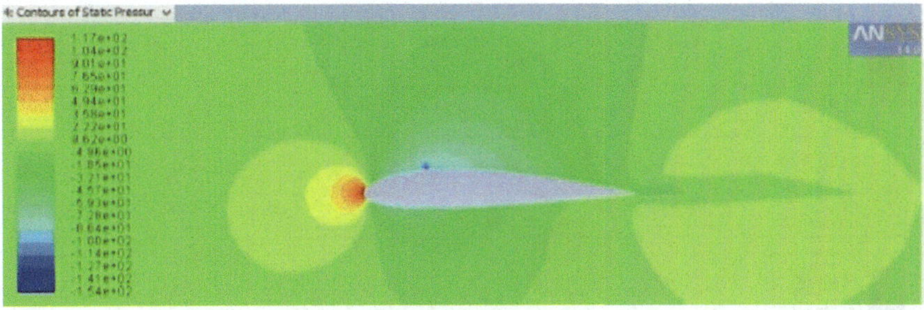

Fig. 21 Pressure contour for Dimpled model at 0°

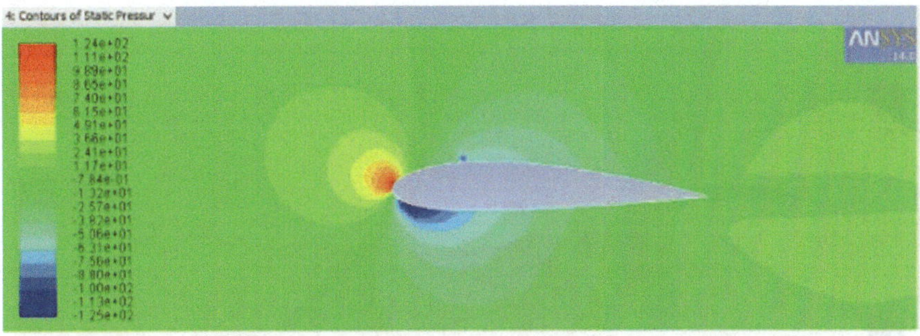

Fig. 22 Pressure contour for Dimpled model at −5°

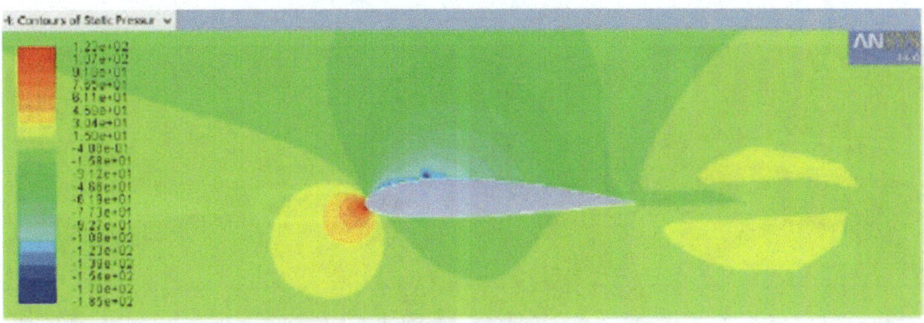

Fig. 23 Pressure contour for Dimpled model at 5°

of L/D was over 1. Within this range, the aerofoil is utilized efficiently. The research showed that the Vortex generators can be used in different ways with different shapes and sizes on the upper surface for delayed separation. The research can be further taken ahead with Positioning of the Vortex Generators with a combination of other moving elements.

Fig. 24 Pressure contour for Dimpled model at 10°

Fig. 25 Pressure contour for Dimpled model at 18°

Fig. 26 The plain model of the aerofoil (by Wirecut M/C)

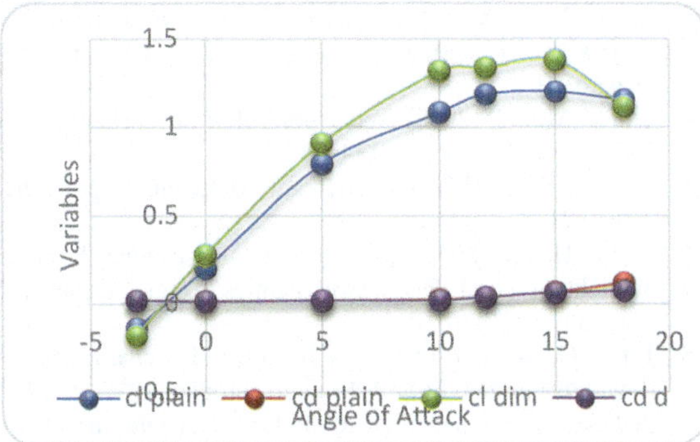

Fig. 27 Coefficient of lift and coefficient of drag versus AOA for both models

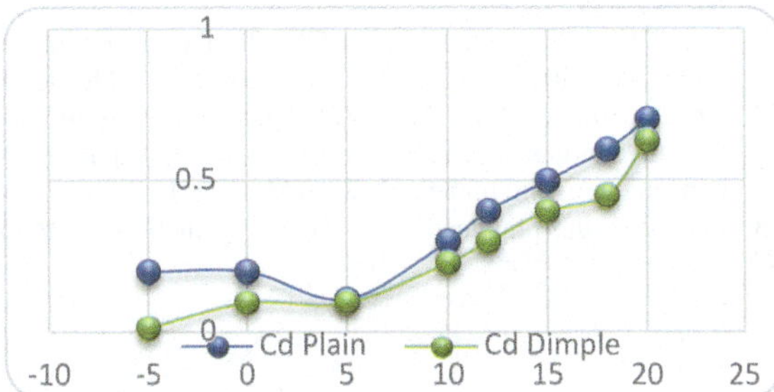

Fig. 28 Comparison of coefficient of drag for both models versus angle of attack

Bibliography

1. Selig MS, Guglielmo JJ (1997) High-lift low Reynolds number airfoil design. J Aircr 34(1):72–79
2. Merabet A, Necib B (2003) Characterisation of wings with NACA 0012 airfoils. Rev. énergies Renouvelables 2:131–137
3. Kunz PJ, Kroo I (2001) Analysis and design of aerofoil's for use at ultra low Reynolds numbers. Prog Astronaut Aeronaut 195:35–60 (Mueller TJ (ed))
4. Rumsey CL, Gatski TB, Ying SX, Bertelrud A (1998) Prediction of high-lift flows using turbulent closure models. AIAA J 36:765–774
5. McCullough ME, Kilpatrick SD, Emmons RA, Larson DB (2004) Is gratitude a moral affect? Psychol Bull 127:249–266
6. Carmichael BH (1981) Low Reynolds number aerofoil survey, vol 1. NASA contractor report 165803, Nov 1981
7. Guilmineau E, Piquet J, Queutey P (1997) Two-dimensional turbulent viscous flow. Comput Fluids 26(2):135–162
8. Loftin LK, Poteat MI. Aerodynamic characteristics of several NACA airfoil sections at seven Reynolds numbers from 0.7 × 10 (exp 6) to 9.0 × 10(exp 6)
9. Srivastav D (2012) Flow control over airfoils using different shaped dimples, vol 33, no. Fdtt, pp 92–97 (2012)
10. Anderson JD Jr (2001) Fundamentals of aerodynamics. McGraw Hill
11. Sathaye SS (2004) Lift distributions on low aspect ratio wings at low Reynolds numbers, May
12. Mirzaei M, Ardekani MA, Doosttalab M (2009) Numerical and experimental study of flow field characteristics of an iced airfoil. Aerosp Sci Technol 13:267–276

13. Emami MR (2007) Aerodynamics forces on an aerofoil. Presented at Aerospace Undergraduate Laboratories-University of Toronto in 2007, for research on lift on aerofoil and forces acting on aerofoil
14. You D, Moin P (2008) Active control of flow separation over an airfoil using synthetic jets in 2008. J Fluids Struct 24:1349–1350
15. Hoerner SF (1965) Fluid dynamic drag. In: Hoerner fluid dynamics, Brick Town, NJ, pp 7.16–7.21
16. Gad-el-Hak M (2001) Micro-air-vehicles: can they be controlled better. J Aircr 38(3):419–429
17. Schlichting H (1979) Aerodynamics of the airplane. Mc-Graw-Hill International Book Company, pp 143–161
18. Torres G, Mueller TJ (2001) Aerodynamic characteristics of low aspect ratio wings at low Reynolds number. Prog Astronaut Aeronaut 195:115–141 (Mueller TJ (ed))
19. Pelletier A, Mueller TJ (2000) Low Reynolds number aerodynamics of low-aspect-ratio, thin/flat/cambered-plate wings. J Aircr 37(5):825–832
20. Spalart PR, Allamaras SR (1992) A one-equation turbulence model for aerodynamic flow. AIAA Pap 92:0439
21. Open∇FOAM, the Open Source CFD Toolbox user guide, version 1.4.1, 1 Aug 2007, copyright 2000, 2001, 2002, 2003, 2004, 2005, 2006, 2007 OpenCFD limited
22. Ghia U, Ghia KN, Shin CT (1982) High-Re solutions for incompressible flows using the Navier-Stokes equations and a multigrid method. J Comput Phys 48:387–411 (in the Langley 8-foot transonic pressure tunnel, NASA TM 81927)
23. Mangler (1939) Induced drag analysis in small aspect ratios. Yearbook D.Lufo, p.I 139

A Review on Vibration Suppression of Flexible Structures Using Piezoelectric Actuators

Aniruddha Mallick, Frank Crasta, and Vijaya Kumar N. Kottur

Abstract Piezoelectric Actuators are devices that convert electrical energy into a mechanical displacement or stress by a phenomenon known as Piezoelectric Effect. These crystals generate energy when mechanical stress is applied. This paper reviews the application of piezoelectric actuators in vibration suppression of flexible lightweight structures. Feedback control strategies are used for suppressing periodic and aperiodic vibrations. Negative velocity feedback and positive position feedback are two such common strategies employed for controller design. A finite element is defined for modelling of a uniform cantilever beam and a thin, flexible plate. The appropriate controller frequency is obtained by modal analysis, for monomodal vibration control.

Keywords Piezoelectric actuators · Piezoelectric effect · Vibration suppression · Feedback control strategies · Negative velocity feedback · Positive position feedback · Finite element · Modal analysis · Monomodal vibration control

1 Introduction

Lightweight flexible structures are used in aircraft, bridges, robot manipulators, large space structures, buildings, etc. Vibrations in any static structures are undesirable, as they induce stress in various structural elements. To reduce the stresses induced by high amplitude vibrations, damping, isolation or any other cancellation method is introduced in these structures. Damping will remove vibration energy, in the form

A. Mallick (✉) · F. Crasta · V. K. N. Kottur
Department of Mechanical Engineering, Dwarkadas J. Sanghvi College of Engineering, Mumbai, India
e-mail: animallick236@gmail.com

F. Crasta
e-mail: prashant.crasta@djsce.ac.in

V. K. N. Kottur
e-mail: vijayakumar.kottur@djsce.ac.in

© Springer Nature Singapore Pte Ltd. 2020

H. Vasudevan et al. (eds.), *Proceedings of International Conference on Intelligent Manufacturing and Automation*, Lecture Notes in Mechanical Engineering,
https://doi.org/10.1007/978-981-15-4485-9_71

of heat. Isolation protects a piece of equipment, by absorbing the mechanical energy from the system.

The two fields to look upon are the applications of smart materials and the adoption of various control techniques for active vibration suppression. Researchers have used smart materials such as Shape Memory Alloys (SMA) [1], Magnetorheological materials [2], Electrorheological materials [3], piezoelectric transducers, etc. Piezoelectric materials are suitable for use in structural vibration control because of (1) fast response, (2) it does not generate a magnetic field during conversion of electrical energy to mechanical motion. The most commonly used material as actuator is PZT (Lead-Zirconate-Titanate) material due to the high electromechanical coupling coefficient.

The piezoelectric materials can also be integrated with laminated composites. However, the challenge faced is the selection of elements for meshing, while accounting for the composite mechanical and electrical properties of the laminate plies. Based on the first-order shear theory (FSDT), a finite element is established to study the effect of stretching-bending coupling of piezoelectric actuator pairs on system stability of smart composite plates. A simple FSDT can be adopted which is similar to classic plate theory and uses a smaller number of unknowns [4]. Alternatively, the analysis can be performed by using a conventional beam or plane stress element, and the effects of the coupling terms can be translated as equivalent concentrated nodal loads.

Velocity Feedback and Positive Position Feedback (PPF) are popularly known control strategies in the classical control theory, due to their easy implementation. The PPF strategy is implemented by using an idealized modal model (analysis of dynamic properties of the system under the frequency domain) of an actuated structure. Lagrange's equations of motion are used to describe the behaviour of the system.

2 Need of Damping in Non-collocated Systems

Collocated control stands for the use of actuator and sensor on a structure when they are physically located at the same place and are energetically conjugated [8]. Studies conducted show how the zeroes migrate when the sensor moves away from the actuator. With the displacement of the sensor along beams with specific boundary conditions, pair of imaginary zeroes will reach infinity and then move towards origin along the real axis. This is known as non-minimum phase and it signifies non-collocated control systems.

Lead compensators produce sinusoidal output with phase lead and this compensates the undesirable phase lag caused by the poles, in the form of imaginary zeroes. However, they provide damping to all flexible modes. Notch filter is used along with lead compensator to introduce two zeroes alongside the flexible poles. It reduces the effect of flexible modes in non-collocated control. In case of systems with large uncertainties, notch filter is not useful as they are tuned to a single frequency and it is not subjected to parameter uncertainty. Therefore, damping is critical for non-collocated systems.

3 Direct Velocity Feedback

If a load is to be moved by means of a motor, the common control method used is Proportional (P) control. It will produce system output proportional to the displacement required and will oscillate about the steady-state value for some time. By using Proportional-Derivative (PD) control, faster response with fewer oscillations is obtained, as compared to P control. The alternative to this method is to use a second feedback loop which returns the rate at which displacement is changing. This is known as velocity feedback.

The governing equation of motion is obtained by using Hamilton's variational principle

$$M\ddot{x} + C\dot{x} + Kx = f. \tag{1}$$

where M, C and K denote the mass, damping and stiffness matrices, respectively, and f stands for the perturbation. Let $\dot{y}$ be a set of velocity measurements

$$\dot{y} = B^T\dot{x}. \tag{2}$$

where B is the influence matrix of dimensions $(n \times m)$. The damping matrix $C = BGB^T$. The velocity distributions which belong to the null space of B^T will remain undamped.

4 Positive Position Feedback (PPF)

This control strategy is appropriate for structures equipped with strain actuators and sensors. It uses a second-order filter. The position response received from the second-order auxiliary system is passed through a second-order filter and force feedback is given to the structure.

For designing the PPF controller, an idealized modal model of a flexible structure is considered. Either Lagrange's [3] or Hamilton's [4, 7] equation can be used to formulate the equation of motion for the structure. The Lagrange equation is

$$\frac{\mathrm{d}}{\mathrm{d}t}\left(\frac{\partial(T - V)}{\partial\dot{q}}\right) + \frac{\partial(T - V)}{\partial q} = Q. \tag{3}$$

where ψ is the displacement of the structure; ζ_s and ω_s are the damping ratio and the natural frequency of the structure; η is the displacement of the compensator, ζ_c and ω_c are the damping ratio and frequency of the compensator; K_{PPF} is the feedback gain. Shan et al. [5] described the state-space form of the equations describing the PPF technique.

When a general case of Multiple Input Multiple Output (MIMO) system with m collocated actuator/sensor pairs and an array of l second-order filters.

$$M\ddot{x} + Kx = Bu \tag{4}$$

$$y = B^T x \tag{5}$$

$$\ddot{v} + \beta_f \dot{v} + \Omega_f^2 v = Ey \tag{6}$$

The structure, sensor and controller equations are (4), (5) and (6) respectively, where $\beta_f = \text{diag}(2\xi_f \omega_f)$, $\Omega_f^2 = \text{diag}(\omega_f^2)$ and $G = \text{diag}(g_i)$, $(g_i) > 0$. G is the diagonal positive gain matrix. E is a rectangular matrix ($l \times m$) which allows the number of filters to be greater than actuators (l modes are dampened by m actuators).

There are three situations regarding the comparison of structural frequency and compensator natural frequency [5],

Active Flexibility: $\zeta_s < \zeta_c$ (due to decrease in stiffness term).
Active Damping: $\zeta_s = \zeta_c$ (due to increase in damping term).
Active Stiffness: $\zeta_s > \zeta_c$ (due to increase in stiffness term).

The robustness of the controller increases by increasing ζ_c as this ensures larger region of active damping.

5 Structure Modelling

5.1 Uniform Cantilever Beam

For system modelling, the flexible structure (for example, robot-arm manipulator [5]) bonded with piezoelectric (PZT) actuators is modelled as a uniform cantilever beam. It is assumed as a Euler-Bernoulli beam. The Euler-Bernoulli equation gives a relation between the deflection of the beam and the applied load

$$\frac{d^2}{dx^2}\left(EI\frac{d^2 w}{dx^2}\right) = q. \tag{7}$$

where $w(x)$ is the curve that defines the deflection of the beam. q is the force per unit length. E is the elastic modulus and I is the area moment of inertia.

Let $w(x, t)$ be the deflection of the beam in the z-direction at point x and at time t. The deflection can be represented as

$$w(x,t) = \sum_{k=1}^{m} \phi_k(x) q_k(t) = \phi(x) q(t). \quad (8)$$

where $\phi_k(x)$ is the shape function, $q_k(t)$ is generalized modal coordinate and k represents the mode number ($k = 1, 2, 3, \ldots, m$). A shape function is a function that is used in finite element simulations to combine the solutions obtained at the mesh nodes. Generalized modal coordinates are obtained by a coordinate transformation,

$$\{q(t)\} = [\psi]\{p(t)\}. \quad (9)$$

where $[\psi]$ is the modal transformation matrix and $p(t)$ denotes the principal coordinates. Such a transformation decouples the coupled set of equations of motion (of higher DOF) so that they can be solved using the single DOF approach.

5.2 Plate

A smart isotropic finite element [7] is used for vibration control of composite plate with bonded piezoelectric PZT actuators. Piezoelectric materials have two constitutive laws, one which is used for sensing and the other for actuation.

$$\{\sigma\}_{3\times1} = [C]^{(E)}_{3\times3}\{\varepsilon\}_{3\times1} - [e]_{3\times2}\{E\}_{2\times1}. \quad (10)$$

$$\{D\}_{2\times1} = [e]^{T}_{2\times3}\{\varepsilon\}_{3\times1} + [\mu]^{(\sigma)}_{2\times2}\{E\}_{2\times1}. \quad (11)$$

where $[C]$ is the mechanical constitutive matrix measure at constant electric field. The above equation can be combined into a single equation

$$\{\varepsilon\} = [S]\{\sigma\} + [d]\{E\}. \quad (12)$$

where $[S]$ is the compliance matrix and $[d]$ is the electromechanical coupling matrix. This law is extensively used for vibrational control, noise control and shape control.

PPF strategy and strain actuators/sensor pairs are relied on for active damping. Nearly collocated configuration of actuator and sensor is preferred as they guarantee alternating poles and zeroes at low frequencies. Precise tuning of the controller natural frequency on the targeted mode is necessary.

Taking an example, a 4-noded isoparametric FE element can be formulated, assuming 2-D plane stress.

At each node, there are three DOFs: (1) $u(x, y, t)$, (2) $w(x, y, t)$ (Displacement components), (3) $E_z(x, y, t)$ (Electrical degree of freedom in z-direction).

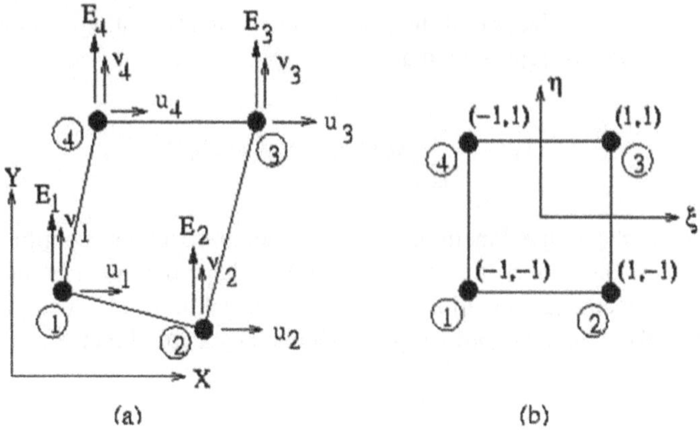

Fig. 1 **a** 4-noded isoparametric element. **b** Coordinate transformation into square of order 2

$$u(x, y, t) = \sum_{i=1}^{4} N_i(\zeta, \eta)u_i(t). \tag{13}$$

$$w(x, y, t) = \sum_{i=1}^{4} N_i(\zeta, \eta)w_i(t). \tag{14}$$

$$E_z(x, y, t) = \sum_{i=1}^{4} N_i(\zeta, \eta)E_{zi}(t). \tag{15}$$

where N_i stands for shape function at the node; ζ and η are the isoparametric coordinates and are the nodal mechanical DOFs. Through a Jacobian transformation, the actual geometry is mapped to a square of order 2 defined in the generalized coordinate system (Fig. 1b). First-order Shear Theory is used to obtain the following linear strain displacement relations

$$u(x, y, z, t) = u_0(x, y, t) + z\beta_x(x, y, t). \tag{16}$$

$$v(x, y, z, t) = v_0(x, y, t) + z\beta_y(x, y, t). \tag{17}$$

$$w(x, y, z, t) = w_0(x, y, t). \tag{18}$$

The nodal displacement vector is used with shape function to obtain reduced strain vector. The governing equation is thus found.

6 Modal Analysis

Modes are inherent properties of a structure. They are characterized by their natural frequency and modal shape. Modal analysis is needed to find out the modal parameters of the structure [6]. This method is an important tool for vibration diagnosis, analysis and control.

Modal analysis is carried out in Ansys workbench. The cross-section of a cantilever beam is created and then extruded up to a given depth. The parameters given in Table 1 are taken into account. The beam is automatically meshed and fixed support constraint is applied to one end of the beam. The displacement should be zero at the fixed support.

From the above observations (Fig. 2; Table 2), it can be concluded that the convenient mode shape to control is the first mode. Thus, a PPF control system can be developed using the first mode frequency, 9.8491 Hz as the controller frequency [6]. An accurate value of the controller is obtained, by reducing the mesh size and defining the mesh by commands.

Table 1 Aluminum alloy beam parameters

Density	2700 kg/m^3
Young's modulus	68.9×10^9 (Pa)
Size	$0.5 \times 0.04 \times 0.003$ m^3

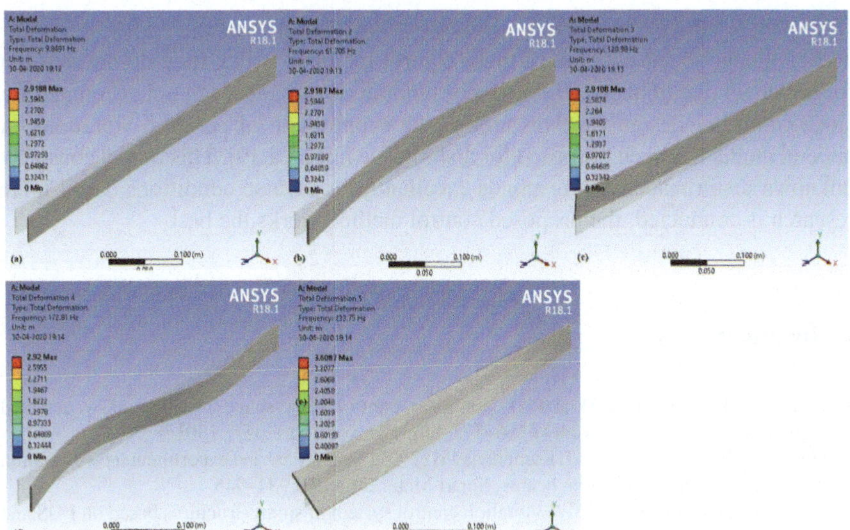

Fig. 2 Mode shapes: **a** first mode, **b** second mode, **c** third mode, **d** fourth mode, **e** fifth mode

Table 2 List of mode frequencies

Tabular Data

	Mode	Frequency [Hz]
1	1.	9.8491
2	2.	61.705
3	3.	129.98
4	4.	172.81
5	5.	233.75

7 Conclusion

Control laws have guaranteed stability due to the interlacing of the poles and zeroes of the structure along the imaginary axis. Stability is ensured for collocated pairs of sensors and actuators. If they are not collocated or the location of the actuator-sensor pair is such that controllability and observability are weak, the system becomes unstable [8]. It has been studied that two such control methods viz., negative velocity feedback and positive position feedback have been developed in the field of vibration suppression.

8 Scope for Future Work

The challenge faced by each of these control methods is that they do not account for unknown disturbances. Unknown or random disturbances are not accounted for since controller design revolves around the suppression of periodic vibrations. A general model has been proposed for disturbance rejection [9]. This model considers unknown disturbances and responds excellently in adverse conditions. Until more research is conducted, the proposed control method works the best.

References

1. Elahinia M, Ashrafiuon H (2002) Nonlinear control of a shape memory alloy actuated manipulator. J Vib Acoust 124(4):566–575. https://doi.org/10.1115/1.1501285
2. Giurgiutiu V, Jichi F, Berman J, Kamphaus JM (2001) Theoretical and experimental investigation of magnetostrictive composite beams. Smart Mater Struct 10:934–945
3. Leng J, Asundi A (1999) Active vibration control system of smart structures based on FOS and ER actuator. Smart Mater Struct 8:252–256
4. Thai HT, Choi DH (2013) A simple first-order shear deformation theory for laminated composite plates. Compos Struct 106:754–763
5. Shan J, Liu HT, Sun D (2005) Slewing and vibration control of a single-link flexible manipulator by positive position feedback (PPF). Mechatronics 15:487–503

6. Huertas VV, Ilkiv BR (2012) Vibration suppression of a flexible structure. Procedia Eng 48:233–241
7. Wang SY, Quek ST, Ang KK (2001) Vibration control of smart piezoelectric composite plates. Smart Mater Struct 10:637–644
8. Preumont A (2011) Vibration control of active structures—an introduction, 3rd edn. Springer, Berlin
9. Zhang XY, Wang RN, Zhang SQ, Wang ZX, Qin XS, Schmidt R (2018) Generalized-disturbance rejection control for vibration suppression of piezoelectric laminated flexible structures. Shock Vib 3:1–17. https://doi.org/10.1155/2018/1538936

Topology Optimization of Wheel Hub Used in Automobiles

Jash H. Patel, Rohan Poojari, Monil K. Shah, Aagam H. Shah, and Vinayak H. Khatawate

Abstract The wheel hub is a vital element of a vehicle that attaches the wheel to the motor shaft. Its main function is to keep the wheel running freely on the bearing while keeping it attached to the vehicle. It experiences a lot of shearing and bending forces when the vehicle in motion. The focus of this paper was to reduce the load on the components due to these forces on the wheel hub by optimizing its design and topology. A solid model was created in accordance with the optimized design and FEA was performed to determine its strength.

Keywords Topology · Wheel hub · Mass reduction · Automobiles · FEA

1 Introduction

The exponential increase in performance-based competition in the automobile industry demands top-notch development and design to meet industry standards. The wheel hub is one of the most essential components which contributes to safe steering, handling and efficiency of a vehicle. Its primary function is to keep the wheel attached to the axle and to allow the wheel to turn freely for safe steering. In order to increase the efficiency of the wheel hub, accurate and specific solutions are required for different working conditions.

In the past, studies have been conducted on fatigue-based design and analysis of the wheel hub by simulation approach and its performance is also checked under non-constant rotational loading [1, 2]. Some other studies have focused on optimizing the design by analyzing strength using different materials and also optimizing material using finite element analysis [3, 4].

Topology deals with the properties of a geometric design that are observed under constant deformation. The optimization of topology is a very useful tool which helps to maximize the performance of the overall product life cycle from raw material to

J. H. Patel (✉) · R. Poojari · M. K. Shah · A. H. Shah · V. H. Khatawate
Department of Mechanical Engineering, Dwarkadas J. Sanghvi College of Engineering, Mumbai, Maharashtra 400056, India
e-mail: jash.ravariya98@gmail.com

© Springer Nature Singapore Pte Ltd. 2020

H. Vasudevan et al. (eds.), *Proceedings of International Conference on Intelligent Manufacturing and Automation*, Lecture Notes in Mechanical Engineering,
https://doi.org/10.1007/978-981-15-4485-9_72

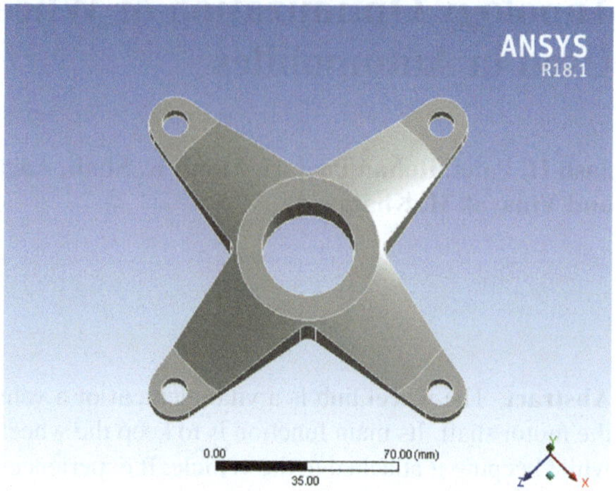

Fig. 1 Basic design of a four-flanged wheel hub

being purchased by the customer. Topology optimization tool attempts to enhance material layout of a design, for a given set of loads, constraints and boundary conditions. Though other optimization tools exist, topology optimization remains one of the most powerful tools for developing novel shapes and concept designs.

This paper aims to optimize the topology of a wheel hub by incorporating changes in the existing basic design shown in Fig. 1. This is done in order to reduce the overall unsprung mass of the wheel assembly and consequently reducing the forces on components. An Over-designed FEA model is analyzed for the strength under shearing and bending loads due to bump force, drive torque and camber thrust.

2 Analytical Calculations

Following assumptions are made for the analytical calculations:

1. The driver is assumed to be divided into three parts during this calculation – head, torso and legs and their weight and centre of gravity location are approximated.
2. The Centre of gravity of the miscellaneous parts which include fasteners, body panels, etc. which are distributed in the whole vehicle is not considered but their overall mass is considered.
3. The vehicle's left:right bias is considered to be 50:50 (the vehicle is symmetrical on both sides and the centre of gravity is on the mid-plane).
4. Forces when the rear wheel hits a bump are more than the forces when the rear wheel hits the ground after a 5 ft drop (as front-wheel takes the major load in a 5 ft drop).
5. For this calculation, the wheel's dynamic rolling radius is taken as 97% of its original radius of 11.5 in.

6. Assumption is made that the momentum is taken as an average momentum when applying the Impulse-Momentum equation, which is found using velocity at an angle that is the mean of the initial and final angles when a bump is encountered.
7. Bump is considered as semi-circular with a radius of 12 in (worst case).
8. Vehicle goes about a corner with a turning radius of 1.64 m.
9. The inertial losses in engine and gearbox due to rotating components at high rpm are neglected, also a homogeneous and isotropic material is used.

2.1 Calculation to Find the Weight Biasing of Vehicle

Centre of gravity of individual components was found out through their CAD model. These points were plotted on a plane (Fig. 2) with a fixed reference point. In this case, the centre of rear wheel has been chosen as the origin or reference point.

Table 1 shows the values of x and y coordinates of the centre of gravity of each part of the vehicle in side-view taking the centre of rear wheel as the origin (reference point).

Referring to Table 1,

$$\mathrm{CG}_x = \frac{\sum_1^n (m_i x_i)}{\sum_1^n m_i} = \frac{-130{,}745}{214.49} = -609.56\ \mathrm{mm}$$

$$\mathrm{CG}_y = \frac{\sum_1^n (m_i y_i)}{\sum_1^n m_i} = \frac{52{,}162.68}{214.49} = 243.19\ \mathrm{mm}$$

This gives the exact location of centre of gravity of vehicle from reference point and thus calculating the weight biasing of the vehicle.

Wheelbase = 54 in

Distance from CG to front tyre = 24.8022 in

$$\text{Front Weight Biasing} = \frac{\text{Distance from CG to front}}{\text{Wheelhase}}$$

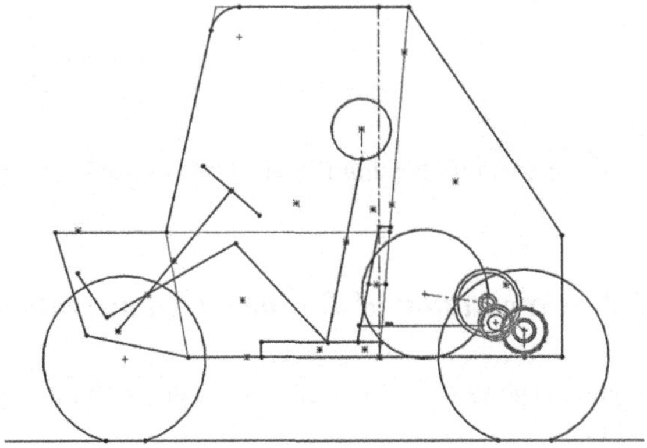

Fig. 2 Centre of gravity layout

Table 1 Coordinates of centre of gravity of vehicle parts in XZ Plane

Parts	mass (m_i)	x_i	y_i	$m_i x_i$	$m_i y_i$
Front Assembly	24.9	−1371.6	−6.6	−34152.84	−164.34
Rear assembly	24.2	0	0	0	0
Front Afco	6	−1290.86	209.58	−7745.16	1257.48
Rear Afco	6	−61.51	240.54	−369.06	1443.24
Steering column	0.6	−1199.3	322.6	−719.58	193.56
Steering assembly	1.1	−1395.22	89.11	−1534.74	98.02
Fuel tank	2.39	−231.92	585.81	−554.29	1400.09
Roll cage	30	−776.65	512.81	−23299.5	15384.3
Steering wheel	0.5	−1003.38	556.09	−501.69	278.05
Brake light	0.3	−404.98	1016.65	−121.49	305
Seat upper	0.75	−501.99	243.15	−376.49	182.36
Seat lower	0.75	−697.12	25.47	−522.84	19.10
Primary	2.8	−337.61	210.18	−945.31	588.50
Secondary	2.1	−123.81	180.15	−260.	378.32
Gear box	4.2	−61.9	133.9	−259.98	562.38
Head	4.74	−553.23	762.07	−2622.31	3612.21
Torso	37.94	−606.54	386.21	−23012.13	14652.81
Leg	21.02	−963.9	189.26	−20261.18	3978.25
Engine	26	−337.61	210.18	−8777.86	5464.68
Fire extinguisher	2.2	−514.57	492.69	−1132.05	1083.92
Shoulder belt	1	−449.48	508.07	−449.48	508.07
Lap belt	1	−544.72	25.47	−544.73	25.47
ASM belt	0.5	−951.12	0.078	−475.56	0.04
Brake assembly	1	−1529.95	423.24	−1529.95	423.24
CVT casing	2.5	−230.71	195.17	−576.78	487.93
Miscellaneous	10				
Total	214.49 Kg			−130745	52162.68

$$= \frac{24.8022}{54} * 100 = 45.93\%$$

Hence, weight biasing is considered as 46:54 (Front:Rear).

2.2 *Calculation of Torque Given by Output Shaft*

Figure 3 shows the vehicle climbing a grade and the system of forces that are acting on it.

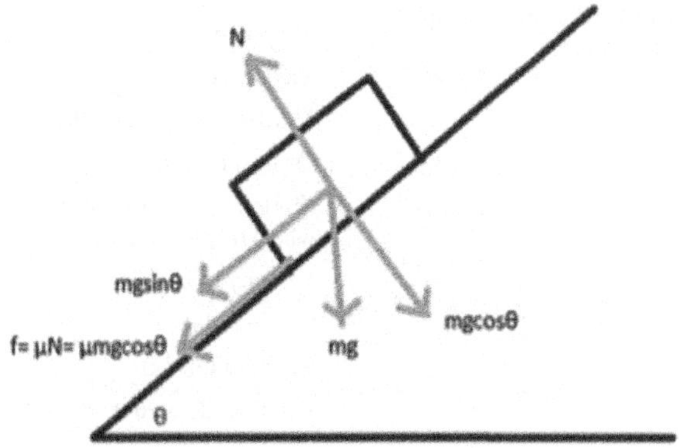

Fig. 3 F.B.D. of vehicle climbing a grade

Table 2 Data for Initial torque calculation

Mass of vehicle (m)	215 kg
Grade of the slope (θ)	45°
Coefficient of rolling resistance (μ)	0.1 (Power limiting condition)
Tyre radius	0.97 * 0.2921 = 0.283337
Engine torque	18.98 Nm
CVT reduction	3.9

Given:
Table 2 shows the available data for calculation of torque on one wheel
Total resistance on wheel $= mg\ \sin\theta + \mu mg\ \cos\theta = 1640.54117$ N
Torque on wheel = Total resistance on wheel * tyre radius = 464.82 Nm
Hence, Torque on single wheel is 464.82 Nm ≈ 465 Nm
Since PCD of rims is 144 mm, Radius of wheel stud points is 72 mm
Therefore, Force due to drive torque on 4 flanges = 465 * (1000/72) = 6458.33 N
Force due to drive torque on 1 flange = (6458.33/4) = 1614.58 N
A force of 1615 N will be applied on each flange anticlockwise in XZ plane.

2.3 *Calculation of Force on the Wheel When Hitting a Bump*

Table 3 shows the available data for calculation of force when the wheel hits a bump.
To find (compression of strut):

Figure 4 shows the rear wheel of the vehicle when hitting the bump and just before leaving contact from the bump.
To find final angle () on the bump:

$$\text{Strut travel}\left(x_f\right) = \text{wheel travel}(x) * \text{Motion ratio}$$

Table 3 Data for bump force calculation

Mass of vehicle	215 kg
Weight bias	46:54 (front: rear)
Mass on single rear wheel (m)	$\frac{215*0.54}{2} = 58.05$ N
Front spring rate (k)	30.39 N/mm
Motion ratio (strut displacement by wheel displacement)	$\frac{18.2-12.9}{7+3} = 0.53$
Velocity of car (v)	15 kmph = 4.1667 m/s
Wheel radius	11.5 in
Bump radius (r)	12 in
Initial angle on the bump (θ)	29.3°

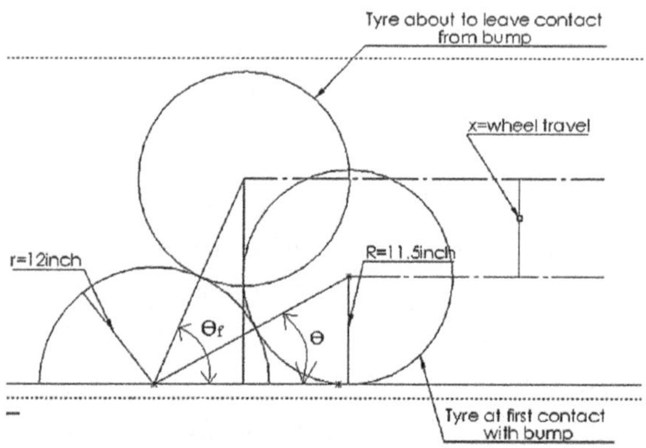

Fig. 4 Tire motion over the bump before leaving contact

$$= 0.0254(23\sin\theta_f - 11.5) \times 0.53$$

Since,

$$N + \frac{mv^2}{r} = mg\ \sin\theta + k(x_5)\cos 45$$

When rear tyre is about to leave contact at that point Normal Reaction (N) tends to 0

$$\frac{mv^2}{r} = mg\ \sin\theta + k\left(x_f\right)\cos 45$$

$$\frac{58.05 \times 4.1667^2 \times \sin^2\theta_f}{58.05 \times 9.81} = 5.8.05 \times 9.81 \times \sin\theta_f$$
$$+ 30.39 \times 1000 \times 0.53 \times \cos(45)$$
$$\times 0.0254 \times \left(23\sin\theta_f - 11.5\right)$$

$\sin\theta_f = 0.66$ Thus, $\theta_f = 41.3°$

Before leaving the bump, tyre makes 41.3° angle with bump.

To find Impact time (t):

$$t = \int_{\theta}^{\theta f} \frac{d\theta}{\omega} = \int_{29.3}^{41.3} \frac{\mathrm{r}d\theta}{v\,\sin\theta} \quad \frac{r}{v}\int_{29.3}^{41.3} \mathrm{cosec}\theta\, d\theta$$

$$t = \frac{23 \times 0.0254 \times 0.3658}{4.1667} = 0.0513\ \mathrm{s}$$

Tyre leaves the contact of bump in 0.0513 s

To Find Normal reaction on tyre [5]:

$$N \times t = \mathrm{mv}\cos\theta_{\mathrm{avg}} = \frac{58.05 \times 4.1667 \times \cos\left(\frac{41.3+29.3}{2}\right)}{0.0513} = 3847\ \mathrm{N}$$

Adding 340 N extra (standard value lies between 300 and 350 N) considering undesired vibrations, forces due to toe changes, forces due to geometric stiffness, forces due to anti-squat property, enhanced safety of the part, etc.

Therefore, $N = 3847 + 340$ N = 4187 N

Force of 4187 N acts on the wheel when it hits a bump.

Taking Bump force on 1 wheel (4 flanges) as 4200 N

Bump force on 1 flange = (4200/4) = 1050 N

A force of 1050 N will be applied on each flange in direction of Z-axis.

2.4 *Calculation of Camber Thrust Force on One Flange*

Table 4 shows the available data for calculation of camber thrust force on one flange.

$$\text{Cornering force} = \frac{mv^2}{R} = \frac{58.05 \times 4.1667^2}{1.64} = 614.52\ \mathrm{N}$$

Camber thrust on 1 flange in plus '+' condition = $\frac{1}{2}[614.52 \times 11.15 \times (25.4/72)] = 1208.6$ N

A force couple of 1209 N will be applied on each flange in plus '+' condition in the direction of positive and negative Y-axis.

Table 4 Given data for cornering force calculation

Mass on single rear wheel (m)	58.05 kg
Turning radius (R)	1.64 m
Velocity of car (v)	15 kmph = 4.1667 m/s
Rolling radius	0.97 × 11.5 = 11.15 in

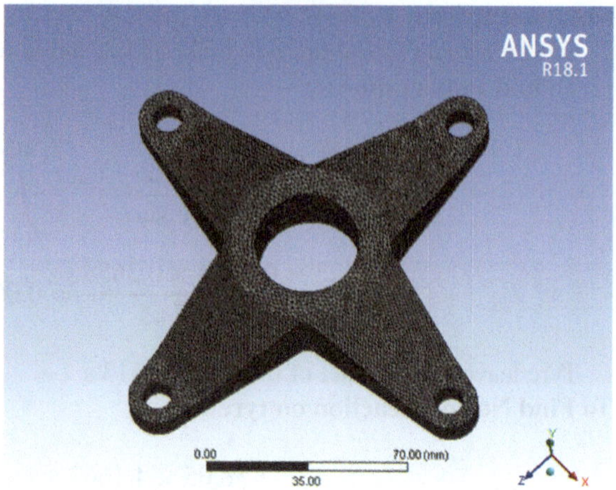

Fig. 5 Meshed model of wheel hub

3 Methodology

The 3D CAD model is made in SolidWorks 2017 [6] and material Aluminum 7075-T6 is assigned. To design the wheel hub, basic considerations like PCD of wheel studs with respect to DWT rims and inner diameter with respect to the tripod housing were taken into account. Static Analysis is done in ANSYS Workbench 18.1 [7] with impact force acting on four-wheel stud points, drive torque on four-wheel stud points acting in anticlockwise direction and camber force couple acting on two flange in plus (+) condition of wheel hub."Topology optimization feature" was used with constraints of 65 percent mass retention and 143.5 MPa maximum stress constraint with desired FOS of 3.5. Design was optimized according to the results and similar static analysis is carried out on new model. Figure 5 shows the meshed model and Fig. 6 shows the input conditions for the analysis.

4 Results and Discussions

Static analysis results for maximum stress come out to be 73.411 MPa (Fig. 7) with a minimum factor of safety of 6.8382 (Fig. 8) and Fatigue factor of safety of 2.4598 (Fig. 9) for 100,000 cycles with reversible loads. These values indicate over-designed condition of the component and hence require some material removal to reduce mass and also keep the F.O.S. in optimal range of 3–3.5. Results for topology optimization are shown in Figs. 10 and 11 which reduce the mass of the component by 35% while maximum stress of 143.5 MPa is kept in check. An optimized design is made based on these results and static analysis is carried out again.

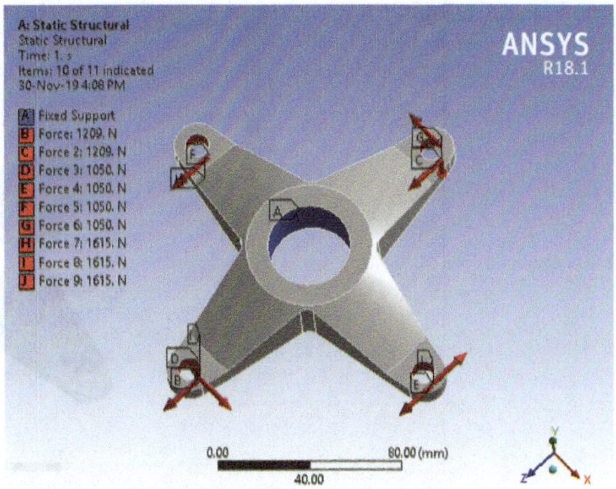

Fig. 6 Input conditions of FEA model

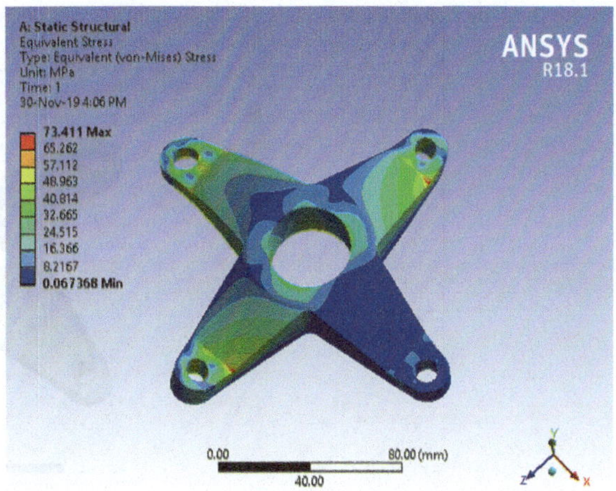

Fig. 7 Maximum stress results

5 Optimized Model

Based on the results of topology optimization a CAD model (Fig. 12) is made with 25.72% reduced weight in new design. The minimum factor of safety comes out to be 3.4188 (Fig. 14) with maximum stress induced as 146.84 MPa (Fig. 13). Fatigue factor of safety also comes out to be 1.2298 (Fig. 15) which is safe for 100,000 cycles of reversible loads.

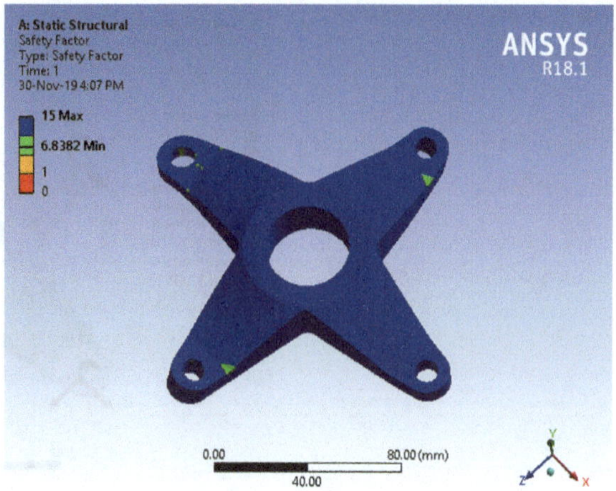

Fig. 8 F.O.S. results

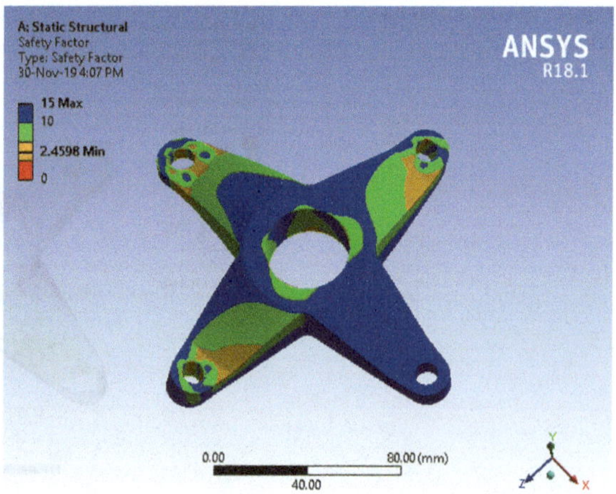

Fig. 9 Fatigue F.O.S. results

6 Conclusion

The basic design had excess material in areas of low-stress concentration, which can be eliminated. The wheel hub was over-designed with FOS of more than 6 and mass of 362.52 gm. The new design with a FOS lying between 3 and 3.5 was optimal, considering the severity of the component's working conditions. The new design has a mass of 269.31 gm, which is 25.71% reduction in overall mass. To sum up, Topology Optimization helps to get an estimate about areas, where excess material is present and an optimal design can be made which is symmetrical, feasible to manufacture and has reduced weight for enhanced performance.

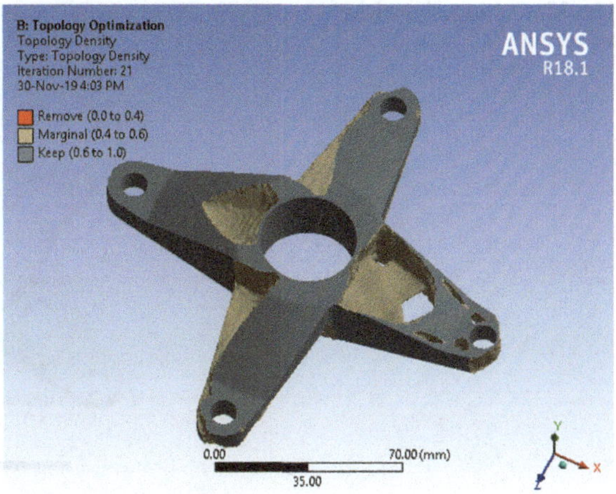

Fig. 10 Topology optimization results view 1

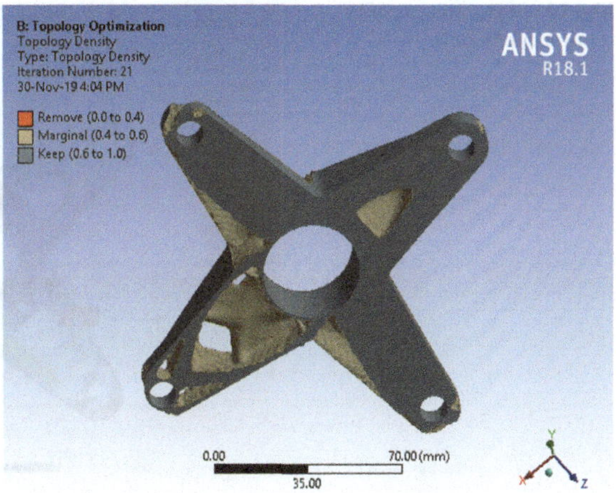

Fig. 11 Topology optimization results view 2

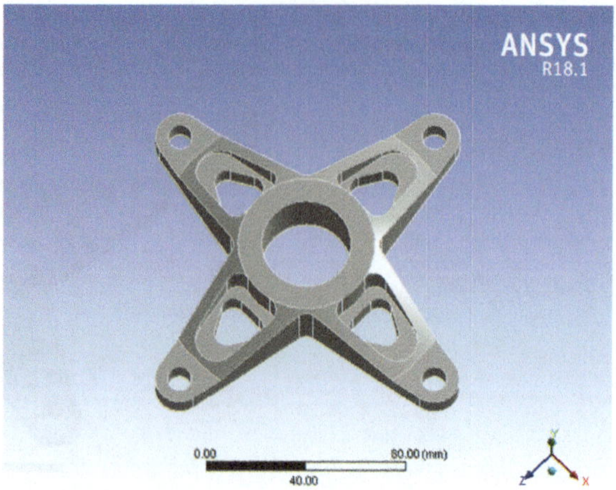

Fig. 12 Optimized model of wheel hub

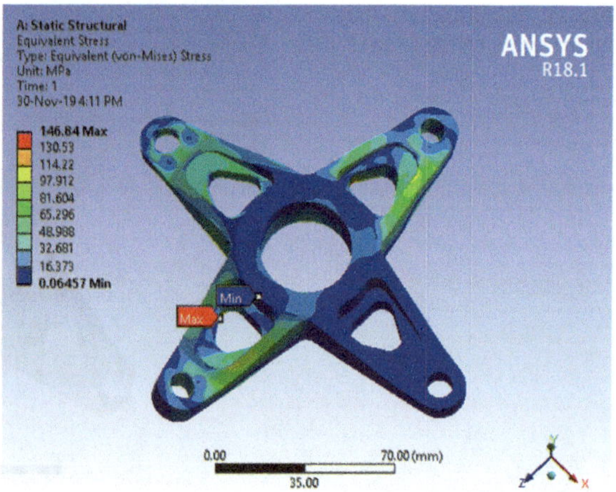

Fig. 13 Max. stress results (optimized model)

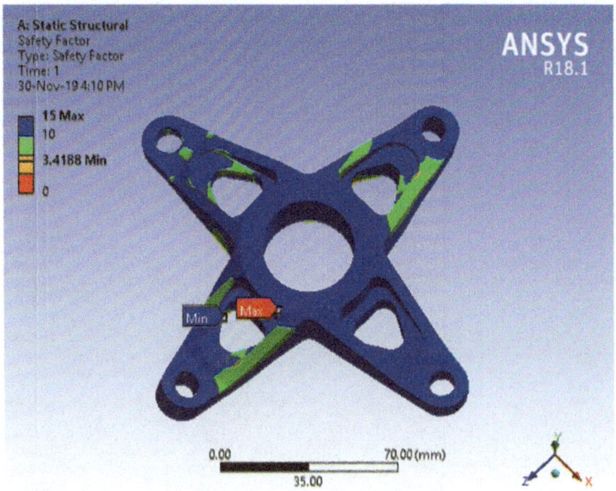

Fig. 14 F.O.S. results (optimized model)

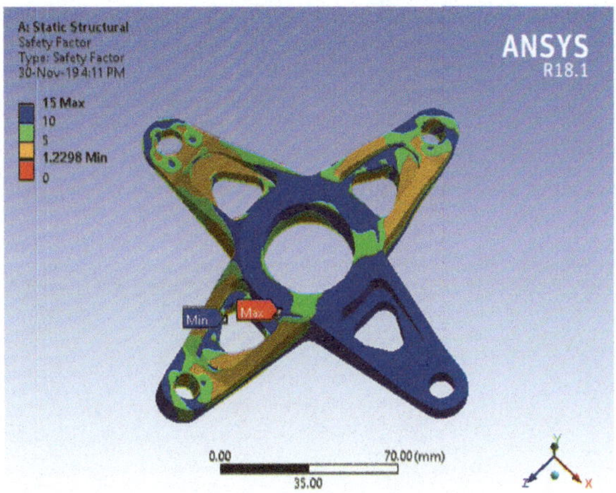

Fig. 15 Fatigue F.O.S. results (optimized model)

References

1. Gowtham V, Ranganathan AS, Satish S, Alexis SJ, Kumar SS (2016) Fatigue based design and analysis of wheel hub for student formula car by simulation approach. In: IOP Conference series: materials science and engineering, vol 149, p 012128
2. Ceyhan A, Duruş M, Akarsu C, Aydın R, Aray A, Hatık G, Tutuk E, Özkardeşler BC (2015) Wheel hub fatigue performance under non-constant rotational loading and comparison to eurocycle test. Procedia Eng 101(2015):77–84
3. Rangababu D, Depti K, Ramana B (2015) Optimal design and strength analysis of a wheel hub by using different materials. J Technolo Adv Sci Res 1(4) (2015). eISSN: 2454-1788, pISSN: 2395-5600
4. Kokate SB, Kulkarni GR (2019) Int Res J Eng Technol 6(5) (2019). e-ISSN: 2395-0056, p-ISSN: 2395-0072
5. Nelson A (2009). Engineering mechanics, India: Mc.Graw Hill Education

6. Solidworks (2017) Introducing solidworks, dassault systemes
7. Autodesk ANSYS 2018, ANSYS theory reference, 11th edn, SAS IP. Inc. pp 77–84

Modelling, Investigation and Refinement of Three Stage Helical Gearcase Housing Utilizing Numerical Approach Contemplating Various Relevant Substances

Ronak D. Gandhi, Ghanshyam V. Patel, and Sanket K. Patel

Abstract The endeavour was undertaken to model and investigate ternary reduction helically coiled gear case body. To initiate the process, examinations were conducted for interrogating and recognizing constructional functioning on the differential gearbox. Afterwards, through a review of various research papers, it was established that majorities of the organizations at present are overlaying the trouble of their prevailing gearbox integrated weight and stability. Thus, drafting and inspection of gearbox housing have emerged as an influential domain for exploration of the causes and optimizing gearbox. This necessitates the accomplishment of various types of analysis for prevailing helically shaped gear housing (S.S.A36) and transformed model (H.C.S and S.S.A36) for forecasting performance of enclosures exposed to discrete charging and discharging environment. Hence, from the viewpoint of drafting and analyzing, reformed drawing having carbon tool steel material is ideally suitable because of flexibility, weight, reverberation, permanence and costing, in comparison with the existing design.

Keywords Tripled reduction gearing dwelling · Robustness · Heaviness · Numerical method and optimization

R. D. Gandhi (✉)
Assistant Professor, Mechanical Engineering Department, ITM Vocational University, Vadodara 391760, Gujarat, India
e-mail: ronakgandhi401@gmail.com

G. V. Patel · S. K. Patel
Assistant Professor, Mechanical Engineering Department, Swaminarayan College of Engineering and Technology, Gandhinagar 382721, Gujarat, India
e-mail: gvpatel22@gmail.com

S. K. Patel
e-mail: er.sanket4u@gmail.com

© Springer Nature Singapore Pte Ltd. 2020
H. Vasudevan et al. (eds.), *Proceedings of International Conference on Intelligent Manufacturing and Automation*, Lecture Notes in Mechanical Engineering,
https://doi.org/10.1007/978-981-15-4485-9_73

1 Introduction

Toothed wheels are one of the principal noteworthy elements in automatic power imparting mechanisms and the foremost industrial revolving equipment. These gear arrangements are confined to the interior in a firm secured box named as gear case cover or else housing. Gearbox shells are enthralled by various diversified forces which have to be intimidated promptly during drafting of gear shell covering. While designing of gear case housing, main causes for demolishing gear housings are either deformed forces, cogwheel tooth's exterior surface toughness or reverberations. Therefore, investigations of stresses are definitely required which would aid the creator to the acknowledged extent of acting forces and also to forecast responses of casing and its constituents exposed to substantial stationary and compelling loads. There are a few researchers who undertook this study and refinement of the helical gear case housing design. The overview of a few scientists and their efforts are explained further.

Emanuel et al. [1] reviewed repercussions of instinctive characteristics for apparent component over pivotal revolving rates in a gearcase manipulating numerical approach. An estimation of trembling performance conceded that ultimate influential meshing loads corresponding to a vibrant way of excitation having extreme energy potential admissible along the rigidity of mesh. Moreover, numerical pretensions revealed that pragmatic predictions of crucial revolving speeds should take into account the complete elements in the gearcase.

Sekar et al. [2] examined the corresponding thermic and von misses pressure intervening on the inside and outside surface of gearbox cover through analytical techniques. Further, the discussion also covers the reaction of convection connecting the surface interior to housing and moving oil. These causes were found to be minor, hence ignored. Also, taking into consideration both force and heat together, the outcomes revealed that cover of gearcase enthralled to thermic loads was found around 86.688 MPa besides directional distortion 0.16343 mm, respectively.

Chabra et al. [3] explored research domain on drafting and examination for compound material gear shell. In such an article, analog-based facility to model and interpret force and twisting on housing was adopted. Additionally, exploitation of blended material gear shell was done to rehabilitate the currently used metallic gear case for certain minimization of weight to gain upgraded power capability, erosion resistance, clam our depreciation, notable intrinsic prevalence and more stimulated design.

Harzy et al. [4] diagnosed gearbox casing vibrations by using the numerical method. The major initiative of this research was to execute analysis for fluctuations and cyclic reaction in pursuance of intercepting reverberation of gearbox cover. Further; it was realized that the critical element was the drafting of the secured linking inside the topmost and bottom-most housing and drawing of the support connections. Moreover, the solutions obtained manifested the frequency range which was acceptable for the gearbox shell to intercept the utmost magnitude.

After contemplating different research papers [1–4], it was culminated that most research survey till now was performed on two-stage helically coiled gear housing. Within this survey [5–7], it was understood that few researchers had worked on the domain of triple gear shell casing. Moreover, it has been concluded that most institutions were encountering concerns for consolidated mass and stability of the current gearbox [8–11]. Hence, the current research was undertaken to achieve targets of reducing load and enhancing the ability of the housing assisting numerical approach.

2 Drafting and Analyzing

2.1 Drafting

2.1.1 Three-Dimensional Computer Aided Design

Computerized assisted drawing defined by manipulating computational equipment through drafting, refining, examination in addition to optimizing models. This mostly includes entire information related to sketch namely geometrical shape, magnitudes, allowances and processing relevant details. Software which could be utilized for drafting are CREO4.0, CATIA V5, N-X, SolidWorks. Hence, according to industrial measurements, three-dimensional designs were drawn by CATIA V5 drafting package.

2.1.2 Intermeshing

ANSYS workbench intermeshing properties aid the designer with diverse impulsive mesh generating technologies. In general, meshing should have been much compact and poor towards loading plus little loading regions accordingly. The current and altered sketch following intermeshing of grid area 6 mm is depicted beneath (Figs. 1 and 2).

2.1.3 Implementing Material Characteristics

This engineering working environment is exposed to a modeller to generate, retain and restore various features of materials. The material selected with regard to existence gear case housing was Structural Mild Steel with elasticity (E) as 29×10^6 psi and Poison rate (μ) as 0.26. While materials opted for the reformed model were Carbon Tool Steel having elasticity modulus (E) 30.45×10^6 psi and Poison rate (μ) as 0.26 along with Structural Mild Steel having elasticity coefficient (E) 29×10^6 psi and Poison rate (μ) as 0.26.

Fig. 1 Meshed model of existing helical gearbox casing

Fig. 2 Reticulated design of refined sketch helix gearcase housing

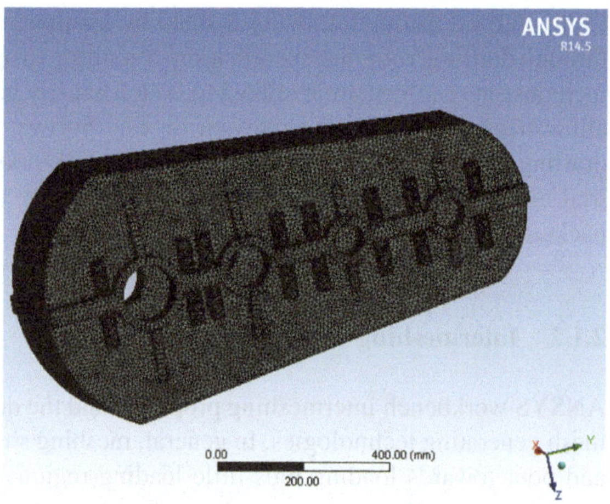

2.1.4 Enforcement of Payload and Marginal Requirements

Support for housing is stubbornly fixed with bottom using six screwed couplings. Lounging cover face have been restricted with the entire extent of directions. Numerous forces are applicable to the gearbox housing. An inert transmission of pressure on gear and driving handle trailing to the carriage opening cleaves towards twice portion notably, radial plus pivotal forces. This thrust force is used to ensure precise influence on stress along with distortions in casing of gearcase. Further, charging and bounded situations given by Finstern Engineering Pvt. Ltd. were employed on housing based on prescribed application.

2.2 *Finite Element Analyzing*

This exploration stage usually relies on problem definition stated previously or the kind of test which has to be performed. Various types of analysis that can be performed are:

2.2.1 Mode Shape Extraction Analysis

This approach is employed to assess an oscillation trait (intrinsic repetitions as well as layouts of various types of modes) for a drawing or apart while designing. Thus, congenital fluctuations along with mode contours investigations are definitely crucial benchmarks in the drafting of an element. Further, such scrutiny can additionally serve in a form of commencement spot for in-depth scrutiny such as transitory vigorous analysis, recurrence respond analysis and range inspection. Entire distortions gained later analyzing mode shape approaching some prevalence with respect to the changed design pondering material H.C.S are delineated into Figs. 3 and 4 accordingly.

Likewise, the outcomes which were captured after executing modal analysis for refined model anticipating S.S.A36 as well as H.C.S substances were tabulated in Table 1.

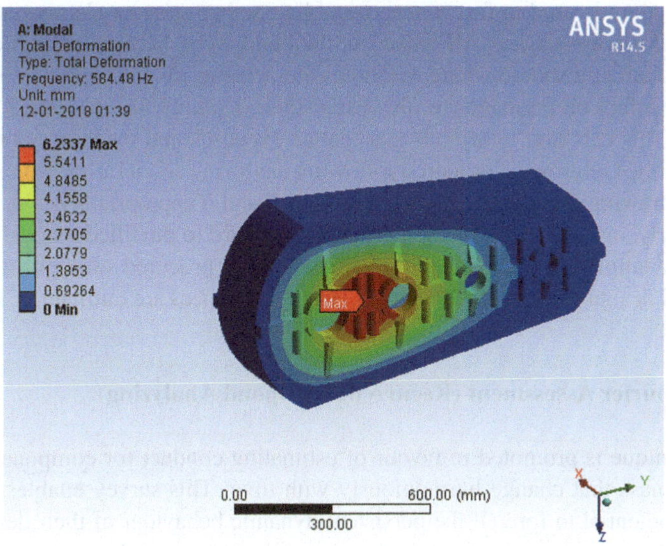

Fig. 3 Entire distortion on changed model (H.C.S) for periodicity 584.48 Hz

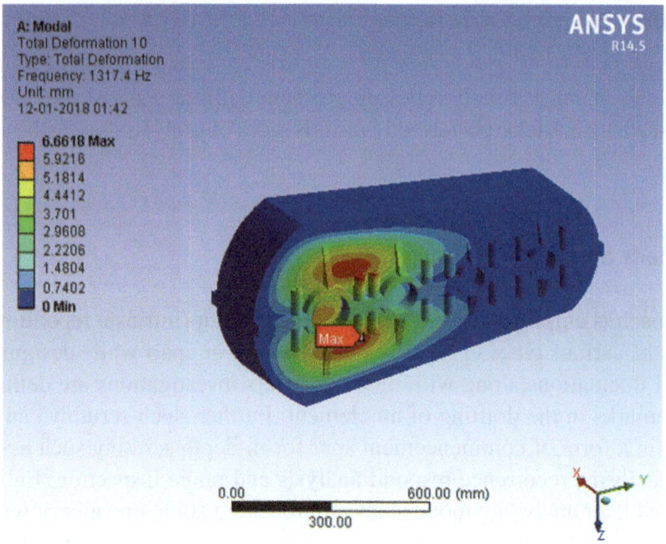

Fig. 4 Entire distortion on changed model (H.C.S) for periodicity 1317.4 Hz

2.2.2 Oscillations Investigation

This procedure is used in forecasting trembling acting upon an element at the time of payload circumstances. All these oscillations maybe because of bad design of goods and an atmosphere where machines are running. At times, this could have a persistent effect on permanence plus exhaustion, leading into shortened servicing lifespan. This can ease enormous sustenance expenses and machinery breakdown considerably. Outgrowths procured following analyzing of vibration with respect to changed drawing (H.C.S) are depicted in Figs. 5 and 6 appropriately.

Similarly, investigations of oscillations with regard to modified model have been completed employing substance S.S.A36. Outcomes procured subsequently for the Reformed design enacting S.S.A36 and H.C.S substances are outlined in Table 2.

2.2.3 Fourier Assessment (Recurrence Respond Analyzing)

This technique is promoted in favour of estimating conduct for components fascinated to mass that change harmoniously with time. This survey enables a creator with the potential to foretell the persistent dynamic behaviour of their design, thus granting the drafter to establish whether or not their design will positively vanquish resonance, exhaust and other adverse results of compelled vibrations. The total directional distortions along with equivalent stress acquired after managing analysis for the altered design deliberating material high carbon steel at diverse frequencies are revealed individually in Figs. 7, 8, 9, 10, 11 and 12.

Table 1 Correlation between mode shape testing outcomes for actual sketch and reformed model

Title	Actual sketch (S.S.A36)			Reformed model (H.C.S)			Reformed model (S.S.A36)		
	Frq. (Hz)	Min. perks (mm)	Max. perks (mm)	Frq. (Hz)	Min. perks (mm)	Max. perks (mm)	Frq. (Hz)	Min. perks (mm)	Max. perks (mm)
Total deformation	731.5	0	6.1273	584.48	0	6.2337	573.13	0	6.2338
Total deformation	734.5	0	6.2085	585.76	0	6.2842	574.39	0	6.2842
Total deformation	1103.3	0	4.5778	777.71	0	4.9307	762.60	0	4.9307
Total deformation	1181.8	0	5.5360	787.19	0	5.0578	768.96	0	5.0578
Total deformation	1357.0	0	4.7364	996.87	0	4.3449	977.51	0	4.3449
Total deformation	1507.0	0	6.6006	1031.2	0	4.7512	1011.2	0	4.7511
Total deformation	1512.0	0	6.6346	1214.8	0	4.3691	1191.2	0	4.3691
Total deformation	1587.2	0	5.1187	1315.7	0	6.0744	1290.2	0	6.0717
Total deformation	1715.4	0	5.3091	1315.9	0	6.6768	1290.3	0	6.6785
Total deformation	1918.4	0	4.6402	1317.4	0	6.6618	1291.8	0	6.6618
Total deformation	2078.4	0	6.8004	1502.3	0	4.2751	1473.2	0	4.2751
Total deformation	2109.6	0	6.9264	1659.0	0	6.7492	1626.8	0	7.0496

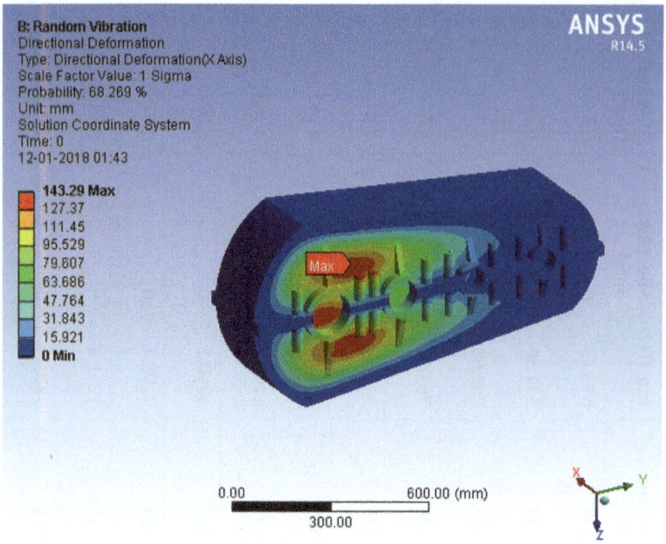

Fig. 5 X orientation distortion gained on altered model (H.C.S)

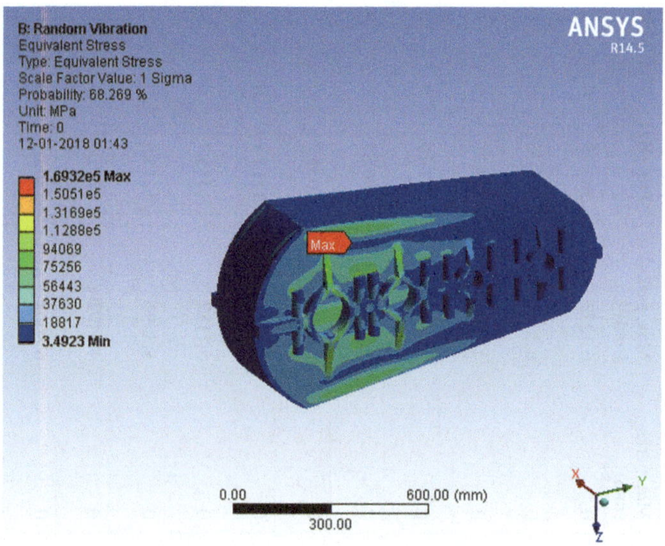

Fig. 6 Identical stresses gained on altered model (H.C.S)

Table 2 Correlation between vibration testing outcomes for actual sketch and reformed model

Title	Actual sketch (S.S.A36)		Reformed model (H.C.S)		Reformed model (S.S.A36)	
	Minimum perks	Maximum perks	Minimum perks	Maximum perks	Minimum perks	Maximum perks
X direction deformation (mm)	0	12.135	0	25.346	0	141.88
Y direction deformation (mm)	0	91.416	0	4.5838	0	23.551
Z direction deformation (mm)	0	41.625	0	17.949	0	55.696
Stress (MPa)	0.55757	2.685 e^5	0.34923	1.6092 e^5	3.7002	1.6121 e^5
Strain	2.7878 e^{-6}	1.3425	1.663 e^{-6}	0.78057	1.8501 e^{-5}	0.80605

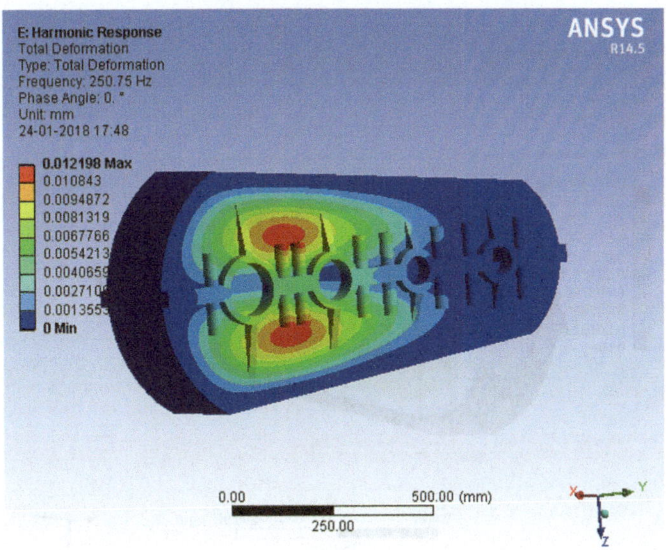

Fig. 7 Entire distortion on refined model (H.C.S) for periodicity 250.0 Hz

Similarly, investigation of frequency response was conducted for a reworked design by use of substance S.S.A36. Outgrowths validating symphonic study for refined drawing anticipating pair of substances approaching peculiar individual recurrences were summed up in Table 3.

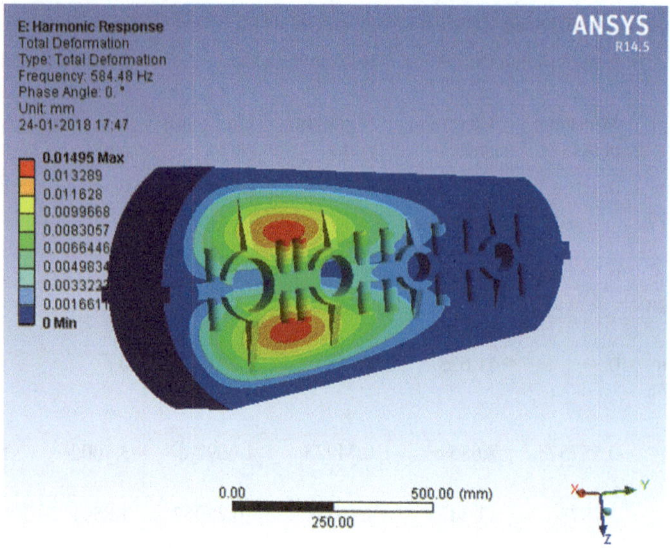

Fig. 8 Entire distortion on refined model (H.C.S) for resonance periodicity 584.4 Hz

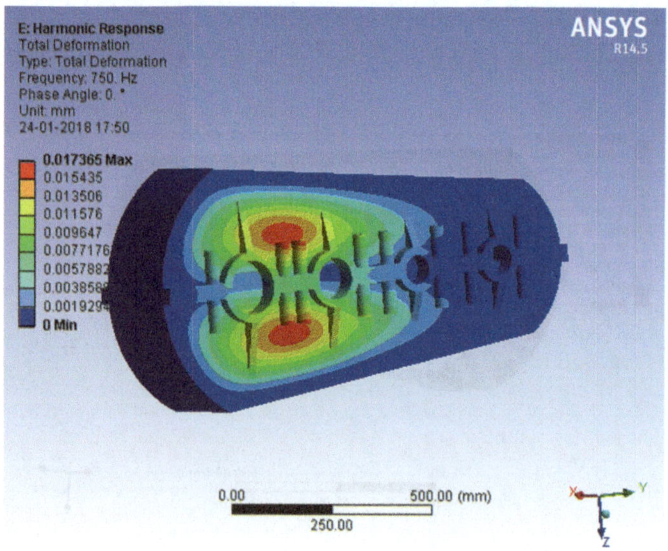

Fig. 9 Entire distortion on improved model (H.C.S) for periodicity 750.0 Hz

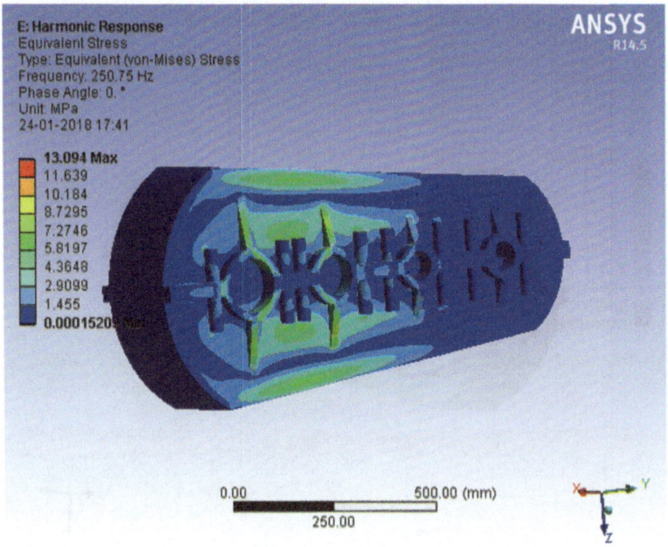

Fig. 10 Identical stress on improved model (H.C.S) for periodicity 250.0 Hz

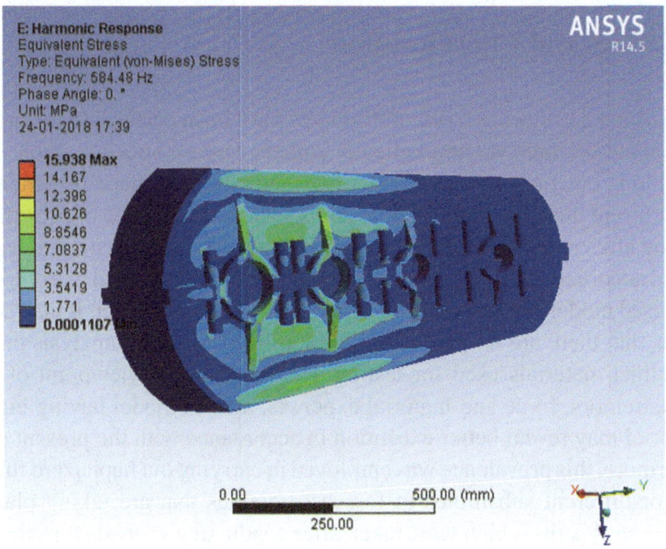

Fig. 11 Identical stress on revised model (H.C.S) for resonance periodicity 584.4 Hz

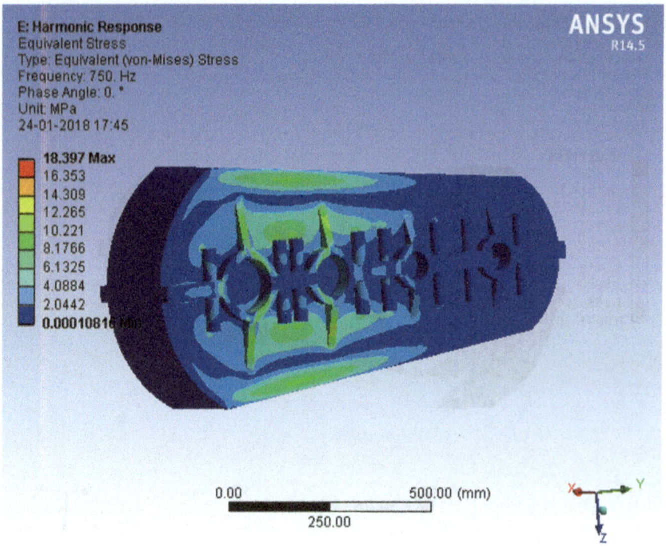

Fig. 12 Identical stress on revised model (H.C.S) for periodicity 750.0 Hz

3 Outcomes and Consideration

In this existing task, thorough research studies have been performed for drafting and testing conduct on three staging helically coiled gearcase housing. Table 1 exhibits relation among outcomes procured after carrying outmode shape investigation with regard to current design and numerical perks obtained in favour of reformed model anticipating alike materials. After collating findings for modal testing, it was evident that ethical accordance was settled up within the outgrowths. Total distortions gained for the revised model privileging pair of materials are nearly closer. It may be because of chances that there are no payload criteria utilized in modal analysis or probably due to distinct materials used for testing. Hence, from the viewpoint of firmness, heaviness, tremors, force and material expenses, altered model having high carbon steel material may reveal better execution in accordance with the present design.

Furthermore, this prevalence was employed in carrying out haphazard fluctuations analysis for different substitutes to forecast trembles that are taking place on the gearbox cover. Results which were taken after conducting analysis for vibrations are exhibited in Table 2. After matching outputs for vibration testing, it surely conveys that fine consent was noticed in the outcomes. It was also found that values of deformations in two different directions are lower in disparity with another direction. This may be because of reality that conditions of loading are utilized in that specific direction where deformity is large. Besides, the value of stress for amended model in which H.C.S material exists is least among the likeness over present layout.

Further, harmonic analysis was attempted for all alternatives to forecast the structure's behaviour enthralled to loads that change sinusoidally with time. This analysis

Table 3 Correlation between harmonized testing outcomes for actual sketch and reformed model

Title	Actual sketch (S.S.A36)			Reformed model (H.C.S)			Reformed model (S.S.A36)		
	Frq. (Hz)	Min. perks	Max. perks	Frq. (Hz)	min. perks	Max. perks	Frq. (Hz)	Min. Perks	Max. perks
Stress (MPa)	500	2.2 e^{-5}	14.648	250.75	1.5 e^{-5}	13.094	250.75	1.6 e^{-4}	13.113
Total deformation (mm)	500	0	0.0720	250.75	0	0.0121	250.75	0	0.0127
Stress (MPa)	731.5	2.5 e^{-5}	16.947	584.48	1.1 e^{-5}	15.938	573.13	9.5 e^{-4}	15.940
Total deformation (mm)	731.5	0	0.0837	584.48	0	0.0149	573.13	0	0.0155
Stress (MPa)	1000	3.3 e^{-5}	22.634	750.00	1.0 e^{-5}	18.397	750.00	1.2 e^{-4}	18.744
Total deformation (mm)	1000	0	0.1115	750.00	0	0.0173	750.00	0	0.0184

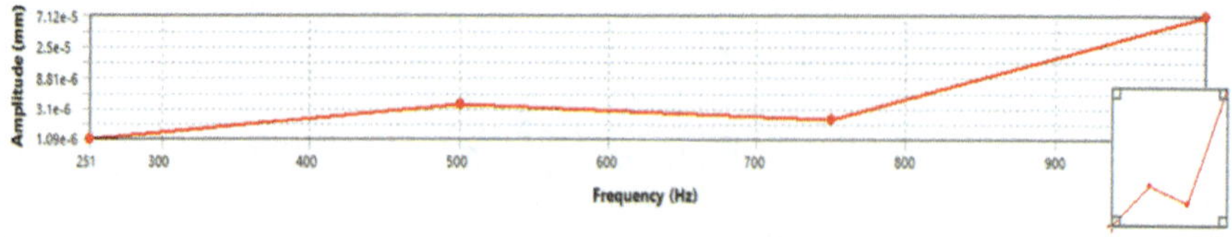

Fig. 13 Recurrence respond distortion plot procured in favour of refined model (H.C.S)

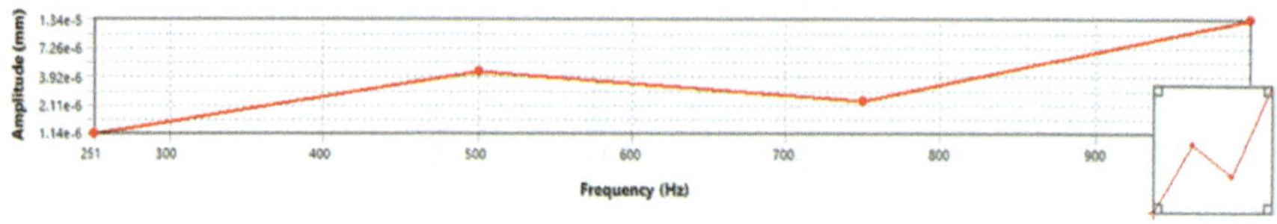

Fig. 14 Recurrence respond distortion plot procured in favour of refined model (S.S.A36)

of frequency response was achieved ideally at the resonance frequency, frequency larger or smaller than reverberation rates for certifying either model will have the capability of resisting vibrancy, exhaust and some adverse impact for strained oscillation. Plots fetched at a time of running harmonized survey deliberating H.C.S and S.S.A36 materials were expressed into respective Figs. 13 and 14. The outcomes gathered after adopting recurrence respond test for specific typical rates were depicted in Table 3.

4 Conclusion

After correlating several numerical assessment outcomes of model alteration (High Carbon Steel plus Structural Steel A36) over various outgrowths recovered for existence drawing (Structural Steel A36), the following consequences have been outlined:

- After comparing mode shape extraction testing results, it was noted in favour of transformed model possessing H.C.S along with S.S.A36 materials that; utmost amount of deformities are decreased and improved by 6.61 and 1.74% individually in resemblance with the current model.
- After equating oscillation exploration findings, it was stated with regard to refined model comprising H.C.S as well as S.S.A36 materials that the; maximum force values are curtailed near 40.06 and 39.95% severally in connectivity over the current design.
- After reviewing harmonic analysis outputs, it was resolved that; the ultimate stress value for adapted model with H.C.S plus S.S.A36 materials approaching concerned frequencies of reverberance are separately amortized by 5.97 and 5.94%,

in connectivity with present design. Furthermore, after undertaking similar analysis for few lowest and highest interval of time than plangency, it is conceived that substantial amount of tension is expediently diminished near 10.6 and 18.7% for material H.C.S and decreased in proportion by 10.4 and 17.1% for material S.S.A36 in settlement with the effective model.

Hence, on the basis of modelling and investigation, it was securely terminated that reformed model possessing H.C.S material is better adapted for housing against actual casing design in terms of permanence, heaviness, oscillations, sustainability and expenses.

References

1. Emanuel B, Sabot S (2010) Study to analyze the impact of the mechanical properties for different elements on the crucial rotational speeds of gearbox. Recent Trends Eng Mater Sci 5(3):388–391
2. Sekar P, Utpat A (2011) Steady state thermal stress analysis of gearbox casing by finite element method. Finite Elem Anal Des 97:1097–1106
3. Chabra P, Bhatia A (2012) Modelling and analysis of composite material gearbox. Adv Manuf Mater Eng 5:40–49
4. Harzy S, Kumar A, Patil P (2014) Study about the vibration analysis for gearbox cover utilizing finite element analysis. In: Proceeding of international conference on advances in manufacturing and material engineering (AMME 2014), pp 140–149, Germany (2014)
5. Ramesh, Renuka D, Jyothirmai S (2013) Study on the vibration analysis of gearbox top casing. In: Proceeding of 2nd international conference on materials processing and characterisation (ICMPC 2013), pp 907–918, Russia (2013)
6. Patel M, Patil A (2015) Study about stress and deformation of 3 stage helical gearbox casing. Int J Adv Res Eng Sci Technol (IJAREST) 02(07):65–71
7. Setty R, Gulam I, Totar D, Naik S (2015) Designing and dynamic analysis of gearbox housing using finite element analysis. J Comput Des Eng 192:953–960
8. Babu M, Reddy Y (2016) Stress analysis of gearbox casing using ansys workbench. J Appl Res Technol 8:1397–1405
9. Zdziennicki Z, Maciejczyk I (2016) Design calculation and modification of industrial gearbox. J Mech Sci Technol 26(3):575–589
10. Hessen J, Vanhollebeke F, Mirant S (2017) Multibody modelling of varying complexity for modal behaviour analysis of gearboxes. J Mech Transm 36(11):184–196
11. Xue Q, Wang D, Zhang X (2017) Study on modal analysis and structure optimization method of gearbox. ASME J Mech Des 117:241–247

Lateral Force Modelling Using Magic Formula Tire Model

Aditya H. Bhatt and Prasad S. Shirodkar

Abstract The aim of this paper is to explain one of the methods that can be used to fit the Pacejka Magic Formula Model for the lateral behaviour of the tire. The content gives the readers an idea in order to utilize data and model tire behaviour. The tool used for the purpose of explanation is MATLAB.

Keywords Tire modelling · Lateral force · Magic formula · Pacejka model

1 Introduction: The Tire Tests [1]

The procedure of the test involves starting with the measurement of spring rate at 0 and 25 mph. Following which, the 'cold-to-hot' test is carried out by making 12 sweeps from $-12°$ Slip Angle (SA) to $+12°$ SA at 8°/s. After this procedure, a warm-up run is done at 250 lbs load and 12 psi pressure for about 1 min. Once these tests are completed, the tires are tested to gather the data for pure cornering conditions. This is done by sweeping the tire from -12-degree SA to $+12$-degree SA at five different loads, five inclination angles and four pressures. The test values of these parameters are provided in Table 1.

A. H. Bhatt (✉)
Dwarkadas J. Sanghvi College of Engineering, Vile Parle (West), Mumbai 400056, India
e-mail: adityabhatt97.ab@gmail.com

P. S. Shirodkar
Department of Mechanical Engineering, Dwarkadas J. Sanghvi College of Engineering, Vile Parle (West), Mumbai 400056, India
e-mail: prasad.shirodkar@djsce.ac.in

© Springer Nature Singapore Pte Ltd. 2020

H. Vasudevan et al. (eds.), *Proceedings of International Conference on Intelligent Manufacturing and Automation*, Lecture Notes in Mechanical Engineering,
https://doi.org/10.1007/978-981-15-4485-9_74

Table 1 The tire test data parameters

Loads (in lbs)	Inclination angles (in degrees)	Pressure (in psi)
50	0	8
100	1	10
150	2	12
250	3	14
350	4	–

2 Data Analysis and Interpretation

In order to understand the test procedure, it is easier to look at the graphs. As shown (Fig. 1) is the diagram of SA with respect to Elapsed Time (ET) for Hoosier R25B tires. From the graph, it is clearly evident that there are five differentiable parts or sweeps. As stated in the procedure, the first two sweeps are cold-to-hot and warm-up (see Fig. 2). After these, the remaining three sweeps are pertaining to 10, 12 and 14 psi pressures with all the load and inclination angle values.

The significance of the first two sweeps is to understand the generation of the temperature in the tires. By understanding the temperature range corresponding to the lateral force values, it will be easier to generate the performance when the tires are initially fitted on the vehicle (i.e. when they are cold and not at the required temperature) by getting the tires to the required temperature.

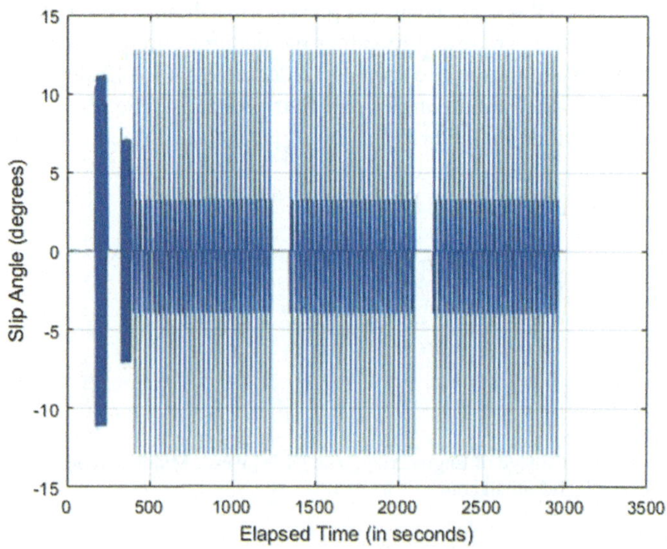

Fig. 1 Elapsed time versus slip angle showing five differentiable sweeps

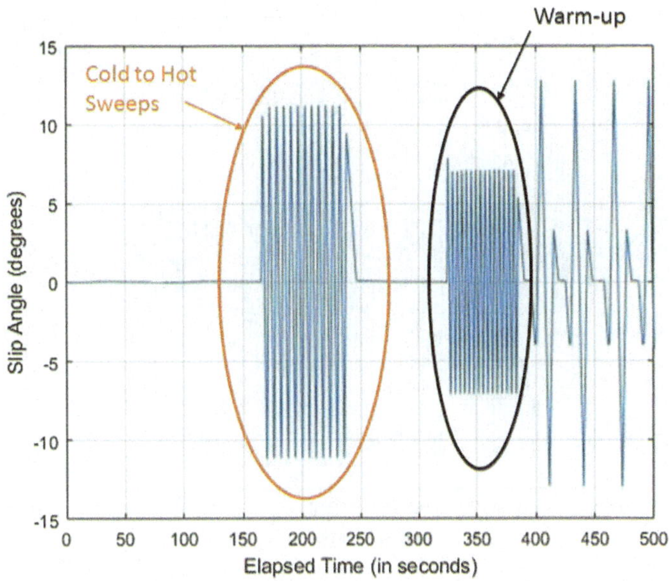

Fig. 2 Cold-to-hot and warm-up sweeps

3 Plotting Lateral Force Versus Slip Angle

Further examining the data, we can see (Fig. 3.) that the first parameter of differentiation is the tire pressure [1]. The order of the test is 82.73 kPa (12 psi), 68.95 kPa (10 psi), 96.53 kPa (14 psi). Now, we will first focus on the 12 psi data. This data is also significant because the warm-up procedure is also carried out on the same pressure; hence, it is easier to compare the effect of temperature at later stages.

Again, we can see that the sweeps can again be differentiated based on the inclination angle and normal loads. As seen in (Fig. 4.), the order of the inclination angles is 0 deg, 2 deg, 4 deg followed by 1 deg and 3 deg.

In order to simplify the understanding of the curve fitting process, we will be using the data at the 0-degree inclination angle so as to eliminate the effect of the inclination angle on the curve.

Now, the process of curve fitting is divided into the following three major parts:

i. Plotting of RAW data
ii. Creating groups based on Slip Angle
iii. Using the groups to fit the Magic Formula and find the coefficients.

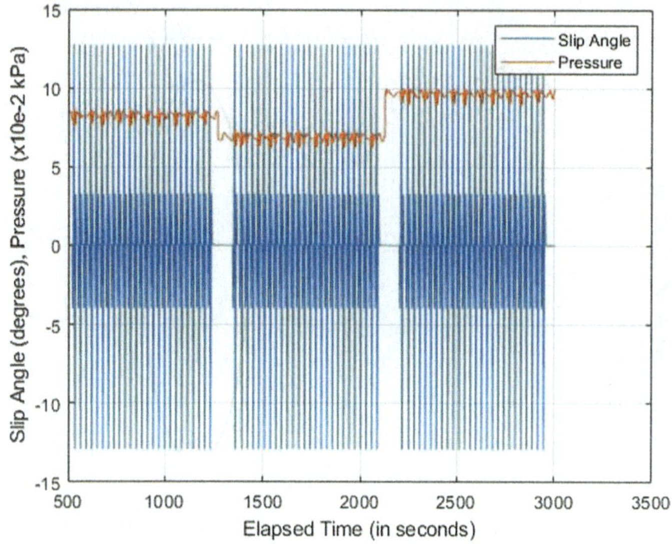

Fig. 3 The distinction of the sweeps based on tire pressure

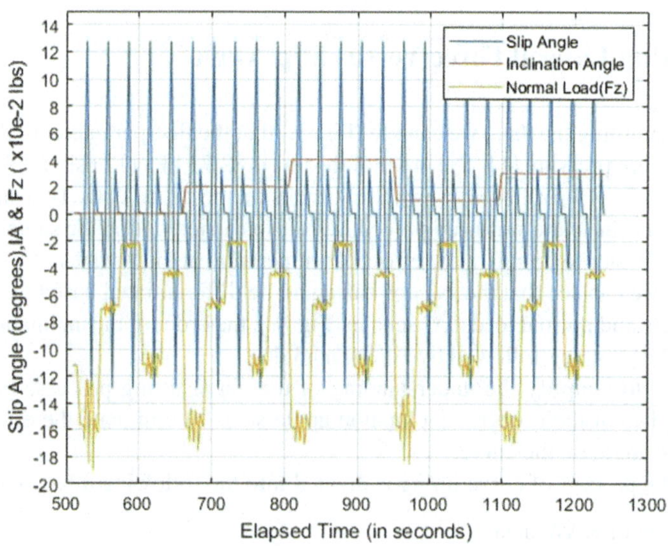

Fig. 4 SA, IA and F_z for 12 psi

3.1 Plotting the RAW Data

RAW data is nothing but the unfiltered data that is recorded by the sensors during the test. Plotting these values will help us visualize the behaviour of the tire in its organic

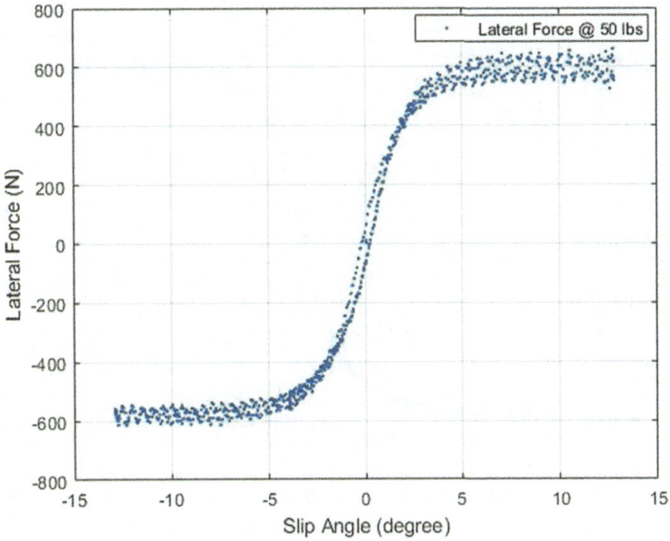

Fig. 5 RAW data plot

state (i.e. without approximations of curve fitting). We can observe that the amount of data points is vast. If we are to use this data to produce a curve, there is ought to be a large amount of deviation because there is a lot of variation for each of the slip angle values. Hence, it is important to filter this data based on certain tolerances. In order to filter the data, the primary parameter will be the slip angles. Since, the Magic Formula tire model is dependent directly on slip angle, forming groups-based slip angle values and their corresponding lateral force values is convenient (Fig. 5).

3.2 Creating Groups Based on Slip Angle

As explained earlier, the slip angle values vary from -12 to $+12°$. Owing to the fact that the control of the machine[1] at low loads is weak, we will need to have a tolerance of $\pm 0.5°$. Hence, for the purpose of filtering the range of slip angle will vary from -12.5 to $+12.5°$. Now, using MATLAB we will create matrices for SA and FY for values of SA starting from $-12.5°$ with an increment of $0.2°$. After finding the corresponding values of lateral force for the slip angles, the mean values of the forces will be calculated and plotted against these slip angle values (Fig. 6).

Hence, now we can see a clear path for the Magic Formula curve to pass through. The key to filtering the data is to identify the key parameter and define the tolerance such that there is the elimination of unwanted data while not missing out on key data values.

[1] Calspan Tire Research Facility (TIRF).

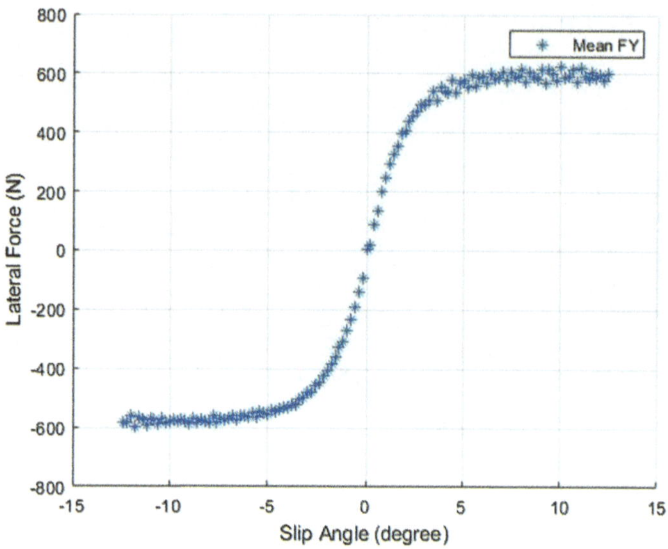

Fig. 6 Filtered data

3.3 Using the Filtered Data to Fit the Magic Formula and Find the Coefficients

The next step is to find the values of the coefficients and fit the Magic Formula [2–4]. In order to find the coefficients, we will need to understand the meaning of each of them.

i. **B—Slope Factor**: The slope of the curve at near 0 slip angle. It is also important to find the value of the cornering stiffness of the tire.
ii. **C—Shape Factor**: The shape of the curve at the peak during the transition phase is controlled by this coefficient.
iii. **D—Peak Factor**: The maximum value of the lateral force.
iv. **E—Curvature Factor**: The curve drops off after reaching the peak. This drop is controlled by increasing the influence of 'sine' function and the coefficient responsible for that is *E*.
v. **SHY & SVY—Horizontal and Vertical Shifts**: The properties of the tire that are described as ply steer and conicity give rise to these coefficients.

Once we have understood the meaning of each of the coefficients, using the equations provided for the Magic Formula we can find the values. Hence, using the filtered data, we can assign *D* as the maximum value of FY. Next, we need to define X_m and $Y_a$which are necessary to find *C* and *E*. X_m is nothing but the value of slip angle for the corresponding value of *D*. Y_a is defined as the minimum value of FY after the peak has been reached [4]. Once the variables have been defined, all that is needed is to plot the RAW data and Magic Formula curve and compare.

As we can see (Fig. 7), the curve is not a perfect fit. The reason for this deviation lies in the value of *B*. The coefficient value generated by the code is not accurate. By defining the value of this variable manually, we can get rid of the error (see Fig. 8).

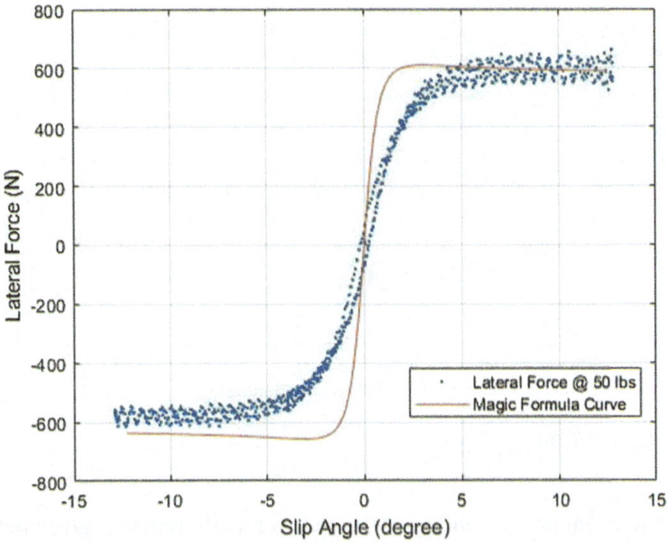

Fig. 7 Magic formula curve

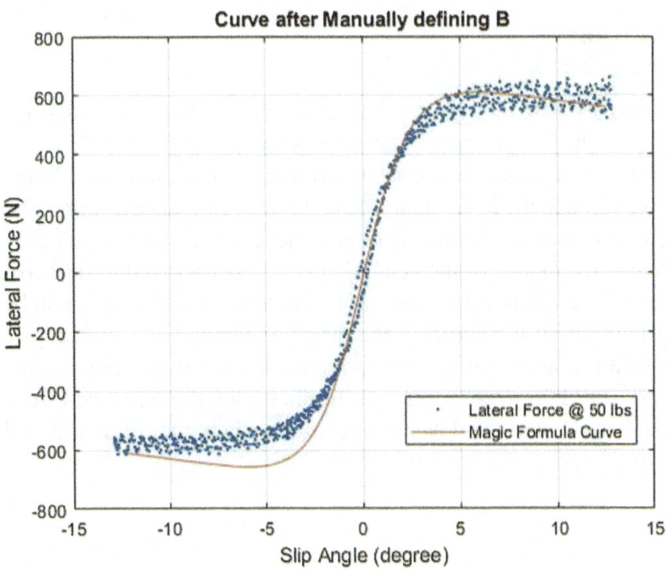

Fig. 8 Eliminating the error in B

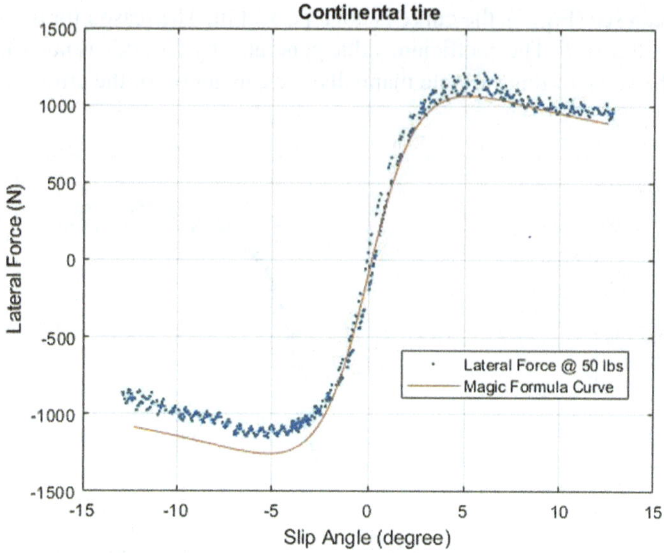

Fig. 9 Continental tire fit

In order to examine the validity of the code for a different tire, given below is the graph for the radial-ply continental tire (Fig. 9).

4 Conclusion

The study was carried to look into the lateral force modelling aspect of Magic Formula tyre model in vehicle dynamics. The parameters to judge the effectiveness of the approach will depend upon the error. The error is the deviation of the Magic Formula curve passing through the RAW data. Using the above-described approach, the Magic Formula curve with considerable error is achieved. Since the tire model will affect the complete performance of the vehicle, it is necessary to reduce the error as far as possible. In order to reduce the error, refined regression techniques will be required. Yet, using the current estimations, the effect of temperature on the generation of lateral force can be analyzed by comparing the FY versus SA curves during warm-up procedure and the sweeps after that. Furthermore, the outcome of the approach provides the base to start the development of the mathematical model of the vehicle to analyze and improve the performance.

5 Future Work

i. Utilizing the list of equations for the lateral dynamics, generating a more accurate curve [4].
ii. Understanding the effects of temperature.
iii. Generating a relation between tire wear and the decrease in the lateral force.
iv. The change in behaviour after re-treading the same tire.

References

1. The Formula SAE Tire Test Consortium (FSAE TTC). http://www.millikenresearch.com/fsaettc.html
2. Bakker E, Nyborg L, Pacejka HB (1987) Tyre modelling for use in vehicle dynamicsstudies. SAE paper 870421:1987
3. Bakker E, Pacejka HB, Lidner L (1989) A new tyre model with an application in vehicledynamics studies. SAE paper 890087:1989
4. Pacejka HB (2006) Tyre and vehicle dynamics, 2nd edn, Butterworth-Heinemann, Oxford, United Kingdom. ISBN-13: 980-0-7506-6918-4

Static Analysis of Tripod Housing Using FEA and Its Validation

Jash H. Patel, Vinayak H. Khatawate, Gaurav Jain, and Param Shah

Abstract Tripod housing plays a major role in transmitting power from the transmission system to the drive wheel and hence it undergoes severe torsion and bending type of deformation. In this paper, a design of Tripod housing for an off-road vehicle was analyzed using a finite element analysis (FEA) and verified mathematically using the principle of impulse-momentum and maximum shear stress theory. For any mechanical component, Factor of Safety (F.O.S.) gives the degree of safeness of the component under consideration. Thus, the focus of concern of this paper was on the value of F.O.S. by FEA and its verification using analytical calculations.

Keywords Tripod housing · Impulse-Momentum · Factor of safety · Shear force · Bending moment diagram · FEA

1 Introduction

With rising competitiveness in the automotive market with regard to speed, performance and durability, the elements providing power to the given automotive plays a huge role in rise or fall of a firm. The Tripod housing works as a shaft or axle used for transmitting power from the engine, through gearbox and straight to the wheels via Tripod type of constant velocity joint (CV-Joint). Tripod roller bearings (Fig. 1) are three cylindrical rollers with convexed curved surface mounted on an adaptor (also known as Spider).

This bearing is housed inside a Tripod housing which is then connected to the wheel in order to transmit power between misaligned shafts. This joint has an added advantage over other joints like more misalignment angle, lightweight, less complexity, more reliability, etc.

Studies were conducted on the wear life model focusing on adhesive wear phenomena [1, 2] and friction behavior of Tripod type CV-Joint [3, 4] in the past. Some

J. H. Patel (✉) · V. H. Khatawate · G. Jain · P. Shah
Department of Mechanical Engineering, Dwarkadas J. Sanghvi College of Engineering, Mumbai, India
e-mail: jash.ravariya98@gmail.com

© Springer Nature Singapore Pte Ltd. 2020
H. Vasudevan et al. (eds.), *Proceedings of International Conference on Intelligent Manufacturing and Automation*, Lecture Notes in Mechanical Engineering,
https://doi.org/10.1007/978-981-15-4485-9_75

Fig. 1 Tripod roller bearing

studies also focused on the changes in lubricant properties under thermal conditions due to the reciprocating motion of Tripod joint [5, 6]. Considering Tripod housing as a stepped shaft under bending and torsional moments with bearing supports, this paper evaluates the safety factor using FEA and compares it with analytical value of F.O.S. Analytical value of F.O.S. is found using basic principles of Machine Design and Engineering Mechanics. The worst-case scenario, that is, force on the tire when hitting a bump and max initial torque on single drive tire when the other tire is not loaded is considered as input condition.

The results will help to determine the closeness or correctness of the Finite element approach to the analytical approach for F.O.S. calculation by getting percentage error between both the values (FEA and analytical model) under 10% which is considered acceptable for practical purposes.

2 Background Theory

2.1 Constant Velocity (CV-Joint)

A Constant velocity joint is a component that helps in the transmission of power from driveshaft to the axle at a variable angle and without any appreciable loss of power or appreciable increase in friction. The different types are:

1. Tripod Joints
2. Tracta Joints
3. Rzeppa Joints
4. Weiss Joints
5. Malpezzi joints
6. Double Cardan joints.

2.2 Maximum Shear Stress Theory

The theory states that the failure of a mechanical component subjected to bi-axial or tri-axial stresses occurs when the maximum shear stress at any point in the component becomes equal to the maximum shear stress in the standard specimen of the tension test when yielding starts [7].

$$\tau_{\max} = 0.5 \times \text{Syt}$$

2.3 Principle of Impulse-Momentum

For a particle or a system of particles acted upon by forces during a time interval the total impulse acting on the system is equal to the difference between the final momentum and initial momentum during that period [8].

$$\int_{t1}^{t2} \Sigma F = m(v_2 - v_1)$$

3 Finite Element Model

The 3D CAD model of Tripod housing was made using SOLIDWORKS 2017 [9] and its finite element analysis (Static Structural) was done using ANSYS Workbench 2018 [10]. Aluminium 7075 T-6 was used for the analysis due to its high strength to weight ratio. Its Yield Strength is 503 MPa and Ultimate Strength is 572 MPa which are used as input parameters in material properties (under Aluminium alloy) in the Static Structural model.

Tetrahedral three-dimensional elements with eight nodes are used for creating the mesh of the model. Cylindrical supports where support bearings are positioned, one-dimensional force in X-direction and torsional moment in the inner profile where Tripod bearing makes contact with the Tripod housing are the boundary conditions applied for the analysis. The required evaluations total deformations, safety factor and equivalent stress (Von-Mises) were found out after the results were processed. Figure 2 shows the meshed CAD model of Tripod housing and Fig. 3 shows the input conditions for the analysis.

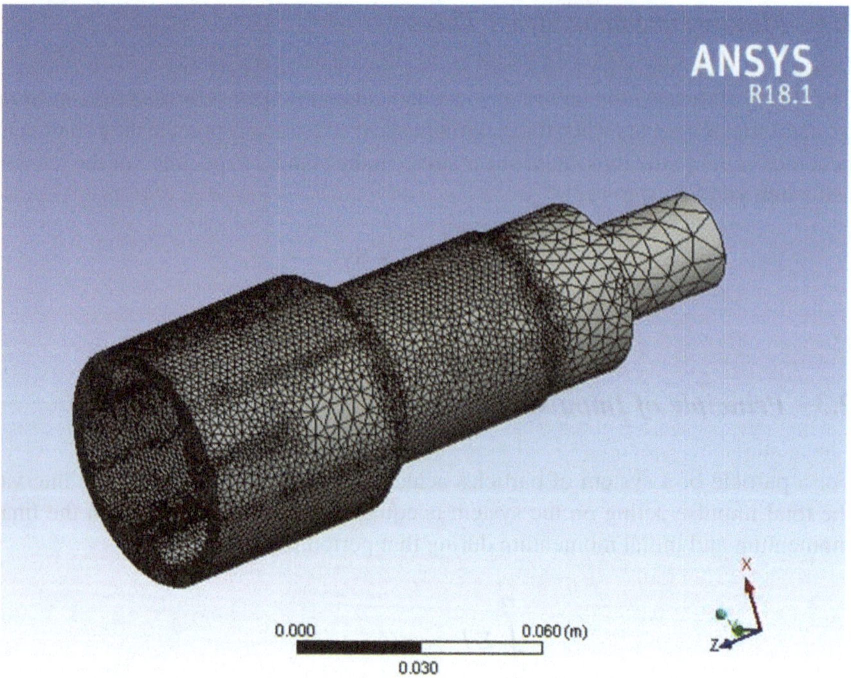

Fig. 2 Meshed model

4 Analytical Calculations

For the analytical model, the following assumptions were made:

1. During this calculation, the driver is assumed to be divided into three parts—head, torso and legs and their weight and its location of centre of gravity are approximated.
2. Centre of gravity of all the parts of the vehicle are found using their CAD model.
3. The Centre of gravity of the miscellaneous parts which include fasteners, body panels, etc. which are distributed in the whole vehicle is not considered but their overall mass is considered.
4. The Left: Right biasing of the vehicle is considered as 50:50 (vehicle is symmetrical on both sides and Centre of gravity lies on the mid-plane).
5. Forces, when the rear wheel hits a bump, are more than the forces when the rear wheel hits the ground after a 5-feet drop (as front-wheel takes the major load in a 5-feet drop).
6. It is assumed that while applying the Impulse-Momentum equation, the momentum is taken as average momentum which is found using velocity at an angle which is the mean of initial and final angles on encountering a bump.

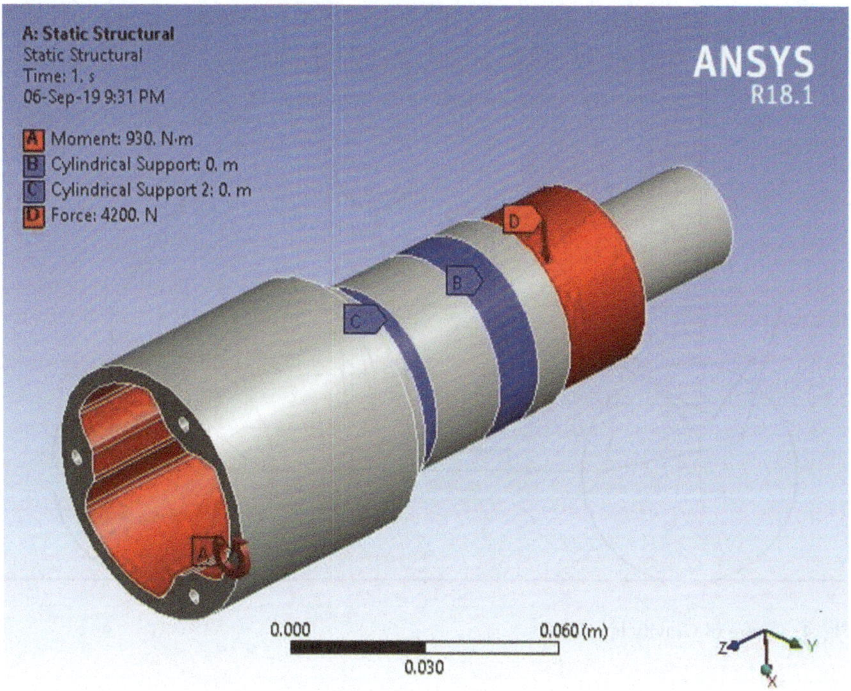

Fig. 3 Input conditions

7. The Tripod housing is subjected to a combination of bending and torsional moments without any axial force. (This is the worst-case scenario which is improbable).
8. Bump is considered as semi-circular with a radius of 12 in. (worst-case scenario).
9. The vehicle has open spool differential, thus, when there is no load on any one of the drive tire, the torque on the other tire is twice the torque on a single tire when both tires are under load.
10. The inertial losses due to rotating components at high rpm in engine and gearbox are neglected and the material used is homogeneous and isotropic.

4.1 Calculation to Find the Weight Biasing of Vehicle

Centre of gravity of individual components was found out through their CAD model. These points were plotted on a plane (Fig. 4) with a fixed reference point. In this case, the centre of rear wheel has been chosen as the origin or reference point.

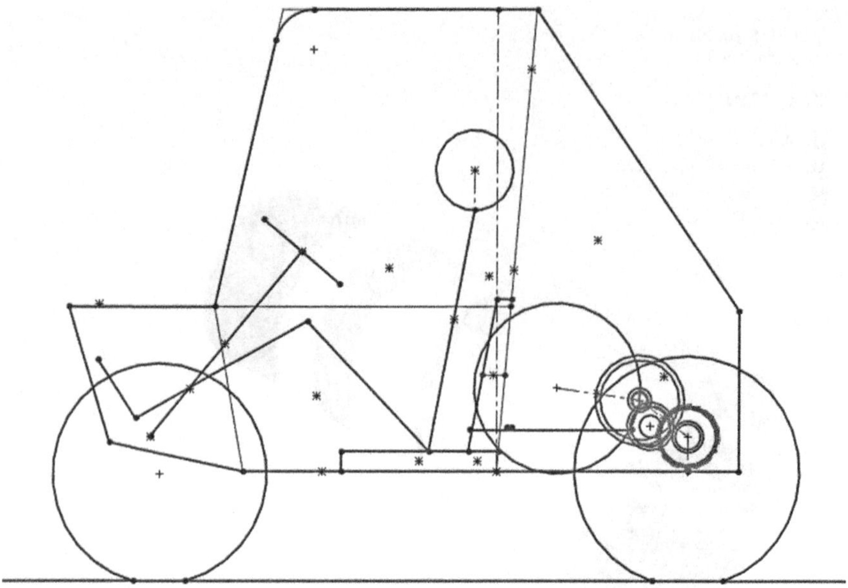

Fig. 4 Centre of Gravity layout

Table 1 shows the values of *x* and *y* coordinates of the centre of gravity of each part of the vehicle in side-view taking centre of rear wheel as the origin (reference point).

Refer to Table 1,

$$\mathrm{CG}_x = \frac{\sum_1^n (m_i x_i)}{\sum_1^n m_i} = \frac{-130{,}745}{214.49} = -609.56\ \text{mm}$$

$$\mathrm{CG}_y = \frac{\sum_1^n (m_i y_i)}{\sum_1^n m_i} = \frac{52{,}162.68}{214.49} = 243.19\ \text{mm}$$

This gives the exact location of centre of gravity of vehicle from reference point and thus calculating the weight biasing of the vehicle.

Wheelbase = 54 in.

Distance from CG to front tyre = 24.8022 in.

$$\text{Front Weight Biasing} = \frac{\text{Distance from CG to front tire}}{\text{Wheelbase}} = \frac{24.8022}{54} \times 100 = 45.93\%$$

Hence, weight biasing is considered as 46:54 (Front: Rear).

Table 1 Co-ordinates of centre of gravity of vehicle parts in XZ-Plane

Parts	Mass (m_i)	x_i	y_i	$m_i x_i$	$m_i y_i$
Front assembly	24.9	−1371.6	−6.6	−34152.84	−164.34
Rear assembly	24.2	0	0	0	0
Front Afco	6	−1290.86	209.58	−7745.16	1257.48
Rear Afco	6	−61.51	240.54	−369.06	1443.24
Steering column	0.6	−1199.3	322.6	−719.58	193.56
Steering assembly	1.1	−1395.22	89.11	−1534.74	98.02
Fuel tank	2.39	−231.92	585.81	−554.29	1400.09
Roll cage	30	−776.65	512.81	−23299.5	15384.3
Steering wheel	0.5	−1003.38	556.09	−501.69	278.05
Brake light	0.3	−404.98	1016.65	−121.49	305
Seat upper	0.75	−501.99	243.15	−376.49	182.36
Seat lower	0.75	−697.12	25.47	−522.84	19.10
Primary	2.8	−337.61	210.18	−945.31	588.50
Secondary	2.1	−123.81	180.15	−260.	378.32
Gear box	4.2	−61.9	133.9	−259.98	562.38
Head	4.74	−553.23	762.07	−2622.31	3612.21
Torso	37.94	−606.54	386.21	−23012.13	14652.81
Leg	21.02	−963.9	189.26	−20261.18	3978.25
Engine	26	−337.61	210.18	−8777.86	5464.68
Fire extinguisher	2.2	−514.57	492.69	−1132.05	1083.92
Shoulder belt	1	−449.48	508.07	−449.48	508.07
Lap belt	1	−544.72	25.47	−544.73	25.47
ASM belt	0.5	−951.12	0.078	−475.56	0.04
Brake assembly	1	−1529.95	423.24	−1529.95	423.24
CVT casing	2.5	−230.71	195.17	−576.78	487.93
Miscellaneous	10				
Total	214.49 Kg			−'130745	52162.68

4.2 Calculation of Torque Given by Output Shaft

Figure 5 shows the vehicle climbing a grade and the system of forces that are acting on it.

Given:

Table 2 shows the available data for calculation of torque on one wheel.

Total resistance on wheel = $mg \sin\theta + \mu mg \cos\theta = 1640.54117$ N

Torque on wheel = Total resistance on wheel*tyre radius = 464.82 Nm

Hence, Torque on single wheel is 464.82 Nm

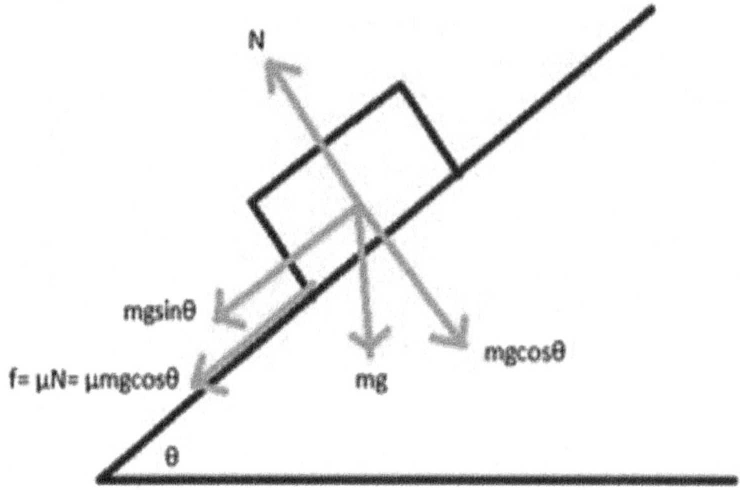

Fig. 5 F.B.D. of vehicle climbing a grade

Table 2 Data for initial torque calculation

Mass of vehicle (m)	215 kg
Grade of the slope (θ)	45 °
Coefficient of rolling resistance (μ)	0.1 (Power limiting condition)
Tyre radius	0.97 * 0.2921 = 0.283337
Engine torque	18.98 Nm
CVT reduction	3.9

4.3 Calculation of Force on Wheel When Hitting a Bump

Table 3 shows the available data for calculation of force when the wheel hits a bump.

Table 3 Data for bump force calculation

Mass of vehicle	215 kg
Weight bias	46:54 (front: rear)
Mass on single rear wheel (m)	$\frac{215\times0.54}{2} = 58.05$ N
Front spring rate (k)	30.39 N/mm
Motion ratio (strut displacement by wheel displacement)	$\frac{18.2-12.9}{7+3} = 0.53$
Velocity of car (v)	15 kmph = 4.1667 m/s
Wheel radius	11.5 in.
Bump radius (r)	12 in.
Initial angle on the bump (θ)	29.3°

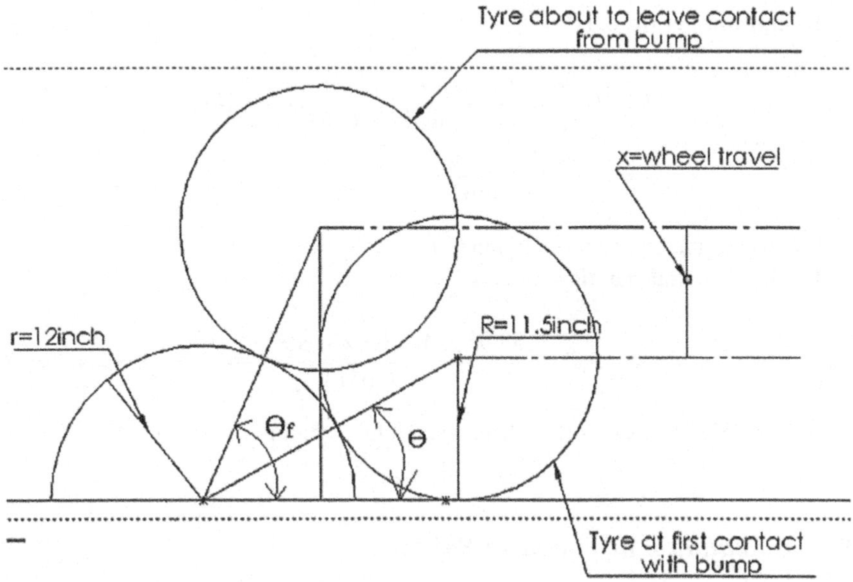

Fig. 6 Tire motion over the bump before leaving contact

To find x_f (compression of strut):

Figure 6 shows the rear wheel of the vehicle when hitting the bump and just before leaving contact from the bump.

To find final angle (θ_f) on the bump

Strut travel (x_f) = wheel travel (x) * Motion ratio = 0.0254 (23 sinθ_f − 11.5) × 0.53

Since, $N + \frac{mv^2}{r} = mg\ \sin\theta + k(x_f)\cos 45$

When rear tyre is about to leave contact at that point Normal Reaction (N) tends to 0

$$\frac{mv^2}{r} = mg\ \sin\theta + k(x_f)\cos 45$$

$$\frac{58.05 \times 4.1667^2 \times \sin^2\theta_f}{12 \times 0.0254} = 58.05 \times 9.81 \times \sin\theta_f + 30.39 \times 1000 \times 0.53 \times \cos(45) \times 0.0254 \times \left(23\sin\theta_f - 11.5\right)$$

sinθ_f = 0 .66 Thus, $\theta_f = 41.3°$

Before leaving the bump, tyre makes 41.3° angle with bump.

To find Impact time (t)

$$t = \int_{\theta}^{\theta_f} \frac{d\theta}{\omega} = \int_{29.3}^{41.3} \frac{r d\theta}{v \sin\omega\theta} = \frac{r}{v} \int_{29.3}^{41.3} \operatorname{cosec}\theta d\theta$$

$$t = \frac{23 \times 0.0254 \times 0.3658}{4.1667} = 0.0513 \text{ s}$$

Tyre leaves the contact of bump in 0.0513 s.

To Find Normal reaction on tyre

$$N \times t = mv \cos\theta_{avg} = \frac{58.05 \times 4.1667 \times \cos\left(\frac{41.3+29.3}{2}\right)}{0.0513} = 4187 \text{ N}$$

Force of 4187 N acts on the wheel when it hits a bump.

4.4 Calculation of Factor of Safety

Figure 7 shows the Shear Force Diagram and Bending Moment Diagram of the Tripod housing under uniformly distributed loading.

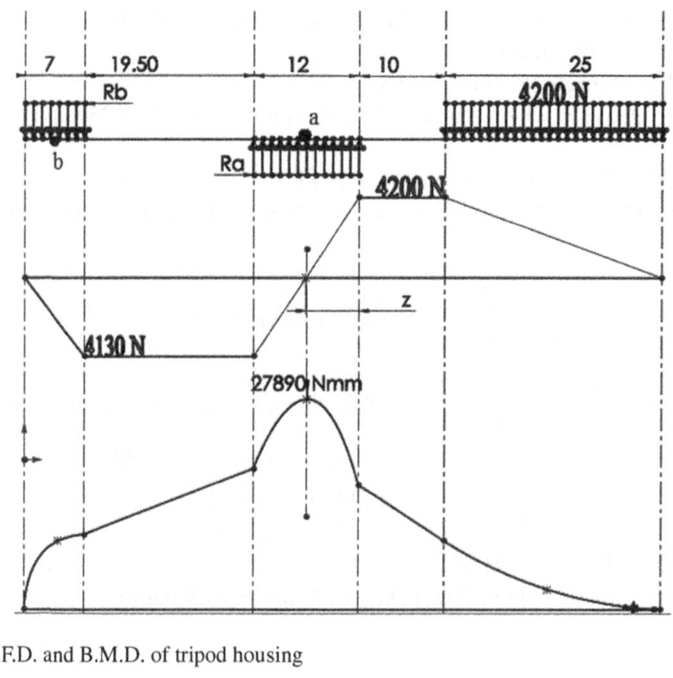

Fig. 7 S.F.D. and B.M.D. of tripod housing

To find distance 'z' where force is zero in SFD
$\frac{12-z}{4130} = \frac{z}{4200}$ Therefore, z = 6.05 mm

Given:

Torsional moment$(M_t) = (2 \times 465{,}000)$ Nmm Yield stress $= S_{yt} = 503$ MPa
Inner diameter $= d_i = 30$ mm Outer diameter $= d_o = 45$ mm
UDL 7 mm in length $= R_b$ UDL 12 mm in length $= R_a$

$$\sum M_b = 0(\text{since in equilibrium})\ R_a = 8330\ \text{N}$$
$$\sum F_y = 0(\text{since in equilibrium})\ R_b = 4130\ \text{N}$$

Therefore Maximum Bending moment is $(M_b) = -R_b(28.95) + R_a \times \frac{(12-z)^2}{2}$
$M_b = 27{,}890$ Nmm
Maximum bending moment $= M_b = 27{,}890$ Nmm

$$c = \frac{d_i}{d_o} = \frac{30}{45} = \frac{2}{3}$$

According to Maximum Shear Stress theory,

$$\frac{0.5 \times S_{yt}}{\text{FOS}} = \frac{16}{\pi\, d_0^3\left(1 - c^4\right)} \times \sqrt{M_b^2 + M_t^2}$$

FOS = 3.88.
Factor of Safety by Analytical calculations is 3.88.

5 Results and Discussion

Figures 8 and 9 shows the value of F.O.S. found out using FEA on ANSYS Workbench. Based on the input boundary conditions, using FEA, a F.O.S. of 3.5188 was obtained which was sufficient considering the Design Failure Modes and Effect Analysis for Tripod housing. Table 4 shows the error percentage between the F.O.S. values of FEA and analytical model.

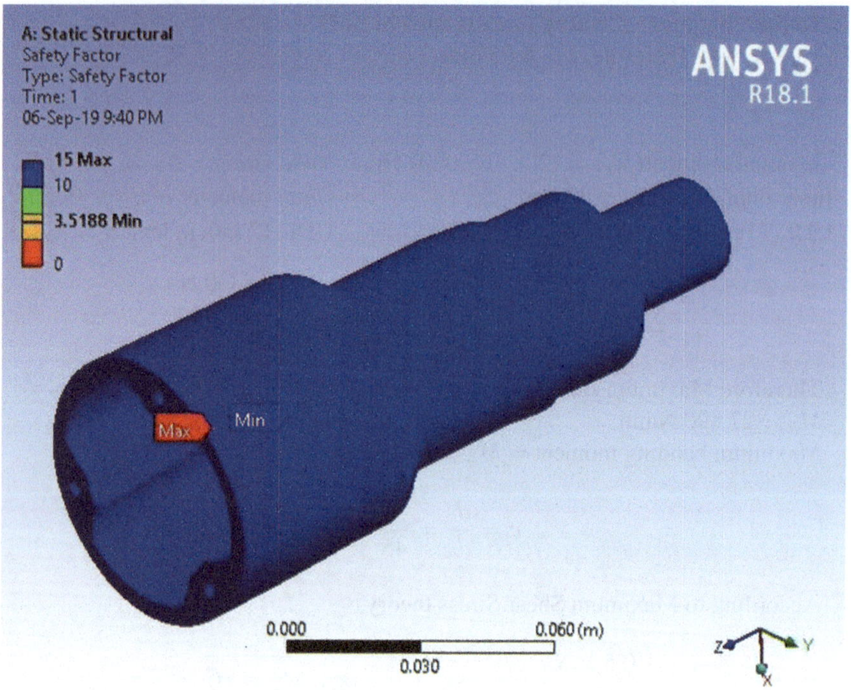

Fig. 8 F.O.S. results (isometric view 1)

6 Conclusion

The safety of Tripod housing plays an important role in transmitting power from transmission system to the drive wheel of automobiles. In this paper, the F.O.S. of Tripod housing is determined by using FEA and it was validated with the analytical calculations. The analytical model gives a F.O.S. of 3.88 while the Finite element model gives a F.O.S. of 3.5188 with an error of 9.3% (<10%) which is practically accepted. Since both values by analytical model and by Finite element model for Factor of safety are close, thus, it is validated.

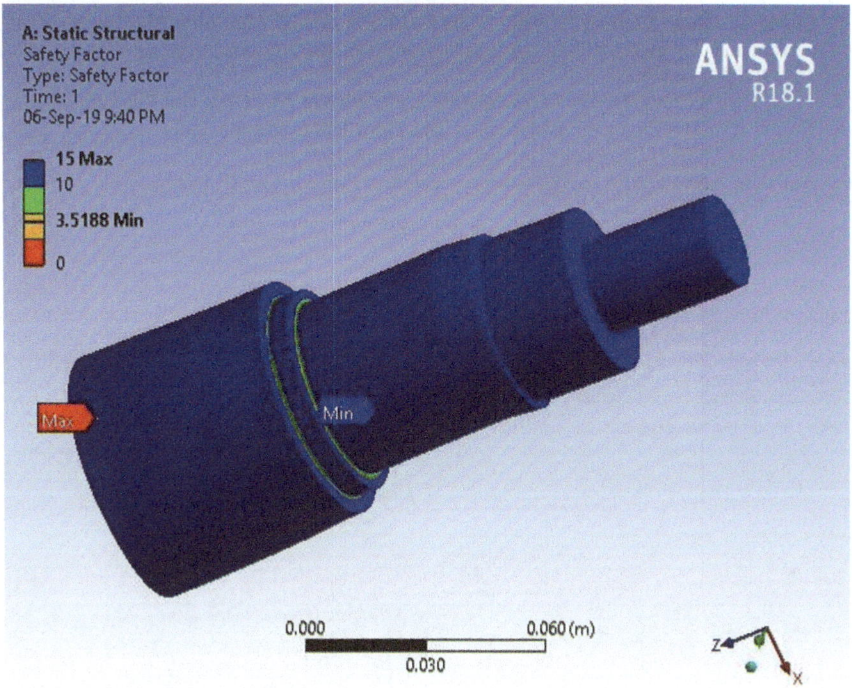

Fig. 9 F.O.S. results (isometric view 2)

Table 4 Comparison of F.O.S. for FEA and analytical method

F.O.S. (Analytical model)	F.O.S. (Finite element model)	Error percentage (practically accepted below 10%)
3.88	3.5188	$= \frac{3.88-3.5188}{3.88} \times 100 = 9.30\% < 10\%$

References

1. Kim DW, Kim SH, Chu CN, Cho JH (2019) A new life model of tripod type constant velocity joint using accelerated life test. Tribol Int 132(2019):130–141
2. Kim DW, Kim SH, Chu CN (2019) A real time wear measurement system for tripod type constant velocity joints. Precis Eng 56(2019):549–558
3. Lee C-H, Polycarpou AA (2010) A phenomenological friction model of tripod constant velocity (CV) joints. Tribol Int 43(2010):844–858
4. Lee C-H (2008) Development of Semi-empiricalfriction model in automotive driveshaftjoints. Int J Automot Technol 9(3):317–322
5. Wang XF, Hu RF, Wang KS, Cui HR (2014) Effects of reciprocating motion parameters on lubricating properties of tripod sliding universal joints. Ind Lubr Tribol 66(1):111–123

6. Chang D, Wang X, Yang F, Wang J (2009) Effects of Lubricant properties on lubricating performance of tripod sliding universal joints under thermal condition. Appl Mech Mater 16–19(2009):302–306
7. Bhandari VB (2017) Design of machine elements, 4th edn. Mc. Graw Hill Education, India
8. Nelson A (2009) Engineering Mechanics. Mc. Graw Hill Education, India
9. Solidworks (2017) Introducing solidworks, Dassault Systemes
10. Autodesk ANSYS 2018, ANSYS theory reference, 11th ed, SAS IP. Inc.

Design and Performance Evaluation of a Cost-Effective Radiant Cooling System

Rohit A. Rawool, Siddharth Saini, Aksheshkumar A. Shah, Tejas P. Shah, and Vinit Katira

Abstract Air conditioning or simply cooling is essential to maintain human comfort in a hot and humid climate. Thus, a system is to be designed that at a minimum energy input gives more efficient cooling along with being highly eco-friendly as compared to conventional Air Conditioners (AC). Hence an idea of cooling by the exchange of sensible heat to the surrounding from one's body can be put in use. During design, various factors were considered so as to accurately find the total cooling load required to cool the space. After successful fabrication and installation of the system, readings were taken to compare both conventional AC system and our designed radiant cooling system and comparisons with respect to human comfort and the cost of operation incurred while running the systems were made. It was concluded that radiant cooling systems save 34% of energy while maintaining superior human comfort.

Keywords Air conditioning · Radiant cooling system · Human comfort

1 Introduction

Due to rising concern for energy reduction and growing need for thermal comfort, the cooling systems which provide better human comfort, are eco-friendly and consume less amount of power are need of the hour. Hence the idea of developing a radiant cooling system was proposed that shows a large potential for energy saving [1]. Moving water is more efficient than moving air because of its physical and thermal properties. Water can carry 3400 times the energy that air can carry for the same volume. This property of water is used to achieve maximum advantage in a radiant cooling system. Also, the natural manner in which the human body dissipates heat is mainly through radiation. This is the primary principle used in radiant cooling. Coldwater flows through pipes embedded in the panels/slabs and cools the entire panel resulting in the panel surface being maintained at about 20 °C. Cooling inside a room is achieved when the cold panel absorbs the heat (radiation) generated by

R. A. Rawool · S. Saini · A. A. Shah · T. P. Shah (✉) · V. Katira
Dwarkadas J. Sanghvi College of Engineering, Mumbai, India
e-mail: tejass121@gmail.com

© Springer Nature Singapore Pte Ltd. 2020

H. Vasudevan et al. (eds.), *Proceedings of International Conference on Intelligent Manufacturing and Automation*, Lecture Notes in Mechanical Engineering,
https://doi.org/10.1007/978-981-15-4485-9_76

people, computers, lighting and other equipment that are exposed to the panel. The air near the cold panels is cooled first which moves down and replaces the hot air at the bottom of the room. This hot air now rises up to the ceiling to replace the cold air which is moving downwards. This hot air now cools as it comes in contact with the cold panels and the cold air present at the bottom gets heated up by absorbing the heat emitted from the equipment and human bodies. Fresh air is supplied through an air system to maintain a healthy indoor environment and also to control the moisture content of air inside the office space. In other words, the sensible heat load is addressed by the cooled panel and the latent heat load is addressed by Dedicated Outdoor Air System (DOAS) [2, 3].

The various heat loads acting taken in consideration are Solar heat gain through glass Solar and transmission heat gain for walls and roof, Transmission heat gain except for walls and roof, Heat gain due to infiltration, Sensible internal Heat Gain and Room latent heat [4, 5].

The relevance to social benefits by this R&D in the proposed area includes energy efficiency, superior thermal comfort, greater architectural flexibility, reduced operating and maintenance costs, effective control of ventilation and absence of harmful chemicals thereby making the system eco-friendly [6].

2 Calculations and Design Methodology

The calculations are one of the most critical stages as the whole designing process to obtain a better human comfort which comprises calculating the net sensible heat load, latent heat load and finally the net room total heat load of the room over a period of time. Since the aim was human comfort around the year so a sample calculation was needed for designing at different points of time as the season changes. Hence, during summer May was chosen and during monsoon, July was chosen and appropriate reading of dry bulb temperature, wet bulb temperature and relative humidity were taken as shown in the following table. Assuming suitable quantities as per the guidelines of ASHRAE handbook. The layout of the room used for calculations is shown in Fig. 1.

2.1 Calculations

A. Figure 2 below shows the general temperature trend in and out of the room for two seasons which are summer and monsoon. The location of the experiment is an office space located in Vasai, Maharashtra, India.

Here, amount of infiltrated air is calculated as,

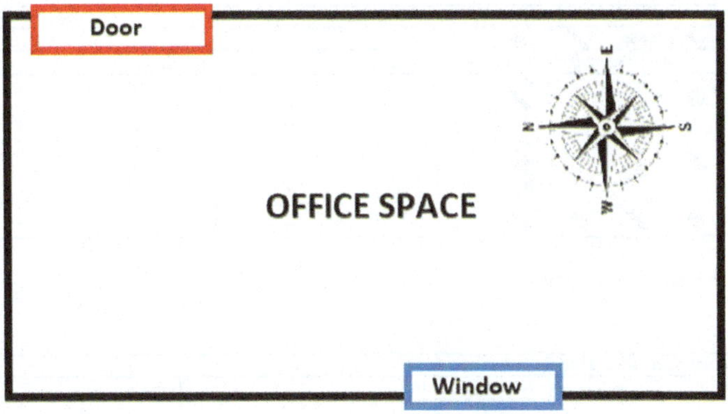

Fig. 1 Layout of the room

			Area:	Vasai								
Project:	OFFICE SPACE COOLING LOAD		City:	Mumbai/Maharashtra								
Space:	OFFICE SPACE		Month:	May for Summer and July for Monsoon								
			Time:	1PM TO 5PM								
Length(m)	3.63		SUMMER					MONSOON				
Width(m)	2.37	Condition	DBT	WBT	%RH	DPT	kg/kg	DBT	WBT	%RH	DPT	kg/kg
Height(m)	2.7	Outside	35	25	45	22	0.016	26	23	75	21	0.017
Area(m^2)	8.603	Inside	25.5	19	52	15	0.011	20	14	50	10	0.007
Volume(n	23.22	Difference	9.5	6	7	7	0.005	6	9	25	11	0.01
BFP	0.12	No. of Air changes/hr		1.5			Infiltrated Air(m^3/min)			0.581		

Fig. 2 General temperature trend in and out of room

$$V_{\text{infiltated}} = (L * B * H * (\text{Number of air changes/hr}))/60 \tag{1}$$

Its unit is m^3/min.

B. Heat gain is calculated using the following formula:

$$\text{Solar Heat Gain for glass} = \text{Area} * \text{Factor} * \text{Heat Flux Density} \tag{2}$$

$$\text{Solar and transmission Heat Gain} = \text{Area} * \text{Factor} * \text{Temperature Difference} \tag{3}$$

$$\text{Transmission Heat Gain except for walls and roof} = \text{Area} * \text{Factor} * \text{Temperature Difference} \tag{4}$$

$$U_{\text{exposed wall}} = 1/((1/h_0) + (2x_1/k_1) + (2x_2/k_2) + (1/k_a) + (1/h_i)) \tag{5}$$

			SUMMER		MONSOON	
SOLAR HEAT GAIN FOR GLASS						
Item	Area(m²)	Factor(W/m²-°C)	Heat Flux Density(W/m²)	W	Heat Flux Density(W/m²)	W
Glass(N)	0	0.48	129	0	136	0
Glass(N-E)	0	0.13	527	0	521	0
Glass(E)	0	0.11	631	0	618	0
Glass(S-E)	0	0.14	372	0	360	0
Glass(S)	0	0.22	129	0	129	0
Glass(S-W)	0	0.52	372	0	360	0
Glass(W)	1.05	0.52	631	344.526	618	337.428
Glass(N-W)	0	0.47	527	0	521	0

Fig. 3 Solar heat gain for glass

SOLAR AND TRANSMSSION HEAT GAIN FOR WALLS AND ROOF						
Item	Area(m²)	Factor(W/m²-°C)	Temp. Diff (°C)	W	Temp. Diff (°C)	W
Glass(N)		1.1675	10	0	5	0
Glass(N-E)		1.1675	10	0	5	0
Glass(E)	8.751	1.1675	10	102.167925	5	51.0839625
Glass(S-E)		1.1675	10	0	5	0
Glass(S)		1.1675	10	0	5	0
Glass(S-W)		1.1675	10	0	5	0
Glass(W)		1.1675	10	0	5	0
Glass(N-W)		1.1675	10	0	5	0

Fig. 4 Solar and transmission heat gain for walls and roof

$$U_{\text{partition}} = 1/((2x_1/k_1) + (x_2/k_2) + (2/h_i)) \tag{6}$$

$$U_{\text{roof}} = 1/((1/h_1) + (x_3/k_3) + (x_2/k_2) + (1/h_2)) \tag{7}$$

$$U_{\text{floor}} = 1/((1/h_1) + (x_3/k_3)) \tag{8}$$

$$U_{\text{glass}} = 1/((1/h_O) + (x_4/k_4) + (1/h_3)) \tag{9}$$

$$\text{Area of door} = 2.07 * 0.72 = 1.49\ \text{m}^2 \tag{10}$$

$$\text{Partition wall area} = 2 * \text{H} * \text{B} + (\text{H} * \text{L} - \text{Area of door}) \tag{11}$$

Length of plaster (x_1) = 0.015 m, Length of brick (x_2) = 0.2 m, Thermal conductance of air gap (k_a) = 5.8 W/m^{2}K, Thermal conductivity of plaster (k_1) = 8.65 W/m^{2}K, Thermal conductivity of brick (k_2) = 0.77 W/m^{2}K, Thermal conductivity of concrete (k_3) = 1.73 W/m^{2}K, Thermal conductivity of glass (k_4) = 0.78 W/m^{2}K. Figures 3, 4 and 5 show the heat gains through glass, roof and walls.

TRANSMISSION HEAT GAIN EXCEPT FOR WALLS AND ROOF						
Item	Area(m^2)	Factor(W/m^2-°C)	Temp. Diff (°C)	W	Temp. Diff (°C)	W
Glass	1.05	5.75	10	60.375	5	30.1875
Partition wall	21.109	2.01	10	424.2909	5	212.14545
Ceiling	8.603	2.615	10	224.96845	5	112.484225
Floor	8.603	4.5	10	387.135	5	193.5675

Fig. 5 Transmission heat gain except for walls and roof

C. Infiltration is calculated using the following formula:

$$\text{Heat Gain due to Infiltration} = \text{Outside air CFM} * \text{BPF} * \text{Factor} * \text{Temperature Difference}\sqrt{a^2 + b^2} \quad (12)$$

Infiltration refers to the outside air which enters through the gaps, door openings. This air adds both sensible and latent heat into the room which needs to be accounted for. The calculations for heat gain due to infiltration is shown in Fig. 6.

D. The room sensible heat was obtained at the end of each respective period [5, 7] (Fig. 7):

The total sensible heat was calculated as shown in Fig. 8, using the following formula:

HEAT GAIN DUE TO INFILTRATION						
Outside Air CFM	Bypass	Factor(W/m^2-°C)	Temp. Diff (°C)	W	Temp. Diff (°C)	W
0.581	1	20.44	10	118.7564	5	59.3782

Fig. 6 Infiltration heat gain

SENSIBLE INTERNAL HEAT GAIN						
Item	Amount	Factor(W/m^2-°C)	Temp. Diff (°C)	W	Temp. Diff (°C)	W
Number of people	5	70	NIL	350	NIL	350
Lights	9	12		108		108
Non-continuous equipments	2	30		60		60
Continuous equipmments	2	125		250		250
Inkjet Printer	1	160		160		160
ROOM SENSIBLE HEAT SUBTOTAL			2590.219675		1924.274838	
Accounting for excess heat gain, leak loss & safety factor (5%)			129.5109838		96.2137419	
ROOM SENSIBLE HEAT (R.S.H.)			2719.730659		2020.48858	

Fig. 7 Room sensible heat gain

ROOM LATENT HEAT CALCULATION						
Outside Air CFM	Bypass	Factor	Diff kg/kg	W	Diff kg/kg	W
0.581	1	50000	0.005	145.25	0.01	290.5
Item		Factor	Diff kg/kg	W	Diff kg/kg	W
Number of people	5	45	NIL	225	NIL	225
ROOM LATENT HEAT SUBTOTAL			370.25		515.5	
Accounting for excess heat gain, leak loss & safety factor (5%)			18.5125		25.775	
ROOM LATENT HEAT (R.L.H.)			388.7625		541.275	
ROOM TOTAL HEAT(R.T.H.)			3108.493159		2561.76358	

Fig. 8 Room latent heat gain

i. **Net room sensible heat (RSH)**

$$\text{Net Room Sensible Heat} = \text{Room Latent Heat Subtotal} * (1 + (\text{Safety Factor}/100)) \quad (13)$$

E. The room latent heat was calculated at the end of each respective period:

i. Room Latent Heat (RLH) calculations are shown in Fig. 8.

$$\text{Net Room Latent Heat} = \text{Room Latent Heat Subtotal} * (1 + (\text{Safety Factor}/100)) \quad (14)$$

F. The outside room sensible heat and latent heat were calculated at the end of each respective period and grand total heat, cooling capacity in tons and sensible heat factor is also calculated. The following formulae are used:

i. **Outside air sensible heat**

$$\text{Outside Air Sensible Heat (OASH)} = \text{Outside air} * (1 - \text{BPF}) * \text{Factor} * \text{Temperature Difference} \quad (15)$$

ii. **Outside air latent heat**

$$\text{Outside Air Latent Heat (OALH)} = \text{Outside air} * (1 - \text{BPF}) * \text{Factor} * \text{Difference kg/kg} \quad (16)$$

iii. **Grand total heat**

$$\text{Grand Total Heat (W)} = (\text{RSH} + \text{RLH} + \text{OASH} + \text{OALH}) * (1 + (\text{Safety Fector}/100)) \quad (17)$$

iv. **Cooling capacity**
v. **Sensible heat factor**

$$\text{Sensible Heat Factor} = \text{Net Room Sensible Heat}/\text{Grand Total Heat} \quad (19)$$

$$\text{Total outside air ventilation} = \text{Panel area} * \text{Ventilation required} \quad (20)$$

$$\text{Panel Area} = 3.63 * 2.37 = 8.603\ \text{m}^2 \quad (21)$$

OUTSIDE AIR HEAT						
			OUTSIDE AIR SENSIBLE HEAT (OASH)			
Outside Air	1-BPF	Factor	Temp. Diff (°C)	W	Temp. Diff (°C)	W
0.172	0.88	20.44	10	30.937984	5	15.468992
			OUTSIDE AIR LATENT HEAT (OALH)			
Outside Air	1-BPF	Factor	Diff kg/kg	W	Diff kg/kg	W
0.172	0.88	50000	0.005	37.84	0.01	75.68
SUBTOTAL			3177.271143		2652.912572	
Accounting for excess heat gain, leak loss & safety factor (5%)			158.8635571		132.6456286	
GRAND TOTAL (W)			3336.1347		2785.5582	
TONS = {(W)/3500}			0.953181343		0.795873772	
SENSIBLE HEAT FACTOR= (RSH/RTH)			0.874935385		0.78871001	
INDICATED ADP			7		11	
SELECTED ADP			7		11	
DEHUMIDIFIED AIR QUANTITY						
Room rise = (1-By-pass factor)*(Room temperature - ADP)			16.28		7.92	
DEHUMIDIFIED AIR = R.S.H/(20.44*Dehumid. Rise)			8.173171369		12.48102712	
Safety factor(5%)			0.408658568		0.624051356	
TOTAL DEHUMIDIFIED AIR			8.581829937		13.10507848	

Fig. 9 Outside air heat and capacity of heat to be removed

The Grand Total Heat (GTH) includes the room heat gain as well as outside air heat gain which gives an indication of the required capacity of the chiller, which is shown in Fig. 9.

G. Design Methodology:

Various equipments were selected to carry out the experimental work. They are:

A. **PEX Pipe**: Diameter of the pipe was taken as ¾ in. The PEX pipe was a preferred choice as it can be bent in any direction without any pinching in the pipe and can turn 90-degree corners without the need for elbow or fittings.
B. **Water Pump**: A water pump was needed to pump the cold water from the water storage tank into the pipes laid out in the room. The selected pump was 1 H.P. having flow rates up to 2500 LPH.
C. **Custom made chiller/out unit**: A chiller was required to cool the water which was to be circulated inside the room to absorb heat loads. The chiller has 1 Ton capacity.
D. **Water Storage Tank**: Capacity: 60L.
E. **Temperature control unit and thermostat**: Range 16–30 °C.

3 Fabrication and Assembly

To fabricate the apparatus for the experiment, dimensions of the room whole room were considered. The dimensions of the room are Height = 2.7 m, Breadth = 2.37 m, Length = 3.63 m. So to cover the ceiling with a hydronic circuit, panels are required. One panel cannot be sufficient as the circuit must cover the whole ceiling and it is difficult to fit the panel on the ceiling. Thus, a 6-panel system was decided based on no. of loops, inlet and outlet of water and loop design.

A. **Dimension of one panel**: Length = 1.2 m, Breadth = 0.9 m, Area = 1.08 m^2, Thickness = 0.005 m

To make the assembly of the panels on the ceiling to be feasible and easy, two panels were joined together to form a mega panel. The pipe was fastened to the aluminum sheet at proper intervals so as to keep the pipe in position and retain the shape of the loops. Figure 10 shows the mega panel and loops of pipe.

Assembly: In our case, the mega panels were mounted to the ceiling via the false ceiling structure. The false ceiling structure was fixed to the ceiling of the given office space and then the mega panels were bolted to the false ceiling structure. The assembled system as rigid as the false ceiling structure was bolted to the ceiling as shown in Figs. 11 and 12.

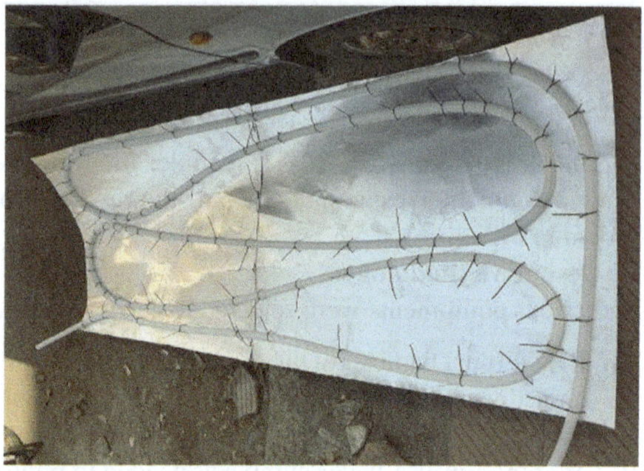

Fig. 10 Mega panel

Fig. 11 False ceiling structure

Fig. 12 Final assembly of panels

4 Result and Discussion

A comparison was made between the readings obtained by performing the experiment with the assembled system switched ON, OFF and the air conditioning system. The parameters on which the comparision were made are shown in Figs. 13, 14 and 15.

1. Room and wall temperature (°C) without switching on Radiant Cooling System shown in Fig. 13.
2. Room and Wall Temperature after switching on the Radiant Cooling system shown in Fig. 14.
3. Room and Wall Temperature After, Switching on the Air Conditioner shown in Fig. 15.
4. **Power consumption**:

$$\text{Power Consumption} = (\text{Number of hours of use} * \text{Number of days of use} * \text{Capacity of the appliance used})/1000 \quad (22)$$

Time	Wall Temperature				Air temperature	Outside Air temperature	Floor temperature	Humidity(%)	Mean Radiant temperature
	W1	W2	W3	W4					
1 pm to 2 pm	29.8	29.5	29.8	29.6	31.1	31.7	29.7	62	29.70125
2pm to 3 pm	30.2	30.3	30.6	30.5	31.4	32.1	30.3	63	30.4
3pm to 4 pm	30.7	30.9	31.1	30.8	31.7	32.5	30.6	63	30.88025
4pm to 5pm	30.5	30.2	30.8	30.6	30.9	31.9	30.1	61	30.55125
5pm to 6pm	30.3	30.5	30.6	30.1	30.8	31.6	29.9	60	30.39075

Fig. 13 Room and wall temperature with no cooling

Time	Wall Temperature				Air temperature	Outside Air temperature	Floor temperature	Humidity(%)	Mean Radiant temperature
	W1	W2	W3	W4					
1 pm to 2 pm	25.2	25.4	25.5	25.1	25	31.7	25.3	55	25.30008425
2pm to 3 pm	25.1	25.3	25.4	25.2	25	32.1	25.1	55	25.24995788
3pm to 4 pm	25.3	25.4	25.4	25.1	25	32.5	25.2	55	25.30008425
4pm to 5pm	25.1	25.1	25.4	25.3	25	31.9	25.1	55	25.22518955
5pm to 6pm	25	25.1	25.3	25.3	25	31.6	25.3	55	25.17497894

Fig. 14 Room and wall temperature with radiant cooling system

Time	Wall Temperature				Air temperature	Outside Air temperature	Floor temperature	Humidity(%)	Mean Radiant temperature
	W1	W2	W3	W4					
1 pm to 2 pm	27.6	27.4	27.9	28	25	31.7	27.3	50	27.72460474
2pm to 3 pm	27.5	27.3	27.6	27.1	25	32.1	27.5	50	27.37579051
3pm to 4 pm	27.3	27.3	27.5	27.1	25	32.5	27.4	50	27.30079051
4pm to 5pm	27.1	27.2	27.4	27.2	25	31.9	27.3	50	27.22549407
5pm to 6pm	27	27.1	27.2	26.9	25	31.6	27.1	50	27.0506917

Fig. 15 Room and wall temperature with air conditioning unit

Unit of Power consumption is kWh.
A normal air conditioning unit consumes from 1950 to 1975 Watts.

i. **Power consumption by radiant cooling system**:

The two main appliances consuming power in a radiant cooling system are the chiller/outdoor unit and the water pump. The power consumed by both the appliances has been mentioned above.

$$\begin{aligned} \text{Power Consumption} &= (10 * 160 * (703.4 + 750))/1000 \\ &= 2325.44\ \text{kWh} \end{aligned} \tag{23}$$

ii. **Power consumed by air conditioning unit**:

$$\text{Power Consumption} = (10 * 160 * 1950)/1000 = 3120\ \text{kWh} \tag{24}$$

5. **Comparison between Radiant cooling system and air conditioning unit**:

The main basis of deciding the better one between the two systems was on the basis of various factors that are:

i. **% Saving of power**:

$$\begin{aligned} \%\ \text{Saving of power} &= ((3120 - 2325.44)/2325.44) * 100 \\ &= 34.17\% \end{aligned} \tag{25}$$

Here, it can be seen that the radiant cooling system saves 34.17% more energy as compared to the conventional air conditioning unit. The following graph shows the energy consumption of both the systems with respect to time:

From Fig. 16, it can be concluded that the conventional air conditioning system consumes more power to cool the room during peak office hours as compared to a radiant cooling system.

ii. **Cost-effectiveness**: The radiant cooling system is a cost-effective system as it does not require expensive appliances but can be made with rather simple and readily available appliances and other auxiliaries. An air conditioning unit costs in the range of Rs. 25,000–30,000 but the whole radiant cooling system was manufactured and assembled for a cost less than Rs. 20,000. Hence, there is

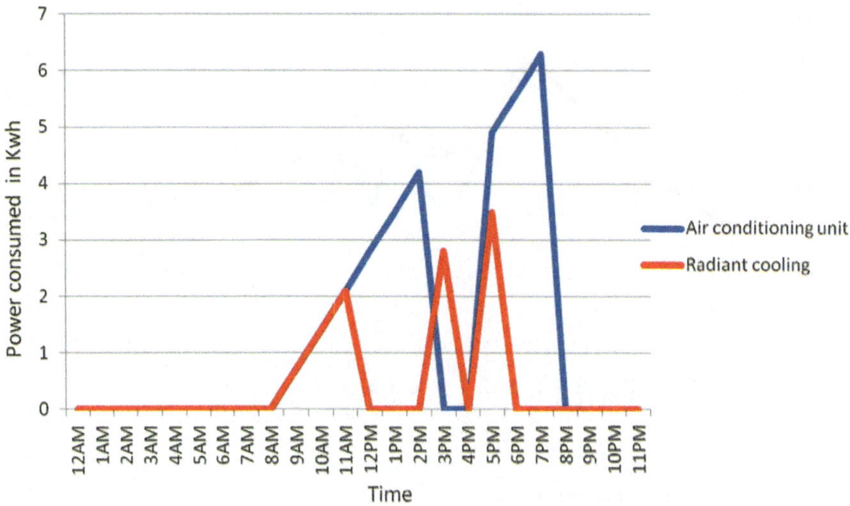

Fig. 16 Graph of power consumed versus time

saving of Rs. 5000 from the start of installation of radiant cooling system and the radiant cooling system exhibits lower overall operating costs.

iii. **Human Comfort**: The quality of human comfort achieved in a radiant cooling system is superior as compared to the conventional air conditioning unit. Figures 17 and 18 represent the mean room temperature and relative humidity obtained with respect to time. From Fig. 18, it can be seen that radiant cooling

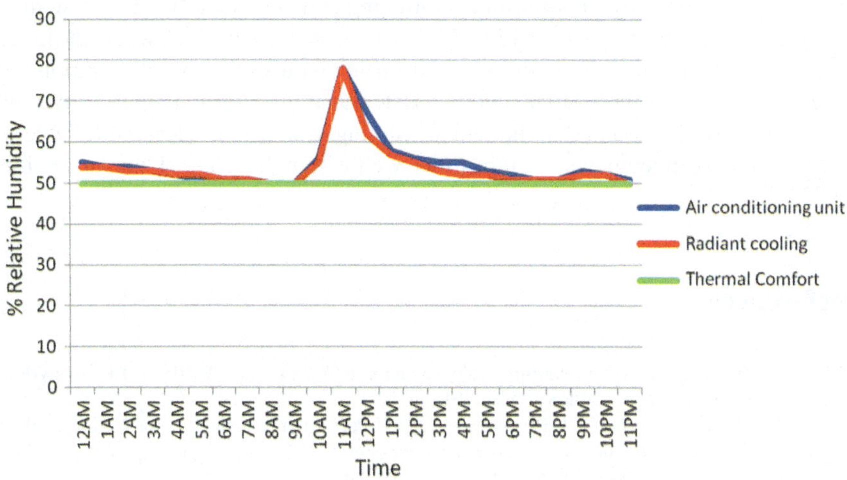

Fig. 17 Graph of relative humidity versus time

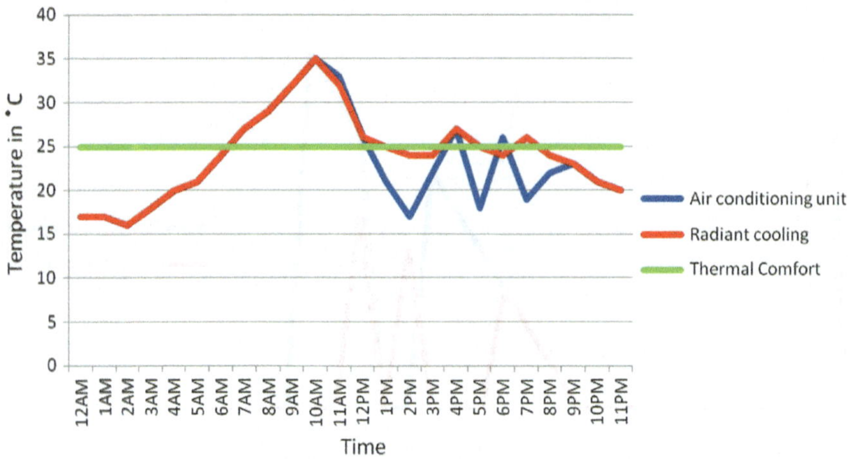

Fig. 18 Graph of temperature versus time

system is able to maintain the temperature in the room closer to the human comfort temperature which ensures superior human comfort.

5 Conclusion

The objective was successfully achieved by a graphical comparison made between the power consumed by a normal air conditioning unit and radiant cooling system. It was proved that the latter was 34.17% efficient than the former which was in the range of 30–50% for dry climatic conditions. This further reduces the cost of operation and investment required for a radiant cooling system making it even more effective and desirable. It is also found that the radiant cooling system was successfully able to maintain the room temperature of the office well within the range of human comfort of 25 °C.

References

1. Uponor Inc. (2013) Radiant cooling design manual, RCDMManual_RC03_0513_, Copyright © 2013 Uponor, MN 55124 USA, May 2013
2. Khan Y, Khare VR, Mathur J, Bhandari M (2015) Performance evaluation of radiant cooling system integrated with air system under different operational strategies. Energy and Buildings (2015). http://dx.doi.org/10.1016/j.enbuild.2015.03.030
3. Košičanová D, Design and calculation of ceiling radiant cooling systems. Technical University of Kosice, Slovakia

4. Bourdakis E, Kazanci OB, Olesen BW (2015) Load calculations of radiant cooling systems for sizing the plant. In: 6th International building physics conference, IBPC 2015. Energy Procedia 78(2015):2639–2644
5. Sahu SK (2014) Cooling load estimation for a multi-story office building. Master of technology thesis. National Institute of technology, Rourkela
6. Caleffi Hydronic Solutions (2013) Idronics. J Des Innov Hydronic Prof
7. Zmrhal V[1], Hensen J[2], Drkal F[1], Modelling and simulation of a room with a radiant cooling ceiling, [1] Czech Technical University in Prague, [2] Technische Universiteit of Eindhoven

Design and Analysis of 'Kangaroo' Boots

Mihir Sanghvi, Hamza Neemuchwala, Md Husain Thekiya, Dinesh Papal, and Kartik Ajugia

Abstract Superman is a hypothetical concept with exceptionally supreme powers. How would it be to have one of his powers? Ever imagined yourself of running at a smeared lightning speed? Well, now you can! 'Kangaroo boots' is a solution to this. Inspired from kangaroos who have a unique back feet structure, these boots are a replica of the same feature which would store and release the elastic energy and increase the running speed of a human up to 25–30 mph. The main perspective of this study is to provide more agility while running off-road and climbing steep roads. It will also make humans more active and travel short distances in minimal time reducing the use of fossil fuels. Keeping in mind the future scope, these boots can be used in human enhancement, military, sports and robotic applications.

Keywords Kangaroo boots · Bio-enhancement boots · Ultra-high-speed · Future scope

1 Introduction

A kangaroo has a unique back feet structure that makes it capable of running to a speed up to 45 mph. They store elastic energy in their back feet and release of this energy generates 80–85% of more power than in humans. The kangaroo boots allow you to run as fast as a vehicle. Such spring shoes also mimic the ostrich move and allow you to go up to 25 miles (40 km/h) an hour. Shoes on the back have movements close to the Achilles ligament of an ostrich. The spring gives the user increasing power down while running in their big, springy walk. Ostriches will reach 40 mph (70 km/h) levels, concealed in a solitary walk up to 5 m (16 ft).

M. Sanghvi (✉) · H. Neemuchwala · M. H. Thekiya · D. Papal · K. Ajugia
Department of Mechanical Engineering, Dwarkadas. J. Sanghvi College of Engineering, Mumbai, India
e-mail: m.sanghvi1998@gmail.com

K. Ajugia
e-mail: Kartik.Ajugia@djsce.ac.in

© Springer Nature Singapore Pte Ltd. 2020
H. Vasudevan et al. (eds.), *Proceedings of International Conference on Intelligent Manufacturing and Automation*, Lecture Notes in Mechanical Engineering,
https://doi.org/10.1007/978-981-15-4485-9_77

We have been taking a shot at the boots for a while and have just given dozens of models. In this present structure, the gadget can arrive at velocities of up to 25 mph (40 km/hr)—or a similar speed as a moderate moving vehicle. Highlighting a spring-stacked sole, it assists individuals with recouping from sport wounds by removing the effect of running on their joints.

1.1 Objectives

1. Improve human potential for greater reach and performance improvement.
2. Reducing fossil fuel reliance and encouraging fitness of fellow human beings.
3. Introducing a new mode of transport between destinations.
4. To move on to new biomechanical engineering directions to facilitate bio-suits completion in the future.

1.2 Methodology

The procedure for designing and analysis of kangaroo boots was generally divided into six major steps. They are as follows:
Step 1: Designing of model
Step 2: Material Survey and selection
Step 3: Virtual analysis of components
Step 4: Manufacturing of components, sub-assemblies and assemblies
Step 5: Testing of components for failure
Step 6: Results validation.

2 Material Survey

Traditional model building materials have always provided for lightweight design while at the same time providing acceptable strength levels. Kangaroo boots are always prone to crash disruption and maneuvering or cyclic pressure cycles (fluctuating loads) because they are capable of operating at substantial speeds. The materials used to make kangaroo boots must, therefore, have certain properties requirements in order to deal with the aforementioned problems. The components of the Kangaroo boots include Leg support, Strut member, Bushing, Extension springs and Footpad. Figure 1 shows the exploded view and the complete assembly of kangaroo boots.

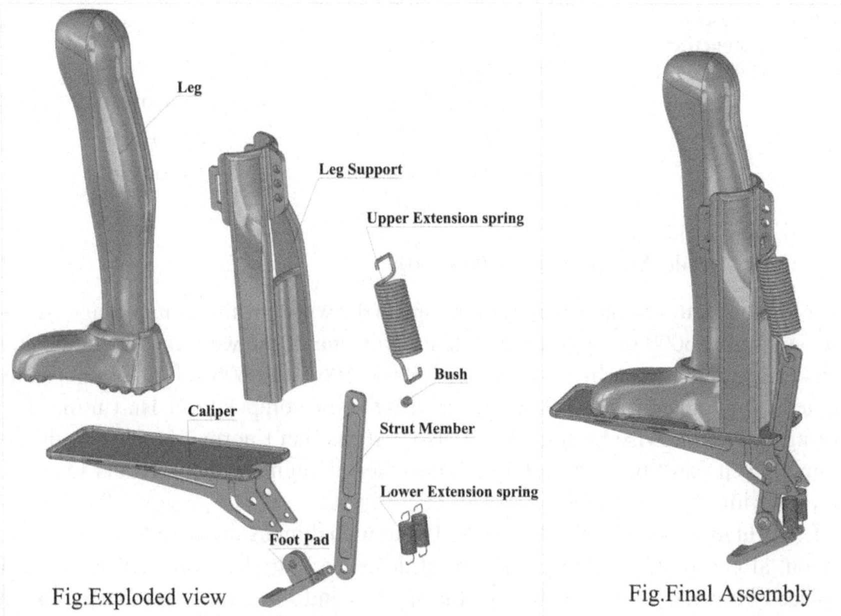

Fig. 1 Assembly of Kangaroo boots

1. Fibre-Reinforced Plastic (FRP)

Fibre-reinforced plastic is a composite material made of a polymer framework strengthened with the help of filaments. The filaments are generally glass, carbon, aramid, or basalt. Once in a while, different strands, for example, paper, wood, or asbestos have been utilized. The polymer is generally an epoxy, vinyl ester, or polyester thermosetting plastic; however, phenol-formaldehyde tars are still being used [1].

FRP enables thermoplastic glass fibres to be matched according to specific design programmes. Glass-reinforced polymers are the strongest and most resistant to deformation when the polymers are parallel to the force and are weakest once the fibres are perpendicular to each other. This ability is, therefore, an advantage or a disadvantage in the usage sense [1].

The leg support for boots are made using FRP moulds. The simplest moulding technique, hand layup, is used in the manufacture of large items with low volume, e.g. components of wind turbines, concrete types and, radomes. For a high-quality layer, a pigmented gel coat is sprayed onto the mould. When the gel coat is healed, the glass mat and/or woven roving is placed in the mould and the catalyzed resin is poured, brushed or sprayed on. Additional layers are applied for thickness of mat or woven roving and resin. Curing in the resin systems is triggered by a catalyst or accelerator which strengthens the material without external heat [1].

Table 1 Represents the mechanical properties of aluminium 6082T6 [2]

Density	2.71 g/cm^3
Young's modulus	71 GPa
Ultimate tensile strength	140–330 MPa
Yield strength	90–280 MPa
Thermal expansion (α)	23.1 μm/m-K

2. Aircraft-Grade Aluminium (Al-6082T6)

6082T6 aluminium composite is a compound in the wrought aluminium-magnesium-silicon family (6000 or 6xxx arrangement). It is one of the well-known composites in its arrangement (nearby combinations 6005, 6061 and 6063) [1]. It is normally formed by expulsion and rolling, yet as a wrought composite, it isn't utilized in casting. It can likewise be forged and clad, yet that isn't normal practice with this compound. It can't be work solidified, however, is regularly heat treated to create tempers with a higher quality yet lower ductility [1].

The strut member is made of Al-6082T6 bar with dimensions to reduce bar weight without affecting its strength, therefore, it drills circular shape slots. This member is placed between the rubber foot of the springs and the tyre. Using bush and pin arrangement, strut is rotated to the sole base clamping plates. At the bottom, there is another pivot to clamp the base of the rubber foot. Since the strut component holds the entire weight of the human body to measure its strength, FEA analysis is carried out to validate its proportions [1].
Mechanical Properties:
See Table 1.

3. Carbon Fibre

Carbon Fibres are fibres concerning 5–10 mm in diameter and composed principally of carbon atoms. They possess many advantages like high stiffness, high tensile strength, high durability, low weight, high chemical resistance, heat tolerance and low thermal expansion. These properties have created carbon fibre very hip in part, military, motorsports, etc. to produce a carbon fibre, The carbon atoms are secure along in crystals that are a lot of or less aligned parallel to the long axis of the fibre as this crystal alignment offers fibre high strength-to-volume ratio (in different words, it's robust for its size). Several thousand carbon fibres are bundled along to create a tow, which can be utilized by itself or woven into a cloth.
Mechanical Properties:

- Tensile Strength: 600 MP
- Compressive Strength: 570 MPa
- Poisson's Ratio: 0.10
- Thermal Coefficient of Expansion : 2.15×10^{-6}.

4. Mild Steel

Mild steel additionally referred to as Plain steel or low carbon steel may be a variety of steel containing a tiny low proportion of carbon and it's robust and hard however not promptly tempered. Mild Steel is currently the foremost widespread and customary variety of steel thanks to its comparatively low price and it provides material properties that are acceptable for several applications. Mild Steel contains roughly 0.05–0.25% carbon that makes it malleable and ductile [2]. It is often used, for example as structural steel, where large quantities of steel are required. The mild steel density is about 7.85 g/cm^3 (7850 kg/m^3 or 0.284 lb/in^3) and the Young module is about 200 GPa (29,000,000 psi) [1].

The research method played an essential role in the liberal design stage. KANGAROO BOOTS concepts and key component aspects were built accordingly to make sure conceptual models can be constructed and quality components obtained.

3 Analysis for Manufacturing

The analysis of caliper or the pivot end and the strut member was done for total deformation under applied load, Von-Mises stress, Factor of safety, Strain analysis, fatigue life and fatigue factor of safety under cyclic loading on ANSYS 17.2 software and the results obtained were in the acceptable range and the components were safe under maximum load conditions with factor of safety greater than one for both the components. The fatigue life calculations were done under fully reversed constant amplitude loading. Carbon fibre was preferred for the manufacturing of the pivot end with an aim to make it light in weight and highly durable with high tensile and shear strength.

A. *Analysis of components*:

1. Caliper analysis:

The ANSYS results for total deformation and Von-Mises stress of the pivot end or caliper obtained were 1.59 mm and 229.3 MPa as pictured in Figs. 2 and 3 respectively. The equivalent stress was quite in the limits with maximum value of 600 MPa for carbon fibre. The factor of safety obtained was 2.68 which is greater than usual value one. The fatigue life was obtained as 10^8 cycles with fatigue factor of safety as 2.575 as indicated in Figs. 4 and 5. Thus, it is safe for maximum load and fluctuating load cycles. The weight after optimization of caliper design was obtained as 0.56 kg.

2. Strut Member analysis:

The ANSYS results for total deformation and Von- Mises stress of the strut member obtained were 0.78 mm and 101.92 MPa as shown in Figs. 6 and 7 respectively. The equivalent stress was quite in the limits with maximum value 330 MPa for aircraft-grade aluminium. The factor of safety obtained was 3.26, hence safe under maximum

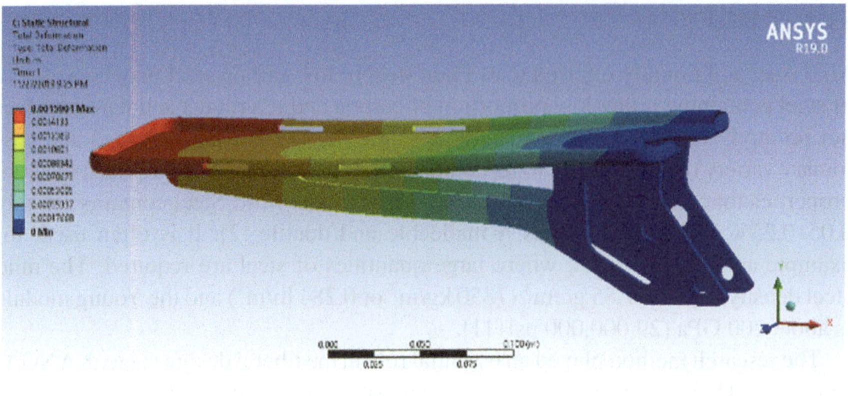

Fig. 2 Deformation in caliper

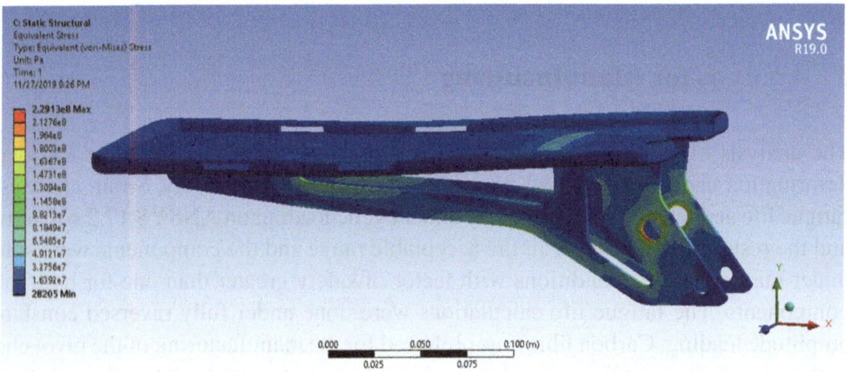

Fig. 3 Von-Mises equivalent stress

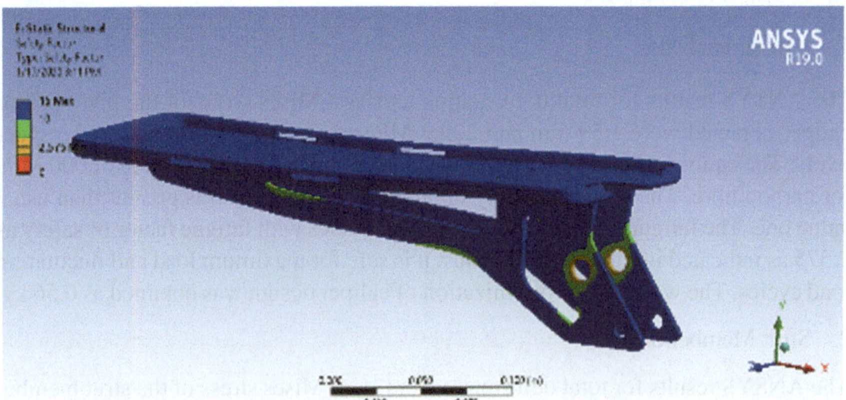

Fig. 4 Fatigue factor of safety

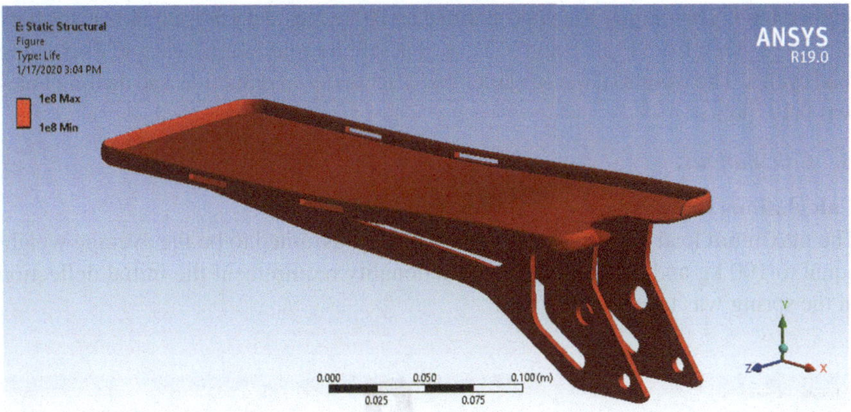

Fig. 5 Fatigue life- 10^8 cycles

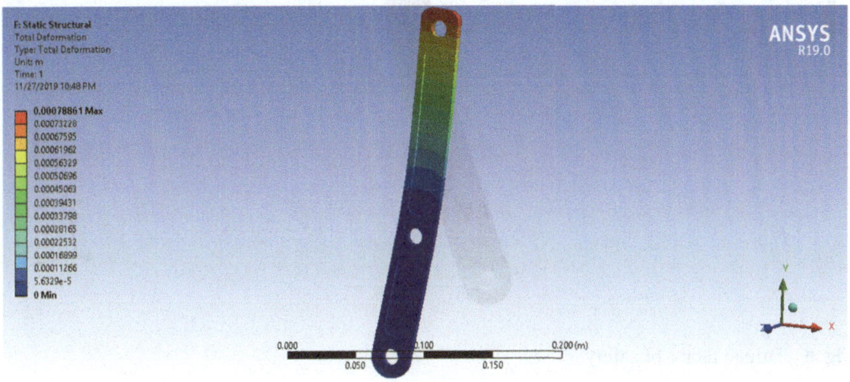

Fig. 6 Deformation of strut member

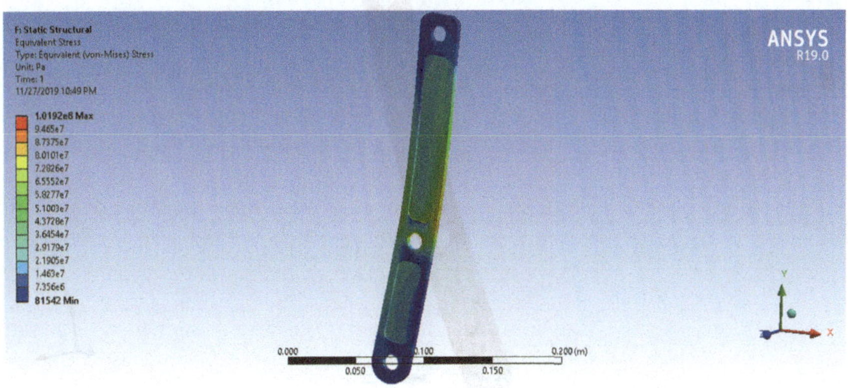

Fig. 7 Von-Mises equivalent stress

applied load. The fatigue life was obtained as 10^8 cycles with fatigue factor of safety as 5.788 as seen from Figs. 8 and 9. Thus, it is safe for maximum load and fluctuating load cycles. The weight of the strut member after material reduction and drilling holes was 0.197 kg.

B. *Calculations*:

Calculations for spring:
The maximum load on the spring was at first was assumed to be the average weight equal to 100 kg and considering the functionality requirement the initial deflection in the spring was taken as 150 mm.

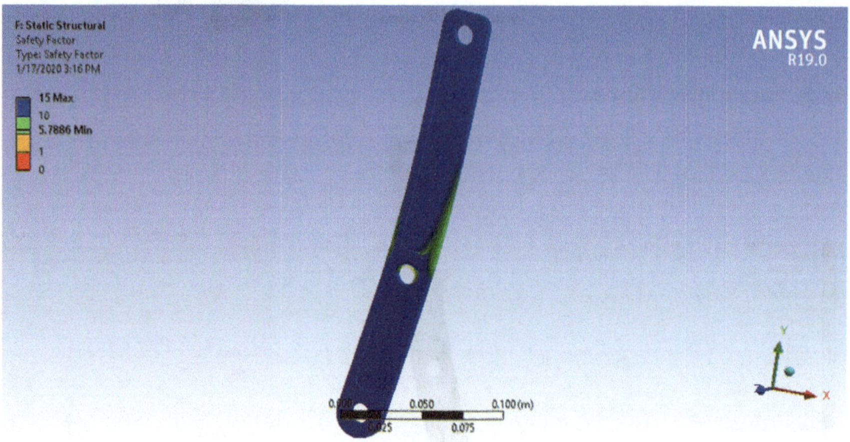

Fig. 8 Fatigue factor of safety

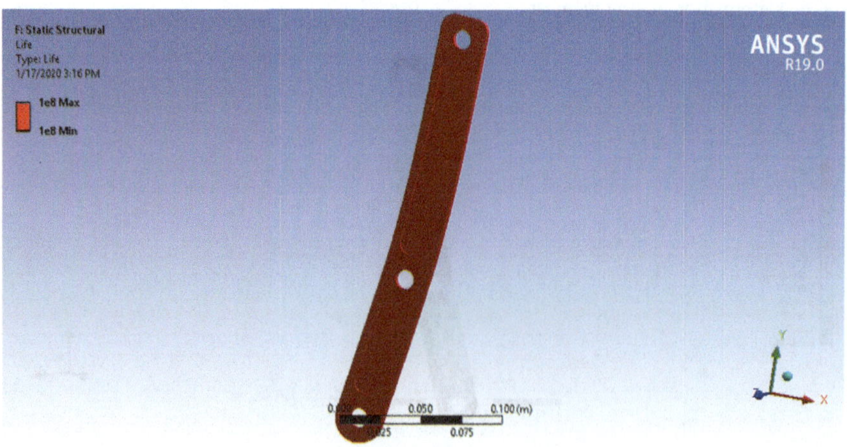

Fig. 9 Fatigue life- 10^8 cycles

Table 2 Indicates the upper and lower spring specifications

	Upper extension spring	Lower extension spring
Wire diameter	4.88 mm	1 mm
Mean diameter	38.01 mm	20 mm
Pitch	4.88 mm	1.05 mm
No. of revolutions	20	40

$$\begin{aligned}\text{Maximum weight} &= 100\ \text{kg} = 100 * 9.81\\ &= 981{-}1000\ \text{N}\end{aligned}$$

The strut member is 210 mm long and 1000 N load as calculated above is allowed to act on it. Thus, the bending force obtained on the strut member was calculated.

$$F_b = 703.64\ \text{N}\ (F_b = \text{Bending Force on Strut member})$$

Stiffness calculation (K):

Bending force on the strut member is equal to the spring force as per geometry [2].

$$K = F_b/\delta = 703.64/150 = 4.69\ \text{N/mm} = 4.7\ \text{N/mm}$$

where δ is the spring deflection.

Rubber pad supportive spring:

Selection of standard tension springm of stiffness greater than 4.7 N/mm (upper extension spring) for additional stability. Therefore, using two springs of stiffness 4.7 N/mm (Table 2).

C. *Working Mechanism*:

Since we realize that Newton's third law of motion says that, 'Every action has the same and opposite reaction'. Thus, following this principle, if a man runs on the hard and rugged (i.e. rigid) road with the usual foot, he must move on the road by pushing it and the road will also respond in the opposite direction with the same strength. But most energies are lost due to the rigidity of the lane. And in the 2nd phase, if a man runs on the rigid ground with the KANGAROO BOOTS, the man's effort is given to both the road then spring as well. The rigid path will behave as normal, but the spring has the property that it can store the energy and then unleash the energy to restore its original form. Therefore the energy given by the human being is first stored as potential energy in the spring and then, in order to regain its original position, it releases the energy as kinetic energy resulting in the generation of a sudden thrust that helps a man to run faster with a kangaroo boot than with the normal shoe [3]. When a man presses the rigid road with the bionic boot, the road reacts and the two springs are stretched after recovering their original position by relieving the pressure that will snap the lever back to its original position, propelling boot and man with more energy than a standard muscle could handle running faster. The spring rate can be taken as 1.3–1.6 N/mm of the weight of the human body [3, 4].

D. *Validation of Results*:

Thus, we have successfully developed an element through virtual and theoretical modeling which will increase human body speed and agility by 50–60% for a fraction of energy needs. The results were obtained with ANSYS 17.2 software by means of static structural analysis to obtain the respective deformation, Von-Mises stress, strain equivalent, safety factor, fatigue life, etc.

Initially, the strut was tested as the working material for mild steel but it soon proved strong and heavy for the necessary application. The mechanical characteristics and wearability of the material chosen have changed into Aluminium 6082T6.

The pivot end has been chosen as being of carbon fibre because it has a large amount of shear loads acting on it and carbon fibre has very high shear strength. It also decreases weight of the caliper by almost 60–65% than if manufactured using mild steel. This would make it more stable and provide the consumer as a whole with adequate protection. This was also validated in practical testing as large forces would not permit the use of any other material in such a place.

4 Conclusion

A new conceptual model of a kangaroo boot was designed, analyzed and studied which showed a large sequential scope for the future. Kangaroo boots have the ability to reduce time, use of oil and make flexibility more realistic. The model was tested for future improvements as compared to different mobility systems and the results compared to normal training. As the findings of the results were very good, offering almost 50 percent boost in length and energy of running.

For its reliable working span, the model is pretty stable and rigid. By improvising materials and manufacturing techniques, it is promising for better and comfortable applications such as amputees, armed forces and other future needs.

5 Future Scope

Future improvements to the kangaroo boots are still a lot to do in order to extend the distance and speed. An onboard electronic input management system to help manage energy and propulsion to achieve the most optimal synchronization of power output throughout the running cadence, thus achieving maximum efficiency and power output can be designed. 3D printing using titanium or carbon fibre, in particular, can be done–even a 20% reduction in weight might produce incredible results [1]. Considering the business model of the kangaroo boots, a pneumatic system like Festo's 'fluidic muscle' can be used and eliminating the use of springs. A GPS system could also be incorporated into the design of the kangaroo boots. This would enable to track the live location and the activities of the user. We see a design sight that involves a full-powered protective suit with onboard readings of speed, range, inputs from device control and more [1]. This bionic idea that in terms of mobility, security, performance and productivity could be groundbreaking for different military, sports and robotic applications.

References

1. Agarwal Rushabh K, Varghese Kevin O, Patil Pramod D, Nimbalkar PADK (2017) Design and analysis of bio-enhancing boots. Int Res J Eng Technol 4:331–339
2. Bhandari V (2010) Design of machine elements
3. Yadav AK, Kari V, Singh S, Chauhan V, Gupta A (2017) Research article fabrication and testing of bionic boot. 9:60317–60321
4. Ratan SS (2009) Theory of machines

Optimization of Brake Rotor Slotting Using Finite Element Analysis

Ujwal Sutaria, Vaibhav Shah, Karan Shah, Chaitanya Shah, and Vinayak H. Khatawate

Abstract The disc braking system in a vehicle is used for either retarding the motion of a vehicle or to bring it to a halt. Due to the application of brakes for number of times, there is a generation of heat, which may result in the failure of brake rotor. In this study, the thermal analysis of the brake rotor was done using Finite Element Analysis (FEA). The primary aim of the project was to compare the effect of different types of slotting on the rotor. The two designs of the rotors with different slotting were considered for analysis. The two designs were compared by their thermal values obtained from the thermal analysis by FEA and the best suitable design was selected for better performance disc brakes.

Keywords Brake rotor · Brake torque · Convection · Finite element analysis · Thermal flux

1 Introduction

In today's advancing automotive industry, there is an ever-growing competition for a better performance vehicle. The disc braking system is a system employed for decelerating the motion of a vehicle or bringing it to a stop. Various materials used for the rotor are typically cast iron or ceramic composites including silica, Kevlar, aluminum, and carbon which are connected to the wheel via its hub to stop the moving vehicle [1–3]. Brake pads, which face the brake disc, are composed of friction material which is forced either electromagnetically, pneumatically, hydraulically or mechanically against each of the faces of the rotor. Deceleration of a vehicle is due to this frictional force only. Basically, two types of braking methods are employed in a vehicle viz. regenerative braking system and friction braking system. In the latter case, friction is produced between two or more surfaces that slide against each other, to retard the motion. As per design configurations, friction braking system

U. Sutaria (✉) · V. Shah · K. Shah · C. Shah · V. H. Khatawate
Department of Mechanical Engineering, Dwarkadas J. Sanghvi College of Engineering, Vile Parle, Mumbai 400061, India
e-mail: sutariaujwal.us@gmail.com

© Springer Nature Singapore Pte Ltd. 2020

H. Vasudevan et al. (eds.), *Proceedings of International Conference on Intelligent Manufacturing and Automation*, Lecture Notes in Mechanical Engineering,
https://doi.org/10.1007/978-981-15-4485-9_78

is categorized into disc brakes and drum brakes. In the present work, modeling and analysis of disc brake are considered. Considerable amount of heat energy is generated upon repetitious braking, which needs to be dissipated for desired braking results. The rise in temperature of the brake components has a major impact on the efficiency of the braking system. Also, it may result in reduction in coefficient of friction, wear, braking system fade and thermal cracking.

While braking, thermal energy is generated at compensation of kinetic and potential energy of the moving vehicle because of friction between brake pads and rotor. This heat needs dispersion through the passage of the surrounding air. The heat transference occurs via the following three modes—conduction, convection and in some cases by radiation. In order to ensure sufficient cooling of rotor and brake pads, study of heat transference between them and the air medium needs to be done. The study of the thermal behavior of the disc braking system is also necessary so as to analyze temperature change while braking. In order to study the cooling potential of the system, a model representing convective heat transfer has been developed [4]. The rotors are distributed with slots in order to enlarge the contact area with ambient air and increase rate of heat transference [5–7]. This rate improves with the increase in slots in the rotor [8–10]. Since more area is open to the atmosphere, there is an increase in the rate of heat transference through convection as well as conduction. However, the strength of the rotor is compensated with a larger number of slots. In this paper, two different designs of rotor are studied for the rate of heat transference.

2 Background Theory

2.1 Braking System

It is a system that produces an artificial frictional force between moving members of a machine, so as to retard the motion of a machine or bring it to halt. In this process, the brakes soak up kinetic energy of the member in motion or the potential energy stored in the machine member. This absorbed energy has to be dispersed in the form of thermal energy into the surrounding atmosphere to slow down or bring the vehicle at halt and hence, the braking system has the following two requirements:

i. In case of any emergency, the brakes should be able to stop a vehicle within a specific minimum distance.
ii. The vehicle should not skid and also the driver should have sufficient control over the vehicle while braking.

2.2 *Heat Transference*

Whenever, there is a difference in temperature between a system and its surrounding, nature tries to reach thermal equilibrium. To achieve this, according to the second law of thermodynamics, heat energy always flows to a system at lower temperature from a system at a higher temperature. The heat can be transferred from one system to another by a combination of the following three modes—Conduction, Convection, and Radiation [4, 8].

2.2.1 Conduction

Conduction is the transference of thermal energy within a system or from a system at higher temperature to that at a lower temperature in physical contact. Conduction of heat can take place through all three mediums—solid, liquid, and gas. It occurs via collision of faster, energetic moving molecules at high temperatures with the slower ones at lower temperature. Heat conduction is governed by Fourier's law of heat conduction which states that the rate at which heat is conducted is directly proportional to the area normal to direction of heat flow and the thermal gradient, where thermal conductivity is the constant of proportionality.

$$q = -kA(\delta T/\delta x) \tag{1}$$

Here, q is the thermal flux normal to the surface of area A, [W]; A is the surface area across which the heat flows, [m^2]; k is the thermal conductivity, [$Wm^{-1}K^{-1}$]; x is the distance normal to the surface traveled by heat (or thermal) flux; T is the temperature [K or °C] [4].

2.2.2 Convection

In convection, heat is transferred by the movement of fluid, hence, it cannot take place in solids. This phenomenon basically consists of heat transfer between a hotter body and a colder fluid by conduction. This fluid is then replaced either by buoyant force (called free convection) or by external forces such as fans or blowers (called forced convection). This phenomenon is governed by Newton's law of cooling whose mathematical form is

$$q = -hA(T_\infty - T_s).$$

Here, h is the convective heat transfer coefficient [$Wm^{-2}K^{-1}$]; T_∞ is the cooling fluid temperature; and T_s is the body surface temperature [4].

2.2.3 Radiation

This is the transfer of heat in the form of electromagnetic waves which is emitted in the wavelength range of 0.1–100 μm, completely due to the temperature of the surface. A body at a higher temperature will emit thermal radiation towards a body at lower temperature. At the same time, the colder body will also emit radiations towards the hotter one. However, as the hotter body emits more than it receives, it will be cooled and the colder body will be heated. The equation that represents thermal radiation is given by the Stefan- Boltzmann law which is of the form:

$$q = -\sigma \varepsilon A\left(T_1^4 - T_2^4\right) \tag{3}$$

here ε is the emission coefficient; σ is the Stefan-Boltzmann constant of proportionality (5.669×10^{-8} [$\mathrm{Wm^{-2}K^{-4}}$]); A is the surface area that radiates heat; T_1 is the radiator's temperature, and T_2 is the ambient temperature.

2.3 *Factors Governing Braking*

There are four major factors that govern the brake power of the system. Three of the elements deal with friction generation: pressure, coefficient of friction, and frictional contact surface. The last element is the effect of this friction which is the dissipation of heat [11].

2.3.1 Pressure

The pressure that is exerted between the relatively moving surfaces largely influences the amount of frictional force produced between them. Usually, a hydraulic system is used to apply pressure in a braking system.

2.3.2 Coefficient of Friction (COF)

Coefficient of friction is an expression for the frictional force produced between two surfaces. The COF of the chosen rotor material SS420 provides reliable and safe braking [9].

2.3.3 Frictional Contact Surface

Frictional contact surface is the area that comes in contact with brake pads. In most of the cases, weight of the vehicle and its velocity decides the area of frictional contact [9].

Table 1 SS420 properties [14]

Properties	
Thermal conductivity ($Wm^{-1}K^{-1}$)	30
Density (Kgm^{-3})	7800
Specific heat capacity ($Jkg^{-1}K^{-1}$)	460
Thermal expansion ($10^{-6}/K$)	9
Elastic modulus (GPa)	210

2.3.4 Heat Dissipation

While designing the brake system, heat dissipation is one of the most crucial parameters that need to be considered. Generation of undue heat while continuous or hard braking may result in brake fading. Also, temperature of the pads may rise drastically, generating gas. Hence, friction material having characteristics of high heat dissipation should be selected to eliminate this generation of gas and prevent brake fading [7].

3 Methodology

The model of the brake rotor is made with the help of Solidworks [12] and its Finite Element Analysis is performed using ANSYS Workbench [13]. Material used of the rotor is SS420 and various thermal properties as shown in Table 1 are input for analysis purposes. The elements used for the mesh of the model are tetrahedral three- dimensional elements with eight nodes [14]. The various constants and material properties used in the calculations are as mentioned in Table 1.

Figures 1 and 2 represents the CAD models of brake rotors with two different slots: Fig. 1 consisting of circular slots and Fig. 2 consisting of cylindrical slots. Figures 3 and 4 are the meshed models of the two respective rotor models.

4 Design Calculations

The following assumptions are made for the analysis:

i. Braking torque is distributed between the front and rear wheels as 65:35 for analysis (due to weight ratio after weight transfer).
ii. Application of brakes is limited to the rear wheels only.
iii. Thermal loading is only considered for analysis and life of rotor isn't determined.
iv. Cooling of rotor is by natural convection only. There is no forced convection.

Fig. 1 Rotor model-1

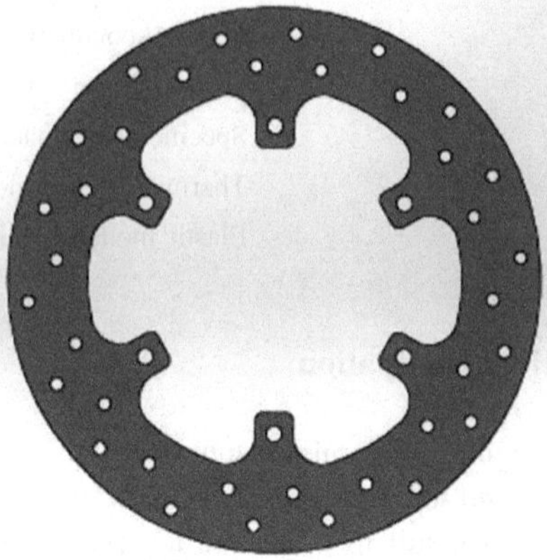

Fig. 2 Rotor model-2

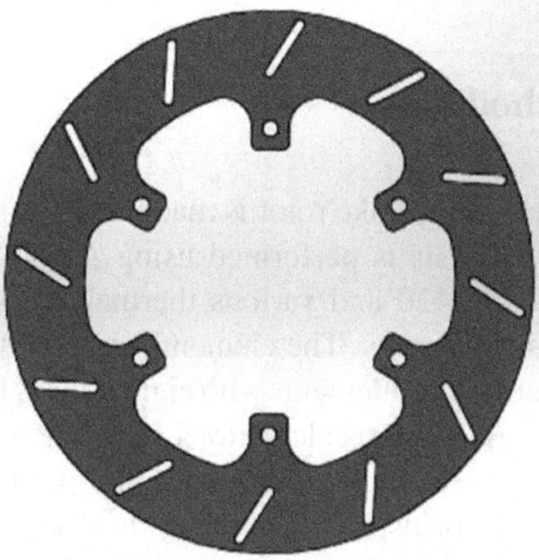

v. Loss of kinetic energy through heat loss between tires and road surface is ignored whereas the loss of energy via brake rotor is considered. The deceleration is uniform.
vi. The material used for both the rotor models is homogeneous in nature.
vii. Material properties like thermal conductivity are considered consistent throughout.
viii. The specific heat capacity of the material is consistent throughout and doesn't vary with change in temperature.
ix. Thermal flux is distributed only on both sides of the rotor of each front wheel.

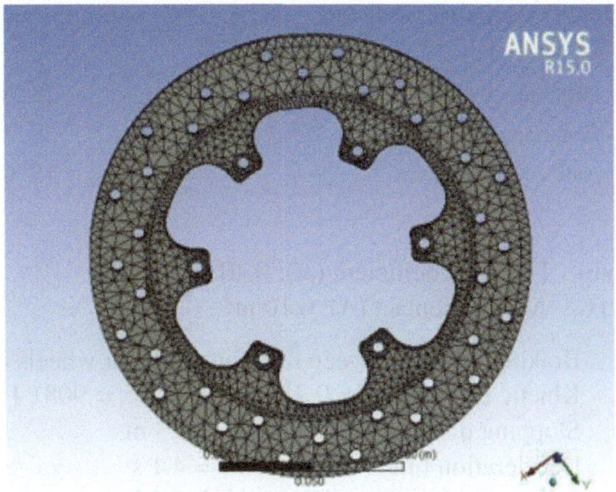

Fig. 3 Meshed rotor model-2

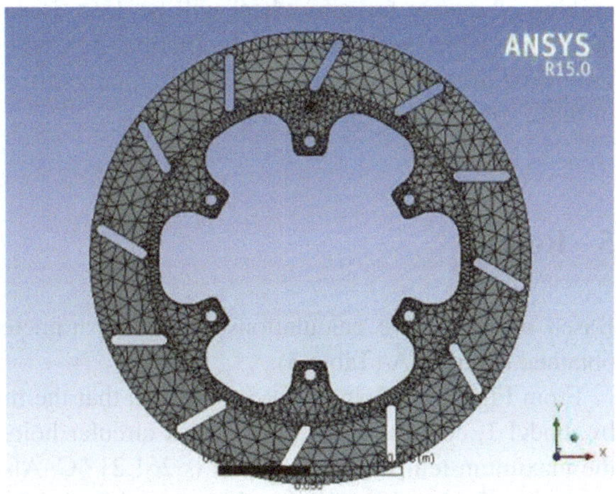

Fig. 4 Meshed rotor model-2

4.1 Calculations for Input Parameters

Heat flux for the two rotor models is calculated for a car advancing with an initial velocity of 27.78 ms^{-1} (100 kmph) (considering the maximum speed that the car can achieve). Further calculations are as shown below [11, 15].

i. Mass of vehicle (m): 150 kg
ii. Initial velocity of the vehicle (u): 100 km/h = 27.78 m/s
iii. Final velocity of the vehicle (v): 0 m/s
iv. Diameter of brake rotor: 0.23 m
v. Weight distributed on axle (γ): 35%
vi. Kinetic energy absorbed by rotor (f): 0.90 (or 90%)
vii. Acceleration due to gravity (g): 9.81 m/s^2

Table 2 Boundary conditions

Properties	
Heat Flux (0.35Pb/A)	22,202 W/m^2
Heat transfer coefficient for SS	28,584.5 W/m^{2}K
Ambient temperature	15 °C

viii. Friction coefficient (μ): 0.40
ix. Area of contact (A): 0.10 m^2

Braking torque between the front and rear wheels is distributed as 65:35.
Kinetic Energy (KE): $0.25 \times \gamma m(v - u)^2 = 9081$ J
Stopping distance (x): $u^2/2\mu g = 98.3$ m
Deceleration time (t): $(v - u)/a = 4.1$ s
Braking power (Pb): KE/t = 22,210.2 W

Since braking torque distribution between both the wheels is taken as 65:35, braking efficiency for rear wheel will be 35%. Because the application of brakes is limited to rear wheels alone, the thermal flux will be 0.35 × total thermal flux. Following are the boundary conditions employed for steady-state thermal analysis of brake rotor (Table 2).

5 Results

Based on the above calculations of input parameters, the following results are obtained using FEA (Table 3).

From Figs. 5, 6, 7 and 8, it can be seen that the maximum temperature attained by model 1, consisting of a number of circular holes, is 225.32 °C. For model-2, the maximum temperature attained is 261.21 °C. Also the maximum heat flux for model-1 is 2.61×10^5 W/m^2 and for model-2, it is 2.99×10^5 W/m^2.

Table 3 FEA results

Results	Temp distribution °C		Heat flux W/m^2	
	Min.	Max.	Min.	Max.
Model 1	72.11	225.32	503.14	2.61e5
Model 2	48.399	261.21	449.5	2.99e5

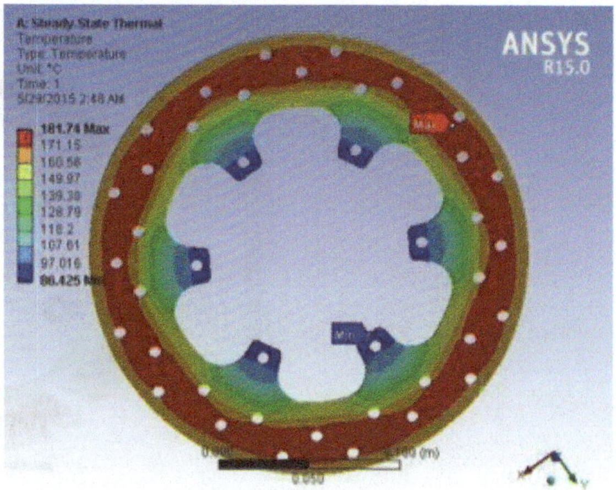

Fig. 5 Temperature distribution of rotor model-1

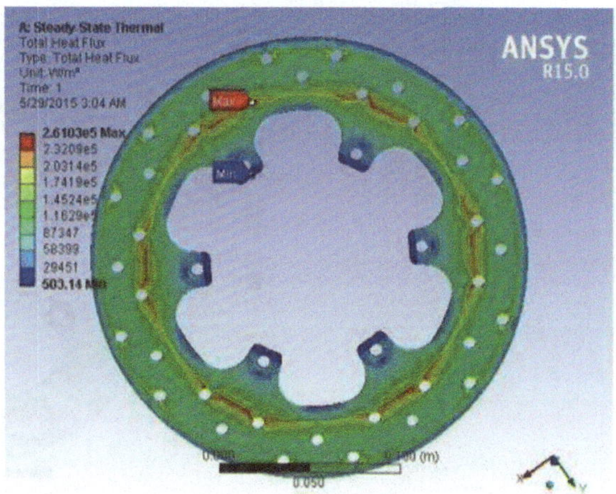

Fig. 6 Distribution of heat flux of rotor model-1

6 Conclusion

In this paper, temperature and heat flux distribution of two brake rotors having different cut patterns were studied in order to determine the most efficient one of the two. It can be concluded that, apart from the material of the rotor, dissipation of heat from brake rotor also depends on the design of cut pattern used. Among the two models made of stainless steel, model-1, consisting of a number of circular holes, showed better heat dissipation as compared to that of model-2 consisting cylindrical cut pattern.

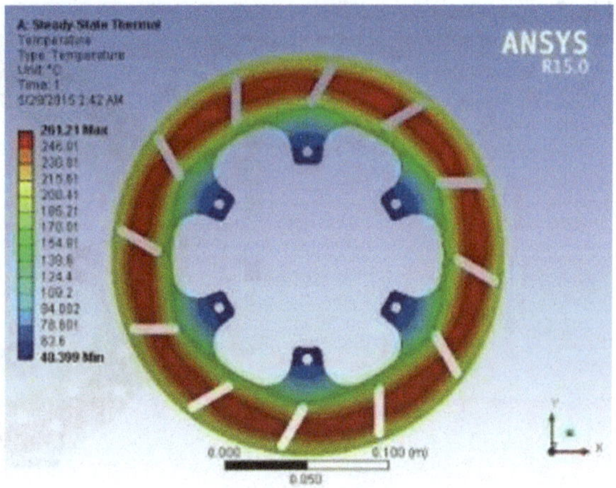

Fig. 7 Temperature distribution of rotor model-2

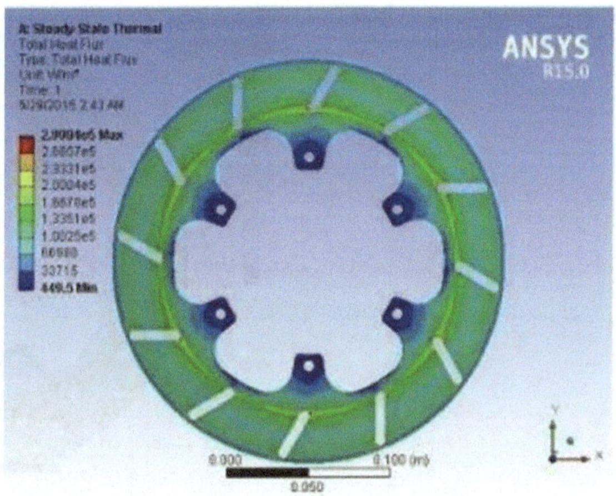

Fig. 8 Distribution of heat flux of rotor model-2

References

1. Shaik AFB, Srinivas CL (2012) Structural and thermal analysis of disc brake with and without cross drilled rotor of race car. Int J Adv Eng Res Stud 1:39–43
2. Sarip S (2013) Design development of lightweight disc brake for regenerative braking and finite element analysis. Int J Appl Phys Math 3:52–58
3. Nathil GM, Charyulu TN (2012) Coupled structural/ thermal analysis of disc brake. IJRET 2:539–553
4. Hasan MI (2011) Influence of wall axial heat conduction on the forced convection heat transfer in rectangular channels. Basrah J Eng Sci 1:31–43
5. Reddy VC, Reddy MG (2013) Modeling and analysis of FSAE car disc brake using FEM. Int J Emerg Technol Adv Eng 3:383–389

6. Kuciej M, Grzes P (2011) The comparable analysis of temperature distributions assessment in disc brake obtained using analytical method and FE model. J Kones Powertrain Transp 18:236–250
7. Papinniemi A, Lai JCS (2007) Disc brake squeal, progress and challenges. ICSV14, Australia, pp 1–8
8. Talati F, Jalalifar S (2008) Investigation of heat transfer phenomenon in a ventilated disk brake rotor with straight radial vanes. J Appl Sci 8:3583–3592
9. Ali B, Mostefa B (2011) Thermal behavior of dry contacts in the brake discs. Int J Automot Eng 3:9–17
10. Mahmod MI, Munisamy KM (2011) Experimental analysis of ventilated brake disc with different blade configuration. Dept Mech Eng 1:1–9
11. Limpert R (1992) Brake design and safety. Soc Automot Eng. Warrandale, Inc, 2nd edn, USA, pp 11–157
12. SOLIDWORKS (2015) Dassault systèmes
13. ANSYS Workbench, Version 15.0, ANSYS Inc
14. Babukanth G, Teja MV (2012) Transient analysis of disk brake by using ANSYS software. Int J Mech Ind Eng 2:21–25
15. Spulber C, Voloaca S, Aspects regarding the disc brake's thermal stress simulation by using Infrared Thermography. In: International conference on optimization of the robots and manipulators romania, ISBN 978, 26-2011

Evaluation of Piping Isometric Drawings Using Six Sigma Process

Abrar Khulli, Prasad Shirodkar, Vijaya Kumar N. Kottur, and Rajendra Khavekar

Abstract Six Sigma is a business strategy that helps organizations to improve its organizational efficiency and customer satisfaction; it decreases operating cost and increases profits. Six Sigma is a widely used method to improve processes from various industrial sectors. The target failure rate for Six Sigma projects is 3.4 parts per million. Numerous practitioner studies claim that Six Sigma improves organizational performance. However, its implementation to design process is limited. So, in this study, we have implemented the Six Sigma philosophy in piping design process. Piping isometric drawings, which feature their intrinsically topological relation rather than just geometrical shape, are important industrial artworks in the field of Computer-Aided Design (CAD). This study suggests that the link between Six Sigma and piping design process performance can be explained and developed by integrating organizational knowledge creation processes. We have quantified the performance of a design process in terms of sigma level and tried to find the possible ways to improve it. The method used in this study is based on five main steps summarized in the acronym (DMAIC): Define Measure, Analyze, Improve and Control. Application of the method on the design process during the five phases of the method will help to reduce costs and losses of man-hours to strive for optimum results in terms of profit and quality.

Keywords Piping engineering · Six sigma · Piping isometrics drawing · DMAIC philosophy · Quality · Design · Computer-Aided design

A. Khulli (✉) · P. Shirodkar · V. K. N. Kottur
Department of Mechanical Engineering, Dwarkadas J. Sanghvi College of Engineering, Vile Parle, Mumbai, India
e-mail: abrarkhulli@gmail.com

R. Khavekar
Department of Production Engineering, Dwarkadas J. Sanghvi College of Engineering, Vile Parle, Mumbai, India

© Springer Nature Singapore Pte Ltd. 2020
H. Vasudevan et al. (eds.), *Proceedings of International Conference on Intelligent Manufacturing and Automation*, Lecture Notes in Mechanical Engineering,
https://doi.org/10.1007/978-981-15-4485-9_79

1 Introduction

Piping engineering is the major branch in Plant Design. It plays a vital role in design of Process plants, Petrochemical Refineries, and Chemical Plants. The global engineering cost of these plants runs into billions of dollars. The cost of piping engineering in these plants is about 3–5% of the total engineering cost. Any Savings in this cost can run into millions of dollars. The improvement in piping design will not only save the cost but also save the time and effort required to complete the project. In this study, it has been tried to improve the quality of piping design by using Six Sigma techniques [1]. Six Sigma is widely recognized as a business strategy that employs statistical and non-statistical tools and techniques, change management tools, project management skills, teamwork skills, and a powerful roadmap DMAIC to maximize an organization's return on investment through the elimination of defects in processes [2]. The approach adopted here is Define, Measure, Analyze, Improve, Control (DMAIC) philosophy [3, 4] to improve the quality of the deliverables [5]. Each step of DMAIC philosophy is used to systematically improve the piping isometric generation in the project. The study involves using actual data from the ongoing project and measuring the current quality of the process in terms of sigma levels. As of now, there is no technique available to quantify the quality of piping design. This paper gives a method to assess the quality of design and improve it by selecting the best practices [6, 7].

2 Research Methodology

Six Sigma is the technique, which gives us a measure of the performance of even routine processes that come across in our daily life [8]. Six Sigma is quantitative defect measurement and analysis of the process, which further gives the remedial action to correct the process and achieve the quality target [9].

2.1 DMAIC Philosophy Application on Piping Isometric Drawing

2.1.1 Design

The aim of the study was to improve the piping design process by minimizing the time and efforts required to produce piping Isometrics. It was observed that in project a lot of rework is done during detail engineering phase and later at construction site because of changes made in piping Isometrics at later stages of the project. Generally, in all projects, the Isometrics are being revised three times before final releasing the final copy to the site. Any changes in the final drawing also increase the work of updating the Isometrics, which increases the complexity. The duration of the revision should be minimum to reduce the changes at the site. The process starts with the inputs P&IDs, Line list, and Equipment layout. The Area Engineer develops piping routing, decides equipment layout and decides requirements of civil structure. The stress analysis of critical piping is done and loading data is provided to the civil engineer. General Arrangement drawings created and supporting are done by area engineer and this input is given to the 3D modeler. A modeler creates a 3D design model of the plant, which is true scale model. The piping Isometrics are extracted from this model.

Piping isometric drawing is 2D CAD drawings with detailed orthographic drawings for plants in the petrochemical industry. These plants usually consist of a large number of piping, which is comprised of various components. Piping isometric drawings represent the details of 3D structure of these piping in the form of 2D diagrams. Hundreds of such drawings are required to create a design project. The sigma level of the project is calculated by the count of the number of defects in the Isometrics, which must be reworked. For the quality check of Isometrics, the Isometric checklist and department SOPs are considered. The number of observations is the total number of points in the Isometric into no. of Isometrics. The DPMO is calculated by number of defeats captured at site in form of query log.

The terminology defective or non-defective is being often used to identify these two classifications [10, 11]. Quality characteristics of this type are called attributes. The control chart that is studied is called the control chart for nonconformities, or the C chart Figs. 1 and 2. The chart selection criteria are given in Table 1. A nonconforming item is a unit of error that does not satisfy one or more of the specifications. Each specific point at which a specification is not satisfied results in a defect or nonconformity. Consequently, a nonconforming item will contain at least one nonconformity. However, depending on their nature and severity, it is quite possible for a unit to contain several nonconformities and not be classified as nonconforming the inspection unit must be the same for each sample [12]. That is, each inspection unit must always represent an identical area of opportunity for the occurrence of nonconformities [13]. In addition, one can count nonconformities of several different types on one unit, as long as the above conditions are satisfied

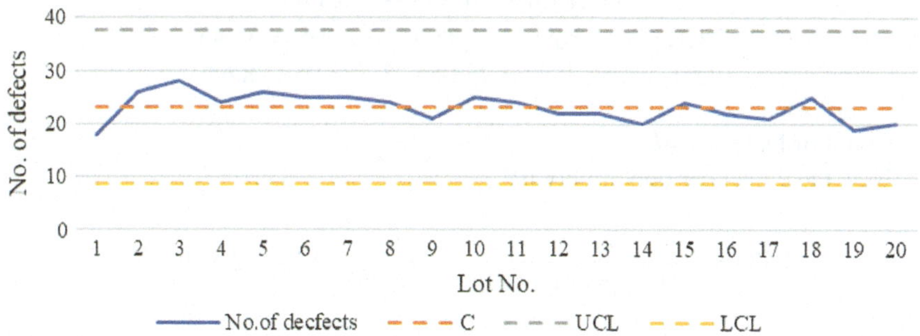

Fig. 1 C-chart for project-1

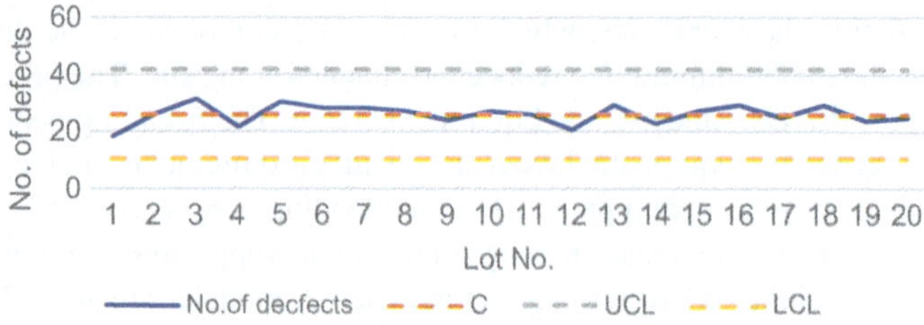

Fig. 2 C-chart for project-2

Table 1 Control chart selection criteria

Lot size	Defects	Defectives
Constant	C chart	NP chart
Variable	U chart	P chart

for each class of nonconformity [14]. Each error in the checking of the Isometrics considered as a defect. The isometrics are divided into 20 equal groups for the analysis of total isometrics in the system.

2.1.2 Measure

A completed project was selected to calculate the sigma level and define the current quality level of the piping design process. The checklist used for the quality check in this project has 72 number of points to checked in single Isometrics. The total number of sheets of Isometrics in the projects were 4305. The calculation of six sigma of the project are as follows:

The opportunities related to Isometrics = 72 × 4305 = 309,960

Now the defects in this Isometrics are revisions of ISOs, the release of Revision Isometrics and mark-up released to site which is divided into groups as shown in Table 2.

$$\text{Centre Line} = c = \tfrac{461}{20} = 23.05$$
$$\text{UCL} = c + 3\sqrt{c} = 23.05 + 3\sqrt{23.05} = 37.45$$
$$\text{LCL} = c - 3\sqrt{c} = 23.05 - 3\sqrt{23.05} = 8.64$$

Count of Defect = 461

Now the defect per opportunity is:

$$\text{DPO} = \frac{\text{Defects}}{\text{Total opportunities}} = \frac{461}{309{,}960} = 1.4872 \times 10^{-03}$$

So, Defects per million Opportunities = $1.4872 \times 10^{-03} \times 10^{6} = 1487.09$

Table 2 Sample lot defects details for project-1

Lot No.	No. of isometrics	No. of defects
1	200	18
2	200	26
3	200	28
4	200	24
5	200	26
6	200	25
7	200	25
8	200	24
9	200	21
10	200	25
11	200	24
12	200	22
13	200	22
14	200	20
15	200	24
16	200	22
17	200	21
18	200	25
19	200	19
20	200	20
Total	4000	461

$$\mathbf{DPMO = 1487}$$

$$P\text{ Value} = \frac{\text{DPMO}}{1{,}000{,}000} = \frac{1487}{1{,}000{,}000} = 0.001487$$

$$\begin{aligned}\text{Sigma Level} &= \text{Normsinv}(1 - \text{P}) + 1.5 \\ &= \text{Normsinv}(1 - 0.001487) + 1.5 = 4.47\end{aligned}$$

The current sigma level for project comes to be **4.47**

$$\text{Process capability index (Cpk)} = \frac{\text{Sigma Level}}{3} = \frac{4.47}{3} = 1.49$$

The process capability for project is **1.49**

$$\text{Process Yield} = \frac{(1 - \text{P})}{100}\% = \frac{(1 - 0.001487)}{100} = 99.85\%$$

The current process can produce Isometrics with **99.85%** accuracy.

2.1.3 Improve

The logs of the project were studied carefully and the defects were grouped in five categories. For each defective or mismatching Isometric, the company maintains the log of error. The logs contain information about the defect and the corrective action required. It also has the status of corrective action and person responsible. Most of the defects can be broadly identified as human errors. These errors were in terms of data entry, reading inputs for 3D model and other documents, misinterpretation, etc. One major reason affecting the quantum of this type of error is individual, who is checking the Isometrics. The more experienced person trends to make less error. The human errors were simple in complexity and can be corrected by the automation of process. It consists of almost 60% of errors in total.

The second biggest reason for the errors in piping Isometrics is that the points to be checked are not covered in the Isometrics or do not check the actual correctness of the activity. There are few Isometrics, which required certain special points to be checked like in terms of specialty items. If the checking does not cover this point it is even unnoticed and might result in mismatch. This type of error increases with an increase in the complexity of the project. Not referring to the latest documents causes the errors in checking the Isometrics. Sometimes the last-minute changes cause misunderstanding and lead to defective Isometrics. A simple system was derived to check the changes and made easily available to the team with a simple dashboard. The frequent changes should be avoided and proper revision summary must be maintained.

2.1.4 Analysis

To improve the current process study was conducted to list the causes of defects by studying the Isometric logs maintained by the department. These defects were analyzed with the help of Fishbone diagram and Pareto analysis. The few common reasons were found with the study of root cause diagram and Pareto chart. The major reasons for defects in piping Isometrics were:

- Changes in Inputs—This change was due to changes in the information provided in Inputs like line size change, additional/Deletion of line, change in the process conditions.
- Change in line list—The main reason for the changes in line list are change in line design and operating conditions like Temperature, Pressure and fluid viscosity.
- Change in equipment layout—Changes in the Isometrics are because of changes in the layout. There can be many reasons for layout change like accessibility, HAZOP and Safety study.

- Change in piping layout—As in the design process, multiple options are available to route the pipes. The piping layout changes when the most optimum design is selected.
- Inputs required not received on time—The inputs required to produce the Isometrics are delayed due to reasons like delay in Equipment GAs, Vendor items details, Interdisciplinary information delay, etc.
- Change in client specifications—Any change in the specification leads to major changes in all the respective Isometrics Changes in the assumption made.

2.1.5 Control

The ideas generated in the improvement phase were applied and a regular check of the quality level of the project was done. The checklist used for the quality check in this project had 72 no. of points to checked-in single Isometrics. The total number of sheets of Isometrics in the projects was 10,346. The opportunities related to Isometrics was 744,912. Now the defects in this Isometrics were revisions of Iso. Release of revision Isometrics and mark-up released to the site are segregated in Table 3.

$$\text{Centre Line} = c = \tfrac{531}{20} = 26.4$$
$$\text{UCL} = c + 3\sqrt{c} = 26.4 + 3\sqrt{26.4} = 41.81$$
$$\text{LCL} = c - 3\sqrt{c} = 26.4 - 3\sqrt{26.4} = 10.98$$

Count of Defect = 531
Now the defect per opportunity is

$$\text{DPO} = \frac{\text{Defects}}{\text{Opportunities}} = \frac{531}{744{,}912} = 7.12 \times 10^{-04}$$

So, Defects per million Opportunities = $7.127822 \times 10^{-04} \times 10^6 = 712.78$
DPMO = 712

$$P\ \text{Value} = \frac{\text{DPMO}}{1{,}000{,}000} = \frac{712}{1{,}000{,}000} = 0.000712$$

$$\begin{aligned}\text{Sigma Level} &= \text{Normsinv}(1 - \text{P}) + 1.5\\ &= \text{Normsinv}(1 - 0.000712) + 1.5 = 4.69\end{aligned}$$

The current sigma level for project comes to be **4.69**

$$\text{Process capability index (Cpk)} = \frac{\text{Sigma Level}}{3} = \frac{4.69}{3} = 1.56$$

The process capability for project is **1.56**

Table 3 Sample lot defects details for project-2

Lot No.	No. of isometrics	No. of defects
1	500	20
2	500	26
3	500	32
4	500	22
5	500	31
6	500	29
7	500	29
8	500	28
9	500	24
10	500	28
11	500	26
12	500	22
13	500	30
14	500	24
15	500	28
16	500	30
17	500	25
18	500	30
19	500	24
20	500	25
Total	10,000	531

$$\text{Process Yield} = \frac{(1 - \text{P})}{100} = \frac{(1 - 0.000712)}{100} = 99.93\%$$

The current process can produce Isometrics with 99.93% accuracy.

The suggestion generated in this study was implemented and the sigma level was observed to be better than the past project. Major errors have occurred due to mistakes in input of data into 3D model.

3 Conclusion

Most important finding of this exercise is that almost care must be taken at the modeling stage itself. Each parameter should be checked thoroughly. It was still observed that piping Isometrics issued to the site has some defects, which further leads to rework and delay in project and cost escalates. After the implementation of the above point, the sigma levels have increased by 0.22 and Process capability by 0.08%.

The topmost priority was given to correctness of plant 3D model to ensure the defect-free Isometrics. A comprehensive checklist should be prepared, which checks each and every aspect of Piping Isometrics for correctness. Frequently changes should be avoided especially at the later stages of the project to reduce the defects in Isometrics and avoid rework at the site. The same approach can be used to improve the quality of other deliverables like Stress report, Bill of Quantity (BOQ), etc. The DMAIC analysis improves the whole process with systematic study of each activity. It strives to reduce the defects in the Piping Isometrics thereby reducing the efforts and time required for the rework. It has an impact on the cost of the project and precious efforts and also saving in the material.

References

1. Narula V, Grover S (2015) Six Sigma: literature review and implications for future research. Int J Ind Eng Prod Res 26(1):13–26
2. Dowaisana TA, Nourelfathb M, Hassan J (2015) Six Sigma performance for non-normal processes. Eur J Oper Res 24(7):968–977
3. Ganguly K (2012) Improvement process for rolling mill through the Dmaic Six Sigma approach. Int J Qual Res UDK- 378.014.3(497.11) short scientific paper (1.03)
4. Subramaniyam P, Srinivasan K, Prabaharan M (2011) An innovative Lean Six Sigma approach for engineering design. Int J Innov Manag Technol 2(2):166–170
5. Sin AB, Zailani S, Iranmanesh M, Ramayah T (2015) Structural equation modelling on knowledge creation in Six Sigma DMAIC project and its impact on organizational performance. Int J Production Econ 168:105–117
6. Toussouf A, Rachid C, Ion V (2014) Contribution to the optimization of strategy of maintenance by Lean Six Sigma. Phys Procedia 55:512–518
7. Erdogen A, Canatan H (2015) Literature search consisting of the areas of Six Sigma's usage. Procedia Soc Behav Sci 195:695–704
8. Nakhai B (2009) The challenges of six sigma in improving service quality. Int J Qual Reliab Manag 26(7):663–684
9. Erbiyika H, Saru M (2015) Six Sigma implementations in supply chain. Procedia Soc Behav Sci 195:2556–2565
10. Gupta N (2013) An application of DMAIC methodology for increasing the yarn quality in textile industry. IOSR J Mech Civil Eng (IOSR-JMCE) 6(1):50–65. e-ISSN: 2278-1684
11. Dabhole S, Raje A, Khandave S, Ghule P, Shinde A (2017) Review on implementation of Six Sigma for controlling defects. Int J Res Appl Sci Eng Technol (IJRASET) 5(11):1874–1879
12. Dambhare S, Aphale S, Kakade K, Thote T, Borade A (2013) Productivity improvement of a special purpose machine using DMAIC principles: a case study. Int J Innov Manag Technol 2(2):1–13
13. Valles A, Sanchez J, Noriega S, Nuñez BG (2009) Implementation of Six Sigma in a manufacturing process: a case study. Int J Ind Eng 16(3):171–181
14. Kabir ME, Boby SMMI, Lutfi M (2013) Productivity improvement by using Six-Sigma. Int J Eng Technol 3(12):1056–1086

Design of an Accumulator Container for a Formula Student Electric Race Car

Harshal Mehta, Ratan Soni, Parvez Shaikh, Raunak Bhanushali, Jobin Abraham, and Dhaval Birajdar

Abstract Growing concerns over carbon emission have lately sparked worldwide interest in more efficient and cleaner transportation systems. Industrialization is increasing exponentially, making energy resources the primary concern of every nation of the globe. Non-renewable sources are on the verge of being extinct, and carbon emissions are becoming a trigger for global extinction. One such effort of reducing the concern is a shift of the automobile industry from a combustion powertrain to the electric counterpart. Challenges accompanied by the electric powertrain include humongous size, reduced safety and sophisticated electronics involved in the electric powertrain relative to the combustion powertrain for the same amount of energy. This paper explores the use of electric powertrain, particularly the battery pack and the accumulator of a Formula Student electric race car. It has also highlighted the intricate design of an accumulator container for a Formula Student electric car.

Keywords Formula student electric car · Electric · Battery pack · FSAE and accumulator

1 Introduction

An accumulator container of a Formula Student electric race car is a component that houses batteries which power the drivetrain. The drivetrain generally consists of motors and the transmission components.

Several constraints impose boundaries on the accumulator design. These constraints come from the FSAE competition rules provided by the authorities, performance goals and integration of the container with the rest of the car. Therefore, before beginning the design, it is of considerable significance to identify these constraints so that the further designing process flows smoothly. The designing process

H. Mehta (✉) · R. Soni · P. Shaikh · R. Bhanushali · J. Abraham · D. Birajdar
Department of Mechanical Engineering, Dwarkadas J. Sanghvi College of Engineering, Mumbai, India
e-mail: harshalwork157@gmail.com

© Springer Nature Singapore Pte Ltd. 2020
H. Vasudevan et al. (eds.), *Proceedings of International Conference on Intelligent Manufacturing and Automation*, Lecture Notes in Mechanical Engineering,
https://doi.org/10.1007/978-981-15-4485-9_80

of an accumulator is iterative and prolonged. To meet, the design requirements and competition regulations, the accumulator container is redesigned.

The accumulator container houses high voltage (HV) batteries and various electronic components. Various rules are being imposed by the competition authorities to ensure the safety of the vehicle. The rulebook [1] is issued by the authorities every year, and every participating team has to follow the rules to compete in the dynamic events of the competition. Every car goes through rigorous technical inspection, and therefore it is of paramount importance to follow every rule listed in the rulebook.

2 Design Process

2.1 Motor Selection

The first step in the design procedure is to select the motors. Motor selection is a critical step since for every change in motor specifications, the powertrain and drivetrain changes accordingly.

Motors are selected based on the torque required on wheels to achieve the performance goals for the season. Current rating is a crucial aspect of motor selection since with the increase in current, the efficiency of the powertrain system decreases. Weight of the motor and its type have a profound impact on the performance of the vehicle. Apart from the technical criteria, availability, logistics and cost also affect motor selection.

Hence two Brushed DC Saietta Agni 119 R motors are used (Table 1).

Table 1 The specifications of the selected motor [2]

Specifications	Single motor	Dual motors connected in series
Voltage	72 V	144 V
Continuous current	200 A	200 A
Peak current	400 A	400 A
Speed	3751 rpm	3751 rpm
Peak torque	68.7 Nm	137.4 Nm
Weight	11 kg	22 kg

Fig. 1 A123 pouch cell

2.2 *Battery Pack*

2.2.1 Cells

Cells form the fundamental part of the accumulator container. They are selected based on their geometry, chemistry and technical specification.

According to geometry, there are two types of cells, cylindrical and pouch cells. Cylindrical cells have their electrodes in the form of a cylinder, and the terminals are at the circular ends of the cell. In pouch cells, the terminals lay on the same edge of the cell, thus forming a flat planar rectangular shape. The required electrical aspects such as nominal voltage, nominal current and peak current are derived based on the motor selected. Another critical specification to be taken care of is the operating temperature. According to the FS Rule *EV 5.8.4* [1], the maximum operating temperature of the cells in the accumulator container is supposed to be below 60 °C or the limit stated in the cell datasheet, whichever is lower. Therefore, it was required that the maximum operating range of the cell is below 60 °C for both charging and discharging or else it would add up complexity to the cooling design of the accumulator.

After thorough discussions and analysis, the team decided to proceed with A123 Pouch Cells, which satisfied all technical criteria listed below [3, 4] (Fig. 1 and Table 2).

2.2.2 Energy Calculations of the Battery Pack

To determine the energy consumed by the motors, the simulation was carried out on Optimum Lap from Optimum G. The vehicle was stimulated on the tracks of FSAE autocross Germany 2012; FSAE endurance Germany 2012 obtained from

Table 2 The Specification of the selected cell [5]

Specifications	Values
Chemistry	Nano phosphate
Nominal capacity	20 Ah
Voltage range	2.0–3.6 V
Nominal voltage	3.3 V
Maximum discharge continuous current	200 A
Pulse 10 s discharge current	600 A
Operating temperature range	−30 to +60 °C
Weight	495 gm

Table 3 Timings of various FSAE events on the formula student germany track

Track	Energy spent per lap (in kJ)	No. of laps	Total energy spent (in kJ)	Lap time (in a sec)
FSAE autocross	840.27	4	3361.08	79.58
FSAE endurance	1037.22	17	17,632.74	90.59
FSAE acceleration	45.15	4	180.6	3.62
FSAE Skid-pad	144	16	2304	5

the Optimum Lap database provided the data regarding the simulation. The vehicle was also simulated on a straight strip of 75 m, which acts as the track of FSAE Acceleration event. Since the acceleration, autocross and skid-pad events take place on the same day and the endurance event on another, the total energy of the battery pack must be such that all the events on a particular day can be completed without the need to charge the accumulator between the events (Table 3).

Since,

$$\begin{aligned} &\text{Req. Energy of Battery Pack} = \max\left[E(\text{accln} + \text{skidpad} + \text{autox}), E(\text{endu})\right] \\ &\therefore \text{Req. Energy of Battery Pack} = 17{,}632.74\ \text{kJ} \end{aligned} \tag{1}$$

Let,

Vm—Voltage across motors
s—number of cells in series
Vn—nominal voltage of a cell
Vd—voltage drop across motor and accumulator
Ah—current capacity of a single cell
p—no. of cells in parallel
TEc—Energy consumed by motors in the event
Ic—continuous current rating of the motor
t—lap time

$$\therefore Vm + Vd = s * Vn \tag{2}$$

From (2) we obtain no. of cells in series (s)

$$TEc = \text{Req. Energy of battery pack} \tag{3}$$

$$\therefore TEc = s * Vn * p * Ah * 3600 \tag{4}$$

Since the software does not consider heat losses,

$$TEc_2 = \text{Eperlap} * \text{total laps} + \text{Heatlosses} \tag{5}$$

However,

$$\text{Heatlosses} = I^2 * R * t \tag{6}$$

$$\therefore \text{Heatlosses} = Ic^2 * R_{\text{pack}} * t * \text{total laps} \tag{7}$$

$$\therefore TEc_2 = \left(\text{Eperlap} + Ic^2 * R_{\text{pack}} * t\right) * \text{total laps} \tag{8}$$

To proceed further, we need to compare Eq. (4) with Eq. (8).
$\therefore$ Proceed further if, $TEc > TEc_2$.

2.2.3 Battery Configuration

From Table 3, the battery pack consists of 90 cells with an overall configuration of 45s2p. The battery pack was divided into segments and modules. After considering the Formula Student rules, the battery pack was divided into five segments with each segment having an equal number of cells. The segment was further divided into what we call modules to simplify the connections. Thus, each segment consists of 9 modules and 18 cells. The battery configurations follow the *EV 5.3.2* rule [1].

Single Module Configuration—1s2p. Single Segment Configuration—9s2p.

Weight of a single segment—10 kg approximately.

2.2.4 Segment Design

The segments were made by connecting nine modules in series with each other. They were geometrically designed in two shapes, Square and L shape. To provide cushioning among the cells in dynamic conditions and to ease assembly, Neoprene, a synthetic rubber polymer is used. It was selected on the basis of its fire-resistant properties and compressibility.

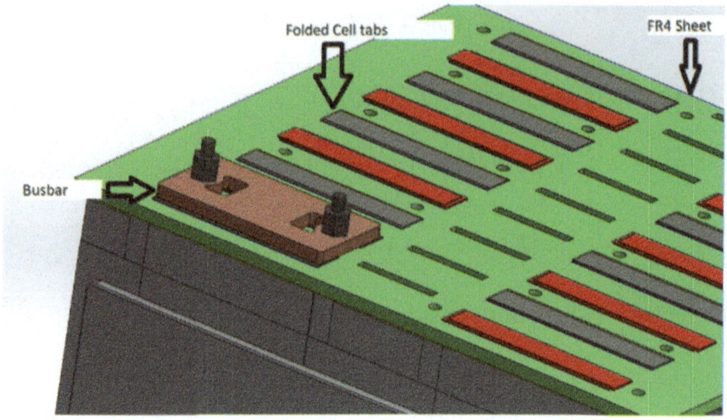

Fig. 2 Detailed view of a segment

The cells are supported on a glass-reinforced epoxy laminate material called FR4. To form a module by connecting two cells in parallel, the cell tabs (*Refer* Fig. 9) of the pouch cells are folded upon each other on an FR4 sheet thus making physical contact. To ensure the connection prevails even during dynamic conditions, bus bars are used, which further presses the cell tabs from above in a way, which forms a sandwich layer of cell tabs between the FR4 sheet and bus bars. Allen bolts were bonded to the FR4 by using UL94 V0 Araldite bond so that the assembly remains compact by eliminating the space required to insert a tool (Fig. 2).

The cells were mounted and connected with each other in series with the help of copper bus bars. Copper is selected due to its superior electrical conductivity. Bus bars were also designed in three shapes to realize various requirements in design.

Box Bus bar was used for intra segment series connections between two modules of parallel-connected cells. Box busbars were made of 3 mm thick copper sheet and were fastened to FR4 sheet using two pairs of M4 Allen bolts and metal locknuts. Apart from the required cross-section, excess material was cut down to reduce weight.

I bus bar was used for intra segment series connections between two modules of parallel-connected cells. I bus bar was exclusively used in L-shaped segments only. I bus bars were made of 6 mm thick copper sheet, fastened to the FR4 sheet (Figs. 3, 4, 6, 7 and 8).

Fig. 3 Box Bus bar

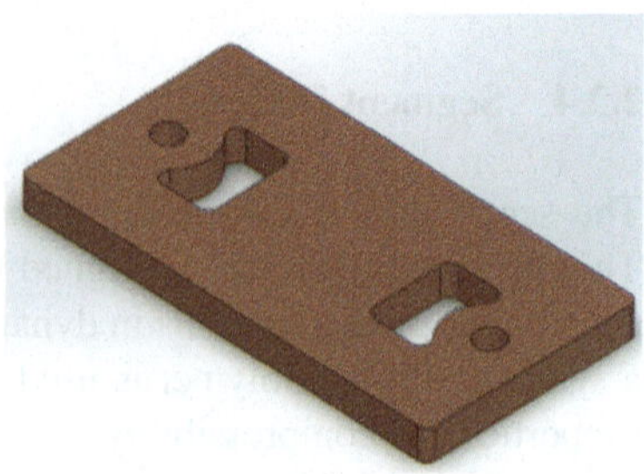

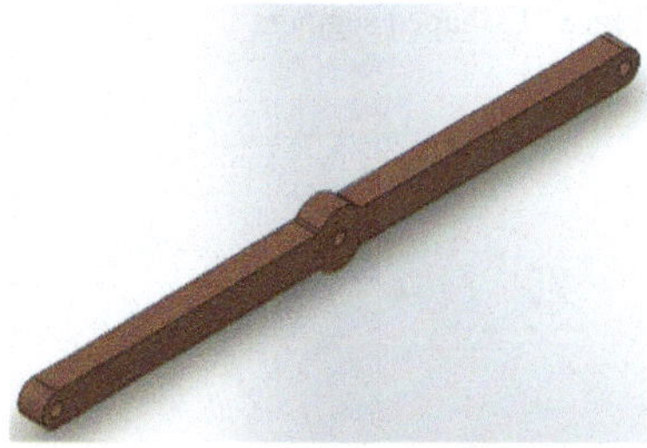

Fig. 4 I Bus bar

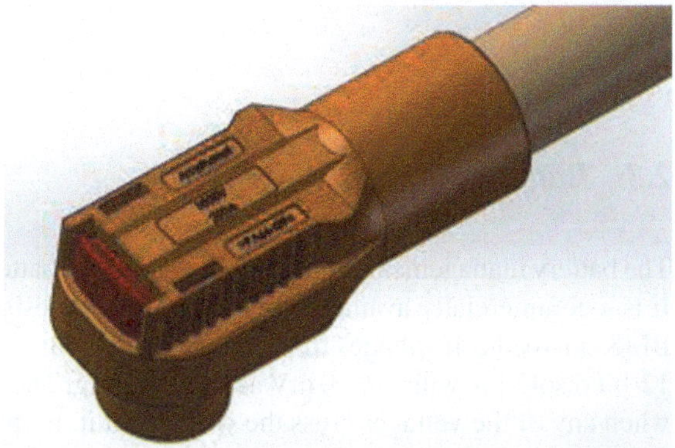

Fig. 5 Radlok connector

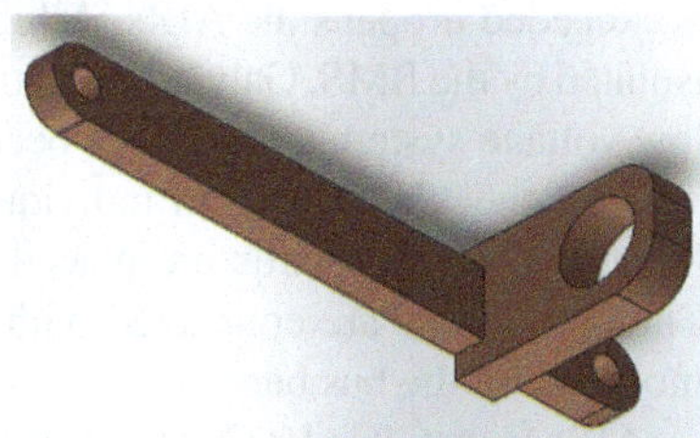

Fig. 6 T Bus bar

Fig. 7 Square-shaped segment

Fig. 8 L-shaped segment

2.3 Battery Management System

The battery management system employed for the battery pack is Elithion HD master. It is a commercially available BMS specifically designed for electric vehicles. The BMS can read cell voltages from 2.09 to 4.545 volts. The ADC within the AMS has a 12-bit resolution with a $\pm$ 10 mV accuracy rating. The AMS is programmed such that when any of the voltages cross the voltage limit, it opens the Accumulator Isolation Relays (AIRs). The temperature limit is set to be 60 °C, and if this temperature is exceeded it opens the AIRs. All the sense wires are electrically and magnetically isolated by the BMS. Galvanic isolation between the tractive system and the grounded low voltage system connections occurs within the BMS.

To gather the data from individual cells, the BMS uses slave PCBs called cell boards. The cell boards are placed in the slots and are bonded to the FR4 sheet. The cell boards are connected with individual modules via ring terminals that are mounted on the bus bars.

AIRs from GIGAVAC are used and Insulation Monitoring Device was bought from Bender that fulfilled all the necessary criterias. High Voltage fuse from Bussman is used.

2.4 Accumulator Design

According to the competition rule, *EV 5.5.4* [1], Aluminum 6061-T6 was selected as the material for the container walls. The metallic conductive walls were made non-conductive to comply with the rule *EV 5.4.1* [1] by bonding 0.25 mm thick NOMEX paper using UL94V0 Araldite. The container is developed to hold five battery segments and necessary high and low voltage components. Partitions are made to isolate the segments. The sheet thickness of the walls was 3 mm and the base thickness was 4 mm. The wall heights were determined such that it follows the FS Rule *EV 5.5.7* [1]. To provide more support to the walls in the axial direction,

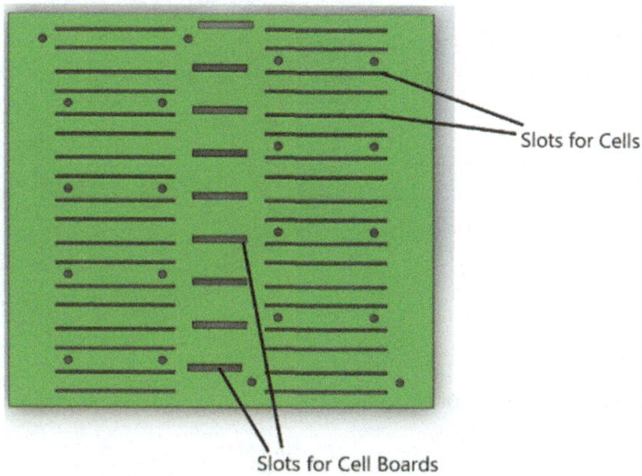

Fig. 9 FR4 Sheet

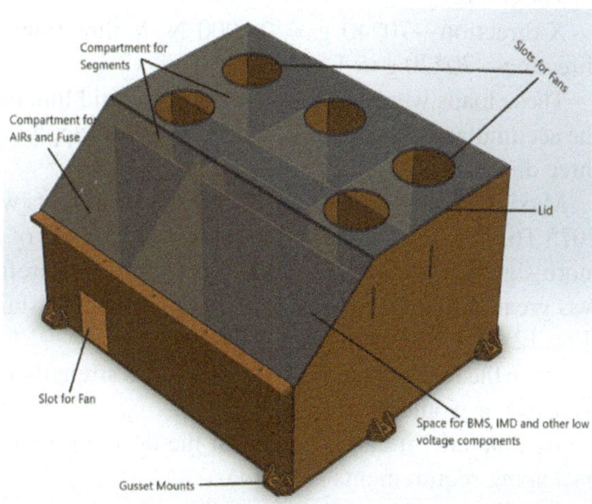

Fig. 10 Accumulator container

10 gusset mounts made of Aluminum 7075-T6 were implemented. All the walls and mounts were welded together to form a rigid structure (*EV 5.5.6*) [1] (Figs. 10 and 11).

2.5 *Structural Analysis*

The design of the accumulator satisfies the structural requirements specified by FS Rule *EV 5.5.9* [1]. The analysis was performed on Ansys Workbench, a CAE software. The chassis bolting points on the gusset mounts were provided fixed support. *EV 5.5.9* [1] mentions the acceleration of the load to be sustained by the accumulator

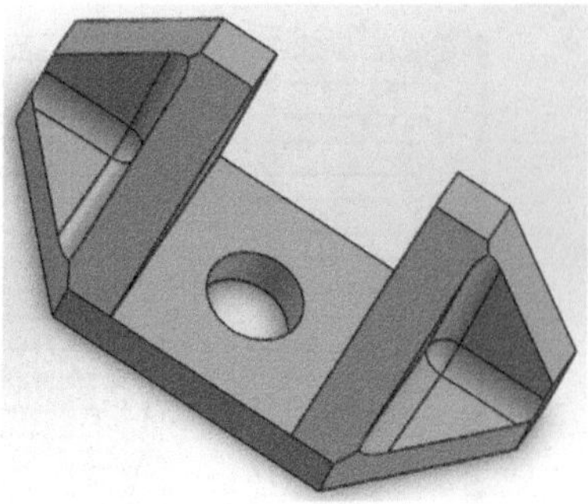

Fig. 11 Gusset mount

in all the three directions. The weight of the accumulator was estimated to be around 70 kilograms.

X-direction—70*40 g = 28,000 N, Y direction—70*40 g = 28,000 N and Z direction—70*20 g = 14,000 N

These loads were applied by creating a rigid link between the center of gravity of the accumulator with all of it's surfaces and by applying these loads in the mentioned three directions.

Maximum stress was acting on the gusset mounts which were made of Aluminum 7075 T6. The yield tensile and yield shear strength of both the aluminum alloys was more significant than 200 MPa, respectively. Therefore, since the factor of safety was greater than 1, it can be said that the design sustains the loads mentioned above (Fig. 12).

Since the loads were lesser than the yield strength of the material, any deformation whatsoever would be elastic (Fig. 13).

To conclude, it can be said that the design satisfies all structural, electrical and packaging requirements (Fig. 14).

3 Conclusion

The container and the battery pack were designed using various softwares such as Optimum Lap, Solidworks and Ansys. The accumulator design was optimal considering the FSAE design rules, packaging and performance requirements of the car, while also contemplating the manufacturing feasibility. The parameters obtained after designing procedures were nominal voltage 148.5 V, maximum voltage 162 V, maximum discharge continuous current 400 A, nominal capacity 40 Ah, weight of all the cells 44.55 kg. The values obtained after the designing of battery pack satisfies all the structural, electrical and packaging requirement.

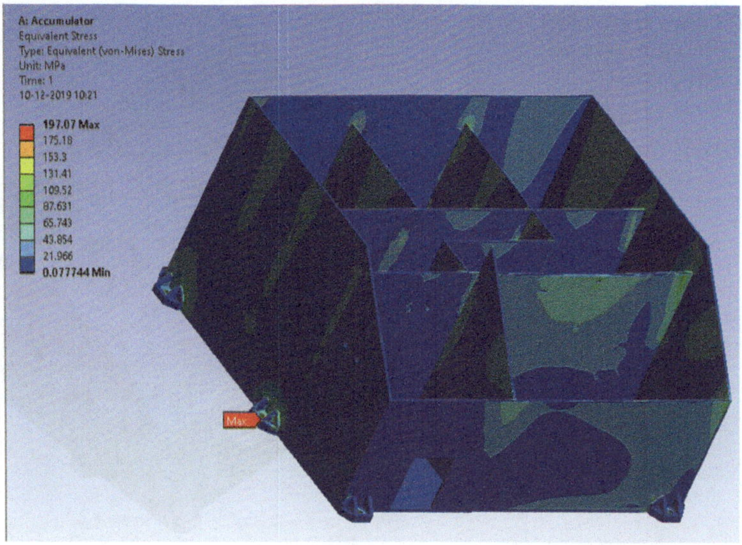

Fig. 12 Stress due to loads (max stress = 198 MPa)

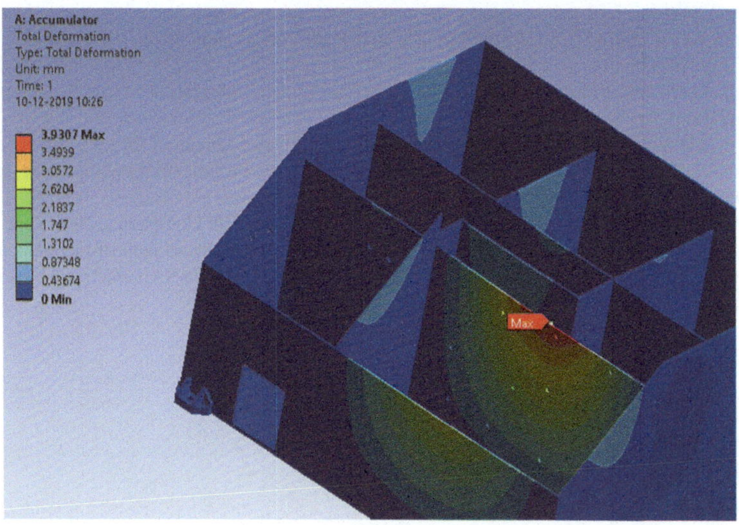

Fig. 13 Deformation due to loads (max deformation = 4 mm)

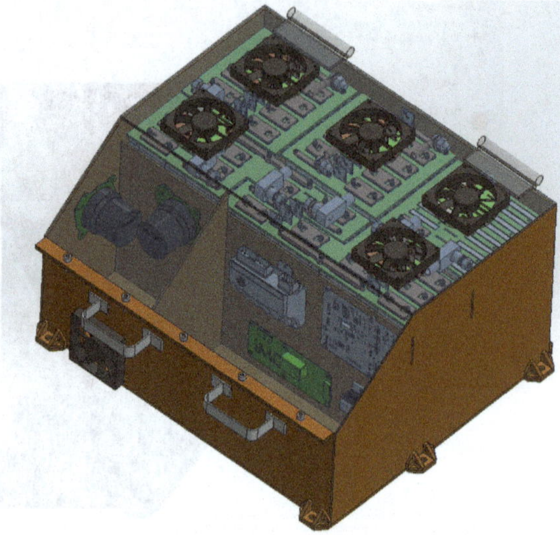

Fig. 14 Complete assembly of the accumulator container with all electronic

References

1. Appendix 1- https://www.formulastudent.de/fileadmin/user_upload/all/2019/rules/FS-Rules_2019_V1.0.pdf
2. Panthakkalakath N, Lee T, Mountford A (2017) Physical integration of engine systems and generator
3. Kalmakov VA, Andreev AA, Salimonenko GN (2016) Development of formula student electric car battery design procedure. In: International conference on industrial engineering, pp1391–1395. https://doi.org/10.1016/j.proeng.2016.07.334
4. Reddy TB (2011) Linden's handbook of batteries. McGraw Hill Companies, New York
5. West L, Shepherd B, Karabon N, Howell J, Pyrtko M (2016) Department of mechanical engineering, design report of the high voltage battery pack for formula SAE electric, University of Wisconsin-Madison

Design and Fabrication of Small Size Parabolic Reflector

Harshal Patil and Nishikant Kale

Abstract This study presents the design and development aspects of 3.2 m^2 small size parabolic reflector, used in domestic solar cooker for single family having four members. The study focused on design, proper material selection, and fabrication of reflector. The proposed reflector is based on Scheffler reflector. It was found that glass mirror has highest reflectivity and hence, is the best suitable material for reflector. Material for supporting frame is taken from scraps which reduces the cost of fabrication. Also, fabrication is simple and model is easy to transport from one place to another. The proposed model of the reflector is suitable for rural area applications and also doesn't require skilled labor to fabricate.

Keywords Parabolic reflector · Glass mirror · Aluminum

1 Introduction

In the current worldwide scenario, the energy demand is met by burning fossils fuel, which is limited [1]. To reduce dependency on conventional fuel, many researchers are making effort to use other sources energy—which is present in abundance like sun, wind, water, etc. These nonconventional sources of energy have the capacity to solve world energy needs. Among the various nonconventional energy sources, solar energy contributes a major portion [2].

Even though the use of solar energy in daily life has been known since 1455 BC [3], still this technology is not popular among society. The total amount of energy received by the earth from the sun, in an hour is more than the total world energy demand in one year [4].

Among the three necessary needs, one is food; this can be achieved with the help of solar energy. Human beings have a habit of eating cooked food. There are various fossil fuels like LPG, Electricity, Wood, Coal, etc., which are used for the cooking process across the world. There are various ways through which heat from the sun

H. Patil (✉) · N. Kale
Department of Mechnical Engineering, PRMIT&R, Badnera, Amravati, India
e-mail: hdpatil1986@gmail.com

© Springer Nature Singapore Pte Ltd. 2020

H. Vasudevan et al. (eds.), *Proceedings of International Conference on Intelligent Manufacturing and Automation*, Lecture Notes in Mechanical Engineering,
https://doi.org/10.1007/978-981-15-4485-9_81

can directed towards the cooking pot. Such cooking is known as solar cooking. Depending on the use, there can be different ways to concentrate solar power to a point (cooking pot). For fast cooking, the temperature at the focus point needs to be high which can be achieved by the use of parabolic through, central tower, parabolic dish and concentrators, etc. Parabolic dish is widely used because of high-temperature generation at the concentrated point. But, these are practically difficult as almost all parabolic concentrators have a rigid structure and the focus. It required skilled labor to operate [5].

This motivated the Scheffler to develop a design where the hot focus is available at a fixed place so that solar energy can be constantly used [6]. He developed the Scheffler reflector in village workshops in India and Kenya so that this technology is in the reach of everyone [7]. Scheffler [8] developed a special Scheffler reflector with 50 m^2 surface area to heat a 2 m long cremation chamber at 700 °C. Then, different researchers started developing the Scheffler dish which finds its application in the households for boiling water, cooking food, and in medical applications [9–13].

This paper attempted a design of a parabolic reflector which is used in solar cookers. This reflector is based on Scheffler technology [14]. In the first section, the details about the parabolic reflector and its characteristics are studied. The later section explains the calculations for various components of the reflector and its material required.

2 Parabolic Dish Type Reflector

A parabolic dish reflector is a concentrating solar collector that is similar in appearance to a large satellite dish but has mirror-like reflectors and an absorber at the focal point. A parabolic dish system uses a dual-axis tracking system to follow the sun across the sky and concentrate the sun's rays onto the receiver located at the focal point in front of the dish. The temperature at the focal depends on the surface area of the reflector [15].The parabolic reflector has mainly two components—Frame and Reflecting surface.

2.1 Frame

Frame is the supporting structure for a reflector. The material is used for fabricating the frame mainly aluminum, cast iron, alloy metals, steels, etc. Following Table 1 shows the density of various materials.

The aluminum has a density around one-third that of steel making; it is one of the lightest commercially available metals. Hence, the aluminum frame has a weight equal to one-third of the weight of the steel frame having the same dimensions. Aluminum is a very desirable metal because it is more malleable and elastic compared to steel [17].

Table 1 Suitable materials for fabrication of reflector [16, 18]

S. No.	Name of the materials	Density (Kg/m^3)	Thermal conductivity ($Wm^{-1}k^{-1}$)	Reflectivity
1	Polished aluminum	2705	215	0.95
2	Stainless steel	7982	16.2	0.90
3	Mild steel	7850	45	0.65
4	Iron	7860	79.5	0.55
5	Glass mirror	8000	1.05	0.99

The high strength to weight ratio makes it an important structural material allowing increased payloads or fuel savings for transport industries in particular. Aluminum shows great tensile strength at low temperature otherwise steel is brittle at low temperature.

When aluminum is exposed to air, a layer of aluminum oxide forms almost instantaneously on the surface of the aluminum and this layer has excellent resistance to corrosion. It is nontoxic and can be easily cleaned. Aluminum is 100% recyclable and recycled aluminum is identical to the virgin products. Aluminum is the world's most abundant metal and does the third most common element comprise 8% of the earth's crust. The versatility of aluminum makes it the most widely used metals after steel. Cost is always an essential factor to consider when making any product. The cost of aluminum per kg is Rs. 220 and that of steel is Rs. 250 per kg. Therefore, considering all these advantages of aluminum, in present research work, aluminum is used for manufacturing of supporting frames of parabolic reflector.

2.2 *Reflecting Surface*

To achieve maximum collector efficiency, the selection of most appropriate material is very important. The material should possess properties like high reflectivity and low absorption [18]. Table 1 shows some suitable material for reflecting surface and their properties.

Though the glass mirror has the highest value of reflectivity, almost all radiations are reflected out of incident radiation on the reflector. Also, the thermal conductivity of glass mirror is very less; hence the heat loss from reflector is negligible through conduction. The present research work is based on glass mirror material for reflector fabrication.

3 Design of Reflector

3.1 Frame

The material considered for frame is aluminum having cross-section as follows: Cross-sectional details (l, b, t): 89.5, 3, 2 cm.

There are 16 links of same cross-section used to form complete parabolic reflector. All links are connected centrally with the help of screw and nut for simple assemble and disassemble the reflector. Due to which transportation of this reflector is easy. [19] At the center of reflector, a circular cast iron plate is mounted, on which links are attached, it provides strength to the reflector and center plate is a connection between reflector and holding stand of reflector.

3.2 *Reflector*

Reflecting Area of Dish: 3.12 m^2 [20]

Reflecting Material: Highly Reflective Mirrors

Calculation of Parabola as below,

Radius of dish (d):	96 cm
Arc Length of curvature:	199 cm
Rim Angle (0):	28 °
Focus Length (F):	6 ft

Mirrors Size

"a" and "b" varies with decrease in cross-section and height "c" is constant for all pieces

Shape of mirror: Trapezoidal as shown in Fig. 1.

Number of mirrors:	160
Total height of single slot:	80 cm
Length of top portion (b):	38 cm
Length of bottom portion (a):	4.5 cm

The following are the parameters which we have to take account while designing Scheffler dish [21].

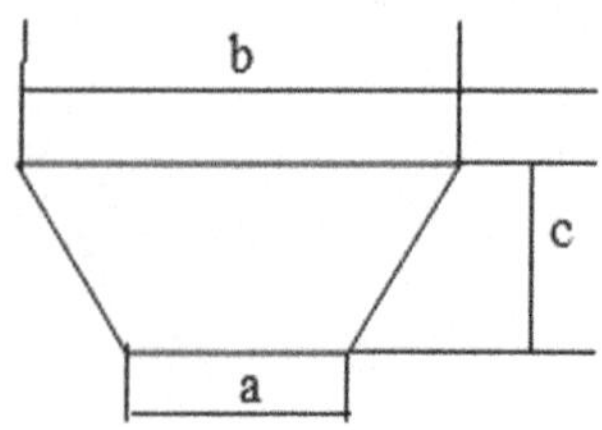

Fig. 1 Dimension of mirror section

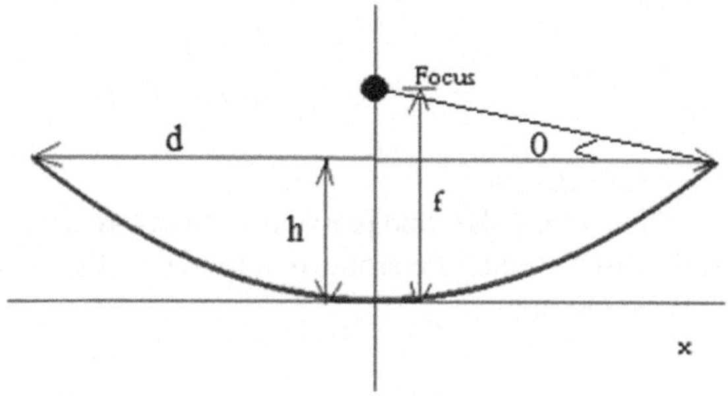

Fig. 2 Parabolic dish

3.3 Mathematical Calculations

Let,

d—Diameter of the Scheffler dish = 6 ft = 1.8288 m

f—focal length of the Scheffler dish = 6 ft = 1.8288 mas shown in Fig. 2.

Height of the curve from center (h) is calculated from Eq. (1)

$$h = \frac{d^2}{16 \times f}$$
$$h = 0.1143\ \text{m} \tag{1}$$

Rim Angles (θ)

$$\theta = \tan^{-1} \frac{1}{\frac{d}{h \times 8} - \frac{2 \times h}{d}}$$
$$\theta = 28^\circ \tag{2}$$

The rim angle of proposed reflector is 28 ° from Eq. (2)

Focal Length to Disc Diameter Ratio

$$\frac{f}{d} = \frac{1.8288}{1.8288} = 1$$

Arc Length (S) shown in Eq. (3)

$$S = \left(\frac{d}{2} \times \sqrt{\left(\frac{4 \times h}{d}\right)^2 + 1}\right) + 2f \times \log\left(\frac{4 \times h}{d} + \sqrt{\left(\frac{4 \times h}{d}\right)^2 + 1}\right)$$
$$S = 1.81\ \text{m} \tag{3}$$

Area of Space (A) available on dish from Eq. (4)

$$A = \frac{2}{3} \times d \times h$$
$$A = 0.386\ \mathrm{m}^2 \tag{4}$$

Force Calculation
The three forces like wind pressure, normal force, and frictional force are considered on the dish, which is determined in following Eqs. (5), (6) and (7), respectively,

Wind pressure

$$A = \frac{1}{2} \times \rho \times V^2$$
$$A = 508.03\ \mathrm{N/m}^2 \tag{5}$$

Normal Pressure/Force
Range = 28° − 75° = 51°

$$P_n = 2 \times P \times \cos(90 - \theta)$$
$$P_n = 789.62\ \mathrm{N} \tag{6}$$

Frictional Force

$$F_f = f \times \mu$$
$$F_f = 276.37\ \mathrm{N} \tag{7}$$

4 Computer-Aided Design of Reflector

CAD model is developed for supporting frame and reflector with a given dimension for physical appearance which is shown in Figs. 3 and 4, respectively. The software tool is used in modeling is CATIA V5.

5 Fabrication of Reflector

According to design parameters, fabrication was performed in two phases. In the first phase, the frame was build up and the second phase reflecting surface was mounting on a frame. Entire fabrication is performed in a college workshop.

Before the actual fabrication of the reflector, the drawing with the exact dimension is carried out on a wooden sheet. The possibility of the focus point is achieved through it and found results are similar to theoretical results. The assignments while performing the drawing on a wooden sheet are shown in Fig. 5. The link required

Fig. 3 CAD model for links of frame

Fig. 4 CAD model for reflector

for the fabrication of the frame is taken from an old satellite dish [22]. The small trapezoidal pieces of mirrors which is used as a reflector are shown in Fig. 6, then the links with proper dimensions are prepared and connected to the central cast iron plate with the help of screw and nuts are shown in Fig. 7. The complete solar Parabolic reflector shown in Fig. 8.

Fig. 5 Actual drawing of a parabola on wooden sheet

Fig. 6 Small trapezoidal pieces of mirror

Fig. 7 Mounting of small mirrors on frame

Fig. 8 Solar parabolic reflector

Table 2 Focus and atmospheric temperature at different times

Time	Atmosphere temperature (°C)	Focus temperature (°C)	Time	Atmosphere temperature (°C)	Focus temperature (°C)
8:30	30	47	12:30	38	125
9:00	31	47	13:00	40	137
9:30	29	64	13:30	42	136
10:00	34	96	14:00	42	140
10:30	35	109	14:30	41	137
11:00	34	110	15:00	40	78
11:30	36	115	15:30	39	76
12:00	37	120	16:00	37	60

6 Experimentation

The experimentation is carried out for the whole day from morning to evening and the focused temperature throughout the day is observed. The readings obtained during testing are shown in Table 2.

7 Conclusion

The design and fabrication of parabolic-type solar reflector are studied as a part of this work. A 3.2 m^2 surface area of the reflector was developed. This small size of the solar reflector is suitable for a family of four members. Though all the links are connected by nuts and screws, it provides easy to assemble and disassemble function but precaution should be considered for handling the mirrors. If any damage occurs to the mirror while working, then there is need to replace only damaged piece of mirror. Since most of the portion is manufactured from scrap material, its cost will be low. Glass mirrors have high value of reflectivity equal to 0.99, which is the best-reflecting surface among all other surfaces; therefore it is preferably used in all type of solar reflector. The peak temperature at focus is found 140 °C for a small size dish. So, it can be concluded that a small parabolic reflector contributes to a large number of functions and is very useful for application requiring medium temperature. Its fabrication is very simple; it doesn't require any skilled labor. This model is suitable in the rural area where insufficient convention fuel supply problem occurs.

References

1. Kannan N, Vakeesan D (2016) Solar energy for future world: a review. Renew Sustain Energy Rev V 62:1092–1105
2. Regin AF, Solanki SC, Saini JS (2008) Heat transfer characteristics of thermal energy storage system using PCM capsules: a review. Renew Sustain Energy Rev 12(9):2438–2458
3. Solar energy solar PV (for electricity) and thermal (for hot water) systems. http://www.yesweb.org/docs/solarenergy.pdf. Accessed on 12 Dec 2015
4. Sansaniwal SK, Sharma V, Mathur J (2018) Energy and exergy analyses of various typical solar energy applications: a comprehensive review. Rene Sustain Energy Rev 82:1576–1601
5. Gadhia D (2009) "Parabolic solar concentrators for cooking and food processing. In: Proceeding of international solar food processing conference, Indore, India. http://www.solarae-bruecke.org
6. Oelher U, Scheffler W (1994) The use of indigenous material for solar conversion. Solar Energy Mater Solar Cells 33:379–387
7. Scheffler W, Bruecke S, von Werdenbergstr G (2006) Introduction to the revolutionary design of scheffler reflectors. In: Proceedings of international conference on solar cookers and food processing, Granada, Spain, pp 12–6
8. Scheffler W, Bruecke S, von Werdenbergstr G (2006) Development of a solar crematorium. In: Proceedings of 6th international conference on solar cooker, Granada, Spain
9. Munir A (2010) Design, development and modeling of a solar distillation system for the processing of medicinal and aromatic plants (Doctoral dissertation). Univ, Kassel
10. Munir A, Hensel O (2009) Biomass energy utilization in solar distillation system for essential oils extraction from herbs. In: Conference of international research on food security, natural resource management and rural development, University of Hamburg; 2009 Oct 6
11. Munir A, Hensel O, Scheffler W (2010) Design principle and calculations of a scheffler fixed focus concentrator for medium temperature applications. Sol Energy 84:1490–1502
12. Tyroller M, Brucke S, Werdenbergstr G (2006) Solar steam sterilizer for rural hospitals Proceedings of international solar cooker conference; 2006 Jul (No. 34)
13. Dupont F, Essential oil extraction from herbs using solar energy. Spain
14. Dafle V, Shinde N (2012) Design, development & performance evaluation of concentrating monoaxial scheffler technology for water heating and low temperature industrial steam application. Int J Eng Res Appl (IJERA) 2(6):848–852
15. Ashutosh CV, Harshit P, Rahul A, Rounak A (2016) Mathematical modeling and performance study of scheffler reflector (UG disseratation). Mesra, India: BIT
16. https://www.theengineeringmindset.com. Last accessed 2019/11/5
17. https://www.wenzelmetalspinning.com. Last accessed 2019/11/6
18. https://www.azom.com. Last accessed 2019/11/4
19. Chandak A, Somani S, Patil M (2011) Comparative analysis of SK-14 and PRINCE-15 solar concentrators. In: Proceeding of the world congress on engineering, vol III, London, U.K
20. Phate M, Gadkari D, Avachat S, Tajne A (2014) Experimental analysis of 2.7 m^2 scheffler reflector and formulation of a model. Int J Eng Trends Technol (IJETT), 12:1–5. ISSN 2231-5381

21. Reddy S, Khan M, Alam MZ, Rashid H (2018) Design charts for scheffler reflector. Solar Energy 163:104–112
22. Chandak A, Somani S (2009) Design, development and testing of multieffect distiller/evaporator using scheffler solar concentrators. J Eng Sci Technol 4(3):315–321

Analysis of Brake Hub Used in Automobiles

Dhairya K. Vora, Rishikesh K. Patil, and Vinayak H. Khatawate

Abstract This paper presents the design and analysis of a brake hub used in an automobile vehicle. While the vehicle is being driven, the power is transmitted from the hub to the rotor. In vehicles, when the brake is applied, the brake rotor receives the braking torque and communicates it to the shaft through the hub. Brakes in automobiles are associated with safety. They play a vital role in the overall performance of the vehicle and also provide comfort and serviceability. The present work consists of the weight optimization of the brake hub and weight optimized part was considered for the structural analysis. It was found that stresses induced in the brake hub were within the permissible values.

Keywords Brake hub · Braking torque · Optimization · Structural analysis

1 Introduction

A brake is a device for slowing or stopping a vehicle typically by applying pressure to the wheels. It is a mechanical device that inhibits motion by absorbing energy from a moving system. The braking system is an important part of automobile and the braking performance directly affects the safety of automobile driving. The structural analysis of the brake drum has very important significance to improve the braking efficiency and stability of the automobile braking system [1]. Braking systems have had tremendous transformations from lever-type brakes on horse carriages to multilayered carbon-ceramic disk-type brakes to air brakes used on land speed record-breaking cars [2]. For a Go-kart which is a type of open-wheel car, hydraulic brakes are more suitable and convenient to use as compared to mechanical system. Brake hub of hydraulic disk-type brake was used in the present work. A hub is a part which is used to assemble two or more components. A brake hub is used to connect the brake rotor to the shaft. The brake hub is rigidly mounted on the shaft with the help

D. K. Vora (✉) · R. K. Patil · V. H. Khatawate
Department of Mechanical Engineering, Dwarkadas J. Sanghvi College of Engineering, Vile Parle, Mumbai, India
e-mail: dhairyakvora2018@gmail.com

© Springer Nature Singapore Pte Ltd. 2020

H. Vasudevan et al. (eds.), *Proceedings of International Conference on Intelligent Manufacturing and Automation*, Lecture Notes in Mechanical Engineering,
https://doi.org/10.1007/978-981-15-4485-9_82

of a key slot which is made on the hub as well as on the shaft, and the brake rotor is attached to the hub with the help of bolts and locknuts. So, the torque from the shaft is transferred to the brake rotor with the help of brake hub. When the driver applies the brakes, forcing the caliper pads against the rotor, the rotor slows down which in turn stops the shaft through the brake hub. During this process, forces are experienced by the brake hub at the bolting points. The material used for brake hub is Aluminum 6061 T6. It has an ultimate tensile strength of 310 MPa and has a yield stress of 276 MPa. Computation of the stresses and structural analysis of brake drum considerably help to avoid failure and possess the ancillary benefits of having optimal weight and cost [3]. The present article deals with the static structural analysis of brake hub.

2 Objectives of the Research

The objectives of the research are as follows:

1. To design a lightweight brake hub to be used in Go-Karts using empirical relations.
2. To assess the structural integrity of the brake hub using FEA.

3 Weight Optimization of Brake Hub

Weight reduction reduces the energy consumption because the energy required moving a vehicle is, except for aerodynamic resistance, directly proportional to its mass. Keeping braking power constant, lightweighting shortens braking distance. Keeping the braking performance constant, lightweighting allows downsizing of the brakes. [4, 5]. In the initial stage, a simple circular hub as shown in Fig. 1 was developed.

The weight of this hub was found to be 0.288 kg; hence, it is decided to make a better design from the prospect of weight, without compromising its strength.

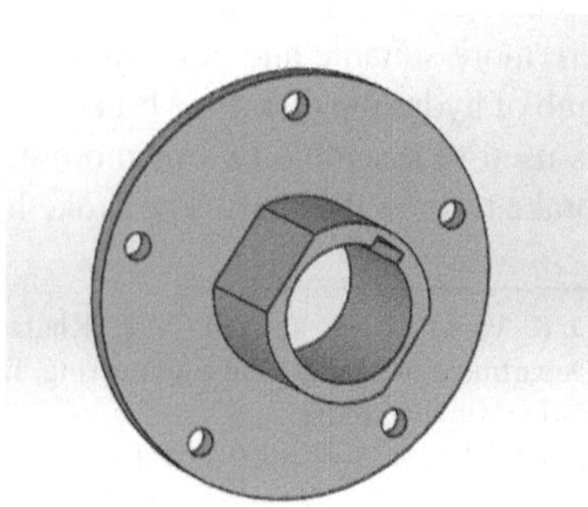

Fig. 1 Brake hub initial design

Fig. 2 Brake hub first iteration

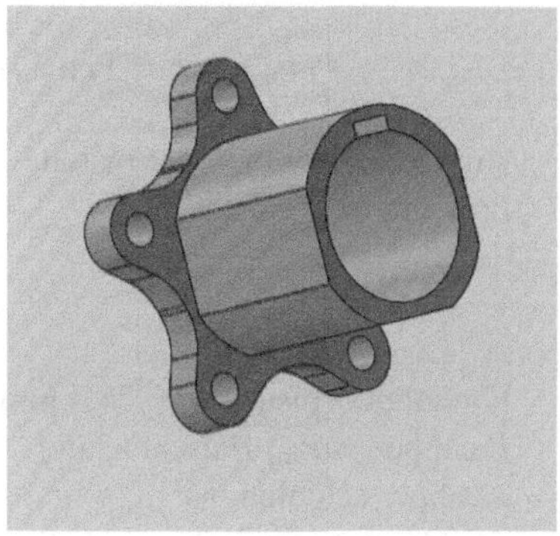

Fig. 3 Brake hub second iteration

Keeping this in mind, the authors came up with the following two iterations as shown in Figs. 2 and 3.

The weight was 0.213 kg, in the first iteration and 0.191 kg in the second iteration as shown in Figs. 2 and 3, respectively.

The second iteration (Fig. 3) was considered for further analysis because of its lightweight and aesthetic appeal [4].

4 Design Calculations

The brake hub experiences frictional force and torque which are exhibited by brake rotor and shaft, respectively. The empirical equations used for designing the hub are described below. [6–8].

4.1 Consideration

Max driver force = 24 kg = 24 * 9.81 = 235.44 N
Pedal ratio = 5:1
Mass of the kart with driver = 180 kg
Coefficient of friction (μ) of rotor and pad = 0.5

4.2 Calculation

For master cylinder (piston diameter (Φ) = 12.7 mm):

$$F_{\mathrm{mc}} = F \ast \text{Pedal ratio} = 235.44 \times 5 = 1177.2\ \mathrm{N}$$

Area of master cylinder piston = $A_{\mathrm{mc}} = 126.677\ \mathrm{m}^2$
Pressure in master cylinder

$$P_{\mathrm{mc}} = 9.293\ \mathrm{MPa}$$

For caliper (piston $\Phi = 25.4$ mm):-
P caliper = P_{mc} [Pascal's law]
Acaliper calculations:

$$A\ \text{caliper} = 506.707 \times 10^{-6}\mathrm{m}^2$$

Now,

$$\begin{aligned} F\ \text{caliper} &= P_{\mathrm{mc}} \ast A\ \text{caliper} \\ &= 9.293 \ast 10^6 \ast 506.707 \ast 10^{-6} \\ F\ \text{caliper} &= 4.7088\ \mathrm{KN} \end{aligned}$$

Since it is a two-piston caliper

$$F\ \text{caliper} = 4.7088 \ast 2 = 9.4177\ \mathrm{KN}$$

$$\begin{aligned} \text{Total frictional force} &= F\ \text{caliper} \ast \mu\ \text{rotor} \\ &= 9.4177 \ast 10^3 \ast 0.5 \\ &= 4.7088\ \mathrm{KN} \end{aligned}$$

$$\text{Disk effective radius} = D + \frac{d}{4}$$

($D = 200$ mm, $d = 138$ mm)
Disk effective radius = 84.5 mm

$$\text{Torque on rotor} = F \text{ frictional} * \text{Re}$$
$$= 4.7088 * 10^3 * 84.5 * 10^3$$
$$\text{Torque on rotor} = 397.89 \text{ Nm.}$$

4.3 Brake Hub

$$s_{yt} = 276 \text{ Mpa} = 276 \text{ N/mm}^2$$
$$\sigma_t = \frac{s_{yt}}{N_f} = \frac{276}{3} = 92 \text{ N/mm}^2$$
$$\sigma_c = \frac{s_{yc}}{N_f} = \frac{1.2 s_{yt}}{N_f} = 110 \text{ N/mm}^2$$
$$\tau = \frac{0.5 s_{yt}}{N_f} = 46 \text{ N/mm}^2$$

Length of the hub = length of the key = 60 mm
Outer diameter of hub = 1.5 D_{shaft}

$$D_h = 60 \text{ mm}$$

Shear stress induced in hub is

$$\tau_{\text{hub}} = \frac{16 T_{\max}}{\pi D h^3 (1 - k^4)}$$

$D_{\text{shaft}} = 40$ mm, $d_{\text{shaft}} = 30$ mm

$$k = \frac{d_{\text{shaft}}}{D_{\text{shaft}}} = 0.375$$

$$\tau_{\text{hub}} = 7.8025 \text{ N/mm}^2$$

Now $\tau_{\text{hub}} < \tau_{\text{allowable}} (46 \text{ N/mm}^2)$

Since, the shear stress induced in the brake hub is below permissible value; therefore, the design of hub is safe.

The dimensions of the brake hub are shown in Fig. 4.

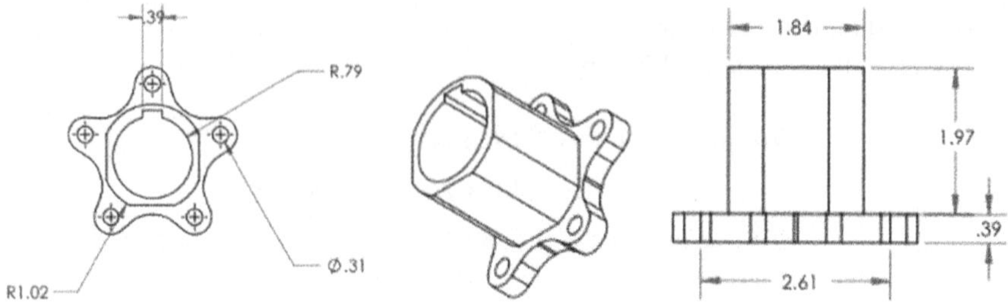

Fig. 4 Different views of brake hub

Table 1 Composition of Al 6061 T6 [11]

S. No.	Elements	Composition (wt%)
1	Silicon	0.4–0.81
2	Iron	0.7
3	Copper	0.15–0.4
4	Manganese	0.15
5	Magnesium	0.8–1.2
6	Chromium	0.04–0.35
7	Zinc	0.25
8	Titanium	0.15
9	Other elements	0.15

5 Stress Analysis of Brake Hub Using FEA

5.1 Material Selection

Aluminum alloy is a panel material which has lightweight and is extensively designed for exterior and interior applications. The material used for analysis is Al 6061 T6 because of its low density, high strength, good workability and high resistance to corrosion [9, 10]. The composition and properties of Al 6061 T6 are tabulated in Tables 1 and 2, respectively.

5.2 Meshing

FEA is a numerical technique for solving problems of engineering and mathematical physics. The most distinctive feature of FEA is, it allows the given domain to divide into simple subdomains. Each subdomain is called element and elements are connected using nodes. The collection of all elements is called meshing. [13].

Table 2 Aluminum Al 6061 properties [12]

S. No.	Property	Value
1	Elastic modulus	70–80 GPa
2	Density	2.7 g/cc
3	Poisson's ratio	0.33
4	Hardness (HB 500)	30
5	Tensile strength	115 MPa
6	Ultimate tensile strength	572 MPa
7	Tensile yield strength	503 MPa
8	Elongation at brake	11%
9	Modulus of rigidity	71.17 GPa
10	Fatigue strength	159 MPa
11	Shear modulus	26.9 GPa
12	Shear strength	331 MPa
13	Machinability	70%

Meshing is performed by taking element size 1 mm and tetrahedron elements are used. The main purpose of selecting tetrahedron type of element is that it covers each and every corner of the brake hub and selecting 1 mm element size results in more accurate and precise results [14]. The meshed diagram of hub is shown in Fig. 5.

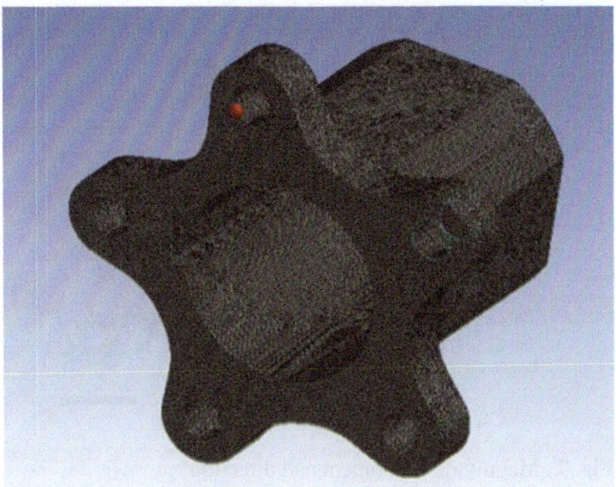

Fig. 5 Mesh image of hub

5.3 Forces and Constraints

A moment of 397.89 Nm is applied on the bolting points of the brake hub in the clockwise direction, whereas the key slot is constrained in all directions. Figures 6, 7 and 8 show applications of boundary conditions.

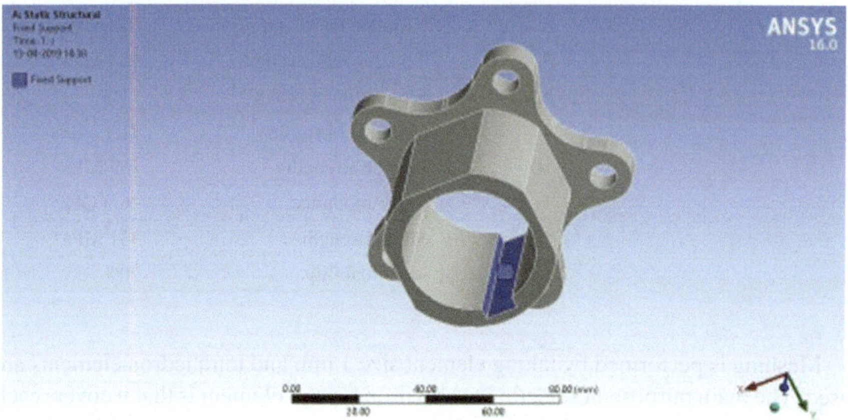

Fig. 6 Fixed constraints

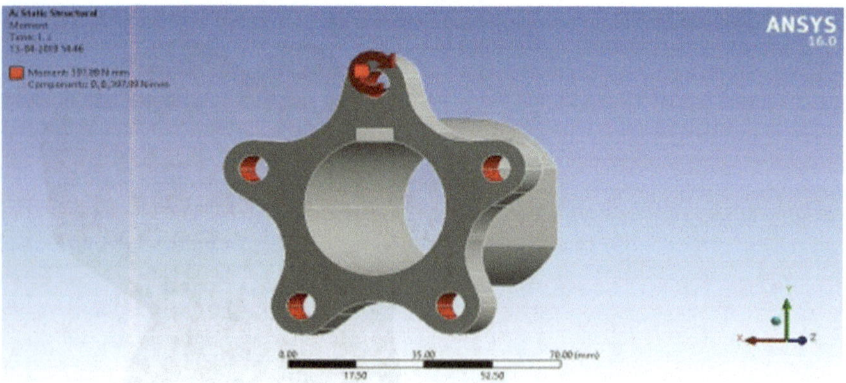

Fig. 7 Magnitude of moment and direction

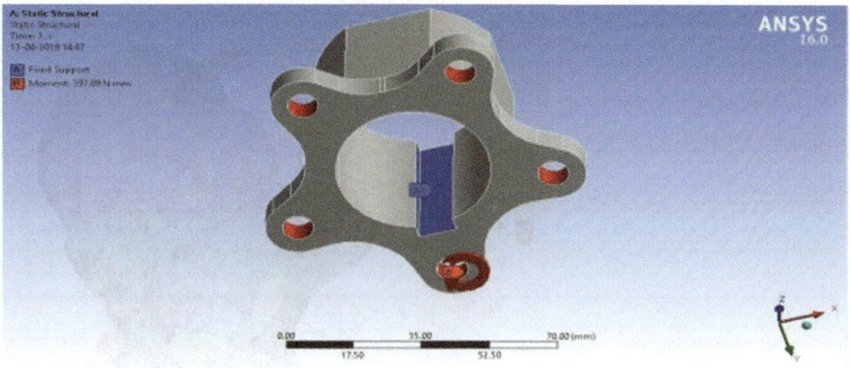

Fig. 8 Application of moment and constraint

5.4 *Results and Discussions*

In Figs. 9, 10 and 11, total deformation, factor of safety and equivalent stress are shown, respectively, as obtained through FEA. Figure 9 depicts the total deformation experienced by the hub. The maximum deformation of the hub is found to be 0.006 mm. The variation of the deformation across the hub is shown by different colors in Fig. 9.

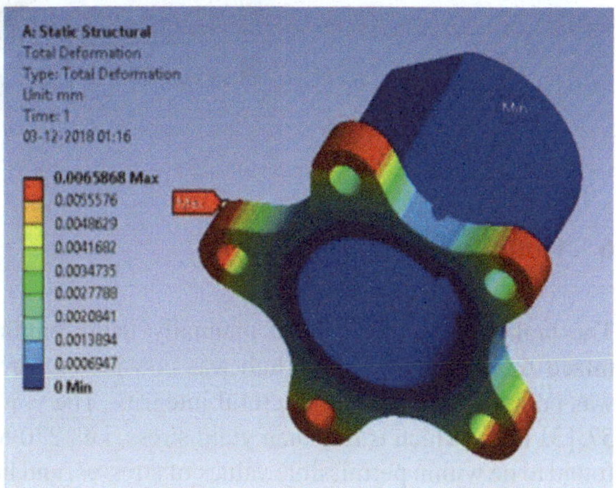

Fig. 9 Total deformation

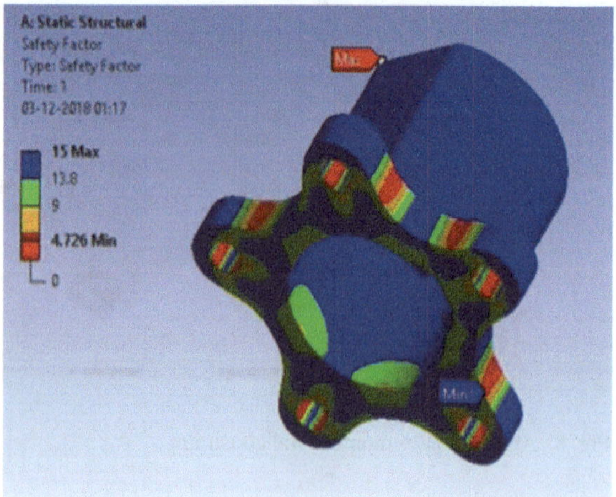

Fig. 10 Factor of safety

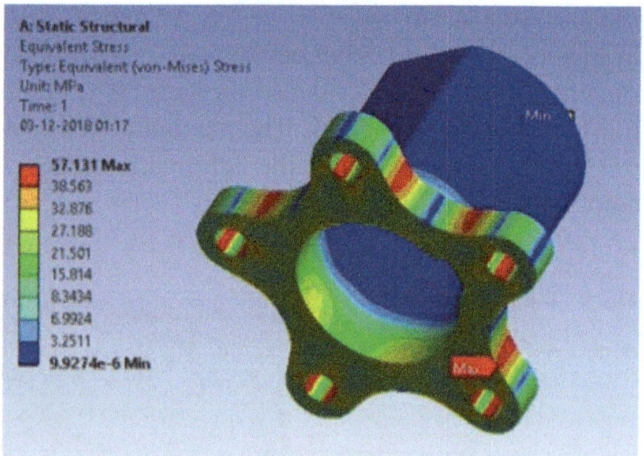

Fig. 11 Equivalent stress of brake

6 Conclusion

The brake hub was designed manually using empirical equations and the optimized solid model of brake hub was considered for FEA. The structural analysis was performed to ensure structural integrity. The von Mises stress was found to be 57.131 MPa, which is less than yield stress, i.e., 270 MPa. The stresses in hub were found to be within permissible values of stresses, and hence, the design was safe. The regions of maximum stress and minimum stresses were identified from the results obtained from FEA, so that if needed further design changes could be made to reduce stress concentration.

Appendix

S_{yt} Yield stress
σ_t Tensile stress
σ_c Compressive stress
τ Shear stress
N_f Factor of safety
M Coefficient of friction
D Outer diameter of shaft
d Inner diameter of shaft
D_h Outer diameter of hub
d_h Inner diameter of hub
Ø Piston diameter of master cylinder

References

1. Bin Z, Jingdon Z (2017) Modal Analysis of the brake drum of the heavy—scale automobiles. In: 2017 IEEE 3rd information technology and mechatronics engineering conference (ITOEC)
2. Sharma A, Marwah AK (2013) Braking systems: past, present & future. Indian J Res 2(3)
3. Bhardwaj M, Mittal S, Kumar V, Sharma R, Heera J (2019) Comparative model analysis of brake rotors. In: Prasad A, Gupta S, Tyagi R (eds) Advances in engineering design. Lecture Notes in Mechanical Engineering. Springer, Singapore
4. Reddy MS, Krishna NN (2015) Trends in weight reduction of automobiles "Alu—maximized". Int J Eng Res Gen Sci 3(3) Part-2
5. Kushal M, Sharma m (2015) Optimization of design of brake drum of two-wheeler through approach of reverse engineering by using ansys software. IOSR J Mech Civil Eng (IOSR-JMCE) 12(4) Ver. IV
6. Shigley JE, Mischke CR (2004) Standard handbook of machine design, 3rd Edition, McGraw Hill Publications
7. Design Data book (2018) Design Data Book for Engineers. P.S.G. College of Technology, Coimbatore, Revised Edition
8. Bhandari VB (2016) Design of mechanical elements, 4th edn, McGraw Hill Publications
9. Rao SS (2010) The finite element method in engineering, 5th edn, Butter Worth Heinemann
10. Gowthami K, Balaji K (2016) Designing and analysis of brake drum. Int J Res Appl Sci Eng Technol (IJRASET) 4(IX)
11. Liu H, Fujii H, Maeda M, Nogi K (2003) Tensile properties and fracture locations of friction-stir welded joints of 6061-T6 aluminum alloy. J Mater Sci Lett (2003)
12. Lang AM (1994) An investigation into heavy vehicle drum brake squeal. A doctoral thesis submitted in partial fulfillment of the requirements for the award of Doctor of Philosophy of the Lough borough University of Technology
13. Reddy JN (2005) An introduction to the finite element method, 3rd ed, McGraw-Hill Education
14. Baba SNN, Hamid MNA, Soid SNM, Zahelem MN, Omar MS (2018) Analysis drum brake system for improvement of braking performance. In: Öchsner A (eds) Engineering applications for new materials and technologies. Advanced structured materials, vol 85, Springer, Cham, pp 345–357

The manufacturer's authorised representative in the EU is Springer Nature Customer Service Centre GmbH, Europaplatz 3, 69115 Heidelberg, Germany. If you have any concerns regarding our products, please contact ProductSafety@springernature.com

Printed and bound by CPI Group (UK) Ltd, Croydon, CR0 4YY
30/09/2026
02225509-0018